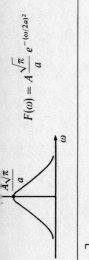

$f(t) = Ae^{-a^2t^2}$

$F(\omega) = A\dfrac{\sqrt{\pi}}{a} e^{-(\omega/2a)^2}$

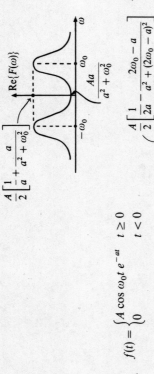

$f(t) = \begin{cases} A\cos\omega_0 t\, e^{-at} & t \geq 0 \\ 0 & t < 0 \end{cases}$

$F(\omega) = A\dfrac{a + j\omega}{(a + j\omega)^2 + \omega_0^2}$

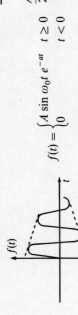

$f(t) = \begin{cases} A\sin\omega_0 t\, e^{-at} & t \geq 0 \\ 0 & t < 0 \end{cases}$

$F(\omega) = \dfrac{A\omega_0}{(a + j\omega)^2 + \omega_0^2}$

SIGNALS AND SYSTEMS

SIGNALS AND SYSTEMS

ALEXANDER D. POULARIKAS
UNIVERSITY OF ALABAMA IN HUNTSVILLE

SAMUEL SEELY
UNIVERSITY OF RHODE ISLAND

PWS-KENT Publishing Company
Boston

PWS-KENT
Publishing Company

Copyright © 1985 by PWS Publishers

All rights reserved. No part of this book may be reproduced, stored in a retrieval system, or transcribed in any form or by any means, electronic, mechanical, photocopying, recording, or otherwise, without the prior written permission of the publisher, PWS-Kent Publishing Company, 20 Park Plaza, Boston, Massachusetts 02116.

PWS-Kent Publishing Company is a division of Wadsworth, Inc.

Library of Congress Cataloging in Publication Data
Poularikas, Alexander D., 1933–
   Signals and systems.
   Includes index.
   1. Signal processing.   I. Seely, Samuel, 1909–
II. Title.
TK5102.5.P658 1984   621.38′043   84-9457
ISBN 0-534-03402-0

ISBN 0-534-03402-0

Printed in the United States of America

    88 89—10 9 8 7 6 5

Sponsoring Editor: John E. Block
Signing Representative: Rich Giggey
Editorial Assistants: Suzi Shepherd and Gaby Bert
Manuscript Editor: Sandy Spiker
Permissions Editor: Mary Kay Hancharick
Text and Cover Design: Nancy Benedict
Text and Cover Illustration: ANCO/Boston
Typesetting: Syntax International Pte Ltd
Cover Printing: Lehigh Autoscreen
Printing and Binding: The Maple-Vail Book Manufacturing Group

Τό βιβλίο αυτό τό αφιερώνω
στήν μητέρα μου Αγγελική
A. D. P.

PREFACE

This book presents the basic concepts and applications of signals and systems analysis to the undergraduate engineering student. It is written for the student who has already studied ac circuits and mathematics through differential equations.

An understanding of the concepts of signals and systems is needed in such diverse fields as electronics, biomedical engineering, industrial processes, chemical process control, and electro-optical systems. In practice, an almost unlimited number and diversity of systems is found: when we turn the ignition key in a car, the input signal is the current to the starter relay and the output is a mechanical torque; when we are trying to cross a street, the input to our eyes is the image of the set of oncoming cars and the output is the brain stimulus signal to the leg muscles. Another system is a computer program that processes reflected radar signals from an airplane and produces signals indicating the location and flight path of the airplane. We will discuss the physical characteristics of electrical, mechanical, and optical systems. We will find that, although their input-output signals can be drastically different, these systems possess two features in common: the input signals to the systems are functions of one or more independent variables, and the systems produce output signals when excited by input signals.

By its nature, the modeling of signals and signal systems is inherently mathematical and involves mathematical forms not ordinarily included within mathematics programs in electrical engineering at the undergraduate level. Consequently, the text provides the mathematical background needed to carry out the intended studies, including convolution, Fourier transforms, discrete Fourier transforms, fast Fourier transforms, Hilbert transforms, orthonormal functions, difference equations, and Z-transforms, as well as a review of forms previously studied. In addition, general mathematical needs are provided for through a detailed appendix on Complex Variable Theory and an appendix on Matrix Theory.

It is our opinion that a solid background in continuous time signals and systems should be achieved before introducing the reader to discrete time systems. Therefore the first half of the book is devoted to continuous time signals and the second half to discrete time signals. The book does not cover all aspects of continuous time and discrete time problems in an exhaustive manner. However, it does include the material necessary for study of these areas and their application to practical problems. We have included simple examples to illustrate and explain the underlying theory. Sections that are mathematically involved are designated by a dagger (†); these can be omitted on first reading.

The book comprises twelve chapters plus two book appendices and four chapter appendices. Chapter 1 presents the mathematical description of continuous time and discrete time signals. In addition, it explains the notion of signal representation by orthogonal functions. Chapter 2 discusses convolution and the impulse response properties of systems, and presents the modeling of electrical, mechanical, and optical systems. Chapter 3 develops the classical field of the Fourier series, including a discussion of the Gibbs phenomenon and window functions. Chapter 4 presents the Fourier transform and its applications. It explores the Fourier spectra of spatial signals and the spectra of modulated signals, as well as such advanced topics as Hilbert transforms and the use of complex variable theory in evaluating Fourier integrals. Chapter 5 uses knowledge acquired in previous chapters to study the effect of ideal linear filters on the spectra of input signals. A variety of electrical, mechanical, and optical systems are studied. Optical systems are introduced because of their growing importance and also because of the analogy between electrical and optical signal processing. This allows students to verify their mathematical formulations with an optical system. Chapter 6 discusses the Laplace transform and its inverse, both by the use of a table and by the use of complex variable theory. Applications of Laplace transform theory to the determination of the response of continuous time systems are included. Chapter 7 introduces the principles of sampling and the effect of sampling rate and values on the reconstruction of signals from their sampled values. Chapter 8 presents the fundamentals of discrete time signals and systems and their block diagram and signal flow graph representations. It includes a discussion of classical procedures for solving difference equations. Chapter 9 contains a fairly detailed development of the Z-transform—its properties, its applications to discrete systems, and its use in solving difference equations. Chapter 10 discusses the features and uses of the discrete Fourier transform, the fast Fourier transform, and the Walsh-Hadamard transform. Chapter 11 is devoted to digital filter design and shows how to develop digital filters to approximate prescribed analog filter behavior. The concluding chapter develops methods for representing continuous time and discrete time systems in state equation form and also methods for the solution of such equations. Such properties as observability, controllability, and stability, which are of special interest in control theory, are also considered.

We have emphasized theoretical content in this text. We do not believe that highlighting the use of the digital computer within the text would serve well in developing the theoretical content. Therefore we leave it to the instructor to define the use of the computer with this text. The instructor can decide when computers may be used to solve problem exercises and whether students should use available software or prepare their own computer programs. We encourage the use of the computer in studying this text, but recommend restraint since an understanding of the theoretical content is most important.

Complete study of the contents of this book would require two courses. Instructors can be selective in their choice of material and tailor the courses to their objectives and the background of their students. A reasonable approach is to divide the material into a one-term course in continuous time systems (which would include selected portions of Chapter 1, Chapters 2–6, and Chapter 8) and a second course in discrete time systems (which would include Chapter 1 for the necessary background and Chapters 7, 9, 10, and 12). A solutions manual accompanies this text.

We believe that this book is an ideal vehicle for the practicing engineer to get a comprehensive and unified overview of analog as well as discrete systems. It also provides an introduction to optical systems, which is useful in light of recent developments in signal processing. Please note that throughout the text the symbol $\doteq$ means approximately equal to and the symbol $\triangleq$ means equal by definition.

We are grateful to the manuscript reviewers for their valuable comments: Buck F. Brown, Rose-Hulman Institute of Technology; and J. W. Howze, Texas A & M University. We extend our special thanks to Doran J. Baker, Utah State University, for his valuable assistance during the review and proofreading processes. We also wish to acknowledge the following publishers and publications for their permission to reproduce figures: Academic Press, Addison-Wesley, *American Journal of Physiology*, *Applied Physics*, Dover, W. H. Freeman, *IEE International Conference on Acoustics, Speech, and Signal Processing*, *IEEE Proceedings*, *IEEE Transactions on Information Theory*, Matrix, Mc-Graw-Hill, *Nature*, Norton, *Optics Letters*, Pergamon Press, Prentice-Hall, *Science*, Wiley.

CONTENTS

CHAPTER 1
SIGNALS AND THEIR FUNCTIONAL REPRESENTATIONS 1

- 1-1 Introduction 1
- 1-2 Periodic Signals 3
- 1-3 Nonperiodic Signals 7
- 1-4 Signal Conditioning 15
- 1-5 Representation of Signals by Orthogonal Functions 19
- 1-6 Orthonormal Functions 27
- 1-7 Representation of Signals by Interpolation 37
- References 41
- Problems 42

CHAPTER 2
CONVOLUTION, IMPULSE RESPONSE, AND SYSTEM REPRESENTATION 46

A. Convolution and Its Properties 46
- 2-1 Introduction 46
- 2-2 Superposition Integral 48
- 2-3 Convolution Integral 52
- 2-4 Properties of Convolution 56
- 2-5 Periodic Convolution 63
- 2-6 Correlation 64

B. Modeling Simple Electrical, Mechanical, and Optical Systems 66
- 2-7 Modeling Simple Systems 66

C. Impulse Response 76
- 2-8 The Impulse Response of Systems 76

D. System Description and Block Diagram Representation 87
- 2-9 Operators of Systems 87
- 2-10 System Classification 90

2-11 Interconnected Linear Systems   101
2-12 State-Space Representation of Systems   104
References   116
Problems   116

CHAPTER 3
FOURIER SERIES   128

3-1 Introduction   128
3-2 Fourier Series and Periodic Functions   130
3-3 Features of Periodic Functions   134
3-4 Choice of Origin   145
3-5 Systems with Periodic Inputs   148
3-6 Gibbs Phenomenon   152
References   155
Problems   156

CHAPTER 4
SPECTRA OF TEMPORAL AND SPATIAL SIGNALS   163

A. The Fourier Transform—Temporal Signals   163
    4-1 The Fourier Transform   163
    4-2 Properties of Fourier Transforms   170
    4-3 Some Special Fourier Transform Pairs   190
    4-4 Gibbs Phenomenon   195

†B. Spectra of Spatial Signals   203
    4-5 Fourier Transforms and Optical Systems   203

†C. Transforms in the Complex Domain—The Hilbert Transform   207
    4-6 Fourier Transforms and Complex Function Theory   207
    4-7 Hilbert Transforms   212

D. Spectra of Modulated Signals   219
    4-8 Amplitude Modulated Signals   219
    4-9 Frequency Modulation   223
    4-10 Modulation of Light   231
References   234
Problems   235

CHAPTER 5
THE RESPONSE AND APPLICATIONS OF LINEAR FILTERS   242

A. Electrical and Mechanical Systems   242
    5-1 Linear Time-Invariant Systems (Filters)   242
    5-2 Filters with Varying Amplitude and Linear Phase Shift   244
    5-3 Filters with Constant Amplitude and Varying Phase   255
    5-4 Symmetrical Bandpass Filters with Linear Phase   259

†B. Optical Systems   261
    5-5 Optical System Functions   261

CONTENTS

5-6 Processing of Signals with Coherent Optics—Optical Filters 273
5-7 Incoherent Optical Processing 282
References 284
Problems 284

CHAPTER 6
THE LAPLACE TRANSFORM 290

6-1 Introduction 290
6-2 The Bilateral Laplace Transform 292
6-3 The One-Sided Laplace Transform 293
6-4 Properties of the Laplace Transform 296
6-5 Systems Analysis—Transfer Function of LTI Systems in Block Diagram Representation 306
6-6 System Analysis—Signal Flow Graph (SFG) Representation of LTI Systems 316
6-7 The Inverse Laplace Transform 328
6-8 Initial Conditions 334
6-9 Problem Solving by Laplace Transforms 336
6-10 Stability of LTI Systems 344
†6-11 The Inversion Integral 350
6-12 Complex Integration and the Bilateral Laplace Transform 356
References 358
Problems 359

CHAPTER 7
SAMPLING OF SIGNALS 366

7-1 Introduction 366
7-2 The Sampling Theorem 370
7-3 Telephone Transmission 383
7-4 Additional Pulse Modulation Techniques 385
References 388
Problems 389

CHAPTER 8
DISCRETE SIGNALS, DIFFERENCE EQUATIONS, AND SYSTEM DESCRIPTION 392

8-1 Introduction 392
8-2 Linear Difference Equations 393
8-3 First-Order Linear Difference Equations with Constant Coefficients 395
8-4 Delay Operations and Signals 406
8-5 Higher-Order Linear Difference Equations with Constant Coefficients 410
8-6 Frequency Response of Discrete Time Systems 423
References 430
Problems 430

CHAPTER 9
THE Z-TRANSFORM   438

9-1 The Z-Transform   438
9-2 Convergence of the Z-Transform   440
9-3 Properties of the Z-Transform   449
9-4 Z-Transform Pairs   461
9-5 The Inverse Z-Transform   465
9-6 The Transfer Function   470
9-7 Transfer Function and Pole Locations   477
9-8 Frequency Response of Systems   480
9-9 Solution of Difference Equations   485
†9-10 The Two-Sided Z-Transform—Region of Convergence   489
†9-11 Relationship of the Z-Transform to the Fourier and Laplace Transforms   497
†9-12 The Z-Transform, Taylor Series, and Fourier Series   498
†9-13 Stability of Time-Invariant Discrete Systems   499
References   501
Problems   502

CHAPTER 10
DISCRETE TRANSFORMS   510

10-1 Introduction   510
10-2 Properties of the DFT   512
10-3 The Fast Fourier Transform—Decimation in Time   531
10-4 The FFT-Matrix Approach   534
10-5 The Base 2 FFT Algorithm   537
10-6 The Sande-Tukey FFT   540
10-7 FFT and Convolution   542
10-8 Walsh Transforms   546
10-9 Hadamard Transforms   550
10-10 Fast Hadamard and Walsh Transformations   552
References   555
Problems   555

CHAPTER 11
ELEMENTS OF DIGITAL FILTER DESIGN   560

11-1 Introduction   560
11-2 The Butterworth Filter   562
11-3 The Chebyshev Low-Pass Filter   568
11-4 Elliptic Filters   576
11-5 Phase Characteristics   577
11-6 Low-Pass to High-Pass Transformation   577
11-7 Low-Pass to Bandpass Transformation   578
11-8 Digital Filters   579

CONTENTS

 11-9 The Invariant Impulse Response Method   579
 11-10 The Bilinear Transformation   588
 †11-11 Prewarping   595
 11-12 Finite Impulse Response (FIR) Filters   599
 11-13 Use of Window Functions for FIR Filters   602
 11-14 The DFT as a Filter   606
 11-15 Frequency Transformation of IIR Filters   609
 11-16 Recursive versus Nonrecursive Designs: General Remarks   611
 References   611
 Problems   612

CHAPTER 12
STATE VARIABLES AND STATE EQUATIONS   615

A. Continuous-Time (Analog) Systems   615
 12-1 Introduction   615
 12-2 State Equations for Linear Systems   616
 12-3 Differential Equations in Normal Form   620
 12-4 State Variable Transformations   627
 12-5 Solution of the Continuous-Time State Equations: Force-Free Equations   628
 12-6 Properties of the Fundamental Matrix   634
 12-7 Complete Solution of the Continuous-Time State Equations   636
 12-8 Laplace Transform Solution to the State Equations   640
 12-9 State Response to Periodic Inputs   644
 12-10 Initial State Vectors and Initial Conditions   647

B. Discrete Time Systems   648
 12-11 Representation of Discrete Time Systems   648
 12-12 Solution to Discrete Time State Equations   658
 12-13 Continuous-Time Systems with Sampled Inputs   665
 12-14 Z-Transform Solution to Discrete Time State Equations   666

†C. Additional Topics   669
 12-15 Controllability and Observability of Linear Systems   666
 12-16 Controllability and Observability Using **A**, **B**, and **C** Matrices   672
 12-17 Stability of Systems   673
 12-18 Stability in the Sense of Liapunov   676
 12-19 Generating Liapunov Functions   678
 References   680
 Problems   681

CHAPTER APPENDIX 1-1
THE DELTA FUNCTION AND ITS PROPERTIES   689

 A1-1 Delta Function as a Limit of Special Sequences   689
 A1-2 Properties of Delta Functions   690
 Problems   694

CHAPTER APPENDIX 1-2
BINARY AND GRAY CODE RELATIONSHIPS   695

CHAPTER APPENDIX 4-1
BESSEL FUNCTIONS   697

         A4-1   Bessel Functions   697

CHAPTER APPENDIX 7-1
THE ERROR FUNCTION   698

         References   698

APPENDIX 1
FUNCTIONS OF A COMPLEX VARIABLE   703

         A1-1   Basic Concepts   703
         A1-2   Sequences and Series   711
         A1-3   Power Series   713
         A1-4   Analytic Continuation   718
         A1-5   Singularities of a Complex Function   719
         A1-6   Theory of Residues   720
         A1-7   Aids to Integration   724
         A1-8   Evaluation of Definite Integrals   725
         A1-9   Principal Value of an Integral   728
         A1-10   Branch Points and Branch Cuts   730
         A1-11   Integral of Logarithmic Derivative   733
         References   736
         Problems   736

APPENDIX 2
MATRICES   738

         A2-1   Introduction   738
         A2-2   Definitions   739
         A2-3   Matrix Algebra   740
         A2-4   Functions of a Matrix   743
         A2-5   Cayley-Hamilton Theorem   746
         References   752

INDEX   753

CHAPTER 1
SIGNALS AND THEIR FUNCTIONAL REPRESENTATIONS

1-1 INTRODUCTION

A broad range of signals are of practical importance in describing human experiences. Signals which are transmitted from one point to another provide the basis for us to see, hear, feel, and act. In engineering systems the signals may be information carrying or they may be energy carrying. For example, signals may be the high-energy microwave pulses in radar or the high energy necessary for performing controlled machine tool operations. They may be telephone or radio signals or the pulses that dictate the operation of a digital computer. The signals with which we are concerned may be the cause of an event or the consequence of an action. The characteristics of any signal may be one of a broad range of shapes, amplitudes, time durations, and perhaps other physical properties. In many cases the signal will be expressed in analytic form; in other cases the signal may be given only in graphical form (see Figure 1.1c as an example), it may be indications on instruments that are read periodically, or it may be data provided in chart or tabular form.

The electrical engineer who is concerned with signal detection and signal design will be concerned with signals of many different amplitudes and shapes and with time durations that can range over widely differing time spans. When concerned with a computer, the engineer must deal with millions of pulses per second; when concerned with radar signals, which are originally transmitted in megawatt-strength pulses, the reflected signal pulse trains are often hidden in noise, can change almost continuously, and recur at rates of several thousand per second. A cardiologist, on the other hand, is interested in signal shape and recurrence rate, since he bases his judgement of a patient's heart on the cardiogram (the heart signal).

Signals in electrical engineering are often encountered as variations of current and voltage versus time. Often many of these time-varying voltage or current signals have been transformed from other types of signals—such as pressure versus time, electrical conductivity versus depth in a well, light intensity versus the sun's position above the horizon, and acceleration of machine parts. Appropriate forms of transducers play important roles in such transformations. Additionally, some signals are continuous functions of time while others are discrete functions of time. Continuous-time signals are a familiar form to the electrical engineer, while discrete time signals are somewhat less common. Discrete time signals are found in many fields of endeavor; for example, the daily chart that gives the temperature, blood pressure, and pulse rate of a patient in a hospital is listed as discrete time information. Likewise the geographic distribution of towns, the crime distribution in a city, the population distribution of animal species in a geographic area, and so on are discrete time signals.

The most common discrete signals the engineer will use are the result of discretizing continuous signals. Accomplishing such discretization is often effected with an electronic instrument known as an analog-to-digital (A/D) converter. When continuous mathematical functions are to be manipulated in a digital computer, we must discretize the functions; that is, the words or data to be entered into the computer must be written in a discrete form deduced from the continuous function. We shall present a theorem in Chapter 7 that provides a guide in finding the appropriate number of discrete samples needed to uniquely represent a continuous function in its discrete form.

Mathematically, signals are presented as functions of one or more independent variables. For example, heat loss from a surface can be represented mathematically by temperature of the surface versus time; a barometric chart can be presented as a function of two spatial variables—height and location. In our present studies, we will consider functions of one independent variable, usually time. It is necessary to understand the important classes of signals and the different mathematical means for their description. A given signal waveform can be described in more than one way; the choice depends upon the signal and the required form in which the analysis is to be undertaken. Suppose, for example, that a recurring square wave is the signal under consideration. An immediate mathematical circumstance is that no exact continuous-function representation exists for this signal over the entire period; such signals are described over a period by piecewise continuous functions. To obtain an approximate functional representation of this signal, we can proceed in several different ways. One procedure is to replace this piecewise continuous function by a sequence of step functions, with amplitudes and displacements chosen appropriate to the function being represented. Another method is to express the function over a designated range by a polynomial approximation obtained by using interpolation formulas. A third method is to express the square wave by a series representation in terms of orthogonal functions. The most common series representation is in terms of trigonometric sine and cosine functions—a Fourier series expansion. Other orthogonal sets may also be used, including Legendre polynomials, Walsh functions (orthogonal, but also nonanalytic*), and Bessel functions. These methods are discussed in the material to follow.

To find the response of a system to an impulse signal, we must use a delta function as the input signal. However, the delta function is a nonanalytic function and requires that the mathematics of continuous functions plus special properties of the delta function be used in such studies.

---

* A function is analytic in a region if it and all its derivatives exist in the region.

SIGNALS AND THEIR FUNCTIONAL REPRESENTATIONS

The presentation and the mathematical description of classes of signals in this chapter will be of importance in our subsequent studies. In the interest of completeness, certain functions of a general class are included although they are not fully explored in this book.

■ ■ ■

1-2 PERIODIC SIGNALS

1-2.a Continuous Signals Any function whose mathematical representation obeys the condition

$$f(t) = f(t + nT) \qquad n = 1, 2, 3, \ldots \tag{1.1}$$

where T is a constant known as the period, is classed as a **periodic** signal. A number of typical periodic signals are shown in Figure 1.1.

Figure 1.1
A number of different types of periodic signals;
(a) sine wave,
(b) rectangular wave,
(c) electrocardiogram (for different electrodes).

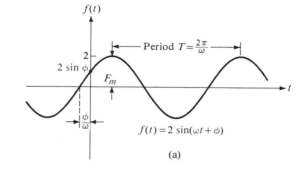

(a)

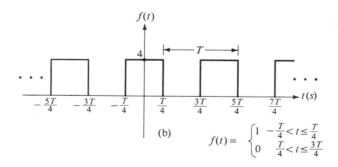

(b)

$$f(t) = \begin{cases} 1 & -\frac{T}{4} < t \leq \frac{T}{4} \\ 0 & \frac{T}{4} < t \leq \frac{3T}{4} \end{cases}$$

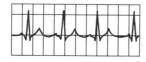 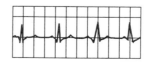

(c)

CHAPTER 1

The real-valued sinusoidal signal $f(t)$ is expressed mathematically by the relation

$$f(t) = F_m \sin(2\pi f_0 t + \varphi) = F_m \sin(\omega_0 t + \varphi) \tag{1.2}$$

where F_m = the peak value of the wave, $f_0 = 1/T$ = frequency (Hz), $T = 1/f_0$ = period (s), $\omega_0 = 2\pi f_0$ = angular frequency (rad/s) and φ = phase shift (rad) relative to a specified time base. This sinusoidal signal can be represented in an equivalent complex form

$$\begin{aligned} f(t) &= \operatorname{Im}\bigl[F_m e^{j(2\pi f_0 t + \varphi)}\bigr] = \operatorname{Im}\bigl[F_m e^{j(\omega_0 t + \varphi)}\bigr] \\ &= \operatorname{Im}\bigl[F_m \cos(\omega_0 t + \varphi) + jF_m \sin(\omega_0 t + \varphi)\bigr] \\ &= \frac{F_m}{2j}\bigl[e^{j(2\pi f_0 t + \varphi)} - e^{-j(2\pi f_0 t + \varphi)}\bigr] \end{aligned} \tag{1.3}$$

where Im indicates that the imaginary part of the Euler expansion of the complex quantity in brackets is the entity of interest.

Sinusoidal signals are key tools enabling us to study the behavior of many systems. For example, if we are interested in determining how well our hi-fi amplifier will reproduce varying frequency inputs, we proceed experimentally as follows: we use as inputs sine voltages of constant amplitude but of different frequencies and plot the ratio (V_0 = output)/(V_i = input) versus frequency, usually on a logarithmic (deciBel = dB) scale. One possible curve is that shown in Figure 1.2. Since our ears do not respond well to frequencies above about 13 kHz, we conclude that our amplifier is of acceptable quality since it amplifies all the frequencies in the audio band equally well. We can perform similar tests on many systems since, as we shall find, any function (periodic or nonperiodic) can be decomposed into the sum of an infinite number of sinusoidal functions. Any linear system affects not only the amplitude of the sinusoidal signals, but it can introduce frequency-dependent phase shifts, which also must be taken into consideration. Matters regarding the response of systems to input signals will receive detailed consideration throughout this book.

As already noted, signals that are periodic but nonsinusoidal and perhaps nonanalytic can be described by means of a Fourier series that consists of the sum of an infinite number of sinusoidal functions. Mathematically, the Fourier series expansion of such functions will be of the form

$$\begin{aligned} f(t) &= A_0 + \sum_{k=1}^{\infty} (A_k \cos k\omega_0 t + B_k \sin k\omega_0 t) \\ &= A_0 + \sum_{k=1}^{\infty} C_k \cos(k\omega_0 t + \varphi_k) \end{aligned} \tag{1.4}$$

Although the number of frequencies necessary to synthesize the original signal is infinite, they do not constitute a continuum; they are all integral multiples of the **fundamental frequency** $\omega_0/2\pi = 1/T = f_0$. The set of coefficients C_k con-

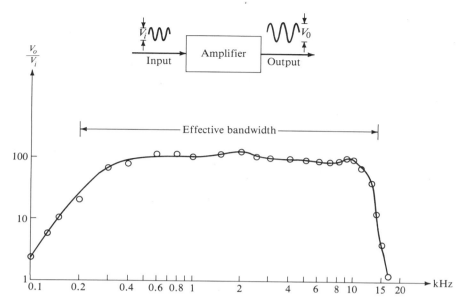

Figure 1.2
Frequency characteristics of an audio amplifier.

stitutes the **amplitude spectrum** of the function $f(t)$ and the phase factors φ_k constitute the **phase spectrum.** Details of the Fourier series expansion and its applications will be considered in Chapter 3.

1-2.b Discrete Time Signals Converting the representation of continuous signals into an equivalent discrete form will be studied in considerable detail in Chapter 7. Here we will examine certain features of typical discrete signals. Refer to Figure 1.3, which shows several discrete signals. In these figures t_k ($k = 1, 2, 3, \ldots$) specifies discrete times and $f(t_k)$ specifies the value of the function $f(t)$ at the particular time t_k. Observe that the function $f(t)$ as shown is defined only at the discrete times t_k. For convenience, we will write $f(k)$ instead of $f(t_k)$ as the value of the function at the time t_k.

The definition of a discrete periodic signal is

$$f(k) = f(k + nT) \qquad n = 1, 2, 3, \ldots \tag{1.5}$$

where T is the period of the function. A cosinusoidal discrete function

$$f(k) = \cos k(2\pi/N)n = \cos k\omega_s \quad \text{(a)} \tag{1.6}$$

can also be written in the form

$$f(k) = \mathrm{Re}\{e^{jk(2\pi/N)n}\} = \mathrm{Re}\{e^{jk\omega_s}\} \quad \text{(b)}$$

where k, n, and N are integers. Since for each n and N the function $f(k)$ is pe-

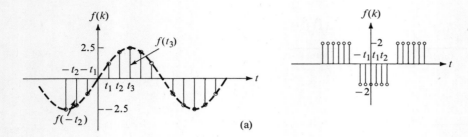

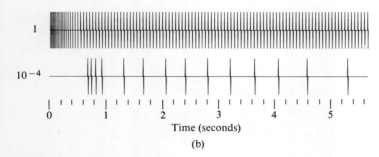

Figure 1.3
(a) Typical discrete periodic signals. (b) Firings of Limulus optical nerve at two different light intensity levels. (From Miller, Ratliff, and Hartline, "How Cells Receive Stimuli," in *Perception: Mechanisms and Models*, 133, W. H. Freeman & Co./Scientific American. Reprinted by permission.)

riodic, we must have

$$\exp\left(jk\frac{2\pi}{N}n\right) = \exp\left[j(k+N_0)\frac{2\pi}{N}n\right]$$

$$= [\exp(jk\omega_s)]\exp\left[j\frac{2\pi}{N}n\frac{N}{\gcd(n,N)}\right] = \exp[jk\omega_s]$$

where N_0 is the fundamental period and is equal to N divided by the **greatest common divisor** (gcd) of n and N; that is,

$$N_0 = \text{fundamental period} = \frac{N}{\gcd(n,N)} \tag{1.7}$$

We can conclude from this development that the function $\exp(jk\omega_s)$ is the same at frequencies ω_s that are multiples of 2π. Thus we usually need only consider the intervals $0 \leq \omega_s \leq 2\pi$ or $-\pi \leq \omega_s \leq \pi$ in which to choose ω_s. Furthermore, if no gcd exists, the discrete exponential is not periodic.

■ ■ ■

SIGNALS AND THEIR FUNCTIONAL REPRESENTATIONS

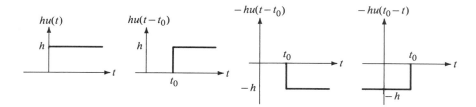

Figure 1.4
The step function and some of its shifted and reflected forms.

1-3 NONPERIODIC SIGNALS

1-3.a Continuous Signals This section introduces a number of special functions for describing particular classes of signals. Many of these have features that make them particularly useful directly in the solution of engineering problems or indirectly in the description of other functions. At this point, we do little more than present a catalog of these signals and their mathematical descriptions.

Step Function The unit step function $u(t)$ is an important signal for analytic studies and it also has many practical applications. Note that while the step function is continuous after its application, there is a discontinuity at the instant of application; therefore this is not a regular function mathematically. When we turn the key in the ignition of our car, we actually introduce a step voltage function (the battery voltage) to the starting motor. Likewise any constant force when applied at a particular time to a body is also described by a step function.

The unit step function is defined by the relation

$$u(t) = \begin{cases} 1 & t > 0 \\ 0 & t < 0 \end{cases} \quad \text{or} \quad u(t - t_0) = \begin{cases} 1 & t - t_0 > 0 \\ 0 & t - t_0 < 0 \end{cases} \quad (1.8)$$

From this definition, the value of the function is unity for positive arguments and zero for negative arguments. The function is not defined at $t = 0$; hence $u(0)$ is not specified. The step function of height h and some of its shifted forms are shown in Figure 1.4.

Ramp Function The ramp function shown in Figure 1.5 is defined by

$$r(at) = ar(t) = \begin{cases} at & t \geq 0 \\ 0 & t < 0 \end{cases} \quad (1.9)$$

As a practical matter, the ramp function is obtained by applying the unit step function to an integrator, since

$$\int_0^t u(t)\,dt = \int_0^t dt = t$$

Figure 1.5
(a) The ramp function.
(b) Block diagram representation.

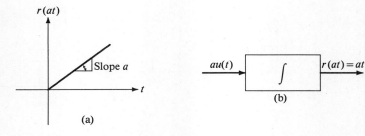

Figure 1.6
The rectangular pulse function.

Figure 1.7
The triangular pulse function.

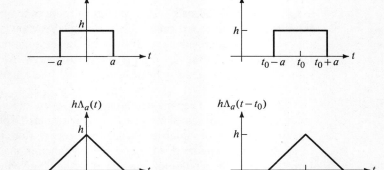

Ramp functions play an important role as the linear sweep waveform of a cathode ray tube.

Rectangular Pulse Function This is another important function with considerable practical utility. The pulse function is the result of an on-off switching operation of a constant voltage source to an electric circuit. For example, it is used to modulate electrical energy in radar transmitters and it is the waveshape of the recurring timing signals (clock pulses) that control the total operation of any digital computer. It is also useful in approximating continuous functions, as we discuss below.

The rectangular pulse functions of height h and its shifted form are shown in Figure 1.6. The mathematical representation is developed from the appropriate choice of step functions. The unit rectangular pulse functions are specified by:

$$p_a(t) = [u(t + a) - u(t - a)] = \begin{cases} 0 & |t| > a \\ 1 & |t| < a \end{cases}$$

$$p_a(t - t_0) = [u(t - t_0 + a) - u(t - t_0 - a)] = \begin{cases} 0 & |t - t_0| > a \\ 1 & |t - t_0| < a \end{cases}$$

(1.10)

Figure 1.8
The sgn function.

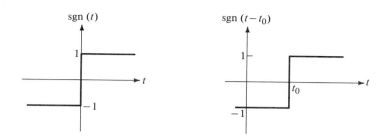

Triangular Pulse Function The unit triangular function is defined by the expression

$$\Lambda_a(t) = \begin{cases} 1 - \dfrac{|t|}{a} & |t| < a \\ 0 & |t| > a \end{cases} \qquad (1.11)$$

This function is illustrated in Figure 1.7 with amplitude h.

Signum Function The signum function (written sgn) is shown in Figure 1.8. The unit sgn function is defined by

$$\text{sgn}(t) = \begin{cases} 1 & t > 0 \\ 0 & t = 0 \\ -1 & t < 0 \end{cases} \qquad (1.12)$$

This function can be expressed in terms of the step function

$$\text{sgn}(t) = -1 + 2u(t)$$

Observe also that for $-1 < t < 1$ sgn(t) is the negative derivative of the unit slope triangle pulse function. The sgn function is often used in communications and in control theory.

Sinc Function The sinc function is defined by the expression

$$\text{sinc}_a(t) = \frac{\sin at}{t} \qquad -\infty < t < \infty \qquad (1.13)$$

This function and its shifted form are shown in Figure 1.9. This function plays an important role in the reconstruction of special band-limited signals. It also appears in many physical optics problems. More will be said about these signals in Chapters 4 and 7.

Figure 1.9
The sinc function.

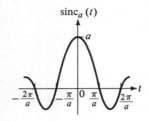
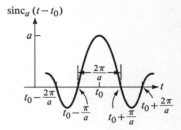

Figure 1.10
The Gaussian function.

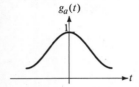
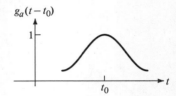

Gaussian Function The Gaussian function is defined by the expression

$$g_a(t) = e^{-at^2} \qquad -\infty < t < \infty \tag{1.14}$$

This function is extensively used in probability theory. It is also used to define the modulating pulses in radars. The Gaussian function is shown in Figure 1.10.

1-3.b Discrete Signals

Delta Function The **delta** function $\delta(t)$, often called the **impulse** or **Dirac delta** function, occupies a central place in signal analysis. Many physical phenomena such as point sources, point charges, concentrated loads on structures, and voltage or current sources acting for very short times can be modeled as delta functions. The delta function $\delta(t)$ has very peculiar properties: it exists only at time t and is of infinite height. It is characterized as follows:

$$\begin{aligned} \delta(t - t_0) &= 0 \qquad t \neq t_0 & \text{(a)} \\ \int_{t_1}^{t_2} f(t)\delta(t - t_0)\,dt &= f(t_0) \qquad t_1 < t_0 < t_2 & \text{(b)} \end{aligned} \tag{1.15}$$

By (1.15) the delta function exists only at the point $t = t_0$ (the argument is zero). Further, these equations show that the impulse function can be used to extract from a function $f(t)$ only the specific value that exists at time t_0, where $f(t)$ is a continuous function at t_0. Several delta functions are shown in Figure 1.11; Figure 1.11b illustrates graphically (1.15b). Observe that the definition of the

Figure 1.11
(a) The representation of several delta functions.
(b) Illustrating the effect of multiplying a well-behaved function and a delta function.

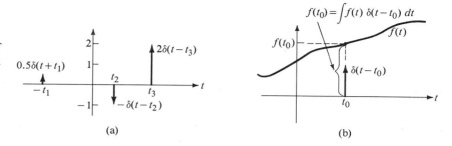

Figure 1.12
The voltage on a capacitor due to an impulse current source.

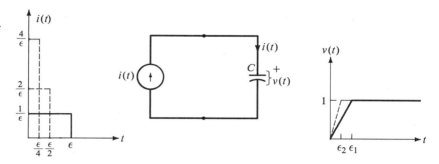

delta function was not presented in functional form; it was presented only by its behavior under the integral sign. Despite the fact that the delta function is not a regular analytic function, a rigorous mathematical definition for this type of function was given by Laurent Schwartz in his work on the **theory of generalized functions**.

If the function $f(t)$ contained in (1.15) is a constant equal to unity, then

$$\int_{t_1}^{t_2} \delta(t - t_0) \, dt = 1 \qquad t_1 < t_0 < t_2 \tag{1.16}$$

This integral shows that the area under the delta function is unity. The height of the spikes shown in Figure 1.11 denotes the area of the delta function.

To see how the delta function can be created, let a current of very short duration ϵ and of amplitude $1/\epsilon$ be applied to the circuit shown in Figure 1.12. From the well known expression

$$v(t) = \frac{1}{C} \int_0^t i(t) \, dt$$

we obtain, for $C = 1F$

$$v(t) = \int_0^\epsilon \frac{1}{\epsilon} \, dt = \int_0^{\epsilon/2} \frac{2}{\epsilon} \, dt = \int_0^{\epsilon/4} \frac{4}{\epsilon} \, dt = 1$$

Figure 1.13
Approximation of a signal by pulse functions.

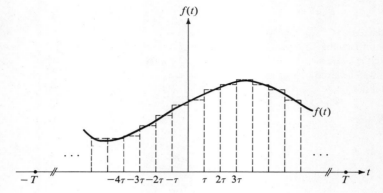

In this example the area under the function will always be equal to unity for any $\epsilon \to 0$, and it will be zero everywhere except at $t = 0$, where it is undefined.

Observe further that in this particular example, which involves the integral of the delta function in the limit as $\epsilon \to 0$, the current function will produce a step voltage of 1 V in the circuit. This is a general result; the integral of the delta function is a step function.

Despite its peculiar properties, the importance of the delta function will become apparent when we study the response of linear systems (those with linear elements) to an arbitrary input. If we know the output of a linear and time-invariant system (a system composed of elements that do not vary with time and are linear) to a delta function input, we can always find the output of the system when the input is any well-behaved function—that is, a function that has a finite number of discontinuities of finite amplitudes on a finite time interval and has finite derivatives at both sides of each discontinuity.

The definition given in (1.15) is less than satisfying, but it does indicate the unusual behavior of this particular function. A more satisfying definition is possible by considering a function that behaves like a delta function upon taking some appropriate limit. Details of such a development are contained in Chapter Appendix 1-1 at the end of the book.

An interesting feature of the delta function is that it can be used to represent any continuous function as an integration of the function itself and the delta function. To explore this matter, refer to Figure 1.13, which shows a function extending from $-T$ to T and divided into segments of length τ. Each segment constitutes a pulse, with the pulse whose center is at $t = k\tau + \tau/2$ written as

$$f(k\tau)[u(t - k\tau) - u(t - k\tau - \tau)]$$

The approximation to $f(t)$ becomes

$$f_a(t) = \sum_{k=-n}^{n} f(k\tau)[u(t - k\tau) - u(t - k\tau - \tau)]$$

$$= \sum_{k=-n}^{n} f(k\tau) \frac{[u(t - k\tau) - u(t - k\tau - \tau)]}{\tau} \tau$$

SIGNALS AND THEIR FUNCTIONAL REPRESENTATIONS

Note that as $\tau \to 0$, the magnitude of k will increase in order that the product $2n\tau$ remain constant and equal to the domain $2T$ of the function. Clearly, the product $k\tau$ will take all possible values in the domain $-T \leq t \leq T$ and can thus be considered as a continuous variable ζ. The above approximate expression $f_a(t)$ thus becomes

$$f(t) = \int_{-T}^{T} f(\zeta)\delta(t - \zeta)\, d\zeta \tag{1.17}$$

We can let $T \to \infty$ with the following relationship

$$f(t) = \int_{-\infty}^{\infty} f(\zeta)\delta(t - \zeta)\, d\zeta \tag{1.18}$$

If we compare (1.18) with (1.15b), we see that they are similar. Notice, however, that the delta function in this expression has been "folded over" since the argument of δ is $(t - \zeta)$ instead of $(\zeta - t)$. In this form the integral is known as the **convolution integral.** It shows that any function can be represented as the convolution of itself with the delta function. A more detailed discussion on convolution integrals will be given in Chapter 2.

EXAMPLE 1.1

Find the velocity of a free body with mass M when an impulse function force is applied to it at $t = 0$.

Solution: Use Newton's force law $f = Ma = M\, dv/dt$ to write

$$v = \frac{1}{M}\int_{0-}^{t} f(t)\, dt = \frac{1}{M}\int_{0-}^{0+} \delta(t)\, dt = \frac{1}{M} \qquad t > 0$$

because $\delta(t) = 0$ at any point $t \neq 0$. This result indicates that the velocity is a step function; it is zero until time $t = 0$ and then acquires the value $1/M$ (m/s) for times $t > 0$. We conclude also that the derivative of the step function $u(t)$ is equal to the delta function $\delta(t)$. Thus we may write, in general,

$$\frac{du(t - t_0)}{dt} = \delta(t - t_0) \qquad \blacksquare$$

EXAMPLE 1.2

Find the derivative of the function $f(t) = 2p_2(t - 2)$.

Solution: The function $f(t)$, which is shown in Figure 1.14, can be written in the form

$$f(t) = 2[u(t) - u(t - 4)]$$

Figure 1.14
A pulse function and its derivative.

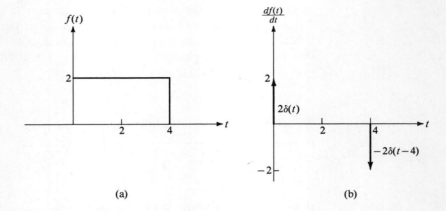

Figure 1.15
The comb function.

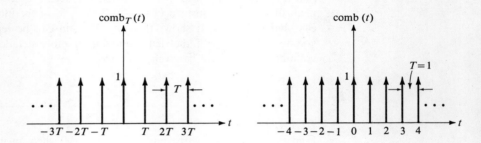

The derivative is given by

$$\frac{df(t)}{dt} = 2\delta(t) - 2\delta(t-4)$$

This result is shown in Figure 1.14b. ∎

Comb Function The comb function which is illustrated in Figure 1.15 is an array of delta functions that are spaced T units apart and extend from $-\infty$ to ∞. The mathematical description of the comb function is

$$\text{comb}_T(t) = \sum_{n=-\infty}^{\infty} \delta(t - nT) \qquad n = 0, \pm 1, \pm 2, \ldots \tag{1.19}$$

Arbitrary Sampled Function The comb function can be used in the representation of any continuous function in its sampled or discrete form. This follows from the properties of the delta function, which permits writing

$$f_s(t) = f(t)\,\text{comb}(t) = \sum_{k=-\infty}^{\infty} f(k)\,\delta(t-k) \quad \text{(a)} \tag{1.20}$$

Figure 1.16 Representation of the discrete form of an arbitrary function and operational form representation.

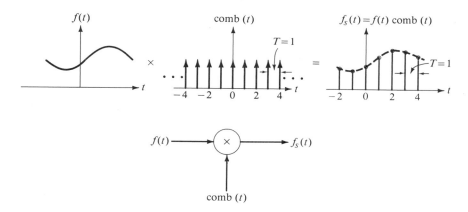

where $f_s(t)$ is the sampled version of the function $f(t)$. Figure 1.16 shows graphically the functions $f(t)$, comb(t), and $f_s(t)$. When the sampling period is T units, (1.20) becomes

$$f_s(t) = f(t)\,\text{comb}_T(t) = \sum_{k=-\infty}^{\infty} f(kT)\,\delta(t-kT) \quad \text{(b)}$$

Basic Discrete Functions A summary of the discrete signals that are of interest in engineering analysis includes:

$$\text{Unit impulse} \quad \delta(k) = \begin{cases} 1 & \text{for } k = 0 \\ 0 & \text{for } k = \pm 1, \pm 2, \dots \end{cases} \quad \text{(a)}$$

$$\text{Unit step} \quad u(k) = \begin{cases} 1 & \text{for } k = 0, 1, 2, \dots \\ 0 & \text{for } k = \text{negative} \end{cases} \quad \text{(b)}$$

$$\text{Unit alternating sequence} \quad u_\pm(k) = \begin{cases} (-1)^k & \text{for } k = 0, 1, 2, \dots \\ 0 & \text{for } k = \text{negative} \end{cases} \quad \text{(c)} \quad (1.21)$$

$$\text{Unit ramp} \quad r(k) = \begin{cases} k & \text{for } k = 0, 1, 2, \dots \\ 0 & \text{for } k = \text{negative} \end{cases} \quad \text{(d)}$$

These functions are illustrated in Figure 1.17.

■ ■ ■

1-4 SIGNAL CONDITIONING

In the area of communications it is usually desired to transmit slowly varying signals, such as voice and low-frequency audio information, over long distances without wires. This would call for a radio transmission channel. If we wish to directly transmit these frequencies, which would be in a range from about 50 Hz to 10 kHz, we would need an antenna system appropriate to signals with wave-

Figure 1.17
Useful elementary discrete functions:
(a) Unit pulse $\delta(k)$.
(b) Unit step $u(k)$.
(c) Unit alternating sequence $u_{\pm}(k)$.
(d) Unit ramp $r(k)$.

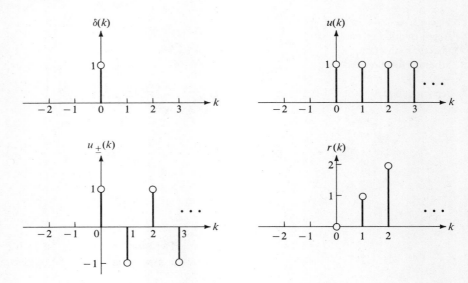

Figure 1.18
Illustrating the usefulness of modulation (multiplication) of signals.

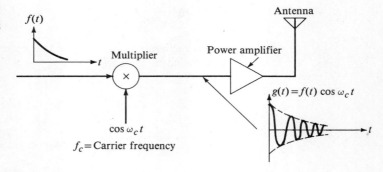

lengths in excess of 3×10^4 m. It is impractical to build an efficient antenna system approximately 10 km in length. To circumvent this problem, one resorts to the radio scheme known as **amplitude modulation,** the elements of which are shown in Figure 1.18. This approach dictates that the signals $f(t)$ must be multiplied with the constant frequency of a high frequency oscillator. What is achieved in this method is that the signal $f(t)$—for example, a decaying exponential signal—will appear at the antenna as a high frequency signal (dictated by the radio frequency oscillator) the amplitude of which is the decaying exponential variation of the original signal. Now if the AM radio station is in the 500–1100 kHz range, the antenna length would be roughly 75 m for a radiated frequency of 1000 kHz. At the receiving point a circuit (a demodulator) would be needed to extract the original signal. The most simple form would be an RC circuit that is acting as an envelope detector. Actually, the circuits are more complicated than those discussed here for both the modulation and the demodulation, but the essential concepts are those that have been discussed.

Figure 1.19
Variable inductance transducer.

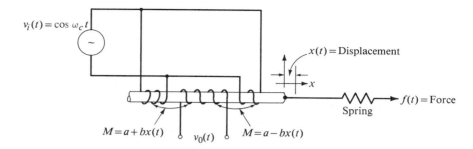

In other applications it may be required to sense the movement of a particular piece of equipment. One of the usual methods is to use a variable inductance transducer, the basic form of which is shown in Figure 1.19. This device is extremely sensitive, with outputs on the order of 1 volt per 0.0025 cm translation and with a linearity better than 0.05%. From Figure 1.19 we find that the output signal, when the self-inductance can be neglected, is given by

$$v_0(t) = A[a + bx(t)] \cos \omega_c t - A[a - bx(t)] \cos \omega_c t = 2Abx(t) \cos \omega_c t$$

where A is a constant. Observe that the output is a modulated signal. With an appropriate processing of the product signal, we can extract the desired displacement $x(t)$.

Most of the signals that we encounter in nature are analog (continuous) signals. A basic reason for this is that physical systems cannot respond instantaneously to changing inputs. Moreover, in many cases the signal is not available in electrical form, thus requiring the use of a transducer (mechanical, electrical, thermal, optical, and so on) to provide an electrical signal that is representative of the system signal. These transducers generally cannot respond instantaneously to changes and tend to smooth out the signals. What we really observe is some close replica of the desired signals, not a signal identical to the original one. The closeness depends, of course, on the sensitivity of the detector.

Depending on the signal processing requirements, detected signals become useful if we are able to analyze them and if we are able to perform mathematical operations on them. The digital computer is of tremendous value in such signal processing; however, digital computers are discrete systems that will accept only discrete information. It is necessary, therefore, to change the analog signals into equivalent digital signals. We have discussed this in general terms in Section 1-3, but the practical realization of very rapid analog to digital conversion is accomplished with an analog/digital (A/D) converter. In many cases the processed digital signals are converted back to analog form because humans are better able to observe analog signals. Such a transformation is accomplished using a digital/analog (D/A) converter.

The elements of an A/D converter are shown in Figure 1.20. A representative sweep-timing system is shown in block form in Figure 1.20a. The input signal is sampled at a regular rate and the sampled values are compared in a comparator with the value achieved by a repetitive linear ramp. Two pulses are

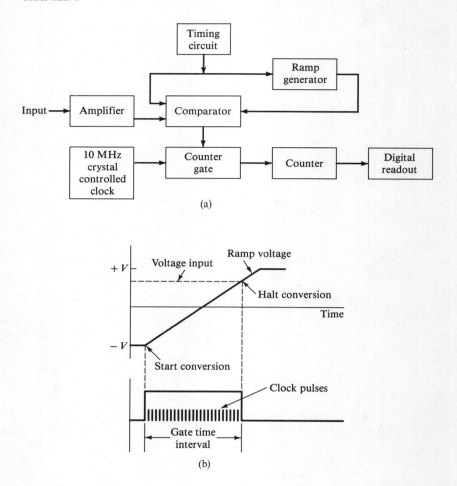

Figure 1.20 (a) Block diagram of a sweep-timing A/D converter. (b) Features of operation. (From Seely and Poularikas, *Electrical Engineering*. Copyright © 1982 by Matrix Publishers, Inc. Used with permission.)

produced—one at the start of the sampling instant and the other at the point of equality with the ramp. The time difference between the two pulses specifies a time proportional to the instantaneous value of the input voltage. These two pulses open and close a gate, thereby allowing a certain number of clock pulses to reach an electronic counter. The subsequent digital readout of the counter is proportional to the input voltage and is available for subsequent processing.

In another case when an optical image, which is a two-dimensional signal, is to be processed by a computer, a self-scanning array of photodiodes can be used to provide equivalent digitized data corresponding to the optical image. Such an array is comprised of a matrix array of light diodes together with a series of electronic switches and associated electronic control circuitry, as depicted schematically in Figure 1.21. When the applied control circuit switching or commutation closes a particular switch of the array of photodiodes, a current I proportional to the light intensity impinging on the diode connected to the circuit will result. By means of appropriate electronic circuitry, the current I

Figure 1.21
A self-scanning photodiode array.

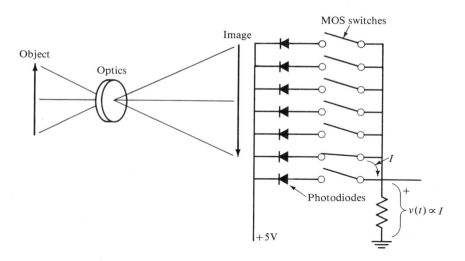

can be digitized and the digital output will give an indication of the radiance of the image. The entire array can be scanned, thereby giving the radiance of the image at each point of the diode array. These sampled and digitized signals can be inputted to a computer for further processing.

■ ■ ■

1-5 REPRESENTATION OF SIGNALS BY ORTHOGONAL FUNCTIONS

We noted in the previous sections that a given signal waveform, whether it is expressed nonanalytically, given graphically by an experimentally determined output from some transducer, or given in chart or tabular form, must often be represented in mathematical form. One approach is to represent a specified function by a sum of elementary functions, called **basis functions.** Such representations make possible straightforward mathematical operations and often result in a better understanding of the system behavior.

When we consider the effects of a physical system on an arbitrary signal $f(t)$, we are actually studying the output of the system when the input is $f(t)$. Because a system is usually described by a mathematical operator, the output of the system is equivalent to the effect of the operator on the signal. However, a mathematical operation (for example, integration) is difficult to perform unless $f(t)$ possesses certain special properties. Therefore we seek a set of simple basis functions $\{\varphi_i\}$ with special properties that will permit the function $f(t)$ to be represented by a linear combination of these basis functions. For linear systems it is relatively easy to determine the effect of the operator on $f(t)$ by finding the effect of the operator on each basis function and then adding these effects; this is the principle of superposition.

To establish certain ideas that will be helpful in understanding the development involving basis functions, consider a vector in space. Let $\mathbf{a}_i$ ($i = 1, 2, 3$), known as a **basis set,** denote the three orthogonal unit vectors along the Cartesian coordinate axes. An arbitrary vector $\mathbf{F}$ can be expanded in the form

$$\mathbf{F} = F_1\mathbf{a}_1 + F_2\mathbf{a}_2 + F_3\mathbf{a}_3 = \sum_{i=1}^{3} F_i\mathbf{a}_i \qquad (1.22)$$

where the F_i's are the projections of $\mathbf{F}$ on each coordinate axis. By taking into consideration the **orthogonality** of the vector set $\mathbf{a}_i$, which is defined by

$$\mathbf{a}_i \cdot \mathbf{a}_j = \begin{cases} 1 & i = j \\ 0 & i \neq j \end{cases} \qquad (1.23)$$

the components of $\mathbf{F}$ are specified by

$$F_i = \frac{\mathbf{F} \cdot \mathbf{a}_i}{\mathbf{a}_i \cdot \mathbf{a}_i} \qquad (1.24)$$

where $\mathbf{F} \cdot \mathbf{a}_i = F_i$ is the dot (inner) product and F_i is the projection of the vector $\mathbf{F}$ on the a_i axis. Note that since **any** three-dimensional vector $\mathbf{F}$ can be represented as a linear combination involving the three orthogonal unit vectors $\mathbf{a}_1$, $\mathbf{a}_2$, $\mathbf{a}_3$, the system is called **complete;** that is, $\mathbf{F}$ is contained in the space **spanned** by the basis set $\{\mathbf{a}_i\}$. On the other hand, a three-dimensional vector $\mathbf{F}$ cannot be represented as a linear combination of only two orthogonal vectors, say $\mathbf{a}_1$ and $\mathbf{a}_2$. Therefore a two-dimensional orthogonal system is **incomplete** to represent three dimensional vectors, and the basis set does not span the space containing $\mathbf{F}$.

Now let us apply these same concepts to functions. It may be possible to express an arbitrary function $f(t)$ as the sum of its components along a set of mutually orthogonal functions that form a complete set over an interval $t = a$ to $t = b$. Thus if we write $\varphi_i(t) \triangleq \mathbf{a}_i$ and $c_i \triangleq F_i$, (the symbol $\triangleq$ indicates equivalence by definition), then (1.22) takes the equivalent form

$$f(t) = \sum_{i=1}^{N} c_i\varphi_i(t) \qquad (1.25)$$

Multiply (1.25) by $\varphi_j(t)\,dt$ or by $\varphi_j^*(t)\,dt$, if the φ_i's are complex functions (the asterisk means complex conjugate function—the function φ_j with all complex components being replaced by their negative values), and integrate over the range of definition $[a, b]$ of $f(t)$. The result is

$$\int_a^b f(t)\varphi_j(t)\,dt = \sum_{i=1}^{N} c_i \int_a^b \varphi_i(t)\varphi_j(t)\,dt$$

or

$$\int_a^b f(t)\varphi_i(t)\,dt = c_i \int_a^b \varphi_i^2(t)\,dt \triangleq c_i\|\varphi_i(t)\|^2 \quad \text{(a)} \qquad (1.26)$$

from which it is found that

$$c_i = \frac{\int_a^b f(t)\varphi_i(t)\,dt}{\|\varphi_i(t)\|^2} = \frac{\int_a^b f(t)\varphi_i(t)\,dt}{\int_a^b \varphi_i^2(t)\,dt} \qquad i = 0, 1, 2, \ldots \quad \text{(b)}$$

where

$$\left[\int_a^b \varphi_i^2(t)\,dt\right]^{1/2} = \|\varphi_i(t)\| \quad \text{(c)}$$

is called the **norm** of the function φ_i and is a real number. Equation (1.26) is the result of the orthogonality property assumed over the interval $t = a$ to $t = b$

$$\int_a^b \varphi_i(t)\varphi_j(t)\,dt = \begin{cases} \|\varphi_i(t)\|^2 & i = j \\ 0 & i \neq j \end{cases} \qquad (1.27)$$

If the value of $\|\varphi_i(t)\|^2 = 1$, the basis functions φ_i are called **orthonormal**. The functions under consideration and the functions that constitute an orthogonal set must be **square integrable**; that is, they must obey the relations

$$\int_a^b |f(t)|\,dt < \infty \quad \text{and} \quad \int_a^b f^2(t)\,dt < \infty \qquad (1.28)$$

including the case when the interval $[a, b] = (-\infty, \infty)$.

EXAMPLE 1.3
Show that the set of functions

$$\{1, \cos t, \cos 2t, \ldots, \cos nt, \ldots\}$$

constitutes orthogonal basis functions over the interval $[0, \pi]$.

Solution: Form the relations

$$\int_0^\pi 1 \cos nt\,dt = 0$$

$$\int_0^\pi \cos nt \cos mt\,dt = 0 \quad m \neq n$$

$$\int_0^\pi \cos^2 nt\,dt = \frac{\pi}{2}$$

Observe that this set constitutes basis functions on the interval $0 \leq t \leq \pi$. The function will become orthonormal if each is multiplied by $\sqrt{2/\pi}$. ∎

In this example we have presented the trigonometric function $\cos nt$ (similarly $\sin nt$), which is one of the most important basis functions and one which we will frequently use in our study of linear systems. The complex represen-

Figure 1.22
Detection scheme using orthogonal functions.

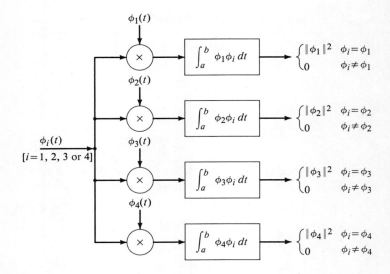

tation of these functions is

$$\varphi_n(t) = \exp(j\omega_0 n t) \qquad n = 0, \pm 1, \pm 2, \ldots$$

which forms the basis for the Fourier series expansion of functions. We will study this matter in some detail in Chapter 3. The main reason for learning about basis functions is that they are the **eigenfunctions of linear time-invariant** operations. This means that if an operator representing a linear physical time-invariant system—that is, one for which any time shift of the input results in the same time shift in the output—operates on the eigenfunction of the system, the output is equal to the input eigenfunction multiplied by a complex constant. Hence the output has the same form as the input but with the possible difference that its magnitude as well as its phase may be changed. This matter will receive more attention in Chapter 2.

Orthogonal functions are also important in detection theory. For example, suppose that we send four messages corresponding to four orthogonal functions $\varphi_1, \varphi_2, \varphi_3, \varphi_4$, respectively. If we construct a receiver of the form shown in Figure 1.22, we observe that each channel will respond only when the incoming signal is equal to the input signal of that channel. Because of the orthogonality of the functions, the remaining channels will produce no output.

An important extension of the concept of orthogonal sets states that a set $\{\Phi_n(t)\}$ $(n = 1, 2, \ldots)$ is orthogonal in the interval $[a, b]$ with respect to a given weight function $w(t)$, where it is usually supposed that $w(t) \geq 0$ in $[a, b]$, if

$$\int_a^b w(t) \Phi_n(t) \Phi_m(t) \, dt = 0 \qquad \text{when } n \neq m \quad (n, m = 1, 2, \ldots)$$

where $\varphi_n(t) = \sqrt{w(t)} \, \Phi_n(t)$. $\Phi_n(t)$ includes a number of well-known functions, including Legendre polynomials, Chebyshev polynomials, and others that we consider below. The integral represents the inner product (φ_m, φ_n) with respect to the weight function. The norm of $\varphi_n(t)$ in this case is

$$U = \int_a^b w(t)[\Phi_n(t)]^2 \, dt \qquad n = 1, 2, \ldots$$

Approximation in the Mean—Completeness Suppose that $f(t)$ is an arbitrary square integrable function and $f_a(t)$ is the approximating function

$$f_a(t) = \alpha_0 \varphi_0(t) + \alpha_1 \varphi_1(t) + \cdots + \alpha_n \varphi_n(t) \tag{1.29}$$

where the α_i's are constants. Since the φ_i's are square integrable, then the linear combination of the φ_i's and $[f(t) - f_a(t)]$ are also square integrable. The important question here is how closely does $f_a(t)$ approximate the function $f(t)$ on $[a, b]$. One criterion that is often used is to require that the approximation $f_a(t)$ be so chosen as to minimize the **mean square error (MSE)** between the true value of $f(t)$ and the approximation function, where the MSE is defined by the relation

$$\text{MSE} = \int_a^b [f(t) - f_a(t)]^2 \, dt \tag{1.30}$$

The MSE is a measure of the closeness of the approximate function $f_a(t)$ to $f(t)$. This quantity is zero only when $f(t) = f_a(t)$. An important question is how to choose the constants α_i in $f_a(t)$ relative to the c_i in the expansion for $f(t)$ given by (1.25) so that the MSE is minimized. We proceed by expanding the integrand in (1.30). Thus we write

$$\text{MSE} = \int_a^b [f^2(t) + f_a^2(t) - 2f(t)f_a(t)] \, dt$$

$$= \int_a^b \left[f^2(t) + \left(\sum_{n=0}^n \alpha_n \varphi_n(t) \right)^2 - 2f(t) \sum_{n=0}^n \alpha_n \varphi_n(t) \right] dt$$

$$= \int_a^b \left[f^2(t) + \sum_{n=0}^n \alpha_n^2 \varphi_n^2(t) + \sum_{i \neq j}^n \alpha_i \alpha_j \varphi_i(t) \varphi_j(t) - 2f(t) \sum_{n=0}^n \alpha_n \varphi_n(t) \right] dt$$

$$= \int_a^b f^2(t) \, dt + \sum_{n=0}^n \alpha_n^2 \|\varphi_n(t)\|^2 - 2 \sum_{n=0}^n c_n \alpha_n \|\varphi_n(t)\|^2$$

$$= \int_a^b f^2(t) \, dt + \sum_{n=0}^n (c_n - \alpha_n)^2 \|\varphi_n(t)\|^2 - \sum_{n=0}^n c_n^2 \|\varphi_n(t)\|^2 \tag{1.31}$$

where (1.26a) and (1.27) were used. However, it is noted that

$$\int_a^b f^2(t) \, dt = \text{positive constant} \qquad \sum_{n=0}^n c_n^2 \|\varphi_n(t)\|^2 = \text{positive constant}$$

and they do not depend on the α_n's. Thus MSE is a minimum when

$$\sum_{n=0}^n (c_n - \alpha_n)^2 \|\varphi_n(t)\|^2 = 0$$

or when

$$\alpha_n = c_n \qquad n = 0, 1, 2, \ldots, n \qquad (1.32)$$

since $\|\varphi_n\| \neq 0$. This shows that the coefficients of the orthogonal basis functions minimize the mean square error. In summary, this development shows that, given n mutually orthogonal functions $\{\varphi_i(t)\}$ on an interval $[a, b]$, the best approximation of an arbitrary function $f(t)$ on $[a, b]$ of the form given by (1.25) is given by choosing the c_i's according to (1.26). The criterion for this approximation is minimization of the mean square error between $f(t)$ and $\sum_{i=1}^{n} c_i \varphi_i(t)$.

To determine the minimum error, introduce the results of (1.32) into (1.31). Thus

$$\text{MSE}_{\min} = \int_a^b f^2(t)\,dt - \sum_{n=0}^{n} c_n^2 \|\varphi_n(t)\|^2$$

But from (1.30) (positive integrand), $\text{MSE}_{\min} \geq 0$, we obtain the inequality

$$\boxed{\int_a^b f^2(t)\,dt \geq \sum_{n=0}^{n} c_n^2 \|\varphi_n(t)\|^2 \qquad \text{Bessel inequality}} \qquad (1.33)$$

But this requires, in light of (1.28), that the sum on the right has a finite value as $n \to \infty$; that is, the series converges. The expression in (1.33) is known as the **Bessel inequality.** When the basis functions are orthonormal $\|\varphi_n(t)\|^2 = 1$, the Bessel inequality assumes the form

$$\int_a^b f^2(t)\,dt \geq \sum_{n=0}^{n} c_n^2 \qquad (1.34)$$

Equation (1.31) suggests that by increasing n, the number of orthogonal functions in the description of $f(t)$, the minimum value of the mean square error decreases. This seems entirely reasonable since it suggests that as we fill more of the space containing $f(t)$, the error should decrease. If, as n increases without limit, the sum $\sum_{i=0}^{\infty} c_i \varphi_i(t)$ converges to the function $f(t)$, the MSE is zero and

$$\boxed{\int_a^b f^2(t)\,dt = \sum_{n=0}^{\infty} c_n^2 \|\varphi_n(t)\|^2 \qquad \text{Parseval's theorem}} \qquad (1.35)$$

This equation is known as **Parseval's theorem.** If this equation is valid, then the sum $\sum_{i=1}^{\infty} c_i \varphi_i(t)$ is said to converge in the mean to $f(t)$ and the set $\{\varphi_i(t)\}$ constitutes a **complete** set. To review, a complete set of functions $\varphi_i(t)$ is one in which all orthogonal elements of the set are included within the set; that is, no function outside of the set exists that is orthogonal to each member of the set. For every complete set, Parseval's theorem holds. No general tests of a practical nature exist for showing that a set of functions is complete.

SIGNALS AND THEIR FUNCTIONAL REPRESENTATIONS

In our study we deal only with orthogonal polynomials and the question of completeness will not be a serious consideration. It has been shown, in fact, that a set of orthogonal polynomials is complete if it contains polynomials of all degrees and the interval $[a, b]$ of its orthogonality is finite. There are some polynomials, specifically Hermite and Laguerre polynomials, whose interval is infinite. The completeness of these polynomials can be established separately. In general, the completeness of polynomials with infinite intervals of orthogonality must be checked.

Signal Energy If we consider any signal $f(t)$ as denoting a voltage that exists across a 1-ohm resistor, the integral

$$E = \int_a^b \frac{f^2(t)}{1} dt \qquad \text{joule} \tag{1.36}$$

represents the energy dissipated in the resistor during the time interval $[a, b]$. The energy dissipated is proportional to the time integral of the square of the signal. A signal is called an **energy signal** if

$$\int_{-\infty}^{\infty} f^2(t) dt < \infty \tag{1.37}$$

A **power signal,** on the other hand, is defined by the relation

$$0 \leq \lim_{T \to \infty} \frac{1}{2T} \int_{-T}^{T} f^2(t) dt < \infty \tag{1.38}$$

We may represent the energy in a finite interval in terms of the coefficients of the basis functions φ_i; that is, write (1.36) in the form

$$E = \int_a^b f^2(t) dt = \int_a^b f(t) \sum_{n=0}^{n} c_n \varphi_n(t) dt = \sum_{n=0}^{n} c_n \int_a^b f(t) \varphi_n(t) dt$$

$$= \sum_{n=0}^{n} c_n^2 \|\varphi_n(t)\|^2 \tag{1.39}$$

where use has been made of (1.26). Since $\|\varphi_n(t)\|^2$ is the energy associated with the nth orthogonal function, (1.39) shows that the energy of any signal is the sum of the energies of its individual orthogonal components weighted by the c_n's. Note that this is the Parseval theorem discussed in (1.35). This equation shows that if $\{\varphi_n(t)\}$ forms an orthogonal set, the signal energy can be calculated from its representation.

Observe from our above analysis that the study of vectors gives us an insight into the properties of orthogonal functions. From Figure 1.23a we observe that the error $\epsilon = \mathbf{F} - \mathbf{F}_i$ between the vector $\mathbf{F}$ and any projection of the $\mathbf{a}_1, \mathbf{a}_2$ plane becomes a minimum if ϵ is perpendicular to the plane. Since the $\mathbf{a}_1, \mathbf{a}_2$ plane is a subspace of the three-dimensional space, we may conjecture that if our approximate function $f_a(t)$ belongs to a subspace S_a of the space S, then the

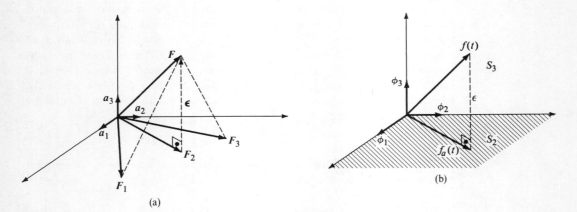

Figure 1.23
Illustrating the projection theorem for a three-dimensional space.

minimum error $\epsilon(t) = f(t) - f_a(t)$ is orthogonal to $f_a(t)$. To show this, we must show that $\int_a^b f_a(t)\epsilon(t)\,dt = 0$. This requires that we examine

$$\int_a^b f_a(t)\epsilon(t)\,dt = \int_a^b f_a(t)f(t)\,dt - \int_a^b f_a^2(t)\,dt$$

$$= \int_a^b f(t) \sum_{n=0}^n c_n \varphi_n(t)\,dt - \int_a^b f_a^2(t)\,dt$$

$$= \sum_{n=0}^n c_n \int_a^b f(t)\varphi_n(t)\,dt - \int_a^b \left[\sum_{n=0}^n c_n \varphi_n(t)\right]^2 dt$$

$$= \sum_{n=0}^n c_n^2 - \sum_{n=0}^n c_n^2 = 0$$

It has been assumed in this development that the φ_i's are orthonormal functions. This proves our assertion. In Figure 1.23b we call $f_a(t)$ the **orthogonal projection** of $f(t)$. The student can extend these ideas from a three-dimensional space and a two-dimensional subspace to an infinite-dimensional space and an n-dimensional subspace.

The ideas discussed above serve to clarify a useful theorem known as the projection theorem.

Projection Theorem. If $f(t)$ is an element of the linear (vector) space S spanned by the orthogonal basis $\{\varphi_i\}$ and $f_a(t)$ is an element in a lower-dimensional subspace $S_a \subset S$, the error ϵ is a minimum in the mean square sense, if and only if $f_a(t)$ is the orthogonal projection of $f(t)$ into S_a. The minimum error and the projection are orthogonal.

∎ ∎ ∎

1-6 ORTHONORMAL FUNCTIONS

We wish now to explore some of the orthonormal function sets that can be used to approximate a given function $f(t)$ on an interval $[a, b]$. The trigonometric functions sine and cosine are perhaps the best known and together are used in the Fourier series. A feature of the Fourier series is that the interval $[a, b]$ can be extended to $[0, \infty]$ if the function $f(t)$ is periodic and satisfies the Dirichlet conditions. This expansion method is of such importance that it will receive detailed consideration in Chapter 3. We wish to examine here a number of other orthogonal functions.

Walsh Functions Walsh functions, introduced in 1923, form a complete set of orthogonal functions on the interval $[0, 1]$. Walsh function expansion is useful in such applications as speech and biomedical signal analysis and filtering, communications signal multiplexing, pattern recognition, and image enhancement. The advantage of Walsh function expansion over other orthogonal function expansions is that Walsh functions assume only the values ± 1 and multiplication by Walsh functions involves only algebraic sign assignment. Since the functions assume only the values ± 1, they can be generated easily by digital circuits. Further, the product of Walsh functions, as discussed below, can be accomplished using Exclusive OR digital logic circuits.

Walsh functions may be presented in three forms and are specified as (a) Walsh (sequency)-ordered, (b) Paley (dyadic)-ordered, and (c) Hadamard (natural)-ordered. Walsh functions are not sinusoidal nor do they possess the notion of frequency in any conventional sense. However, the functions cross the zero axis, and as an analog to frequency, the term **sequency** has been introduced. As we shall see, the Walsh functions comprise block waves over the interval.

1-6.a Walsh-Ordered Functions

The Walsh-ordered function set is denoted by

$$\Phi_w = \{\text{wal}_w(i, t) \quad i = 0, 1, \ldots, N - 1\} \quad \textbf{(a)} \tag{1.40}$$

where

$$N = 2^n \quad n = 1, 2, \ldots \quad \textbf{(b)}$$

The subscript w denotes that the function set is Walsh-ordered, and i denotes the ith member of the basis set Φ_w. If s_i represents the sequency, or the number of zero axis crossings from $+$ to $-$, as shown in Figure 1.24, then s_i is given by

$$s_i = \begin{cases} 0 & i = 0 \\ \dfrac{i}{2} & i = \text{even} \\ \dfrac{i+1}{2} & i = \text{odd} \end{cases} \tag{1.41}$$

Figure 1.24
Walsh functions; sequency ordered, $N = 8$.

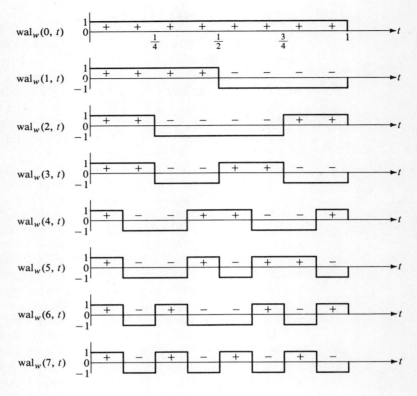

The function $\text{wal}_w(i, t)$ is defined by

$$\text{wal}_w(2i + p, t) \triangleq \text{wal}_w[i, 2(t + \tfrac{1}{4})] + (-1)^{i+p} \text{wal}_w[i, 2(t - \tfrac{1}{4})] \quad (1.42)$$

where

$p = 0$ or 1 $\quad i = 0, 1, 2, \ldots \quad \text{wal}_w(0, t) = 1, \quad$ for $0 \leq t \leq 1$

$\text{wal}_w[i, 2(t + \tfrac{1}{4})] =$ length of $\text{wal}_w(i, t)$ considered as a block reduced by a factor 2, with the center of the block shifted to the left by its length

$\text{wal}_w[i, 2(t - \tfrac{1}{4})] =$ length of $\text{wal}_w(i, t)$ considered as a block reduced by a factor 2, with the center of the block shifted to the right by its length

---

EXAMPLE 1.4
Evaluate and illustrate $\text{wal}_w(1, t)$; $\text{wal}_w(2, t)$.

Solution: For this we set $i = 0$ and $p = 1$. Equation (1.42) becomes

$$\text{wal}_w(1, t) = \text{wal}_w[0, 2(t + \tfrac{1}{4})] + (-1)^{0+1} \text{wal}_w[0, 2(t - \tfrac{1}{4})]$$

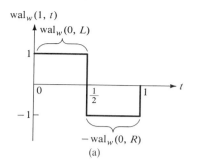
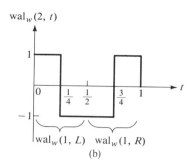

Figure 1.25
The construction of Walsh functions.

This function is shown in Figure 1.25a.

To evaluate $\text{wal}_w(2, t)$ we set $i = 1$ and $p = 0$ in (1.42). This yields

$$\text{wal}_w(2, t) = \text{wal}_w[1, 2(t + \tfrac{1}{4})] + (-1)^{1+0} \text{wal}_w[1, 2(t - \tfrac{1}{4})]$$

Observe that this requires $\text{wal}_w(1, t)$ in its representation; it comprises two $\text{wal}_w(1, t)$ blocks, one shifted to the left and one to the right and then combined. The result is represented, by writing $L = 2t + \tfrac{1}{2}$ and $R = 2t - \tfrac{1}{2}$, as

$$\text{wal}_w(2, t) = \text{wal}_w(1, L) - \text{wal}_w(1, R)$$

This function is shown in Figure 1.25b. ∎

By employing the same procedure used in this example, we can form successive forms of the Walsh functions. Figure 1.24 shows eight Walsh-ordered functions. The orthogonality property of these functions can be proved by showing that they satisfy (1.27).

An alternate designation of $\text{wal}_w(i, t)$ functions often found in the literature is:

$$\begin{aligned} \text{cal}(i, t) &= \text{wal}_w(i, t) \quad i \text{ even} \quad \textbf{(a)} \\ \text{sal}(i, t) &= \text{wal}_w(i, t) \quad i \text{ odd} \quad \textbf{(b)} \end{aligned} \quad (1.43)$$

The product of two Walsh functions is obtained by using the formula

$$\text{wal}_w(k, t) \, \text{wal}_w(l, t) = \text{wal}_w(k \oplus l, t) \quad (1.44)$$

where the symbol $\oplus$ denotes the logic operation Exclusive OR with $1 \oplus 0 = 1$, $0 \oplus 1 = 1$, $0 \oplus 0 = 0$, $1 \oplus 1 = 0$. To find the number $k \oplus l$ analytically, it is first necessary to write k and l in binary form and then apply the Exclusive OR operation. If graphs of Walsh functions are available, the prescribed Exclusive OR operation can be accomplished graphically.

EXAMPLE 1.5

Find the product of the Walsh functions $\text{wal}_w(2, t)\,\text{wal}_w(3, t)$.

Solution: Apply (1.44) to write

$$\text{wal}_w(2, t)\,\text{wal}_w(3, t) = \text{wal}_w(10_b \oplus 11_b, t) = \text{wal}_w(01_b, t) = \text{wal}_w(1, t)$$

The reader should verify the result using Figure 1.24. ∎

1-6.b Paley-Ordered Walsh Functions. The Paley order of the Walsh functions can be effected using the Gray code (refer to Chapter Appendix 1-2 at the end of the book). The Paley-ordered set is denoted by

$$\Phi_p = \{\text{wal}_p(i, t);\ i = 0, 1, \ldots, N - 1\} \tag{1.45}$$

with the sets of functions specified in (1.40) and (1.45) being related by the expression

$$\text{wal}_p(i, t) = \text{wal}_w(Gb(i), t) \tag{1.46}$$

where $Gb(i)$ represents the Gray-binary conversion of i. Table 1.1 illustrates the relation specified by (1.46). The first six Paley-ordered Walsh functions are shown in Figure 1.26a.

Table 1.1 Paley-Ordered Walsh Functions.

i_{dec}	$i_b[i_G]$	$Gb(i)$	$Gb(i)_{\text{dec}}$	Equation (1.46)
0	000	000	0	$\text{wal}_p(0, t) = \text{wal}_w(0, t)$
1	001	001	1	$\text{wal}_p(1, t) = \text{wal}_w(1, t)$
2	010	011	3	$\text{wal}_p(2, t) = \text{wal}_w(3, t)$
3	011	010	2	$\text{wal}_p(3, t) = \text{wal}_w(2, t)$
4	100	111	7	$\text{wal}_p(4, t) = \text{wal}_w(7, t)$
5	101	110	6	$\text{wal}_p(5, t) = \text{wal}_w(6, t)$

Table 1.2 Hadamard-Ordered Walsh Functions.

i_{dec}	i_b	$i_{br}[i_G]$	$Gb(i_r)$	$Gb(i_r)_{\text{dec}}$	Equation (1.48)
0	000	000	000	0	$\text{wal}_h(0, t) = \text{wal}_w(0, t)$
1	001	100	111	7	$\text{wal}_h(1, t) = \text{wal}_w(7, t)$
2	010	010	011	3	$\text{wal}_h(2, t) = \text{wal}_w(3, t)$
3	011	110	100	4	$\text{wal}_h(3, t) = \text{wal}_w(4, t)$
4	100	001	001	1	$\text{wal}_h(4, t) = \text{wal}_w(1, t)$
5	101	101	110	6	$\text{wal}_h(5, t) = \text{wal}_w(6, t)$

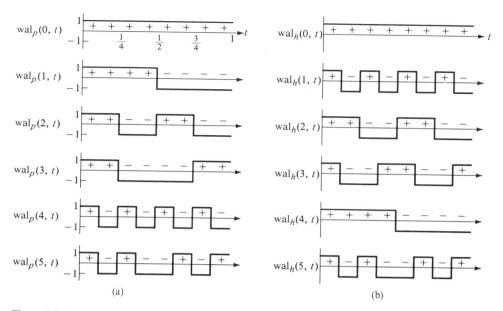

Figure 1.26
(a) Paley-ordered Walsh functions. (b) Hadamard-ordered Walsh functions.

1-6.c Hadamard-Ordered Walsh Functions. The Hadamard-ordered Walsh functions are specified by

$$\Phi_h = \{\text{wal}_h(i, t); i = 0, 1, \ldots, N - 1\} \tag{1.47}$$

with the sets of functions specified in this expression related to the Walsh-ordered functions by the relation

$$\text{wal}_h(i, t) = \text{wal}_w(Gb(i_r), t) \tag{1.48}$$

where i_r denotes the bit reversal of the Gray-binary conversion of i. Table 1.2 displays the relation specified by (1.48). Figure 1.26b displays the Hadamard-ordered Walsh functions.

EXAMPLE 1.6
Find a four-term Walsh approximation for the function shown in Figure 1.27a.

Solution: The approximating function is written

$$f_a(t) = c_0 \, \text{wal}_w(0, t) + c_1 \, \text{wal}_w(1, t) + c_2 \, \text{wal}_w(2, t) + c_3 \, \text{wal}_w(3, t)$$

From the figure, the function $f(t)$ is

$$f(t) = \begin{cases} t & 0 \leq t \leq 0.5 \\ 1 & 0.5 \leq t \leq 1 \end{cases}$$

Figure 1.27
Approximation of $f(t)$ with Walsh functions.

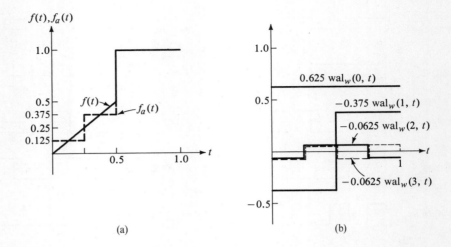

The coefficients c_i are evaluated using (1.26) and Figure 1.24. Because the Walsh functions are orthonormal, we obtain

$$c_0 = \int_0^1 f(t)\,\text{wal}_w(0, t)\,dt = \int_0^{0.5} t\,dt + \int_{0.5}^1 dt = 0.625$$

$$c_1 = \int_0^1 f(t)\,\text{wal}_w(1, t)\,dt = \int_0^{0.5} t\,dt + \int_{0.5}^1 (-1)\,dt = -0.375$$

$$c_2 = \int_0^1 f(t)\,\text{wal}_w(2, t)\,dt = \int_0^{0.25} t\,dt - \int_{0.25}^{0.5} t\,dt - \int_{0.5}^{0.75} dt$$
$$+ \int_{0.75}^1 dt = -0.0625$$

$$c_3 = \int_0^1 f(t)\,\text{wal}_w(3, t)\,dt = \int_0^{0.25} t\,dt - \int_{0.25}^{0.5} t\,dt + \int_{0.5}^{0.75} dt$$
$$- \int_{0.75}^1 dt = -0.0625$$

The resulting function $f_a(t)$ is shown in Figure 1.27a and the individual functions comprising $f_a(t)$ are shown in Figure 1.27b. ∎

†**Classical Orthogonal Polynomials** Among the often used orthogonal functions, in addition to the trigonometric sine and cosine functions, are the following:

a. Legendre Functions. These functions are orthonormal in the range $-1 \leq t \leq 1$ and in one form are given by the Rodrigues formula

$$\begin{aligned}
\varphi_n(t) &= \sqrt{n + \tfrac{1}{2}}\, P_n(t) \quad & -1 \leq t \leq 1 & \quad \text{(a)} \\
P_n(t) &= \frac{1}{2^n}\frac{1}{n!}\frac{d^n}{dt^n}(t^2 - 1)^n \quad & n = 0, 1, 2, \ldots & \quad \text{(b)}
\end{aligned} \quad (1.49)$$

Figure 1.28
Approximations of a triangular wave with three terms of a Legendre function series.

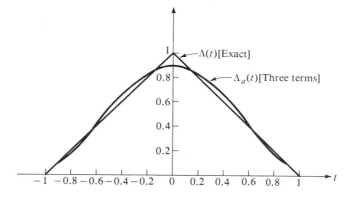

The functions $P_n(t)$ are **Legendre polynomials,** which possess the following properties:

$$\int_{-1}^{1} P_n^2(t)\, dt = \frac{1}{n + \frac{1}{2}} \quad \text{(a)}$$

$$(n + 1)P_{n+1}(t) = (2n + 1)t\, P_n(t) - n P_{n-1}(t) \quad \text{(b)} \qquad (1.50)$$

$$P_n(-t) = (-1)^n P_n(t) \quad \text{(c)}$$

EXAMPLE 1.7

Approximate the $\Lambda(t)$ function by three terms using Legendre basis functions.

Solution: Since $\Lambda(t)$ is an even function, $c_n = 0$ for n odd, whereas for n even

$$c_n = 2\sqrt{n + \tfrac{1}{2}} \int_0^1 \Lambda(t) P_n(t)\, dt = 2\sqrt{n + \tfrac{1}{2}} \int_0^1 (1 - t) P_n(t)\, dt$$

The expansion constants are given by

$$c_0 = 2\frac{1}{\sqrt{2}} \int_0^1 (1 - t)\, dt = \frac{1}{\sqrt{2}}$$

$$c_2 = 2\frac{\sqrt{5}}{\sqrt{2}} \int_0^1 (1 - t)\left(\frac{3}{2}t^2 - \frac{1}{2}\right) dt = -\frac{\sqrt{5}}{4\sqrt{2}}$$

$$c_4 = 2\frac{\sqrt{9}}{\sqrt{2}} \int_0^1 (1 - t)\left(\frac{35}{8}t^4 - \frac{15}{4}t^2 + \frac{3}{8}\right) dt = \frac{1}{24}\frac{\sqrt{9}}{\sqrt{2}}$$

The approximate function is given as

$$\Lambda_a(t) \doteq \frac{1}{\sqrt{2}}\frac{1}{\sqrt{2}} - \frac{1}{4}\left(\sqrt{\frac{5}{2}}\right)^2\left(\frac{3}{2}t^2 - \frac{1}{2}\right) + \frac{1}{24}\left(\sqrt{\frac{9}{2}}\right)^2\left(\frac{35}{8}t^4 - \frac{15}{4}t^2 + \frac{3}{8}\right)$$

Figure 1.28 shows the function and the approximate function for three terms.

b. Laguerre Functions. These functions are orthonormal in the range $0 \le t < \infty$. They are given by

$$\varphi_n(t) = e^{-t/2} L_n(t) \qquad \text{(a)}$$

$$L_n(t) = \frac{e^t}{n!} \frac{d^n}{dt^n} (t^n e^{-t}) \qquad n = 0, 1, 2, \ldots \qquad \text{(b)}$$

(1.51)

The functions $L_n(t)$ are **Laguerre polynomials.** These do not constitute an orthogonal set. However, the related set of functions $\varphi_n(t)$ is orthonormal in the interval $0 \le t < \infty$. The Laguerre polynomials possess the following properties:

$$\int_0^\infty e^{-t} L_n(t) L_m(t)\, dt = \delta_{m,n} = \begin{cases} 0 & \text{for } m \ne n \\ 1 & \text{for } m = n \end{cases} \qquad \text{(a)}$$

$$(n+1) L_{n+1}(t) = (2n+1-t) L_n(t) - n L_{n-1}(t) \qquad \text{(b)}$$

$$L_n(0) = 1 \qquad \text{(c)}$$

(1.52)

c. Hermite Functions. These functions are orthonormal in the range $-\infty < t < \infty$. They are given by

$$\varphi_n(t) = (2^n n! \sqrt{\pi})^{-1/2} e^{-t^2/2} H_n(t) \qquad -\infty < t < \infty \qquad \text{(a)}$$

$$H_n(t) = (-1)^n e^{t^2} \frac{d^n}{dt^n} (e^{-t^2}) \qquad n = 0, 1, 2, \ldots \qquad \text{(b)}$$

(1.53)

$H_n(t)$ are the **Hermite polynomials** with the following properties:

$$\int_{-\infty}^\infty e^{-t^2} H_n^2(t)\, dt = 2^n n! \sqrt{\pi} \qquad \text{(a)}$$

$$H_{2n}(0) = (-1)^n \frac{(2n)!}{n!} \qquad H_{2n+1}(0) = 0 \qquad \text{(b)}$$

$$H_n(t) = (-1)^n H_n(-t) \qquad \text{(c)}$$

$$H_{n+1}(t) = 2t H_n(t) - 2n H_{n-1}(t) \qquad \text{(d)}$$

(1.54)

d. Chebyshev Functions. These functions are orthonormal in the range $-1 \le t \le 1$. They are given by

SIGNALS AND THEIR FUNCTIONAL REPRESENTATIONS

$$\varphi_0(t) = \frac{1}{\sqrt{\pi}} \frac{1}{(1-t^2)^{1/4}} C_0(t) \qquad n = 0 \qquad \text{(a)}$$

$$\varphi_n(t) = \sqrt{\frac{2}{\pi}} \frac{1}{(1-t^2)^{1/4}} C_n(t) \qquad n \geq 1 \qquad \text{(b)} \qquad (1.55)$$

$$C_n(t) = \frac{(-2)^n n!}{(2n)!} \sqrt{1-t^2} \frac{d^n}{dt^n} (1-t^2)^{n-1/2} \qquad \text{(c)}$$

where $C_n(t)$ are the **Chebyshev polynomials.** They have the following properties:

$$C_{n+1}(t) - 2tC_n(t) + C_{n-1}(t) = 0 \qquad n \geq 1 \qquad \text{(a)}$$
$$C_1(t) - tC_0(t) = 0 \qquad \text{(b)} \qquad (1.56)$$

The general features of the polynomials mentioned above are shown in Figure 1.29.

EXAMPLE 1.8

Find an orthonormal set of polynomials in the interval $0 \leq t < \infty$ that satisfies the weighted orthonormal relationship of (1.27) and is developed from the orthogonal set $\{1, t, t^2, \ldots\} \triangleq \{m_1, m_2, \ldots\}$. The weighting function is $w(t) = e^{-t}$.

Solution: Use the results of Problem 1-5.2 which defines and uses the Gram-Schmidt orthogonalization procedure to express the elements of the set of polynomials $\{\varphi_1, \varphi_2, \varphi_3, \ldots\}$ appropriate to the orthogonal set $\{m_1, m_2, \ldots\}$. We obtain

$$\tilde{\varphi}_1(t) \triangleq m_1 = 1 \qquad \|\tilde{\varphi}_1(t)\| = \left[\int_0^\infty e^{-t} \cdot 1 \, dt\right]^{1/2} = 1$$

$$\varphi_1(t) = \frac{\tilde{\varphi}_1(t)}{\|\tilde{\varphi}_1(t)\|} = 1$$

$$\tilde{\varphi}_2(t) \triangleq m_2 - \langle m_2, \varphi_1 \rangle_w \varphi_1 = t - \left(\int_0^\infty t \cdot 1 e^{-t} dt\right) \cdot 1 = t - 1$$

$$\|\tilde{\varphi}_2(t)\| = \left[\int_0^\infty e^{-t}(t-1)^2 \, dt\right]^{1/2} = 1 \qquad \varphi_2(t) = \frac{\tilde{\varphi}_2(t)}{\|\tilde{\varphi}_2(t)\|} = t - 1$$

$$\tilde{\varphi}_3(t) \triangleq m_3 - \langle m_3, \varphi_2 \rangle_w \varphi_2 - \langle m_3, \varphi_1 \rangle_w \varphi_1$$

$$= t^2 - \left[\int_0^\infty e^{-t} t^2 (t-1) \, dt\right] \cdot (t-1) - \left[\int_0^\infty t^2 \cdot 1 e^{-t} dt\right] \cdot 1$$

$$= t^2 - 4t + 2$$

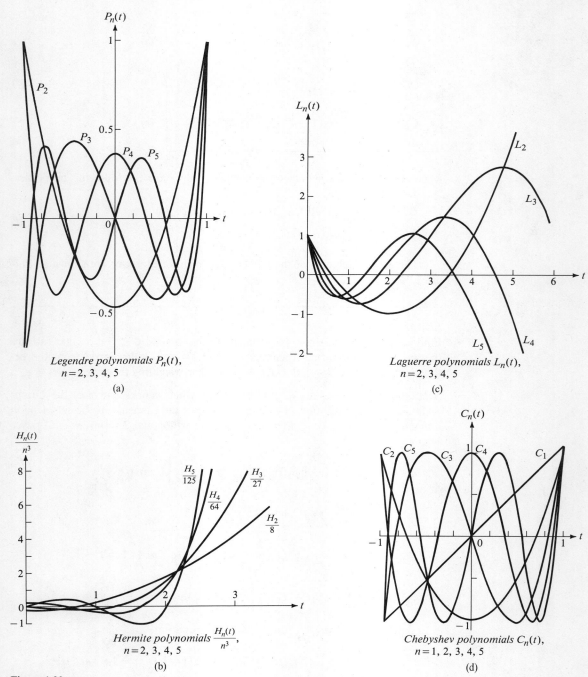

Figure 1.29
(a) Legendre Polynomials $P_n(x)$, $n = 2, 3, 4, 5$. (b) Hermite Polynomials $H_n(x)/n^3$, $n = 2, 3, 4, 5$.
(c) Laguerre Polynomials $L_n(x)$, $n = 2, 3, 4, 5$. (d) Chebyshev Polynomials $C_n(x)$, $n = 1, 2, 3, 4, 5$.

$$\|\tilde{\varphi}_3(t)\| = \left[\int_0^\infty e^{-t}(t^2 - 4t + 2)^2 \, dt \right]^{1/2} = 2$$

$$\varphi_3(t) = \frac{\tilde{\varphi}_3(t)}{\|\tilde{\varphi}_3(t)\|} = \frac{t^2 - 4t + 2}{2}$$

and so on. These are the Laguerre orthonormalized polynomials.

■ ■ ■

1-7 REPRESENTATION OF SIGNALS BY INTERPOLATION

Polynomial Interpolation It is known from the calculus that a function that is continuous and has continuous derivatives can be written in the form of a Taylor expansion

$$f(t) = f(t_0) + (t - t_0) \frac{df}{dt}\bigg|_{t=t_0} + \frac{(t - t_0)^2}{2!} \frac{d^2 f}{dt^2}\bigg|_{t=t_0} + \cdots$$

$$+ \frac{(t - t_0)^{n-1}}{(n - 1)!} \frac{d^{n-1} f}{dt^{n-1}}\bigg|_{t=t_0} + R_n \tag{1.57}$$

where R_n is the **remainder** that vanishes as $n \to \infty$. This shows that a function $f(t)$ can be written in terms of its value at a particular instant of time, if the values of all of its derivatives at the same instant are also known. In many cases these quantities are not known and one must then resort to other methods.

Another approach is to approximate the function by polynomials. The characteristic feature of most numerical methods is that values of a function $f(t)$ are given for a set of distinct values of t, but not for intermediate values. These intermediate values can be estimated on the hypothesis that $f(t)$ can be replaced by a polynomial agreeing with $f(t)$ at the points where its values are given. The simplest case is that of linear interpolation; this involves taking only two adjacent values of $f(t)$ from a table and calculating intermediate values based on the supposition that the first derivative $f^{(1)}(t)$ is constant in the interval. This approach is accurate provided that $f^{(1)}(t)$ changes little in the interval. Often, however, allowance must be made for higher derivatives. The use of a polynomial for curve-fitting can never be mathematically exact unless $f(t)$ is itself a polynomial. It can, in suitable circumstances, be as accurate as the tabulated values that are given.

Suppose therefore that a function $f(t)$ is approximated by a polynomial

$$f_a(t) = \sum_{n=0}^{N} \alpha_n t^n \tag{1.58}$$

It is assumed that the function $f(t)$ is known (collocates) at $N + 1$ points with values $f(t_0), f(t_1), \ldots, f(t_N)$, where $t_0 < t_1 < t_2 < \cdots < t_N$. This leads to the set of equations:

$$\alpha_0 + \alpha_1 t_0 + \alpha_2 t_0^2 + \cdots + \alpha_N t_0^N = f(t_0)$$
$$\alpha_0 + \alpha_1 t_1 + \alpha_2 t_1^2 + \cdots + \alpha_N t_1^N = f(t_1)$$
$$\vdots \qquad \vdots \qquad \vdots \qquad \vdots \qquad \vdots$$
$$\alpha_0 + \alpha_1 t_N + \alpha_2 t_N^2 + \cdots + \alpha_N t_N^N = f(t_N)$$
(1.59)

This set of equations can be written in compact matrix form

$$\begin{bmatrix} 1 & t_0 & t_0^2 & \cdots & t_0^N \\ 1 & t_1 & t_1^2 & \cdots & t_1^N \\ \vdots & \vdots & \vdots & & \vdots \\ 1 & t_N & t_N^2 & \cdots & t_N^N \end{bmatrix} \begin{bmatrix} \alpha_0 \\ \alpha_1 \\ \vdots \\ \alpha_N \end{bmatrix} = \begin{bmatrix} f(t_0) \\ f(t_1) \\ \vdots \\ f(t_N) \end{bmatrix} \quad \text{(a)}$$
(1.60)

or using matrix notation

$$\mathbf{T}\boldsymbol{\alpha} = \mathbf{f} \quad \text{(b)}$$

The matrix $\mathbf{T}$ is known as the Vandermonde matrix. The determinant of this matrix is found to be

$$\det \mathbf{T} = |\mathbf{T}| = \prod_{i>j} (t_i - t_j)$$
(1.61)

where the symbol $\prod$ denotes the continuous product. That is, the value of det $\mathbf{T}$ is the product of all binomials $t_i - t_j$ with $i > j$, and is

$$\det \mathbf{T} = [(t_1 - t_0)][(t_2 - t_1)(t_2 - t_0)] \cdots [(t_N - t_{N-1}) \cdots (t_N - t_0)]$$

The Vandermonde matrix is nonsingular if and only if $t_0, t_1, \ldots, t_N$ are distinct. In the present case, since our sampling gives distinct values to t, the matrix $\mathbf{T}$ is nonsingular, and we can solve (1.60) for the unknown α_n's. This involves inverting the matrix $\mathbf{T}$, with the result that

$$\boldsymbol{\alpha} = \mathbf{T}^{-1}\mathbf{f}$$
(1.62)

The inverse matrix $\mathbf{T}^{-1}$ is given by (see Appendix 2 at the end of the book)

$$\mathbf{T}^{-1} = [T_{ij}]^{-1} = \frac{\text{transpose cofactor } [T_{ij}]}{\det \mathbf{T}} = \frac{\text{cofactor } [T_{ji}]}{\det \mathbf{T}}$$

EXAMPLE 1.9

Find a three-term (parabolic) approximation for the function $f(t) = \sin(\pi/2)t$ over the interval $[-1, 1]$.

Solution: Refer to Figure 1.30, which shows a sketch of the function and arbitrarily select the three points: $t_0 = -1$, $t_1 = 0$, $t_2 = 1$. The resulting set of equations is:

$$f_a(-1) = \alpha_0 - \alpha_1 + \alpha_2 = f(-1) = -1$$

Figure 1.30
Polynomial approximation to $\sin(\pi/2)t$.

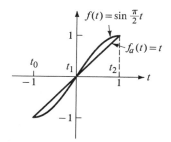

$$f_a(0) = \alpha_0 + 0\alpha_1 + 0\alpha_2 = \alpha_0 = f(0) = 0$$
$$f_a(1) = \alpha_0 + \alpha_1 + \alpha_2 = f(1) = 1$$

Solving this set of equations yields for the values of α_i: $\alpha_0 = 0$, $\alpha_1 = 1$, $\alpha_2 = 0$. The approximate function is simply

$$f_a(t) = t$$

This is shown in Figure 1.30. Note that a better approximation is possible by using more terms in the polynomial expansion. ∎

Lagrange Interpolation Formula This method avoids the need for matrix inversion discussed above. The result in this case is expressed as follows: let $f(t)$ be given for $t = t_0, t_1, \ldots, t_K$; then the function $f_a(t)$

$$f_a(t) = f(t_0)\frac{(t-t_1)(t-t_2)\cdots(t-t_K)}{(t_0-t_1)(t_0-t_2)\cdots(t_0-t_K)} + f(t_1)\frac{(t-t_0)(t-t_2)\cdots(t-t_K)}{(t_1-t_0)(t_1-t_2)\cdots(t_1-t_K)}$$
$$+ \cdots + f(t_K)\frac{(t-t_0)(t-t_1)\cdots(t-t_{K-1})}{(t_K-t_0)(t_K-t_1)\cdots(t_K-t_{K-1})} \quad (1.63)$$

This expression can be written in the compact form

$$f_a(t) = \sum_{i=0}^{K} f(t_i)\lambda_{K,i}(t) \quad \text{(a)}$$

where

$$\lambda_{K,i}(t) = \prod_{\substack{k=0 \\ k \neq i}}^{K} \frac{t - t_k}{t_i - t_k} \quad i = 0, 1, 2, \ldots, K \quad \text{(b)}$$

(1.64)

As chosen, the value of f_a tends to $f(t_0)$ for $t = t_0$, to $f(t_1)$ for $t = t_1$, and so on. Each polynomial $\lambda_{K,i}(t)$ is of degree K. Observe that it is symmetrical in the sense that it is unaltered by any interchange of the suffixes; that is, the tabulated values can be taken in any order. Observe also that each polynomial is zero at times t_n when $n = i$.

EXAMPLE 1.10

Find an approximation of order 3 for the function of Figure 1.27 using the Lagrange interpolation formula.

Solution: By (1.64) we write

$$f_a(t) = f(t_0)\lambda_{3,0}(t) + f(t_1)\lambda_{3,1}(t) + f(t_2)\lambda_{3,2}(t) + f(t_3)\lambda_{3,3}(t)$$

where, for $t_0 = 0$, $t_1 = 0.4$, $t_2 = 0.7$ and $t_3 = 1$,

$$f(0) = 0 \qquad f(0.4) = 0.4 \qquad f(0.7) = 1 \qquad f(1) = 1$$

The several values of λ are

$$\lambda_{3,0}(t) = \frac{(t-t_1)(t-t_2)(t-t_3)}{(t_0-t_1)(t_0-t_2)(t_0-t_3)} = \frac{(t-0.4)(t-0.7)(t-1)}{(0-0.4)(0-0.7)(0-1)}$$

$$= -\frac{(t-0.4)(t-0.7)(t-1)}{0.28}$$

$$\lambda_{3,1}(t) = \frac{(t-t_0)(t-t_2)(t-t_3)}{(t_1-t_0)(t_1-t_2)(t_1-t_3)} = \frac{t(t-0.7)(t-1)}{0.4(0.4-0.7)(0.4-1)}$$

$$= \frac{t(t-0.7)(t-1)}{0.072}$$

$$\lambda_{3,2}(t) = \frac{(t-t_0)(t-t_1)(t-t_3)}{(t_2-t_0)(t_2-t_1)(t_2-t_3)} = \frac{t(t-0.4)(t-1)}{(0.7-0)(0.7-0.4)(0.7-1)}$$

$$= -\frac{t(t-0.4)(t-1)}{0.063}$$

$$\lambda_{3,3}(t) = \frac{(t-t_0)(t-t_1)(t-t_2)}{(t_3-t_0)(t_3-t_1)(t_3-t_2)} = \frac{t(t-0.4)(t-0.7)}{0.18}$$

The final function is

$$f_a(t) = 0.4\,\frac{t(t-0.7)(t-1)}{0.072} - 0.7\,\frac{t(t-0.4)(t-1)}{0.063} + \frac{t(t-0.4)(t-0.7)}{0.18}$$

The exact function $f(t)$ and its approximation $f_a(t)$ are shown in Figure 1.31.   ∎

A number of other interpolation formulas are available, most of which can be derived from (1.63). Such interpolation formulas are not usually convenient because in practice $f_a(t)$ will usually be determined mainly by the adjacent tabulated values so that linear interpolation will need only a small correction. However, since all arguments appear symmetrically in (1.63), the contributions from all terms must be taken into account. Computations are made easier by

Figure 1.31
Approximation by Lagrange polynomials (refer to Figure 1.27).

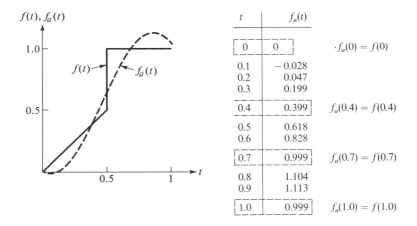

developing forms that make the special dependence on neighboring values explicit and therefore abandon the symmetry. For further details, reference should be made to works on numerical methods.

REFERENCES

1. Challifour, J. L. *Generalized Functions and Fourier Analysis*. Reading, Mass.: Benjamin, 1972.
2. Courant, R., and D. Hilbert. *Methods of Mathematical Physics*. Vol. 1. New York: Interscience, 1953.
3. Gel'fand, I. M., and G. E. Shilov. *Generalized Functions*. Vol. 1. New York: Academic Press, 1964.
4. Harmuth, E. H. *Transmission of Information by Orthogonal Functions*. 2d ed. New York: Springer-Verlag, 1972.
5. Lackey, R. B., and D. Meltzer. "A Simplified Definition of Walsh Functions." *IEEE Transactions Computers,* C-20 (1971): 211–213.
6. Lighthill, M. J. *Introduction to Fourier Analysis and Generalized Functions*. Cambridge University Press, 1964.
7. Miller, W. H., F. Ratliff, and K. Hartline. "How Cells Receive Stimuli." In *Perception: Mechanisms and Models*. San Francisco: Scientific American/W. H. Freeman, 1972.
8. Papoulis, A. *Systems and Transforms with Applications in Optics*. New York: McGraw-Hill, 1968.
9. Perlis, S. *Theory of Matrices*. 3d ed. Reading, Mass.: Addison-Wesley, 1958.
10. Schwartz, L. *Methods of Mathematical Physics*. Reading, Mass.: Addison-Wesley, 1964.

11. Seely, S., and A. D. Poularikas. *Electromagnetics—Classical and Modern Theory and Applications*. New York: Marcel Dekker, 1979.
12. Seely, S., and A. D. Poularikas. *Electrical Engineering: Introduction and Concepts*. Beaverton, Ore.: Matrix, 1982.
13. Stakgold, I. *Boundary Value Problems of Mathematical Physics*. Vol. 1. New York: Macmillan, 1967.
14. Yuen, C. *Walsh Functions and Gray Code*. Proc. 1971 Walsh Functions Symposium, 68–73, National Technical Information Service, Springfield, Va.

PROBLEMS

1-2.1 Describe three periodic phenomena that are observed in nature.

1-2.2 Add the two periodic functions shown in Figure P1-2.2 with $T_1 = 1$ for $f_1(t)$ and $T_2 = 2$ for $f_2(t)$, and indicate if the new function is periodic.

Figure P1-2.2

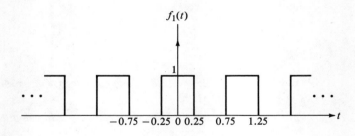

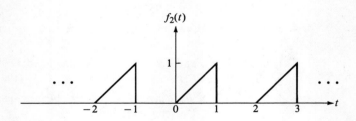

1-2.3 Determine which of the following signals are periodic. For those signals that are periodic, determine their fundamental frequencies ω_0.
  a. $f(t) = -4 \sin(-3.4 \times 10^{-6} t + \pi/6)$
  b. $f(t) = 2.5 \cos(2\pi \times 10^2 t + 30°) u(-t)$
  c. $f(t) = 2.5 \cos^2(2\pi \times 10^{-3} t) - 2.5 \sin^2(2\pi \times 10^{-3} t)$
  d. $f(t) = \sum_{k=0}^{\infty} e^{-jk3t}$
  e. $f(k) = 2 \cos(2\pi k/8 + 30°)$
  f. $f(k) = \sin^2(k/12)$
  g. $f(k) = 2 \sin(2k/4) \sin(k/4)$

1-2.4 Plot the following signals:
  a. $f(t) = \sin(3t - 20°)\,u(t - 2)$
  b. $f(t) = \text{Re}\{e^{-j2t}e^{j30°}u(t + 1)\}$
  c. $f(t) = \sin(0t)$ $f(t) = \sin[(\pi/8)t]$ $f(t) = \sin[(\pi/4)t]$ $f(t) = \sin[(\pi/2)t]$ $f(t) = \sin(\pi t)$
     $f(t) = \sin[(3\pi/2)t]$ $f(t) = \sin[(7\pi/4)t]$ $f(t) = \sin[(15\pi/8)t]$ $f(t) = \sin(2\pi t)$
  d. The following values of $f(k)$: $\sin(0k)$, $\sin[(\pi/8)k]$, $\sin[(\pi/4)k]$, $\sin[(\pi/2)k]$, $\sin(\pi k)$, $\sin[(3\pi/2)k]$, $\sin[(7\pi/4)k]$, $\sin[(15\pi/8)k]$, $\sin(2\pi k)$.
  Draw an important conclusion about the difference between continuous and discrete sinusoidal functions from parts (c) and (d).

1-2.5 If $z = x + jy = re^{j\theta}$, prove the following relations:
  a. $(zz^*)^{1/2} = r$ b. $(z/z_1)^* = (r/r_1)e^{-j(\theta - \theta_1)}$ c. $z + z^* = 2\,\text{Re}\,z$
  d. $z - z^* = 2j\,\text{Im}\,z$ e. $|z|^2 = |z^*|^2$ f. $|z_1 z_2|^2 = |z_1|^2\,|z_2|^2$

1-3.1 If $f(t) = u(t) - u(1 - t)$, sketch the following functions:
  a. $f(-t)$ b. $f(2 - t)$ c. $f(t - 1)$

1-3.2 If $f(t) = p(t - 2) + p(t + 2)$, sketch the following functions:
  a. $f(t - 2)$ b. $f(t)\,\text{sgn}(t)$ c. $f(t + 1) - f(t - 1)$

1-3.3 Sketch the following functions:
  a. $f(t) = 2u(t - 2) - u(t - 3) + p_2(t - 1)$
  b. $f(t) = 2\Lambda(t) - 3\Lambda_2(t - 3)$
  c. $f(t) = 2\Lambda_2(t)\,\text{sgn}(t)$

1-3.4 For $f(t) = [\Lambda(t) + p(t - 2)]u(t)$, sketch the following functions:
  a. $f(t)$ b. $f(t/2)$ c. $f(t + 3)$ d. $f(-t - 5)$ e. $f(\frac{t - 5}{2})$

1-3.5 Write the mathematical representation of each of the functions shown in Figure P1-3.5. Also, write $\Lambda_a(t)$ in terms of pulse and ramp functions.

Figure P1-3.5

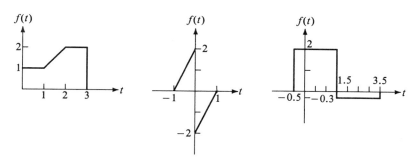

1-3.6 For the signal $f(t)$ shown in Figure P1-3.6, sketch each of the following signals:
  a. $f(t - 1)$ b. $f(2 - t)$ c. $f(2t + 1)$ d. $f[(t/2) - 1]$ e. $f(t - 1)u(t - 2)$
  f. $f(t)[\delta(t + 0.5) - \delta(t - 1.5)]$

Figure P1-3.6

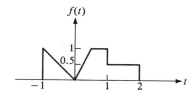

1-3.7 Find the following signals, given that $f(t)$ is the signal shown in Figure P1-3.6 and $g(t)$ is the signal shown in Figure P1-3.7:
  a. $f(t)g(t)$ **b.** $f(t)g(t-1)$ **c.** $f(t+1)g(t-1)$

Figure P1-3.7

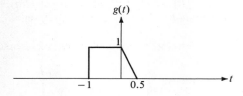

1-3.8 Two discrete time signals are shown in Figure P1-3.8. Find the following signals:
  a. $f(k)g(k)$ **b.** $f(-k)g(k)$ **c.** $f(k-2)g(1-k)$ **d.** $f(k+1)g(k+2)$ **e.** $(-1)^k f(k) + (0.8)^k g(k)$

Figure P1-3.8

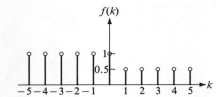 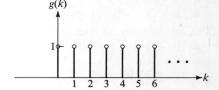

1-3.9 Sketch the functions:
  a. $f(t) = p_4(t)\,\text{comb}(t-1)$ **b.** $f(t) = \Lambda_2(t)\,\text{comb}_{0.5}(t)$ **c.** $f(t) = \text{sinc}_4(t)\,\text{comb}_{0.5}(t)$
  d. $f(t) = \text{sgn}(t)\,\text{comb}(t)$

1-3.10 Sketch the following discrete signals:
  a. $f(k) = u(k) + \cos(k + 30°)$ **b.** $f(k) = 2(0.8)^k$ **c.** $f(k) = \delta(k+2) - 2\delta(k-2)$
  d. $f(k) = u(k) - u(k-4)$ **e.** $f(k) = -r(-k)$ **f.** $f(k) = -u(3-k)$
  g. $f(k) = r(k-2)$ **h.** $f(k) = 2r(k)\,\delta(k-3)$ **i.** $f(k) = r(k)[u(k-2) - u(k-5)]$

1-3.11 Sketch the following discrete signals:
  a. $f(k) = (-1)^k 2(0.9)^k \quad k \geq 0$ **b.** $f(k) = 0.5(1.1)^k \quad k \geq 0$
  c. $f(k) = (0.9)^k \sin(\pi/8)k \quad k \geq 0$ **d.** $f(k) = (1.1)^2 \sin(\pi/8)k \quad k \geq 0$

1-5.1 If the basis functions are complex functions, show that

$$\int_a^b |f(t)|^2 \, dt = \int_a^b f(t) f^*(t) \, dt = \sum_{n=0}^{\infty} |c_n|^2 |\varphi_n|^2$$

1-5.2 Given a linear independent system of functions $\varphi_1, \varphi_2, \varphi_3, \ldots, \varphi_n, \ldots$ defined on the interval $[a, b]$, define a new system $\psi_1, \psi_2, \ldots, \psi_n, \ldots$ as follows

$$\psi_1 = \varphi_1 \qquad \psi_2 = \varphi_2 - \frac{\int_a^b \psi_1(t)\varphi_2(t)\,dt}{\int_a^b \psi_1^2(t)\,dt}\psi_1(t)$$

$$\psi_3 = \varphi_3 - \frac{\int_a^b \psi_1(t)\varphi_3(t)\,dt}{\int_a^b \psi_1^2(t)\,dt}\psi_1(t) - \frac{\int_a^b \psi_2(t)\varphi_3(t)\,dt}{\int_a^b \psi_2^2(t)\,dt}\psi_2(t) \quad \cdots$$

This procedure is called the **Gram-Schmidt orthogonalization process.** Apply this process to the set of functions $1, t, t^2, t^3, \ldots$ in the interval $-1 \leq t \leq 1$ and show that they generate the Legendre polynomials within a constant.

1-5.3 Are the basis functions $\{\varphi_n(t)\} = \{\exp(jn\omega_0 t)\}$ orthonormal in the range $0 \leq t \leq 2\pi/\omega_0$? If not, find the appropriate constant that will make them orthonormal ($\omega_0 = 2\pi/T$).

1-5.4 a. Show that the functions $\varphi_0(t) = \sqrt{\frac{1}{2}}$, $\varphi_1(t) = \sqrt{\frac{3}{2}} t$ are orthonormal in the interval $-1 \leq t \leq 1$.
   b. Apply the mean square error minimization procedure to find the constants c_0 and c_1 that will represent $f(t) = t^3$ in the approximate form $f_a(t) = c_0 \varphi_0(t) + c_1 \varphi_1(t)$.

1-6.1 Find the functions:
   a. $\text{wal}_w(8, t)$ **b.** $\text{wal}_p(6, t)$ **c.** $\text{wal}_h(6, t)$

1-6.2 Find an approximation to the function shown in Figure 1.27 using the first three $\text{wal}_p(i, t)$ functions.

1-6.3 Approximate the function shown in Figure P1-6.3 by the first three Legendre functions.

Figure P1-6.3

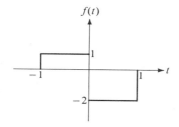

1-7.1 Use a 3 × 3 matrix and show that the formula given by (1.61) is satisfied.

1-7.2 Approximate with a third-order polynomial the function given in Figure 1.27.

1-7.3 Approximate the function shown in Figure 1.27 with a fourth-order Lagrange polynomial. Compare these results with those shown in Figure 1.31.

CHAPTER 2
CONVOLUTION, IMPULSE RESPONSE, AND SYSTEM REPRESENTATION

A. CONVOLUTION AND ITS PROPERTIES

2-1 INTRODUCTION

Every **physical system** is broadly characterized by its ability to accept an **input** such as voltage, pressure, and displacement and to produce an output response to this input. For example, a modern video-disc player is an optoelectronics system; the input is the reflection (or transmission) of laser light from the grooves of the record and the output is a video signal that is viewed on a television screen. A telescope is an optical system that accepts as its input the radiance from the stars and produces as its output their images on a film. A mass on a spring is an elementary seismograph with the input being the force of the undulating earth and the output the oscillations of the mass.

When studying the behavior of systems, the procedure is to model mathematically each element that comprises the system and then to consider the interconnected array of elements. The resulting interconnected system is described mathematically; the specific form of the description is dictated by the domain of description, whether a time-domain or frequency-domain description. For a system that consists of elements described by proportional, continuous time derivative or integral functions that relate the input-output properties, the resulting description is given as an integrodifferential equation. Such a description results when we apply the Kirchhoff voltage law to a series *RLC* circuit with a voltage-source input and with the circuit current as the output. Systems that are described in discrete time form are described mathematically by difference equations. More will be said about these systems in Chapters 8 and 9.

The analysis of most systems can be reduced to the study of the relationship between certain input excitations and the resulting outputs. We can represent the system in any of the forms shown in Figure 2.1. The major categories of systems are shown in Figure 2.2 and are described as follows.

Lumped-parameter systems are those comprised of a finite number of physically discrete elements, each of which is able to store or dissipate energy or, if it is a source, to deliver energy. In our study we shall consider: electrical sys-

Figure 2.1
Operator representation of systems. (a) Operator formalism. (b) Impulse response formalism. (c) System function formalism. (d) Multi-input multi-output formalism.

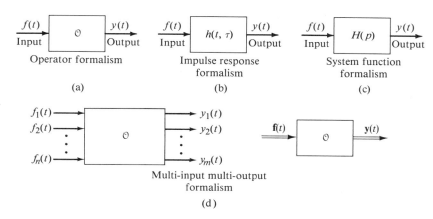

Figure 2.2
Major categories of systems. Paths discussed in our studies are illustrated by a dotted line.

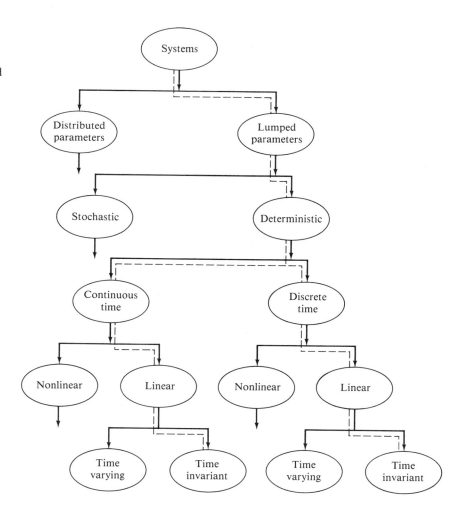

tems (resistors, inductors, capacitors, sources) and mechanical systems (masses, springs, dampers, sources). Such systems assume that no propagation delays occur in signals traversing each element, and they are described by ordinary differential equations. Conversely, **distributed-parameter** systems consist of elements that cannot be described by simple lumped elements because of propagation time delays in signals traversing the elements. However, each infinitesimal part of distributed elements can be modeled in lumped-parameter form. Distributed-parameter systems are described in terms of partial differential equations. Optical systems (films, lenses, apertures, sources) constitute a class of distributed-parameter systems.

Systems may be **linear** or **nonlinear.** A linear system is one for which a linear relation exists between cause and effect or between excitation and response. We will find that superposition applies for linear systems. A casual inspection will not show linearity, since both lumped and distributed elements can be linear or nonlinear. We will find that systems analysis can often be carried completely in closed mathematical form for systems composed of linear elements. This is rarely possible for nonlinear systems. To deduce a closed-form solution, one often assumes linearity for nonlinear systems. Extreme care is necessary in such cases since the assumptions of linearity may completely negate important features of the system. Although we confine ourselves to linear systems in this book, be aware that physical systems, when analyzed in detail, might actually be nonlinear. However, if their nonlinearities are small, we can approximate physical systems as linear systems for small variations around their operating points—a common approach in electronics and in control theory.

Systems whose parameters vary with time are called **time-varying** systems. We will not consider them in this book. We will confine our attention to systems with constant parameters, the so-called **time-invariant** systems. That is, we will concentrate our efforts on a study of those systems included by the dotted line in Figure 2.2. Even within these limitations, the number of specific systems is almost limitless. Therefore we will confine our attention to a study of some basic electrical, mechanical, and optical systems. Three such simple systems are illustrated in Figure 2.3. Essentially, for electrical and mechanical systems, we will treat only linear time-invariant systems (LTI), and optical systems will be limited to linear space-invariant systems (LSI).

Whether systems are continuous time or discrete time, the input-output relation is given in operator form or by means of a system function operator. For linear time-invariant systems, the input-output relation is often given by the convolution of the system impulse (or delta function) response function with the input function. All of these methods receive detailed attention in this text.

■ ■ ■

2-2 SUPERPOSITION INTEGRAL

A feature of the superposition integral is that it makes possible a determination of the time response of a linear time-invariant system to a general excitation

Figure 2.3
Three elementary systems.

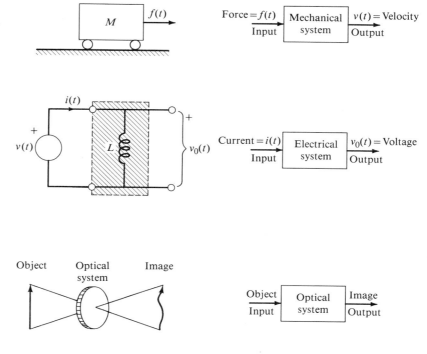

function from knowledge of the system response to a unit step excitation. In practice, finding the response of a system to a step function may be a complicated problem in itself, and then carrying out the subsequent steps required by the superposition integral can be quite involved mathematically. Analytically, however, the process has interest.

Consider a causal and linear system to which a unit step excitation is applied. Causal functions are those that start at some finite time, say $t = 0$, and have zero value for $t < 0$. Thus, in general, causal functions are those for which $f(t) = 0$ for $t < 0$. Also, a system is causal or nonanticipatory if the output of the system at time t depends only on the applied input after $t = 0$ and does not depend on the input applied after time t. Hence, past inputs affect the future, but future inputs do not affect the past.

If a system is noncausal, it is anticipatory. Thus the output of a noncausal system depends both on past inputs and on the future value of the inputs. This implies that a noncausal system is able to predict the input that will be applied in the future, which is impossible for a real physical system.

A system is linear if the parameter values of the elements thereof do not depend on through-variables (in electrical systems, currents; in mechanical systems, forces and torques) or across-variables (in electrical systems, voltages; in mechanical systems, velocities and angular velocities); that is, parameter values are independent of magnitudes of through- or across-variables. Other definitions of linear systems will be given later, all of which imply the existence of

Figure 2.4
Step-wise approximation to a given waveform; representation in the τ-domain.

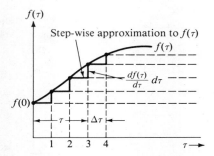

superposition; namely, the consequence of two or more excitations applied simultaneously to the system is the sum of the consequences of each excitation acting separately.

The response of the system to unit step excitation, which can be obtained by classical network analysis techniques, is termed the **indicial** response, and is written $y_u(t)$. Knowing this response, we can find the system response to a general excitation function $f(t)$. The procedure is to represent the given excitation function $f(t)$ by a step-wise approximation with time intervals $\Delta\tau$, as shown in Figure 2.4. Clearly, the given waveform $f(t)$ is approximated by a number of incremental step functions that are delayed in accordance with the sampling time intervals. The system response will then be the superposition of the response of the system to each incremental step function, with due account taken of the time delay associated with each step. We can formalize this process mathematically.

The components of response corresponding to the successive voltage steps are the following:

$$g_0(t) = f(0)y_u(t) \qquad\qquad t \geq 0$$

$$g_1(t) = \left.\frac{df}{d\tau}\right|_{\Delta\tau} y_u(t - \Delta\tau)\,\Delta\tau \qquad t \geq \Delta\tau$$

$$g_2(t) = \left.\frac{df}{d\tau}\right|_{2\,\Delta\tau} y_u(t - 2\,\Delta\tau)\,\Delta\tau \qquad t \geq 2\,\Delta\tau \qquad (2.1)$$

$$g_3(t) = \left.\frac{df}{d\tau}\right|_{3\,\Delta\tau} y_u(t - 3\,\Delta\tau)\,\Delta\tau \qquad t \geq 3\,\Delta\tau$$

and so on. The response at any time t, say the interval between $n\,\Delta\tau$ and $(n+1)\,\Delta\tau$, is given approximately by the expression

$$g(t) = \sum_i g_i(t) = f(0)y_u(t) + \left[\left.\frac{df}{d\tau}\right|_{\Delta\tau} y_u(t - \Delta\tau) + \cdots + \left.\frac{df}{d\tau}\right|_{n\,\Delta\tau} y_u(t - n\,\Delta\tau)\right]\Delta\tau$$

which is

$$g(t) = f(0)y_u(t) + \sum_{n=1}^{N} \frac{df}{d\tau} y_u(t - n\,\Delta\tau)\,\Delta\tau \qquad (2.2)$$

This form is suitable for numerical calculations using a digital computer.

To improve the approximation, $\Delta\tau$ is made progressively smaller and n is permitted to approach infinity. Actually, this involves a limiting procedure, a complicated process, since at the same time that $\Delta\tau$ approaches zero, n approaches infinity. This requires consideration of the two limits:

$$\lim_{\Delta\tau \to 0} \Delta\tau \qquad \lim_{\substack{\Delta\tau \to 0 \\ n \to \infty}} (n\,\Delta\tau)$$

The limiting process cannot be given rigorous treatment without the use of advanced mathematics, but an outline of the processes involved follows. As $\Delta\tau$ is allowed to approach zero, one limit approaches the condition of an increment of a continuous variable, which is denoted $d\tau$,

$$\lim_{\Delta\tau \to 0} \Delta\tau \to d\tau$$

Further, for each value of $\Delta\tau$ the summation in (2.2) takes on all positive integer values, and so the product $n\,\Delta\tau$ varies from 0 to infinity. It is represented by the variable τ, thus

$$\lim_{\substack{\Delta\tau \to 0 \\ n \to \infty}} (n\,\Delta\tau) \to \tau$$

For each finite value of $\Delta\tau$, the quantity $n\,\Delta\tau$ varies with n in a step-like fashion, but the variation becomes smooth as $\Delta\tau$ approaches zero. In the limit as $\Delta\tau \to 0$, the sum becomes an integral and the integrand becomes $y_u(t-\tau)$, and (2.2) leads to

$$g(t) = f(0)y_u(t) + \int_0^t y_u(t-\tau)\frac{df}{d\tau}\,d\tau \tag{2.3}$$

This is one of many forms of the superposition (Duhamel) integral that relates the response of the system to a general waveform $f(t)$ from a knowledge of the indicial response to the unit-step excitation.

Another form of interest is obtained by carrying out the integration in (2.3).

Let

$$u = y_u(t-\tau) \qquad du = -\frac{dy_u(t-\tau)}{d(t-\tau)}d\tau \qquad dv = \frac{df}{d\tau}d\tau \qquad v = f(\tau)$$

Equation (2.3) can be written

$$g(t) = f(0)y_u(t) + y_u(t-\tau)f(\tau)\Big|_0^t + \int_0^t f(\tau)\frac{dy_u(t-\tau)}{d(t-\tau)}\,d\tau$$

which becomes

$$g(t) = f(t)y_u(0) + \int_0^t f(\tau)\frac{dy_u(t-\tau)}{d(t-\tau)}\,d\tau \tag{2.4}$$

This form of the superposition integral warrants special attention. Recall that $y_u(t)$ is the response of the initially relaxed network to a unit step. Then $dy_u(t)/dt$, in accordance with the discussion in Section 1-3 and (1.18), is the response of the same initially relaxed network to the unit impulse. We can interpret (2.4) to show that the superposition integral in the form given resolves the excitation function $f(t)$ into a set of successive and appropriately summed impulse functions with amplitudes $f(\tau)\,d\tau$ at each instant.

A third form of the superposition integral follows directly from (2.4) by a simple change of variable. If we write $t - \tau = \tau'$ then the integral in (2.4) becomes

$$\int_0^t f(\tau)\frac{dy_u(t-\tau)}{d(t-\tau)}\,d\tau = -\int_t^0 f(t-\tau')\frac{dy_u(\tau')}{d\tau'}\,d\tau' = \int_0^t f(t-\tau)\frac{dy_u(\tau)}{d\tau}\,d\tau$$

The final step is valid because the definite integral has a value that is not dependent on the variable used. Hence (2.4) becomes

$$g(t) = f(t)y_u(0) + \int_0^t f(t-\tau)\frac{dy_u(\tau)}{d\tau}\,d\tau \tag{2.5}$$

Other forms are possible (see Problem 2-2.1).

■ ■ ■

2-3 CONVOLUTION INTEGRAL

The superposition integral in the forms given in (2.4) and (2.5) is frequently written in forms that reflect special properties of the system input. First, consider the term $f(t)y_u(0)$, which, as is evident from Figure 2.4, is a measure of the response of the system to an initial excitation at time $t = 0$. This might be considered as the response to an excitation applied at a negative time with the value of the response being precisely $f(t)y_u(0)$ at $t = 0$. This factor might then be written

$$f(t)y_u(0) = \int_{-\infty}^0 f(\tau)\frac{dy_u(t-\tau)}{d(t-\tau)}\,d\tau = \int_{-\infty}^0 f(t-\tau)\frac{dy_u(\tau)}{d\tau}\,d\tau \tag{2.6}$$

If this expression is combined with (2.5) then we can write

$$g(t) = \int_{-\infty}^t f(\tau)\frac{dy_u(t-\tau)}{d(t-\tau)}\,d\tau \tag{2.7}$$

In this form the expression contains the response at $t = 0$, but it is not shown explicitly.

For convenience, we now designate the impulse response of the system

$$h(t-\tau) = \frac{dy_u(t-\tau)}{d(t-\tau)} \tag{2.8}$$

Using this notation (2.7) assumes the form

$$g(t) = \int_{-\infty}^t f(\tau)h(t-\tau)\,d\tau \tag{2.9}$$

CONVOLUTION, IMPULSE RESPONSE, AND SYSTEM REPRESENTATION

Correspondingly, if we proceed from (2.5), the equivalent form will be

$$g(t) = \int_{-\infty}^{t} f(t - \tau)h(\tau)\,d\tau \tag{2.10}$$

The integrals in (2.9) and (2.10) are **convolution** integrals. These integrals can be interpreted to show that if the impulse response of a system $h(t)$ is known, the response to any other input excitation $f(t)$ can be determined by a simple convolution operation of these two functions, which is designated symbolically by $f(t) * h(t)$. The upper limit is t and not ∞ because we are interested in the impulse response at any time t resulting from the application of a delta-function excitation. Further, if the system is causal and relaxed at $t = 0$, the lower limits in (2.9) and (2.10) will be zero.

A reasonable interpretation of convolution can be given. Let us specify $f(t)$ as an infinite sum of weighted impulses. Refer to (1.18) and write

$$f(t) = \int_{-\infty}^{\infty} f(\tau)\,\delta(t - \tau)\,d\tau \tag{2.11}$$

But we know that the response of a linear system to a unit impulse is $h(t)$. It then follows that the response to $f(\tau)\,\delta(t - \tau)$ will be $f(\tau)h(t - \tau)$, and by superposition, the response to $f(t)$ will be

$$g(t) = \int_{-\infty}^{\infty} f(\tau)h(t - \tau)\,d\tau \tag{2.12}$$

which is (2.9) where $h(t - \tau) = 0$ for $\tau > t$.

Convolution is a general mathematical process involving real-valued functions. For the case of two real-valued functions, say $h(t)$ and $f(t)$, the convolution of these two functions will be given by

$$\boxed{g(t) \triangleq f(t) * h(t) = \int_{-\infty}^{\infty} f(\tau)h(t - \tau)\,d\tau = \int_{-\infty}^{\infty} f(t - \tau)h(\tau)\,d\tau} \tag{2.13}$$

Suppose that $f(t)$ denotes the input to an integrator. The output from the integrator will be

$$g(t) \triangleq \text{output} = \int_{-\infty}^{t} f(t)\,dt = \int_{-\infty}^{t} \int_{-\infty}^{\infty} f(\tau)\delta(t - \tau)\,d\tau\,dt$$

$$= \int_{-\infty}^{\infty} f(\tau) \int_{-\infty}^{t} \delta(t - \tau)\,dt\,d\tau = \int_{-\infty}^{\infty} f(\tau)u(t - \tau)\,d\tau$$

Further, we shall designate any linear operator (system) by the symbol $\mathcal{O}^{-1}$, and for the particular system under review (the integrator) we must write the identity

$$\mathcal{O}^{-1} \equiv \int_{-\infty}^{t} dt$$

Figure 2.5
The convolved functions in the t and τ domains.

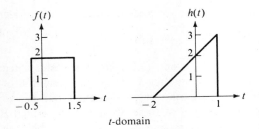

t-domain

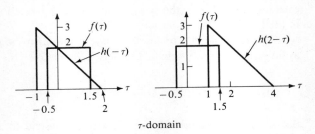

τ-domain

Hence, the output of any system can be written as

$$g(t) \triangleq f(t) * h(t) = \int_{-\infty}^{\infty} f(\tau)\mathcal{O}^{-1}\{\delta(t-\tau)\}\, d\tau$$

$$= \int_{-\infty}^{\infty} f(\tau)h(t-\tau)\, d\tau \qquad (2.14)$$

which again shows that the output of this LTI system when excited by any input is given by the convolution of its input and its impulse response $h(t)$.

The integral of (2.14) has a graphical representation, as shown by the illustration in Figure 2.5. Observe that the integrand function is equal to the product of one of the functions and the other function folded over and shifted by t. This integral is often referred to as the Faltung integral in mathematical literature.

The following are a number of important observations relative to convolution:

1. If $f(t) = 0$ for $t < 0$, then $f(t) * h(t) = \int_0^{\infty} f(\tau)h(t-\tau)\, d\tau$ (see Figure 2.6a) for the special case when $f(t) = e^{-t}u(t)$ and $h(t) = \operatorname{sinc} t \triangleq (\sin t)/t$.

2. For $h(t - \tau) = 0$ for $\tau > t$, the convolution integral is given by $f(t) * h(t) = \int_{-\infty}^{t} f(\tau)h(t-\tau)\, d\tau$ (see Figure 2.6b).

3. For causal systems $h(t) = 0$ for $t < 0$ and $f(t) = 0$ for $t < 0$. This implies that $h(t - \tau) = 0$ for $\tau > t$ and $h(\tau) = 0$ for $\tau < 0$. Their convolution is given by $f(t) * h(t) = \int_0^t f(\tau)h(t-\tau)\, d\tau$. (see Figure 2.6c).

For a causal system, the response can never precede the input. In the manner discussed in connection with (2.6), we split the convolution integral into two parts

Figure 2.6
Convolution properties of different functions.

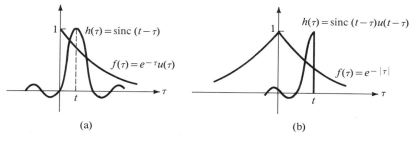

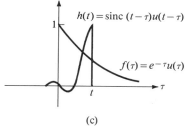

$$g(t) \triangleq f(t) * h(t) = \underbrace{\int_{-\infty}^{t_0-} f(\tau)h(t-\tau)\,d\tau}_{g_{zi}(t)} + \underbrace{\int_{t_0-}^{t} f(\tau)h(t-\tau)\,d\tau}_{g_{zs}(t)}$$

$$= g_{zi}(t) + g_{zs}(t) \tag{2.15}$$

The function $g_{zi}(t)$ is the **zero input** response of the system and $g_{zs}(t)$ is the **zero state** response of the system, where t_0 is arbitrarily taken as the initial time. The symbol t_{0-} indicates that functions with finite discontinuity are included in the definition of the zero-state response.

The second convolution integral $g_{zs}(t)$ given by (2.15) represents the zero-state response of the system. By this we mean that at time $t = t_0$ the system was relaxed; that is, there were no charges on capacitors, no currents through inductors, no deformation of springs, no velocities of masses, and so on. However, if the initial state of the system was not zero, we must add a zero-input response produced by the initial state at $t = t_0$. That is, the first integral is the response at time t caused by the initial state; this is precisely the term given by (2.6), which specifies that the state at $t = t_0$ depends on the inputs prior to that time. Thus if a force (input) is applied to a body (system) at time $t = t_0$, the motion of the particle (output) for $t \geq t_0$ is uniquely determined if we know its position and velocity at $t = t_0$, and it is immaterial how the body attained these initial values. This means that the state of this system at $t = t_0$ is its velocity and position at this instant. More is said about the state of a system later in this chapter, and Chapter 12 is devoted to the analysis of systems by a state-space method.

■ ■ ■

2-4 PROPERTIES OF CONVOLUTION

We now review many important properties of convolved functions.

It is important to know that the convolution $g(t) = f(t) * h(t)$ does not exist for all possible functions. The sufficient conditions are:

a. Both $f(t)$ and $h(t)$ must be absolutely integrable on the interval $(-\infty, 0]$.

b. Both $f(t)$ and $h(t)$ must be absolutely integrable on the interval $[0, \infty)$.

c. Either $f(t)$ or $h(t)$ (or both) must be absolutely integrable on the interval $(-\infty, \infty)$.

For example, the convolution $\cos \omega_0 t * \cos \omega_0 t$ does not exist. The important properties of convolution are the following:

1. Commutative. This states that

$$g(t) = \int_{-\infty}^{\infty} f(\tau) h(t - \tau) \, d\tau = \int_{-\infty}^{\infty} f(t - \tau) h(\tau) \, d\tau \qquad (2.16)$$

This is easily proved by setting $t - \tau = \tau'$ in the first integral and then renaming the dummy variable τ' to τ.

2. Distributive. This specifies that

$$g(t) = f(t) * [h_1(t) + h_2(t)] = f(t) * h_1(t) + f(t) * h_2(t) \qquad (2.17)$$

This property follows directly as a result of the linear property of integration.

3. Associative. This specifies that

$$[f(t) * h_1(t)] * h_2(t) = f(t) * [h_1(t) * h_2(t)] \qquad (2.18)$$

EXAMPLE 2.1
Verify (2.18).

Solution: Expand the expression

$$[f(t) * h_1(t)] * h_2(t) = \int_{-\infty}^{\infty} \left[\int_{-\infty}^{\infty} f(\lambda) h_1(\tau - \lambda) \, d\lambda \right] h_2(t - \tau) \, d\tau$$

$$= \int_{-\infty}^{\infty} f(\lambda) \left[\int_{-\infty}^{\infty} h_1(\tau - \lambda) h_2(t - \tau) \, d\tau \right] d\lambda$$

By a change of variable, with $\tau - \lambda = \mu$, then $d\tau = d\mu$; when $\tau = -\infty$, $\mu = -\infty$ and when $\tau = \infty$, $\mu = \infty$, the expression becomes

$$= \int_{-\infty}^{\infty} f(\lambda) \left[\int_{-\infty}^{\infty} h_1(\mu) h_2(t - \lambda - \mu) \, d\mu \right] d\lambda$$

Now set $t - \lambda = v$. Then we have $d\lambda = -dv$; when $\lambda = -\infty$, $v = \infty$ and when $\lambda = \infty$, $v = -\infty$, the expression becomes

$$= \int_{-\infty}^{\infty} f(t - v) \left[\int_{-\infty}^{\infty} h_1(\mu) h_2(v - \mu) \, d\mu \right] dv$$

CONVOLUTION, IMPULSE RESPONSE, AND SYSTEM REPRESENTATION

which shows that

$$[f(t) * h_1(t)] * h_2(t) = [h_1(v) * h_2(v)] * f(v) = f(t) * [h_1(t) * h_2(t)]$$ ∎

4. Shift Invariance. This property specifies that if

$$g(t) = f(t) * h(t)$$

then

$$g(t - t_0) = f(t - t_0) * h(t) = \int_{-\infty}^{\infty} f(\tau - t_0) h(t - \tau) d\tau \quad (2.19)$$

This property is proved by writing $g(t)$ in its integral form, substituting $t - t_0$ for t, setting $\tau + t_0 = \tau'$, and then renaming the dummy variable.

EXAMPLE 2.2

Suppose that the input to a linear time-invariant system (LTI) is the function $f(t) = \cos \omega t \triangleq \text{Re}\{e^{j\omega t}\} \triangleq \text{Re}\{\tilde{f}(t)\}$, where $\tilde{f}(t) = \cos \omega t + j \sin \omega t$ is a complex valued function. Find the corresponding response for a given $h(t)$.

Solution: We write, by (2.13),

$$g(t) = \int_{-\infty}^{\infty} h(\tau) \text{Re}\{e^{j\omega(t-\tau)}\} d\tau = \text{Re}\left\{e^{j\omega t} \int_{-\infty}^{\infty} h(\tau) e^{-j\omega \tau} d\tau\right\}$$

Observe that the last integral is a function of ω, and we write this

$$H(\omega) = \int_{-\infty}^{\infty} h(\tau) e^{-j\omega \tau} d\tau$$

$H(\omega)$ is a complex valued function in general and is known as the **system function**. In fact, we find in Chapter 4 that this integral is the Fourier transform of $h(t)$. Since $H(\omega)$ is frequently a complex function, we can represent it in the following general form:

$$H(\omega) = H_r(\omega) + jH_i(\omega) = |H(\omega)| e^{j\varphi(\omega)}$$

where

$H_r(\omega)$ and $H_i(\omega)$ are real-valued functions

$|H(\omega)| = [H_r^2(\omega) + H_i^2(\omega)]^{1/2}$

$\varphi(\omega) = \tan^{-1}(H_i(\omega)/H_r(\omega))$

The output is now written in the form

$$g(t) = \text{Re}\{e^{j\omega t} H(\omega)\} \triangleq \text{Re}\{\tilde{g}(t)\} = |H(\omega)| \cos[\omega t + \varphi(\omega)]$$

These results show that if

$$\tilde{f}(t) = e^{j\omega t} \quad \text{then} \quad \tilde{g}(t) = H(\omega) e^{j\omega t} = |H(\omega)| e^{j[\omega t + \varphi(\omega)]}$$

Further, we see that the output is identical in form with the input, but its amplitude has been changed and it has been shifted in phase. ∎

CHAPTER 2

For the case when the impulse response of a system is the pulse function $h(t) = p_1(t-1)$, the system function is

$$H(\omega) = \int_{-\infty}^{\infty} p_1(\tau-1)e^{-j\omega\tau}\,d\tau = \int_0^2 e^{-j\omega\tau}\,d\tau = \frac{1}{-j\omega}e^{-j\omega\tau}\Big|_0^2$$

$$= \frac{e^{-j\omega}}{-j\omega}(e^{-j\omega} - e^{j\omega})$$

$$= 2\frac{\sin\omega}{\omega}e^{-j\omega} = 2e^{-j\omega}\,\text{sinc}\,\omega$$

The output of a system with this value of $h(t)$ to a $\cos\omega t$ input is

$$g(t) = \text{Re}\{2e^{-j\omega}e^{+j\omega t}\,\text{sinc}\,\omega\} = 2\,\text{sinc}\,\omega\,\cos\omega(t-1)$$

If $\omega = \pi$, the output signal is zero even though the input is the signal $\cos\pi(t-1)$, which is complete frequency elimination.

5. The Area Property. Consider the integrals

$$A_f = \int_{-\infty}^{\infty} f(t)\,dt = \text{area} \qquad m_f = \int_{-\infty}^{\infty} tf(t)\,dt = \text{first moment}$$

Also let $K_f = m_f/A_f = $ center of gravity. Then the convolution $g(t) = f(t) * h(t)$ leads to

$$\begin{aligned} A_g &= A_f A_h \quad &\text{(a)} \\ K_g &= K_f + K_h \quad &\text{(b)} \end{aligned} \tag{2.20}$$

EXAMPLE 2.3
Prove the validity of (2.20b).

Solution: The first moment of $g(t)$ is written

$$m_g = \int_{-\infty}^{\infty} tg(t)\,dt = \int_{-\infty}^{\infty} t\left[\int_{-\infty}^{\infty} f(\tau)h(t-\tau)\,d\tau\right]dt$$

$$= \int_{-\infty}^{\infty} f(\tau)\left[\int_{-\infty}^{\infty} th(t-\tau)\,dt\right]d\tau$$

Write $t - \tau = \lambda$, then

$$= \int_{-\infty}^{\infty} f(\tau)\left[\int_{-\infty}^{\infty}(\lambda + \tau)h(\lambda)\,d\lambda\right]d\tau$$

which is written

$$= \int_{-\infty}^{\infty} f(\tau)\,d\tau\int_{-\infty}^{\infty}\lambda h(\lambda)\,d\lambda + \int_{-\infty}^{\infty}\tau f(\tau)\,d\tau\int_{-\infty}^{\infty} h(\lambda)\,d\lambda$$

$$= A_f m_h + m_f A_h$$

Hence, the quantity

$$\frac{m_g}{A_f A_h} = \frac{m_g}{A_g} \triangleq K_g = \frac{A_f m_h + m_f A_h}{A_f A_h} = K_h + K_f$$

6. Scaling Property. This property states that if
$$g(t) = f(t) * h(t)$$
then

$$|a|g\left(\frac{t}{a}\right) = f\left(\frac{t}{a}\right) * h\left(\frac{t}{a}\right) \quad (2.21)$$

7. Complex-Valued Functions. If both $f(t)$ and $h(t)$ are complex functions, then $g(t)$ can be written

$$\begin{aligned} g(t) = f(t) * h(t) &= [f_r(t) + jf_i(t)] * [h_r(t) + jh_i(t)] \\ &= [f_r(t) * h_r(t) - f_i(t) * h_i(t)] + j[f_r(t) * h_i(t) + f_i(t) * h_r(t)] \end{aligned} \quad (2.22)$$

8. Derivative. In addition to (1.18), which proved that $f(t) = f(t) * \delta(t)$, we rewrite the expression

$$g(t) = f(t) * \frac{d\,\delta(t)}{dt} = \int_{-\infty}^{\infty} f(\tau) \frac{d\,\delta(t-\tau)}{dt}\,d\tau$$

$$= \frac{d}{dt} \int_{-\infty}^{\infty} f(\tau) \delta(t-\tau)\,d\tau = \frac{df(t)}{dt} \quad (2.23)$$

EXAMPLE 2.4
Discuss the convolution of the pulse functions $p_a(t)$ and $p_{2a}(t)$.

Solution: Refer to Figure 2.7a, which shows the overlapping of the two pulse functions for different values of t. Figure 2.7b shows the resulting function $g(t)$. The points on the curve represent the values of the integrals at the values of t shown in Figure 2.7a. Since the two rectangular pulses have unit amplitudes, the value of $g(t)$ is equal to the area of overlap. Observe that the resulting function is smoother than either of the convolving functions. ∎

EXAMPLE 2.5
Determine the convolution

$$g(t) = p_a(t-a) * [\delta(t+2a) - \delta(t-2a)]$$

Solution: The essentials of the evaluation are shown in Figure 2.8. Figure 2.8a shows the respective functions in the τ-plane. The convolution is shown in Figure 2.8b. ∎

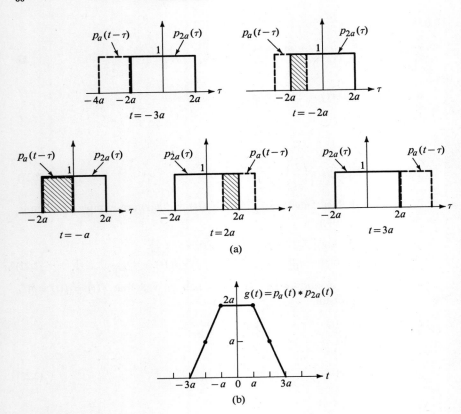

Figure 2.7
Graphical representation of the convolution of two unit pulse functions.

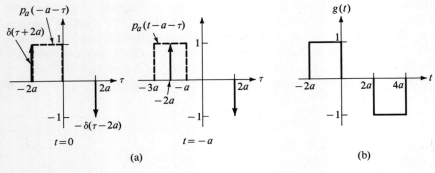

Figure 2.8
Convolution with delta functions.

CONVOLUTION, IMPULSE RESPONSE, AND SYSTEM REPRESENTATION

EXAMPLE 2.6
Determine the convolution given by the functions

$$g(t) = p_{1/2}(t - \tfrac{1}{2}) * 2p_{1/4}(t - \tfrac{1}{4})$$

Solution: Figure 2.9 includes the details of the solution. ■

EXAMPLE 2.7
Determine the convolution of the functions

$$g(t) = p_{1/2}(t - 0.5) * \left[e^{-t}u(t)\right]$$

Solution: The essential details of the evaluation are contained in Figure 2.10. The student will remember from basic circuit analysis studies that the application of a delta function voltage to a series RL circuit yields a current that is an exponential function. Further, when the input voltage is a pulse, the current has the shape shown in Figure 2.10. ■

The properties of the convolution operation are summarized in Table 2.1.

Table 2.1 Convolution Properties

1. Causal system	$g(t) = \int_{-\infty}^{t} f(\tau)h(t-\tau)\,d\tau$	$h(t) = 0$ for $t < 0$
2. Delta function convolution	$g(t) = f(t) * \delta(t) = f(t)$	
3. Commutative	$g(t) = \int_{-\infty}^{\infty} f(\tau)h(t-\tau)\,d\tau = \int_{-\infty}^{\infty} f(t-\tau)h(\tau)\,d\tau$	
4. Distributive	$g(t) = f(t) * [h_1(t) + h_2(t)] = f(t) * h_1(t) + f(t) * h_2(t)$	
5. Associative	$[f(t) * h_1(t)] * h_2(t) = f(t) * [h_1(t) * h_2(t)]$	
6. Shift invariance	$g(t - t_0) = f(t - t_0) * h(t)$	
7. Area	$A_g = A_f A_h \quad A_g =$ area of $g(t)$ $A_f =$ area of $f(t) \quad A_h =$ area of $h(t)$	
8. Center of gravity	$K_g = K_f + K_h \quad K_g = \dfrac{m_g}{A_g} \quad m_g = \int_{-\infty}^{\infty} t g(t)\,dt$	
9. Scaling	$f\left(\dfrac{t}{a}\right) * h\left(\dfrac{t}{a}\right) = |a| g\left(\dfrac{t}{a}\right)$	
10. Derivative	$f(t) * \dfrac{d\,\delta(t)}{dt} = \dfrac{df(t)}{dt}$	

■ ■ ■

Figure 2.9
Development of the convolution of two rectangular pulses.

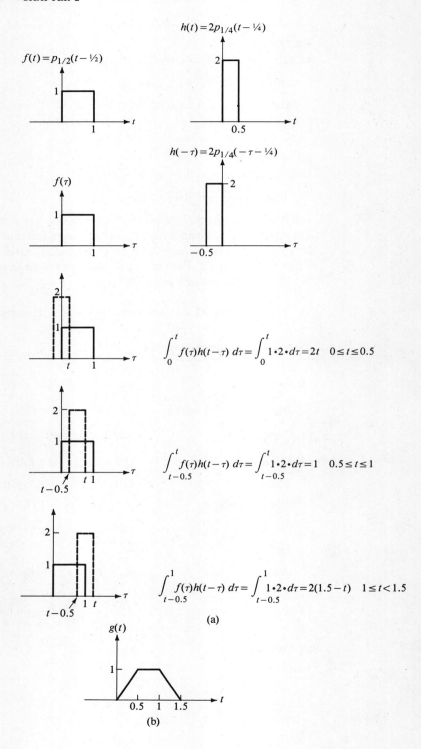

Figure 2.10
The convolution of a pulse and an exponential function.

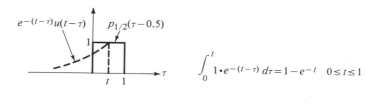

$$\int_0^t 1 \cdot e^{-(t-\tau)}\,d\tau = 1 - e^{-t} \quad 0 \le t \le 1$$

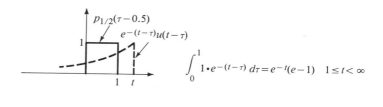

$$\int_0^1 1 \cdot e^{-(t-\tau)}\,d\tau = e^{-t}(e-1) \quad 1 \le t < \infty$$

$g(t) = p_{1/2}(t-0.5) * [e^{-t}u(t)]$

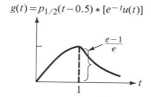

2-5 PERIODIC CONVOLUTION

Periodic or circular convolution, which is a special case of general convolution, attains a special form when $f(t)$ and $h(t)$ are periodic functions with the same period. Periodic convolution is defined by the integral

$$g(t) = \frac{1}{T}\int_0^T f(\tau)h(t-\tau)\,d\tau \tag{2.24}$$

where the integral is taken over one period T. We wish to show that $g(t)$ is also periodic. For this, consider the integral

$$g_p(t) = \frac{1}{T}\int_c^{c+T} f(\tau)h(t-\tau)\,d\tau \tag{2.25}$$

and we choose $c = kT + a$ with $0 \le a < T$. Equation (2.25) then becomes

$$g_p(t) = \frac{1}{T}\int_{kT+a}^{kT+a+T} f(\tau)h(t-\tau)\,d\tau$$

$$= \frac{1}{T}\int_0^T f(\tau' + kT + a)h[t - (\tau' + kT + a)]\,d\tau'$$

$$= \frac{1}{T}\int_0^T f(\tau')h(t-\tau')\,d\tau' = g(t)$$

since $f(t)$ and $h(t)$ are periodic. This result shows that $g(t)$ itself is **periodic**.

We substitute $t - \tau = \tau'$ in (2.24) and obtain

$$-\int_{t}^{t-T} f(t-\tau')h(\tau')\,d\tau' = -\int_{T}^{0} f(t-\tau')h(\tau')\,d\tau' = \int_{0}^{T} f(t-\tau')h(\tau')\,d\tau'$$

This relationship shows that the periodic convolution is **commutative**; that is,

$$f(t) * h(t) = h(t) * f(t) \tag{2.26}$$

In a similar manner, we can also show that the periodic convolution is **associative**. Hence we write

$$f_1(t) * [f_2(t) * f_3(t)] = [f_1(t) * f_2(t)] * f_3(t) \tag{2.27}$$

This topic will receive further attention in Chapter 3.

■ ■ ■

2-6 CORRELATION

The **cross-correlation** of two different functions is defined by the relation

$$\boxed{R_{fh}(t) \triangleq f(t) \star h(t) = \int_{-\infty}^{\infty} f(\tau)h(\tau-t)\,d\tau = \int_{-\infty}^{\infty} f(\tau+t)h(\tau)\,d\tau} \tag{2.28}$$

When $f(t) = h(t)$, the correlation operation is called **autocorrelation**.

Equation (2.28) shows that cross-correlation does not obey the commutative rule; that is

$$f(t) \star h(t) \neq h(t) \star f(t) \tag{2.29}$$

The cross-correlation and autocorrelation of complex functions are defined as follows:

$$R_{fh}(t) \triangleq f(t) \star h^*(t) = \int_{-\infty}^{\infty} f(\tau)h^*(\tau-t)\,d\tau \tag{2.30a}$$

$$R_f(t) \triangleq f(t) \star f^*(t) = \int_{-\infty}^{\infty} f(\tau)f^*(\tau-t)\,d\tau \tag{2.30b}$$

Now apply the Schwarz inequality, which is given by

$$\left| \int f_1(t)f_2(t)\,dt \right| \leq \left[\int |f_1(t)|^2\,dt \right]^{1/2} \left[\int |f_2(t)|^2\,dt \right]^{1/2} \tag{2.31}$$

to (2.30b). We obtain the relation

$$\left| \int_{-\infty}^{\infty} f(\tau)f^*(\tau-t)\,d\tau \right| \leq \left[\int_{-\infty}^{\infty} |f(\tau)|^2\,d\tau \right]^{1/2} \left[\int_{-\infty}^{\infty} |f(\tau-t)|^2\,d\tau \right]^{1/2} \tag{2.32}$$

Since the value of the integral does not change when the function is shifted, (2.32) becomes

Figure 2.11
Illustration of the correlation principle.

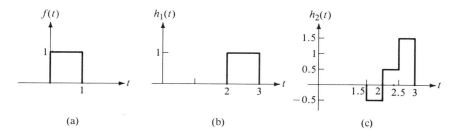

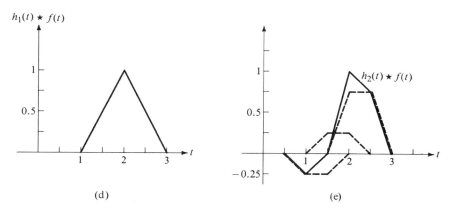

$$\left| \int_{-\infty}^{\infty} f(\tau) f^*(\tau - t) \, d\tau \right| \leq \int_{-\infty}^{\infty} |f(\tau)|^2 \, d\tau = \int_{-\infty}^{\infty} f(\tau) f^*(\tau) \, d\tau$$

from which it follows that

$$|R_f(t)| \triangleq |f(t) \star f^*(t)| \leq R_f(0) \tag{2.33}$$

This equation indicates that there is a time $t = 0$ at which the absolute value of the autocorrelation function is equal to or larger than at any other time. This fact is of great importance in signal detection. It is routinely used in radar signal detection when correlation between the emitted signal and the signal returned from a target is performed. A large peak indicates a resemblance between the returned signal and the emitted signal, from which we assume that a target is present.

EXAMPLE 2.8

Find the correlation between a pulse transmitted by a radar, as shown in Figure 2.11a and each of the two possible received pulses shown in Figures 2.11b and 2.11c.

Solution: Employ (2.28) plus the graphical approach discussed in Section 2-4 in connection with convolution. The correlation between $f(t)$ and $h_1(t)$ is shown in Figure 2.11d. Also the correlation between $f(t)$ and $h_2(t)$ can be accomplished

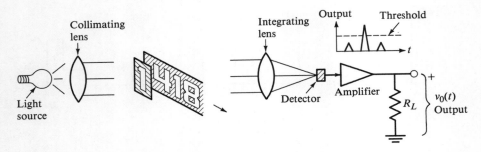

Figure 2.12
Simple optical correlating device.

by constructing three independent correlations between $f(t)$ and each of the three pulses of the function $h_2(t)$. The first pulse extends between 1.5 and 2, the second is between 2 and 2.5, and the third is between 2.5 and 3. The dotted lines in Figure 2.11e are the results of the three correlation operations, and the solid line is their total sum.

It is interesting to compare the resulting functions in Figures 2.11d and 2.11e; the first results from the correlation of two similar pulses, and the second from a pulse that might be considered to be distorted by noise. We see that if the detected signal is distorted by noise, the resulting output signal of the correlation operation is also distorted. However, if the noise is not severe, we can always assign a threshold value to specify whether or not the correlated output indicates the presence of a signal. This is essentially a go/no-go situation; if the output of the correlator is larger than some threshold value, we say that the signal is present. ∎

The idea of matching signals may take a completely different format. For example, Figure 2.12 illustrates a practical way of correlating shapes—in this case, that of recognizing numbers.

Correlation techniques are also used in medicine where acoustic (ultrasonic) waves are used as probes to detect anomalies inside the human body. This technique also finds important applications in many areas of physics and technology.

■ ■ ■

B. MODELING SIMPLE ELECTRICAL, MECHANICAL, AND OPTICAL SYSTEMS

2-7 MODELING SIMPLE SYSTEMS

As already noted, an essential requirement in systems analysis is a mathematical input-output description of the elements comprising the interconnected system.

Figure 2.13
(a) A capacitor across a voltage source. (b) Linear charge-voltage relationship.

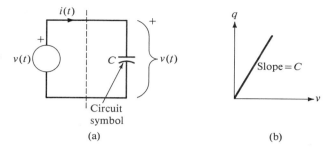

This requirement exists because we wish to perform mathematical studies on systems comprising hardware components or to carry out designs mathematically that will ultimately be realized by hardware. We must stress that the analysis or design can be no better than the quality of the models used. We here review the essential ideas in the modeling process.

2-7.a Electrical Elements

The Capacitor The linear capacitor is an idealized circuit element in which energy may be stored in electric form. In its most elementary configuration, the capacitor consists of two closely spaced metallic plates that are separated by a single or multiple layers of nonconducting (insulating) dielectric material (air, glass, paper). The schematic representation of the capacitor is shown to the right of the broken line in Figure 2.13a. The terminal properties of a linear time-invariant capacitor are described graphically by a charge-voltage relationship of the form shown in Figure 2.13b. The capacitors used in most electronic circuits are time-invariant, although the capacitor microphone used in radio studios is an example of a time-varying and linear capacitor.

By definition, the capacitance

$$C = \frac{q}{v} \qquad \frac{\text{coulomb}}{\text{volt}} = \text{farad (F)} \qquad (2.34)$$

and since the current i is defined by

$$i = \frac{dq}{dt} \qquad \frac{\text{coulomb}}{\text{second}} = \text{ampere (A)} \qquad (2.35)$$

we obtain the relations

$$\boxed{\begin{aligned} v(t) &= \frac{1}{C} \int_{-\infty}^{t} i(t')\, dt' \quad &\text{(a)} \\ i(t) &= C \frac{dv(t)}{dt} \quad &\text{(b)} \end{aligned}} \qquad (2.36)$$

Figure 2.14
(a) An inductor across a voltage source. (b) Linear flux-current relationship. (c) Illustration of flux linkages.

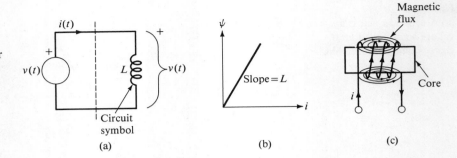

Here $i(t)$ is denoted as a **through variable** and $v(t)$ is an **across variable,** which is descriptive of their inherent properties.

The Inductor Another important electrical element that stores energy in the magnetic field instead of the electric field, is the inductor (sometimes called a coil or solenoid). The terminal properties of the linear inductor are described graphically by the flux linkage-current ψ, i relationship shown in Figure 2.14b. Figure 2.14a shows the circuit representation of the inductor, and Figure 2.14c gives a rough illustration of the flux linkages. The simple telephone receiver is an example of a time-varying inductor.

By definition, the inductance of an element is written

$$L = \frac{\psi}{i} \qquad \frac{\text{weber}}{\text{ampere}} = \text{henry (H)} \tag{2.37}$$

and using Faraday's law

$$v(t) = \frac{d\psi}{dt} \qquad \text{volt (V)} \tag{2.38}$$

we obtain the relations

$$
\begin{aligned}
i(t) &= \frac{1}{L} \int_{-\infty}^{t} v(t')\, dt' \quad &\text{(a)} \\
v(t) &= L \frac{di(t)}{dt} \quad &\text{(b)}
\end{aligned}
\tag{2.39}
$$

The Resistor Unlike the capacitor and inductor, which store energy, the resistor dissipates energy. For a linear resistor, the (v, i) characteristic is a straight line with a factor of proportionality

CONVOLUTION, IMPULSE RESPONSE, AND SYSTEM REPRESENTATION

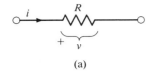

(a)

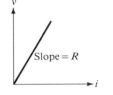

(b)

Figure 2.15
Network representation of the resistor.

$$R = \frac{v(t)}{i(t)} \quad \frac{\text{volt}}{\text{ampere}} = \text{ohm } (\Omega) \quad \text{(a)}$$

$$i(t) = \frac{1}{R} v(t) = Gv(t) \quad \text{(b)}$$

(2.40)

where R = resistance (ohm) and $G = 1/R$ = conductance (mho). The network representation is shown in Figure 2.15. Ordinary resistors are usually assumed to be time-invariant but the microphone in the ordinary telephone set is an example of a time-varying resistor.

2-7.b Translational Mechanical Elements

The Ideal Mass Element   The dynamics of a mass element are described by Newton's second law of motion

$$f(t) = M \frac{dv(t)}{dt} = M \frac{d^2 x(t)}{dt^2} = Ma \qquad \text{newton} = \text{kg} \cdot \text{m} \cdot \text{s}^{-2} \text{ (N)} \quad (2.41)$$

which relates the force with the acceleration a. The motional variable v enters in a relative form: it is the velocity of the mass relative to the velocity of the ground, which is zero, and is an **across** variable. The force is transmitted through an element; hence it is a **through** variable. The integral form of (2.41) is given by

$$v(t) = \frac{1}{M} \int_{-\infty}^{t} f(t') \, dt' \qquad (2.42)$$

It is interesting to compare (2.36) and (2.42), which shows an **analogy** between the mass in a mechanical system and the capacitor in the electrical system. The schematic representation of the mass system is shown in Figure 2.16.

Figure 2.16
Schematic representation of the mass element.

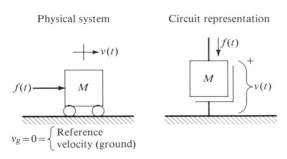

Figure 2.17
Schematic representation of the spring.

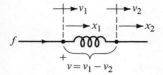

The Spring A spring element is one that stores energy during the variation of its shape due to elastic deformation resulting from the application of a force. Over its linear region, the spring satisfies Hooke's law which relates the linear force to the displacement by the expression

$$f(t) = Kx(t) \quad \text{newton} \tag{2.43}$$

where K is the spring constant, with units newton/m.

By differentiating (2.43) with respect to time, we obtain the relation

$$v(t) = \frac{1}{K}\frac{df(t)}{dt} \tag{2.44}$$

Refer to the schematic representation of the spring in Figure 2.17, and use (2.43) and (2.44). We have

$$f(t) = K[x_1(t) - x_2(t)] = K\int_{-\infty}^{t}[v_1(t') - v_2(t')]\,dt'$$
$$= K\int_{-\infty}^{t} v(t')\,dt' \tag{2.45}$$

If $x_1 > x_2$, there is a compressive force and $f > 0$. If $x_1 < x_2$, there is a negative or extensive force. Attention is called to the **analogy** between the spring and the inductor.

The Damper We shall limit our considerations only to viscous friction, which, for a linear dependence between force and velocity, is given by

$$f(t) = Dv(t) \quad \text{(a)} \tag{2.46}$$

from which

$$v(t) = \frac{1}{D}f(t) \quad \text{(b)}$$

where D is the damping constant (newton-second/m). A mechanical damper and its schematic representation are shown in Figure 2.18. Observe the **analogy** between the damper and the resistor.

Figure 2.18
Physical and diagrammatic representations of a dash pot (damper).

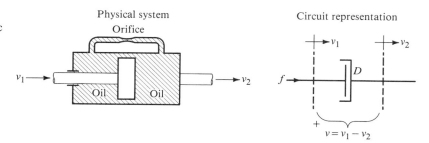

By maintaining the parallelism between through and across variables, we found that mass and capacitor, spring and inductor, and damper and resistor were analogous entities. In this case, there would be a direct parallelism in the resulting topology, with a parallel connection of mechanical elements leading to an analogous parallel connection of electrical elements. From a purely mathematical point of view, we could have chosen a parallelism between velocity and current, and force and voltage, in which case we would obtain an analogy between mass and inductor, spring and capacitor, and damper and resistor. Now, however, the topology would be dually related; that is, a parallel connection of mechanical elements would lead to an analogous series connection of electrical elements. While some of the older writings (particularly in acoustics) employed this dual relationship, we will confine ourselves only to the through- and across-variable parallelism and thereby maintain the parallel topological structure.

2-7.c Rotational Mechanical Elements A set of rotational mechanical elements and rotational variables exist that bear a one-to-one correspondence to the translational mechanical elements and the translational variables discussed in the foregoing sections. In the rotational system, **torque** is the through variable and **angular velocity** is the motional or across variable. The corresponding fundamental entities are:

J = polar moment of inertia	corresponds to M in translation
K = rotational spring constant	corresponds to K in translation
D = rotational damper	corresponds to D in translation
$\mathcal{T}$ = torque	corresponds to f in translation
$\omega = d\theta/dt$ angular velocity	corresponds to $v = dx/dt$ in translation
$\alpha = d\omega/dt$ angular acceleration	corresponds to $a = dv/dt$ in translation

The Inertial Element In this rotational set, J is the rotational parameter, the assumed proportionality factor between torque and angular acceleration. When the motion is considered on one axis only,

$$\mathcal{T} = \frac{d(J\omega)}{dt} \quad \text{(a)} \tag{2.47}$$

Figure 2.19
Moment of inertia of a homogeneous solid disk.

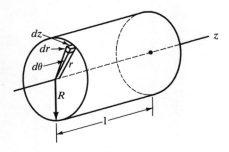

and for systems with constant J

$$\boxed{\mathcal{T} = J\frac{d\omega(t)}{dt} = J\frac{d^2\theta(t)}{dt^2}} \qquad \text{newton-meter} \quad \textbf{(b)}$$

J, the moment of inertia of a rotational body, depends upon the mass and the square of a characteristic distance of the body called the **radius of gyration** k. This relation is given by

$$J = Mk^2 \qquad \text{kg-m}^2 \tag{2.48}$$

For a simple point mass rotating about an axis at a distance r from the center of mass, $k = r$ and

$$J = Mr^2 \tag{2.49}$$

For a simple disk of radius r rotating about its center, $k = r/\sqrt{2}$, with

$$J = \tfrac{1}{2}Mr^2 \tag{2.50}$$

In general, moment of inertia J is related to the mass by the integral

$$J = \int r^2\, dm \tag{2.51}$$

where the integral is taken over the body and the rotation occurs on an axis that is perpendicular to its center.

EXAMPLE 2.9
Find the moment of inertia and the radius of gyration of a homogeneous disk about its axis, as shown in Figure 2.19.

Solution: Use (2.51), with $\rho =$ density of the material in kg/m^3,

$$J = \iiint_V r^2\, dm = \iiint_V r^2 \rho\, dV = \rho \int_0^l dz \int_0^{2\pi} d\theta \int_0^R r^2 r\, dr$$

$$= \rho\, \frac{l2\pi R^4}{4} = \frac{M}{V} l2\pi R^4 = \frac{M}{\pi R^2 l}\, \frac{l2\pi R^4}{4} = \frac{M}{2} R^2$$

Figure 2.20
Schematic representation of rotational inertia.

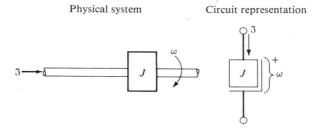

Figure 2.21
Schematic representation of a rotational spring.

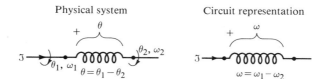

By (2.48) we find that

$$k = \sqrt{\frac{J}{M}} = \sqrt{\frac{MR^2}{2M}} = \frac{R}{\sqrt{2}}$$

The essential features of the rotational system are given schematically in Figure 2.20. ∎

Spring Element A rotational spring is one that will twist under the action of a torque. A linear spring element is described by the pair of equations

$$\boxed{\begin{aligned} \mathcal{T}(t) &= K\theta(t) = K \int_{-\infty}^{t} \omega(t')\,dt' \quad &\text{(a)} \\ \omega(t) &= \frac{1}{K}\frac{d\mathcal{T}(t)}{dt} \quad &\text{(b)} \end{aligned}} \quad (2.52)$$

The schematic representation of the rotational spring is shown in Figure 2.21.

The Damper The rotational damper differs from the translational damper principally in the character of the motion. The schematic representation is given in Figure 2.22, and the equations that describe a rotational damper are:

$$\boxed{\begin{aligned} \mathcal{T}(t) &= D\omega(t) \quad &\text{(a)} \\ \omega(t) &= \frac{1}{D}\mathcal{T}(t) \quad &\text{(b)} \end{aligned}} \quad (2.53)$$

Figure 2.22
Schematic representation of a rotational viscous damper.

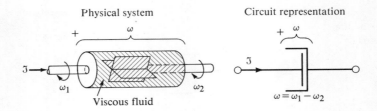

†2-7.d Optical Systems
The fields of geometrical and physical optics have long been important concerns of electrical engineers through their work in television cameras and their extensive work on antennas for radio, radar, and radio astronomy. More recently interest has expanded in the field of electro-optics involving optical signal processing and pattern recognition. We therefore consider the basics of optical systems in our studies.

An optical system involves lenses and apertures as a means of modifying an electromagnetic field (the optical field). A description of an optical system relates the output field to the input field, usually under the simplifying assumption that the field can be described by a scalar function instead of the more complicated vector function. Note, however, that in optical systems we observe the intensity of the field (the field amplitude squared) and not the field itself—that is, the squared value of the amplitude of the electric field component of the electromagnetic wave E^2. E^2 is proportional to the power that the field carries with it. This is what exposes a photographic film.

Although optical systems are two-dimensional and usually involve circular lenses, for simplicity we will treat only the one-dimensional case. This simplified discussion will still allow a presentation of the basic principles of optical systems. Moreover, this simplification is not completely artificial since cylindrical lenses exist and are often used in signal processing. Their index of refraction varies only in one dimension.

The elements that comprise optical systems include: lenses (spherical, cylindrical, and so on); thin films (often introducing known attenuation and phase shifts in the light as it passes through the film); space distances (which exist between the several elements of the system); light sources (these might be point sources, sets of point sources, or distributed sources of various sizes and shapes); and detectors. The light sources might be special sources plus apertures having shapes for particular needs of the problem. Moreover, the light might be coherent (constant phase) or noncoherent (random phase). We will discuss some of these factors in much the same manner as the discussion in Section 2-7a on electrical elements R, L, C. In our discussion of optical systems, we assume a monochromatic (single-frequency) electromagnetic field. Although pure monochromaticity is not physically realizable, it is a very useful simplifying approximation.

The Film The relationship between the input complex field $\Phi_0(y)$ to an optical element and the resulting output field $\Phi_1(y)$ is

Figure 2.23
Schematic and operational representation of a film optical element.

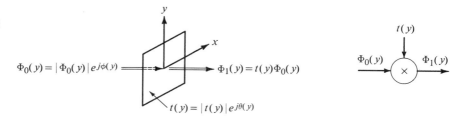

Figure 2.24
Schematic and operational representation of the space distance d.

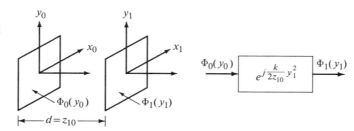

$$\Phi_1(y) = t(y)\Phi_0(y) = |t(y)|e^{j\theta(y)}\Phi_0(y) \tag{2.54}$$

where $t(y)$ denotes the transmission properties of the element, here assumed to be very thin, and $\Phi_0(\cdot)$ is the complex form representation of the electric field E of the electromagnetic wave. We specify $\Phi_0(\cdot)$ in complex form because we know from electromagnetic field studies that the representation of the field in complex form facilitates computations. In addition, the complex representation provides us with information of amplitude and phase of the field in a compact manner. A schematic representation of the optical thin film is shown in Figure 2.23.

The Space Distance When the field traverses an optical path distance $d = z_{10}$ (the distance from one plane to another), its length is ordinarily many wavelengths and a substantial phase change is involved. The relationship between the distributed initial optical field and the final optical field at a point is given by the Fresnel integral in the form

$$\Phi_1(y_1) = \int_{-\infty}^{\infty} \Phi_0(y_0) e^{j(k/2z_{10})(y_1 - y_0)^2} \, dy_0 \tag{2.55}$$

where $k = 2\pi/\lambda$ and $\lambda =$ optical wavelength. For simplicity, the proportionality constant in front of the integral has been set equal to unity. The optical path factor system and its operational form are shown in Figure 2.24. The quantity within the rectangle is the impulse response of the space distance system.

Figure 2.25
Schematic and operational representation of a thin cylindrical lens.

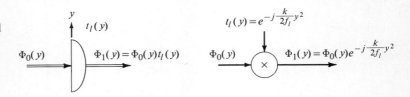

The Lens The transmittance factor $t_l(y)$ for a thin lens differs from that of a thin film owing to the curvature of the lens. The evaluation of the factor $t_l(y)$ involves the use of the paraxial ray approximation, which replaces the spherical wavefront by a parabolic wavefront. We are assuming that the amplitude remains unchanged (no attenuation), but there is a phase shift only, given by

$$t_l(y) = e^{-j(k/2f_l)y^2} \tag{2.56}$$

where f_l = focal length of the lens. This expression assumes that there is no amplitude reduction in passing through the lens, only a phase shift. Figure 2.25 shows the schematic and operational form of a simple one-dimensional (cylindrical) lens.

■ ■ ■

C. IMPULSE RESPONSE

2-8 THE IMPULSE RESPONSE OF SYSTEMS

In this section we will ascertain the impulse response of a system using the classical solution of ordinary differential equations. In Section 2-3 we found the impulse response using a system operator. In Chapter 6 we employ Laplace transform techniques to find system response. Recall the assertion that a knowledge of the impulse response of a system permits finding, by means of the convolution integral, the response of the system to any input. Recall also that the impulse response of a physical system is the time derivative of the step function response.

EXAMPLE 2.10
Find the impulse response of the system shown in Figure 2.26. Use this result to find its output when the input is the function $p_1(t-1)$.

Solution: Initially we wish to represent the system in its network representation. To construct the equivalent circuit representation of the mechanical system shown in Figure 2.26a, we first create a diagram of velocity levels, as shown in

Figure 2.26b. Next, we connect the source (force) from the ground velocity to the node marked v_1. Since the mass element has a velocity v with respect to the ground, we insert the mass between v and v_g. Similarly, the relative velocity of the damper with respect to the ground is v, and we insert the damper element between v and v_g. Furthermore, v_1 and v are the same; hence we connect the two points. The resulting circuit is that shown in Figure 2.26c. Note specifically that even though the system appears to be connected in series, it is actually a parallel combination of elements.

For mechanical circuits **D'Alembert's principle** parallels the Kirchhoff current law in electrical circuits. This is usually written

$$\sum_{\text{forces}} f_i(t) - M_i \frac{dv_i}{dt} = 0 \qquad i = \text{node number} \tag{2.57}$$

The term $M(dv/dt)$ is often termed the **kinetic reaction.** This principle essentially requires that each node be isolated in the analysis and that the $M(dv/dt)$ force be distinguished from the other forces. This yields a "free-body" diagram for each portion of the system. The appropriate free-body diagram for this example is shown in Figure 2.26d. When using the equivalent network representation for mechanical circuits, it is more convenient to write the mechanical equilibrium equation by including f_M with other forces and considering (2.57) as a point law.

The equilibrium equation is seen, from an examination of Figure 2.26c, to be

$$-f(t) + f_M(t) + f_D(t) = 0 \tag{2.58}$$

by an application of the point law at node 1 (note that the **algebraic sum** of the forces = 0 has been written). Use the relationships for mechanical elements (see Section 2-7) to write this expression in the form ($M = 1$, $D = 2$)

$$\frac{dv}{dt} + 2v = f(t) \tag{2.59}$$

We set $f(t) = u(t)$ and solve the first-order differential equation, for relaxed initial conditions $v(0) = 0$. The step response solution is easily obtained as

$$v_u(t) = \tfrac{1}{2}(1 - e^{-2t})u(t) \tag{2.60}$$

By (2.8) the impulse response of the system is

$$h(t) = \frac{dv_u(t)}{dt} = e^{-2t}u(t) + \tfrac{1}{2}(1 - e^{-2t})\delta(t)$$

from which

$$h(t) = e^{-2t}u(t) \qquad t \geq 0+ \tag{2.61}$$

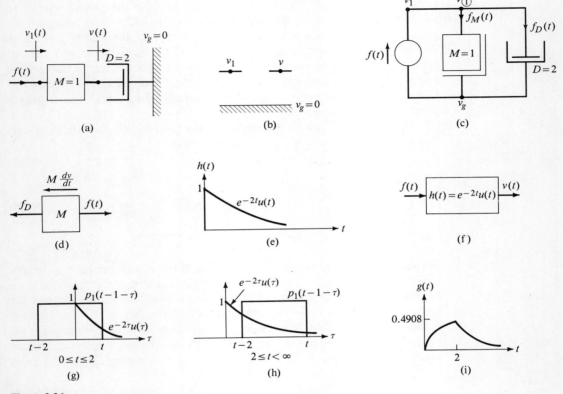

Figure 2.26
The impulse response of a mechanical system and its output to a pulse input.

The input pulse to the system and the impulse response of the system are shown in Figures 2.26e, 2.26g, and 2.26h. The output signal of the system, which is given by the convolution of these signals, is

$$g(t) = \begin{cases} \int_0^t 1 \cdot e^{-2\tau} d\tau = -\tfrac{1}{2} e^{-2\tau} \Big|_0^t = \tfrac{1}{2}(1 - e^{-2t}) & 0 \le t \le 2 \\ \int_{t-2}^t 1 \cdot e^{-2\tau} d\tau = -\tfrac{1}{2} e^{-2\tau} \Big|_{t-2}^t = \tfrac{1}{2} e^{-2t}(e^4 - 1) & 2 \le t < \infty \end{cases}$$

The output signal is shown in Figure 2.26i. ∎

The following example presents a different technique for finding the impulse response of a system.

EXAMPLE 2.11
Find the impulse response of the relaxed system shown in Figure 2.27a; also find

Figure 2.27
A simple electrical system.

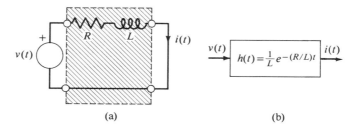

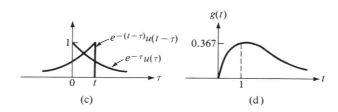

its output if the input is the function $e^{-t}u(t)$, with $R = L = 1$.

Solution: Apply Kirchhoff's voltage law around the circuit, which yields the equation

$$L\frac{di(t)}{dt} + Ri(t) = v(t) \qquad (2.62)$$

If $v(t) = \delta(t)$, we write this equation in the form

$$\frac{dh(t)}{dt} + \frac{R}{L}h(t) = \frac{1}{L}\delta(t) \qquad (2.63)$$

Integrate this equation in the range $(0-) \leq t \leq (0+)$ to find

$$\int_{0-}^{0+} \frac{dh(t)}{dt}\,dt + \frac{R}{L}\int_{0-}^{0+} h(t)\,dt = \frac{1}{L}\int_{0-}^{0+} \delta(t)\,dt \qquad (2.64)$$

Assume that $h(t)$ does not possess an impulse function in the origin; hence the second integral vanishes. The first integral yields

$$\int_{0-}^{0+} \frac{dh(t)}{dt}\,dt = h(0+) - h(0-) = h(0+) \qquad (2.65)$$

since the initially relaxed (zero initial conditions) system specifies that $h(0-) = 0$. The third integral is equal to one, as specified by the definition of the delta function. Upon combining these results, we find that the initial condition is $h(0+) = 1/L$.

Now we use the fact that at $t > (0+)$ the delta function is equal to zero, which permits (2.63) to be written

$$\frac{dh(t)}{dt} + \frac{R}{L} h(t) = 0$$

This equation is a simple first-order ordinary differential equation, with the solution

$$h(t) = Ke^{-Rt/L} \qquad t > 0 \qquad (2.66)$$

where K is an unknown constant that must be determined from the initial conditions. We have, therefore

$$h(0+) = K = \frac{1}{L}$$

Thus the impulse response of the system becomes

$$h(t) = \frac{1}{L} e^{-Rt/L} \qquad t > 0 \qquad (2.67)$$

Correspondingly, its shifted form is given by

$$h(t - t_0) = \frac{1}{L} e^{-R(t-t_0)/L} \qquad t > t_0$$

The input pulse to the system and the impulse response of the system are shown in Figure 2.27c. The output signal is then given by

$$g(t) = \int_0^t e^{-\tau} e^{-(t-\tau)} d\tau = t e^{-t} \qquad t > 0$$

The output is shown in Figure 2.27d. ∎

EXAMPLE 2.12
Find the impulse response of the rotational mechanical system shown in Figure 2.28a.

Solution: Use will be made of the analogous D'Alembert's principle for rotating mechanical systems, corresponding to (2.57): for any system, the algebraic sum of externally applied torques and the torques resisting rotation about any axis is zero. Hence for each node, with due account of sign, $\sum_{\text{node}} \mathcal{T}_i = 0$ or basically

$$\mathcal{T} + \mathcal{T}_I + \mathcal{T}_D + \mathcal{T}_K = 0 \qquad (2.68)$$

where

$\mathcal{T}$ = externally applied torque

$\mathcal{T}_I$ = inertial torque = $J \dfrac{d\omega}{dt} = J \dfrac{d^2\theta}{dt^2}$

$\mathcal{T}_D$ = damping torque = $D\omega = D \dfrac{d\theta}{dt}$

Figure 2.28
A rotational mechanical system.

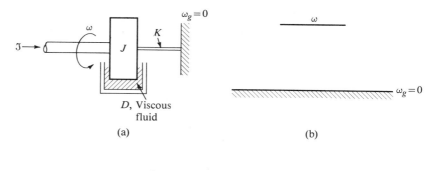

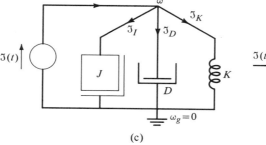

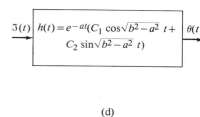

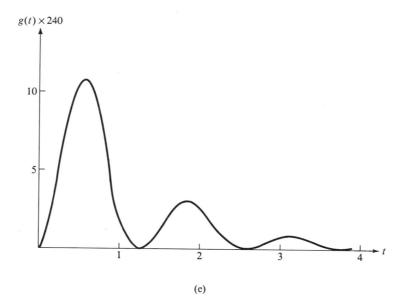

$$\mathcal{T}_K = \text{spring torque} = K \int \omega \, dt = K\theta$$

To determine the network diagram of the rotational mechanical system shown, proceed in a manner that parallels that for a mechanical system in translational

CHAPTER 2

motion. This involves:

    a. The angular velocity ω is identified and one node is identified (see Figures 2.28b and 2.28c).

    b. Elements J, D, and K are located from ω to the reference node.

    c. Source $\mathcal{T}$ is then inserted.

With the help of Figure 2.28c and (2.68), we obtain

$$\mathcal{T} + \mathcal{T}_I + \mathcal{T}_D + \mathcal{T}_K = -\mathcal{T}(t) + J\frac{d^2\theta(t)}{dt^2} + D\frac{d\theta(t)}{dt} + K\theta(t) = 0 \quad (2.69)$$

If the input torque is a delta function, this equation becomes

$$J\frac{d^2h(t)}{dt^2} + D\frac{dh(t)}{dt} + Kh(t) = \delta(t) \quad (2.70)$$

If the system is initially relaxed, we have the initial conditions

$$h(0-) = \frac{dh(0-)}{dt} = 0 \quad (2.71)$$

Next, we integrate (2.70) between the limits $t = 0--$ and $t = 0+$. This is written

$$J\int_{0-}^{0+} \frac{d^2h}{dt^2} dt + D\int_{0-}^{0+} \frac{dh}{dt} dt + K\int_{0-}^{0+} h\, dt = \int_{0-}^{0+} \delta(t)\, dt \quad (2.72)$$

We assume that both h and dh/dt do not include impulse functions, and (2.72) gives

$$J\left[\frac{dh(0+)}{dt} - \frac{dh(0-)}{dt}\right] + D[h(0+) - h(0-)] = 1$$

or

$$J\frac{dh(0+)}{dt} + Dh(0+) = 1 \quad (2.73)$$

Integrate (2.70) twice to find

$$J\int_{0-}^{0+} \frac{dh}{dt} dt + D\int_{0-}^{0+} h\, dt = \int_{0-}^{0+} \int \delta(t)\, dt\, dt$$

From this

$$Jh(0+) - Jh(0-) = 0$$

from which

$$h(0+) = h(0-) \quad (2.74)$$

Combining (2.74), (2.71), and (2.73), we obtain

$$\frac{dh(0+)}{dt} = \frac{1}{J} \quad (2.75)$$

The above equation indicates that the impulse torque forces the angular velocity $d\theta/dt \triangleq dh/dt$ to jump from 0 to $1/J$ instantaneously while the angle $\theta(t)$ remains at zero.

We are now ready to solve our problem, which becomes

$$\frac{d^2 h(t)}{dt^2} + \frac{D}{J}\frac{dh(t)}{dt} + \frac{K}{J} h(t) = 0 \qquad t > 0 \quad \text{(a)}$$

$$h(0+) = 0 \qquad \frac{dh(0+)}{dt} = \frac{1}{J} \qquad \text{(b)}$$

(2.76)

Set $D/J = 2a$ and $b^2 = K/J$, and with $D > 0$, $J > 0$, $K > 0$, the solution is

$$h(t) = e^{-at}(C_1 \cos \sqrt{b^2 - a^2}\, t + C_2 \sin \sqrt{b^2 - a^2}\, t) \qquad t > 0 \qquad (2.77)$$

It is here assumed that $b > a$, which is the **underdamped** case for this system. The constants are easily found using the initial conditions given by (2.76b). These lead to $C_1 = 0$ and $C_2 = 1/(J\sqrt{b^2 - a^2})$.

To investigate the output of this system to an input $f(t) = e^{-t}u(t)$, we introduce the following constants: $J = 1$, $b = 5$, and $a = 1$. The impulse response is then

$$h(t) = \frac{1}{\sqrt{24}} e^{-t} \sin \sqrt{24}\, t \qquad t \geq 0$$

For the function $f(t) = e^{-t}u(t)$ we follow similar steps to those in the previous two examples. The output is given by

$$g(t) \triangleq f(t) * h(t) = \int_0^t e^{-(t-\tau)} \frac{1}{\sqrt{24}} e^{-\tau} \sin \sqrt{24}\, \tau \, d\tau$$

$$= \frac{1}{24}(1 - \cos \sqrt{24}\, t) e^{-t} \qquad t \geq 0 \qquad (2.78)$$

The output of the system is plotted in Figure 2.28e. ∎

EXAMPLE 2.13

Find the impulse response of the system shown in Figure 2.29. The system is in a relaxed state at $t = 0-$.

Solution: The integrodifferential equation that describes the system for an impulse function input is

$$\frac{dh}{dt} + 5h + 4\int h \, dt = \delta(t) \qquad (2.79)$$

We consider the equation at two specific instants of time $t = 0-$ and $t = 0+$ and subtract the two. The result is

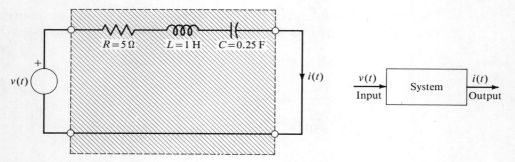

Figure 2.29
Illustrating a second-order system.

$$\frac{dh(0+)}{dt} - \frac{dh(0-)}{dt} + 5h(0+) - 5h(0-) + 4\int_{0-}^{0+} h\,dt = \delta(t)\Big|_{0-}^{0+}$$

The delta function difference over this time interval is zero. Further, the area under the $h(t)$ curve from $0-$ to $0+$ is zero. From initial conditions, $h(0-) = 0$ and $dh(0-)/dt = 0$. Therefore we have

$$\frac{dh(0+)}{dt} + 5h(0+) = 0 \tag{2.80}$$

Now we integrate (2.79) between the limits $0-$ and $0+$. This gives

$$\int_{0-}^{0+} \frac{dh}{dt}\,dt + 5\int_{0-}^{0+} h\,dt + 4\int_{0-}^{0+}\left(\int h\,dt\right)dt = \int_{0-}^{0+} \delta(t)\,dt$$

From this, remembering that the area under the δ-function is unity and that $h(t)$ is a smooth function, then

$$h(0+) - h(0-) = 1$$

Therefore, we have

$$h(0+) = 1$$

This result together with (2.80) gives

$$\frac{dh(0+)}{dt} = -5$$

To find the solution for $h(t)$, differentiate (2.79) to express the result as a differential equation

$$\frac{d^2h}{dt^2} + 5\frac{dh}{dt} + 4h = \frac{d\delta}{dt}$$

Since $\delta(t) = 0$ for $t > 0$, the solution to this second-order equation is

$$h = Ae^{-4t} + Be^{-t}$$

CONVOLUTION, IMPULSE RESPONSE, AND SYSTEM REPRESENTATION

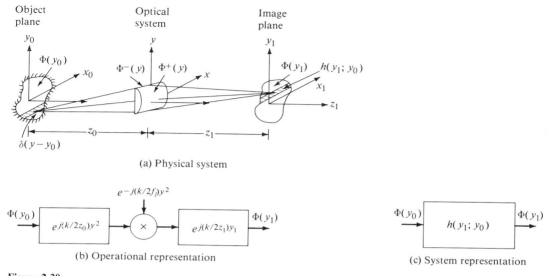

Figure 2.30
(a) Simple optical system. (b) Operational representation. (c) Impulse response representation.

The constants A and B are obtained by using the initial conditions $dh(0+)/dt$ and $h(0+)$, the resulting expression for $h(t)$ is

$$h(t) \triangleq i(t) = \frac{4}{3}e^{-4t} - \frac{1}{3}e^{-t} \qquad t \geq 0 \qquad \blacksquare$$

EXAMPLE 2.14
Find the impulse response of the optical system shown in Figure 2.30.

Solution: We proceed to build up the requisite function. With the use of (2.55), we write the relationship between the field $\Phi(y_1)$ and $\Phi^+(y)$ in the form

$$\Phi(y_1) = \int_{-\infty}^{\infty} \Phi^+(y) P(y) e^{j(k/2z_1)(y_1 - y)^2} \, dy$$

$$= e^{j(k/2z_1)y_1^2} \int_{-\infty}^{\infty} \Phi^+(y) P(y) e^{-j(k/2z_1)(2yy_1 - y^2)} \, dy$$

where we have included the pupil function $P(y)$ (also called the **aperture function** or the **window function**), where

$$P(y) = \begin{cases} P(y) & \text{for } y \text{ inside the aperture of the lens} \\ 0 & \text{elsewhere} \end{cases}$$

This assumes a sharp light discontinuity at the edge of the aperture, with total darkness outside of the aperture. This assumption ignores the diffraction of light into the dark zone, which is very small. $\Phi(y_1)$ is then written in the form

$$\Phi(y_1) = \psi(y_1; z_1) \int_{-\infty}^{\infty} \Phi^+(y)P(y)\psi(y; z_1)e^{-j(k/z_1)yy_1} \, dy \quad \text{(a)} \tag{2.81}$$

where we have defined the phase function

$$\psi(y; z) = e^{j(k/2z)y^2} \quad \text{(b)}$$

The function $\Phi^+(y)$ is related to $\Phi^-(y)$ through the transmittance function $t_l(y)$ of (2.56), and we write

$$\Phi^+(y) = \Phi^-(y)e^{-j(k/2f_l)y^2} = \Phi^-(y)\psi(y; -f_l) \tag{2.82}$$

Next we relate $\Phi^-(y)$ to $\Phi(y_0)$ through another expression that is similar to (2.81a). We now write the total expression relating to the field at the image plane in terms of the field at the object plane. The resulting expression is

$$\Phi(y_1) = \psi(y_1; z_1) \int_{-\infty}^{\infty} P(y)\psi(y; -f_l)$$
$$\times \left[\psi(y; z_0) \int_{-\infty}^{\infty} \Phi(y_0)\psi(y_0; z_0)e^{-j(k/z_0)y_0y} \, dy_0 \right]$$
$$\times \psi(y; z_1)e^{-j(k/z_1)yy_1} \, dy \tag{2.83}$$

Now include in (2.83) the quantity that is defined by

$$\frac{1}{z_0} + \frac{1}{z_1} - \frac{1}{f_l} = \frac{1}{w} \tag{2.84}$$

We can then write

$$\Phi(y_1) = \int_{-\infty}^{\infty} \Phi(y_0)\psi(y_0; z_0)\psi(y_1; z_1)$$
$$\times \int_{-\infty}^{\infty} \psi(y; w)P(y)e^{-jk[(y_0/z_0)+y_1/z_1)y]} \, dy \, dy_0 \tag{2.85}$$

But the input-output relationship of a system in general form can be written as an integral of its impulse response—this is known in optics as the **spread function**—and the input function. We can thus write

$$\Phi(y_1) = \int_{-\infty}^{\infty} h(y_1; y_0)\Phi(y_0) \, dy_0 \tag{2.86}$$

A comparison of (2.86) and (2.85) shows that $h(y_1; y_0)$ has the form

$$h(y_1; y_0) = \psi(y_1; z_1)\psi(y_0; z_0)$$
$$\times \int_{-\infty}^{\infty} P(y)\psi(y; w) \exp\left\{ -jk\left[\left(\frac{y_0}{z_0} + \frac{y_1}{z_1}\right)y \right] \right\} dy \tag{2.87}$$

Based on this, the following observations are made:

- Since at the image plane we observe only the intensity of the field—that is, $\Phi(y_1)\Phi^*(y_1) = |\Phi(y_1)|^2$—each of the phase factors $\psi(y_1; z_1)$ and $\psi(y_0; z_0)$ gives $\psi\psi^* = 1$.

CONVOLUTION, IMPULSE RESPONSE, AND SYSTEM REPRESENTATION

- For each object located at a distance z_0, there exists an image plane at $z = z_1$ such that $1/z_0 + 1/z_1 - 1/f_l = 0$ or $w = \infty$.
- When dealing with nearly ideal optical systems, the point (x_1, y_1) accepts light from a small region centered at the geometric object point (x_0, y_0). Therefore we may use the approximation

$$\exp[jk(y_0^2/2z_0)] \doteq \exp[jk(y_1^2/2M^2 z_0)]$$

where the constant M is equal to one if no magnification takes place. With this simplification we have removed the dependence on x_0 and y_0 and have expressed the results as a function of x_1 and y_1. This phase factor does not contribute to the intensity during the recording of the image, and so it can be ignored.

Upon introducing these simplifications into (2.87), the approximate form of the impulse function becomes

$$h(y_1; y_0) = \int_{-\infty}^{\infty} P(y)\psi(y; w) e^{-jk[(y_0/z_0 + y_1/z_1)y]} \, dy \tag{2.88}$$

If we set $1/w = 0$ (the **lens law** condition) and write $z_1/z_0 = M = $ **magnification,** the impulse (spread) function becomes

$$h(y_1; y_0) = \int_{-\infty}^{\infty} P(y) e^{-j(k/z_1)[(y_1 + My_0)y]} \, dy \tag{2.89}$$

It is observed that if we set $My_0 = -y_0'$—that is, if we describe the relationship in terms of a reflected and magnified coordinate system—then $h(\cdot)$ is a function of the difference $(y_1 - y_0')$. This means that the image is the **convolution** of the impulse response $h(\cdot)$ given by (2.89) and the object function. The result is a smooth version of the object, since $h(\cdot)$ has a finite width. In physical terms the smoothing process changes a high-contrast picture into one with low contrast; that is, it softens the crispness of a picture. The resulting image is like that obtained with a camera that is slightly out of focus. ∎

In the foregoing we have developed certain details for ascertaining the impulse response of simple systems. We will develop some general properties of specific systems in the following section. These general properties are applicable, of course, to the simple systems above.

■ ■ ■

D. SYSTEM DESCRIPTION AND BLOCK DIAGRAM REPRESENTATION

2-9 OPERATORS OF SYSTEMS

We can generalize the system descriptions discussed above: for each system introduce an operator $\mathcal{O}$ that maps, according to some specified rule, the input

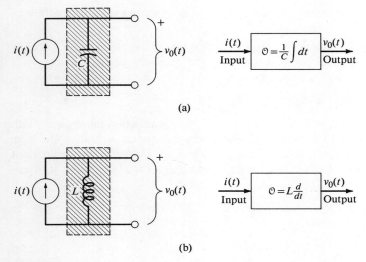

Figure 2.31
Simple electrical systems.

functions $f(t)$ belonging to a set S_1 (the domain of the operator) into the output functions $y(t)$ belonging to another set S_2 (the range of the operator). Symbolically, the operation is given by

$$\mathcal{O}\{f(t)\} = y(t) \tag{2.90}$$

This indicates that once the operator of a system is known, the response to any appropriate excitation can be found.

To relate this concept to the systems with which we are familiar, consider the circuit shown in Figure 2.31a. By application of the Kirchhoff voltage law, we write

$$v_0(t) = \frac{1}{C} \int i(t)\, dt = \left[\frac{1}{C} \int dt\right] i(t) \triangleq \mathcal{O}\{i(t)\} \tag{2.91}$$

A comparison of this expression with (2.90) shows that the input function to this system is $i(t)\,(= f(t))$ and the output function is $v_0(t)\,(= y(t))$. The limits of integration have been left undefined deliberately since initial conditions play an important role in studying systems. In practice, because any system at $t = -\infty$ is relaxed, any input $f(t)$ in the range $(-\infty, \infty)$ will create an output $y(t)$ that is the result **only** and **uniquely** of the input. This observation is very important because, if an input is applied in the time range (t_1, ∞), we cannot find the output unless we know if an input was applied earlier. Thus in developing our input-output relationships, we must assume that at a specified time the system is relaxed (zero initial conditions, which means no energy is present in the system) and that the output is excited solely by the input applied after the designated starting

CONVOLUTION, IMPULSE RESPONSE, AND SYSTEM REPRESENTATION

time. Under these assumptions the operator $\mathcal{O}$ in (2.90) will uniquely specify the output $y(\cdot)$ in terms of the input $f(\cdot)$.

In a similar manner for the system shown in Figure 2.31b, we obtain

$$v_0(t) = L\frac{di(t)}{dt} = L\frac{d}{dt}i(t) \triangleq \mathcal{O}\{i(t)\} \tag{2.92}$$

It was easy to find the operators in these simple examples because each involved only first-order differential operators. For more complicated circuits, deducing the operators is somewhat more complicated.

For example, consider the second-order system shown in Figure 2.32a. An application of the Kirchhoff voltage law yields

$$L\frac{di(t)}{dt} + \frac{1}{C}\int i(t)\,dt = v(t) \quad \text{(a)} \tag{2.93}$$

which can be written

$$\frac{d^2q(t)}{dt^2} + \frac{1}{LC}q(t) = \frac{1}{L}v(t) \quad \text{(b)}$$

where $i(t) = dq(t)/dt$. This equation is written

$$\mathcal{O}^{-1}\{q(t)\} \triangleq \frac{1}{L}v(t) \quad \text{(c)}$$

where

$$\mathcal{O}^{-1} \triangleq \frac{d^2}{dt^2} + \frac{1}{LC}$$

Observe that we here have indicated the operator by the symbol $\mathcal{O}^{-1}$. This representation characterizes operators that operate on the output function of the system. Note that the operator $\mathcal{O}$, which operates on the input, and the operator $\mathcal{O}^{-1}$, which operates on the output, may have a simple interrelationship: they are reciprocal to each other. Most often the two operators have different mathematical forms.

If we introduce $h(t)$, the impulse response of the system (this is called the **Green's function** by physicists and the spread function by optical engineers), then the input is a delta function, and for the system of Figure 2.32, we write

$$\frac{d^2h(t,\tau)}{dt^2} + \frac{1}{LC}h(t,\tau) = \frac{\delta(t-\tau)}{L} \quad \text{(a)}$$
$$\mathcal{O}^{-1}\{h(t,\tau)\} = \frac{\delta(t-\tau)}{L} \quad \text{(b)} \tag{2.94}$$

If (2.93c) is multiplied by $h(t,\tau)$, (2.94b) by $q(t)$, the two resulting equations subtracted, and the difference integrated, we find the relation

$$\int [h(t,\tau)\mathcal{O}^{-1}\{q(t)\} - q(t)\mathcal{O}^{-1}\{h(t,\tau)\}]\,dt$$
$$= \int h(t,\tau)\frac{1}{L}v(t)\,dt - \int q(t)\frac{\delta(t-\tau)}{L}\,dt \tag{2.95}$$

Figure 2.32
An *LC* circuit and its system representation.

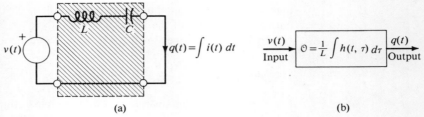

The left hand side of this equation gives

$$\left[h(t,\tau)\frac{dq(t)}{dt} - q(t)\frac{dh(t,\tau)}{dt} \right]_{-\infty}^{\infty} = 0 \tag{2.96}$$

because it is assumed that both the functions and their derivatives are bounded functions. Using the property of the delta function, it is noted also that the second term on the right-hand side is equal to $q(\tau)/L$. Therefore (2.95) takes the form

$$q(\tau) = \int h(t,\tau)\frac{1}{L}v(t)\,dt = \mathcal{O}\{v(t)\} \quad \text{(a)} \tag{2.97}$$

where

$$\mathcal{O} = \frac{1}{L}\int h(t,\tau)\,dt \quad \text{(b)}$$

The system representation of the circuit is shown in Figure 2.32b. It is noted that since t is a dummy variable and τ is a parameter, (2.97) can also be written in the form

$$g(t) = \int h(t,\tau)\frac{1}{L}v(\tau)\,d\tau = \mathcal{O}\{v(\tau)\} \quad \text{(a)} \tag{2.98}$$

where

$$\mathcal{O} = \frac{1}{L}\int h(t,\tau)\,d\tau \quad \text{(b)}$$

■ ■ ■

2-10 SYSTEM CLASSIFICATION

We will now address features of the general representation of a system, as shown in Figure 2.1a, and will study some properties useful in subsequent analyses.

2-10.a Linear System A **linear system** is characterized by the relation

$$\mathcal{O}\{a_1 f_1(t) + a_2 f_2(t)\} = a_1 \mathcal{O}\{f_1(t)\} + a_2 \mathcal{O}\{f_2(t)\}$$
$$= a_1 y_1(t) + a_2 y_2(t) \tag{2.99}$$

CONVOLUTION, IMPULSE RESPONSE, AND SYSTEM REPRESENTATION

where a_1 and a_2 are constants, and $y_1(t)$ and $y_2(t)$ are the outputs of the system when the inputs are $f_1(t)$ and $f_2(t)$, respectively. If this relation does not hold, the system is **nonlinear**. Observe that for linear systems the additivity and homogeneity properties are satisfied. In addition, the additivity of the zero-input and zero-state response is also satisfied for linear systems.

Suppose that a system is characterized by an input-output relation of the form

$$y(t) = \mathcal{O}\{f^2(t)\} = af^2(t) + b$$

When two inputs $f_1(t)$ and $f_2(t)$ are applied to the system specified, the outputs are

$$\mathcal{O}\{f_1^2(t)\} = af_1^2(t) + b$$
$$\mathcal{O}\{f_2^2(t)\} = af_2^2(t) + b$$

Correspondingly,

$$\mathcal{O}\{f_1^2(t) + f_2^2(t)\} = a[f_1^2(t) + f_2^2(t)] + b$$

These relations indicate that $\mathcal{O}\{f_1^2(t)\} + \mathcal{O}\{f_2^2(t)\} \neq \mathcal{O}\{f_1^2(t) + f_2^2(t)\}$, which means that the system is nonlinear. However, if b in the defining equation is zero so that $\mathcal{O}\{f^2(t)\} = af^2(t)$, then the system is linear.

2-10.b Invariant Systems The response of an **invariant** system follows the input response; that is, if a shift occurs in the input signal, an equal shift occurs in the output. This requires that if

$$\mathcal{O}\{f(t)\} = g(t) \quad \textbf{(a)} \tag{2.100}$$

then

$$\mathcal{O}\{f(t - \tau)\} = g(t - \tau) \quad \textbf{(b)}$$

A system that does not meet these conditions is said to be **time-varying**. For a **linear and shift-invariant** (LSI), also called **linear time-invariant** (LTI) system,

$$\mathcal{O}\{af_1(t - \tau_1) + bf_2(t - \tau_2)\} = ag_1(t - \tau_1) + bg_2(t - \tau_2) \tag{2.101}$$

where τ_1 and τ_2 are constants.

EXAMPLE 2.15
The node equation describing the relaxed system shown in Figure 2.33 is

$$\frac{dv(t)}{dt} + \frac{v(t)}{RC} = \frac{1}{C}i(t)$$

Find the general form of the solution and also the solution to an impulse function current.

Solution: From the fact that the impulse response of this circuit is given by $h(t) = (1/C)e^{-t/RC}$ for $t \geq 0$, we can write

$$v(t) = \int_0^t h(t-\tau)i(\tau)\,d\tau = \frac{1}{C}\int_0^t e^{-(t-\tau)/RC}i(\tau)\,d\tau \qquad (2.102)$$

Since the input signal is part of the integrand, this indicates that the system is linear.

We will write $v_s(t)$ for the solution when $i = i(t - t_0)$. In this case (2.102) becomes

$$v_s(t) = \frac{1}{C}\int_0^t e^{-(t-\tau)/RC}i(\tau - t_0)\,d\tau \qquad (2.103)$$

Set $\tau - t_0 = x$, then $d\tau = dx$. Also, since $i(t - t_0) = 0$ for $t < t_0$, (2.103) becomes

$$v_s(t) = \frac{1}{C}\int_0^{t-t_0} e^{-[t-(x+t_0)]/RC}i(x)\,dx$$

or

$$v_s(t) = \frac{1}{C}\int_0^{t-t_0} e^{-[(t-t_0)-\tau]/RC}i(\tau)\,d\tau \triangleq v(t - t_0) \qquad (2.104)$$

Equation (2.104) indicates that the system is time-invariant; that is, when the input is shifted by the quantity $t - t_0$, the output is also shifted by the same amount.

The shifted impulse response of the system is readily obtained from (2.103) by setting $i(\tau - t_0) = \delta(\tau - t_0)$. The result is

$$h(t) \triangleq v(t) = \begin{cases} \dfrac{1}{C}e^{-(t-t_0)/RC} & t \geq t_0 \\ 0 & t < t_0 \end{cases} \qquad (2.105)$$

■

For an invariant system, the impulse response due to an input impulse $\delta(t - \tau)$ takes the form $h(t - \tau)$, and thus

$$\boxed{\begin{aligned} y(t) &= \int f(\tau)h(t-\tau)\,d\tau \triangleq \mathcal{O}\{f(t)\} \quad &\text{(a)} \\ \text{where}& \\ \mathcal{O} &= \int h(t-\tau)\,d\tau \quad &\text{(b)} \end{aligned}} \qquad (2.106)$$

An optical system is **space-invariant** or **isoplanatic** if its impulse response $h(y_1; y_0)$ depends on the distance $y_1 - y_0$. Hence for one-dimensional linear space-invariant (LSI) systems, we have

$$h(y_1; y_0) = h(y_1 - y_0) \qquad (2.107)$$

Therefore (2.86), which represents the input-output relationship of optical sys-

Figure 2.33
A low-pass filter.

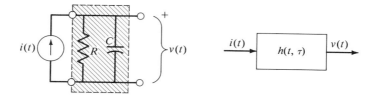

tems for one- and two-dimensional input functions, becomes

$$\Phi(y_1) = \int_{-\infty}^{\infty} h(y_1 - y_0)\Phi(y_0)\,dy_0 \qquad \text{(a)} \qquad (2.108)$$

$$\Phi(y_1) = \iint_{-\infty}^{\infty} h(y_1 - y_0)\Phi(x_0, y_0)\,dx_0\,dy_0$$

$$= \int_{-\infty}^{\infty} h(y_1 - y_0)\varphi(y_0)\,dy_0 \qquad \text{(b)}$$

where

$$\varphi(y_0) = \int_{-\infty}^{\infty} \Phi(x_0, y_0)\,dx_0 \qquad \text{(c)}$$

2-10.c Causality Consider a system for which

$$f_1(t) = f_2(t) \quad \text{for} \quad t \leq t_0 \quad \text{(a)} \qquad (2.109)$$

This implies that

$$\mathcal{O}\{f_1(t)\} = \mathcal{O}\{f_2(t)\} \quad \text{for} \quad t \leq t_0 \quad \text{(b)}$$

Because of these properties, when an impulse is applied to a causal system at $t = 0$, its response $h(t)$ is zero for $t < 0$. The convolution integral then assumes the form

$$g(t) = \int_{-\infty}^{t} f(\tau)h(t - \tau)\,d\tau = \int_{0}^{\infty} h(\tau)f(t - \tau)\,d\tau \qquad (2.110)$$

Furthermore, suppose that a system responds instantaneously to its input; this means that its present output depends only on the input at the present time. This system is called **memoryless** (see 1-10.f *Memory*); otherwise the system is called **dynamical** or having **memory.** For example, the voltage across a resistor, which is given by $v(t) = Ri(t)$, implies that the resistor is a memoryless system. However, the voltage across a capacitor depends on the previous input current, and the capacitor is a system with memory.

2-10.d Stability A system is stable if for any bounded input the output is bounded. This requires that the impulse response must be **absolutely integrable,** which means that

$$\int_{-\infty}^{\infty} |h(t)|\,dt < \infty \qquad (2.111)$$

This relation is found by using the following definition of stability:

Definition: A system is **bounded-input bounded-output** (bibo) **stable** if for each bounded input, the output is bounded. Hence we obtain

$$|g(t)| = \left|\int_{-\infty}^{\infty} h(\tau)f(t-\tau)\,d\tau\right| \leq \int_{-\infty}^{\infty} |h(\tau)||f(t-\tau)|\,d\tau \leq M \int_{-\infty}^{\infty} |h(\tau)|\,d\tau$$

which verifies (2.111). M is the maximum value of $f(t)$ that is attained in the interval of integration. Intuitively we know that a cube lying on one of its surfaces is stable and lying on one of its apexes is unstable. That is, a system is characterized as stable if small inputs produce outputs that do not diverge. More will be said about stability in Chapter 12.

EXAMPLE 2.16
Investigate the stability of an **integrator,** a system that is defined by the relationship

$$g(t) = \int_{-\infty}^{t} f(t')\,dt'$$

Solution: The impulse response of this system, which is that for $f(t) = \delta(t)$, is given by

$$h(t) = \int_{-\infty}^{t} \delta(t')\,dt' = \int_{-\infty}^{t} \frac{du(t')}{dt'}\,dt' = u(t)$$

But since

$$\int_{-\infty}^{\infty} u(t)\,dt = \infty$$

then we know the system is unstable. ∎

The stability of an analog system can also be studied through its representation in the complex plane. We will study this in later chapters.

2-10.e Homogeneity A system is called **homogeneous** if it satisfies the following property:

$$\mathcal{O}\{af(t)\} = a\mathcal{O}\{f(t)\} \tag{2.112}$$

It is easily shown that a system is linear if and only if it is additive and homogeneous. The additivity property is found from (2.99) if $a_1 = a_2 = 1$.

2-10.f Memory A system is **memoryless** if its output at time t depends on the input at that time and not on past or future times. Memoryless systems obey

Figure 2.34
A simple linear memoryless system.

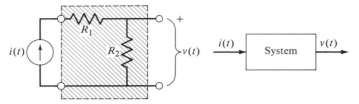

the relation

$$g(t) \triangleq \text{output} = \mathcal{O}\{f(t)\} \triangleq \mathcal{O}\{\text{input}\}$$

where $f(t)$ is the input. This relation is also valid for systems with several inputs and several outputs. In addition, since we talk about the values that $f(t)$ and $g(t)$ will take during the operation of the system, we assume the existence of two sets F and G that will have as subsets all the possible values of f and g. For discrete systems the sets F and G will have a finite number of elements.

If the output of the system at time t is determined by the input at time $t - t_0$, the system does not satisfy our definition and we say that the system has memory. Recall that we have mentioned signal detection by using a cross-correlation operation. For example, each time a pulse is transmitted from a radar antenna, a replica of the pulse is retained and is delayed appropriately by being transmitted through a delay line. The delayed pulse is then cross-correlated with the received pulse for signal detection. The relation between the input and output signals of the delay line is

$$g(t) \triangleq \text{output} = \mathcal{O}\{f(t)\} = f(t - t_0)$$

Therefore a delay line is a system with memory.

If the output of the system at time t is determined by the input in the interval $t - t_0$, $t > t_0$ for $t_0 > 0$, the system is said to have memory of length t_0. Obviously the output of an integrator is a function of the integration length $t - t_0$, and the integrator has memory of length t_0.

In addition to the foregoing properties, systems are also characterized by other properties (such as observability and controllability), but these properties are not important to our present development. However, some reference to these topics is made in Chapter 12.

The foregoing characterizations of systems apply equally well to discrete systems and to systems in which the operator is given in a matrix representation.

EXAMPLE 2.17

Refer to the system shown in Figure 2.34. Is the system linear, time-invariant, and memoryless?

Solution: An application of the system property definitions given above shows that this system is linear, time-invariant, and memoryless. ∎

Figure 2.35
Two commonly used systems.

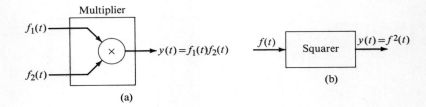

EXAMPLE 2.18
Refer to Figure 2.31a. Is this system linear, time-invariant, and memoryless?

Solution: By an application of the system property definitions, it is established that this system is linear and time-invariant, but it does possess memory. This is so because its output depends on the values of $i(t)$ for all times from $-\infty$ to t. ∎

EXAMPLE 2.19
Shown in Figure 2.35a is a **multiplier** circuit. Are this circuit and the corresponding squaring circuit shown in Figure 2.35b linear?

Solution: By the fundamental nature of the multiplying process, the devices shown are nonlinear systems. ∎

2-10.g Exponential Complex Periodic Input Function The output of a linear, invariant system with an input function $\exp(j\omega t)$ is

$$y(t) = \int_{-\infty}^{\infty} h(\tau)f(t-\tau)\,d\tau = \int_{-\infty}^{\infty} h(\tau)e^{j\omega(t-\tau)}\,d\tau = e^{j\omega t}\int_{-\infty}^{\infty} h(\tau)e^{-j\omega\tau}\,d\tau$$
$$= e^{j\omega t}H(\omega) \quad \text{(a)} \tag{2.113}$$

where

$$H(\omega) = \int_{-\infty}^{\infty} h(\tau)e^{-j\omega\tau}\,d\tau \quad \text{(b)}$$

As already discussed in Example 2.2, we write

$$\mathcal{O}\{e^{j\omega t}\} = H(\omega)e^{j\omega t} \quad \text{(c)}$$

In this context $\exp(j\omega t)$ is the **eigenfunction** of the system, and $H(\omega)$ is its **eigenvalue**. This equation is of fundamental importance because it tells us that if we know the **system function** (eigenvalue) of a system, we can find its output to any input function by appropriately modifying the phases and amplitudes of the sine and cosine functions, which when added construct the input function. This is another indication of the importance of the analysis of signals in its

Figure 2.36
A simple linear and time-invariant system.

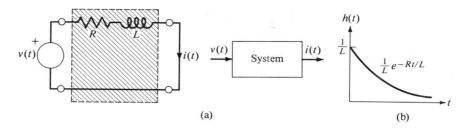

Fourier series representation, a matter that will be discussed in detail in Chapter 3.

In a linear-invariant system that has a real-valued impulse response, the eigenvalues $H(\omega)$ are **Hermetian;** that is, the real part of $H(\omega)$ is even and the imaginary part is odd. This property is expressed by

$$H(\omega) = H^*(-\omega) \tag{2.114}$$

In such a system, if the input is $\sin \omega_0 t$, then

$$\mathcal{O}\{\sin \omega_0 t\} = \mathcal{O}\left\{\frac{1}{2j}[e^{j\omega_0 t} - e^{-j\omega_0 t}]\right\} = \frac{1}{2j}\mathcal{O}\{e^{j\omega_0 t}\} - \frac{1}{2j}\mathcal{O}\{e^{-j\omega_0 t}\}$$

$$= \frac{1}{2j}H(\omega_0)e^{j\omega_0 t} - \frac{1}{2j}H(-\omega_0)e^{-j\omega_0 t}$$

$$= \frac{1}{2j}H(\omega_0)e^{j\omega_0 t} + \left[\frac{1}{2j}H(\omega_0)e^{j\omega_0 t}\right]^*$$

or

$$\mathcal{O}\{\sin \omega_0 t\} = 2\,\text{Re}\left\{\frac{1}{2j}H(\omega_0)e^{j\omega_0 t}\right\} = |H(\omega_0)|\sin(\omega_0 t + \varphi(\omega_0))$$

where

$$H(\omega_0) = |H(\omega_0)|e^{j\varphi(\omega_0)}$$

We have used the property given by (2.114) in this development.

EXAMPLE 2.20
Find the impulse response of the system shown in Figure 2.36a.

Solution: By an application of the Kirchhoff voltage law, we write

$$\frac{di(t)}{dt} + \frac{R}{L}i(t) = \frac{1}{L}v(t) \tag{2.115}$$

If we set $v(t) = \exp(j\omega t)$ we have, from (2.113a),

$$i(t) = H(\omega)e^{j\omega t}$$

Combine this expression with (2.115) to obtain

$$j\omega H(\omega)e^{j\omega t} + \frac{R}{L}H(\omega)e^{j\omega t} = \frac{1}{L}e^{j\omega t}$$

We find from this that

$$H(\omega) = \frac{1}{L}\frac{1}{\frac{R}{L} + j\omega} \tag{2.116}$$

Note, however, that the integral [see (2.113b)]

$$\frac{1}{L}\int_0^\infty e^{-Rt/L}e^{-j\omega t}\,dt = \frac{1}{L}\frac{1}{\frac{R}{L} + j\omega} \triangleq \int_0^\infty h(\tau)e^{-j\omega \tau}\,d\tau$$

Therefore the impulse response of the system is

$$h(t) = \frac{1}{L}e^{-Rt/L} \qquad t > 0 \tag{2.117}$$

and it is shown in Figure 2.36b. The solution to (2.115) is then given by

$$i(t) = h(t) * v(t) = \frac{1}{L}e^{-Rt/L}\int_{-\infty}^{t} v(\tau)e^{R\tau/L}\,d\tau \tag{2.118}$$

∎

2-10.h Zero-State and Zero-Input Response We have defined the zero-state and zero-input response of systems in Section 2-3 in relation to their convolution properties. We will investigate these definitions further because they are important when linear systems are described in their state-space representation (later in this chapter and in Chapter 12). When the input-output relationship of a system is written in matrix form with no equation in this description containing derivatives of higher than the first order, this description is called state-space. We will find that the response of a linear system represented in this form can always be decomposed into the zero-input response plus the zero-state response. Here we examine certain features of such response characteristics.

1. Zero-input Response: If initial energy is stored in a system, the system will have an output even when no inputs are present. For example, if we charge an ideal capacitor and then connect it across an ideal inductor (no losses), there will be a continuing current in this circuit.

EXAMPLE 2.21
Find the zero-input response of the system shown in Figure 2.37. The current at $t = 0$ is assumed to be $i(0) = 1$.

Figure 2.37
Illustrating Example 2.21 for computing zero-input response.

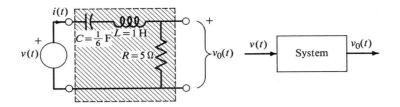

Solution: The equation that describes the system is

$$\frac{di}{dt} + 6\int i\,dt + 5i = v(t) \tag{2.119}$$

which we write in Heaviside operational form with $p = d/dt$, $1/p = \int dt$

$$\left(p + \frac{6}{p} + 5\right)i = v(t) \tag{2.120}$$

With no input excitation, the characteristic equation becomes

$$(p^2 + 5p + 6)i = 0 \tag{2.121}$$

For an assumed solution of the form $i = A\exp(\lambda t)$, this equation yields two roots: $\lambda_1 = -3$ and $\lambda_2 = -2$; the resulting solution is

$$i(t) = Ae^{-3t} + Be^{-2t} \tag{2.122}$$

We set $t = 0$ in (2.119) to obtain the zero-input condition, which is

$$\frac{di(0)}{dt} + 6\int_0^0 i(t)\,dt + 5i(0) = 0$$

or, with $i(0) = 1$,

$$\frac{di(0)}{dt} = -5 \tag{2.123}$$

The lower limit of the integral is taken to be zero because we assume that no initial charge exists on the capacitor. If we introduce the initial conditions into (2.122), we finally obtain

$$i_{zi}(t) = 3e^{-3t} - 2e^{-2t} \tag{2.124}$$

∎

2. Zero-state Response: the zero-state response of a system is due solely to applied input sources. The following example will clarify the difference between the zero-state and zero-input responses and will also specify the natural (transient) and forced (steady-state) responses.

Figure 2.38
Illustrating Example 2.22 for computing zero-input and zero-state responses.

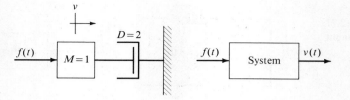

EXAMPLE 2.22

Find the zero-input and zero-state response of the system shown in Figure 2.38. The initial condition for the velocity is $v(0) = 2$ and the input force is $f(t) = u(t)$.

Solution: The equation that describes the system is

$$\frac{dv}{dt} + 2v = f(t) \tag{2.125}$$

The zero-input response is found from the characteristic equation $(p + 2)v = 0$ that has the root $\lambda = -2$. The zero-input response is

$$v_{zi}(t) = Ae^{-2t} \tag{2.126}$$

Using the initial condition, we find that $A = 2$, and so

$$v_{zi}(t) = 2e^{-2t} \tag{2.127}$$

The impulse response of (2.125), setting $f(t) = \delta(t)$, is found to be

$$h(t) = e^{-2t} u(t)$$

The zero-state response is given by [see (2.15)]

$$v_{zs}(t) = h(t) * f(t) = \int_0^t e^{-2\tau} d\tau = \tfrac{1}{2}(1 - e^{-2t}) \tag{2.128}$$

The total response is given by

$$v(t) = v_{zi}(t) + v_{zs}(t) = \underbrace{2e^{-2t}}_{\text{zero-input}} + \underbrace{\tfrac{1}{2}(1 - e^{-2t})}_{\text{zero-state}} = \underbrace{\tfrac{3}{2}e^{-2t}}_{\text{transient}} + \underbrace{\tfrac{1}{2}}_{\text{steady state}} \quad \blacksquare$$

2-10.i Superposition The ideas of superposition were employed in Section 2-2. We use these same ideas in a slightly different context. Here we note that for any linear time-invariant (LTI) system with input consisting of a linear combination of signals

$$f(t) = a_1 f_1(t) + a_2 f_2(t) + \cdots + a_n f_n(t) = \sum_{i=1}^{n} a_i f_i(t) \tag{2.129}$$

the output is of the form

Figure 2.39
Systems in cascade and in parallel.

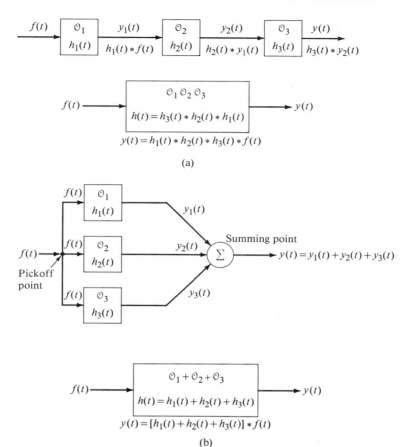

(a)

(b)

$$y(t) = a_1 y_1(t) + a_2 y_2(t) + \cdots + a_n y_n(t) = \sum_{i=1}^{n} a_i y_i(t) \qquad (2.130)$$

where each response $y_k(t)$ is that produced by $f_k(t)$. Therefore, if we can decompose any arbitrary input signal applied to a system as a linear combination of other (basic) signals, as was done in Chapter 1, the output to the system is given by the superposition of the responses to the individual input components of the excitation.

■ ■ ■

2-11 INTERCONNECTED LINEAR SYSTEMS

When systems are connected in **cascade,** as shown in Figure 2.39a, the combined operator of the system is equal to the product of the individual operators, provided of course that we can ignore any loading of one system on the other. In

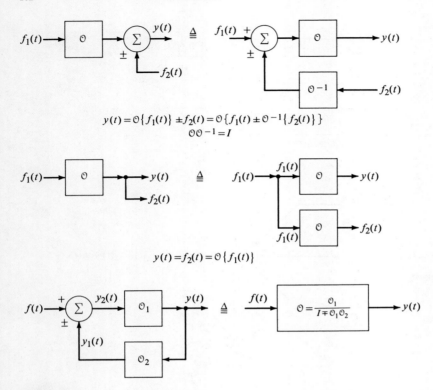

Figure 2.40
Basic system configurations.

certain electronics problems when we connect two amplifiers in cascade, we might adjust them so that the input impedance of the second is equal to the output impedance of the first. While this provides for optimum power transfer and in the case of high frequency signals could result in distortion of the signal due to reflection caused by the impedance mismatch, it requires attention to the loading effect. If we use a buffer amplifier between the two amplifier stages to avoid or reduce the mutual loading effects between the two amplifiers, we write for a **cascade** configuration

$$\mathcal{O}_3\{\mathcal{O}_2\{\underbrace{\underbrace{\mathcal{O}_1\{f(t)\}}_{y_1(t)}}_{y_2(t)}\}\} = \mathcal{O}_3\mathcal{O}_2\mathcal{O}_1\{f(t)\} \qquad (2.131)$$

$$y_3(t)$$

This scheme is applicable to any finite number of operators.
If the systems are connected in **parallel,** then

CONVOLUTION, IMPULSE RESPONSE, AND SYSTEM REPRESENTATION

$$y(t) = y_1(t) + y_2(t) + y_3(t) = \mathcal{O}_1\{f(t)\} + \mathcal{O}_2\{f(t)\} + \mathcal{O}_3\{f(t)\}$$
$$= [\mathcal{O}_1 + \mathcal{O}_2 + \mathcal{O}_3]\{f(t)\} \qquad (2.132)$$

Such a configuration is shown in Figure 2.39b.

A number of additional fundamental system configurations are shown in Figure 2.40.

EXAMPLE 2.23

Prove the operator for the third system configuration shown in Figure 2.40.

Solution: From the figure shown on the left, we obtain the following relations:

$$y_2(t) = f(t) \pm y_1(t)$$
$$y(t) = \mathcal{O}_1\{y_2(t)\}$$
$$y_1(t) = \mathcal{O}_2\{y(t)\}$$

It follows from these relations that

$$\mathcal{O}_1\{y_2(t)\} = y(t) = \mathcal{O}_1\{f(t)\} \pm \mathcal{O}_1\{y_1(t)\} = \mathcal{O}_1\{f(t)\} \pm \mathcal{O}_1\mathcal{O}_2\{y(t)\}$$

or

$$[I \mp \mathcal{O}_1\mathcal{O}_2]\{y(t)\} = \mathcal{O}_1\{f(t)\}$$

so that finally

$$y(t) = [[I \mp \mathcal{O}_1\mathcal{O}_2]^{-1}\mathcal{O}_1]\{f(t)\} = \mathcal{O}\{f(t)\}$$

The symbol I in these equations may assume different meanings depending on the character of the operator. For example, if $\mathcal{O} = d/dt$, we know that its inverse is $\mathcal{O}^{-1} = \int dt$, and in this case $\mathcal{O}^{-1}\mathcal{O} = I = 1$. If $\mathcal{O}$ denotes a matrix operator, I stands for the unit matrix, which has ones along the diagonal and zeros for all other elements. ∎

EXAMPLE 2.24

Find the resultant operator $\mathcal{O}$ of the system shown in Figure 2.41.

Solution: The given system can be split into two elementary systems, as shown in Figures 2.41b and 2.41c. Figure 2.41b leads to the relation

$$\mathcal{O}_1\{f(t) - \mathcal{O}_2\{f_1(t)\}\} = f_1(t)$$

or

$$f_1(t) = [\mathcal{O}_1\mathcal{O}_2 + I]^{-1}[\mathcal{O}_1\{f(t)\}]$$

Also from Figure 2.41c, we obtain

$$f_1(t) - \mathcal{O}_3\{f_1(t)\} = y(t)$$

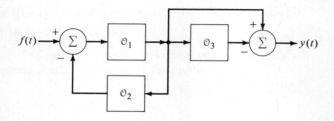

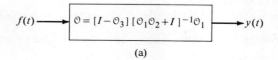

(a)

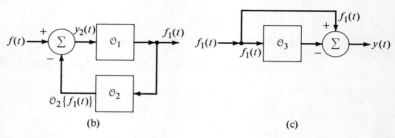

(b)　　　　　　　　　　　　　　　(c)

Figure 2.41
A system with feedback and feed-forward paths.

or

$$[I - \mathcal{O}_3]\{f_1(t)\} = y(t)$$

For the total system, by combining the above equations

$$[I - \mathcal{O}_3][\mathcal{O}_1\mathcal{O}_2 + I]^{-1}[\mathcal{O}_1\{f(t)\}] = y(t)$$

and the total system operator is

$$\mathcal{O} = [I - \mathcal{O}_3][\mathcal{O}_1\mathcal{O}_2 + I]^{-1}\mathcal{O}_1 \qquad \blacksquare$$

■ ■ ■

2-12 STATE-SPACE REPRESENTATION OF SYSTEMS

In addition to the representation of linear time-invariant systems in the form of input-output relationships, we will also study a state-space representation of such

CONVOLUTION, IMPULSE RESPONSE, AND SYSTEM REPRESENTATION

systems. In this representation a system is represented by a set of equations that describe unique relations among the input, output, and state of the system. Here, if we know the state of the system [a set of variables called **state variables**— for example, $x_1(t), x_2(t), \ldots, x_n(t)$ at some time t_0] and the input functions $w_1(t)$, $w_2(t), \ldots, w_p(t)$ for $t_0 \le t \le \infty$, we can find the behavior of the system at any time $t_0 \le t \le \infty$, and this behavior is unique. The state equations for a linear time-invariant system are given by the set

$$\frac{dx_1(t)}{dt} = a_{11}x_1(t) + a_{12}x_2(t) + \cdots + a_{1n}x_n(t) + b_{11}w_1(t) + b_{12}w_2(t)$$
$$+ \cdots + b_{1p}w_p(t)$$

$$\frac{dx_2(t)}{dt} = a_{21}x_1(t) + a_{22}x_2(t) + \cdots + a_{2n}x_n(t) + b_{21}w_1(t) + b_{22}w_2(t)$$
$$+ \cdots + b_{2p}w_p(t) \qquad \text{(a)} \quad (2.133)$$

$$\vdots \quad \vdots$$

$$\frac{dx_n(t)}{dt} = a_{n1}x_1(t) + a_{n2}x_2(t) + \cdots + a_{nn}x_n(t) + b_{n1}w_1(t) + b_{n2}w_2(t)$$
$$+ \cdots + b_{np}w_p(t)$$

and

$$y_1(t) = c_{11}x_1(t) + c_{12}x_2(t) + \cdots + c_{1n}x_n(t) + d_{11}w_1(t) + d_{12}w_2(t)$$
$$+ \cdots + d_{1p}w_p(t)$$

$$y_2(t) = c_{21}x_1(t) + c_{22}x_2(t) + \cdots + c_{2n}x_n(t) + d_{21}w_1(t) + d_{22}w_2(t)$$
$$+ \cdots + d_{2p}w_p(t) \qquad \text{(b)}$$

$$\vdots \quad \vdots$$

$$y_q(t) = c_{q1}x_1(t) + c_{q2}x_2(t) + \cdots + c_{qn}x_n(t) + d_{q1}w_1(t) + d_{q2}w_2(t)$$
$$+ \cdots + d_{qp}w_p(t)$$

These equations, which represent an n-dimensional dynamic system, can be written in matrix notation as follows:

$$\frac{d\mathbf{x}(t)}{dt} = \mathbf{A}\mathbf{x}(t) + \mathbf{B}\mathbf{w}(t) \qquad \text{state equation} \quad \text{(a)}$$
$$\mathbf{y}(t) = \mathbf{C}\mathbf{x}(t) + \mathbf{D}\mathbf{w}(t) \qquad \text{output equation} \quad \text{(b)} \qquad (2.134)$$

where the matrix factors are of an order $\mathbf{A} = \mathbf{A}_{n \times n}$, $\mathbf{B} = \mathbf{B}_{n \times p}$, $\mathbf{C} = \mathbf{C}_{q \times n}$, $\mathbf{D} = \mathbf{D}_{q \times p}$; the vector $\mathbf{x}(t) = [x_1(t), x_2(t), \ldots, x_n(t)]^T$ is the state vector; the $x_i(t)$'s are the state variables; and the exponent T signifies the transpose of the row vector. In matrix form (2.133) is also written as follows:

$$\begin{bmatrix} \dfrac{dx_1(t)}{dt} \\ \vdots \\ \dfrac{dx_n(t)}{dt} \end{bmatrix} = \begin{bmatrix} a_{11} & a_{12} & \cdots & a_{1n} \\ a_{21} & a_{22} & \cdots & a_{2n} \\ \vdots & \vdots & & \vdots \\ a_{n1} & a_{n2} & \cdots & a_{nn} \end{bmatrix} \begin{bmatrix} x_1(t) \\ x_2(t) \\ \vdots \\ x_n(t) \end{bmatrix}$$

$$+ \begin{bmatrix} b_{11} & b_{12} & \cdots & b_{1p} \\ b_{21} & b_{22} & \cdots & b_{2p} \\ \vdots & \vdots & & \vdots \\ b_{n1} & b_{n2} & \cdots & b_{np} \end{bmatrix} \begin{bmatrix} w_1(t) \\ w_2(t) \\ \vdots \\ w_p(t) \end{bmatrix} \quad \text{(a)}$$

$$\begin{bmatrix} y_1(t) \\ y_2(t) \\ \vdots \\ y_q(t) \end{bmatrix} = \begin{bmatrix} c_{11} & c_{12} & \cdots & c_{1n} \\ c_{21} & c_{22} & \cdots & c_{2n} \\ \vdots & \vdots & & \vdots \\ c_{q1} & c_{q2} & \cdots & c_{qn} \end{bmatrix} \begin{bmatrix} x_1(t) \\ x_2(t) \\ \vdots \\ x_n(t) \end{bmatrix}$$

$$+ \begin{bmatrix} d_{11} & d_{12} & \cdots & d_{1p} \\ d_{21} & d_{22} & \cdots & d_{2p} \\ \vdots & \vdots & & \vdots \\ d_{q1} & d_{q2} & \cdots & d_{qp} \end{bmatrix} \begin{bmatrix} w_1(t) \\ w_2(t) \\ \vdots \\ w_p(t) \end{bmatrix} \quad \text{(b)}$$

(2.135)

The representation of systems in this form has many advantages, some of which are: (a) this form of system differential equations has been extensively investigated and used, (b) it can be extended to time-varying and nonlinear systems, (c) the state variables are part of the solution and may be important to know, (d) this form is compact in its representation and is very suitable for analog and digital computer solution, and (e) the form of the solution is common for all systems. Additional studies of such equations and their solution will be found in Chapter 12.

The examples to follow will give some idea of how to develop the state-variable descriptions of systems. A critical element in the description is the selection of the state variables. In one-dimensional rigid-body mechanics (Newtonian), we know that if an external force (input to the system) is applied to a body (system) at time t_0, the motion (output of the system) of the body for $t > t_0$ is uniquely defined if we know the position and its velocity at $t = t_0$. Hence, the **position** and **velocity** may be used as state variables. For electrical circuits the situation is a little more complicated. However, if we assume that we do not have loops made up exclusively of voltage sources and capacitors or nodes made up exclusively of current sources and inductors, we can proceed in the selection of state variables as follows:

Figure 2.42
An *RL* electrical circuit and its block diagram representation.

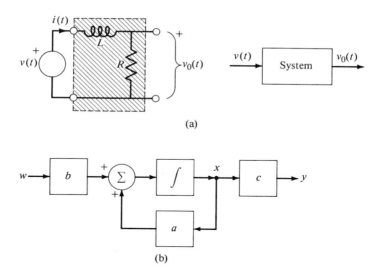

1. Choose all **capacitor voltages** and **inductor currents** as state variables.
2. Write relationships among mesh currents and state variables.
3. Write mesh equations.
4. Eliminate from the two sets of equations all variables except the state variables.

A similar set of procedures is employed for node equations.

EXAMPLE 2.25

Develop the state model representation for the system shown in Figure 2.42a.

Solution: The Kirchhoff voltage law around the loop and Ohm's law for the output voltage yield the expressions

$$\frac{di}{dt} = -\frac{R}{L}i + \frac{1}{L}v$$

$$v_0 = Ri$$
(2.136)

If we set the current through the inductor as the state variable with $x = i$, $y = v_0$, $a = -R/L$, $b = 1/L$, $c = R$, and $v = w$ in these equations, the resulting equations are in standard form of (2.134). Here, of course, since this system is of order one, the vector functions become simple functions and the matrices become constant numbers. The transformed equations are

$$\frac{dx}{dt} = ax + bw$$

$$y = cx$$
(2.137)

Figure 2.43
The circuit and block diagram representation of a mechanical system.

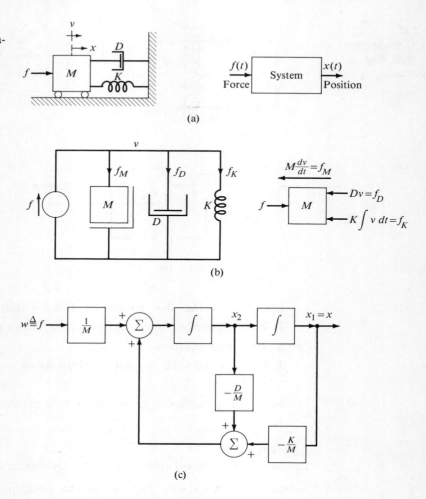

(a)

(b)

(c)

which are the state-variable description of the system. Figure 2.42b shows the implicit feedback structure of the state equations, and this is apparent from the block-diagram representation. ∎

EXAMPLE 2.26

Develop the state model representation for the system shown in Figure 2.43a.

Solution: Since the algebraic sum of the forces at the node must be equal to zero (see Figure 2.43b), we write

$$\frac{dv}{dt} = -\frac{D}{M}v - \frac{K}{M}\int v\,dt + \frac{1}{M}f \qquad (2.138)$$

CONVOLUTION, IMPULSE RESPONSE, AND SYSTEM REPRESENTATION

This expression is written in terms of x, the displacement, with $v = dx/dt$,

$$\frac{d^2x}{dt^2} = -\frac{D}{M}\frac{dx}{dt} - \frac{K}{M}x + \frac{1}{M}f \qquad (2.139)$$

Now make the following substitutions:

$$x = x_1$$

$$\frac{dx}{dt} = \frac{dx_1}{dt} = x_2$$

$$f = w$$

and write (2.139) in terms of two first-order state equations

$$\dot{x}_1 = \frac{dx_1}{dt} = x_2$$

$$\dot{x}_2 = \frac{dx_2}{dt} = \frac{d^2x}{dt^2} = -\frac{D}{M}x_2 - \frac{K}{M}x_1 + \frac{1}{M}w \qquad (2.140)$$

$$y = x_1$$

This set of equations is written in matrix form

$$\begin{bmatrix} \dot{x}_1 \\ \dot{x}_2 \end{bmatrix} = \begin{bmatrix} 0 & 1 \\ -K/M & -D/M \end{bmatrix} \begin{bmatrix} x_1 \\ x_2 \end{bmatrix} + \begin{bmatrix} 0 \\ 1/M \end{bmatrix} w \quad \text{(a)}$$

$$y = \begin{bmatrix} 1 & 0 \end{bmatrix} \begin{bmatrix} x_1 \\ x_2 \end{bmatrix} \quad \text{(b)}$$

(2.141)

In this example we used the position and the velocity as state variables. Figure 2.43b shows the circuit representation of the physical system, and Figure 2.43c shows its block diagram representation. ■

EXAMPLE 2.27
Develop the state equations for the electric circuit shown in Figure 2.44a.

Solution: We follow the guides given above for the selection of state variables for an electric circuit.

1. We choose the current through the inductor $i_2 = x_1$ as one state variable, and the voltage across the capacitor $v_C = x_2$ as a second state variable.

2. Next write the relationships between the mesh currents i_1 and i_2 and the state variables, which are

$$x_1 = i_2 \quad \text{(a)}$$

$$C\frac{dx_2}{dt} = i_1 - i_2 \quad \text{(b)}$$

(2.142)

Figure 2.44
(a) A two-loop electric circuit; (b) its state-space block diagram.

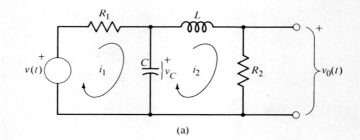

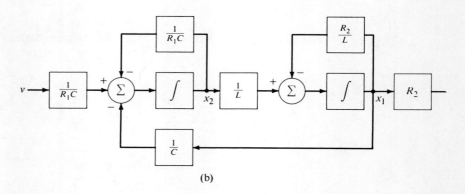

3. The loop equations yield the following equations

$$v - R_1 i_1 - x_2 = 0 \quad \text{(a)}$$

$$-L\frac{dx_1}{dt} - R_2 x_1 + x_2 = 0 \quad \text{(b)}$$

(2.143)

4. Eliminate i_1 from (2.142) and (2.143) to obtain

$$\dot{x}_1 = -\frac{R_2}{L} x_1 + \frac{1}{L} x_2$$

$$\dot{x}_2 = -\frac{1}{C} x_1 - \frac{1}{R_1 C} x_2 + \frac{1}{R_1 C} v \quad \text{(b)}$$

(2.144)

which, in matrix form, is written

$$\begin{bmatrix} \dot{x}_1 \\ \dot{x}_2 \end{bmatrix} = \begin{bmatrix} -\dfrac{R_2}{L} & \dfrac{1}{L} \\ -\dfrac{1}{C} & -\dfrac{1}{R_1 C} \end{bmatrix} \begin{bmatrix} x_1 \\ x_2 \end{bmatrix} + \begin{bmatrix} 0 \\ \dfrac{1}{R_1 C} \end{bmatrix} v \quad \text{(c)}$$

$$\mathbf{y} = [R_2 \ 0] \begin{bmatrix} x_1 \\ x_2 \end{bmatrix} \triangleq v_0 \quad \text{(d)}$$

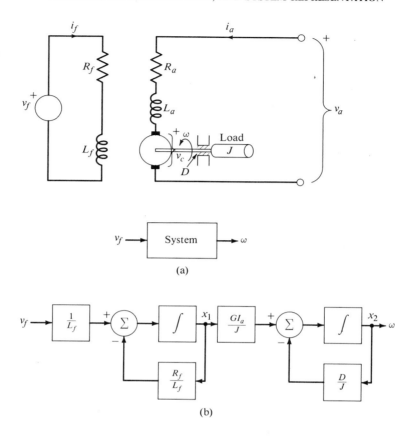

Figure 2.45
(a) Field-controlled *DC* motor; (b) its state-space block diagram.

The procedure outlined above is considerably facilitated when only one loop current passes through each of the inductors or capacitors. The block diagram representation is given in Figure 2.44b. ∎

EXAMPLE 2.28

Find the state equation for a dc motor operating with constant armature current. In this example v's define voltages and ω's define rotational velocities. (For more details see Seely and Poularikas, *Electrical Engineering: Introduction and Concepts*, Matrix, 1982).

Solution: For the field-controlled dc motor shown schematically in Figure 2.45a, the elements R_a and L_a denote the resistance and inductance that are present in the armature winding of the motor. The voltages applied to the field and armature circuits are v_f and v_a, respectively. The currents corresponding to these voltages are i_f and i_a. The motor drives a load having inertia J and damping D. The torque generated by the motor is $\mathcal{T}_e$, and ω is the motor angular speed.

CHAPTER 2

The equation for the field circuit is given by

$$v_f = R_f i_f + L_f \frac{di_f}{dt} \tag{2.145}$$

The equation for the armature circuit is

$$v_a = R_a i_a + L_a \frac{di_a}{dt} + v_c \tag{2.146}$$

where

$$v_c = G\omega i_f \tag{2.147}$$

G is an electromechanical constant for the motor and relates the emf (electromotive force) induced in the armature windings, which are rotating in the magnetic field produced by the field current i_f. Finally, the electric torque produced by the motor must balance the torques due to the inertia and the damping, in the absence of other torques. Hence we have

$$\mathscr{T}_e = D\omega + J\frac{d\omega}{dt} \tag{2.148}$$

In accordance with Ampere's law the electric torque produced by the motor depends on the two currents i_f and i_a or, equivalently, on the magnetic field produced by the field current i_f and the armature current i_a. The electrically produced torque is given by the expression

$$\mathscr{T}_e = G i_f i_a \tag{2.149}$$

In our problem we have assumed that the armature current is to be maintained constant, so that

$$i_a = I_a = \text{constant}$$

The two important equations, when modified to take this constraint into account, become

$$\begin{aligned} v_f &= R_f i_f + L_f \frac{di_f}{dt} \quad \textbf{(a)} \\ J\frac{d\omega}{dt} &= -D\omega + GI_a i_f \quad \textbf{(b)} \end{aligned} \tag{2.150}$$

These equations completely describe the operation of the dc field-controlled motor under constant armature current. It is clear from these equations that the state variables are $x_1 = i_f$ and $x_2 = \omega$; the input variable is v_f, and the output variable is the motor speed ω.

The state equations are now written

CONVOLUTION, IMPULSE RESPONSE, AND SYSTEM REPRESENTATION

$$\frac{dx_1}{dt} = -\frac{R_f}{L_f} x_1 + \frac{1}{L_f} v_f \quad \text{(a)}$$

$$\frac{dx_2}{dt} = \frac{GI_a}{J} x_1 - \frac{D}{J} x_2 \quad \text{(b)}$$

(2.151)

which, in matrix form, become

$$\begin{bmatrix} \dot{x}_1 \\ \dot{x}_2 \end{bmatrix} = \begin{bmatrix} -\frac{R_f}{L_f} & 0 \\ \frac{GI_a}{J} & -D/J \end{bmatrix} \begin{bmatrix} x_1 \\ x_2 \end{bmatrix} + \begin{bmatrix} 1/L_f \\ 0 \end{bmatrix} v_f \quad \text{(a)}$$

$$y = \begin{bmatrix} 0 & 1 \end{bmatrix} \begin{bmatrix} x_1 \\ x_2 \end{bmatrix} \quad \text{(b)}$$

(2.152)

A block diagram and signal flow representation of the field-controlled motor are shown in Figure 2.45b. Observe from the figure that there is no feedback from the second state variable to the first state variable. This is a consequence of the fact that no coupling exists between the system of equations given in (2.151); that is, x_1 and x_2 do not appear in both equations. Physically, it means that the motion of the shaft does not affect the current in the field. ∎

The following examples illustrate a second approach in establishing the state-space representation of systems when they are described by differential equations. A third approach is discussed in Chapter 12. Although the different approaches lead to different state-space representations, they are all equivalent and can be derived from each other. Clearly the individual states in these forms possess mathematical significance; they have no physical significance in themselves.

EXAMPLE 2.29

Find a state-space representation of the system described by the following differential equation

$$\frac{d^2 y(t)}{dt^2} + 3 \frac{dy(t)}{dt} + 2y(t) = 4 \frac{dw(t)}{dt} \quad (2.153)$$

Solution: Write this equation in the form

$$\frac{d^2 y(t)}{dt^2} = -3 \frac{dy(t)}{dt} - 2y(t) + 4 \frac{dw(t)}{dt} \quad (2.154)$$

Integrate this equation twice, writing the results

$$y(t) = \int \left[-3y(t) - \int 2y(t)\,dt + 4w(t) \right] dt \quad (2.155)$$

The diagram that simulates this equation is shown in Figure 2.46.

Figure 2.46
Diagrammatic representation of (2.155).

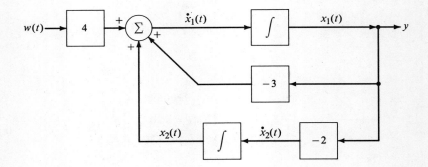

We choose as state variables the outputs of the integrators so that

$$\frac{dx_1}{dt} = -3x_1(t) + x_2(t) + 4w(t)$$

$$\frac{dx_2}{dt} = -2x_1(t) \qquad (2.156)$$

$$y(t) = x_1(t)$$

The matrix representation of this set of equations is

$$\dot{\mathbf{x}}(t) = \begin{bmatrix} -3 & 1 \\ -2 & 0 \end{bmatrix}\mathbf{x}(t) + \begin{bmatrix} 4 \\ 0 \end{bmatrix} w(t) \qquad y(t) = \begin{bmatrix} 1 & 0 \end{bmatrix}\mathbf{x}(t) \qquad (2.157)$$

This form is known as the **first canonical** form.

Suppose that we select the state variables to be the output of the integrators, but label them the reverse of (2.156); that is, we set $x_2 = x_1$ and $x_1 = x_2$. This substitution leads to

$$\frac{dx_1}{dt} = -2x_2(t)$$

$$\frac{dx_2}{dt} = x_1(t) - 3x_2(t) + 4w(t)$$

$$y(t) = x_2(t)$$

This set has the matrix representation

$$\dot{\mathbf{x}}(t) = \begin{bmatrix} 0 & -2 \\ 1 & -3 \end{bmatrix}\mathbf{x}(t) + \begin{bmatrix} 0 \\ 4 \end{bmatrix} w(t) \qquad y(t) = \begin{bmatrix} 0 & 1 \end{bmatrix}\mathbf{x}(t) \qquad (2.158)$$

■

EXAMPLE 2.30
Find the state-space representation of the system described by the differential equation

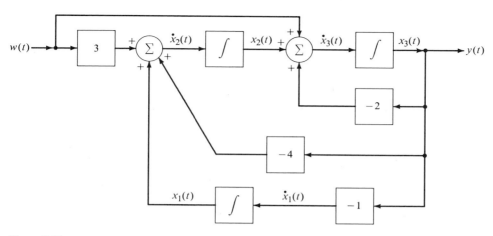

Figure 2.47
Diagrammatic representation of (2.160).

$$\frac{d^3y(t)}{dt^3} + 2\frac{d^2y(t)}{dt^2} + 4\frac{dy(t)}{dt} + y(t) = \frac{d^2w(t)}{dt^2} + 3\frac{dw(t)}{dt} \quad (2.159)$$

Solution: We follow the same general procedure as in Example 2.29 by integrating three times. We arrange the terms in the following way

$$y(t) = \iint \left[3w(t) - \int y(t)\,dt - 4y(t) \right] dt\,dt + \int [w(t) - 2y(t)]\,dt \quad (2.160)$$

The schematic representation of this equation is shown in Figure 2.47. We select the integrator outputs as state variables, and from the figure we write

$$\dot{x}_1(t) = -x_3(t)$$
$$\dot{x}_2(t) = x_1(t) - 4x_3(t) + 3w(t)$$
$$\dot{x}_3(t) = x_2(t) - 2x_3(t) + w(t)$$
$$y(t) = x_3(t)$$

These equations, in matrix form, yield

$$\dot{\mathbf{x}}(t) = \begin{bmatrix} 0 & 0 & -1 \\ 1 & 0 & -4 \\ 0 & 1 & -2 \end{bmatrix} \mathbf{x}(t) + \begin{bmatrix} 0 \\ 3 \\ 1 \end{bmatrix} \mathbf{w}(t)$$

$$\mathbf{y}(t) = \begin{bmatrix} 0 & 0 & 1 \end{bmatrix} \mathbf{x}(t)$$

∎

REFERENCES

1. Finizio, N., and G. Ladas. *Ordinary Differential Equations with Modern Applications*. Belmont, Calif.: Wadsworth, 1978.
2. Papoulis, A. *Signal Analysis*. New York: McGraw-Hill, 1977.
3. Seely, S., and A. D. Poularikas. *Electrical Engineering: Introduction and Concepts*. Beaverton, Or.: Matrix, 1982.
4. Seely, S., and A. D. Poularikas. *Electromagnetics: Classical and Modern Theory and Applications*. New York: Marcel Dekker, 1979.
5. Swisher, G. M. *Introduction to Linear Systems Analysis*. Champaign, Ill.: Matrix, 1976.

PROBLEMS

2-2.1 Prove the following two forms of the superposition integral

$$g(t) = \frac{d}{dt} \int_{-\infty}^{t} y_u(t - \tau) f(\tau) \, d\tau$$

$$\dot{g}(t) = \frac{d}{dt} \int_{-\infty}^{t} y_u(\tau) f(t - \tau) \, d\tau$$

2-2.2 The indicial response of a system is $3 \exp(-t)$. Find the response to the excitation shown in Figure P2-2.2.

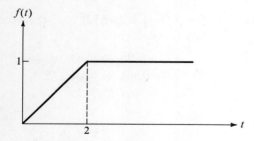

Figure P2-2.2

2-2.3 Consider the circuit shown in Figure P2-2.3. Find the output $v_2(t)$ employing the superposition integral. There is zero initial charge on the capacitor.

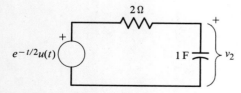

Figure P2-2.3

2-4.1 Graphically determine the convolution of the pairs of functions shown in Figure P2-4.1.

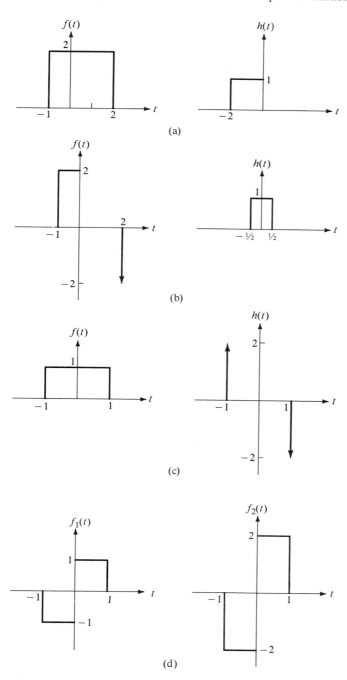

Figure P2-4.1

2-4.2 Evaluate the following operations and sketch your results.

    a. $[\Lambda_a(t)\,\text{sgn}(t)] * [2\Lambda_a(t)\,\text{sgn}(t)]$ **b.** $u(t) * p_a(t)$

    c. $p_a(t) * p_a(t + 3a)$ **d.** $p_a(t) * \delta\left(\dfrac{t-a}{2}\right)$

    e. $p_a(t) * p_{2a}(t) * p_{3a}(t)$ **f.** $[p_a(t) - 2p_a(t - 3a)] * \delta(t)$

    g. $\Lambda_a(t) * \dfrac{d\delta(t)}{dt}$ **h.** $[\delta(t) + \delta(t - 1)] * [\delta(t) + \delta(t - 1)]$

    i. $[2\Lambda_a(t) + 2\delta(t - 3a)] * [2\Lambda_a(t) + 2\delta(t - 3a)]$ **j.** $u(t) * \dfrac{d\delta(t)}{dt}$

2-4.3 Prove the following relations:

    a. $f(t) * \dfrac{d^n\delta(t)}{dt^n} = \dfrac{d^n f(t)}{dt^n}$

    b. $\dfrac{d^n f(t)}{dt^n} * \dfrac{d^m h(t)}{dt^m} = \dfrac{d^{n+m}}{dt^{n+m}}[f(t) * h(t)]$

2-4.4 Show by an example that the repeated convolution of one function with a second function produces a smoothing effect. This indicates that the resulting function under a convolution operation is smoother than either of the two functions involved in the convolution.

2-4.5 Find the convolution of the two functions shown in Figure P2-4.5 for the following values of a: 0.25, 0.5, 0.75.

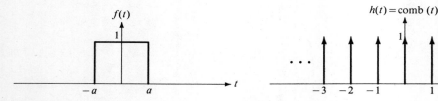

Figure P2-4.5

2-4.6 Find the convolution of the functions shown in Figure P2-4.6.

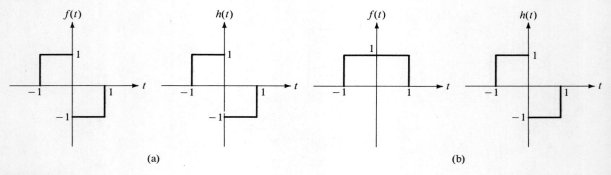

Figure 2-4.6

2-4.7 Prove (2.20a) and (2.21).

2-5.1 Find the periodic convolution of the two signals shown in Figure P2-5.1.

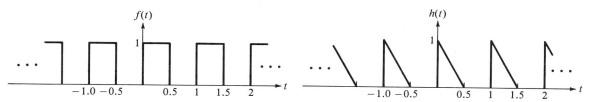

Figure P2-5.1

2-6.1 Verify the following relations:
  a. $f(t) \star h(t) = f(t) * h(-t)$ $f(t)$ and $h(t)$ are real functions
  b. $f(t) \star h(t) = f(t) * h^*(-t)$ $f(t)$ and $h(t)$ are complex functions
  c. $R_{fh}(t) = R_{hf}^*(-t)$ $f(t)$ and $h(t)$ are complex functions

2-6.2 Find the correlation between the functions given
  a. $f(t) = p_1(t)$, $h(t) = e^{-t}u(t)$ b. $f(t) = p_1(t)$, $h(t) = \delta(t-1) + p_1(t-3)$
  c. $f(t) = e^{-t}u(t)$, $h(t) = e^{-t}u(t)$ d. $f(t) = tp_1(t-1)$, $h(t) = e^{-t}u(t)$

2-6.3 A manufacturer needs to separate three different-shaped items as they ride along a conveyor belt. Sketch the elements of a correlation mechanism that will effect this separation.

2-7.1 If the current through a capacitor ($C = 1F$) is that shown in Figure P2-7.1, sketch the voltage across the capacitor.

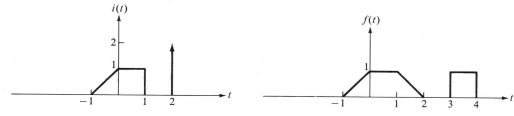

Figure P2-7.1

Figure P2-7.2

2-7.2 A force is applied to a translational spring with $K = 1$. For the $f(t)$ shown in Figure P2-7.2, sketch the waveform of its relative velocity.

2-7.3 Find the operational representation of the optical system shown in Figure P2-7.3.

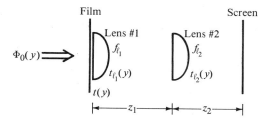

Figure P2-7.3

2-7.4 The electric field amplitude varies over the aperture of an optical system as indicated in Figure P2-7.4. If He-Ne laser light ($\lambda = 0.63 \ \mu m$) impinges on the aperture, what is the phase and amplitude of the field at $y_1 = 15$ mm?

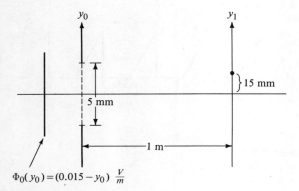

$\Phi_0(y_0) = (0.015 - y_0) \ \frac{V}{m}$

Figure P2-7.4

2-8.1 Find the impulse response of the initially relaxed system shown in Figure P2-8.1.

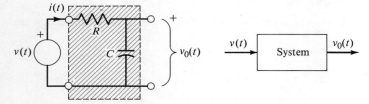

Figure P2-8.1

2-8.2 Find the impulse response of the initially relaxed system shown in Figure P2-8.2. Use this result to find the output of this system when the input is the function $p_a(t - a)$.

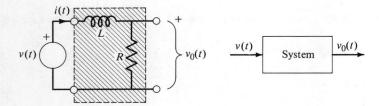

Figure P2-8.2

2-8.3 Find the impulse response of the initially relaxed system shown in Figure P2-8.3. Assume the underdamped case for the system.

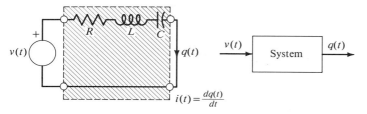

Figure P2-8.3

2-8.4 Set up the circuit diagrams and write the differential equations that describe the systems shown in Figure P2-8.4.

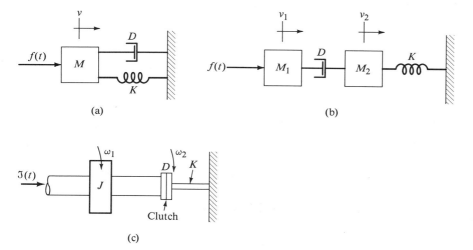

Figure P2-8.4

2-8.5 Find the output of the systems shown in Figure P2-8.5 if the input is the function $f(t) = tp_1(t-1)$, where $f(t)$ is current $i(t)$ for the electrical system and torque $\mathcal{T}$ for the mechanical system. The systems are relaxed at $t = 0$.

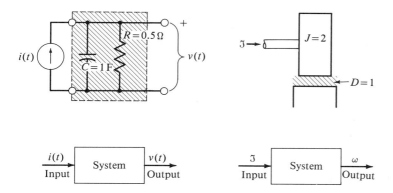

Figure P2-8.5

2-8.6 The systems shown in Figure P2-8.6 are initially relaxed. The input to system (a) is $p_1(t-1)$, and the input to system (b) is $p_1(t)$. Determine the corresponding outputs.

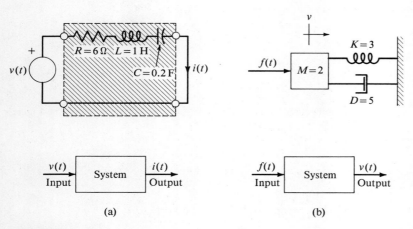

Figure P2-8.6

2-9.1 Find the operators for the systems shown in Figure P2-9.1.

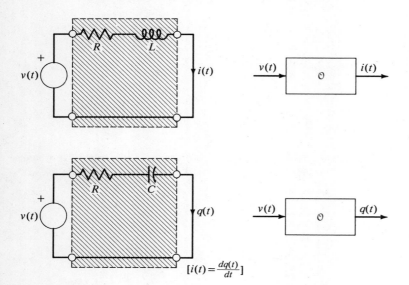

Figure P2-9.1

2-9.2 Verify (2.96).

2-10.1 The low-pass circuit shown in Figure P2-8.1 is initially relaxed. Show that the system is linear.

2-10.2 Consider the systems described by the following equations. Determine whether each is linear, nonlinear, time invariant, or time variant; $y(t)$ denotes the output and $f(t)$ denotes the input.

a. $y(t) = \int f(t)\,dt$ **b.** $\dfrac{dy(t)}{dt} + 2y(t) = e^{-t}$ for $t \geq 0$

2-10.3 An optical system has a pupil function equal to 1 in the range $-1 \leq y \leq 1$. Find the field at $y_1 = 1$ m if the input field to the optical system is an He-Ne laser with light having $\lambda \doteq 0.63\ \mu$m. The magnification M is equal to 1.

2-10.4 Find the transfer functions $H(\omega)$ for the systems shown in Figure P2-10.4.

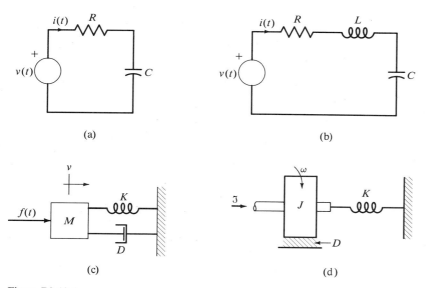

Figure P2-10.4

2-10.5 Find the zero-input response, the zero-state response, the transient response and the steady state response for the systems described by the following equations:

a. $\dfrac{dy(t)}{dt} + 2y(t) = e^{-t}u(t)$ $y(0) = 2$

b. $\dfrac{dy(t)}{dt} + y(t) = tu(t)$ $y(0) = 1$

c. $\dfrac{d^2y}{dt^2} + 5\dfrac{dy(t)}{dt} + 4y(t) = u(t)$ $y(0) = 1$ $\dfrac{dy(0)}{dt} = 4$

2-10.6 Find the complete response of the system shown in Figure P2-10.6. The initial conditions are: $i(0) = 2$ and $v_c(0) = 4$.

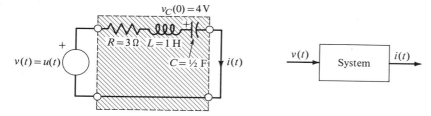

Figure P2-10.6

2-11.1 Give an example to show that $\mathcal{O}_1\mathcal{O}_2\{f(t)\} = \mathcal{O}_2\mathcal{O}_1\{f(t)\}$ is not always true.

2-11.2 Determine the resultant operator for each system shown in Figure P2-11.2.

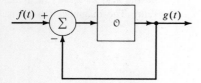

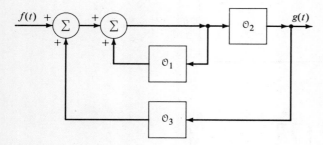

Figure P2-11.2

2-11.3 Find equivalent diagrams for the systems shown in Figure P2-11.3, given that a is a constant (see also Figure 2.40).

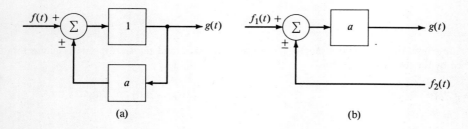

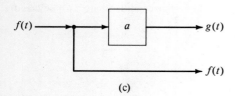

Figure P2-11.3

2-11.4 Find the resultant operator for the system shown in Figure P2-11.4.

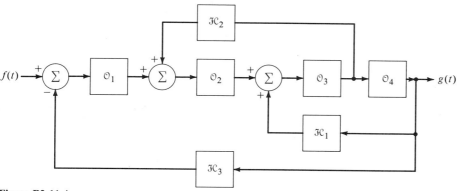

Figure P2-11.4

2-12.1 Develop the state model representations for the systems shown in Figure P2-12.1 and draw their block diagrams.

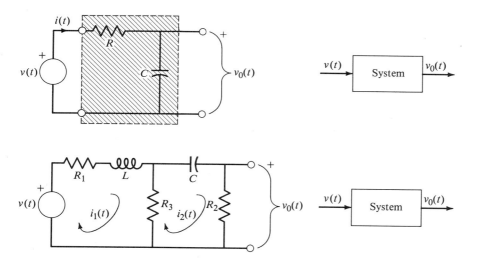

Figure P2-12.1

2-12.2 Develop a state model representation for the system shown in Figure P2-12.2a. Figure P2-12.2b shows the equivalent circuit representation.

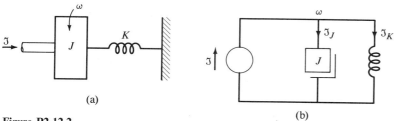

Figure P2-12.2

2-12.3 Develop a state model representation for the dc motor operated under constant field current conditions (see Example 2.28).

2-12.4 Develop a state model for the electrical filter shown in Figure P2-12.4.

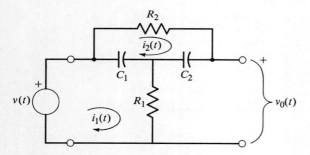

Figure P2-12.4

2-12.5 Develop a state model for the system shown in Figure P2-12.5, where f_a denotes the applied force, f denotes the force produced by the current, e denotes voltages, and v specifies the velocity. Give a diagrammatic representation of the system.

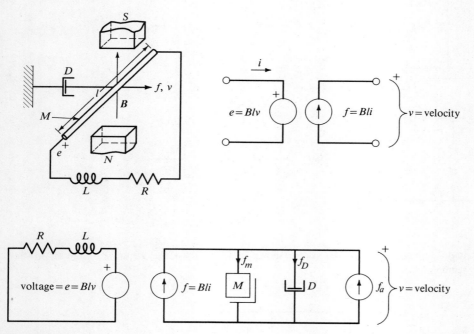

Figure P2-12.5

2-12.6 Find state-space representations of the systems specified by the following differential equations. Show block diagrams for each.

a. $\dfrac{dy(t)}{dt} + ay(t) = w(t)$ **b.** $\dfrac{dy(t)}{dt} + ay(t) = \dfrac{dw(t)}{dt} + bw(t)$

c. $\dfrac{d^2y(t)}{dt^2} + 2\dfrac{dy(t)}{dt} + y(t) = 2\dfrac{dw(t)}{dt} + 5w(t)$ **d.** $\dfrac{d^3y}{dt^3} + \dfrac{d^2y}{dt^2} + \dfrac{dy}{dt} + y = \dfrac{dw}{dt} + w$

e. $\dfrac{d^2y_1}{dt^2} + 3\dfrac{dy_1}{dt} + 4y_1 + 2y_2 = \dfrac{dw_1}{dt} + w_1$ $\dfrac{d^2y_2}{dt^2} + \dfrac{dy_2}{dt} + 5y_2 + 6\dfrac{dy_1}{dt} = w_2$

CHAPTER 3
FOURIER SERIES

3-1 INTRODUCTION

One of the most frequent features of natural phenomena is periodicity. A wide range of periodic phenomena exist—the audible note of a mosquito, the beautiful patterns of crystal structures, acoustic and electromagnetic waves of most types, the periodic vibrations of musical instruments. In all of these phenomena there is a pattern or displacement that repeats itself over and over again in time or in space (see Figure 3.1). In some instances (for example, in wave patterns) the pattern itself will move in time. A periodic pattern might be a simple one or it might be rather complicated, as shown in Figure 3.1. It is important to our subsequent work that we understand the properties of periodic waves.*

Chapter 1 demonstrated that function $f(t)$ (a physical pattern or phenomenon) is periodic, with period T, if it satisfies the condition

$$f(t + T) = f(t) \quad \text{(a)} \tag{3.1}$$

or

$$f(t + nT) = f(t) \qquad n = 1, 2, \ldots \quad \text{(b)}$$

for every t. Refer to Figure 3.2a, which shows a periodic or repetitive function, and to Figure 3.2b, which shows we can create this periodic function by adding a basic functional form having a length equal to its period and appropriately displaced. We can write this latter function as the summation

$$f(t) = \sum_{n=-\infty}^{\infty} f_p(t - nT) \tag{3.2}$$

We mentioned in Chapter 1 that sine and cosine functions, and as a consequence complex exponential functions, constitute a set of orthogonal functions over one period, and that these can be used to expand any periodic function in a series of such functions. The use of complex exponential functions to study linear time-invariant (LTI) system behavior is important since their response to such signals is easily obtained, thereby allowing a ready review of their properties.

* We will often use the term *periodic waves* to mean *periodic functions*.

FOURIER SERIES

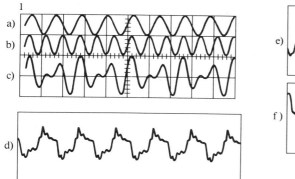
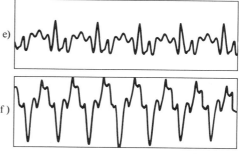

Figure 3.1
Simple and complicated periodic waveforms. (a) Sinusoid of 400 Hz. (b) Sinusoid of 600 Hz. (c) Superposition of (a) and (b) when their zeros coincide at $t = 0$. (d) Clarinet waveform. (e) Oboe waveform; (f) Saxophone waveform. (Reprinted, by permission, from French, *Vibrations and Waves*, 25, 215.)

Figure 3.2
A periodic function and its elementary representation.

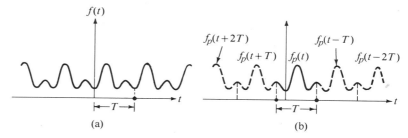

As already discussed in Chapter 2, when the input to an LTI system is $\exp(j\omega t)$, the output is $H(\omega)\exp(j\omega t)$. Therefore if an input signal is the sum of elementary signals, say

$$f_i(t) = a_1 e^{j\omega_1 t} + a_2 e^{j\omega_2 t} + \cdots + a_n e^{j\omega_n t} \tag{3.3}$$

then the output of the system will be

$$f_0(t) = a_1 H(\omega_1) e^{j\omega_1 t} + a_2 H(\omega_2) e^{j\omega_2 t} + \cdots + a_1 H(\omega_n) e^{j\omega_n t}$$

$$= \sum_{n=1}^{N} a_n H(\omega_n) e^{j\omega_n t} \tag{3.4}$$

This relationship results because LTI systems are linear and homogeneous and therefore obey the superposition principle. The relationship indicates that if we know the eigenvalues $H(\omega_i)$'s of the system, its output can be constructed in a simple way. We will expand on this notion in Section 3-5 when the ω_i's are linearly related.

■ ■ ■

3-2 FOURIER SERIES AND PERIODIC FUNCTIONS

An important feature of a general periodic function is that it can be represented as an infinite sum of sine and cosine functions, which are themselves periodic. This series of sine and cosine terms is known as a **Fourier series.** For a function to be Fourier series transformable, it must possess the following properties, which are known as the **Dirichlet conditions.** These conditions are sufficient but not necessary and require that within a period (a) only a finite number of maxima and minima can be present, (b) the number of discontinuities must be finite, and (c) the discontinuities must be bounded; that is, the function must be absolutely integrable, $\int_0^T |f(t)| \, dt < \infty$. Fortunately, all of the periodic functions (phenomena) with which we deal in practice obey these conditions.

A striking feature of the Fourier series is that it permits an arbitrary function defined over a finite interval, even a rough or a discontinuous graph, to be represented as an infinite summation of sine and cosine functions, each of which is an analytic function. In this respect it is quite different from the Taylor series, which allows us to predict the value of an analytic function at any given point in terms of a knowledge of the value an infinitesimal distance away (see Appendix 1). That is, the Fourier series gives a knowledge of the function over the entire range whereas the Taylor series gives a strict prediction at a finite distance from a point. We will find that the chasm between the Taylor series and the Fourier series can be bridged by the Z-transform, which is discussed in some detail in Chapter 9.

Another significant difference between the Fourier and Taylor expansions is that the coefficients of a Fourier series are obtained by integration whereas the coefficients of a Taylor expansion are obtained by differentiation.

Any periodic signal $f(t)$ that satisfies the Dirichlet conditions can be expressed over an interval $t_0 \leq t \leq t_0 + T$, where T is the period, by the expansion

$$f(t) = \sum_{n=-\infty}^{\infty} \alpha_n e^{jn\omega_0 t}$$

$$= \sum_{n=-\infty}^{\infty} |\alpha_n| e^{j(n\omega_0 t + \varphi_n)} \qquad t_0 \leq t \leq t_0 + T \qquad \text{(a)}$$

where (3.5)

$$\alpha_n = \frac{1}{T} \int_{t_0}^{t_0+T} f(t) e^{-jn\omega_0 t} \, dt = \text{constant} = |\alpha_n| e^{j\varphi_n} \qquad \text{(b)}$$

where

$$\omega_0 = \frac{2\pi}{T}$$

This representation is called the complex or exponential form of the Fourier series. Observe that the amplitudes α_n of the component functions in the Fourier series expansion of $f(t)$ are determined from the given $f(t)$. Correspondingly, $f(t)$ is determined from a knowledge of α_n. Often these two related equations are referred to as a **Fourier series transform pair**; that is, knowing one, the second can be found, and vice versa. In addition, any periodic function that satisfies the Dirichlet conditions and is written in Fourier series representation converges at every point $t = t_0$ to $f(t_0)$, provided that the function is continuous at t_0. If $f(t)$ is discontinuous at $t = t_0$, the function $f(t_0)$ will converge to $f(t_0) = [f(t_0+) + f(t_0-)]/2$, the mean value at the point of discontinuity (the arithmetic mean of the left-hand and right-hand limits).

The complex functions $\exp(jn\omega_0 t) = \cos n\omega_0 t + j \sin n\omega_0 t$ constitute an orthogonal set (see Chapter 1) over a full period. This means that

$$\int_{t_0}^{t_0+T} e^{jn\omega_0 t}(e^{jm\omega_0 t})^* \, dt = \begin{cases} 0 & n \neq m \\ T & n = m \end{cases} \tag{3.6}$$

If $f(t)$ is real, (3.5b) becomes

$$\alpha_{-n} = \frac{1}{T}\int_{t_0}^{t_0+T} f(t)e^{jn\omega_0 t} \, dt = \left[\frac{1}{T}\int_{t_0}^{t_0+T} f(t)e^{-jn\omega_0 t} \, dt\right]^* = \alpha_n^*$$

This result, when combined with (3.5a), yields

$$f(t) = \alpha_0 + \sum_{n=1}^{\infty} \left[(\alpha_n + \alpha_n^*)\cos n\omega_0 t + j(\alpha_n - \alpha_n^*)\sin n\omega_0 t\right] \tag{3.7}$$

This shows that the Fourier series expansion of a periodic real function can also be written in the explicit form

$$f(t) = \frac{A_0}{2} + \sum_{n=1}^{\infty} (A_n \cos n\omega_0 t + B_n \sin n\omega_0 t) \quad \text{(a)} \tag{3.8}$$

where

$$A_n = \alpha_n + \alpha_n^* = \frac{1}{T}\int_{t_0}^{t_0+T} f(t)e^{-jn\omega_0 t} \, dt + \frac{1}{T}\int_{t_0}^{t_0+T} f(t)e^{jn\omega_0 t} \, dt$$

$$= \frac{1}{T}\int_{t_0}^{t_0+T} f(t)[e^{-jn\omega_0 t} + e^{jn\omega_0 t}] \, dt$$

$$= \frac{2}{T}\int_{t_0}^{t_0+T} f(t)\cos n\omega_0 t \, dt \qquad n = 0, 1, 2, \ldots \quad \text{(b)}$$

Similarly

$$B_n = \frac{2}{T}\int_{t_0}^{t_0+T} f(t)\sin n\omega_0 t \, dt \qquad n = 0, 1, 2, \ldots \quad \text{(c)}$$

Additionally, we observe that

$$A_0 = 2\alpha_0 \qquad A_n = \alpha_n + \alpha_n^* = 2\,\text{Re}\{\alpha_n\}$$

$$B_n = j(\alpha_n - \alpha_n^*) = -2\,\text{Im}\{\alpha_n\} \quad \text{(d)}$$

Table 3.1 Forms of Fourier Series for Real Functions

Complex Form		Formulas for the Coefficients
$f(t) = \sum_{n=-\infty}^{\infty} \alpha_n e^{jn\omega_0 t}$ $\quad = \sum_{n=-\infty}^{\infty} |\alpha_n| e^{j(n\omega_0 t + \varphi_n)}$	$\omega_0 = \dfrac{2\pi}{T}$ $\alpha_n = |\alpha_n| e^{j\varphi_n}$ $\alpha_n = \dfrac{1}{T} \int_{t_0}^{t_0+T} f(t) e^{-jn\omega_0 t} \, dt$ $t_0 \le t \le t_0 + T$	$A_0 = 2\alpha_0$ $A_n = \alpha_n + \alpha_n^* = 2\,\text{Re}\{\alpha_n\}$ $B_n = j(\alpha_n - \alpha_n^*) = -2\,\text{Im}\{\alpha_n\}$ $C_n = [A_n^2 + B_n^2]^{1/2}$ $\varphi_n = -\tan^{-1}[B_n/A_n]$ $A_n = C_n \cos \varphi_n$ $B_n = -C_n \sin \varphi_n$
Trigonometric Forms	$A_n = \dfrac{2}{T} \int_{t_0}^{t_0+T} f(t) \cos n\omega_0 t \, dt$ $n = 1, 2, \ldots$ $B_n = \dfrac{2}{T} \int_{t_0}^{t_0+T} f(t) \sin n\omega_0 t \, dt$ $n = 1, 2, \ldots$	
$f(t) = \dfrac{A_0}{2} + \sum_{n=1}^{\infty} (A_n \cos n\omega_0 t + B_n \sin n\omega_0 t)$ $f(t) = \dfrac{A_0}{2} + \sum_{n=1}^{\infty} C_n \cos(n\omega_0 t + \varphi_n)$		

It follows from (3.8d) that the coefficients A_0, A_n, and B_n are real, whereas the coefficients α_n are complex, in general.

We can write (3.8a) in a slightly different form, as follows:

$$\begin{aligned} f(t) &= \frac{A_0}{2} + \sum_{n=1}^{\infty} A_n \left(\cos n\omega_0 t + \frac{B_n}{A_n} \sin n\omega_0 t\right) \\ &= \frac{A_0}{2} + \sum_{n=1}^{\infty} A_n (\cos n\omega_0 t - \tan \varphi_n \sin n\omega_0 t) \\ &= \frac{A_0}{2} + \sum_{n=1}^{\infty} C_n \cos(n\omega_0 t + \varphi_n) \quad \text{(a)} \end{aligned} \qquad (3.9)$$

where

$$C_n = \sqrt{A_n^2 + B_n^2} \qquad \varphi_n = -\tan^{-1}\left(\frac{B_n}{A_n}\right) \quad \text{(b)}$$

The Fourier series in the forms given in (3.8) and (3.9) has greater practical use in representing periodic functions than the exponential form. However, the exponential form is usually more convenient for analytic purposes.

FOURIER SERIES

Table 3.2 Examples of Fourier Series

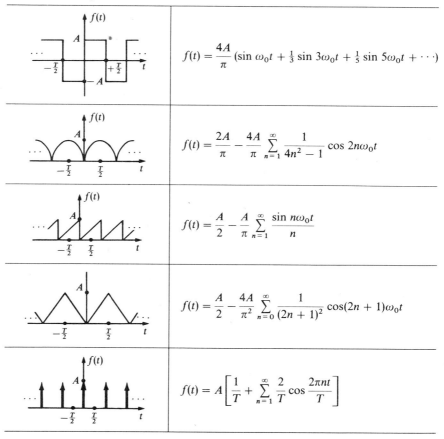

(square wave)	$f(t) = \dfrac{4A}{\pi}(\sin \omega_0 t + \tfrac{1}{3}\sin 3\omega_0 t + \tfrac{1}{5}\sin 5\omega_0 t + \cdots)$
(full-wave rectified)	$f(t) = \dfrac{2A}{\pi} - \dfrac{4A}{\pi}\sum_{n=1}^{\infty}\dfrac{1}{4n^2-1}\cos 2n\omega_0 t$
(sawtooth)	$f(t) = \dfrac{A}{2} - \dfrac{A}{\pi}\sum_{n=1}^{\infty}\dfrac{\sin n\omega_0 t}{n}$
(triangular)	$f(t) = \dfrac{A}{2} - \dfrac{4A}{\pi^2}\sum_{n=0}^{\infty}\dfrac{1}{(2n+1)^2}\cos(2n+1)\omega_0 t$
(impulse train)	$f(t) = A\left[\dfrac{1}{T} + \sum_{n=1}^{\infty}\dfrac{2}{T}\cos\dfrac{2\pi n t}{T}\right]$

Table 3.1 contains a tabulation of the several forms of the Fourier series.

We observe that even though an infinite number of frequencies are used to synthesize the original signal in the Fourier series expansion, they do not constitute a continuum; each frequency term is a multiple of the fundamental frequency $\omega_0/2\pi$. The set of coefficients C_n constitutes the **amplitude spectrum** for the periodic function; correspondingly, the set of angles φ_n constitutes the **phase spectrum**.

It is important to realize that the Fourier series representation of a periodic function $f(t)$ cannot be differentiated term by term to give the representation of df/dt. A simple example shows this; consider the square wave having the Fourier series representation given in Table 3.2. The derivative of this wave in the period is a pair of equally spaced, oppositely directed δ-functions (see Problem P3-2.5b). These are not specified by the derivative of the Fourier series representation of $f(t)$. However, a Fourier series can be integrated term by term to yield a valid representation of $\int f(t)\,dt$.

■ ■ ■

3-3 FEATURES OF PERIODIC FUNCTIONS

Next we develop some important properties of periodic functions.

Parseval's Formula Consider two periodic functions $f_1(t)$ and $f_2(t)$ that have the same period T. These functions are expressed respectively by their series

$$f_1(t) = \sum_{n=-\infty}^{\infty} \alpha_n e^{jn\omega_0 t} \quad \text{(a)}$$

$$f_2(t) = \sum_{m=-\infty}^{\infty} \beta_m e^{jm\omega_0 t} \quad \text{(b)}$$

(3.10)

The mean value of the product of these two functions is given by

$$\frac{1}{T}\int_{t_0}^{t_0+T} f_1(t) f_2^*(t)\, dt = \frac{1}{T} \sum_{n=-\infty}^{\infty} \sum_{m=-\infty}^{\infty} \alpha_n \beta_m^* \int_{t_0}^{t_0+T} e^{j(n-m)\omega_0 t}\, dt$$

$$= \begin{cases} 0 & n \neq m \\ \sum_{n=-\infty}^{\infty} \alpha_n \beta_n^* & n = m \end{cases}$$

(3.11)

by (3.6). If we set $f_1(t) = f_2(t) = f(t)$ in this expression, then as a consequence $\alpha_n = \beta_n$, and the result is

$$\frac{1}{T}\int_{t_0}^{t_0+T} |f(t)|^2\, dt = \sum_{n=-\infty}^{\infty} |\alpha_n|^2$$

(3.12)

This is the **energy relation** discussed in Section 1-5. It can be thought of as the energy dissipated during one period by a one-Ohm resistor when the voltage source $f(t)$ volts is applied to it.

We can also write the energy relation in the following way:

$$\frac{1}{T}\int_{t_0}^{t_0+T} |f(t)|^2\, dt = \alpha_0^2 + \sum_{\substack{n=-\infty \\ n \neq 0}}^{\infty} \alpha_n \alpha_n^* = \frac{A_0^2}{4} + \sum_{n=1}^{\infty} (\alpha_n \alpha_n^* + \alpha_{-n}\alpha_{-n}^*)$$

$$= \frac{A_0^2}{4} + \sum_{n=1}^{\infty} 2\alpha_n \alpha_n^*$$

$$= \frac{A_0^2}{4} + \sum_{n=1}^{\infty} 2[(\text{Re}\{\alpha_n\})^2 + (\text{Im}\{\alpha_n\})^2]$$

$$= \frac{A_0^2}{4} + \sum_{n=1}^{\infty} \left(\frac{A_n^2}{2} + \frac{B_n^2}{2}\right) \quad \text{(a)}$$

(3.13)

Hence we have Parseval's formula

$$\frac{1}{T}\int_{t_0}^{t_0+T} |f(t)|^2\, dt = \sum_{n=-\infty}^{\infty} |\alpha_n|^2 = \frac{A_0^2}{4} + \sum_{n=1}^{\infty} \left(\frac{A_n^2}{2} + \frac{B_n^2}{2}\right)$$

$$= \frac{A_0^2}{4} + \sum_{n=1}^{\infty} \frac{C_n^2}{2}$$

(b)

FOURIER SERIES

If $f(t)$ denotes a current, this result specifies that the rms value squared of the current is the sum of the dc component and the rms values of the ac components. Stated differently, the total power is the power of the dc component plus the power of all the ac components.

Effects of Symmetry The waveforms that occur in electrical engineering applications frequently possess symmetries of various types. The important symmetries are illustrated in Figure 3.3, together with the mathematical conditions that express these symmetries. Symmetry is important because it may lead to simplifications of the formulas for the coefficients.

Figure 3.3
Illustrations of types of symmetry. (a) Zero average value. (b) Zero-axis symmetry, $f(t) = f(-t)$. (c) Zero-point symmetry, $f(t) = -f(-t)$. (d) Half-wave symmetry, $f(t) = -f(t + (T/2))$.

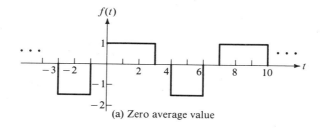
(a) Zero average value

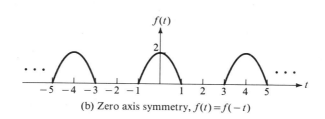
(b) Zero axis symmetry, $f(t) = f(-t)$

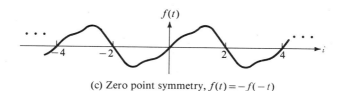
(c) Zero point symmetry, $f(t) = -f(-t)$

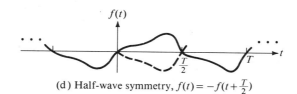
(d) Half-wave symmetry, $f(t) = -f(t + \frac{T}{2})$

a. Average value of function is zero. In this case the area enclosed by the positive part of the function equals that enclosed by the negative part.
b. Zero-axis symmetry* $f(t) = f(-t)$. A wave possesses zero-axis symmetry if it is symmetrical about the axis $t = 0$.
c. Zero-point symmetry* $f(t) = -f(-t)$. A wave possesses zero-point symmetry if it is symmetrical about the origin.
d. Half-wave or mirror symmetry $f(t) = -f(t + T/2)$. A wave possesses half-wave symmetry when the negative portion of the wave is the mirror image of the positive portion of the wave, displaced horizontally by distance $(T/2)$.

The recognition of the existence of one or more of the above symmetries results in the simplification of the computation of the harmonic coefficients. The consequences of these symmetries are given in Table 3.3.

Table 3.3 Symmetries and the Fourier Coefficients

Symmetry	Mathematical Requirements	A_0	A_n	B_n	Coefficient Special Remarks
Zero-axis	$f(t) = f(-t)$	✓	✓	0	Integration required over half cycle
Zero-point	$f(t) = -f(-t)$	0	0	✓	Integration required over half cycle
Half-wave	$f(t) = -f\left(t + \dfrac{T}{2}\right)$	0	✓	✓	Integration required over half cycle and no even terms.

As a specific application of the results contained in Table 3.3, refer to the wave of Figure 3.4, which is seen to exhibit both zero-point and half-wave symmetries. As a result, its Fourier representation would be characterized by

$A_0 = 0 \qquad A_n = 0 \qquad$ no even terms

*In mathematical literature the terms **even** and **odd** are used to describe functions that are designated respectively as having zero-axis and zero-point symmetry in this chapter.

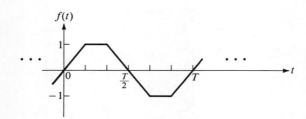

Figure 3.4 Typical wave illustrating effect of symmetries.

Figure 3.5
The construction of a signal from its even and odd parts. (a) Given wave. (b) Even part. (c) Odd part.

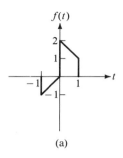

(a)

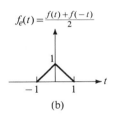

(b)

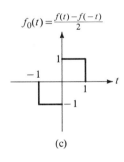

(c)

Thus the Fourier series representation is

$$f(t) = \sum_{n(\text{odd})=1}^{\infty} B_n \sin n\omega_0 t$$

where

$$B_n = \frac{2}{\pi} \int_0^{\pi} f(t) \sin n\omega_0 t \, d(\omega_0 t)$$

Even Function If a periodic function is an even function so that $f(-t) = f(t) \triangleq f_e(t)$, then it follows from (3.8c) that all coefficients B_n are zero since the integrand is (even × odd =) odd. The Fourier series expansion becomes

$$f_e(t) = \frac{A_0}{2} + \sum_{n=1}^{\infty} A_n \cos n\omega_0 t \quad \text{(a)} \tag{3.14}$$

where

$$A_n = \frac{4}{T} \int_{t_0}^{t_0 + T/2} f_e(t) \cos n\omega_0 t \, dt \quad \text{(b)}$$

Odd Function If a periodic function is an odd function so that $f(-t) = -f(t) \triangleq f_0(t)$, then it follows from (3.13b) that $A_n = 0$, and the Fourier series expansion becomes

$$f_0(t) = \sum_{n=1}^{\infty} B_n \sin n\omega_0 t \quad \text{(a)} \tag{3.15}$$

where

$$B_n = \frac{4}{T} \int_{t_0}^{t_0 + T/2} f_0(t) \sin n\omega_0 t \, dt \quad \text{(b)}$$

It is important to realize that any periodic function can always be resolved into even and odd parts. This follows because we can write $f(t)$ in the equivalent form

$$f(t) = \underbrace{\frac{f(t) + f(-t)}{2}}_{\text{even}} + \underbrace{\frac{f(t) - f(-t)}{2}}_{\text{odd}} \quad \text{(a)} \tag{3.16}$$

From (3.14) and (3.15) the result is

$$f(t) = \frac{A_0}{2} + \sum_{n=1}^{\infty} A_n \cos n\omega_0 t + \sum_{n=1}^{\infty} B_n \sin n\omega_0 t \quad \text{(b)}$$

where the A_n's are deduced from (3.14b) and the B_n's from (3.15b). The illustration in Figure 3.5 shows the construction of a function $f(t)$ from its even and odd parts.

CHAPTER 3

Average Value The full-cycle time average value of a periodic signal over the interval $t_0 \leq t \leq t_0 + T$ is given by

$$\langle f(t) \rangle_T = \frac{1}{T} \int_{t_0}^{t_0+T} f(t)\, dt = \frac{1}{T} \int_{-T/2}^{T/2} f(t)\, dt = \frac{A_0}{2} \qquad (3.17)$$

One must exercise care in measuring the average value, say of a voltage or current, because electrical indicating instruments do not always carry out the average over a full cycle. Specifically, a half-wave average reading instrument may read only the average of the positive portion or the negative portion of the wave, and for unsymmetrical waves the readings will differ.

EXAMPLE 3.1
Deduce the Fourier series expansion for the waves shown in Figure 3.6.

Solution: **a.** The function given in Figure 3.6a is an even function, and thus $B_n = 0$. This function is created when a sine voltage or current waveform is rectified by a single diode, a process known as half-wave rectification. As shown, the total period of the wave is the interval $-1 \leq t \leq 1$. Thus from (3.14) we deduce

$$A_n = \frac{4}{2} \int_0^{1/2} \cos \pi t \cdot \cos n\omega_0 t\, dt + \frac{4}{2} \int_{1/2}^1 0 \cdot \cos n\omega_0 t\, dt$$

$$= 2 \int_0^{1/2} \left[\frac{\cos(\pi + n\omega_0)t}{2} + \frac{\cos(\pi - n\omega_0)t}{2} \right] dt$$

$$= \frac{2\pi}{\pi^2 - (n\omega_0)^2} \cos\left(\frac{n\omega_0}{2} \right)$$

Hence the function is expressed by the series

$$f(t) = \frac{1}{\pi} + \sum_{n=1}^{\infty} \frac{2\pi}{\pi^2 - (n\omega_0)^2} \cos\left(\frac{n\omega_0}{2} \right) \cos n\omega_0 t \qquad (3.18)$$

b. The function in Figure 3.6b is neither even nor odd, and the Fourier series expansion will contain both A_n and B_n terms. By (3.8) and (3.9) we obtain

$$A_0 = \frac{2}{9} \int_{-1}^{0.5} 2\, dt = \frac{6}{9}$$

$$A_n = \frac{2}{9} \int_{-1}^{0.5} 2 \cos n \frac{2\pi}{9} t\, dt = \frac{2}{\pi n} \left(\sin \frac{n\pi}{9} + \sin \frac{2n\pi}{9} \right) \qquad n = 1, 2, \ldots$$

$$B_n = \frac{2}{9} \int_{-1}^{0.5} 2 \sin n \frac{2\pi}{9} t\, dt = \frac{2}{\pi n} \left(-\cos \frac{n\pi}{9} + \cos \frac{2n\pi}{9} \right) \qquad n = 1, 2, \ldots$$

$$C_n = \sqrt{A_n^2 + B_n^2} \qquad \varphi_n = -\tan^{-1}\left(\frac{B_n}{A_n} \right)$$

Figure 3.6
Two periodic functions.
(a) Output from a half-wave rectifier. (b) Function with neither odd nor even symmetry.

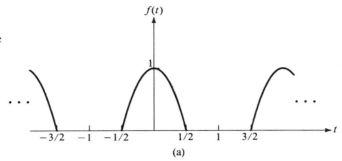

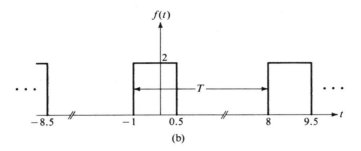

The final result is

$$f(t) = \frac{1}{3} + \sum_{n=1}^{\infty} C_n \cos(n\omega_0 t + \varphi_n) \qquad n = 1, 2, \ldots \tag{3.19}$$

The frequencies present in this series are $f_n = \omega_n/2\pi = n(2\pi/9)/2\pi = n/9$ Hz. The average or dc term has a value = 1/3. The frequency corresponding to $n = 1$ is the **fundamental or first harmonic;** for $n = 2$, it is the second harmonic, and so on. As already noted, the **frequency** or **amplitude spectrum** is the set of absolute values of the amplitudes of the harmonics C_n; the **phase spectrum** is the set of all phases of the harmonics. The amplitude and phase spectrums of the given function are contained in Figure 3.7. ∎

The Fourier series expansion of several important waveforms are contained in Table 3.2.

Finite Signals It can be shown that a continuous function $f(t)$ within an interval $t_0 \leq t \leq t_0 + T$ can be approximated by a trigonometric polynomial of the form given in (3.9) to any degree of accuracy specified in advance.

To approximate the function $f(t)$ shown in Figure 3.8a, the periodic function $f_p(t)$ shown in Figure 3.8b is created. The expansion then proceeds as before for the periodic function, but the final range of applicability is constrained to coincide with the range of the original function. The solution is then specified as

Figure 3.7
Amplitude and phase spectra for the signal shown in Figure 3.6b.

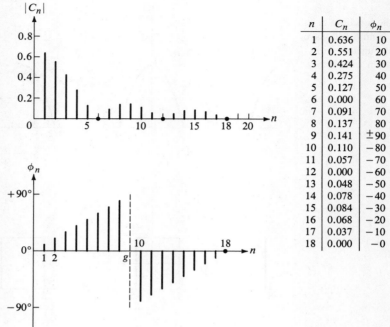

n	C_n	ϕ_n
1	0.636	10
2	0.551	20
3	0.424	30
4	0.275	40
5	0.127	50
6	0.000	60
7	0.091	70
8	0.137	80
9	0.141	±90
10	0.110	−80
11	0.057	−70
12	0.000	−60
13	0.048	−50
14	0.078	−40
15	0.084	−30
16	0.068	−20
17	0.037	−10
18	0.000	−0

Figure 3.8
Representation of a non-periodic function by a Fourier series.

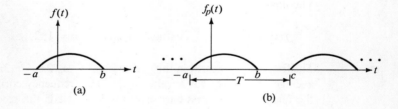

$$f(t) = \begin{cases} \sum_{n=-\infty}^{\infty} \alpha_n e^{jn\omega_0 t} & -a \leq t \leq c \\ 0 & t < -a \quad t > c \end{cases} \quad (3.20)$$

where

$$\alpha_n = \frac{1}{T} \int_{t_0}^{t_0+T} f_p(t) e^{-jn\omega_0 t} \, dt$$

Partial Sums If the summation of the series of a periodic function extends to N, this suggests either that terms greater than N in the exponential are zero (such a series is band-limited) or that the expansion has been truncated to N terms, presumably because the higher-order terms sum to a negligible amount

FOURIER SERIES

for the purposes at hand. In such cases the series is written

$$f_N(t) = \frac{A_0}{2} + \sum_{n=1}^{N} (A_n \cos n\omega_0 t + B_n \sin n\omega_0 t) \tag{3.21}$$

If we combine this expression with the known forms for the factors A_n and B_n, we obtain

$$f_N(t) = \frac{2}{T} \int_{-T/2}^{T/2} f(\tau) \left[\frac{1}{2} + \sum_{n=1}^{N} (\cos n\omega_0 t \cos n\omega_0 \tau + \sin n\omega_0 t \sin n\omega_0 \tau) \right] d\tau$$

$$= \frac{2}{T} \int_{-T/2}^{T/2} f(\tau) \left[\frac{1}{2} + \sum_{n=1}^{N} \cos n\omega_0(t - \tau) \right] d\tau \tag{3.22}$$

Change the variable by writing $t - \tau = v$; this integral becomes

$$f_N(t) = \frac{2}{T} \int_{t-T/2}^{t+T/2} f(t - v) \left[\frac{1}{2} + \sum_{n=1}^{N} \cos n\omega_0 v \right] dv \tag{3.23}$$

It is possible to sum the expression in the bracket, which is now designated S_N. To do this, consider the quantity

$$2S_N \sin\left(\frac{\omega_0 v}{2}\right) = \sin\left(\frac{\omega_0 v}{2}\right)(1 + 2 \cos \omega_0 v$$

$$+ 2 \cos 2\omega_0 v + \cdots + 2 \cos N\omega_0 v)$$

$$= \sin \frac{\omega_0 v}{2} + \left(\sin \frac{3\omega_0 v}{2} - \sin \frac{\omega_0 v}{2} \right)$$

$$+ \left(\sin \frac{5\omega_0 v}{2} - \sin \frac{3\omega_0 v}{2} \right)$$

$$+ \cdots + \left[\sin\left(N + \frac{1}{2}\right)\omega_0 v - \sin\left(N - \frac{1}{2}\right)\omega_0 v \right]$$

$$= \sin\left[\left(N + \frac{1}{2}\right)\omega_0 v\right] \tag{3.24}$$

It follows, therefore, that

$$S_N \triangleq \frac{1}{2} + \sum_{n=1}^{N} \cos n\omega_0 v = \frac{\sin\left(N + \frac{1}{2}\right)\omega_0 v}{2 \sin\left(\frac{\omega_0 v}{2}\right)} \tag{3.25}$$

Since the functions $f(t - v)$ and $S_N(v)$ are now both periodic with period T, then (3.23) gives the partial sum of the Fourier series in the form

$$f_N(t) = \frac{1}{T} \int_{-T/2}^{T/2} f(t - v) \frac{\sin\left[(2N + 1)\omega_0 \frac{v}{2}\right]}{\sin\left(\frac{\omega_0 v}{2}\right)} dv \quad \text{(a)} \tag{3.26}$$

or

$$f_N(t) = \frac{1}{T} \int_{-T/2}^{T/2} f(v) \frac{\sin\left[(2N+1)\omega_0 \frac{(t-v)}{2}\right]}{\sin \omega_0 \left(\frac{t-v}{2}\right)} dv \qquad \text{(b)}$$

The same results can be found by the simple transformation $t - v = \tau$ and then setting $\tau = v$. This formula provides a closed-form analytic expression for the Fourier expansion.

Least-Squares Approximation Property The least-squares approximation property of the Fourier series relates to the difference between the specified function $f(t)$ and its Fourier series approximation. Suppose that a real function $f(t)$ is approximated by a finite or truncated series of exponentials, not necessarily of the Fourier type

$$f_N(t) = \sum_{n=-N}^{N} D_n e^{jn\omega_0 t} \qquad (3.27)$$

We wish to select the coefficients D_n in a manner that minimizes the mean square error resulting from approximating $f(t)$ by $f_N(t)$. We denote the error $\epsilon(t)$ as

$$\epsilon(t) = f(t) - f_N(t)$$

so that

$$\epsilon(t) = \sum_{n=-\infty}^{\infty} C_n e^{jn\omega_0 t} - \sum_{n=-N}^{N} D_n e^{jn\omega_0 t}$$

We observe that $\epsilon(t)$ is a periodic function; we can write it in the form of a Fourier series

$$\epsilon(t) = \sum_{n=-\infty}^{\infty} G_n e^{jn\omega_0 t} \qquad \text{(a)} \qquad (3.28)$$

where

$$G_n = \begin{cases} C_n - D_n & -N \leq n \leq N \\ C_n & |n| > N \end{cases} \qquad \text{(b)}$$

We wish to examine the mean square value of $\langle \epsilon^2(t) \rangle$. Assume that $f(t)$ is real; we can write from (3.28)

$$\langle \epsilon^2 \rangle = \left\langle \left(\sum_{n=-\infty}^{\infty} G_n e^{jn\omega_0 t} \right) \left(\sum_{m=-\infty}^{\infty} G_m^* e^{-jm\omega_0 t} \right) \right\rangle$$

$$= \sum_{n=-\infty}^{\infty} \sum_{m=-\infty}^{\infty} G_n G_m^* \langle e^{j(n-m)\omega_0 t} \rangle$$

$$= \sum_{n=-N}^{N} |C_n - D_n|^2 + \sum_{|n|>N} |C_n|^2 \qquad (3.29)$$

where the time average $\langle \exp[j(n-m)\omega_0 t]\rangle = 0$ for $n \neq m$; and $= 1$ for $n = m$. Each term in this expression is positive. Clearly, to minimize $\langle \epsilon^2 \rangle$ we should set

$$D_n = C_n \qquad (3.30)$$

since this will make the first summation vanish. The result is

$$\langle \epsilon_{\min}^2 \rangle = \sum_{|n|>N} C_n^2 \qquad (3.31)$$

This result shows that the mean square error is minimized by making the coefficient D_n in the finite exponential series (3.27) identical with the Fourier coefficient C_n. That is, if the Fourier series expansion of a function $f(t)$ is truncated at any given value of N, it approximates $f(t)$ with a smaller mean square error than any other exponential series with the same number of terms. Furthermore, since the error is the sum of the positive terms, the error is a monotonically decreasing function of the number of harmonics used in the approximation.

Sum and Difference Functions Suppose that $f(t)$ and $h(t)$ are periodic, both having the same period. Then we can write

$$p(t) = C_1 f(t) \pm C_2 h(t) = \sum_{n=-\infty}^{\infty} [C_1 \beta_n \pm C_2 \gamma_n] e^{jn\omega_0 t}$$

$$= \sum_{n=-\infty}^{\infty} \alpha_n e^{jn\omega_0 t} \qquad (3.32)$$

where C_1 and C_2 are constants and where β_n and γ_n are the appropriate Fourier expansion coefficients for the functions $f(t)$ and $h(t)$, respectively, and $\alpha_n = C_1 \beta_n \pm C_2 \gamma_n$.

Product of Two Functions If $f(t)$ and $h(t)$ are periodic with the same period, the product

$$p(t) \triangleq f(t)h(t) = \sum_{l=-\infty}^{\infty} \beta_l e^{jl\omega_0 t} \sum_{m=-\infty}^{\infty} \gamma_m e^{jm\omega_0 t} = \sum_{l=-\infty}^{\infty} \sum_{m=-\infty}^{\infty} \beta_l \gamma_m e^{j(l+m)\omega_0 t}$$

$$= \sum_{n=-\infty}^{\infty} \sum_{m=-\infty}^{\infty} (\beta_{n-m} \gamma_m) e^{jn\omega_0 t} \qquad (3.33)$$

where $l + m = n$. The sum $\sum_{m=-\infty}^{\infty} \beta_{n-m} \gamma_m$ is the product function of the two sequences of the β_m's and γ_m's. Equation (3.33) indicates that the Fourier coefficients of the new function $p(t) = f(t)h(t)$ are related to the coefficients of the Fourier expansion of (3.5) by

$$\boxed{\frac{1}{T} \int_{-T/2}^{T/2} f(t)h(t) e^{-jn\omega_0 t} \, dt = \sum_{m=-\infty}^{\infty} \beta_{n-m} \gamma_m} \qquad (3.34)$$

The summation in this equation denotes a convolution process. The β_{n-m} sequence is the result of a folded sequence β_m that is shifted by n.

Convolution of Two Functions Periodic convolution was defined by (2.24) as

$$g(t) = \frac{1}{T}\int_{-T/2}^{T/2} f(\tau)h(t-\tau)\,d\tau \tag{3.35}$$

Further, if $f(t)$ and $h(t)$ are each periodic, then $g(t)$ is periodic, as discussed in Section 2-5. Thus we can write $g(t)$ in a Fourier series representation with coefficients

$$\alpha_n = \frac{1}{T}\int_{-T/2}^{T/2} g(t)e^{-jn\omega_0 t}\,dt = \frac{1}{T^2}\iint_{-T/2}^{T/2} f(\tau)h(t-\tau)e^{-jn\omega_0 t}\,dt\,d\tau$$

We write this

$$\alpha_n = \frac{1}{T}\int_{-T/2}^{T/2} f(\tau)e^{-jn\omega_0 \tau}\,d\tau \, \frac{1}{T}\int_{-T/2}^{T/2} h(t-\tau)e^{-jn\omega_0(t-\tau)}\,dt \tag{3.36}$$

By writing $t - \tau = v$ in the second integral, we have

$$\alpha_n = \frac{1}{T}\int_{-T/2}^{T/2} f(\tau)e^{-jn\omega_0 \tau}\,d\tau \, \frac{1}{T}\int_{-T/2-\tau}^{T/2-\tau} h(v)e^{-jn\omega_0 v}\,dv$$

For periodic functions, integration of the second integral is independent of a shift τ. Then

$$\alpha_n = \beta_n \gamma_n \tag{3.37}$$

where the coefficients β_n and γ_n belong to the functions $f(t)$ and $h(t)$, respectively. Therefore we have

$$\boxed{\begin{aligned} g(t) &= \sum_{n=-\infty}^{\infty} \alpha_n e^{jn\omega_0 t} = \sum_{n=-\infty}^{\infty} \beta_n \gamma_n e^{jn\omega_0 t} \\ &= \frac{1}{T}\int_{-T/2}^{T/2} f(\tau)h(t-\tau)\,d\tau \end{aligned}} \tag{3.38}$$

EXAMPLE 3.2

Two functions $f(t)$ and $h(t)$ are explicitly shown in Figure 3.9. Find the Fourier series coefficients of the product function and the convolution of these functions.

Solution: Use the results of Problem 3-2.1a and 3-2.1f, which show the Fourier coefficients of the series expansion for $f(t)$, β_n and those for $h(t)$, γ_n; we have

$$\beta_n = 4j\,\frac{\sin^2\left(\dfrac{n\omega_0}{2}\right)}{n\omega_0}$$

Figure 3.9
Two periodic functions.

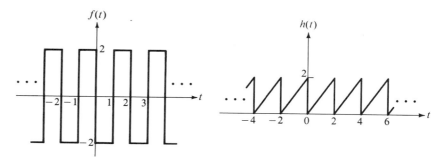

$$\gamma_n = \frac{j}{n\omega_0}$$

The Fourier coefficients for the convolution function follow from (3.38)

$$\alpha_n = \beta_n \gamma_n = \frac{4\sin^2\left(\frac{n\omega_0}{2}\right)}{n^2\omega_0^2}$$

The coefficients for the product function follow from (3.34)

$$\alpha_n = \sum_{m=-\infty}^{\infty} \beta_{n-m}\gamma_m = \sum_{m=-\infty}^{\infty} \frac{4\sin^2\left[\frac{(n-m)\omega_0}{2}\right]}{(n-m)\omega_0 m\omega_0}$$

■ ■ ■

3-4 CHOICE OF ORIGIN

The form of the Fourier series representation of a given wave is intimately related to the origin about which the representation is to be developed. It is possible to use the series representation of a given wave that is specified about one origin to yield the series representation of the wave relative to another origin. It is often easier to find the Fourier series relative to one origin and then shift the origin to a previously specified point in the wave than it is to evaluate the coefficients directly for the wave as specified.

To examine the situation, refer to the particular case of a square wave illustrated in Figure 3.10. The Fourier representation of this wave relative to an axis at point 0 is given by (see Table 3.3)

$$f(t) = \frac{4}{\pi}\left(\sin t + \frac{1}{3}\sin 3t + \frac{1}{5}\sin 5t + \cdots\right)$$

The coefficients are obtained by a straightforward application of (3.8).

Suppose we want to obtain the Fourier series representation of the wave relative to the axis at $0'$ chosen $\pi/2$ to the right of point 0. A direct application

Figure 3.10
A square wave and the reference origins 0 and 0'.

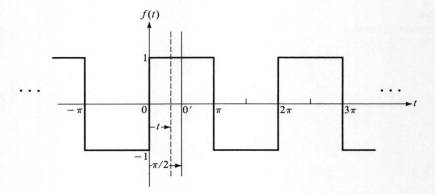

of (3.8) will again yield the required results. If one notes, however, that the Fourier series representation completely describes a given wave over a full period except at points of discontinuity, it is expected that the representation relative to one origin should be related to the representation relative to any other origin. Clearly, in the present case the value of the function $f(t)$ at point 0' is $f(t - \pi/2)$, relative to the function at point 0. Therefore, the Fourier series for the square wave relative to point 0' is obtained from (3.39) by replacing t by $t - \pi/2$. This yields

$$f(t) = \frac{4}{\pi}\left[\sin(t - \pi/2) + \frac{1}{3}\sin 3(t - \pi/2) + \frac{1}{5}\sin 5(t - \pi/2) + \cdots\right]$$

which is written

$$f(t) = \frac{4}{\pi}\left(-\cos t + \frac{1}{3}\cos 3t - \frac{1}{5}\cos 5t - \cdots\right) \tag{3.40}$$

These results may be generalized and expressed in the form of a set of rules. Thus if the Fourier series representation of a complex wave $f(t)$ is known relative to one origin, the Fourier series relative to another origin at an angle θ to the **left** is obtained from the first series by writing $t - \theta$ for t in each term of the series. Since in general

$$\cos(t - \theta) = \cos t \cos \theta + \sin t \sin \theta$$

and

$$\sin(t - \theta) = \sin t \cos \theta - \cos t \sin \theta$$

a representation that might have been expressible in terms of sine or cosine terms alone may now contain both sine and cosine terms.

EXAMPLE 3.3

The Fourier series of the triangular wave shown in Figure 3.11a is

Figure 3.11
Illustrating a shift of origin to simplify calculation of coefficients.

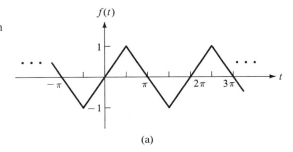

(a)

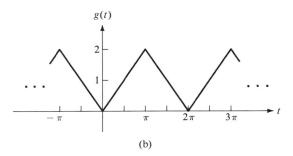

(b)

$$f(t) = \frac{8}{\pi^2}\left(\sin t - \frac{1}{3^2}\sin 3t + \frac{1}{5^2}\sin 5t - \cdots\right) \quad (3.41)$$

Use this information to write the series for $g(t)$ given in Figure 3.11b.

Solution: From an inspection of the two waves, it is evident that t in (3.41) is to be replaced by $t - \pi/2$, and that

$$g(t) = 1 + f\left(t - \frac{\pi}{2}\right) \quad (3.42)$$

Therefore

$$g(t) = 1 + \frac{8}{\pi^2}\left[\sin\left(t - \frac{\pi}{2}\right) - \frac{1}{3^2}\sin 3\left(t - \frac{\pi}{2}\right) \right.$$
$$\left. + \frac{1}{5^2}\sin 5\left(t - \frac{\pi}{2}\right) - \cdots\right]$$

or finally

$$g(t) = 1 - \frac{8}{\pi^2}\left(\cos t + \frac{1}{3^2}\cos 3t + \frac{1}{5^2}\cos 5t + \cdots\right) \quad (3.43)$$

This result should be compared with the fourth entry in Table 3.2. ■

■ ■ ■

3-5 SYSTEMS WITH PERIODIC INPUTS

Suppose that the input signal to a linear time-invariant system is a periodic function. Because of the properties of the Fourier series, we can visualize the input source function as a sum of sources, each of which produces a sine wave and each of which is shifted by an appropriate amount with respect to the time (or space, for optical signals) origin. Recall from (2.113) that when an input signal $\exp(jn\omega_0 t)$ is applied to an LTI system, the output is equal to

$$\mathcal{O}\{e^{jn\omega_0 t}\} = H(n\omega_0)e^{jn\omega_0 t} \tag{3.44}$$

where

$$H(n\omega_0) = \text{transfer function of the system} = |H(n\omega_0)|\exp[j\varphi(n\omega_0)] \tag{3.45}$$

By the superposition theorem for linear systems, which we learned in circuit analysis studies, we obtain

$$y(t) = \mathcal{O}\{f(t)\} = \mathcal{O}\left\{\sum_{n=-\infty}^{\infty} \alpha_n e^{jn\omega_0 t}\right\} = \sum_{n=-\infty}^{\infty} \alpha_n H(n\omega_0)e^{jn\omega_0 t} \tag{3.46}$$

However, $H(n\omega_0)$ is a complex constant for each n; therefore it follows that the output is also periodic. Specifically for an $f(t)$ that is given by

$$f(t) = \frac{A_0}{2} + \sum_{n=1}^{\infty} (A_n \cos n\omega_0 t + B_n \sin n\omega_0 t) \tag{3.47}$$

the output is

$$y(t) = \frac{A_0}{2}H(0) + \sum_{n=1}^{\infty} |H(n\omega_0)|[A_n \cos[n\omega_0 t + \varphi(n\omega_0)]$$
$$+ B_n \sin[n\omega_0 t + \varphi(n\omega_0)]] \tag{3.48}$$

EXAMPLE 3.4
Find the output voltage of the system shown in Figure 3.12a if the input voltage source is the periodic function given by

$$v(t) = f(t) = 10(\sin t - 0.5 \sin 2t + 0.33 \sin 3t)$$

which is shown by Figure 3.12c.

Solution: Apply Kirchhoff's voltage law to the circuit to give

$$\frac{L}{R}\frac{d(Ri)}{dt} + Ri = v$$

or

$$\frac{dv_0}{dt} + \frac{R}{L}v_0 = \frac{R}{L}v$$

If we set $v = \exp(j\omega t)$ in this equation, the voltage v_0 is given by [see (2.113)]

FOURIER SERIES

$$v_0 = H(\omega)e^{j\omega t}$$

Using the relations for v and v_0, we obtain

$$j\omega H(\omega)e^{j\omega t} + \frac{R}{L} H(\omega)e^{j\omega t} = \frac{R}{L} e^{j\omega t}$$

from which

$$H(\omega) = \frac{\frac{R}{L}}{\frac{R}{L} + j\omega} = \frac{\frac{R}{L}}{\sqrt{\left(\frac{R}{L}\right)^2 + \omega^2}} \bigg/ -\tan^{-1}\left(\frac{\omega L}{R}\right)$$

The transfer function (system function) at any frequency $\omega = n\omega_0$ is written as follows:

$$H(n\omega_0) = \frac{\frac{R}{L}}{\sqrt{\left(\frac{R}{L}\right)^2 + (n\omega_0)^2}} \bigg/ -\tan^{-1}\left(\frac{n\omega_0 L}{R}\right) \triangleq |H(n\omega_0)| \bigg/ -\varphi(n\omega_0)$$

Hence the output [see (3.48)] is given by

$$y(t) = v_0(t) = \sum_{n=1}^{3} \frac{\frac{R}{L}}{\sqrt{\left(\frac{R}{L}\right)^2 + (n\omega_0)^2}} B_n \sin[n\omega_0 t - \varphi(n\omega_0)]$$

For the special case of $\omega_0 = 1$ and $R/L = 2$, the output is

$$v_0(t) = \frac{20}{\sqrt{5}} \sin(t - \tan^{-1} 0.5) - \frac{5}{\sqrt{2}} \sin(2t - \tan^{-1} 1)$$

$$+ \frac{6.6}{\sqrt{13}} \sin(3t - \tan^{-1} 1.5)$$

The characteristics of this output are shown in Figure 3.12e. The amplitude and phase characteristics of the transfer function are shown in Figure 3.12d. ∎

Transmission without Distortion Let us consider the case of a filter system that, when placed in a circuit, causes a periodic input function to undergo a change in amplitude and a shift in time in the output, but otherwise causes no changes in the waveform. Such a filter allows distortionless transmission of all frequency components. This statement in mathematical form is given by

$$y(t) = h_0 f(t - t_0) \tag{3.49}$$

where $f(t)$ is the input function to the filter, h_0 is a constant, and $y(t)$ is the output. Since $f(t)$ is a periodic function, this equation can be written

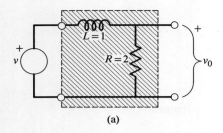
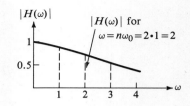
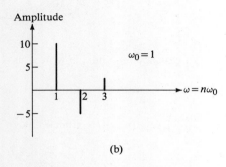
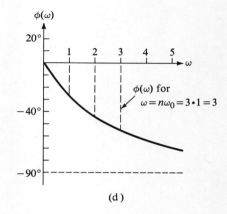
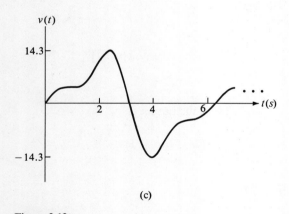
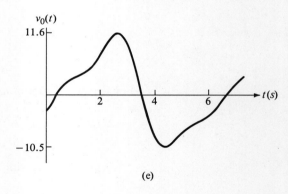

Figure 3.12
(a) System. (b) Amplitude spectrum of $v(t)$. (c) Input signal $v(t)$. (d) Amplitude and phase characteristic of the transfer function $H(n\omega_0)$. (e) Output signal $v_0(t)$.

$$y(t) = \sum_{n=-\infty}^{\infty} h_0 \alpha_n e^{jn\omega_0(t-t_0)} = \sum_{n=-\infty}^{\infty} |\alpha_n| h_0 e^{j[n\omega_0 t + \varphi_n - n\omega_0 t_0]} \quad (3.50)$$

where $\alpha_n = |\alpha_n| \exp[\varphi_n]$. Observe that the coefficients of the output are $h_0 |\alpha_n|$ and the phase angle is $\varphi_n - n\omega_0 t_0$. But since the general form of the transfer function of a system is

$$H(\omega) = |H(\omega)| e^{j\varphi(\omega)}$$

FOURIER SERIES

we see that this filter must possess the transfer function

$$H(\omega) = h_0 e^{jn\omega_0 t_0} \tag{3.51}$$

where

$|H(\omega)| = h_0 =$ constant

$\varphi(\omega) = n\omega_0 t_0 =$ linear function with respect to frequency

A filter with such a transfer function characteristic is called an **ideal filter.** We note, however, that a physical filter system which consists of L's and C's cannot respond instantaneously to a sudden change in excitation; that is, if a step function is applied to its input, its output will not respond instantaneously but will start slowly and will ultimately level off. Since the ideal filter assumes that, except for an amplitude change and a time delay, an input step function will result in an output step function, we conclude that we cannot build ideal filters using inductors and capacitors and that ideal filters are not physically realizable. Ideal filters may be approximated, as we will study in some detail in Chapters 5 and 11.

Observe that the transfer function of the circuit in Figure 3.12 over a limited band at low frequencies has an amplitude response approximately constant and a phase characteristic approximately linear with frequency. Hence we may say that over this frequency range the system approximates an ideal filter.

Band-Limited Signals As already noted in Section 3-3, a periodic function is band-limited if its amplitude spectrum contains only a finite number of coefficients α_n or C_n. The explicit form is given by

$$f(t) = \sum_{n=-N}^{N} \alpha_n e^{jn\omega_0 t} = C_0 + \sum_{n=1}^{N} C_n \cos(n\omega_0 t + \varphi_n) \tag{3.52}$$

Clearly, if we know $\omega_0 = 2\pi/T$, then we need only determine $2N + 1$ terms—namely, $C_0, C_1, C_2, \ldots, C_n$ and the phase functions $\varphi_1, \ldots, \varphi_n$—to completely specify the function $f(t)$. This permits writing the theorem:

Theorem 3.1 (Sampling Theorem). A periodic band-limited function containing only N harmonics in its Fourier series expansion is uniquely specified by its values at $2N + 1$ instants in one period.

Proof: This theorem is easily proved: if $2N + 1$ values of $f(t)$, written $f(t_1), f(t_2), \ldots, f(2N+1)$, are introduced in (3.52), this generates $2N + 1$ equations. These can be solved simultaneously to specify the factors C_0, $C_1, \ldots, C_n$; $\varphi_1, \varphi_2, \ldots, \varphi_n$; or equivalently α_0, Re$\{\alpha_1\}, \ldots,$ Re$\{\alpha_n\}$, Im$\{\alpha_1\}, \ldots,$ Im$\{\alpha_n\}$. A more complete discussion of sampling is given in Section 7-2.

■ ■ ■

3-6 GIBBS PHENOMENON

Begin with a periodic function $f(t)$ expressed by its Fourier series

$$f(t) = \sum_{n=-\infty}^{\infty} \beta_n e^{jn\omega_0 t} \tag{3.53}$$

The truncated form of this expression is written

$$f_N(t) = \sum_{n=-N}^{N} \beta_n e^{jn\omega_0 t} = \sum_{n=-\infty}^{\infty} \beta_n w_n e^{jn\omega_0 t} \quad \text{(a)} \tag{3.54}$$

where w_n denotes the factor

$$w_n = \begin{cases} 1 & |n| \leq N \\ 0 & |n| > N \end{cases} \quad \text{(b)}$$

This factor can be considered as a *window* function that has a constant value in the range $-N$ to N and is zero outside this range. By comparing (3.54) with (3.38), $f_N(t)$ can be considered the convolution of $f(t)$ with a second function $h(t)$ that has $2N+1$ nonzero Fourier coefficients w_n. In Fourier series form, the function $h(t)$ can be written

$$h(t) = \sum_{n=-N}^{N} e^{jn\omega_0 t}$$
$$= (e^{-j\omega_0 t})^N + (e^{-j\omega_0 t})^{(N-1)} + \cdots + e^{-j\omega_0 t}$$
$$+ 1 + e^{j\omega_0 t} + \cdots + (e^{j\omega_0 t})^N$$

Rearrange this in the form

$$h(t) = \{e^{-j\omega_0 t}(1 + e^{-j\omega_0 t} + (e^{-j\omega_0 t})^2 + \cdots + (e^{-j\omega_0 t})^{(N-1)})\}$$
$$+ \{1 + e^{j\omega_0 t} + (e^{j\omega_0 t})^2 + \cdots + (e^{j\omega_0 t})^N\}$$

Now make use of the relation

$$\frac{1-x^N}{1-x} = 1 + x + x^2 + \cdots + x^{N-1} \tag{3.55}$$

With this, $h(t)$ becomes

$$h(t) = \frac{e^{j(N+1/2)\omega_0 t} - e^{-j(N+1/2)\omega_0 t}}{e^{j\omega_0 t/2} - e^{-j\omega_0 t/2}} = \frac{\sin\left[\left(N+\frac{1}{2}\right)\omega_0 t\right]}{\sin\left(\omega_0 \frac{t}{2}\right)} \tag{3.56}$$

This function is known as the **Fourier kernel.**

FOURIER SERIES

This relation when combined with (3.54) gives [see (3.38)]

$$f_N(t) = \sum_{n=-N}^{N} \beta_n e^{jn\omega_0 t} = \frac{1}{T} \int_{-T/2}^{T/2} f(t-\tau) \frac{\sin\left[\left(N+\frac{1}{2}\right)\omega_0 \tau\right]}{\sin\left(\frac{\omega_0 \tau}{2}\right)} d\tau \qquad (3.57)$$

which is seen to be identical with (3.26). This result shows that truncating a Fourier series is the same as convolving the given function $f(t)$ with the Fourier kernel $h(t)$. However, convolution is a shifting process with integration of the product of the given function with $h(t)$. Therefore the resulting function $f_N(t)$ exhibits oscillations on both sides of the points of discontinuity. These oscillations are known as the **Gibbs phenomenon** and appear whenever a discontinuity is being approximated, no matter what value of N is used. The overshoot at the discontinuity remains 9% higher than the function being approximated independent of the number of terms used to reconstruct the function. It is the result of the nonuniform convergence of the approximation process.

If the window function w_n of (3.54b) is modified slightly to the form

$$w_n = \begin{cases} 1 & |n| \leq N-1 \\ \frac{1}{2} & |n| = N \\ 0 & |n| > N \end{cases} \qquad (3.58)$$

then $h(t)$ has the form

$$h(t) = \sum_{n=-N}^{N} w_n e^{jn\omega_0 t} = (e^{-j\omega_0 t})^N + \cdots + (e^{j\omega_0 t})^N - \frac{1}{2} e^{-j\omega_0 tN} - \frac{1}{2} e^{j\omega_0 tN}$$

$$= \frac{\sin\left[\left(N+\frac{1}{2}\right)\omega_0 t\right]}{\sin\left(\frac{\omega_0 t}{2}\right)} - \cos N\omega_0 t \qquad (3.59)$$

By expansion of the sine term in the numerator and some algebraic manipulation, this expression becomes

$$h(t) = \frac{\sin(N\omega_0 t)}{\sin\left(\frac{\omega_0 t}{2}\right)} \cos\left(\omega_0 \frac{t}{2}\right) \qquad (3.60)$$

But the factor $\cos \omega_0 t/2$ decreases to zero at $t = \pm T/2$, and the resulting function is somewhat smoother than that of $h(t)$ given by (3.56).

It is informative to consider $h(t)$ as a convolving **window function** that serves to extract the truncated function from the given function. The form of w_n given by (3.54b), which is a sequence of the form $\{\ldots 0, 0, 1, 1, \ldots, 1, 1, 0, 0, \ldots\}$, is called a **rectangular window**. The form of w_n given by (3.58), which is $\{\ldots 0, 0, \frac{1}{2}, 1, 1, \ldots, 1, 1, \frac{1}{2}, 0, 0, \ldots\}$, is called the **modified rectangular window**.

Figure 3.13
This effect of smoothing on Gibbs phenomenon.
(a) Three coefficients.
(b) Six coefficients.

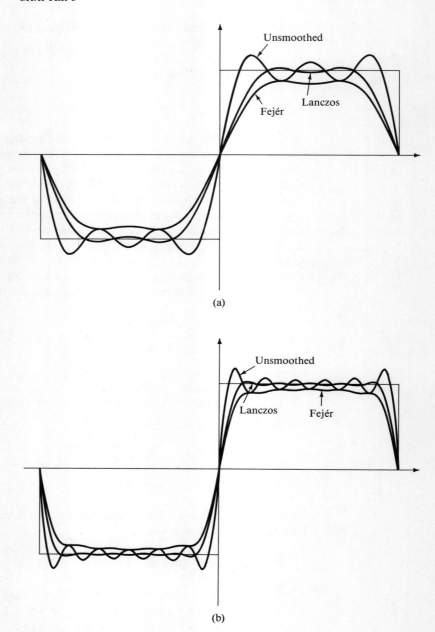

Studies have been undertaken to suggest methods for smoothing the truncated function. Lanczos suggested that the truncated Fourier series

$$f_N(t) = \frac{A_0}{2} + \sum_{n=1}^{N} (A_n \cos n\omega_0 t + B_n \sin n\omega_0 t) \tag{3.61}$$

have imposed on it over the interval T/N a carefully established averaging process. Hence, centered at t, one chooses

$$\begin{aligned} l_N(t) &= \frac{N}{T} \int_{t-T/2N}^{t+T/2N} f_N(t')\, dt' \\ &= \frac{N}{T} \int_{t-T/2N}^{t+T/2N} \frac{A_0}{2}\, dt' + \sum_{n=1}^{N} \left[\frac{N}{T} A_n \int_{t-T/2N}^{t+T/2N} \cos n\omega_0 t'\, dt' \right. \\ &\quad \left. + \frac{N}{T} B_n \int_{t-T/2N}^{t+T/2N} \sin n\omega_0 t'\, dt' \right] \end{aligned}$$

This process leads to the form

$$l_N(t) = \frac{A_0}{2} + \sum_{n=1}^{N} \frac{\sin\left(\frac{n\pi}{N}\right)}{\left(\frac{n\pi}{N}\right)} [A_n \cos n\omega_0 t + B_n \sin \omega_0 t] \tag{3.62}$$

In this calculation the factor $\sin(n\pi/N)/(n\pi/N)$ is called the **Lanczos factor** (and is often referred to as the sigma factor). If we introduce the Fejér factor $(N - n)/N$, the resulting **Fejér smoothing series** of a function $f_N(t)$ is

$$\varphi_N(t) = \frac{A_0}{2} + \sum_{n=1}^{N} \frac{N - n}{N} [A_n \cos n\omega_0 t + B_n \sin n\omega_0 t] \tag{3.63}$$

The Gibbs phenomenon and the effects of the Lanczos and the Fejér smoothing are shown in Figure 3.13. Observe that adding terms in the Fourier expansion does not have any effect in reducing the first overshoot of the Gibbs phenomenon, as already noted. This phenomenon was observed by Michelson using his harmonic analyzer. He found that truncated series often approximated smooth functions very well, but any time he approximated a discontinuous function, he observed unusual overshoots close to the discontinuities. He wrote about his results to Gibbs, who proved that Michelson's analyzer was without defects. Gibbs reported his explanation of this phenomenon in 1899.

■ ■ ■

REFERENCES

1. French, A. P. *Vibration and Waves.* New York: Norton, 1971.
2. Lanczos, C. *Discourse on Fourier Series.* New York: Hafner, 1966.
3. Papoulis, A. *Signal Analysis.* New York: McGraw-Hill, 1977.
4. Stark, H., and F. B. Tuteur. *Modern Electrical Communications—Theory and Systems.* Englewood Cliffs, N.J.: Prentice-Hall, 1979.
5. Tolstov, G. P. *Fourier Series.* New York: Dover, 1976.

PROBLEMS

3-1.1 Find the condition that is required in order that two periodic waves with periods T_1 and T_2, respectively, have a common period.

3-1.2 If $f(t) = \sin t$, find the period of the function $f_1(t) = \sin^3 t$, also $f_2(t) = \sin^4 t$.

3-1.3 The signal $f(t) = a_0 \cos \omega_0 t$, $\omega_0 = 2\pi/T$, is periodic with fundamental period T. Since the function is periodic for $3T$, find the Fourier coefficients of the given function if we regard it as periodic with period $3T$.

3-2.1 Find the Fourier series representations of the functions shown in Figure P3-2.1, and plot the amplitude and phase spectra.

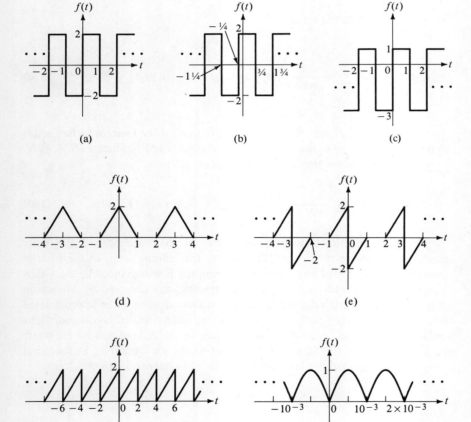

Figure P3-2.1

3-2.2 Plot the given function for $n' = 1$, $n' = 3$ and $n' = 5$

$$f(t) = 1 + \sum_{n=1}^{n'} 2 \cos n\omega_0 t$$

and predict the form of $f(t)$ as $n' \to \infty$. The period is $T = 1$.

3-2.3 Prove (3.9).

3-2.4 a. Show that if a periodic function is absolutely integrable—that is, $\int_0^T |f(t)|\, dt < \infty$—then $|\alpha_n| < \infty$.
  b. Construct a periodic function that has infinite maxima and minima within a period. HINT: use the sine function.
  c. Construct a periodic function that has an infinite discontinuity.

3-2.5 Find the Fourier series for the functions shown in Figure P3-2.5.

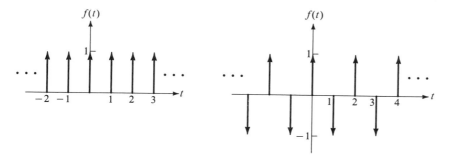

Figure P3-2.5

3-3.1 If the energy of the first four harmonics of the signal shown in Figure P3-2.1a is equal to 1.28 joules, what is the energy contained in the rest of the harmonics?

3-3.2 If we know the expansion coefficients for the periodic function $f_1(t)$ shown in Figure P3-3.2a, find the coefficients of the second periodic function $f_2(t)$ shown in Figure P3-3.2b by appropriately modifying the coefficients found for the first periodic function.

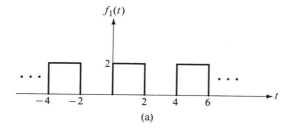
(a)

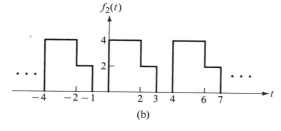
(b)

Figure P3-3.2

3-3.3 Specify which terms are zero in the Fourier series representations of the waves shown in Figure P3-3.3.

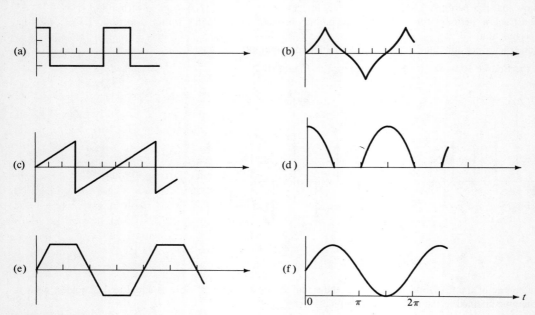

Figure P3-3.3

3-3.4 Decompose the signals shown in Figure P3-3.4 into their even and odd parts.

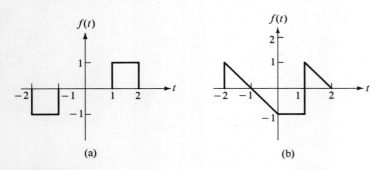

Figure P3-3.4

3-3.5 Show that if a periodic function has half-wave symmetry, then its Fourier series representation contains only odd harmonics.

3-3.6 Consider a function that has the following properties: $f(t) = -f(-t)$ and $f(T/4 - t) = f(T/4 + t)$. Prove that the function has the period $T/2$.

3-3.7 Write the first five terms of the series for the function that is specified as follows:

$$f(x) = \frac{2}{\pi} x + \sin x \qquad 0 \leq x \leq \frac{\pi}{2}$$

$$= 2 - \frac{2}{\pi}x + \sin x \qquad \frac{\pi}{2} \le x \le \pi$$

with the additional knowledge that the wave has half-wave and zero-point symmetry.

3-3.8 Show that $g(t) = f(t)*h_T(t) = f_T(t)*h(t)$, where the subscript T indicates that the length of the function is equal to one period.

3-3.9 Show that the following are identities:

a. $\displaystyle \frac{2}{T}\int_{-T/2}^{T/2} \frac{\sin\left(N+\frac{1}{2}\right)\omega_0 v}{2\sin(\omega_0 v/2)}\, dv = 1$

b. $\displaystyle \sin t + \sin 2t + \cdots + \sin Nt = \frac{\cos\frac{1}{2}t - \cos\left(N+\frac{1}{2}\right)t}{2\sin\frac{t}{2}}$

3-3.10 For the waves of Figure P3-3.10 do the following:
  a. Specify the symmetries that exist.
  b. Write the expressions for the general terms of the series in integral form.

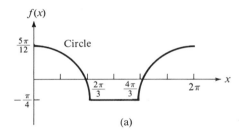

(a)

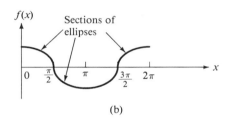

(b)

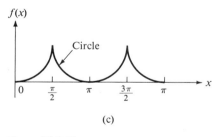

(c)

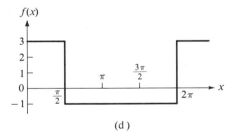

(d)

Figure P3-3.10

3-3.11 Let $f(x)$ be a function having half-wave symmetry. Prove that its Fourier coefficients are given by the formulas

$$A_n = \frac{2}{\pi}\int_A^{A+\pi} f(x)\cos nx\, dx \qquad (n \text{ odd})$$

$$B_n = \frac{2}{\pi}\int_B^{B+\pi} f(x)\sin nx\, dx \qquad (n \text{ odd})$$

where A and B are arbitrary constants.

3-3.12 Prove that if a function $f(x)$ has point symmetry at one point, and axis symmetry at a different point, the formulas for its Fourier coefficients can be written

$$A_n = \frac{4}{\pi} \int_{x_0}^{x_0 + \pi/2} f(x) \cos nx \, dx \quad (n \text{ odd})$$

$$B_n = \frac{4}{\pi} \int_{x_0}^{x_0 + \pi/2} f(x) \sin nx \, dx \quad (n \text{ odd})$$

where x_0 is the location on the x-axis of the points of symmetry.

3-4.1 The Fourier series representation of the triangular wave shown in Figure P3-4.1a is

$$f(x) = \frac{8}{\pi^2} \left(\sin x - \frac{1}{3^2} \sin 3x + \frac{1}{5^2} \sin 5x - \frac{1}{7^2} \sin 7x + \cdots \right)$$

Use this result to obtain four terms of the series for the wave shown in Figure P3-4.1b.

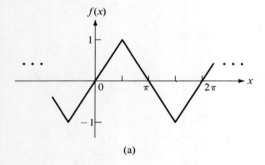

(a)

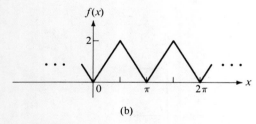

(b)

Figure P3-4.1

3-4.2 Determine the effect of a time displacement of a periodic function on the Fourier coefficients. In particular, what is the relation between the Fourier series for $f(t)$ and $f(t - t_0)$?

3-4.3 The Fourier series for the half-wave rectified wave shown in Figure P3-4.3 with period 2π is

$$f(x) = \frac{1}{\pi} \left(1 + \frac{\pi}{2} \cos x + \frac{2}{3} \cos 2x - \frac{2}{15} \cos 4x + \frac{2}{35} \cos 6x - \cdots \right)$$

a. From this determine the series for a half-wave rectified wave which starts from 0 at $x = 0$.
b. From part (a) derive the series for a full-wave rectified sinusoid having the value 0 at $x = 0$.

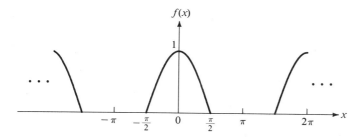

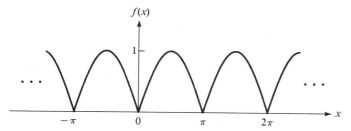

Figure P3-4.3

3-5.1 If the input voltage to the circuit shown in Figure P3-5.1 is

$$v(t) = 1 + 2(\cos t + \cos 2t + \cos 3t)$$

find the output voltage $v_0(t)$.

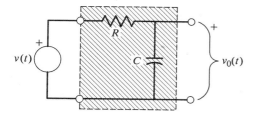

Figure P3-5.1

3-5.2 The triangular wave of Figure P3-4.1a has a peak value of 10 volts. It is the voltage source applied to a series combination of a resistor $R = 200\ \Omega$ and an inductor $L = 0.15$ H. The frequency of the wave is 100 Hz. Obtain an approximate solution for the power dissipated in the resistor.

3-5.3 Compute the energy delivered to the circuit shown in Figure P3-5.3 during each period, if the input voltage is given by

$$v(t) = \frac{2A}{\pi} - \frac{4A}{\pi} \sum_{n=1}^{\infty} \frac{1}{4n^2 - 1} \cos 2n\omega_0 t = \sum_{n=-\infty}^{\infty} \alpha_n e^{jn\omega_0 t}$$

where $\omega_0 = 3$ rad/s. Note that this function describes a full-wave rectified sine wave.

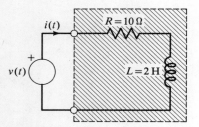

Figure P3-5.3

3-5.4 The input $f(t)$ is applied to three different filters, the resulting outputs being given by the expressions

a. $y(t) = \sum_{n=-\infty}^{\infty} (h_0 - 3)|\alpha_n|e^{j[n\omega_0 t + \varphi - 3n\omega_0 t_0]}$

b. $y(t) = \sum_{n=-\infty}^{\infty} h_0^2|\alpha_n|e^{j[n\omega_0 t + \varphi - n^2\omega_0 t_0]}$

c. $y(t) = \sum_{n=-\infty}^{\infty} h_0|\alpha_n|e^{j[n\omega_0 t + \varphi - [n(n+1)/2]\omega_0 t]}$

Which, if any, is an ideal filter?

3-5.5 Refer to Figure P3-5.5. The input voltage is a square wave of amplitude 10 V (on each side of the axis), with a recurrence frequency of 1000 Hz. Calculate the approximate power dissipated in each resistor.

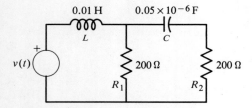

Figure P3-5.5

CHAPTER 4
SPECTRA OF TEMPORAL AND SPATIAL SIGNALS

A. THE FOURIER TRANSFORM—TEMPORAL SIGNALS

The Fourier transform is a mathematical technique that finds extensive application in a variety of physical and engineering problems. It bears a close relationship to the Fourier series, and we will develop it from the Fourier series by applying appropriate conditions. Unlike the Fourier series, which is essentially oriented toward periodic functions, the Fourier integral permits a description of nonperiodic functions. In this chapter we also consider the properties of the Fourier integral transform, with examples of the calculations of some elementary transform pairs. Among the important applications of the Fourier transform that we will consider in subsequent chapters are: analytically representing nonperiodic functions, solving differential equations, aiding in the analysis of linear time-invariant systems, and analyzing and processing signal spectra.

The inception of the ideas behind the Fourier analysis technique stems from Euler's studies in 1748 of the motion of a vibrating string. At about this time, Bernoulli argued, without proof, that all physical motions could be represented by a linear combination of harmonics. Lagrange opposed this notion since he felt that it was not possible to represent discontinuous signals by trigonometric series. A half century later, Jean Baptiste Joseph Fourier presented his ideas on the subject while working on heat propagation and diffusion. His treatise *Theorie Analytique de la Chaleur* was published in 1822.

4-1 THE FOURIER TRANSFORM

The Fourier transform of a function $f(t)$ and its inverse are defined by the integral relations:

$$\mathscr{F}\{f(t)\} \triangleq F(\omega)$$
$$= \int_{-\infty}^{\infty} f(t)e^{-j\omega t}\, dt \quad \text{direct Fourier transform} \quad (4.1a)$$

$$\mathscr{F}^{-1}\{F(\omega)\} \triangleq f(t)$$
$$= \frac{1}{2\pi}\int_{-\infty}^{\infty} F(\omega)e^{j\omega t}\, d\omega \quad \text{inverse Fourier transform} \quad (4.1b)$$

Observe that this transform pair permits the frequency-domain function $F(\omega)$ to be obtained from a knowledge of the time-domain function $f(t)$, and vice versa; a knowledge of $F(\omega)$ permits a determination of the corresponding $f(t)$ from the inverse transform [compare with (3.5) for the Fourier series]. Not all functions $f(t)$ are Fourier transformable. Sufficiency conditions for a function $f(t)$ to be Fourier transformable are the Dirichlet conditions. These are

  a. $\int_{-\infty}^{\infty} |f(t)|\,dt < \infty$.

  b. $f(t)$ has finite maxima and minima within any finite interval.

  c. $f(t)$ has a finite number of discontinuities within any finite interval.

If these conditions are met, $f(t)$ can be transformed uniquely. Some functions exist that do not possess Fourier transforms in the strict sense since they violate one or another of the Dirichlet conditions. Yet in many cases it is still possible to deduce the Fourier transform if the functions under consideration belong to a set known as **generalized functions**. One such function is the delta function $\delta(t)$, and we develop its Fourier transform below.

To deduce the Fourier transform from the Fourier series, begin by writing the Fourier series in the form

$$f(t) = \sum_{n=-\infty}^{\infty} \alpha_n e^{jn\omega_0 t} = \sum_{n=-\infty}^{\infty} e^{jn\omega_0 t} \frac{1}{T} \int_{-T/2}^{T/2} f(t) e^{-jn\omega_0 t}\,dt$$

$$= \sum_{n=-\infty}^{\infty} e^{jn\omega_0 t} \left(\frac{2\pi}{T}\right) \frac{1}{2\pi} \int_{-T/2}^{T/2} f(t) e^{-jn\omega_0 t}\,dt \qquad (4.2)$$

We wish to examine this expression as $T \to \infty$. It is necessary here to consider simultaneously the two limits

$$\lim_{T\to\infty} (\omega_0) = \lim_{T\to\infty} \left(\frac{2\pi}{T}\right) \qquad \lim_{\substack{T\to\infty \\ n\to\infty}} (n\omega_0) = \lim_{\substack{T\to\infty \\ n\to\infty}} n\left(\frac{2\pi}{T}\right)$$

This involves a complicated process, since at the same time that T approaches ∞, n approaches ∞. The limiting processes involved cannot be given rigorous treatment without the use of advanced mathematics. The procedure here parallels that discussed in passing from (2.2) to (2.3). As T is allowed to approach infinity, one limit approaches the condition of an increment of a continuous variable, which may be called $d\omega$:

$$\lim_{T\to\infty} \left(\frac{2\pi}{T}\right) \to d\omega$$

Further, for each value of T, the summation in (4.2) shows that n takes on all integer values, and so the product $n(2\pi/T)$ varies from $-\infty$ to $+\infty$. It may be represented by the variable ω, thus:

$$\lim_{\substack{T\to\infty \\ n\to\infty}} n\left(\frac{2\pi}{T}\right) \to \omega$$

For each finite value of T the quantity $n(2\pi/T)$ varies with n in a steplike fashion, but the variation becomes smooth as T approaches infinity.

As T approaches infinity, the integral in (4.2) attains limits of $-\infty$ and ∞ and the integrand becomes $\exp[-j\omega t]f(t)$. The summation conforms to the definition of an integral, so it becomes an integration with respect to the variable ω between infinite limits. This is seen by considering successive summations for a sequence of increasing but momentarily fixed values of T. Therefore

$$f(t) = \frac{1}{2\pi}\int_{-\infty}^{\infty} e^{j\omega t}\,d\omega \left[\int_{-\infty}^{\infty} f(t)e^{-j\omega t}\,dt\right] = \frac{1}{2\pi}\int_{-\infty}^{\infty} F(\omega)e^{j\omega t}\,d\omega \quad (4.3)$$

This establishes the Fourier transform pair given in (4.1).

In the limit-taking process leading to (4.1), both limits approach infinity in the same way. That is, these two infinite integrals are shorthand notation for the following

$$\int_{-\infty}^{\infty} F(\omega)e^{j\omega t}\,d\omega = \lim_{a\to\infty}\int_{-a}^{a} F(\omega)e^{j\omega t}\,d\omega \quad \textbf{(a)}$$

$$\int_{-\infty}^{\infty} f(t)e^{-j\omega t}\,dt = \lim_{a\to\infty}\int_{-a}^{a} f(t)e^{-j\omega t}\,dt \quad \textbf{(b)} \quad (4.4)$$

When an integral is so defined, it is said to be the **principal value** (see Appendix 1). If the integrals from 0 to ∞ and 0 to $-\infty$ each converge, other values might be obtained for the integral if the two limits were allowed to approach infinity independently. For example, the form

$$\lim_{\substack{a\to\infty \\ b\to\infty}}\int_{-a}^{b} f(t)e^{-j\omega t}\,dt = \lim_{b\to\infty}\int_{0}^{b} f(t)e^{-j\omega t}\,dt + \lim_{a\to\infty}\int_{-a}^{0} f(t)e^{-j\omega t}\,dt$$

would not necessarily be the same as (4.4b).

An indication of the process leading to (4.3) is made evident by reference to Figure 4.1, which shows a train of discrete pulses of width a with a recurrence period T. The amplitude spectrum of the series representation is also shown.

Figure 4.1
The effect of increasing the period T.

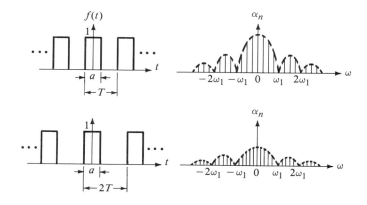

Recall that the spacing of the components in the amplitude spectrum is $\omega_0 = (2\pi/T)$. Observe from the figure that for the recurrence period $2T$ the spacing of the components in the amplitude spectrum has decreased to half of the first value. With increasing recurrence period, this trend continues, and in the limit as $T \to \infty$, the amplitude components become a continuum with the same form of envelope. The decrease of the amplitude spectrum shown in Figure 4.1b occurs because of the equation

$$\alpha_n = \frac{1}{T}\int_{-T/2}^{T/2} f(t)e^{-jn\omega_0 t}\,dt = \Delta f \int_{-1/2\Delta f}^{1/2\Delta f} f(t)e^{-j(n\,d\omega)t}\,dt$$

However, the ratio

$$\lim_{\Delta f \to 0} \frac{\alpha_n}{\Delta f} = \int_{-\infty}^{\infty} f(t)e^{-j(n\,d\omega)t}\,dt$$

remains finite, provided of course that the integral remains finite. This shows, and this is a general characteristic of the Fourier integral, that the amplitude spectrum is a continuous function extending, in general, over all frequencies from $\omega = -\infty$ to $\omega = +\infty$.

The point to be noted is that the spectrum of a finite-length signal is continuous with range $-\infty < \omega < +\infty$, and as a result the interpretation here is different from that applied to periodic signals. Specifically, we found in Chapter 3 that any periodic signal can represented by a sum of discrete exponentials with finite amplitudes. For finite-length signals thereby involving the Fourier integral, the spectrum exists for every value of ω, although at any one frequency ω the amplitude of the frequency component can be zero. However, a contribution that is equal to $F(\omega)\,d\omega/2\pi$ exists for each frequency component, and the function $f(t)$ can be expressed in terms of the continuous sum of such infinitesimal components; this is the inverse transform integral given by (4.1b). In a parallel way, note that similar arguments can be used in interpreting a continuously distributed load on a beam. The loading exists at every point, but at any one point, the loading is zero. However, the loading in any distance dx along the beam is equal to $f(x)\,dx$, where $f(x)$ is the loading density, given in kg/m.

A word is in order about the convergence of the integrals in (4.1). Because the limits are infinite, the integrals are called improper integrals. Those functions $f(t)$ and $F(\omega)$ that approach zero rapidly enough for the integral to have a finite value in spite of the infinite limits are said to converge. If the integral does not have a finite value, it is said to diverge. Only those functions that lead to convergent integrals are admissible.

We will now examine the important properties of Fourier integral signal analysis.

Fourier Transforms of Complex and Real Functions: Complex Functions To examine the important features of Fourier transforms of complex functions, begin by writing $f(t)$ in the form that shows its real and imaginary parts. Also write the corresponding $F(\omega)$ to show its real and imaginary parts. Thus

$$f(t) = f_r(t) + jf_i(t) \quad \text{(a)} \tag{4.5}$$

and

$$F(\omega) = R(\omega) + jX(\omega) \quad \text{(b)}$$

where from 4.1a, it will be found that

$$R(\omega) = \int_{-\infty}^{\infty} \left[f_r(t) \cos \omega t + f_i(t) \sin \omega t \right] dt \quad \text{(a)} \tag{4.6}$$

$$X(\omega) = \int_{-\infty}^{\infty} \left[f_i(t) \cos \omega t - f_r(t) \sin \omega t \right] dt \quad \text{(b)}$$

The inverse pair, given by (4.1b), is

$$f_r(t) = \frac{1}{2\pi} \int_{-\infty}^{\infty} \left[R(\omega) \cos \omega t - X(\omega) \sin \omega t \right] d\omega \quad \text{(a)} \tag{4.7}$$

$$f_i(t) = \frac{1}{2\pi} \int_{-\infty}^{\infty} \left[R(\omega) \sin \omega t + X(\omega) \cos \omega t \right] d\omega \quad \text{(b)}$$

Real Functions If the function $f(t)$ is real, then $f_i(t) = 0$, and (4.5) and (4.6) become

$$F(-\omega) = F^*(\omega) \quad \text{(a)}$$

$$R(\omega) = \int_{-\infty}^{\infty} f(t) \cos \omega t \, dt \quad \text{(b)} \tag{4.8}$$

$$X(\omega) = -\int_{-\infty}^{\infty} f(t) \sin \omega t \, dt \quad \text{(c)}$$

where we have set $f(t) \triangleq f_r(t)$.

a. Even and Real. For $f(t)$ even and real, $f(t) = f_e(t)$ and (4.8b) and (4.8c) become

$$R(\omega) = \int_{-\infty}^{\infty} f_e(t) \cos \omega t \, dt = F(\omega) = 2 \int_{0}^{\infty} f_e(t) \cos \omega t \, dt \quad \text{(a)} \tag{4.9}$$

$$X(\omega) = 0 \quad \text{(b)}$$

b. Odd and Real. For an odd and real function, $f(t) = f_o(t)$, and (4.8b) and (4.8c) become

$$R(\omega) = 0 \quad \text{(a)} \tag{4.10}$$

$$F(\omega) = jX(\omega) = -j \int_{-\infty}^{\infty} f_o(t) \sin \omega t \, dt = -2j \int_{0}^{\infty} f_o(t) \sin \omega t \, dt \quad \text{(b)}$$

General Function For a nonsymmetrical function, we can write for $f(t)$, $f(t) = f_e(t) + f_o(t)$ and then

$$\mathscr{F}\{f(t)\} \triangleq F(\omega) = R(\omega) + jX(\omega) = \int_{-\infty}^{\infty} f_e(t) \cos \omega t \, dt - j \int_{-\infty}^{\infty} f_o(t) \sin \omega t \, dt$$

$$= F_e(\omega) + F_o(\omega) \tag{4.11}$$

Based on the foregoing, the following observations can be made.

Property of $f(t)$	Characteristics of $F(\omega)$
$f(t) = f_e(t)$	Real transform and even
$f(t) = f_0(t)$	Imaginary transform and odd
$f(t) =$ real	Even real part and odd imaginary part
$f(t) =$ complex and no symmetry	Complex transform and no symmetry
$f(t) =$ complex even or odd	Complex and even transform or complex and odd

Causal Functions A causal (positive time) function is one for which $f(t) = 0$ for $t < 0$. This implies that $f(-t) = 0$, but not necessarily for $t = 0$. This means that

$$f_e(t) = \frac{f(t) + f(-t)}{2} = \frac{f(t) + 0}{2} \quad \text{or} \quad f(t) = 2f_e(t) \quad t > 0 \quad \textbf{(a)}$$

$$f_0(t) = \frac{f(t) - f(-t)}{2} = \frac{f(t) - 0}{2} \quad \text{or} \quad f(t) = 2f_0(t) \quad t > 0 \quad \textbf{(b)}$$

(4.12)

and so

$$\mathcal{F}\{f(t)\} = 2\int_{-\infty}^{\infty} f_e(t) \cos \omega t\, dt = 2F_e(\omega) \quad \textbf{(c)}$$

$$\mathcal{F}\{f(t)\} = -j2\int_{-\infty}^{\infty} f_0(t) \sin \omega t\, dt = 2F_0(\omega) \quad \textbf{(d)}$$

By (4.1b), (4.9), and (4.10), we obtain

$$f(t) = \frac{1}{2\pi}\int_{-\infty}^{\infty} 2F_e(\omega)e^{j\omega t}\, d\omega = \frac{1}{\pi}\int_{-\infty}^{\infty} F_e(\omega) \cos \omega t\, d\omega$$

$$= \frac{1}{\pi}\int_{-\infty}^{\infty} R(\omega) \cos \omega t\, d\omega, \quad t > 0 \quad \textbf{(a)} \quad (4.13)$$

and

$$f(t) = \frac{1}{2\pi}\int_{-\infty}^{\infty} 2F_0(\omega)e^{j\omega t}\, d\omega = \frac{j}{\pi}\int_{-\infty}^{\infty} F_0(\omega) \sin \omega t\, d\omega$$

$$= \frac{j}{\pi}\int_{-\infty}^{\infty} jX(\omega) \sin \omega t\, d\omega = -\frac{1}{\pi}\int_{-\infty}^{\infty} X(\omega) \sin \omega t\, d\omega \quad t > 0 \quad \textbf{(b)}$$

Furthermore, the two functions $R(\omega)$ and $X(\omega)$ of a causal signal are related to each other by the equations

SPECTRA OF TEMPORAL AND SPATIAL SIGNALS

$$R(\omega) = \frac{1}{\pi} \dashint_{-\infty}^{\infty} \frac{X(\tau)}{\omega - \tau} d\tau \tag{c}$$

$$X(\omega) = -\frac{1}{\pi} \dashint_{-\infty}^{\infty} \frac{R(\tau)}{\omega - \tau} d\tau \tag{d}$$

These relationships are known as **Hilbert transforms** (see Section 4-7); and the bars on the integral signs denote the principal value of the integrals.

EXAMPLE 4.1
Find the Fourier transform of the causal function

$$f(t) = \begin{cases} e^{-t} & t \geq 0 \\ 0 & t < 0 \end{cases}$$

and show the validity of (4.13c) and (4.13d).

Solution: By (4.1a) we write directly

$$F(\omega) = \int_0^{\infty} e^{-t} e^{-j\omega t} dt = \frac{1}{-(1+j\omega)} e^{-(1+j\omega)t} \Big|_0^{\infty} = \frac{1}{1+j\omega}$$

$$= \frac{1}{1+\omega^2} - j\frac{\omega}{1+\omega^2} = R(\omega) - jX(\omega)$$

This establishes the forms for $R(\omega)$ and $X(\omega)$.

Now introduce the expression for $R(\omega)$ into (4.13d) to get

$$-\frac{1}{\pi} \int_{-\infty}^{\infty} \frac{1}{(1+\tau^2)(\omega - \tau)} d\tau$$

$$= \frac{-1}{\pi(1+\omega^2)} \int_{-\infty}^{\infty} \left(\frac{\tau}{1+\tau^2} + \omega \frac{1}{1+\tau^2} + \frac{1}{\omega - \tau} \right) d\tau$$

$$= \frac{-1}{\pi(1+\omega^2)} \left[\frac{1}{2} \ln(1+\tau^2) + \omega \tan^{-1}\tau - \ln(\omega - \tau) \right]_{-\infty}^{\infty}$$

$$= -\frac{\omega}{(1+\omega^2)} = X(\omega)$$

In this development use was made of the expansion

$$\frac{1}{(1+\tau^2)(\omega - \tau)} = \frac{A\tau + B}{1+\tau^2} + \frac{C}{\omega - \tau}$$

The reader should verify the validity of (4.13c) for this problem. ■

Interpretation of the Fourier Transform To interpret the defining Fourier transform integral of (4.1a), let us assume that the function $f(t)$ is real and even. Let us further assume that $f(t) = p_{0.5}(t)$, the pulse function having limits -0.5

to 0.5. Under these conditions, the integral becomes

$$F(\omega) = \int_{-\infty}^{\infty} p_{0.5}(t) \cos \omega t \, dt = \int_{-0.5}^{0.5} \cos \omega t \, dt = 2\frac{\sin\left(\dfrac{\omega}{2}\right)}{\omega}$$

$$\triangleq 2 \operatorname{sinc}_{1/2}(\omega) \tag{4.14}$$

At any particular frequency $\omega = \omega_i$, this equation takes on the form

$$F(\omega_i) = \int_{-\infty}^{\infty} p_{0.5}(t) \cos \omega_i t \, dt$$

where the number $F(\omega_i)$ is the area under the product function $p_{0.5}(t) \cos \omega_i t$. Figure (4.2a) illustrates this assertion in graphical form for specific values of ω. If we imagine that this process is repeated for an infinite number of ω's, the continuous curve $F(\omega)$ will result. Figure 4.2b shows how the function $f(t)$ is constructed. Each number of $f(t_i)$ is equal to the area under the product function curve $2 \operatorname{sinc}_{1/2}(\omega) \cos \omega t_i$, which is the integrand of the inverse transform of the $2 \operatorname{sinc}_{1/2}(\omega)$ function. This is because $\operatorname{sinc}(\omega)$ is real and symmetric.

■ ■ ■

4-2 PROPERTIES OF FOURIER TRANSFORMS

Fourier transforms possess a number of very important properties. These are developed as follows.

1. Linearity.

$$\mathscr{F}\{af_1(t) + bf_2(t)\} = a\mathscr{F}\{f_1(t)\} + b\mathscr{F}\{f_2(t)\} = aF_1(\omega) + bF_2(\omega) \tag{4.15}$$

where a and b are constants. This property is the direct result of the linear operation of integration.

2. Symmetry. If $\mathscr{F}\{f(t)\} = F(\omega)$, then

$$2\pi f(-\omega) = \int_{-\infty}^{\infty} F(t) e^{-j\omega t} \, dt \tag{4.16}$$

Proof: From (4.1b) it follows that

$$2\pi f(t) = \int_{-\infty}^{\infty} F(\omega) e^{j\omega t} \, d\omega$$

Now interchange the symbols ω and t, and this equation takes the form

$$2\pi f(\omega) = \int_{-\infty}^{\infty} F(t) e^{j\omega t} \, dt$$

By introducing the change from ω to $-\omega$, the result is (4.16). This symmetry property can be used to extend tables of transforms.

EXAMPLE 4.2
Examine the symmetry property of the pulse function $p_a(t)$ and the impulse function (delta function) $\delta(t)$.

SPECTRA OF TEMPORAL AND SPATIAL SIGNALS 171

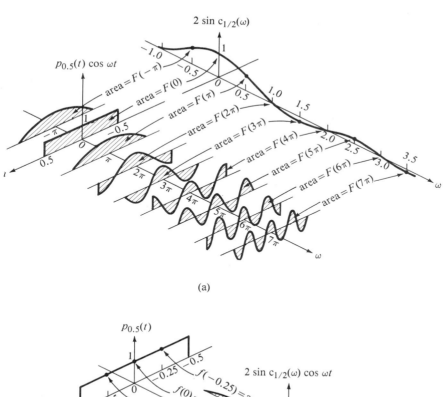

(a)

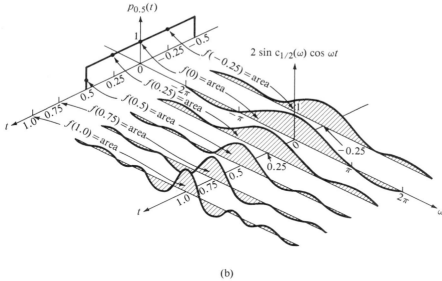

(b)

Figure 4.2
Illustration of the Fourier integral pair. (a) Shows $p_{0.5}(t)$ and successive values of $\cos \omega t$ for $\omega t = \pi, 2\pi, 3\pi, \ldots$ and the corresponding value of the Fourier transform $F(\omega)$. (b) Shows the variation of $F(\omega) \cos \omega t$ for successive values of t. (Reprinted, by permission, from Gaskill, *Linear Systems, Fourier Transforms and Optics*, 189, 190.)

Solution: The Fourier transform of the pulse function is

$$\mathscr{F}\{p_a(t)\} = \int_{-\infty}^{\infty} p_a(t)e^{-j\omega t}\,dt = \int_{-a}^{a} e^{-j\omega t}\,dt = 2\frac{\sin \omega a}{\omega} = 2\operatorname{sinc}_a \omega \tag{4.17}$$

From the symmetry property, we write

$$2\pi p_a(-\omega) = \mathscr{F}\left\{2\frac{\sin ta}{t}\right\} \tag{4.18}$$

The graphical representation of these formulas is shown in Figure 4.3a. To prove (4.18) requires the use of the integral

$$2\int_0^\infty 2\frac{\sin at}{t}\cos \omega t\,dt = 2\pi \qquad -a \le \omega \le a$$

For the delta function (see Chapter 1) we have

$$\mathscr{F}\{\delta(t)\} = \int_{-\infty}^{\infty} \delta(t)e^{-j\omega t}\,dt = e^{-j\omega \cdot 0} = 1 \triangleq \Delta(\omega) \quad \text{(a)} \tag{4.19}$$

Figure 4.3
(a) Illustration of the symmetry property of the pulse function. (b) Illustration of the symmetry property of the delta function.

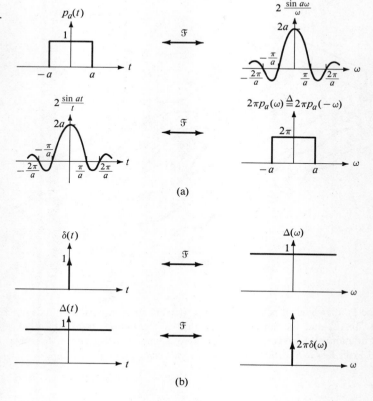

By the symmetry property

$$2\pi \delta(-\omega) = 2\pi \delta(\omega) = \mathscr{F}\{\Delta(t)\} = \mathscr{F}\{1\} \qquad \text{(b)}$$

These relationships are shown in Figure 4.3b.

3. Shifting. For any real time t_0

$$\mathscr{F}\{f(t \pm t_0)\} = e^{\pm j\omega t_0}\mathscr{F}\{f(t)\} = e^{\pm j\omega t_0}F(\omega) \qquad (4.20)$$

EXAMPLE 4.3

Find the Fourier transform of the function $f(t - t_0) = \exp[-(t - t_0)]u(t - t_0)$.

Solution: Use (4.20) to write (see also Example 4.1)

$$\mathscr{F}\{f(t - t_0)\} = e^{-j\omega t_0}\mathscr{F}\{f(t)\} = e^{-j\omega t_0}\mathscr{F}\{e^{-t}u(t)\} = \frac{e^{-j\omega t_0}}{1 + j\omega}$$

The effect of time shifting is shown in Figure 4-4. Observe that only the phase spectrum is modified.

4. Scaling. If $\mathscr{F}\{f(t)\} = F(\omega)$, then

$$\mathscr{F}\{f(at)\} = \frac{1}{|a|} F\left(\frac{\omega}{a}\right) \qquad \text{(a)} \qquad (4.21)$$

and from this we see that

$$\mathscr{F}\{f(-t)\} = F(-\omega) \qquad \text{(b)}$$

Figure 4.4
The effect of shifting the function $\exp(-t)u(t)$ in the time domain.

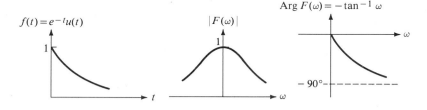
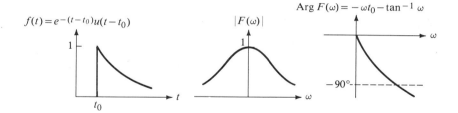

EXAMPLE 4.4

Discuss the Fourier transform of the pulse function $f(t) = p_a(bt)$ where $b > a$.

Solution: The Fourier transform of $f_a(t) = p_a(t)$ is, by Example 4.2, $2(\sin a\omega)/\omega$. By (4.21a) the Fourier transform of $p_{a/b}(bt)$ is

$$\mathscr{F}\{p_{a/b}(bt)\} = \frac{1}{|b|} F_a(\omega)\bigg|_{\omega=\omega/b} = \frac{1}{|b|} \frac{2\sin\dfrac{a}{b}\omega}{\dfrac{\omega}{b}}$$

The respective functions are shown in Figure 4.5. ∎

5. Central Ordinate. By setting $\omega = 0$ and $t = 0$ in (4.1a) and (4.1b), respectively, the resulting expressions are:

$$F(0) = \int_{-\infty}^{\infty} f(t)\,dt \quad \text{(a)}$$

$$f(0) = \frac{1}{2\pi} \int_{-\infty}^{\infty} F(\omega)\,d\omega \quad \text{(b)}$$

(4.22)

The first of these equations shows that the area under the $f(t)$ curve is equal to the central ordinate of the Fourier transform. The second of these equations shows that the area under the $F(\omega)$ curve is 2π times the value of the function at $t = 0$.

6. Frequency Shift. If $\mathscr{F}\{f(t)\} = F(\omega)$, then

$$\mathscr{F}\{e^{\pm j\omega_0 t}f(t)\} = F(\omega \mp \omega_0) \tag{4.23}$$

Figure 4.5
The effect of scaling in time and in frequency domains, $b > a$.

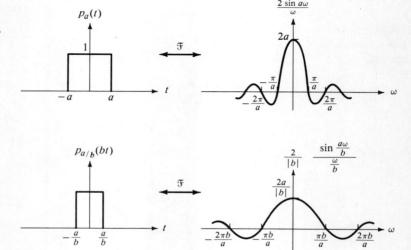

7. Modulation.
If $\mathscr{F}\{f(t)\} = F(\omega)$, then

$$\mathscr{F}\{f(t)\cos\omega_0 t\} = \frac{1}{2}[F(\omega+\omega_0) + F(\omega-\omega_0)] \quad \text{(a)}$$

$$\mathscr{F}\{f(t)\sin\omega_0 t\} = \frac{1}{2j}[F(\omega-\omega_0) - F(\omega+\omega_0)] \quad \text{(b)}$$

(4.24)

Equation (4.24a) is shown graphically in Figure 4.6a for the particular case $f(t) = p_1(t)$.

Figure 4.6
(a) Illustration of the spectrum shift of a modulated signal $f(t) = p_1(t)$. (b) Simultaneously recorded evoked response (AEPs) of the medial geniculate nucleus and the reticular formation in the same cat (A and B). The same AEP's filtered with passband filters of 33–42 Hz are shown below (C and D). (From E. Basar, *Biophysical and Physiological Systems Analysis*, © 1976. Addison-Wesley, Reading, MA. Reprinted with permission.)

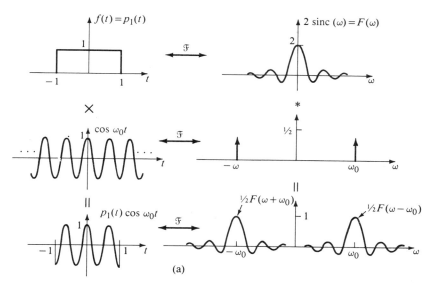

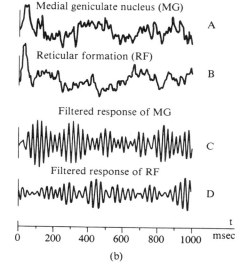

These formulas are easily derived by expanding the cosine and sine terms into their equivalent exponential forms and then using (4.23). The results of (4.24) constitute the fundamental properties of modulation and are fundamental to the field of communications. More will be said about modulation later in this chapter. Figure 4.6b shows that modulation is present in these particular biological signals.

8. Derivatives. If $\mathscr{F}\{f(t)\} = F(\omega)$, then

$$\mathscr{F}\left\{\frac{df(t)}{dt}\right\} = j\omega F(\omega) \quad \text{(a)}$$

$$\mathscr{F}\left\{\frac{d^n f(t)}{dt^n}\right\} = (j\omega)^n F(\omega) \quad \text{(b)}$$
(4.25)

EXAMPLE 4.5

Find the transformed input-output relationship of the system shown in Figure 4.7a.

Solution: It is first necessary to write the differential equation of the system in the time domain. This is readily seen to be

$$\frac{v(t)}{R} + C\frac{dv(t)}{dt} = i(t)$$

Take the Fourier transform of both sides of this equation to get

$$\left(\frac{1}{R} + j\omega C\right) V(\omega) = I(\omega)$$

from which

$$\underbrace{V(\omega)}_{\text{Output}} = \frac{1}{\frac{1}{R} + j\omega C} I(\omega) = \frac{1}{Y(\omega)} I(\omega) = Z(\omega) I(\omega) = \underbrace{H(\omega)}_{\substack{\text{System} \\ \text{function}}} \underbrace{I(\omega)}_{\text{Input}}$$

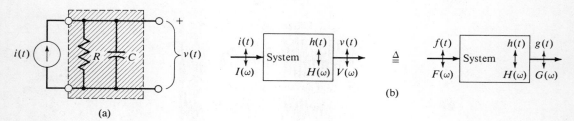

Figure 4.7
An electrical system and its schematic representation.

SPECTRA OF TEMPORAL AND SPATIAL SIGNALS

where, in this case, the system function $H(\omega) = Z(\omega)$. This expression shows that the input-output relationship of any linear time-invariant system is given by

$$\underbrace{G(\omega)}_{\text{Output}} = \underbrace{H(\omega)}_{\substack{\text{System} \\ \text{Function}}} \underbrace{F(\omega)}_{\text{Input}} \qquad (4.26)$$

The system function $H(\omega)$ discussed here is the same as that talked about in Chapter 2. As previously noted, it is the Fourier transform of the impulse response $h(t)$ of the system.

If the input to the system is an impulse function, the impulse response of the system is deduced from the differential equation

$$\frac{h(t)}{R} + C\frac{dh(t)}{dt} = \delta(t)$$

By following the same procedure in the solution as that developed in Chapter 2, we obtain

$$h(t) = \frac{1}{C} e^{-t/RC} \qquad t \geq 0$$

The Fourier transform of this $h(t)$ is given by

$$H(\omega) = \frac{1}{C}\int_0^\infty e^{-t/RC} e^{-j\omega t}\, dt = -\frac{1}{C}\frac{1}{j\omega + \frac{1}{RC}} e^{-[j\omega + (1/RC)]t}\Bigg|_0^\infty$$

$$= \frac{1}{j\omega C + \frac{1}{R}}$$

which is identical with $H(\omega)$ developed above but which was found by a different approach. ∎

The result of this example indicates that the transform of a differential equation describing a physical system of the form

$$\sum_{n=0}^{N} a_n \frac{d^n g(t)}{dt^n} = \sum_{m=0}^{M} b_m \frac{d^m f(t)}{dt^m} \qquad (4.27)$$

will be the algebraic equation

$$\sum_{n=0}^{N} a_n (j\omega)^n G(\omega) = \sum_{m=0}^{M} b_m (j\omega)^m F(\omega)$$

from which it follows that

$$\underbrace{G(\omega)}_{\text{Output}} = \frac{\sum_{m=0}^{M} b_m(j\omega)^m}{\sum_{n=0}^{N} a_n(j\omega)^n} \underbrace{F(\omega)}_{\text{Input}} = H(\omega)\,F(\omega) \quad (4.28)$$

$$\text{System function}$$

The output in the time domain then follows from the inverse transform

$$g(t) = \frac{1}{2\pi} \int_{-\infty}^{\infty} H(\omega)F(\omega)e^{j\omega t}\,d\omega \quad (4.29)$$

To find the inverse Fourier transform, as specified by this equation, we must be sure that $H(\omega)$ exists. We know that this is the case if $h(t)$ is an absolutely integrable function, and this also guarantees the stability of the system (see Section 2-10d). Recall that this is one of the Dirichlet conditions. Further, since we are dealing with signals associated with physical systems, the impulse response always satisfies the other two Dirichlet conditions. Thus the Fourier transform of $h(t)$ will exist.

9. Time Convolution. If $f(t)$ and $h(t)$ have Fourier transforms $F(\omega)$ and $H(\omega)$, respectively, then

$$\mathscr{F}\{f(t) * h(t)\} = \mathscr{F}\left\{\int_{-\infty}^{\infty} f(\tau)h(t-\tau)\,d\tau\right\} = F(\omega)H(\omega) \quad (4.30)$$

Proof: The Fourier transform is

$$\int_{-\infty}^{\infty} [f(t) * h(t)]\, e^{-j\omega t}\,dt = \int_{-\infty}^{\infty} f(\tau)\,d\tau \int_{-\infty}^{\infty} h(t-\tau)e^{-j\omega t}\,dt$$

$$= \int_{-\infty}^{\infty} f(\tau)e^{-j\omega\tau}\,d\tau \int_{-\infty}^{\infty} h(s)e^{-j\omega s}\,ds$$

$$= F(\omega)H(\omega)$$

where we write $t - \tau = s$. This result agrees with (4.26). If (4.30) is written in the form

$$g(t) = \int_{-\infty}^{\infty} f(\tau)h(t-\tau)\,d\tau = \mathscr{F}^{-1}\{F(\omega)H(\omega)\}$$

$$= \frac{1}{2\pi} \int_{-\infty}^{\infty} F(\omega)H(\omega)e^{j\omega t}\,d\omega \quad (4.31)$$

this shows that the output of a linear time-invariant system in the time domain is equal to the convolution of its transfer function and the input signal. Figure 4.8 shows this relationship in graphical form.

Figure 4.8
Graphical representation of linear time-invariant system response.

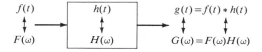

EXAMPLE 4.6
Determine the Fourier transform of the triangular function $2a\Lambda_{2a}(t)$.

Solution: We observe that the triangular function $2a\Lambda_{2a}(t)$ is given as the convolution of the pulse function $p_a(t)$ with itself, that is

$$2a\Lambda_{2a}(t) = p_a(t) * p_a(t) \tag{4.32}$$

This equation is shown graphically in Figure 4.9a. By (4.15) and (4.17) it follows that

$$\mathscr{F}\{2a\Lambda_{2a}(t)\} = \mathscr{F}\{p_a(t) * p_a(t)\} = 4\left(\frac{\sin a\omega}{\omega}\right)^2 = 4(\operatorname{sinc}_a \omega)^2 \tag{4.33}$$

This result is shown in Figure 4.9b. ∎

10. Frequency Convolution. If $f(t)$ and $h(t)$ are Fourier transformable, then

$$\mathscr{F}\{f(t)h(t)\} = \frac{1}{2\pi}\int_{-\infty}^{\infty} F(\tau)H(\omega - \tau)\,d\tau = \frac{1}{2\pi}F(\omega) * H(\omega) \tag{4.34}$$

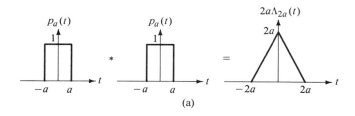

(a)

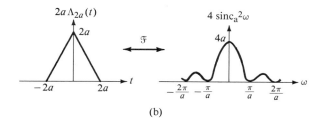

(b)

Figure 4.9
(a) The convolution of two pulse functions. (b) The Fourier transform of the triangle function.

Proof: Proceed by writing

$$\mathscr{F}^{-1}\{\mathscr{F}\{f(t)h(t)\}\} = f(t)h(t) = \frac{1}{2\pi}\frac{1}{2\pi}\iint_{-\infty}^{\infty} F(\tau)H(\omega-\tau)e^{j\omega t}\,d\tau\,d\omega$$

$$= \frac{1}{2\pi}\int_{-\infty}^{\infty} F(\tau)e^{j\tau t}\,d\tau \frac{1}{2\pi}\int_{-\infty}^{\infty} H(s)e^{jst}\,ds = f(t)h(t)$$

where we set $\omega - \tau = s$ (that is, $\omega = \tau + s$ and $d\omega = ds$).

EXAMPLE 4.7

Find the Fourier transform of the function $f(t) = p_1(t)\cos 2\pi t$, which is shown in Figure 4.10a.

Solution: Using the results of Example 4.2, we obtain

$$\mathscr{F}\{\cos 2\pi t\} = \int_{-\infty}^{\infty} \frac{e^{j2\pi t} + e^{-j2\pi t}}{2} e^{-j\omega t}\,dt$$

$$= \frac{1}{2}[2\pi\,\delta(\omega - 2\pi) + 2\pi\delta(\omega + 2\pi)] \qquad (4.35)$$

Therefore we can write from (4.34)

$$\mathscr{F}\{f(t)\} = \mathscr{F}\{p_1(t)\cos 2\pi t\}$$

$$= \frac{1}{2\pi}\int_{-\infty}^{\infty} 2\frac{\sin\tau}{\tau}[\pi\,\delta(\omega - 2\pi - \tau) + \pi\,\delta(\omega + 2\pi - \tau)]\,d\tau$$

$$= \frac{\sin(\omega - 2\pi)}{\omega - 2\pi} + \frac{\sin(\omega + 2\pi)}{\omega + 2\pi} \qquad (4.36)$$

This function is shown in Figure 4.10b.

Equation (4.36) shows that the spectrum function is made up of two parts, each of which is similar to (4.17) and Figure 4.3a but centered around the two frequencies -2π and 2π. It is seen that the two components of $F(\omega)$ are largely independent, since one of them is small where the other is large.

Figure 4.10
The modulated pulse function and its Fourier transform.

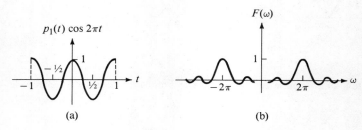

SPECTRA OF TEMPORAL AND SPATIAL SIGNALS

The longer we make the wave train by replacing $p_1(t)$ by $p_n(t)$ with $n > 1$, the narrower will be the spectrum function on the ω scale. This can be seen by reference to Figure 4.3a from which it follows that the width on the ω scale between the first zeros to the right and left of the peak is 2π. This conclusion also follows directly from (4.36). This result shows that the bandwidth required of a bandpass network that is to approximately pass the given wave train must become increasingly wide as the wave train is shortened.

Another view of this matter is that if we truncate a signal in time, the signal is spread out in the frequency domain. This suggests that if a signal consists of a number of lines in its frequency spectrum, each line will be spread out by the same function, which is of the form $\sin \omega / \omega$. This is important to communication engineers who must truncate signals in order to process them and to physicists who like to observe signals for a relatively "long time" in order to resolve adjacent atomic spectral lines. We previously found similar effects when we tried to represent a periodic function by a finite sum. ∎

EXAMPLE 4.8

Find the Fourier transform of the Gaussian modulated cosine signal

$$f(t) = e^{-at^2} \cos \omega_c t = e^{-at^2} \frac{(e^{j\omega_c t} + e^{-j\omega_c t})}{2} \tag{4.37}$$

Solution: The Fourier transform of this function is

$$\mathscr{F}\{f(t)\} = F(\omega) = \frac{1}{2} \int_{-\infty}^{\infty} [e^{-at^2} e^{j\omega_c t} + e^{-at^2} e^{-j\omega_c t}] e^{-j\omega t} \, dt \tag{4.38}$$

We now use the fact that the Fourier transform of $g_a(t) = \exp(-at^2)$ can be found to be

$$G_a(\omega) = \int_{-\infty}^{\infty} e^{-at^2} e^{-j\omega t} \, dt = e^{-\omega^2/4a} \int_{-\infty}^{\infty} e^{-(t\sqrt{a} + j\omega/2\sqrt{a})^2} \, dt$$

$$= \frac{e^{-\omega^2/4a}}{\sqrt{a}} \int_{-\infty}^{\infty} e^{-\beta^2} \, d\beta = \sqrt{\frac{\pi}{a}} e^{-\omega^2/4a} \tag{4.39}$$

Also, the Fourier transforms of $\exp(j\omega_c t)$ and $\exp(-j\omega_c t)$ are, respectively, $2\pi\delta(\omega - \omega_c)$ and $2\pi\delta(\omega + \omega_c)$. Now we use (4.34) to write the Fourier transform of $f(t)$, which is

$$F(\omega) = \frac{1}{2}\left[\frac{1}{2\pi} \int_{-\infty}^{\infty} \sqrt{\frac{\pi}{a}} e^{-\tau^2/4a} 2\pi \delta(\omega - \omega_c - \tau) \, d\tau \right.$$

$$\left. + \frac{1}{2\pi} \int_{-\infty}^{\infty} \sqrt{\frac{\pi}{a}} e^{-\tau^2/4a} 2\pi \delta(\omega + \omega_c - \tau) \, d\tau \right]$$

$$= \frac{1}{2}\sqrt{\frac{\pi}{a}} e^{-(\omega - \omega_c)^2/4a} + \frac{1}{2}\sqrt{\frac{\pi}{a}} e^{-(\omega + \omega_c)^2/4a} = F_1(\omega) + F_2(\omega) \tag{4.40}$$

These results are shown in Figure 4.11. ∎

Figure 4.11
The Fourier transform of a Gaussian modulated carrier.

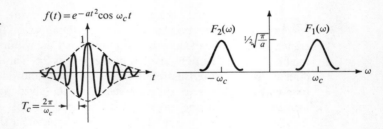

11. Autocorrelation. As defined in (2.30b) autocorrelation is written

$$\mathscr{F}\{r_{f,f}(t)\} = \mathscr{F}\{f(t) \star f^*(t)\} \triangleq R_{f,f}(\omega) = \mathscr{F}\left\{\int_{-\infty}^{\infty} f(\tau)f^*(\tau - t)\,d\tau\right\}$$

$$= \int_{-\infty}^{\infty} f(\tau)\,d\tau \int_{-\infty}^{\infty} f^*(\tau - t)e^{-j\omega t}\,dt$$

$$= \int_{-\infty}^{\infty} f(\tau)e^{-j\omega\tau}\,d\tau \left[\int_{-\infty}^{\infty} f(\beta)e^{-j\omega\beta}\,d\beta\right]^*$$

$$= F(\omega)F^*(\omega) = |F(\omega)|^2 \tag{4.41}$$

If we use two different functions—for example, $f(t)$ and $g(t)$—in this equation, the result would be the Fourier transform of the **cross-correlation** of these two functions, given by $R_{f,g}(\omega)$.

12. Parseval's Theorem. As an extension of the discussion in Section 3-3

$$E = \int_{-\infty}^{\infty} |f(t)|^2\,dt = \frac{1}{2\pi}\int_{-\infty}^{\infty} |F(\omega)|^2\,d\omega \tag{4.42}$$

Proof: Proceed as follows:

$$\int_{-\infty}^{\infty} |f(t)|^2\,dt = \int_{-\infty}^{\infty} f(t)f^*(t)\,dt = \int_{-\infty}^{\infty}\left[\frac{1}{2\pi}\int_{-\infty}^{\infty} F(\omega)e^{j\omega t}\,d\omega\right]f^*(t)\,dt$$

$$= \frac{1}{2\pi}\int_{-\infty}^{\infty} F(\omega)\left[\int_{-\infty}^{\infty} f(t)e^{-j\omega t}\,dt\right]^*\,d\omega$$

$$= \frac{1}{2\pi}\int_{-\infty}^{\infty} F(\omega)F^*(\omega)\,d\omega$$

This is a statement of the conservation of energy: the energy of the time-domain signal is equal to the energy of the frequency-domain transform.

If the **power density** spectrum of a signal is defined by

$$W(\omega) = \frac{1}{2\pi}|F(\omega)|^2 \tag{4.43}$$

then the energy in an infinitesimal band of frequencies $d\omega$ is $W(\omega)\,d\omega$, and within a

SPECTRA OF TEMPORAL AND SPATIAL SIGNALS

band $\omega_1 \leq \omega \leq \omega_2$ the energy contained is

$$\Delta E = \int_{\omega_1}^{\omega_2} \frac{1}{2\pi} |F(\omega)|^2 \, d\omega \tag{4.44}$$

The fraction of the total energy that is contained within the band $\Delta\omega$ is

$$\frac{\Delta E}{E} = \frac{\text{Energy in band}}{\text{Total energy}} = \frac{\frac{1}{2\pi} \int_{\omega_1}^{\omega_2} |F(\omega)|^2 \, d\omega}{\frac{1}{2\pi} \int_{-\infty}^{\infty} |F(\omega)|^2 \, d\omega} = \frac{\int_{\omega_1}^{\omega_2} |F(\omega)|^2 \, d\omega}{\int_{-\infty}^{\infty} |F(\omega)|^2 \, d\omega} \tag{4.45}$$

The interpretation of energy and power of signals in this manner is possible because $f(t)$ may be a voltage so that $f(t)/1$ Ohm = current, and thus $f(t)^2$ is proportional to power.

EXAMPLE 4.9

Determine the total energy associated with the function $f(t) = e^{-t}u(t)$.

Solution: The total energy is, from (4.42),

$$E = \int_{-\infty}^{\infty} (e^{-t})^2 u(t) \, dt = \int_{0}^{\infty} e^{-2t} \, dt = \frac{1}{2}$$

We can also proceed from a frequency viewpoint, and by using the results of Example 4.1, we have

$$E = \frac{1}{2\pi} \int_{-\infty}^{\infty} \left(\frac{1}{1+j\omega}\right)\left(\frac{1}{1-j\omega}\right) d\omega = \frac{1}{2\pi} \int_{-\infty}^{\infty} \frac{1}{1+\omega^2} \, d\omega$$

$$= \frac{1}{\pi} \int_{0}^{\infty} \frac{1}{1+\omega^2} \, d\omega = \frac{1}{\pi} \frac{\pi}{2} = \frac{1}{2} \qquad \blacksquare$$

Another important form follows by using the fact that the spectrum between input and output is specified by

$$G(\omega) = F(\omega)H(\omega)$$

so that

$$|G(\omega)|^2 = |F(\omega)H(\omega)|^2 = |F(\omega)|^2 |H(\omega)|^2 \tag{4.46}$$

In general $H(\omega)$ is a complex quantity that can be written

$$H(\omega) = H_0(\omega)e^{j\theta(\omega)} \tag{4.47}$$

where $H_0(\omega)$ is a real quantity, and (4.46) becomes

$$\boxed{|G(\omega)|^2 = H_0^2(\omega)|F(\omega)|^2} \tag{4.48}$$

This equation shows that the power density spectrum of the response of a linear time-invariant system is the product of the power density spectrum of the input function and the squared amplitude function of the system (network). The phase characteristics of the network do not affect the energy density of the output.

EXAMPLE 4.10

Find the input-output power density spectrum of the system shown in Figure 4.12a. Also find the fractional energy ratio of the input and output signals.

Solution: By an application of the Kirchhoff voltage law

$$v(t) = L\frac{di}{dt} + Ri$$

The Fourier transform of this equation permits writing

$$I(\omega) = \frac{1}{R + j\omega L} V(\omega) = \frac{1}{1 + j\omega} V(\omega) = H(\omega)V(\omega)$$

The transfer function $H(\omega)$ can be written in the form

$$H(\omega) = H_0(\omega)e^{j\theta} = \frac{1}{(1 + \omega^2)^{1/2}} e^{-j\tan^{-1}\omega}$$

Figure 4.12
The fractional energy relations of a linear system.

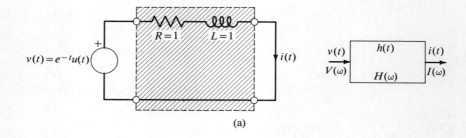

(a)

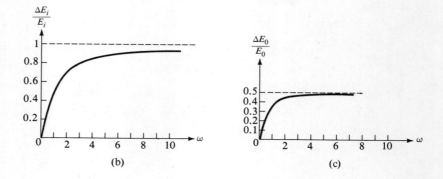

(b)   (c)

so that
$$H_0(\omega) = \frac{1}{(1 + \omega^2)^{1/2}}$$
Furthermore
$$V(\omega) = \mathscr{F}\{v(t)\} = \mathscr{F}\{e^{-t}u(t)\} = \frac{1}{1 + j\omega}$$
and so
$$|V(\omega)|^2 = \frac{1}{1 + \omega^2}$$

The fractional energy of the input signal is given by
$$\frac{\Delta E_i}{E_i} = \frac{\frac{2}{2\pi}\int_0^\omega \frac{1}{1 + \omega'^2}\,d\omega'}{\frac{2}{2\pi}\int_0^\infty \frac{1}{1 + \omega'^2}\,d\omega'} = \frac{2}{\pi}\tan^{-1}\omega$$

The fractional energy of the output is given by
$$\frac{\Delta E_0}{E_0} = \frac{\int_0^\omega \left(\frac{1}{1 + \omega'^2}\right)\left(\frac{1}{1 + \omega'^2}\right)d\omega'}{\frac{\pi}{2}} = \frac{2}{\pi}\left[\frac{\omega}{2(1 + \omega^2)} + \frac{1}{2}\tan^{-1}\omega\right]$$
$$= \frac{\omega}{\pi(1 + \omega^2)} + \frac{1}{\pi}\tan^{-1}\omega$$

These fractional energy relations are plotted in Figures 4.12b and c. ∎

13. Moments. The nth moment m_n is specified by

$$m_n = \int_0^\infty t^n f(t)\,dt = \frac{F^{(n)}(0)}{(-j)^n} \qquad F^{(n)}(0) = \left.\frac{d^n F(\omega)}{d\omega^n}\right|_{\omega = 0} \tag{4.49}$$

Proof: Differentiate the Fourier transform n times with respect to ω. This gives

$$F^{(n)}(\omega) = \int_{-\infty}^\infty (-jt)^n e^{-j\omega t} f(t)\,dt$$

For $\omega = 0$, and dividing by $(-j)^n$, (4.49) results.

It is interesting to examine specific moments. Using the relationship from (4.49), we find

$$n = 0 \qquad m_0 = \int_{-\infty}^\infty f(t)\,dt = F(0) = \text{Area} \quad \text{(a)}$$
$$n = 1 \qquad m_1 = \int_{-\infty}^\infty tf(t)\,dt = \frac{F^{(1)}(0)}{-j} \quad \text{(b)} \tag{4.50}$$
$$n = 2 \qquad m_2 = \int_{-\infty}^\infty t^2 f(t)\,dt = F^{(2)}(0) \quad \text{(c)}$$

CHAPTER 4

But the **centroid** is given by m_1/m_0; this yields

$$\bar{t} = \text{centroid} = \frac{m_1}{m_0} = \frac{F^{(1)}(0)}{-jF(0)} \quad \text{(a)} \tag{4.51}$$

Similarly

$$\overline{t^2} = \text{mean square abscissa} = \frac{m_2}{m_0} = \frac{-F^{(2)}(0)}{F(0)} \quad \text{(b)}$$

$$(\overline{t^2})^{1/2} = \text{radius of gyration} \quad \text{(c)}$$

$$\sigma^2 = \text{variance} = \overline{t^2} - \bar{t}^2 \quad \text{(d)}$$

$$\sigma = \sqrt{\sigma^2} = \text{standard deviation} \quad \text{(e)}$$

A number of important Fourier transform properties are summarized in Table 4.1. Figures 4.13, 4.14, and 4.15 show Fourier spectra of representative nonelectrical signals. This is indicative of the power of Fourier techniques in analyzing and understanding different types of signals and systems.

EXAMPLE 4.11
Find the impulse response of the system shown in Figure 4.16 if the input is the function $f(t) = e^{-t}u(t)$. Also find the Fourier transform of the output.

Solution: The output of the summer (adder) is equal to $f(t) + f(t-2)$ because the impulse response of the subsystem with $H_1(\omega) = e^{-j2\omega}$ is $\delta(t-2)$. Therefore the output is

$$g(t) = [f(t) + f(t-2)] * h_2(t)$$

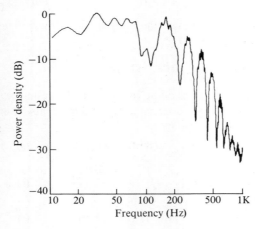

Figure 4.13
Power spectrum of myoelectric signal from the biceps brachii muscle (single motor unit). (Reprinted, by permission, from Lindström and Magnusson, "Interpretation of Myoelectric Power Spectra: A Model and its Applications," 653.)

The Fourier transform of the output is

$$G(\omega) = F(\omega)H_2(\omega) + e^{-j2\omega}F(\omega)H_2(\omega) = F(\omega)H_2(\omega)[1 + e^{-j2\omega}]$$

where, for the specified functions $F(\omega) = H_2(\omega) = 1/(1 + j\omega)$

$$G(\omega) = \frac{1 + e^{-j2\omega}}{(1 + j\omega)^2}$$

∎

Table 4.1 Fourier Transform Properties

Operation	$f(t)$	$F(\omega)$
1. Transform-direct	$f(t)$	$\int_{-\infty}^{\infty} f(t)e^{-j\omega t} dt$
2. Inverse transform	$\frac{1}{2\pi}\int_{-\infty}^{\infty} F(\omega)e^{j\omega t} d\omega$	$F(\omega)$
3. Linearity	$af_1(t) + bf_2(t)$	$aF_1(\omega) + bF_2(\omega)$
4. Symmetry	$F(t)$	$2\pi f(-\omega)$
5. Time shifting	$f(t \pm t_0)$	$e^{\pm j\omega t_0}F(\omega)$
6. Scaling	$f(at)$	$\frac{1}{|a|} F\left(\frac{\omega}{a}\right)$
7. Frequency shifting	$e^{\pm j\omega_0 t}f(t)$	$F(\omega \mp \omega_0)$
8. Modulation	$f(t)\cos\omega_0 t$	$\frac{1}{2}[F(\omega + \omega_0) + F(\omega - \omega_0)]$
	$f(t)\sin\omega_0 t$	$\frac{1}{2j}[F(\omega - \omega_0) - F(\omega + \omega_0)]$
9. Time differentiation	$\frac{d^n}{dt^n} f(t)$	$(j\omega)^n F(\omega)$
10. Time convolution	$f(t) * h(t) = \int_{-\infty}^{\infty} f(\tau)h(t - \tau) d\tau$	$F(\omega)H(\omega)$
11. Frequency convolution	$f(t)h(t)$	$\frac{1}{2\pi} F(\omega) * H(\omega) = \frac{1}{2\pi}\int_{-\infty}^{\infty} F(\tau)H(\omega - \tau) d\tau$
12. Autocorrelation	$f(t) \star f(t) = \int_{-\infty}^{\infty} f(\tau)f^*(\tau - t) d\tau$	$F(\omega)F^*(\omega) = |F(\omega)|^2$
13. Parseval's formula	$E = \int_{-\infty}^{\infty} f(t)^2 dt$	$E = \frac{1}{2\pi}\int_{-\infty}^{\infty} |F(\omega)|^2 d\omega$
14. Moments formula	$m_n = \int_{-\infty}^{\infty} t^n f(t) dt = \frac{F^{(n)}(0)}{(-j)^n};$	$F^{(n)}(0) = \left.\frac{d^n F(\omega)}{d\omega^n}\right|_{\omega=0}$

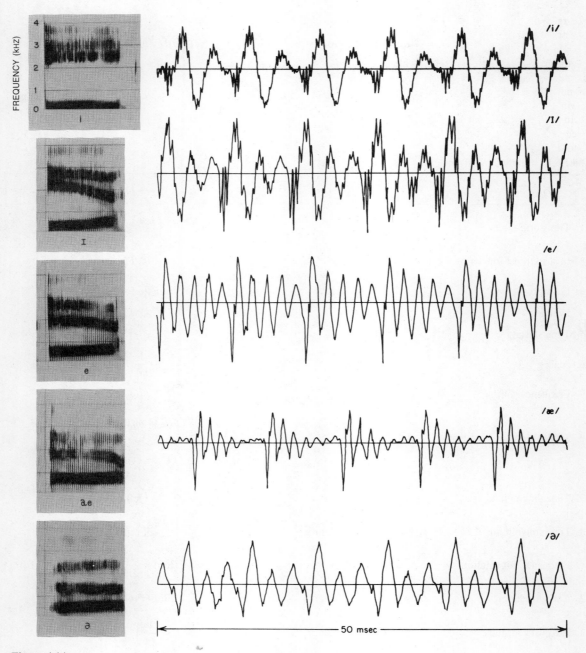

Figure 4.14
The acoustic waveforms for several American English vowels and corresponding spectrograms. (Reprinted, by permission, from Rabiner and Schafer, *Digital Processing of Speech Signals*, 46, 48.)

Figure 4.15
Spectrogram of the word "READ" computed from contiguous 8 msec speech segments. (Reprinted, by permission, from Tufts, Levinson, and Rao, "Measuring Pitch and Formant Frequencies for Speech Understanding System," 314.)

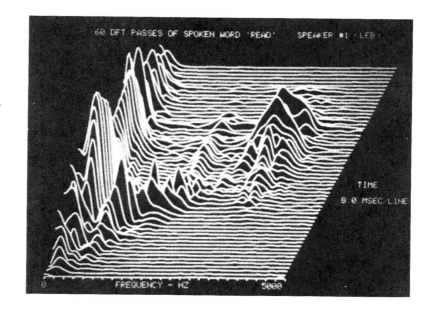

Figure 4.16
Illustrating Example 4.11.

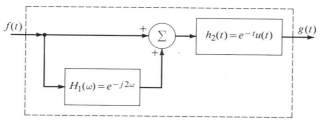

EXAMPLE 4.12

Find the frequency response of an LTI system, given that the input is $f(t) = e^{-t}u(t)$ and the output is $g(t) = 2e^{-2t}$. Determine the impulse response of the system.

Solution: From (4.26) we write the relation

$$H(\omega) = \frac{\text{Output spectrum}}{\text{Input spectrum}} = \frac{\dfrac{2}{2+j\omega}}{\dfrac{1}{1+j\omega}} = 2\frac{1+j\omega}{2+j\omega} = 2 - \frac{2}{2+j\omega}$$

We thus find the impulse response, which is the inverse transform of this expression,

$$h(t) = 2\delta(t) - 2e^{-2t}$$

■

Figure 4.17
A delta-function sequence.

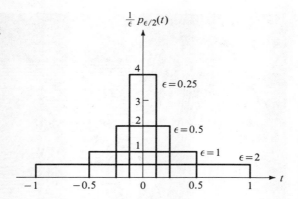

4-3 SOME SPECIAL FOURIER TRANSFORM PAIRS

As has already been noted, we can often employ the notion of generalized functions and limiting processes to find the Fourier transform of a function that does not satisfy the Dirichlet conditions. We will consider a number of such special functions.

EXAMPLE 4.13
Find the Fourier transform of $\delta(t)$ illustrated in Figure 4.17 and defined here by the limiting formula, for different positive values of ϵ.

$$\delta(t) = \lim_{\epsilon \to 0} \frac{1}{\epsilon} p_{\epsilon/2}(t)$$

Solution: The Fourier transform of this equation is

$$\mathscr{F}\{\delta(t)\} = \mathscr{F}\left\{\lim_{\epsilon \to 0} \frac{1}{\epsilon} p_{\epsilon/2}(t)\right\} = \lim_{\epsilon \to 0} \mathscr{F}\left\{\frac{1}{\epsilon} p_{\epsilon/2}(t)\right\}$$

$$= \lim_{\epsilon \to 0} \frac{\sin \epsilon\omega/2}{\epsilon\omega/2} = 1$$

where the limit is found using L'Hospital's rule. ∎

EXAMPLE 4.14
Find the Fourier transform of the function sgn(t) shown in Figure 4.18a.

Solution: We proceed by writing the sgn(t) function as $\lim_{\epsilon \to 0} \exp[-\epsilon|t|] \operatorname{sgn}(t)$, as shown in Figure 4.18b. The procedure is now direct and yields

$$\mathscr{F}\{\operatorname{sgn}(t)\} = \mathscr{F}\left\{\lim_{\epsilon \to 0} e^{-\epsilon|t|} \operatorname{sgn}(t)\right\} = \lim_{\epsilon \to 0} \int_{-\infty}^{\infty} e^{-\epsilon|t|} \operatorname{sgn}(t) e^{-j\omega t} dt$$

SPECTRA OF TEMPORAL AND SPATIAL SIGNALS

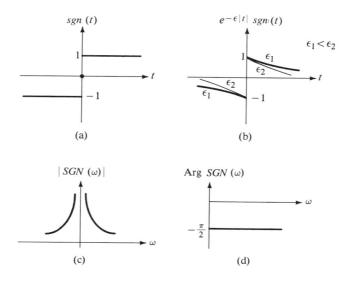

Figure 4.18
The sgn function and its Fourier representation.

$$= \lim_{\epsilon \to 0} \left[\int_{-\infty}^{0} -e^{(\epsilon - j\omega)t} \, dt + \int_{0}^{\infty} e^{-(\epsilon + j\omega)t} \, dt \right]$$

$$= \lim_{\epsilon \to 0} \left(-\frac{1}{\epsilon - j\omega} + \frac{1}{\epsilon + j\omega} \right) = \frac{2}{j\omega}$$

$$= \frac{2}{\omega} e^{-j\pi/2} \triangleq SGN(\omega) \tag{4.53}$$

The Fourier amplitude and phase spectra of sgn(t) are shown in Figures 4.18c and 4.18d. It is noted that a mathematical subtlety exists regarding the interchange of the limit-taking and the integration. Such an interchange requires mathematical justification that is beyond the scope of this text. ■

EXAMPLE 4.15
Find the Fourier transform of $\exp[j\omega_0 t]$.

Solution: This result can be written down on the basis of (4.19b) and the frequency shifting property. The result is

$$\mathscr{F}\{e^{j\omega_0 t}\} = \mathscr{F}\{1 \cdot e^{j\omega_0 t}\} = 2\pi \delta(\omega - \omega_0) \tag{4.54}$$

■

EXAMPLE 4.16
Find the Fourier transform of the unit step function $u(t)$.

Solution: Begin by writing $u(t)$ in its equivalent representation involving the

Figure 4.19
The unit step function and its Fourier transform representation.

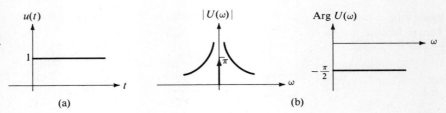

(a)   (b)

sgn(t) function

$$u(t) = \frac{1}{2} + \frac{1}{2}\operatorname{sgn}(t) \tag{4.55}$$

From Example 4.15 and Example 4.14 we find that

$$U(\omega) \triangleq \mathscr{F}\{u(t)\} = \frac{2\pi}{2}\delta(\omega) + \frac{2}{2j\omega} = \pi\delta(\omega) + \frac{1}{j\omega} \tag{4.56}$$

The unit step function and its transform are shown in Figure 4.19. ∎

EXAMPLE 4.17
Find the Fourier transform of the $\operatorname{comb}_T(t)$ function, as shown in Figure 4.20a.

Solution: Carrying out this problem requires that the $\operatorname{comb}_T(t)$ function first be represented in its Fourier series expansion. This is given by

$$\operatorname{comb}_T(t) = \sum_{n=-\infty}^{\infty} \delta(t - nT) \triangleq \sum_{n=-\infty}^{\infty} \alpha_n e^{jn\omega_0 t} = \frac{1}{T}\sum_{n=-\infty}^{\infty} e^{jn\omega_0 t} \tag{4.57}$$

where

$$\alpha_n = \frac{1}{T}\int_{-T/2}^{T/2} \delta(t)e^{-jn\omega_0 t}\,dt = \frac{1}{T} \qquad \omega_0 = \frac{2\pi}{T}$$

The Fourier transform is then

$$\mathscr{F}\{\operatorname{comb}_T(t)\} = \frac{1}{T}\sum_{n=-\infty}^{\infty} \mathscr{F}\{e^{jn\omega_0 t}\} = \frac{1}{T}\sum_{n=-\infty}^{\infty} 2\pi\delta(\omega - n\omega_0)$$

$$= \frac{2\pi}{T}\sum_{n=-\infty}^{\infty} \delta\left(\omega - n\frac{2\pi}{T}\right) \triangleq \frac{2\pi}{T}\operatorname{COMB}_{2\pi/T}(\omega) \tag{4.58}$$

The spectrum of the function is shown in Figure 4.20b.

This result shows that any periodic function of the form

$$f(t) = \sum_{n=-\infty}^{\infty} \alpha_n e^{jn\omega_0 t}$$

has a Fourier transform of the form

$$\mathscr{F}\{f(t)\} = F(\omega) = 2\pi\sum_{n=-\infty}^{\infty} \alpha_n \delta(\omega - n\omega_0) \tag{4.59}$$

∎

Figure 4.20
The Fourier transform of the $\text{comb}_T(t)$ function.

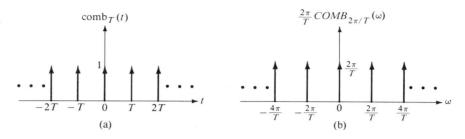

EXAMPLE 4.18
Prove the validity of the **Poisson sum formula**

$$\sum_{n=-\infty}^{\infty} f(t + nT) = \frac{1}{T}\sum_{n=-\infty}^{\infty} e^{jn\omega_0 t}F(n\omega_0) \qquad \omega_0 = \frac{2\pi}{T} \qquad (4.60)$$

Solution: Represent the left hand side by its equivalent form

$$\sum_{n=-\infty}^{\infty} f(t + nT) = f(t) * \text{comb}_T(t)$$

from which

$$\mathscr{F}\left\{\sum_{n=-\infty}^{\infty} f(t + nT)\right\} = \mathscr{F}\{f(t) * \text{comb}_T(t)\}$$

$$= F(\omega)\frac{2\pi}{T}\sum_{n=-\infty}^{\infty} \delta(\omega - n\omega_0)$$

$$= \sum_{n=-\infty}^{\infty} \omega_0 F(n\omega_0)\delta(\omega - n\omega_0)$$

The Fourier transform of the righthand member of (4.60) is

$$\mathscr{F}\left\{\frac{1}{T}\sum_{n=-\infty}^{\infty} e^{jn\omega_0 t}F(n\omega_0)\right\} = \frac{F(n\omega_0)}{T}\sum_{n=-\infty}^{\infty} \mathscr{F}\{e^{jn\omega_0 t}\}$$

$$= \frac{F(n\omega_0)}{T}2\pi\sum_{n=-\infty}^{\infty} \delta(\omega - n\omega_0)$$

$$= \sum_{n=-\infty}^{\infty} \omega_0 F(n\omega_0)\delta(\omega - n\omega_0)$$

This shows the validity of (4.60).
By setting $t = 0$ in (4.60), the following expression results

$$\sum_{n=-\infty}^{\infty} f(nT) = \frac{1}{T}\sum_{n=-\infty}^{\infty} F(n\omega_0) \qquad \omega_0 = \frac{2\pi}{T} \qquad (4.61)$$

For causal functions $f(t) = 0$ for $t < 0$, which means that $f(t)$ may be discontinuous at $t = 0$; in this case $f(t)|_0$ must be equal to $[f(0+) + f(0-)]/2$, and (4.61) becomes

$$\frac{f(0+) + f(0-)}{2} + \sum_{n=1}^{\infty} f(nT) = \frac{1}{T} \sum_{n=-\infty}^{\infty} F(n\omega_0)$$

from which

$$\sum_{n=0}^{\infty} f(nT) = -\frac{f(0+)}{2} + \frac{1}{T} \sum_{n=-\infty}^{\infty} F(n\omega_0) \tag{4.62}$$

since $f(0-) = 0$ and $f(0) = f(0+)$. The factor $f(0+)/2$ arises from the property of the inverse Fourier transform that defines a function at a point of discontinuity as

$$f(t) = \frac{f(t+) + f(t-)}{2} \tag{4.63}$$

which is the arithmetic mean of the values on each side of the discontinuity. ∎

EXAMPLE 4.19
Find the Fourier transform of the periodic function shown in Figure 4.21a.

Solution: The function $f(t)$ is written as a convolution product [see (3.8)]

$$f(t) = p_a(t) * \text{comb}_T(t)$$

The Fourier transform is given by (see Table 4.1)

$$F(\omega) = \mathscr{F}\{f(t)\} = \mathscr{F}\{p_a(t)\}\mathscr{F}\{\text{comb}_T(t)\} \qquad T = 4a$$

By (4.17) and (4.58) the result is

$$F(\omega) = \frac{2 \sin a\omega}{\omega} \frac{2\pi}{T} \sum_{n=-\infty}^{\infty} \delta\left(\omega - n\frac{2\pi}{T}\right)$$

$$= \frac{2\pi}{T} \sum_{n=-\infty}^{\infty} \frac{2 \sin\left(\frac{an2\pi}{T}\right)}{\frac{n2\pi}{T}} \delta\left(\omega - n\frac{2\pi}{T}\right)$$

$$= \sum_{n=-\infty}^{\infty} \frac{2}{n} \sin \frac{n\pi}{2} \delta\left(\omega - \frac{n\pi}{2a}\right) \qquad \omega_0 = \frac{2\pi}{T} = \frac{\pi}{2a} \tag{4.64}$$

This equation shows that the Fourier transform of the periodic function $f(t)$ is a train of impulses weighted by the factor $2\sin(n\pi/2)/n$, as shown in Figure 4.21b. Attention is called to the similarity of this result with the discussion of Figure 4.1. ∎

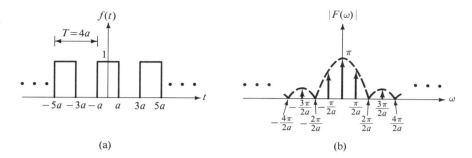

Figure 4.21 The Fourier transform of a periodic function.

By proceeding as in the examples given, a table of transforms can be developed; such a table is given as Table 4.2. Note that owing to the uniqueness property of the Fourier transform, there is a one-to-one correspondence between the direct and the inverse transform; hence corresponding to a function $f(t)$, the transform $F(\omega)$ is unique, and vice versa. Consequently, we can use Table 4.2 to write the $f(t)$ appropriate to a given $F(\omega)$. If the available tables do not include a given $F(\omega)$ and if the $F(\omega)$ cannot be reduced to available forms, recourse to the inversion integral will be necessary.

■ ■ ■

4-4 GIBBS PHENOMENON

The Gibbs phenomenon appears when we try to reconstruct discontinuous finite functions using their Fourier transform spectrums. Similar effects were observed when we tried to reconstruct discontinuous periodic signals using their Fourier series. To study this phenomenon, denote the spectrum of the nonperiodic step function $u(t)$ by $U(\omega)$. For its truncated spectrum, we write

$$U_{\omega_0}(\omega) = \begin{cases} U(\omega) & \text{for } |\omega| \leq \omega_0 \\ 0 & \text{elsewhere} \end{cases} \quad (4.65)$$

which is conveniently written

$$U_{\omega_0}(\omega) = U(\omega) p_{\omega_0}(\omega) \quad (4.66)$$

where $p_{\omega_0}(\omega)$ denotes a unit pulse of width $2\omega_0$ centered about 0. Hence the approximate reconstructed function $u_a(t)$ is given by [see (4.30)]

$$u_a(t) = \mathscr{F}^{-1}\{U(\omega)p_{\omega_0}(\omega)\} = u(t) * \mathscr{F}^{-1}\{p_{\omega_0}(\omega)\}$$

$$= u(t) * \frac{\sin \omega_0 t}{\pi t} = u(t) * \frac{1}{\pi} \operatorname{sinc}_{\omega_0}(t) \quad (4.67)$$

This expression indicates that as the sinc function slides towards the discontinuity, oscillations will be created in the vicinity of the discontinuity and will

Table 4.2 Fourier Transforms of Signals

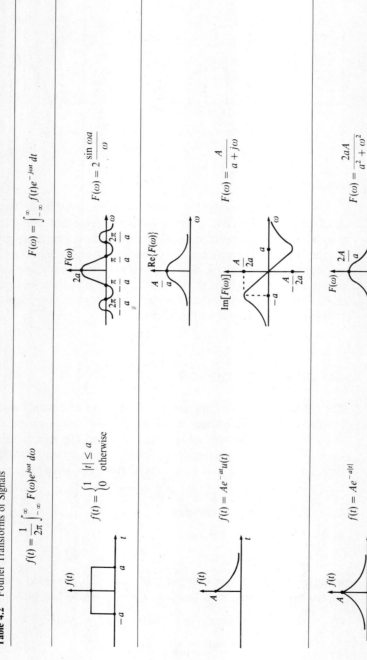

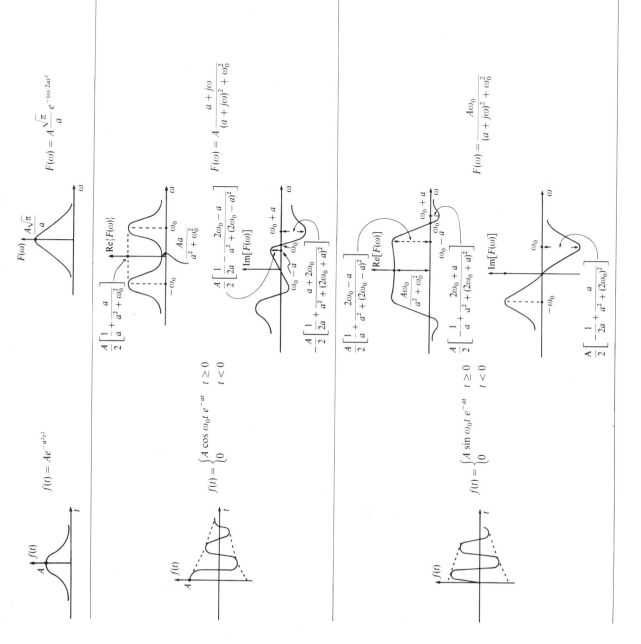

Table 4.2 (continued)

$$f(t) = \frac{1}{2\pi}\int_{-\infty}^{\infty} F(\omega)e^{j\omega t}\,d\omega \qquad F(\omega) = \int_{-\infty}^{\infty} f(t)e^{-j\omega t}\,dt$$

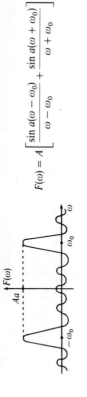

$$f(t) = \begin{cases} A & (T-a) < |t| < (T+a) \\ 0 & \text{otherwise} \end{cases}$$

$$F(\omega) = 4A\,\frac{\cos T\omega \sin a\omega}{\omega}$$

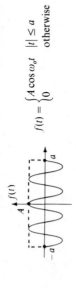

$$f(t) = \begin{cases} A\cos\omega_0 t & |t| \le a \\ 0 & \text{otherwise} \end{cases}$$

$$F(\omega) = A\left[\frac{\sin a(\omega - \omega_0)}{\omega - \omega_0} + \frac{\sin a(\omega + \omega_0)}{\omega + \omega_0}\right]$$

$$f(t) = A\,\delta(t)$$

$$F(\omega) = A$$

$$f(t) = \begin{cases} A & t > 0 \\ 0 & \text{otherwise} \end{cases}$$

$$F(\omega) = A\left[\pi\,\delta(\omega) - j\,\frac{1}{\omega}\right]$$

$$f(t) = \begin{cases} A & t > 0 \\ 0 & t = 0 \\ -A & t < 0 \end{cases}$$

$$F(\omega) = -j2A\,\frac{1}{\omega}$$

Table 4.2 (continued)

$f(t) = \frac{1}{2\pi} \int_{-\infty}^{\infty} F(\omega) e^{j\omega t} d\omega$	$F(\omega) = \int_{-\infty}^{\infty} f(t) e^{-j\omega t} dt$
$f(t) = A$	$F(\omega) = 2\pi A \delta(\omega)$
$f(t) = A \cos \omega_0 t$	$F(\omega) = \pi A[\delta(\omega - \omega_0) + \delta(\omega + \omega_0)]$
$f(t) = A \sin \omega_0 t$	$F(\omega) = j\pi A[\delta(\omega + \omega_0) - \delta(\omega - \omega_0)]$
$f(t) = \sum_{n=-\infty}^{\infty} \delta(t - nT) \triangleq \text{comb}_T(t)$	$F(\omega) = \frac{2\pi A}{T} \sum_{n=-\infty}^{\infty} \delta\left(\omega - \frac{2\pi n}{T}\right) \triangleq \frac{2\pi A}{T} \text{COMB}_{2\pi/T}(\omega)$

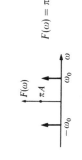

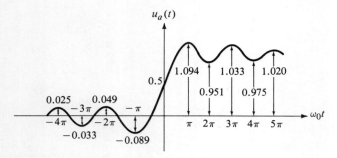

Figure 4.22
Illustration of the function $1/2 + Si(\omega_0 t)/\pi$.

continue as the sinc function slides past the discontinuity. To establish these conclusions mathematically, (4.67) is written

$$u_a(t) = \frac{\omega_0}{\pi} \int_{-\infty}^{t} \frac{\sin \omega_0 \tau}{\omega_0 \tau} d\tau = \frac{1}{\pi} \int_{-\infty}^{\omega_0 t} \frac{\sin x}{x} dx \qquad (4.68)$$

where $x = \omega_0 \tau$.

Use is now to be made of the sine integral $Si(x)$, which is defined by

$$Si(x) = \int_0^x \frac{\sin \tau}{\tau} d\tau = \int_0^x \operatorname{sinc} \tau \, d\tau \qquad (4.69)$$

Properties of the integral $Si(x)$ include: $Si(-x) = Si(x)$, also $Si(\infty) = \pi/2$. These properties permit (4.68) to be written in the form

$$u_a(t) = \frac{1}{2} + \frac{1}{\pi} Si(\omega_0 t) \qquad (4.70)$$

A plot of this expression is given in Figure 4.22.

The maximum occurs at $t = \pi/\omega_0$; this indicates that as ω_0 increases, the time scale changes but the amplitude of the oscillations remain unchanged. That is, an increase of the spectrum bandwidth results only in the compression of the oscillations close to the discontinuity.

The foregoing discussion shows we must be careful in selecting not only the width of the spectrum but also the window function in the frequency domain, in order that its inverse transform is a smoother time function than the $\operatorname{sinc}(t)$ function. This will result in a reconstructed function with smaller overshoots, overshoots that can be completely eliminated when special windows are used. This windowing is often called **apodization.** Window functions that serve better than the unit pulse function $p_{\omega_0}(\omega)$ include

triangle window: $\Lambda_{\omega_0}(\omega)$

SPECTRA OF TEMPORAL AND SPATIAL SIGNALS

$$\text{Hamming window: } W_{hm}(\omega) = \begin{cases} 0.54 + 0.46 \cos \dfrac{\pi\omega}{\omega_0} & |\omega| < \omega_0 \quad \text{(a)} \\ 0 & \text{elsewhere} \quad \text{(b)} \end{cases} \quad (4.71)$$

Examine how the truncated time function affects the spectrum. Specifically, consider the truncated step function, which has been truncated at $t = 2a$. Clearly, this truncated step function is equivalent to the pulse function $u_t(t) = p_a(t - a)$. The Fourier transform of $u_t(t)$ is found to be

$$\mathscr{F}\{u_t(t)\} \triangleq U_t(\omega) = e^{-ja\omega} 2 \frac{\sin a\omega}{\omega} = \frac{\sin 2a\omega}{\omega} - j2\frac{\sin^2 a\omega}{\omega}$$

The real and imaginary parts of the truncated step function $U_t(\omega)$ are plotted in Figure 4.23, together with the real and imaginary parts of the step function $U(\omega)$. The effect of the truncation is obvious; it tends to spread and deform the spectrum.

Exactly similar arguments apply for signals truncated in time. In this case it is desired to approximate the spectrum $F(\omega)$ of $f(t)$ as closely as possible using the truncated form $f_t(t)$. Actually, the truncated signal is the one that appears in practice, since we are forced to collect data within a finite time interval. Therefore multiply the truncated function $f_t(t)$ by a window function $w(t)$ so that the difference $F_w(\omega) - F(\omega)$, where $F_w(\omega)$ is the Fourier transform of $f_w(t) = w(t)f_t(t)$, is reduced in some prescribed sense. The window function is selected so that

$$w(0) = \frac{1}{2\pi} \int_{-\infty}^{\infty} W(\omega)\,d\omega = 1$$

$$w(t) = 0 \quad \text{for } |t| > a$$

A number of available window functions possess good overall properties and have been used extensively. These, with their Fourier spectra, include

$$\text{Hamming: } w_{hm}(t) = \left(0.54 + 0.46 \cos \frac{\pi}{a} t\right) p_a(t) \quad \text{(a)}$$

$$W_{hm}(\omega) = \frac{1.08\pi^2 - 0.16\, a^2\omega^2}{\omega(\pi^2 - a^2\omega^2)} \sin a\omega \quad \text{(b)} \quad (4.72)$$

$$\text{Hanning (or Hann): } w_{hn}(t) = \frac{1}{2}\left(1 + \cos \frac{\pi}{a} t\right) p_a(t) \quad \text{(a)}$$

$$W_{hn}(\omega) = \frac{\pi^2 \sin a\omega}{\omega(\pi^2 - a^2\omega^2)} \quad \text{(b)} \quad (4.73)$$

$$\text{Bartlett: } w_b(t) = \Lambda_a(t) \quad \text{(a)}$$

$$W_b(\omega) = 4\frac{\sin^2\left(\dfrac{a\omega}{2}\right)}{a\omega^2} \quad \text{(b)} \quad (4.75)$$

Figure 4.23
Comparison of Fourier spectra of truncated and nontruncated step functions.

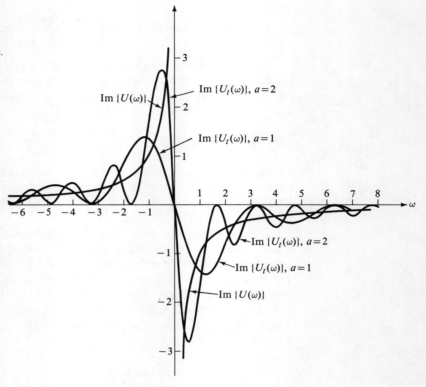

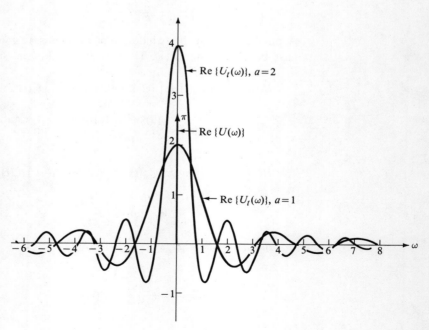

†B. SPECTRA OF SPATIAL SIGNALS

4-5 FOURIER TRANSFORMS AND OPTICAL SYSTEMS

We develop in this section the conditions under which the output field created in an optical system (the simple lens system shown in Figure 2.30) is equal to the Fourier transform of the field entering the system. For this purpose we rewrite (2.85) in the following form

$$\Phi(y_1) = \psi(y_1; z_1) \int_{-\infty}^{\infty} \Phi(y_0)\psi(y_0; z_0)$$
$$\times \int_{-\infty}^{\infty} \psi(y; w)P(y)e^{-jk[y_0/z_0 + y_1/z_1]y} \, dy \, dy_0 \quad \text{(a)} \quad (4.75)$$

where

$$\psi(y_1; z_1) = e^{j(k/2z_1)y_1^2} \quad \text{(b)}$$

$$\psi(y; w) = e^{j(k/2w)y^2} \qquad \frac{1}{w} = \frac{1}{z_0} + \frac{1}{z_1} - \frac{1}{f_l} \quad \text{(c)}$$

$$P(y) = 1, \text{ since } \lambda \to 0 \quad \text{(d)}$$

$$\Phi(y_0) = \text{input field} \qquad \Phi(y_1) = \text{output field} \quad \text{(e)}$$

$$b = \frac{y_0}{z_0} + \frac{y_1}{z_1} \quad \text{(f)}$$

We first focus attention on the inner integral, which can be written in the following form (see Example 4.8)

$$\int_{-\infty}^{\infty} e^{j(k/2w)y^2} e^{-jkby} \, dy = e^{-j(k/2)wb^2} \int_{-\infty}^{\infty} e^{j(k/2w)(y-wb)^2} \, dy$$

$$= je^{j(\pi/4)} \sqrt{\frac{2\pi w}{2k}} e^{-j(k/2)wb^2} \triangleq je^{j(\pi/4)} \sqrt{\frac{2\pi w}{2k}} \psi\left(b; -\frac{1}{w}\right) \quad (4.76)$$

Introduce these results into (4.75) to obtain

$$\Phi(y_1) = \psi\left(y_1; \frac{z_1}{1 - \frac{w}{z_1}}\right) \int_{-\infty}^{\infty} \Phi(y_0)\psi\left(y_0; \frac{z_0}{1 - \frac{w}{z_0}}\right) e^{-j(kw/z_0 z_1)y_0 y_1} \, dy_0 \quad (4.77)$$

We have omitted the constant term $\sqrt{w\pi/k}$ because it adds a constant intensity term in any image-processing technique involving the exposure of a film. This constant term can be subtracted, in much the same way that a dc term in time signals can be removed, when it is required to do so. Remember that we observe $|\Phi(y_1)|^2$ on the film and not $\Phi(y_1)$.

If the object is located in the **front focal plane** ($z_0 = f_l$) and the observation plane is located in the **back focal plane** ($z_1 = f_l$), then $w = f_l$ and the image field is related to the object by the Fourier transform

CHAPTER 4

$$\Phi(y_1) = \int_{-\infty}^{\infty} \Phi(y_0) e^{-j\omega_{y_1} y_0} \, dy_0 \quad \text{(a)}$$

where

$$\omega_{y_1} = \frac{k y_1}{f_l} = 2\pi \frac{y_1}{\lambda f_l} = 2\pi f_{y_1} \quad m^{-1} \quad \text{(b)}$$

(4.78)

These formulas have physical interpretations; in fact, in the following two examples we can view the Fourier spectra of field distributions.

Initially, we must point out that $\Phi(y_1)$ should really have been written in the form $\Phi(\omega_{y_1})$ to emphasize that this entity denotes a frequency spectrum. However, in optics we observe light along a particular direction y_1 of the x_1, y_1 plane; therefore the spectrum of $\Phi(y_0)$, which is $\Phi(y_1)$, is written as a function of the coordinate y_1. Thus if we look at the point $y_1 = 0$, the field is $\Phi(y_1)$ (even though we see the intensity on the film) at the optical center, where $\omega_{y_1} = 0$. This is analogous to the dc term in time signals. Any field distribution away from the optical center indicates that discontinuities exist in the field $\Phi(y_0)$. Since ω_{y_1} is proportional to the distance y_1, the field (light) away from the optical center will contribute to the higher frequency bands of the optical transform. This conclusion means we can always manipulate the amplitude and phase of the light at the (x_1, y_1) plane, known as the Fourier plane, and produce the equivalent of filtering of time-domain signals. The effects of filtering on optical signals will be discussed in the next chapter.

EXAMPLE 4.20
Find the Fourier transform of the field due to the two pinholes shown in Figure 4.24a.

Solution: Assume the incident field to be of unit amplitude, hence the field distribution on the (x_0, y_0) plane is given by

$$\Phi(y_0) = \delta(y_0 - a) + \delta(y_0 + a) \tag{4.79}$$

Introduce this expression into (4.78) to obtain

$$\Phi(y_1) = \int_{-\infty}^{\infty} [\delta(y_0 - a) + \delta(y_0 + a)] e^{-j\omega_{y_1} y_0} \, dy_0$$

$$= e^{-j\omega_{y_1} a} + e^{j\omega_{y_1} a} = 2 \cos \omega_{y_1} a = 2 \cos\left(\frac{2\pi}{\lambda f_l} y_1 a\right) \tag{4.80}$$

The intensity is given by

$$|\Phi(y_1)|^2 = 4 \cos^2\left(\frac{2\pi a}{\lambda f_l} y_1\right)$$

and this is shown in Figure 4.24b.

SPECTRA OF TEMPORAL AND SPATIAL SIGNALS

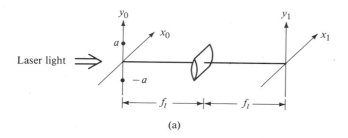

(a)

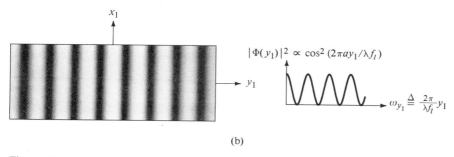

(b)

Figure 4.24
The Fourier transform of two point sources. (Reprinted, by permission, from Born and Wolf, *Principles of Optics*, 260a.)

EXAMPLE 4.21
A slit of height $2a$ along the y_0-axis is illuminated by a constant amplitude field. Find its Fourier transform.

Solution: Note that a slit in optics having an opening of $2a$ meters is equivalent to a pulse function form when dealing with time functions. The field function $\Phi(y_0)$ in this case is given by

$$\Phi(y_0) = \begin{cases} 1 & -a \leq y_0 \leq a \\ 0 & \text{otherwise} \end{cases} \quad (4.81)$$

Equation (4.78a) becomes

$$\Phi(y_1) = \int_{-a}^{a} e^{-j\omega_{y_1} y_0} dy_0 = \frac{2 \sin a\omega_{y_1}}{\omega_{y_1}} = 2 \operatorname{sinc}_a(\omega_{y_1}) \quad (4.82)$$

■

Figure 4.25a shows the field distribution of a square aperture imaged by a circular lens. The intensity distribution along the y_1-axis is proportional to

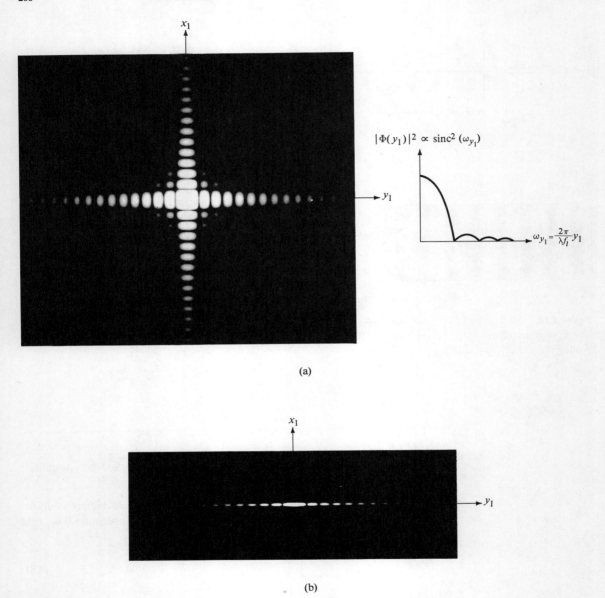

Figure 4.25
(a) The Fourier transform of a square aperture field. (b) The Fourier transform of a slit. (From E. Hecht and A. Zajac, *Optics*, © 1974. Addison-Wesley, Reading, MA. Reprinted with permission.)

sinc$^2(\omega_{y_1})$, which is proportional to $|\Phi(y_1)|^2$, as we would expect from (4.82). The distribution of the field in the x_1-direction is due to the finite aperture in the x_0-direction of the slit. In the limit when the x_0-direction opening becomes

SPECTRA OF TEMPORAL AND SPATIAL SIGNALS

large ($a \to \infty$), the light intensity distribution along the y_1-axis approaches that given by (4.82). This is illustrated in Figure 4.25b. The observation from Figure 4.25b is equivalent to a slit in the y_0-direction imaged by a cylindrical lens, as shown in the previous figure.

■ ■ ■

†C. TRANSFORMS IN THE COMPLEX DOMAIN—THE HILBERT TRANSFORM

4.6 FOURIER TRANSFORMS AND COMPLEX FUNCTION THEORY

NOTE: This section presupposes familiarity with the theory of functions of a complex variable (see Appendix 1).

Often in the evaluation of the direct or inverse Fourier transform, functions are involved that do not lend themselves to simple integration in the real plane, as indicated in (4.1). In many cases these integrals can be evaluated using methods involving complex function theory. This calls for contour integration in the complex plane involving the specific integrand or an integrand from which the given function can be obtained.

The basic considerations involve the **Cauchy Integral Theorem,** which relates the line integral of a complex function $W(z)$ around a closed path in the complex plane to the residues enclosed by the chosen path. Specifically, the theorem states that

$$\oint_C W(z)\,dz = 2\pi j \sum \text{Res} \tag{4.83}$$

where $\sum \text{Res}$ denotes the "residues" in the poles contained within the contour C in the complex plane.

In applying this theorem, the integrand must be written in a form that exhibits the function in complex variable form. When applied in the evaluation of Fourier integrals, a closed path of integration is selected that, includes the real axis from $-\infty$ to $+\infty$. Usually the path to be selected is clearly evident although in some cases special care is required. If poles or zeros of the integrand exist along the chosen path, hooks are formed around them to avoid having the path pass through any roots of $W(z)$.

An important feature of this formula is that the result is independent of the shape of the contour. Consequently, one is free to select the contour to make the integrations tractable. The problem is then carrying out the integrations along the various segments of the path making up the total contour C. Also, the right-hand member must also be evaluated. This process leads to forms from which the desired Fourier integral is extracted. The following examples clarify this procedure.

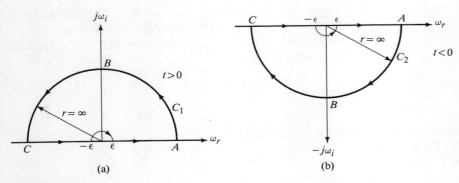

Figure 4.26
Contours for evaluating the integral $1/\pi \int_{-\infty}^{\infty} (\sin \omega t/\omega)\, d\omega$.

EXAMPLE 4.22
Evaluate the inverse Fourier transform of the function $F(\omega) = 2/j\omega$, which is known to be $f(t) = \text{sgn}(t)$.

Solution: We are to evaluate the integral

$$\mathcal{F}^{-1}\left\{\frac{2}{j\omega}\right\} = \frac{1}{2\pi}\int_{-\infty}^{\infty} \frac{2}{j\omega} e^{j\omega t}\, d\omega = \frac{1}{\pi j}\int_{-\infty}^{\infty} \frac{e^{j\omega t}}{\omega}\, d\omega = \frac{1}{\pi}\int_{-\infty}^{\infty} \frac{\sin \omega t}{\omega}\, d\omega \quad (4.84)$$

The latter integral is valid because the function $\cos \omega t/\omega$ is odd and the integral is zero. Instead of considering the exact integral specified, we initially consider the evaluation of the integral $\oint \exp(j\omega t)/\omega\, d\omega$ around the contour shown in Figure 4.26a for $t > 0$. Observe the hook in the curve around the pole at the origin. This corresponds to (4.83) with $z = \omega$; that is, ω is taken to be the complex variable. We write ω_r to be the real part and ω_i the imaginary part of ω. The contour integral is written explicitly

$$\oint \frac{e^{j\omega t}}{\omega}\, d\omega = \int_{-\infty}^{-\epsilon} \frac{e^{j\omega_r t}}{\omega_r}\, d\omega_r + \int_{\pi}^{0} \frac{e^{jt\epsilon(\cos\theta + j\sin\theta)}}{\epsilon e^{j\theta}}\, d(\epsilon e^{j\theta})$$

$$+ \int_{\epsilon}^{\infty} \frac{e^{j\omega_r t}}{\omega_r}\, d\omega_r + \int_{ABC} \frac{e^{j\omega t}}{\omega}\, d\omega = 0 \quad (4.85)$$

The right-hand side is set equal to zero because the path selected does not enclose any singularities of the integrand. We consider the evaluation of each integral in (4.85).

To evaluate the fourth integral, we parameterize it by setting $\omega = re^{j\theta} = r\cos\theta + jr\sin\theta$. Thus we have

$$\int_{ABC} e^{jtre^{j\theta}} \frac{1}{re^{j\theta}} jre^{j\theta}\, d\theta = \int_{ABC} e^{jtre^{j\theta}} e^{j(\pi/2)}\, d\theta$$

However,

$$\left|e^{jtre^{j\theta}}\right| = e^{-tr\sin\theta} \quad \text{and} \quad \left|e^{j(\pi/2)}\right| = 1$$

and so

$$\left|\int_{ABC} \frac{e^{j\omega t}}{\omega} d\omega\right| = \left|\int_0^\pi e^{jtre^{j\theta}} e^{j(\pi/2)} d\theta\right| \leq \int_0^\pi \left|e^{jtre^{j\theta}}\right| \left|e^{j(\pi/2)}\right| d\theta = \int_0^\pi e^{-tr\sin\theta} d\theta$$

Since the function $\exp(-tr\sin\theta)$ is symmetric about $\theta = \pi/2$ in $0 \leq \theta < \pi$, then

$$\int_0^\pi e^{-tr\sin\theta} d\theta = 2\int_0^{\pi/2} e^{-tr\sin\theta} d\theta$$

We note that $\sin\theta \geq (2/\pi)\theta$ in the range $0 \leq \theta \leq \pi/2$, which thus implies that $\exp[-tr\sin\theta] \leq \exp[-t2r\theta/\pi]$, and we obtain

$$\int_0^\pi e^{-tr\sin\theta} d\theta \leq 2\int_0^{\pi/2} e^{-2tr\theta/\pi} d\theta = 2\left(\frac{-\pi}{2tr}\right)(e^{-tr} - 1) < \frac{\pi}{tr}$$

Therefore, as $r \to \infty$, the integration along the semicircle does not contribute to the integral on the left of (4.85) as long as $t > 0$. This result is in accord with the **Jordan Lemma,** which specifies that if $t > 0$ and $P(z)/Q(z)$ is the quotient of two polynomials such that degree $Q \geq 1 +$ degree P, then

$$\lim_{r \to \infty} \int_C e^{jtz} \frac{P(z)}{Q(z)} dz = 0$$

where C is the upper half-circle of radius r.

The remaining integrals, in the limit as $\epsilon \to 0$, become

$$-\int_0^\infty \frac{e^{-j\omega_r t}}{\omega_r} d\omega_r + \int_0^\infty \frac{e^{j\omega_r t}}{\omega_r} d\omega_r - j\int_0^\pi d\theta = 0$$

It follows from this that

$$\int_0^\infty \frac{\sin\omega_r t}{\omega_r} d\omega_r = \frac{\pi}{2} \quad \text{or} \quad \frac{1}{\pi}\int_{-\infty}^\infty \frac{\sin\omega_r t}{\omega_r} d\omega_r = 1 \quad (4.86)$$

For $t < 0$, the contour is that shown in Figure 4.26b. The integration around the hook (the small semicircle) extends from $-\pi$ to 0 and yields the value $-\pi$. The remaining integrals remain unchanged in the integration. Hence it follows that for $t > 0$, $\mathscr{F}^{-1}\{2/j\omega\}$ is $+1$ and for $t < 0$, $\mathscr{F}^{-1}\{2/j\omega\} = -1$, which is the definition of the sgn(t) function. ∎

EXAMPLE 4.23

Find the Fourier transform of the Gaussian function $g_a = \exp[-at^2]$.

Solution: By definition

$$F(\omega) = \int_{-\infty}^\infty e^{-j\omega t} e^{-at^2} dt = e^{-\omega^2/4a} \int_{-\infty}^\infty e^{-(\sqrt{a}t + j(\omega/2\sqrt{a}))^2} dt \quad (4.87)$$

Figure 4.27
Contour for the integral of (4.88).

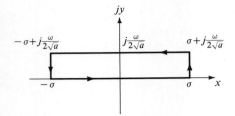

Set $\sqrt{a}\, t + j(\omega/2\sqrt{a}) = z$, and (4.87) becomes

$$F(\omega) = e^{-\omega^2/4a} \int_{-\infty + j(\omega/2\sqrt{a})}^{\infty + j(\omega/2\sqrt{a})} e^{-z^2}\, dz \tag{4.88}$$

where the integration is designated to be along the line $y = j(\omega/2\sqrt{a})$. To proceed, consider the complex integral $\oint \exp[-z^2]\, dz$ around the contour shown in Figure 4.27. The contour integral now becomes

$$\oint_{\sigma \to \infty} e^{-z^2}\, dz = \lim_{\sigma \to \infty} \left[\int_{-\sigma}^{\sigma} e^{-x^2}\, dx + \int_{0}^{j(\omega/2\sqrt{a})} e^{-(\sigma + jy)^2} j\, dy \right.$$

$$\left. + \int_{\sigma}^{-\sigma} e^{-[x + j(\omega/2\sqrt{a})]^2}\, dx + \int_{j(\omega/2\sqrt{a})}^{0} e^{-(-\sigma + jy)^2} j\, dy \right] = 0$$

$$\tag{4.89}$$

The combined contribution from the integrations along the two vertical edges is zero. Hence the remaining terms yield

$$\lim_{\sigma \to \infty} \int_{-\sigma}^{\sigma} e^{-x^2}\, dx - \lim_{\sigma \to \infty} \int_{-\sigma}^{\sigma} e^{-[x + j(\omega/2\sqrt{a})]^2}\, dx = 0$$

or

$$\lim_{\sigma \to \infty} \int_{-\sigma}^{\sigma} e^{-x^2}\, dx = 2 \lim_{\sigma \to \infty} \int_{-\sigma}^{0} e^{-x^2}\, dx = \lim_{\sigma \to \infty} \int_{-\sigma}^{\sigma} e^{-[x + j(\omega/2\sqrt{a})]^2}\, dx$$

The right-hand integral is written in the form

$$\int_{-\infty}^{\infty} e^{-[\sqrt{a}t + j(\omega/2\sqrt{a})]^2}\, d(t\sqrt{a})$$

from which it follows that

$$\frac{1}{\sqrt{a}} \int_{-\infty}^{\infty} e^{-x^2}\, dx = \int_{-\infty}^{\infty} e^{-[\sqrt{a}t + j(\omega/2\sqrt{a})]^2}\, dt \tag{4.90}$$

To evaluate the left-hand integral, consider the related integral $I = \int_{0}^{\infty} \exp[-x^2]\, dx$. For convenience focus on I^2, which is written

$$I^2 = \int_{0}^{\infty} e^{-x^2}\, dx \int_{0}^{\infty} e^{-y^2}\, dy = \int_{0}^{\infty} \int_{0}^{\infty} e^{-(x^2+y^2)}\, dx\, dy$$

$$= \int_{0}^{\pi/2} d\theta \int_{0}^{\infty} r e^{-r^2}\, dr = \frac{\pi}{2}\frac{1}{2}$$

SPECTRA OF TEMPORAL AND SPATIAL SIGNALS

so that

$$I = \frac{\sqrt{\pi}}{2}$$

Relate this result to (4.88) from which it follows that

$$F(\omega) = \sqrt{\frac{\pi}{a}}\, e^{-\omega^2/4a} \tag{4.91}$$

∎

EXAMPLE 4.24
Find the inverse Fourier transform of the function $F(\omega) = 2a/(a^2 + \omega^2)$.

Solution: Here the problem is to evaluate the integral

$$f(t) = \frac{a}{\pi}\int_{-\infty}^{\infty} \frac{1}{\omega^2 + a^2}\, e^{j\omega t}\, d\omega = \frac{a}{\pi}\int_{-\infty}^{\infty} \frac{1}{(\omega + ja)(\omega - ja)}\, e^{j\omega t}\, d\omega \tag{4.92}$$

Consider ω to be a complex quantity, and the integrand is seen to have two poles at $\pm ja$, as shown in Figure 4.28. Carrying out the integration along the upper contour for $t > 0$ requires the residue at the pole $\omega = +ja$. This is given by

$$\text{Res}\, \frac{e^{j\omega t}}{(\omega + ja)(\omega - ja)}\bigg|_{\omega = ja} = \frac{e^{j\omega t}}{(\omega + ja)}\bigg|_{\omega = ja} = \frac{e^{-at}}{2ja} \qquad t > 0$$

The lower contour applies for $t < 0$, and this yields for the residue at the pole at $\omega = -ja$

$$\text{Res}\, \frac{e^{j\omega t}}{(\omega + ja)(\omega - ja)}\bigg|_{\omega = -ja} = \frac{e^{j\omega t}}{(\omega - ja)}\bigg|_{\omega = -ja} = \frac{e^{at}}{-2ja} \qquad t < 0$$

Figure 4.28
Contour for evaluating (4.92)

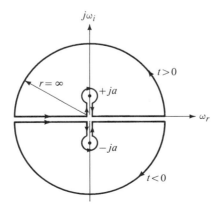

Figure 4.29
Contour for (4.94).

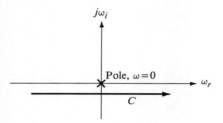

Combine these with (4.92), and recalling (4.83), we get

$$f(t) = \begin{cases} \dfrac{a}{\pi} 2\pi j \dfrac{e^{-at}}{2ja} = e^{-at} & t > 0 \\ -\dfrac{a}{\pi} 2\pi j \dfrac{e^{+at}}{-2ja} = e^{+at} & t < 0 \end{cases} \qquad (4.93)$$

which is conveniently written

$$f(t) = e^{-a|t|}$$

■

EXAMPLE 4.25
Find the inverse transform of $F(\omega) = 2(\sin \omega/\omega)$.

Solution: The integral to be evaluated is

$$f(t) = \frac{1}{\pi} \int_{-\infty}^{\infty} \frac{\sin \omega}{\omega} e^{j\omega t} d\omega = \frac{1}{2\pi j} \int_C \frac{e^{j\omega(t+1)}}{\omega} d\omega - \frac{1}{2\pi j} \int_C \frac{e^{j\omega(t-1)}}{\omega} d\omega \qquad (4.94)$$

where C is the contour shown in Figure 4.29. If $t < -1$, we close the contour in the lower half-plane for both integrals. Since no pole is enclosed, $\sum \text{Res} = 0$ and $f(t) = 0$. For $t > 1$, we can close the contour in the upper half-plane for both integrals. Since a pole is enclosed, we evaluate the residue at this pole, which is given by $\{\exp[j0(t+1)] - \exp[j0(t-1)]\} = 0$, and thus $f(t) = 0$. When t is in the range $-1 < t < 1$, we close the contour in the upper half-plane for the first integral. Since there is a pole, we find the residue, which is equal to 1. Hence $f(t)$ is the unit pulse function $p_1(t)$.

■

■ ■ ■

4-7 HILBERT TRANSFORMS

Hilbert transform techniques are important when dealing with narrow-band signals. Such signals contain a small band of frequencies around a center or carrier frequency [see also (4.24)]. For example, AM transmission of voice sig-

SPECTRA OF TEMPORAL AND SPATIAL SIGNALS

nals usually involves a carrier frequency of several megahertz with bandwidths of 3000 Hz. The function

$$f(t) = A \sin \omega_0 t \, p_T\left(t - \frac{T}{2}\right) \qquad T \gg \frac{2\pi}{\omega_0} \tag{4.95}$$

approximately satisfies the narrow-band condition.

Given a real function $f(t)$ in $-\infty < t < \infty$, its Hilbert transform is given by

$$\hat{f}(t) = \mathcal{H}\{f(t)\} = \frac{1}{\pi} \overline{\int_{-\infty}^{\infty}} \frac{f(\tau)}{t-\tau} d\tau = f(t) * h(t) \tag{4.96}$$

where the bar on the integral sign indicates the principal value of the integral and $h(t) = 1/\pi t$. The inverse transform is defined as follows:

$$f(t) = \mathcal{H}^{-1}\{\hat{f}(t)\} = -\frac{1}{\pi} \overline{\int_{-\infty}^{\infty}} \frac{\hat{f}(\tau)}{t-\tau} d\tau \tag{4.97}$$

Note that since $\mathcal{F}\{1/\pi t\} = -j \, \text{sgn} \, \omega$, the Fourier transform of $\hat{f}(t)$ is given by

$$\hat{F}(\omega) = \mathcal{F}\{f(t) * h(t)\} = -jF(\omega) \, \text{sgn} \, \omega \tag{4.98}$$

This equation shows we can obtain the Hilbert transform of a function by passing it through an ideal 90° phase shifter. Figure 4.30 shows the amplitude and phase characteristics of such a system, which is known as a **quadrature filter** because its output is a sine function when its input is a cosine signal.

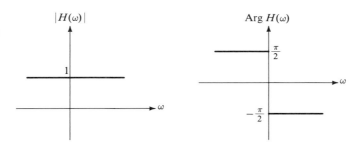
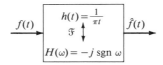

Figure 4.30 The system representation of a Hilbert transform.

Properties of Hilbert Transforms

1. Fourier Transform:

$$\mathscr{F}\{\hat{f}(t)\} = \hat{F}(\omega) = -jF(\omega)\,\text{sgn}\,\omega$$

The validity of this relation has been shown in (4.98).

2. Transform of Cosine Signals

$$\mathscr{H}\{\cos(\omega t + \varphi)\} = \sin(\omega t + \varphi) \tag{4.99}$$

Proof:

$$\mathscr{H}\{\cos(\omega t + \varphi)\} = \frac{1}{\pi}\int_{-\infty}^{\infty} \frac{\cos(\omega(t-\tau) + \varphi)}{\tau}\,d\tau$$

$$= \frac{1}{\pi}\int_{-\infty}^{\infty} \frac{\cos(\omega t + \varphi)\cos\omega\tau}{\tau}\,d\tau$$

$$+ \frac{1}{\pi}\int_{-\infty}^{\infty} \frac{\sin(\omega t + \varphi)\sin\omega\tau}{\tau}\,d\tau$$

Since the integrand of the first integral is odd, the integral yields zero. We thus have

$$\mathscr{H}\{\cos(\omega t + \varphi)\} = \frac{\sin(\omega t + \varphi)}{\pi}\int_{-\infty}^{\infty} \frac{\sin\omega\tau}{\tau}\,d\tau = \sin(\omega t + \varphi)$$

since the value of the integral is equal to π ([see (4.86)].

3. The Transform of a Transform

$$\mathscr{H}\{\hat{f}(t)\} = \mathscr{H}\{\mathscr{H}\{f(t)\}\} = -f(t) \tag{4.100}$$

Proof:

$$\mathscr{H}\{\hat{f}(t)\} = \frac{1}{\pi}\int_{-\infty}^{\infty} \frac{\hat{f}(t)}{t-\tau}\,d\tau = -\left\{-\frac{1}{\pi}\int_{-\infty}^{\infty} \frac{\hat{f}(t)}{t-\tau}\,d\tau\right\} = -f(t)$$

by (4.97). Additionally, owing to (4.100) we also obtain

$$\mathscr{H}\{\sin(\omega t + \varphi)\} = \mathscr{H}\{\mathscr{H}\{\cos(\omega t + \varphi)\}\} = -\cos(\omega t + \varphi) \tag{4.101}$$

4. Transform of Convolution

$$\hat{g}(t) = \mathscr{H}\{g(t)\} \triangleq \mathscr{H}\{f(t) * h(t)\} = f(t) * \hat{h}(t) = \hat{f}(t) * h(t) \tag{4.102}$$

Proof:

$$\hat{g}(t) = \frac{1}{\pi}\int_{-\infty}^{\infty} \frac{g(\tau)}{t-\tau}\,d\tau = \frac{1}{\pi}\int_{-\infty}^{\infty}\int_{-\infty}^{\infty} \frac{f(\xi)h(\tau-\xi)}{t-\tau}\,d\tau\,d\xi$$

$$= \int_{-\infty}^{\infty} f(\xi)\,d\xi \int_{-\infty}^{\infty} \frac{1}{\pi}\frac{h(\tau-\xi)}{t-\tau}\,d\tau = \int_{-\infty}^{\infty} f(\xi)\,d\xi \int_{-\infty}^{\infty} \frac{1}{\pi}\frac{h(\eta)}{t-\xi-\eta}\,d\eta$$

$$= \int_{-\infty}^{\infty} f(\xi)\hat{h}(t - \xi)\,d\xi = f(t) * \hat{h}(t)$$

We can similarly prove the other equality.

5. Modulated Signals

$$\hat{f}(t) = \mathcal{H}\{f(t)\} \triangleq \mathcal{H}\{a(t)\cos\omega_0 t\} = a(t)\sin\omega_0 t$$

$$\mathcal{F}\{a(t)\} = \begin{cases} A(\omega) & |\omega| < B \\ 0 & \text{otherwise} \end{cases} \tag{4.103}$$

Proof: From (4.98) we find

$$\hat{F}(\omega) = \mathcal{F}\{\hat{f}(t)\} = -j(\operatorname{sgn}\omega)F(\omega) = -j\operatorname{sgn}\omega\,\mathcal{F}\{a(t)\cos\omega_0 t\}$$

$$= -j\operatorname{sgn}\omega\left[\frac{A(\omega - \omega_0)}{2} + \frac{A(\omega + \omega_0)}{2}\right]$$

$$= \begin{cases} -\dfrac{j}{2} A(\omega - \omega_0) & \omega > 0 \\ \dfrac{j}{2} A(\omega + \omega_0) & \omega < 0 \end{cases}$$

Figure 4.31 shows some of these functions.

Figure 4.31 Spectra of modulated signals.

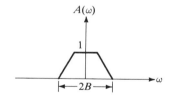
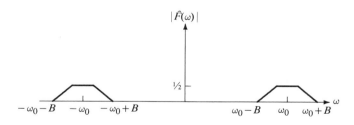
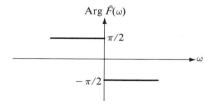

If we now take the inverse Fourier transform $\hat{F}(\omega)$, we shall obtain $\hat{f}(t)$. Hence

$$\hat{f}(t) = \frac{1}{2\pi} \int_{-\infty}^{\infty} -j \operatorname{sgn} \omega F(\omega) e^{j\omega t} d\omega$$

$$= \frac{1}{2\pi} \int_{0}^{\infty} -\frac{j}{2} A(\omega - \omega_0) e^{j\omega t} d\omega + \frac{1}{2\pi} \int_{-\infty}^{0} \frac{j}{2} A(\omega + \omega_0) e^{j\omega t} d\omega$$

and

$$\hat{f}(t) = -\frac{j}{2} \frac{e^{j\omega_0 t}}{2\pi} \int_{-B}^{B} A(\xi) e^{j\xi t} d\xi + \frac{j}{2} \frac{e^{-j\omega_0 t}}{2\pi} \int_{-B}^{B} A(\xi) e^{j\xi t} d\xi$$

$$= \frac{ja(t)}{2} (e^{-j\omega_0 t} - e^{j\omega_0 t}) = a(t) \sin \omega_0 t$$

By using (4.100) we find that

$$\mathscr{H}\{a(t) \sin \omega_0 t\} = -a(t) \cos \omega_0 t \tag{4.104}$$

Let us assume that a modulating signal $f_m(t)$ has the spectrum shown in Figure 4.32a. The single sideband spectrum with suppressed carrier is shown in Figure 4.32b. We can write

$$g_{\text{SSB}}(t) = \frac{1}{2} \int_{\omega_c}^{\infty} F_m(\omega - \omega_c) e^{j\omega t} d\omega + \frac{1}{2} \int_{-\infty}^{-\omega_c} F_m(\omega + \omega_c) e^{j\omega t} d\omega$$

$$= \frac{1}{2} e^{j\omega_c t} \int_{0}^{\infty} F_m(u) e^{jut} du + \frac{1}{2} e^{-j\omega_c t} \int_{-\infty}^{0} F_m(v) e^{jvt} dv$$

$$= \frac{1}{4} e^{j\omega_c t} \int_{-\infty}^{\infty} F_m(u) e^{jut} du + \frac{1}{4} e^{-j\omega_c t} \int_{-\infty}^{\infty} F_m(v) e^{jvt} dv$$

$$+ \frac{1}{4} e^{j\omega_c t} \int_{-\infty}^{\infty} F_m(u) \operatorname{sgn}(u) e^{jut} du$$

$$+ \frac{1}{4} e^{-j\omega_c t} \int_{-\infty}^{\infty} F_m(v) \operatorname{sgn}(-v) e^{jvt} dv$$

$$= \frac{1}{2} [f_m(t) \cos \omega_c t - \hat{f}_m(t) \sin \omega_c t]$$

where we have used (4.98) and the known relationship $\operatorname{sgn}(x) = -\operatorname{sgn}(-x)$. The production of a single sideband signal using a quadrature filter that introduces a $\pi/2$ phase shift to every frequency component of the signal is shown in Figure 4.32c. If the two signals at the summation point are subtracted, the output is only the upper sideband present; if they are added, only the lower sideband is present.

6. The Analytic Signal (Preenvelope)

The **analytic signal** (preenvelope) is defined by the relation

$$f_a(t) = f(t) + j\hat{f}(t) \tag{4.105}$$

Figure 4.32
Single sideband modulation.

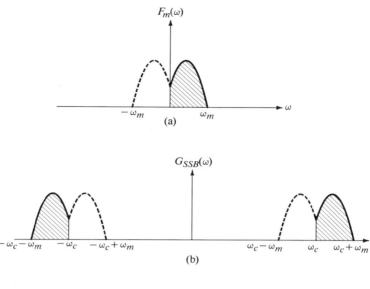

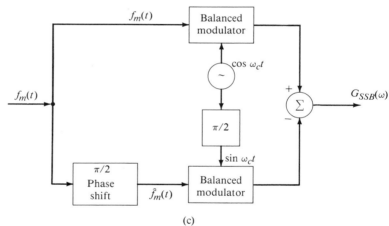

and its Fourier transform is

$$F_a(\omega) = F(\omega) + j\hat{F}(\omega) = F(\omega) + j(-j \,\text{sgn}\, \omega)F(\omega)$$

$$= \begin{cases} 2F(\omega) & \text{for } \omega > 0 \\ F(\omega) & \text{for } \omega = 0 \\ 0 & \text{for } \omega < 0 \end{cases} \quad (4.106)$$

This equation indicates that the analytic signal has only positive frequencies. Thus if we want to find the analytic signal of a signal $f(t)$, we proceed as follows:

a. Find its Hilbert transform $\hat{f}(t)$ and use (4.105).

b. Find $F(\omega) = \mathscr{F}\{f(t)\}$ and use the inverse transform

$$f_a(t) = 2 \int_0^\infty F(\omega)e^{j\omega t}\, d\omega \tag{4.107}$$

The analytic signal simplifies the handling of narrow-band signals and systems analogously to the use of phasors to simplify the handling of sinusoidally-varying signals in circuit analysis.

Let the analytic signal be represented by the relation

$$f_a(t) = \tilde{f}(t)e^{j\omega_0 t} \tag{4.108}$$

where $\tilde{f}(t) = \tilde{f}_r(t) + j\tilde{f}_i(t)$ is called the **complex envelope** of the signal. But shifting the spectrum of $f(t)$ at the origin will result in $F_a(\omega)$, and this implies that the frequency spectrum of $\tilde{f}(t)$ at $\omega = \omega_0$ has a width of $2B$ centered at $\omega = \omega_0$.

From (4.105) we observe that the signal $f(t)$ is equal to the real part of the analytic signal, hence by (4.108) we obtain

$$\begin{aligned} f(t) &= \text{Re}\{f_a(t)\} = \text{Re}\{\tilde{f}(t)e^{j\omega_0 t}\} \\ &= \tilde{f}_r(t)\cos\omega_0 t - \tilde{f}_i(t)\sin\omega_0 t \end{aligned} \tag{4.109}$$

where $\tilde{f}_r(t)$ is the **in-phase component** of $f(t)$ and $\tilde{f}_i(t)$ is the **quadrature component.** This equation shows that $f(t)$ is the projection on the real axis of the phasor $\tilde{f}(t)$ lying on the $\tilde{f}_r, \tilde{f}_i$ plane.

If we were to define the complex envelope by

$$\tilde{f}(t) = a(t)e^{j\varphi(t)} \tag{4.110}$$

then

$$f(t) = \text{Re}\{f_a(t)\} = a(t)\cos[\omega_0 t + \phi(t)] \tag{4.111}$$

where $a(t)$ is the **envelope** of the signal and $\varphi(t)$ is its **phase.**

Suppose that we multiply (4.109) by $\cos\omega_0 t$ and then pass the signal through a low-pass filter; the result is $\tfrac{1}{2}\tilde{f}_r(t)$. If we multiply (4.109) by $-\sin\omega_0 t$ and use a low-pass filter, we obtain $\tfrac{1}{2}\tilde{f}_i(t)$.

EXAMPLE 4.26

Find the complex envelope and the envelope of the narrow-band signal

$$f(t) = x(t)\cos\omega_0 t - y(t)\sin\omega_0 t \tag{4.112}$$

where $x(t)$ and $y(t)$ are both narrow-band signals.

Solution: We wish to use (4.105), but to do so we must find $\hat{f}(t)$. We write [see (4.103) and (4.104)]

$$\mathscr{H}\{f(t)\} = x(t)\sin\omega_0 t + y(t)\cos\omega_0 t$$

and the preenvelope is

SPECTRA OF TEMPORAL AND SPATIAL SIGNALS

$$f_a(t) = x(t)\cos\omega_0 t - y(t)\sin\omega_0 t + jx(t)\sin\omega_0 t + jy(t)\cos\omega_0 t$$
$$= x(t)(\cos\omega_0 t + j\sin\omega_0 t) + jy(t)(\cos\omega_0 t + j\sin\omega_0 t)$$
$$= [x(t) + jy(t)]e^{j\omega_0 t}$$

Therefore the envelope is

$$|f_a(t)| = \sqrt{x(t)^2 + y(t)^2}$$

and the complex envelope is

$$\tilde{f}(t) = x(t) + jy(t)$$ ∎

7. Narrow-band Filters

$$y_a(t) = h(t) * f_a(t) \tag{4.113}$$

where $h(t)$ is the impulse response of a linear time-invariant narrow-band filter, $f_a(t)$ is the analytic input narrow-band signal, and $y_a(t)$ is the system output.

Proof: The analytic signal of the output is [see (4.102)]

$$y_a(t) = h(t) * f(t) + j\mathcal{H}\{h(t) * f(t)\} = h(t) * f(t) + jh(t) * \hat{f}(t)$$
$$= h(t) * [f(t) + j\hat{f}(t)] = h(t) * f_a(t)$$

∎ ∎ ∎

D. SPECTRA OF MODULATED SIGNALS

4-8 AMPLITUDE MODULATED SIGNALS

The concept of modulation plays a central role in a variety of engineering systems; however, communication networks are the most important systems where modulation is the central focus. We will not explore all aspects of modulation in our present study, but we consider certain features important in signal analysis, such as the Fourier spectra of modified signals.

One type of modulation is **amplitude modulation** (AM), which is created by controlling the amplitude of one signal by a second signal. As noted in Section 1-4, the need for modulation in communication systems arises from practical considerations. Suppose that we wish to transmit a low-frequency radio wave, such as voice converted into electrical form with an appropriate transducer, or telegraph pulses, from one point to another without wires. A similar situation would arise if we wanted to use a given **channel** to transmit a number of messages simultaneously without interference. For the wireless case, we would need an antenna as the radiator of the signal. For efficient radiation an antenna approximately $l = \lambda/2$ in length is required, where λ is the wavelength of the signal. If a grounded vertical antenna is used, the physical length is approximately $l = \lambda/4$, since the total radiation results from the antenna and its electrical image in the ground, with each $\lambda/4$ long. For a frequency of 3 kHz, the wavelength is equal to $\lambda_m = c/f_m = 3 \times 10^8/3 \times 10^3 = 10^5$ km, and the required length of the antenna would be $10^5/4 = 25$ km, an impractical length. In the amplitude modulation process, a high-frequency oscillator produces a **carrier**, with the amplitude of the carrier being altered by the information signal. If the carrier frequency is 500 kHz, the required antenna length for a grounded radiator is $l = \lambda_c/4 = c/4f_c = 3 \times 10^8/4 \times 5 \times 10^5 = 150$ m, a manageable length.

In the AM system, the carrier signal is easily produced by a stable oscillator with power amplifiers to establish the desired power level. The carrier signal is of the form $v_c(t) = V_c \cos \omega_c t$, where ω_c is called the **carrier frequency.** The modulating signal (tone modulation), in its simplest form, is $v_m(t) = V_m \cos \omega_m t$ and may denote one component of a complex signal pattern, where ω_m is the modulating frequency and $\omega_m \ll \omega_c$. The general form of the modulation process, which is produced electronically in a communication system, is specified by the equation

$$v(t) = V_c[1 + mv_m(t)] \cos \omega_c t \qquad (4.114)$$

where m is the **modulation index** and is defined as $m = V_m/V_c$. The maximum value of $mv_m(t)$ must be less than one if distortion of the envelope is to be avoided. For the case of simple sinusoidal modulation, this equation can be expanded to

$$v(t) = V_c \cos \omega_c t + V_c m \cos \omega_c t \cos \omega_m t$$

or

$$v(t) = V_c \left\{ \cos \omega_c t + \frac{m}{2} [\cos(\omega_c + \omega_m)t + \cos(\omega_c - \omega_m)t] \right\} \qquad (4.115)$$

The frequency spectra of the three waveforms are given in Figure 4.33a, and Figure 4.33b shows the system of an AM generator in schematic form. We observe from Figure 4.33 that the amplitude modulated signal has twice the bandwidth of the original signal and, in addition, the carrier is also present.

Suppose that we had used a modulated signal of the form

$$v(t) = mv_m(t) \cos \omega_c t = m \cos \omega_m t \cos \omega_c t$$

$$= \frac{m}{2} [\cos(\omega_c + \omega_m)t + \cos(\omega_c - \omega_m)t]$$

In such a system, the carrier frequency is not present. This type of modulation is known as **double-sideband suppressed carrier** (DSBSC). While such modulation does save the energy of the carrier, a frequency component carrying no information, subsequent extraction of the information component is rather

Figure 4.33
(a) The frequency spectrum of an AM signal modulated by $V_m \cos \omega_m t$.
(b) A schematic representation of an ideal generator of an AM signal.

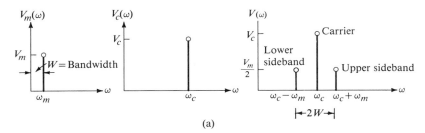

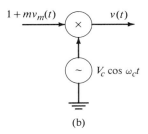

difficult. Moreover, the information is contained redundantly in each sideband; hence it is desirable to remove one or the other sideband to yield **single-sideband modulation** (SSB).

EXAMPLE 4.27

Consider a band-limited modulating signal $v_m(t)$ with a bandwidth ω_b, as shown in Figure 4.34a. Sketch the frequency spectrum of the signal. This signal modulates a carrier $v_c(t) = \cos \omega_c t$, so that the modulated signal is $v(t) = v_m(t) \cos \omega_c t$. Sketch its form in the time domain and in the frequency domain. Finally, multiply the received signal by a signal $\cos \omega_c t$, and show that the modulating signal is recovered in this process (called **demodulation**).

Solution: Suppose that the signal is that shown in Figure 4.34a and that its Fourier spectrum is that shown in Figure 4.34b. The Fourier spectrum of $v(t)$, the modulated signal, is

$$\mathscr{F}\{v(t)\} = \mathscr{F}\left\{v_m(t) \frac{e^{j\omega_c t} + e^{-j\omega_c t}}{2}\right\}$$

$$= \frac{1}{2}\left[\int_{-\infty}^{\infty} v_m(t) e^{-j(\omega + \omega_c)t} \, dt + \int_{-\infty}^{\infty} v_m(t) e^{-j(\omega - \omega_c)t} \, dt\right]$$

$$= \frac{1}{2} V_m(\omega + \omega_c) + \frac{1}{2} V_m(\omega - \omega_c)$$

This is shown in Figure 4.34c.

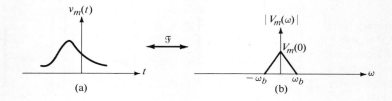
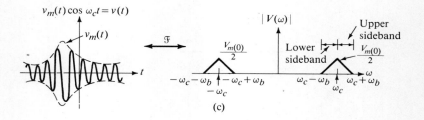
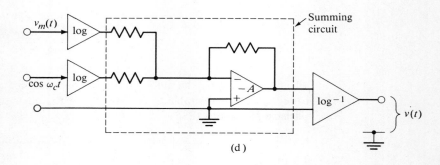
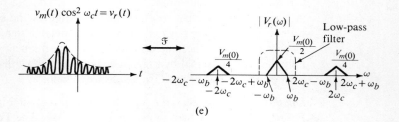
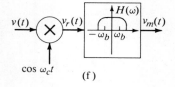

Figure 4.34
(a) Information-carrying signal $v_m(t)$. (b) The Fourier spectra of $v_m(t)$. (c) The modulated signal and its Fourier spectra. (d) One type of multiplier circuit. (e) The received signal multiplied by $\cos \omega_c t$ and its Fourier spectrum. (f) Demodulation scheme.

SPECTRA OF TEMPORAL AND SPATIAL SIGNALS

The Fourier spectrum of the received signal is given by

$$\mathscr{F}\{v_r(t) = v(t)\cos\omega_c t\} = \mathscr{F}\left\{v_m(t)\left[\frac{e^{j\omega_c t} + e^{-j\omega_c t}}{2}\right]^2\right\}$$

$$= \mathscr{F}\left\{\frac{v_m(t)}{2} + \frac{v_m(t)}{4}e^{j2\omega_c t} + \frac{v_m(t)}{4}e^{-j2\omega_c t}\right\}$$

$$= \frac{V_m(\omega)}{2} + \frac{V_m(\omega + 2\omega_c)}{4} + \frac{V_m(\omega - 2\omega_c)}{4}$$

Now use a low-pass filter with cutoff ω_b to pass frequencies equally well up to ω_b and to reject all frequencies higher than ω_b, and $V_m(\omega)$, the desired signal, is recovered. ∎

The ability to shift the frequency spectrum of the information-carrying signal by means of amplitude modulation allows us to transmit many different signals simultaneously through a given channel (transmission line, telephone line, microwave link, radio broadcast, and so on). This scheme is known as **frequency division multiplexing** (FDM). Of course, the ability to extract a given signal from the channel is equally essential, and a variety of demodulation or detection schemes exist. The sum of the frequency bands of the transmitted signals plus the frequency guard bands separating different signals must be less than or equal to the bandwidth of the channel for undistorted transmission.

A schematic representation of frequency division multiplexing for three signals is shown in Figure 4.35a. We present here only the positive part of the total spectrum since the other part appears on the negative ω-axis and is an exact mirror-image reflection of the positive portion of the spectrum relative to the origin. Bear in mind that negative frequencies do not exist physically but they do play an important role in our mathematical formulations. Figure 4.35b shows the schematic representation for detecting the combined signal and separating it into its three components.

For purposes of classification, the frequency spectrum is roughly divided as follows: 3–300 kHz for telephony, navigation, industrial communication, and long-range navigation; 0.3–30 MHz for AM broadcasting, military communication, and amateur and citizen-band radio; 30–300 MHz for FM broadcasting, TV broadcasting, and land transportation; 0.3–3 GHz for UHF TV, radar, and military applications; 3–30 GHz for satellite and space communication, microwave, and radar applications. The frequencies above 30 GHz are used mostly for research purposes and radio astronomy.

■ ■ ■

4-9 FREQUENCY MODULATION

Frequency modulation (FM) transmission proves to be very effective in improving immunity to noise since the modulated signal amplitudes are maintained constant by limiting or clipping circuits. The modulation is effected by

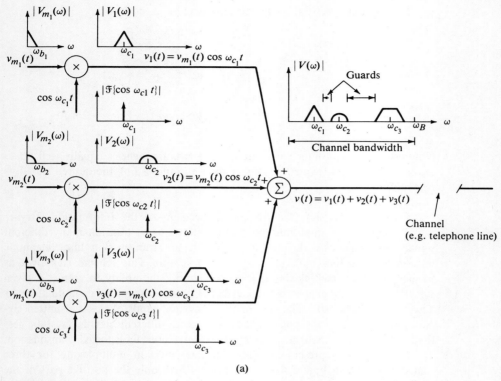

Figure 4.35
Frequency multiplexing, demultiplexing, and demodulating signals.

having the input signal vary the carrier frequency by an amount that depends on the amplitude and a rate that depends upon the frequency of the modulating signal. The fact that FM achieves noise immunity is easily verified by listening to an AM broadcast and an FM broadcast during a thunderstorm. Furthermore, a smaller geographical interference area exists when two nearby FM stations operate simultaneously on the same frequency. It is a characteristic of FM that the receiver will **capture** the stronger signal and almost completely eliminate the weaker signal. However, FM transmission requires a considerably larger bandwidth than AM transmission (up to 20 times as large). This plus the line-of-sight transmission limitation plus the strong nonlinearity of the signal creates some serious drawbacks for frequency modulation schemes.

We define the **instantaneous frequency** of an FM wave by

$$\omega_i = \frac{d\theta}{dt} = \omega_c + kv_m(t) \tag{4.116}$$

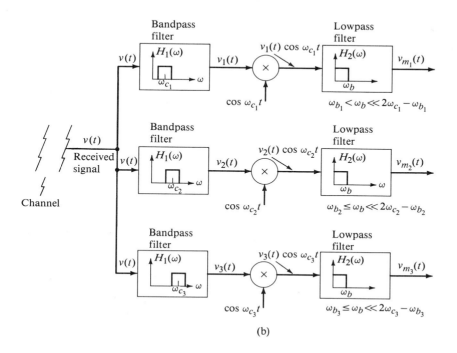

(b)

where θ is the instantaneous phase angle, ω_c = carrier frequency, k = constant, and $v_m(t)$ is the signal to be transmitted. This linear relation between the frequency and the modulating signal is the basis for the name **frequency modulation**. From (4.116) the total instantaneous phase angle is given by

$$\theta = \int_0^t \omega_i(t') \, dt' = \omega_c t + \int_0^t k v_m(t') \, dt' + \theta_0 \qquad (4.117)$$

where θ_0 is some initial constant phase, which can be set equal to zero without loss of generality.

EXAMPLE 4.28

Determine the instantaneous frequency of a signal $f(t) = V_m \cos(10^6 t + 10 t^2)$.

Solution: The instantaneous phase of the given signal is

$$\theta = 10^6 t + 10 t^2$$

and from (4.116) we obtain

$$\omega_i = \frac{d\theta}{dt} = 10^6 + 20t$$

At $t = 0$ the instantaneous frequency is 10^6 rad/s and increases linearly at a rate of 20 Hz/s. ∎

To understand some of the characteristics of FM signals, assume that the modulating signal is a pure tone

$$v_m(t) = V_m \cos \omega_m t \qquad \omega_m \ll \omega_c \tag{4.118}$$

Combine this expression with (4.116) with the result

$$\theta(t) = \int_0^t \omega_i(t')\, dt' = \int_0^t \left(\omega_c + kV_m(t') \cos \omega_m t' \right) dt'$$

$$= \omega_c t + \beta \sin \omega_m t \tag{4.119}$$

where

$$\boxed{\begin{aligned} \beta &= \frac{kV_m}{\omega_m} = \frac{\omega_d}{\omega_m} = \frac{\text{peak frequency deviation}}{\text{modulating frequency}} \\ &= \text{maximum phase deviation} = \text{deviation ratio} \end{aligned}} \tag{4.120}$$

The modulated signal thus has the form

$$v(t) = A \cos \theta(t) = \text{Re}\{Ae^{j\theta(t)}\} = \text{Re}\{Ae^{j\omega_c t} e^{j\beta \sin \omega_m t}\}$$

or

$$v(t) = A \cos(\omega_c t + \beta \sin \omega_m t) \tag{4.121}$$

For small values of β (that is, $\beta \ll 1$), this equation becomes

$$\begin{aligned} v(t) &= A \cos \omega_c t \cos(\beta \sin \omega_m t) - A \sin \omega_c t \sin(\beta \sin \omega_m t) \\ &\doteq A \cos \omega_c t - A\beta \sin \omega_m t \sin \omega_c t \end{aligned} \tag{4.122}$$

since, for small β, $\cos(\beta \sin \omega_m t) \doteq 1$ and $\sin(\beta \sin \omega_m t) \doteq \beta \sin \omega_m t$. The Fourier spectrum of this signal, which is known as the **narrow-band** frequency modulated signal, can be found easily by representing the sine and cosine functions in their exponential form. This is shown in Figure 4.36. From the figure we observe that the narrow-band frequency modulated signal resembles the amplitude modulated signal; that is, it has a carrier plus the two sidebands. However, the sideband spectrum for the FM signal has a phase shift of $\pi/2$ with respect to the carrier, a feature that is evident from (4.121). Figure 4.37 shows the elements of a narrow-band FM generator.

For the wide-band FM wave, we note that $\exp[j\beta \sin \omega_m t]$ is periodic in time with fundamental frequency ω_m. This permits it to be expanded into a

Figure 4.36
The frequency spectrum of a narrow-band frequency modulated signal.

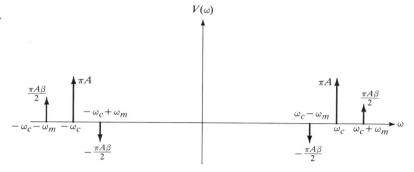

Figure 4.37
Narrow-band FM generating system.

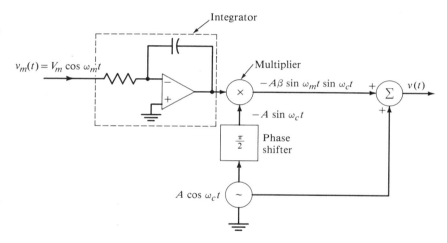

Fourier series

$$e^{j\beta \sin \omega_m t} = \sum_{n=-\infty}^{\infty} \Theta_n e^{jn\omega_m t} \qquad \text{(a)} \qquad (4.123)$$

where

$$\Theta_n = \frac{1}{T} \int_{-T/2}^{T/2} e^{j\beta \sin \omega_m t} e^{-jn\omega_m t} dt \quad \text{(b)}$$

We now set $\eta = \omega_m t = (2\pi/T)t$ in this expression to obtain

$$\Theta_n = \frac{1}{2\pi} \int_{-\pi}^{\pi} e^{j(\beta \sin \eta - n\eta)} d\eta = J_n(\beta) \qquad (4.124)$$

This integral is a function of β and n and is the integral form of the Bessel function of the first kind, $J_n(\beta)$ (see Chapter Appendix 4-1 at the end of this book). A number of Bessel functions are shown in Figure 4.38.

Figure 4.38
Bessel functions of the first kind.

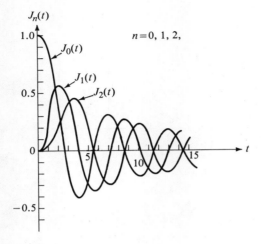

When we combine (4.123a) with (4.121), we obtain the following relation

$$v(t) = \text{Re}\left\{Ae^{j\omega_c t} \sum_{n=-\infty}^{\infty} J_n(\beta)e^{jn\omega_m t}\right\} = A \sum_{n=-\infty}^{\infty} J_n(\beta)\cos(\omega_c + n\omega_m)t$$

$$= A\{J_0(\beta)\cos\omega_c t + J_1(\beta)[\cos(\omega_c + \omega_m)t - \cos(\omega_c - \omega_m)t]$$
$$+ J_2(\beta)[\cos(\omega_c + 2\omega_m)t + \cos(\omega_c - 2\omega_m)t]$$
$$+ J_3(\beta)[\cos(\omega_c + 3\omega_m)t - \cos(\omega_c - 3\omega_m)t] + \cdots\} \quad (4.125)$$

In developing this expression, use has been made of the following properties of Bessel functions:

a. $J_n(\beta)$ are real valued functions
b. $J_n(\beta) = J_{-n}(\beta)$ n even, n = integer
c. $J_n(\beta) = -J_{-n}(\beta)$ n odd, n = integer
d. $\sum_{n=-\infty}^{\infty} J_n^2(\beta) = 1$

Equation (4.125) shows that the frequency modulated signal has an infinite number of sidebands. However, since $J_n(\beta)$ decreases rapidly for large n's, the amplitudes of the sidebands for large n's decrease rapidly, and for all practical purposes, most of the energy contained in the signal is confined within a finite bandwidth. The line spectra for an FM sinusoidally modulated waveform are given in Figure 4.39. A more precise example of the sideband distribution is shown in Figure 4.40. A detailed indication of the sideband composition is given in Figure 4.41, which shows a plot of $J_n(10)$ as a function of n and the corresponding spectral distribution of an FM wave with $\beta = 10$. Notice that $J_n(10)$ falls off toward zero rapidly for n greater than 10, but the amplitudes are significant out to about 14.

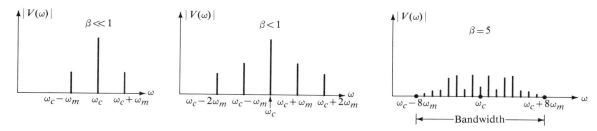

Figure 4.39
Frequency spectra of FM signals for different β.

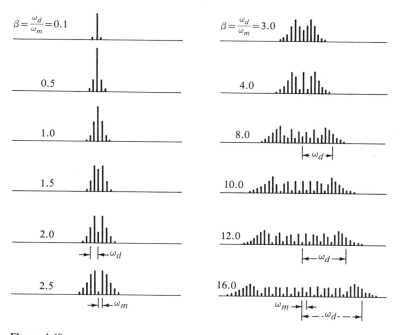

Figure 4.40
The spectral distribution in an FM wave for different values of β, different ω_d, and fixed ω_m. [Reprinted, by permission, from S. Seely, *Electron Tube Circuits*, 606.]

A practical constraint to the allowable bandwidth for an FM signal is the Federal Communications Commission (FCC) specification of ± 75 kHz. Suppose that we wish to provide for high-fidelity transmission that would accept modulating frequencies as high as 15 kHz. For the allowable 75 kHz half bandwidth, we would have $\beta = 75/15 = 5$. But for $\beta = 5$ there are six significant sideband pairs, which would require $6 \times 15 = 90$ kHz. As a practical matter,

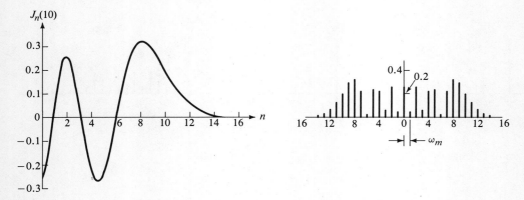

Figure 4.41
(a) A plot of $J_n(10)$ as a function of n. (b) The spectral distribution in an FM wave with $\beta = 10$.

therefore, circuitry is provided in an FM transmitter to predistort the energy content of the higher frequency components before transmission in a prescribed way so that the total bandwidth does not exceed the allowable 150 kHz, and receivers are provided with attenuation circuits that will compensate for this predistortion.

The curves of Figure 4.42 show a series of spectra for constant frequency deviation ω_d for different modulating frequencies ω_m. This plot shows that the total bandwidth required to include all significant sidebands decreases somewhat with increasing ω_m. For a given frequency deviation ω_d, except for small values of β, almost all significant sidebands are contained within the range ω_d.

Had we set the instantaneous phase factor $\theta(t)$ to be

$$\theta(t) = \omega_c t + k v_m(t) + \theta_0 \tag{4.126}$$

instead of (4.117), we would find that the instantaneous frequency is of the form

$$\omega_i(t) = \frac{d\theta}{dt} = \omega_c + k \frac{dv_m}{dt} \tag{4.127}$$

Here, because the phase is linearly related to $v_m(t)$, this modulation is known as **phase modulation** (PM). FM and PM are closely related because any variation in phase will result in a frequency variation, and vice versa. The difference between FM and PM appears in the expression for the deviation ratio, (4.120); specifically:

$$\beta = k_p V_m \quad \text{for PM values}$$

$$\beta = \frac{k_f V_m}{\omega_m} \quad \text{for FM values}$$

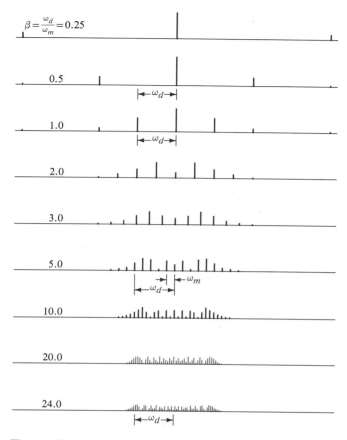

Figure 4.42
The spectral distribution in an FM wave for different values of fixed ω_d and different ω_m. [Reprinted, by permission, from S. Seely, *Electron Tube Circuits*, 607.]

In fact, one method for producing FM waves begins with PM waves and is accomplished by including a circuit that introduces the term $1/\omega_m$ in the β-factor. Figure 4.43 shows the time variations of these two types of modulation.

■ ■ ■

4-10 MODULATION OF LIGHT

Fiber optics communication is a rapidly growing field because of the very large channel capacity of small glass fibers used as conduits at optical frequencies. Modulation of light is used not only for transmitting information through fibers, but it is also the vehicle for space optical communication.

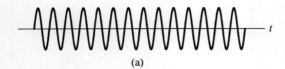
(a)

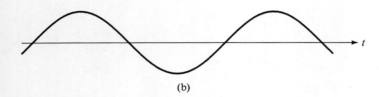
(b)

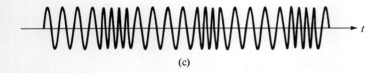
(c)

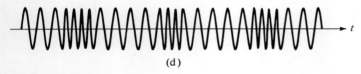
(d)

Figure 4.43
(a) Carrier signal. (b) Modulating signal. (c) Phase-modulated wave. (d) Frequency-modulated signal.

A typical simplified amplitude modulating scheme that involves an electro-optical crystal is shown in Figure 4.44a. As shown in the figure, a polarized incoming optical beam, when passing through the electro-optic crystal, is split into two components along the two optical axes. As a result, the field at the exit is elliptically polarized, the extent of elliptical polarization being dependent upon the voltage across the electro-optic crystal. The changing polarization results in intensity variations through an analyzer (polarizer) at the output.

It can be shown that the ratio of the output irradiance I_0 to the input irradiance I_{in} varies as

$$\frac{I_0}{I_{\text{in}}} = \sin^2\left[\frac{\pi}{2}\frac{v_m(t)}{V_\pi}\right] \quad (4.128)$$

where V_π is the voltage needed to produce total transmission at the modulator.

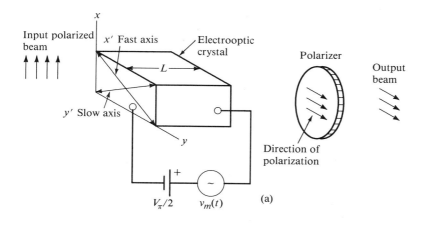

Figure 4.44
Amplitude modulating scheme at optical frequencies. (a) A simplified electro-optic modulator. (b) Input beam polarized in the x-direction. (c) Field components just inside the electro-optic crystal. (d) Field components and their resultant as they leave the crystal. (e) Amplitude of the field permitted to pass through the polarizer. [Reprinted, by permission, from Seely and Poularikas, *Electrical Engineering, Introduction and Concepts*, 568.]

If we set

$$\varphi = \frac{v_m(t)\pi}{V_\pi} = \frac{\pi}{2} + m \sin \omega_m t = \text{phase retardation} \quad (4.129)$$

where $\pi/2$ is the bias retardation and $v_m(t) = V_m \sin \omega_m t$, then (4.128) becomes

$$\frac{I_0}{I_{in}} = \sin^2\left(\frac{\pi}{4} + \frac{m}{2} \sin \omega_m t\right) = \frac{1}{2}[1 + \sin(m \sin \omega_m t)]$$

$$\doteq \frac{1}{2}(1 + m \sin \omega_m t) \quad (4.130)$$

for $m \ll 1$. This equation shows that the output intensity is identical with the

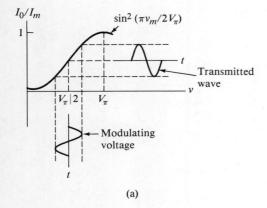

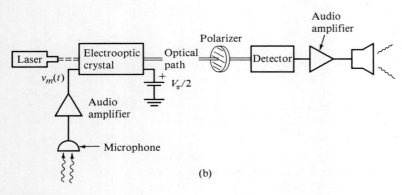

Figure 4.45
(a) Biasing characteristics of an electro-optic system. (b) Typical optical communication system. [Reprinted, by permission, from Seely and Poularikas, *Electrical Engineering, Introduction and Concepts*, 569.]

input but the signal has been amplitude modulated. Figure 4.45a shows the modulating process, and Figure 4.45b shows a typical communication system.

∎ ∎ ∎

REFERENCES

1. Başar, E. *Biophysical and Physiological Systems Analysis*. Reading, Mass.: Addison-Wesley, 1976.
2. Born, M., and E. Wolf. *Principles of Optics*. 3d ed. London: Pergamon Press, 1963.
3. Bracewell, R. *The Fourier Transform and its Applications*. New York: McGraw-Hill, 1965.
4. Gaskill, J. D. *Linear Systems, Fourier Transforms and Optics*. New York: Wiley, 1978.

5. Hecht, E., and A. Zajac. *Optics*. Reading, Mass.: Addison-Wesley, 1974.
6. Jahnke, E., and F. Emde. *Tables of Functions*. New York: Dover, 1945.
7. Lindström, L. H., and R. I. Magnusson. "Interpretation of Myoelectric Power Spectra: A Model and its Applications," *IEEE Proceedings* 65 (1977): 653.
8. Neff, H. P. Jr., *Continuous and Discrete Linear Systems*. New York: Harper and Row, 1984.
9. Papoulis, A. *The Fourier Integral and its Applications*. New York: McGraw-Hill, 1962.
10. Rabiner, L. R., and R. W. Schafer. *Digital Processing of Speech Signals*. Englewood Cliffs, N. J.: Prentice-Hall, 1978.
11. Seely, S., and A. D. Poularikas, *Electromagnetics—Classical and Modern Theory and Applications*. New York: Marcel Dekker, 1979.
12. Seely, S., *Electron Tube Circuits*, 2d ed. New York: McGraw-Hill, 1958.
13. Stark, H. *Applications of Optical Fourier Transforms*. New York: Academic Press, 1982.
14. Tufts, D. W., S. E. Levinson, and R. Rao. "Measuring Pitch and Formant Frequencies for a Speech Understanding System," *Proceedings of the 1976 IEEE International Conference on Acoustics, Speech and Signal Processing*, April 1976, 314.

PROBLEMS

4-1.1 Prove (4.6) and (4.7).

4-1.2 **a.** Obtain the spectrum functions for each of the two time functions shown in Figure P4-1.2a and b.
b. Obtain the spectrum function for the time function shown in Figure P4-1.2c directly from the Fourier integral. Compare the result so obtained by suitably combining the two spectra obtained in part (a).

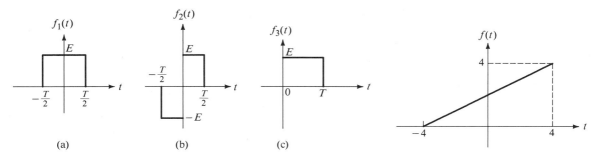

Figure P4-1.2

Figure P4-1.3

4-1.3 Consider the time function shown in Figure P4-1.3.
a. Find the even and odd functions for the given time function, $R(\omega)$ and $X(\omega)$.
b. Break the time function into two parts having the two types of symmetry, and again find $R(\omega)$ and $X(\omega)$. Check these results with those of part (a).

c. Write expressions for the given time function as the sum of two infinite integrals and as a single integral, with ω as the variable of integration. In each case write the integrals in the simplest possible form.

4-1.4 Give a graphical interpretation of the Fourier pair $f(t) = \exp[-t^2]$ and $F(\omega) = \sqrt{\pi} \exp[-\omega^2/4]$ as was done in Figure 4.2.

4-2.1 Verify (4.20), (4.21), (4.23), and (4.24).

4-2.2 If $\mathscr{F}\{f(t)\} = F(\omega)$ and a, b, and ω_0 are constants, prove the following identities:

a. $\mathscr{F}\left\{f\left(\dfrac{t}{a} + b\right)\right\} = |a|e^{jab\omega}F(a\omega)$
b. $\mathscr{F}\{f(t)\delta(t-a)\} = f(a)e^{-j\omega a}$

c. $\mathscr{F}\left\{f(t) * \delta\left(\dfrac{t}{a} - b\right)\right\} = |a|F(\omega)e^{-jab\omega}$
d. $\mathscr{F}\{e^{jat}f(bt)\} = \dfrac{1}{|b|} F\left(\dfrac{\omega - a}{b}\right)$

e. $\mathscr{F}\{f_1(t) * f_2(t) * f_3(t)\} = F_1(\omega)F_2(\omega)F_3(\omega)$
f. $\mathscr{F}^{-1}\{F_1(\omega) * F_2(\omega)\} = 2\pi[f_1(t)f_2(t)]$

g. $\mathscr{F}\{\mathscr{F}\{f(t)\}\} = f(-t)$
h. $\mathscr{F}\{f^*(t)\} = F^*(-\omega)$

i. $\displaystyle\int_{-\infty}^{\infty} f(t)h^*(t)\,dt = \dfrac{1}{2\pi} \int_{-\infty}^{\infty} F(\omega)H^*(\omega)\,d\omega$

4-2.3 Find the Fourier transforms of the following functions and sketch their amplitude and phase spectra.

a. $f(t) = p_2(t-1)$ b. $f(t) = p_1\left(\dfrac{t-2}{3}\right)$ c. $f(t) = 0.5 p_{0.5}(t) + 2\Lambda_1(t-2)$

d. $f(t) = 2\delta(t-1) + p_2(t+3)$ e. $f(t) = p_1(t) * p_2(t-2)$ f. $f(t) = p_2(t+2) * 2p_1(t-3)$

g. $f(t) = 2\,\mathrm{sinc}_2(t)\,\mathrm{sinc}(t)$ h. $f(t) = e^{j2t}p_1(t)$ i. $f(t) = e^{-|t|}\cos\omega_0 t$

j. $f(t) = e^{j3t}p_2(3t)$

4-2.4 Find m_0, $\bar{t}$, and σ^2 for the following functions:

a. $f(t) = p_2(t)$ b. $f(t) = \mathrm{sinc}_2(t)$ c. $f(t) = g_a(t) = e^{-at^2}$ d. $f(t) = \delta(t-2)$

4-2.5 Find the Fourier transforms of the following functions:

a. $\dfrac{1}{1+t^2}$ b. $\dfrac{\sin t}{t}$ c. $\delta\left(\dfrac{t-t_0}{a}\right)$ d. $\dfrac{ne^{-n^2t^2}}{\sqrt{\pi}}$

4-2.6 Based on (4.21a), explain the phenomenon of a recorded voice when it is played back (a) at a higher speed and (b) at a lower speed than the speed at which it was recorded.

4-2.7 Suppose that we represent the Fourier transform of a signal $f(t)$ in the form $F(\omega) = |F(\omega)|\exp[j\varphi(\omega)]$. We will perform operations on this signal so that $\varphi(\omega)$ is (a) substituted by $-\varphi(\omega)$ and (b) substituted by $\varphi(\omega) + a\omega$, where a is a positive constant. Find the new signals, and draw the conclusions implied by these phase functions.

4-2.8 The amplitude modulation process is specified by the equation

$v(t) = V_c[1 + mv_m(t)]\cos\omega_c t$

Find and sketch the Fourier transform of the AM signal, if $v_m(t) = p(t)$.

4-2.9 Find the Fourier transform of the signal $f(t)$ shown in Figure P4-2.9a. Draw a conclusion about the Fourier transform of the signal $f(t)$ shown in Figure P4-2.9b.

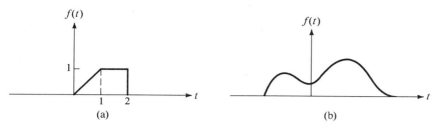

Figure P4-2.9

4-2.10 Compute the convolution of the signals $f(t) = \exp[-2t]u(t)$ and $h(t) = \exp[2t]u(-t)$ by finding their Fourier transforms $F(\omega)$ and $H(\omega)$ and then using the inverse transformation.

4-2.11 Find the transfer function of the systems shown in Figure P4-2.11 and the Fourier transforms of their outputs, if the input signal is $\exp[-2t]u(t)$.

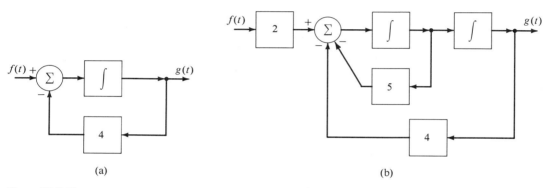

Figure P4-2.11

4-2.12 Find the Fourier transform of the correlation functions $r_{f,g}(t)$; $r_{g,f}(t)$. If $f(t)$ is the input to an LTI system whose impulse response is the real function $h(t)$ and $g(t)$ is the output, find expressions for $R_{f,g}(\omega)$ and $R_{g,g}(\omega)$ in terms of $R_{f,f}(\omega)$ and $H(\omega)$.

4-2.13 Use the Fourier transform method to find the transfer functions of the systems shown in Figure P4-2.13. Also, repeat by first finding the $h(t)$ function for each and then using the Fourier transform to find the transfer functions. Compare your results. The systems are relaxed at $t = 0$.

4-2.14 First find the autocorrelation function and then obtain its Fourier transform for each of the functions given. Verify your results using (4.41).
   a. $f(t) = p_2(t)$ **b.** $f(t) = e^{-t^2}$
   HINT: $\int_{-\infty}^{\infty} \exp[-p^2 x^2 \pm qx]\,dx = (\sqrt{\pi}/p)\exp(q^2/4p^2)$ $p > 0$

4-2.15 Prove the following properties for the cross-correlation function.
   a. $\mathscr{F}\{f(t) \star h(t)\} = F(\omega)H(-\omega)$ $f(t)$ and $h(t)$ are real functions
   b. $\mathscr{F}\{f(t)h^*(t)\} = (1/2\pi)F(\omega) \star H^*(\omega)$ $f(t)$ and $h(t)$ are complex functions

4-2.16 a. A function $f(t)$ and its approximating function $f_a(t)$ are shown in Figure P4-2.16a. Find the Fourier transforms of $f(t)$ and $f_a(t)$ and compare your results.
   b. Repeat part (a) for the approximating function shown in Figure P4-2.16b.

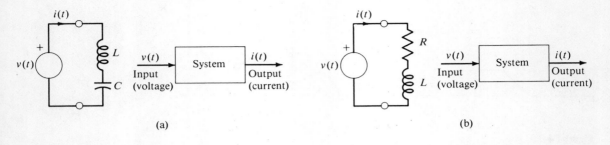

Figure P4-2.13

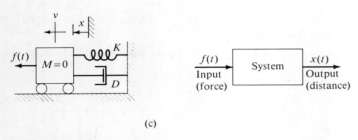

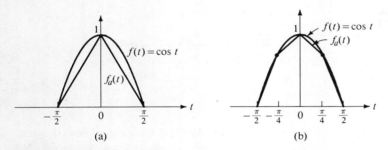

Figure P4-2.16

4-2.17 Find the Fourier spectrum of the signal shown in Figure P4-2.17a. Now deduce the Fourier transform of the signals shown in Figures P4-2.17b, c, and d using the spectrum for the signal in P4-2.17a together with other Fourier transform properties given in Table 4.1.

4-2.18 The impulse response of a system is given as $h(t) = p_1(t) - 0.5p_2(t)$. Find the Fourier transform of $h(t)$ and from its spectrum indicate what effect you think this system has on input signals.

4-3.1 Find the Fourier transforms of the following functions:
  a. $f(t) = 1 + e^{-|t|} \cos \omega_0 t$
  b. $f(t) = \sin \omega_0 t + \cos \omega_0 (t - t_0)$
  c. $f(t) = u(t) \cos \omega_0 t$
  d. $f(t) = e^{-t} u(t) \cos \omega_0 t$
  e. $f(t) = (1 + 2 \cos \omega_m t) \cos \omega_0 t$
  f. $f(t) = a_0/2 + \sum_{n=1}^{\infty} (a_n \cos n\omega_0 t + b_n \sin n\omega_0 t)$

4-3.2 Find the Fourier transforms of the functions shown in Figure P4-3.2 and plot their amplitude spectra.

4-3.3 Use the Poisson sum formula to prove that

$$\sum_{n=-\infty}^{\infty} e^{-|n|} = \sum_{n=-\infty}^{\infty} \frac{2}{1 + (2\pi n)^2}$$

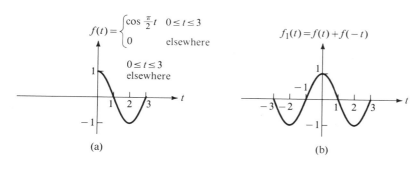

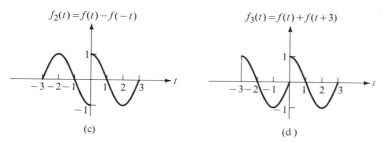

Figure P4-2.17

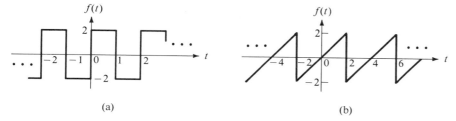

Figure P4-3.2

4-3.4 Show, for periodic functions (T = period), that (4.60) yields

$$\sum_{n=-\infty}^{\infty} \alpha(nT) e^{-jn\omega_0 T} = \frac{1}{T} \sum_{n=-\infty}^{\infty} F(n\omega_0)$$

4-4.1 Show that when we use a window function $\Lambda_{\omega_c}(\omega)$, a discontinuous function can be reproduced approximately without overshoots.

4-4.2 Find and plot the spectrum of the truncated signal $f_t(t) = [(2 \sin t)/t] p_1(t)$.

4-5.1 Verify (4.76), (4.77), and (4.78).

4-5.2 Four pinholes are located along the y_0-axis at the following distances from the x_0-axis: $-2a$, $-a$, a, $2a$. Suppose that this system of pinholes is illuminated with laser light (constant amplitude and zero phase). Find the light intensity variation along the y_1-axis.

4-6.1 Complete the development of the second part of Example 4.22.

4-6.2 Show that the combined contribution from the integrations along the two vertical edges shown in Figure 4.27 is zero.

4-7.1 Determine the complex envelope and the envelope for the function

$$f(t) = p_T\left(t - \frac{T}{2}\right)\cos \omega_0 t \qquad T \gg \frac{2\pi}{\omega_0}$$

4-7.2 Find the Hilbert transform of the pulse $f(t) = p_a(t - T)$ and sketch the result.

4-8.1 Sketch the signal $v(t) = V_c(1 + m \cos \omega_m t) \cos \omega_c t$ for:
  a. $m < 1$,   **b.** $m = 1$,   **c.** $m > 1$.

4-8.2 The equation of a sinusoidally modulated carrier wave is

$$f(t) = (1 - m \sin \omega_m t) \sin \omega_0 t$$

Consider the case where $m = 0.57$, $\omega_0 = 10^7$, $\omega_m = 10^4$.
  a. What is the period of $f(t)$?
  b. Completely specify the Fourier series for this function.

4-8.3 The voltage across a resistor that is connected in series with a diode (a nonlinear element) is given by the relation

$$v_0(t) = a_1 v_i(t) + a_2 v_i^2(t)$$

where the input voltage $v_i(t)$ across the combined circuit is small. If $v_i(t) = v_m(t) + \cos \omega_c t$, find the form of the AM signal. If $v_m = \cos \omega_m t$, find the frequency spectrum of $v_0(t)$.

4-8.4 An envelope detector of an AM signal is shown in Figure P4-8.4. If the RC time constant is large compared with $2\pi/\omega_c$, find the output voltage if $v_i(t) = [1 + mv_m(t)] \cos \omega_c t$.

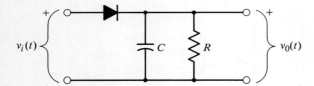

Figure P4-8.4

4-8.5 A periodic signal $f_p(t)$ shown in Figure P4-8.5 is modulated by a signal $v_m(t)$ whose bandwidth is ω_b. Find the Fourier spectrum of the modulated signal and the relationship between the period T of $f(t)$ and the bandwidth ω_b so that no overlapping of the Fourier spectra bands exists.

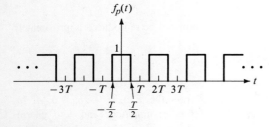

Figure P4-8.5

4-9.1 An FM signal is specified by

$$v(t) = 5 \cos[10^5 \pi t + 16 \sin(100 \pi t)]$$

Determine the following:
a. the carrier frequency.
b. the phase deviation factor β.
c. the peak frequency deviation ω_d.

4-9.2 Suppose that the deviation ratio β is very small compared to one; that is, $\beta \ll 1$. Find the frequency spectrum of a phase-modulated signal if $v_m(t) = \cos \omega_m t$. How does this result compare with an AM signal?

4-9.3 Find the output signal of a linear system with transfer function $H(\omega)$ if the input is an FM modulated signal, $v(t) = A \cos(\omega_c t + \beta \sin \omega_m t)$.

4-9.4 An FM signal is the input to the system shown in Figure P4-9.4. Find the signal at the points indicated on this same figure.

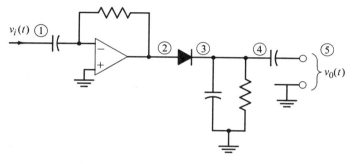

Figure P4-9.4

CHAPTER 5

THE RESPONSE AND APPLICATIONS OF LINEAR FILTERS

A. ELECTRICAL AND MECHANICAL SYSTEMS

The Fourier transform discussed in the foregoing chapter has important uses in the study of the input-output relationships of linear time-invariant (LTI) systems. It is also important in determining the frequency spectra of LTI systems. We have already found that the output of an LTI system is related to the input through a convolution integral. Moreover, we have shown in the previous chapter that this relationship becomes a simple algebraic one in the frequency domain: the frequency spectrum of the output is equal to the product of the spectrum of the input and the spectrum of the impulse response of the system. Consequently, we may regard the output spectrum in the most general terms as a filtered (or an influenced) version of the input.

Filters, which for our purposes are systems that influence a signal in a prescribed way, have many useful applications and are used in diverse fields—for example, filtering of mechanical vibrations, fluctuations of economic data, electrical signals, biological signals, and two-dimensional signals (pictures). Their construction is such that when filters are included in a given system, the detected output will have some desired characteristic. For example, when we purchase an audio amplifier (a hi-fi system), we expect that it will be equipped with both bass and treble controls so that we can modify to our liking the low- and high-frequency response of the program to which we are listening. What is actually provided in the amplifier is essentially two variable filters. In the case of AM communications (see Section 4-8), we may wish to transmit the upper sideband only since the complete signal information is redundantly contained in each band. This might require that we insert a high-pass filter in the system so that only those frequencies above the carrier frequency are passed. Filtering is used extensively to eliminate undesired low- or high-frequency noise. Our purpose in this chapter is to investigate the characteristics and applications of some typical filters. We will not discuss designing filters to meet prescribed specifications at this time, but some aspects of filter design are included in Chapter 11.

■ ■ ■

5-1 LINEAR TIME-INVARIANT SYSTEMS (FILTERS)

As discussed in the previous chapter, the Fourier transform of the output signal of a system $g(t)$ is the convolution of the input signal $f(t)$ and the impulse re-

THE RESPONSE AND APPLICATIONS OF LINEAR FILTERS

sponse of the system $h(t)$, and is given by the relation

$$G(\omega) = F(\omega) H(\omega) \tag{5.1}$$

In general, the system function or transfer function $H(\omega)$ will be of the form

$$H(\omega) \triangleq |H(\omega)|e^{j\theta_h(\omega)} \tag{5.2}$$

where $|H(\omega)|$ is the **amplitude transfer function** and $\theta_h(\omega)$ is the **phase transfer function.** Similarly, we can write the input signal spectrum function in the form

$$F(\omega) \triangleq |F(\omega)|e^{j\theta_f(\omega)} \tag{5.3}$$

where $|F(\omega)|$ is the amplitude spectrum of the input signal and $\theta_f(\omega)$ is its phase spectrum. Upon combining (5.3) and (5.2) with (5.1), we obtain the relation

$$G(\omega) \triangleq |G(\omega)|e^{j\theta_g(\omega)} = |H(\omega)||F(\omega)|e^{j[\theta_h(\omega) + \theta_f(\omega)]} \quad \text{(a)}$$
$$\text{where}$$
$$|G(\omega)| = |H(\omega)||F(\omega)| \quad \text{(b)} \tag{5.4}$$
$$\theta_g(\omega) = \theta_h(\omega) + \theta_f(\omega) \quad \text{(c)}$$

This shows that the output amplitude spectrum is equal to the product of the input amplitude spectrum and the spectrum of the amplitude transfer function. Also, the output phase spectrum is equal to the sum of the input phase spectrum and the transfer phase function. Equation (5.4) clearly shows that the output signal will be modified by the presence of the filter.

In the general case, as well as that for periodic input signals, the filter is **distortionless** when the form of the output is exactly identical with the input, except perhaps for a change in the amplitude and a possible time lag. Under these conditions, the output is specified as

$$g(t) = H_0 f(t - t_0) \tag{5.5}$$

where $f(t)$ is the input signal and H_0 is the constant amplitude response of the filter, which may be less than or equal to unity. It follows from this equation that

$$G(\omega) = H_0 e^{-j\omega t_0} F(\omega) \tag{5.6}$$

Hence the system function of a distortionless filter [see (5.1)] is

$$H(\omega) = H_0 e^{-j\omega t_0} \triangleq \text{Transfer function of distortionless filter} \tag{5.7}$$

A graphical representation of this function is shown in Figure 5.1.

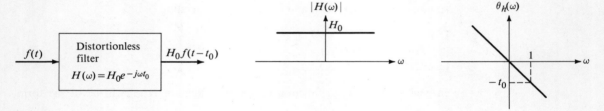

Figure 5.1
The transfer function of a distortionless filter.

If $|H(\omega)|$ is not constant, we say that the filter is **amplitude distorted,** and if $\theta_h(\omega)$ is a nonlinear function, the filter is **phase distorted.** A number of particular cases will warrant our attention.

■ ■ ■

5-2 FILTERS WITH VARYING AMPLITUDE AND LINEAR PHASE SHIFT

5-2.a Ideal Low-Pass Filter An ideal low-pass filter is one that has a transfer function of the form

$$H(\omega) = H_0 p_{\omega_0}(\omega) e^{-j\omega t_0} \triangleq \text{Ideal low-pass filter} \tag{5.8}$$

The characteristics of this filter are shown in Figure 5.2a. Clearly, the ideal low-pass filter is one that permits ideal transmission over a specified band of frequencies and completely excludes all other frequencies. Such ideal filters are not physically realizable, as discussed below, but the concept does help in an understanding of different types of physical filters whose response may be approximated by those of the ideal filters.

The impulse response of the ideal low-pass filter specified by (5.8) is found by taking the inverse Fourier transform of $H(\omega)$. This is

$$h(t) = \frac{1}{2\pi} \int_{-\infty}^{\infty} H_0 p_{\omega_0}(\omega) e^{-j\omega t_0} e^{j\omega t} \, d\omega = \frac{1}{2\pi} H_0 \int_{-\omega_0}^{\omega_0} e^{j\omega(t-t_0)} \, d\omega$$

or

$$h(t) = \frac{H_0}{\pi} \frac{\sin \omega_0(t - t_0)}{t - t_0} = \frac{H_0}{\pi} \text{sinc}_{\omega_0}(t - t_0) \tag{5.9}$$
$$\triangleq \text{Impulse response of ideal low-pass filter}$$

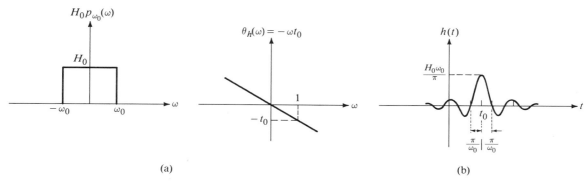

Figure 5.2
The impulse response characteristics of an ideal low-pass filter.

The form dictated by this expression is shown in Figure 5.2b. This figure shows that the impulse response is not identically zero for $t \leq 0$. This would suggest that when a delta source is applied at $t = 0$, the system anticipates the input and a signal occurs at the output of the system even before the source has been applied. Of course, this is not true for physical systems. This is proof that ideal filters are not physically realizable; that is, it is not possible to build an ideal filter with any combination of resistors, capacitors, and inductors.

EXAMPLE 5.1
Find the output waveform of an ideal low-pass filter if the input is a unit step function.

Solution: The Fourier transform of the unit step function was deduced in Example 4.16, and was found to be

$$U(\omega) \triangleq \mathscr{F}\{u(t)\} = \pi \delta(\omega) + \frac{1}{j\omega} \tag{5.10}$$

By an application of (5.1), we can write the output of the filter in the form

$$g(t) \triangleq \mathscr{F}^{-1}\{G(\omega)\} = \mathscr{F}^{-1}\{H(\omega)U(\omega)\}$$

$$= \frac{1}{2\pi} \int_{-\infty}^{\infty} H_0 p_{\omega_0}(\omega) e^{-j\omega t_0} \left[\pi \delta(\omega) + \frac{1}{j\omega}\right] e^{j\omega t} d\omega$$

$$g(t) = \frac{H_0}{2} \int_{-\omega_0}^{\omega_0} \delta(\omega) e^{j\omega(t-t_0)} d\omega + \frac{H_0}{2j\pi} \int_{-\omega_0}^{\omega_0} \frac{e^{j\omega(t-t_0)}}{\omega} d\omega \tag{5.11}$$

The first integral is equal to $H_0/2$, by the properties of the delta function. The integrand of the second integral can be expanded in Euler form to $\cos \omega(t - t_0)/\omega + j \sin \omega(t - t_0)/\omega$. The first factor of this expansion is an odd function; since the integration is symmetric around the origin, the integral will vanish.

Figure 5.3
The response of an ideal low-pass filter to a step function.

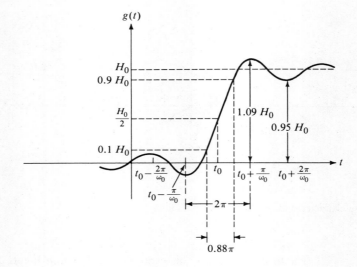

Thus (5.11) becomes

$$g(t) = \frac{H_0}{2} + \frac{H_0}{\pi} \int_0^{\omega_0} \frac{\sin \omega(t - t_0)}{\omega} d\omega$$

which can be written, with $x = \omega(t - t_0)$,

$$g(t) = \frac{H_0}{2} + \frac{H_0}{\pi} \int_0^{\omega_0(t-t_0)} \frac{\sin x}{x} dx = \frac{H_0}{2} + \frac{H_0}{\pi} Si(\omega_0(t - t_0)) \quad (5.12)$$

where $Si(\omega_0(t - t_0))$ is the sine integral, which was discussed in Section 4-4. The graphical representation of the output function $g(t)$ is shown in Figure 5.3. We observe the following for this response function:

- The output signal is shifted from 0 by t_0 in the time domain.
- The response is not identically zero for $t < 0$, as expected for nonrealizable filters.
- The response has a gradual rise, in contrast to the abrupt input rise.
- The **rise time** in the time interval from $t_0 - \pi/\omega_0$ to $t_0 + \pi/\omega_0$ is

$$t_r = \frac{2\pi}{\omega_0} = \frac{1}{f_0}$$

which is equal to the reciprocal of the cutoff frequency of the filter. (This is not the only definition of rise time used in electrical engineering literature. For example, the rise time defined for electronic circuits specifies the time for a signal to rise from $0.1H_0$ to $0.9H_0$ of its height.)

- The response is oscillatory with frequency equal to the cutoff frequency. ∎

EXAMPLE 5.2

Compare the response of the seismic instrument (filter) shown in Figure 5.4a with that of an ideal low-pass filter. The input is a delta force function $f(t) = \delta(t)$ and the output is the displacement x.

Solution: The circuit diagram representation for this system, following the development of Section 2-7, is shown in Figure 5.4b. Since the algebraic sum of the forces must be equal to zero, the equation governing the system is

$$M\frac{d^2x}{dt^2} + D\frac{dx}{dt} + Kx = \delta(t) \tag{5.13}$$

This equation is identical with (2.70), and after applying the same initial condition as in (2.77), the solution is

$$x(t) = \begin{cases} \dfrac{1}{M\sqrt{b^2 - a^2}} e^{-at} \sin\sqrt{b^2 - a^2}\, t & t > 0 \\ 0 & \text{otherwise} \end{cases} \quad \text{(a)} \tag{5.14}$$

where

$$2a = \frac{D}{M} \qquad b^2 = \frac{K}{M} \qquad b > a \qquad \text{(b)}$$

and all three constants K, M, and D are greater than zero.

We could have proceeded in the following manner: first finding the transfer function of the system and then taking the inverse Fourier transform of the resulting expression. However, this approach results in an expression requiring the use of complex function theory in its evaluation.

If we introduce a sinusoidal excitation function $\exp[j\omega t]$ into (5.13) and include the equivalent quantities given by (5.14b), we obtain the following relationship* for the distance traveled by the seismic stylus versus frequency

$$\begin{aligned}
X(\omega) &= \frac{1}{M} \frac{1}{b^2 - \omega^2 + 2ja\omega} \\
&= \frac{1}{K} \frac{1}{\left[\left(1 - \left(\dfrac{\omega}{b}\right)^2\right)^2 + 4\left(\dfrac{a}{b}\right)^2\left(\dfrac{\omega}{b}\right)^2\right]^{1/2}} \\
&\quad \times \exp\left[-j\tan^{-1}\left(2\left(\dfrac{a}{b}\right)\frac{\dfrac{\omega}{b}}{1 - \left(\dfrac{\omega}{b}\right)^2}\right)\right]
\end{aligned} \tag{5.15}$$

---

* We have not used the term transfer function or system function because we have defined these as the ratio of the output to the input when both are written in terms of through- and across-variables. Here we are concerned with displacement that is the integral of velocity, the nominal across-variable in mechanical systems.

This relationship has been plotted for the specific values $K = 1$ and $(a/b) = 0.5$ in Figure 5.4c. Remember that we are restricted in the values of $\tan^{-1}(\theta)$ from $-\pi/2$ to $\pi/2$; hence at the discontinuity we must subtract π, which is equal to the jump across the point of discontinuity.

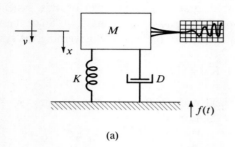

(a)

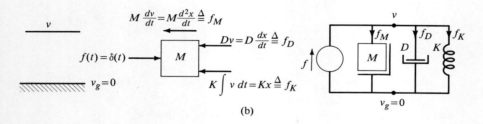

(b)

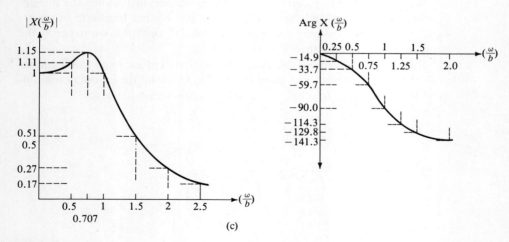

(c)

Figure 5.4
Characteristics and response of a physically realizable low-pass filter. (a) Physical system. (b) Circuit representation of the system. (c) Frequency response characteristics. (d) Time response. (e) Typical average evoked potential responses of the brain cells of a cat when stimulated (a tool for neurologists and psychologists studying brain electrical activity.) (From E. Başar, *Biophysical and Physiological Systems Analysis*, © 1976. Addison-Wesley, Reading, MA. Reprinted with permission.)

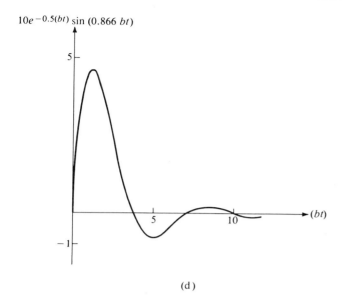

(d)

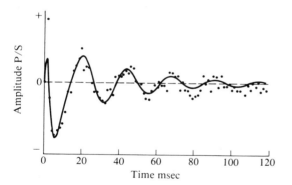
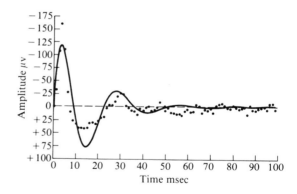
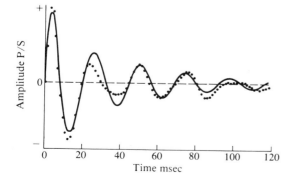
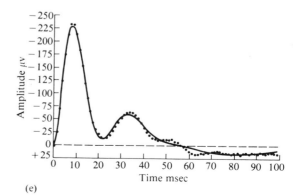

(e)

The response function

$$10e^{(-a/b)(bt)} \sin\left(\sqrt{1-\left(\frac{a}{b}\right)^2}\, bt\right) \tag{5.16}$$

has been graphed in Figure 5.4d. It is interesting to compare Figure 5.3 with the response of this physically realizable filter; we observe some similarities and some differences. In both cases the response reaches a peak that increases with the bandwidth and is oscillatory with the frequency increasing as the bandwidth increases. They differ in the form of frequency cutoff curves; the ideal filter has an abrupt change and also possesses the anticipatory response, and these do not exist for the real filter.

Figure 5.4e shows the response characteristics from a particular area of a cat's brain when the optic nerve is excited by light pulses. Here the input is light and the output is the evoked potential. The system comprises an eye-brain combination. ∎

EXAMPLE 5.3
Assume an ideal amplitude filter having the transfer function of the form ($t_0 = 0$)

$$H(\omega) = p_{\omega_0}(\omega) \tag{5.17}$$

If the input function is the periodic function shown in Figure 5.5a, find the output of the filter for $\omega_0 = 4$ and $\omega_0 = 1$.

Solution: Note that the function $f(t)$ can be written as the convolution of the triangular function $\Lambda_{0.5}(t)$ and the $\text{comb}_2(t)$ function. That is

$$f(t) = \Lambda_{0.5}(t) * \text{comb}_2(t) \tag{5.18}$$

Figure 5.5
Effect of low-pass filter on a periodic input function.

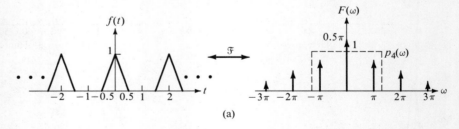

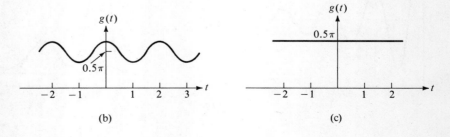

THE RESPONSE AND APPLICATIONS OF LINEAR FILTERS

This has the spectrum function [see Table 4.2 and also, (4.58)]

$$F(\omega) = 8\pi\, SINC_{0.25}^2(\omega)\, COMB_\pi(\omega) \tag{5.19}$$

The output spectrum has the form

$$G(\omega) = 8\pi p_{\omega_0}(\omega)\, SINC_{0.25}^2(\omega)\, COMB_\pi(\omega) \tag{5.20}$$

For $\omega_0 = 4$, as shown in Figure 5.5a,

$$G(\omega) = 8\pi p_4(\omega)\, SINC_{0.25}^2(\omega)\, COMB_\pi(\omega)$$
$$= 0.5\pi\, \delta(\omega) + 0.4\pi\, \delta(\omega - \pi) + 0.4\pi\, \delta(\omega + \pi) \tag{5.21}$$

The inverse Fourier transform of this function is

$$g(t) = 0.5\pi + 0.4 \cos \pi t \tag{5.22}$$

This is shown in Figure 5.5b. For the case $\omega_0 = 1$, only the first term $0.5\pi\, \delta(\omega)$ will be transmitted, and this results in a dc term in the time domain, as shown in Figure 5.5c. ■

EXAMPLE 5.4

Find the transfer function $V_0(\omega)/V_i(\omega) = H(\omega)$ for the relaxed system shown in Figure 5.6a. Plot $|H(\omega)|$ versus frequency on both a linear and a logarithmic scale. The value $20 \log |H(\omega)|$ versus $\log \omega$ is measured in **decibels** (dB) and yields what is known as a **Bode plot**.

Solution: The equation that describes the system is

$$L\frac{di(t)}{dt} + \frac{1}{C} \int i(t)\, dt + Ri(t) = v_i(t) \tag{5.23}$$

The Fourier transform of both sides of this equation yields

$$j\omega L I(\omega) + \frac{1}{j\omega C} I(\omega) + RI(\omega) = V_i(\omega)$$

from which

$$\frac{I(\omega)}{V_i(\omega)} = \frac{j\omega C}{LC(j\omega)^2 + j\omega RC + 1} \tag{5.24}$$

The voltage across the capacitor, which is also the output voltage, is given by

$$v_0(t) = \frac{1}{C} \int i(t)\, dt$$

which is, after being Fourier transformed,

$$V_0(\omega) = \frac{1}{j\omega C} I(\omega) \tag{5.25}$$

Combining this result with (5.24), we have

Figure 5.6
Linear and Bode plots of the transfer function $|H(\omega)|$ and linear plots of the phase function Arg $H(\omega)$.

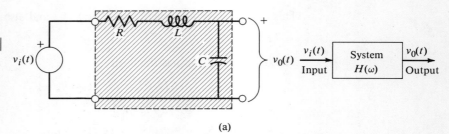

(a)

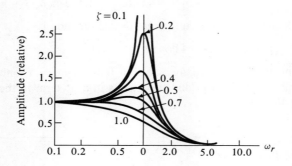

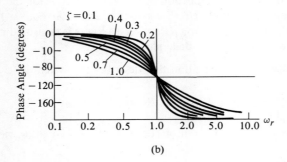

(b)

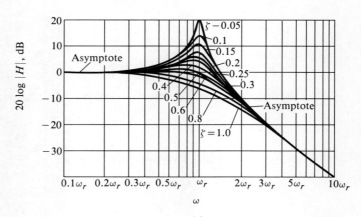

(c)

THE RESPONSE AND APPLICATIONS OF LINEAR FILTERS

$$\frac{\left(\frac{I(\omega)}{j\omega C}\right)}{V_i(\omega)} = \frac{V_0(\omega)}{V_i(\omega)} \triangleq H(\omega) = \frac{1}{LC(j\omega)^2 + RC(j\omega) + 1}$$

$$= \frac{\frac{1}{LC}}{(j\omega)^2 + 2\left(\frac{R}{2L}\right)(j\omega) + \frac{1}{LC}} = \frac{\omega_r^2}{(j\omega)^2 + j2\zeta\omega_r\omega + \omega_r^2} \quad (5.26)$$

where $\omega_r = 1/\sqrt{LC}$ and $\zeta = (R/2L)/\omega_r$.

The amplitude and phase plots of $H(\omega)$ are shown in Figure 5.6b. To plot the magnitude of $H(\omega)$ on a logarithmic scale, we first write (5.26) in the form

$$H(\omega) = \frac{1}{1 + j2\zeta\left(\frac{\omega}{\omega_r}\right) - \left(\frac{\omega}{\omega_r}\right)^2} \quad (5.27)$$

Therefore

$$20 \log |H(\omega)| = 20 \log 1 - 20 \log \left|1 + j2\zeta\left(\frac{\omega}{\omega_r}\right) - \left(\frac{\omega}{\omega_r}\right)^2\right|$$

$$= -20 \log \left|1 + j2\zeta\left(\frac{\omega}{\omega_r}\right) - \left(\frac{\omega}{\omega_r}\right)^2\right| \quad (5.28)$$

For $\omega \ll \omega_r$, this formula becomes

$$20 \log |H(\omega)| \doteq -20 \log 1 = 0 \quad (5.29)$$

For $\omega \gg \omega_r$, (5.28) becomes

$$20 \log |H(\omega)| \doteq -20 \log \left|-\left(\frac{\omega}{\omega_r}\right)^2\right| = -40 \log \left(\frac{\omega}{\omega_r}\right)$$

$$= -40 \log \omega + 40 \log \omega_r \quad (5.30)$$

Equations (5.29) and (5.30) indicate that the asymptotes are zero for $\omega \ll \omega_r$ and $-40 \log \omega + 40 \log \omega_r$ for $\omega \gg \omega_r$. The slope of the second line plotted on a log scale is a straight line with a slope of -40 dB/decade (or -12 dB/octave). The Bode plot of $|H(\omega)|$ with its asymptotes is shown in Figure 5.6c. Observe that the equations describing the asymptotes are independent of ζ. However, the exact value is

$$20 \log |H(\omega)| = -20 \log \left[\left[1 - \left(\frac{\omega}{\omega_r}\right)^2\right]^2 + 4\zeta^2\left(\frac{\omega}{\omega_r}\right)^2\right]^{1/2}$$

$$= -10 \log \left[\left[1 - \left(\frac{\omega}{\omega_r}\right)^2\right]^2 + 4\zeta^2\left(\frac{\omega}{\omega_r}\right)^2\right] \quad (5.31)$$

which is a function of ζ also. The effect of varying ζ is shown in Figures 5.6b and c.

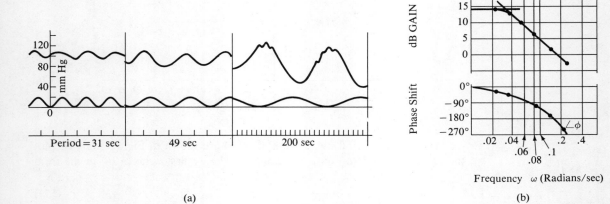

Figure 5.7
(a) Output response of systemic arterial pressure (SAP) (upper curve) of the animal's ischemic reflex to three sinusoidal inputs (lower curve) of different frequencies. (b) Bode diagram of the ischemic reflex system of the animal. (Reprinted, by permission, from Sagawa, Taylor, and Guyton, "Dynamic Performance and Stability of the Cerebral Ischemic Pressor Response," 1164.)

Considering this circuit as a filter, we see that it is not an ideal one, but for $\zeta \doteq 0.5$ (see Figure 5.6b), it approximates one in the range $0 \leq \omega/\omega_r < 1$. ∎

It is interesting that in animals severe ischemia of the brain produces a pressor response that in turn increases the systemic blood pressure to compensate for the decreased cerebral blood flow. Figure 5.7a shows three output responses of the systemic arterial pressure (upper curves) to sinusoidal inputs (lower curves) of cerebral perfusion pressure. Figure 5.7b shows the Bode diagram for one experimental animal.

5-2.b Ideal High-Pass Filter The frequency characteristic of an ideal high-pass filter is given by

$$H(\omega) = [H_0 - H_0 p_{\omega_0}(\omega)]e^{-j\omega t_0} \tag{5.32}$$

The corresponding impulse response function, which is obtained by taking the inverse Fourier transform of this equation, is (Table 4.2)

$$\begin{aligned} h(t) &= H_0\,\delta(t - t_0) - \frac{H_0}{\pi}\frac{\sin \omega_0(t - t_0)}{t - t_0} \\ &= H_0\left[\delta(t - t_0) - \frac{\omega_0}{\pi}\operatorname{sinc}_{\omega_0}(t - t_0)\right] \end{aligned} \tag{5.33}$$

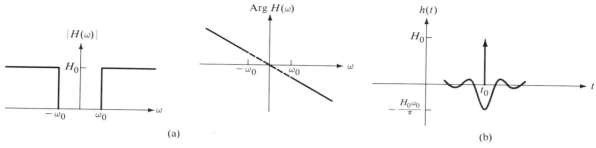

Figure 5.8
Frequency characteristics and impulse response of a high-pass filter.

EXAMPLE 5.5
What is the output of a high-pass filter, if the input is the function $\delta(t - \tau)$, where τ is a constant?

Solution: Since $\mathscr{F}\{\delta(t - \tau)\} = e^{-j\omega\tau}$, the output is given by

$$g(t) = \mathscr{F}^{-1}\{F(\omega)H(\omega)\}$$

$$= \frac{H_0}{2\pi} \int_{-\infty}^{\infty} e^{j\omega(t - (t_0 + \tau))} d\omega - \frac{H_0}{2\pi} \int_{-\omega_0}^{\omega_0} e^{j\omega(t - (t_0 + \tau))} d\omega$$

$$= H_0 \delta[t - (t_0 + \tau)] - \frac{H_0}{\pi} \frac{\sin \omega_0[t - (t_0 + \tau)]}{t - (t_0 + \tau)}$$

This shows that the impulse response $h(t)$ has been shifted by τ, an expected result. ∎

■ ■ ■

5-3 FILTERS WITH CONSTANT AMPLITUDE AND VARYING PHASE

Systems in which the phase of each sinusoidal component of the input changes in a predetermined manner are known as **phase filters**. The general form of the frequency characteristic of such a filter is

$$H(\omega) = H_0 e^{j\theta_h(\omega)} \tag{5.34}$$

This equation indicates that all frequencies of the signal will pass, but each frequency is shifted by an amount dictated by $\theta_h(\omega)$. Phase filters are very important

CHAPTER 5

in optics and have been used successfully in microscopes for viewing transparent objects (see Section 5-6g).

The following examples clarify some of the basic features of phase filters.

EXAMPLE 5.6
Find the output waveform of the phase filter having a frequency characteristic given by

$$H(\omega) = H_0 e^{-j\omega t_0} \qquad t_0 = \text{constant} \tag{5.35}$$

Solution: The impulse response of this function is

$$h(t) \triangleq \mathscr{F}^{-1}\{H(\omega)\} = \frac{1}{2\pi} \int_{-\infty}^{\infty} H_0 e^{j\omega(t-t_0)} d\omega = H_0 \delta(t - t_0) \tag{5.36}$$

The output is then given by

$$g(t) = f(t) * h(t) = H_0 f(t - t_0) \tag{5.37}$$

From this see that we recapture the input function $f(t)$ without any distortion, although there is a time delay by an amount t_0 and amplitude change by H_0. This is, of course, the distortionless filter discussed in Section 5-1. ∎

EXAMPLE 5.7
Determine the unit step function response of a phase filter that has the phase spectral characteristics shown in Figure 5.9a.

Solution: A careful consideration of the characteristic given shows that we can replace the given filter by two filters in parallel, one being a low-pass filter and the second being a high-pass filter, as shown in Figure 5.9b. That is, the transfer function of the specified filter is written

$$\begin{aligned} H(\omega) &= H_{LP}(\omega) + H_{HP}(\omega) \\ &= H_0 p_{\omega_0}(\omega) e^{-j(\pi/\omega_0)\omega} + [H_0 - H_0 p_{\omega_0}(\omega)] e^{-j\pi} \end{aligned} \tag{5.38}$$

From this, the impulse response of the filter is

$$h(t) \triangleq \mathscr{F}^{-1}\{H(\omega)\}$$

$$= \frac{H_0}{\pi} \frac{\sin\left[\omega_0\left(t - \frac{\pi}{\omega_0}\right)\right]}{t - \frac{\pi}{\omega_0}} - H_0 \delta(t) + \frac{H_0}{\pi} \frac{\sin \omega_0 t}{t} \tag{5.39}$$

The output of this filter to a step function excitation is, by the convolution integral,

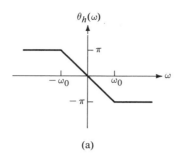

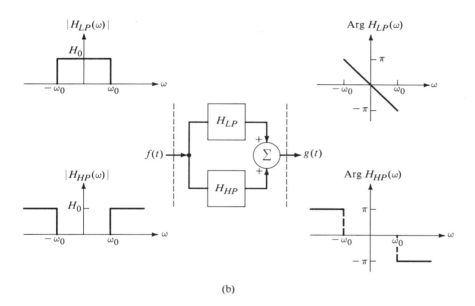

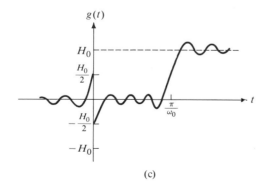

Figure 5.9
The unit pulse response of a phase filter.

CHAPTER 5

$$g(t) = u(t) * h(t) = \frac{H_0}{\pi} \int_{-\infty}^{\infty} u(t-\tau) \frac{\sin\left[\omega_0\left(\tau - \frac{\pi}{\omega_0}\right)\right]}{\tau - \frac{\pi}{\omega_0}} d\tau$$

$$- H_0 \int_{-\infty}^{\infty} u(t-\tau)\delta(\tau) d\tau + \frac{H_0}{\pi} \int_{-\infty}^{\infty} u(t-\tau) \frac{\sin \omega_0 \tau}{\tau} d\tau$$

$$= \frac{H_0}{\pi} \int_{-\infty}^{t} \frac{\sin\left[\omega_0\left(\tau - \frac{\pi}{\omega_0}\right)\right]}{\tau - \frac{\pi}{\omega_0}} d\tau - H_0 u(t) + \frac{H_0}{\pi} \int_{-\infty}^{t} \frac{\sin \omega_0 \tau}{\tau} d\tau$$

Write this

$$g(t) = \frac{H_0}{\pi} \int_{-\infty}^{0} \frac{\sin\left[\omega_0\left(\tau - \frac{\pi}{\omega_0}\right)\right]}{\tau - \frac{\pi}{\omega_0}} d\tau + \frac{H_0}{\pi} \int_{0}^{t} \frac{\sin\left[\omega_0\left(\tau - \frac{\pi}{\omega_0}\right)\right]}{\tau - \frac{\pi}{\omega_0}} d\tau$$

$$- H_0 u(t) + \frac{H_0}{\pi} \int_{-\infty}^{0} \frac{\sin \omega_0 \tau}{\tau} d\tau + \frac{H_0}{\pi} \int_{0}^{t} \frac{\sin \omega_0 \tau}{\tau} d\tau$$

$$= \left[\frac{H_0}{2} + \frac{H_0}{\pi} \operatorname{Si}\left[\omega_0\left(t - \frac{\pi}{\omega_0}\right)\right]\right] - H_0 u(t) + \left[\frac{H_0}{2} + \frac{H_0}{\pi} \operatorname{Si}(\omega_0 t)\right]$$

(5.40)

This expression is plotted in Figure 5.9c. ∎

EXAMPLE 5.8
Find the response of a phase filter to an input $f(t)$ that has a phase of the form

$$\theta_h(\omega) = -\epsilon \sin\left(\frac{\pi}{\omega_0}\omega\right) \qquad \epsilon \ll 1 \tag{5.41}$$

Solution: From the definition of a phase filter given by (5.34), we can write

$$H(\omega) = H_0 e^{-j\epsilon \sin((\pi/\omega_0)\omega)} \doteq H_0\left[1 - j\epsilon \sin\left(\frac{\pi}{\omega_0}\omega\right)\right]$$

$$= H_0 - \frac{\epsilon H_0}{2}(e^{j(\pi/\omega_0)\omega} - e^{-j(\pi/\omega_0)\omega}) \tag{5.42}$$

The impulse response of this filter, which is given by the inverse Fourier transform of (5.42), is

$$h(t) = H_0 \delta(t) - \frac{\epsilon H_0}{2}\left[\delta\left(t + \frac{\pi}{\omega_0}\right) - \delta\left(t - \frac{\pi}{\omega_0}\right)\right]$$

THE RESPONSE AND APPLICATIONS OF LINEAR FILTERS 259

The resulting output to an applied signal $f(t)$ is

$$g(t) = f(t) * h(t) = H_0 f(t) - 0.5\epsilon H_0 \left[f\left(t + \frac{\pi}{\omega_0}\right) - f\left(t - \frac{\pi}{\omega_0}\right) \right] \quad (5.43)$$

∎

■ ■ ■

5-4 SYMMETRICAL BANDPASS FILTERS WITH LINEAR PHASE

The spectral characteristic of an ideal bandpass filter is shown in Figure 5.10. We write $H(\omega)$ for this filter in terms of the two sections $H_1(\omega)$ and $H_2(\omega)$ symmetrically displaced about $\omega = 0$. Because of this symmetric distribution, the transfer functions of the two sections obey the conditions

$$H_1(\omega + \omega_c) = H_1^*(-\omega + \omega_c) \quad \text{(a)}$$
$$H_2(\omega - \omega_c) = H_2^*(-\omega - \omega_c) \quad \text{(b)} \quad (5.44)$$

where ω_c is the center frequency of the filter passband. Further, we write for the sectional transfer functions

$$H_1(\omega) = H_{01} e^{j\theta_1(\omega)} \quad \text{(a)}$$
$$H_2(\omega) = H_{02} e^{j\theta_2(\omega)} \quad \text{(b)} \quad (5.45)$$

In light of (5.44), we write the following equalities:

$$H_{01}(\omega + \omega_c) = H_{01}(-\omega + \omega_c); \quad \theta_1(\omega + \omega_c) = -\theta_1(-\omega + \omega_c) \quad \text{(a)}$$
$$H_{02}(\omega - \omega_c) = H_{02}(-\omega - \omega_c); \quad \theta_2(\omega - \omega_c) = -\theta_2(-\omega - \omega_c) \quad \text{(b)} \quad (5.46)$$

These equations indicate that the amplitude spectra are even functions and that the phase spectra are odd functions.

Observe that if we shift $H_1(\omega)$ or $H_2(\omega)$ to be centered at the origin, the result is a low-pass filter, as shown by the dotted lines in Figure 5.10a. This shows that we can write $H_1(\omega)$ and $H_2(\omega)$ as shifted versions of the low-pass filter

$$H_1(\omega) = H_{LP}(\omega - \omega_c) \quad \text{(a)}$$
$$H_2(\omega) = H_{LP}(\omega + \omega_c) \quad \text{(b)} \quad (5.47)$$

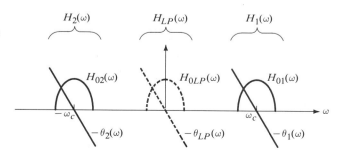

Figure 5.10
The spectral characteristics of a passband filter.

Figure 5.11
The impulse response of a bandpass filter.

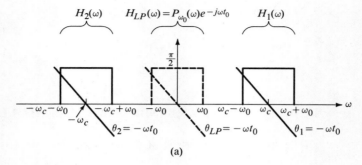

(a)

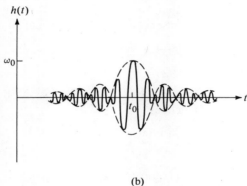

(b)

Taking the inverse Fourier transform of these terms, with due account of the shifting property of Fourier transforms, we obtain

$$\mathscr{F}^{-1}\{H_1(\omega)\} = h_{LP}(t)e^{j\omega_c t} \quad \text{(a)}$$
$$\mathscr{F}^{-1}\{H_2(\omega)\} = h_{LP}(t)e^{-j\omega_c t} \quad \text{(b)}$$
(5.48)

From the fact that the filter system transfer function is given by

$$H(\omega) = H_1(\omega) + H_2(\omega) \tag{5.49}$$

the impulse response is

$$\boxed{\begin{aligned} h(t) &= \mathscr{F}^{-1}\{H(\omega)\} = \mathscr{F}^{-1}\{H_1(\omega)\} + \mathscr{F}^{-1}\{H_2(\omega)\} \\ &= 2h_{LP}(t)\cos\omega_c t \end{aligned}} \tag{5.50}$$

This expression is that of an amplitude modulated signal. In communication theory, as we have already indicated, ω_c is known as the carrier frequency and, in this case, is equal to the center frequency of the bandpass filter.

EXAMPLE 5.9
Determine the impulse response of the bandpass filter shown in Figure 5.11a.

THE RESPONSE AND APPLICATIONS OF LINEAR FILTERS   261

Solution: The desired result is given by (5.50) together with (5.9), the impulse characteristic of the low-pass filter. The result is $\left(H_0 = \dfrac{\pi}{2}\right)$

$$h(t) = 2h_{LP}(t)\cos\omega_c t = \frac{\sin[\omega_0(t - t_0)]}{t - t_0}\cos\omega_c t \qquad (5.51)$$

In general $\omega_0 \ll \omega_c$, and the function $h(t)$ will have the form shown in Figure 5.11b. ∎

■ ■ ■

†B. OPTICAL SYSTEMS

We make certain observations concerning optical waves. Optical waves cannot be accurately characterized by monochromatic harmonic functions because such waves have finite line-width. Thus an optical wave comprises wave packets of slightly different frequencies, and this causes the phase to be a complicated function of frequency. If the phase of each component is proportional to the frequency, the waves are **coherent,** and coherent monochromatic waves can be treated as harmonic waves. There are many sources of waves that closely approximate coherent waves, but not all are optical waves. These include masers, microwave sources derived from high-Q tubes, such as magnetrons, klystrons, and traveling wave tubes, and ultrahigh frequency and radio frequency sources produced with high-Q tank circuits.

Waves with varying or random phase are called **incoherent.** Ordinary light from an incandescent source is a good example of incoherent waves. A coherent wave scattered from a random surface will become incoherent.

In the material to follow, we study general aspects of optical systems of simple types in order to establish methods of analysis. While the details of optical systems differ markedly from electrical systems analysis, the general results obtained with optical signal processors parallel those obtained with electrical signal processors. Often, therefore, optical systems provide a convenient means for performing signal processing. A number of techniques will be studied involving both coherent and incoherent waves, which are employed in present-day signal and image processors. The world of optics is a broad and varied one.

■ ■ ■

5-5 OPTICAL SYSTEM FUNCTIONS

The formation of optical images, which can be considered two-dimensional signals, can easily be interpreted if we borrow concepts from circuit and signal theory. As with many electronic or other systems, optical systems possess **linearity, invariance (space), causality,** and **stability.** The causality condition, in general, can be relaxed in optics since the light distribution exists on both sides of the coordinate axis. The invariance property does not strictly apply in optics unless additional assumptions are made. However, if we consider a working area very close to the optic axis of the system, we can freely invoke the principle of invariance. Since we shall limit our study to linear systems, we will accept the linearity principle as valid here. As far as stability is concerned, optical systems possess this property to a high degree; no further assumptions are needed.

We represent the operation of a linear optical system symbolically by the operator $\mathcal{O}\{\cdot\}$; hence the one-dimensional image (output) is related to the object (input) by the expression

$$\Phi_i(y_1) = \mathcal{O}\{\Phi_0(y_0)\} \tag{5.52}$$

where $\Phi_0(y_0)$ is the amplitude of the wavefront in the object plane and $\Phi_i(y_1)$ is the amplitude of the wavefront on the image plane. Observe specifically that the operator $\mathcal{O}$ operates on the coordinate y_0. Since we can represent any function with the help of the Dirac delta function (see Section 2-3), (5.52) can be written

$$\Phi_i(y_1) = \mathcal{O}\left\{\int_{-\infty}^{\infty} \Phi_0(y_0')\delta(y_0 - y_0')\,dy_0'\right\}$$
$$= \int_{-\infty}^{\infty} \Phi_0(y_0')\mathcal{O}\{\delta(y_0 - y_0')\}\,dy_0' \tag{5.53}$$

The function

$$\mathcal{O}\{\delta(y_0' - y_0)\} = h(y_0', y_1) \tag{5.54}$$

is the response of the optical system to a point source located at y_0', as shown in Figure 5.12. Since we assume space invariance for optical systems, the impulse response of the optical system, known as the **spread function,** is written in the familiar form $h(y_1 - y_0)$; it describes the light (field) distribution at y_1 due to a point source at y_0. That is, point sources are described mathematically by delta functions.

Introduce the invariant form of the spread function into (5.53) to obtain

$$\Phi_i(y_1) = \int_{-\infty}^{\infty} \Phi_0(y_0)h(y_1 - y_0)\,dy_0 = \Phi_0(y_0) * h(y_0) \tag{5.55}$$

where the dummy variable y_0' was changed to y_0. Equation (5.55) is important because the output (image) of an optical system (like that of an electrical system) is the convolution of its input (object) and its spread function. However, any detector of optical fields responds to the field intensity (irradiance), which is the square of the field amplitude; hence we are interested in the quantity

$$I(y_1) = \Phi_i(y_1)\Phi_i^*(y_1)$$
$$= \iint_{-\infty}^{\infty} \Phi_0(y_0)\Phi_0^*(y_0')h(y_1 - y_0)h^*(y_1 - y_0')\,dy_0\,dy_0' \tag{5.56}$$

THE RESPONSE AND APPLICATIONS OF LINEAR FILTERS

Figure 5.12
The impulse response of an optical system.

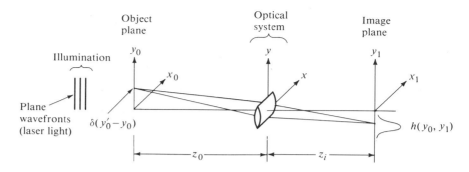

We wish to examine the form of this expression for both incoherent and coherent light sources. To simplify matters we assume that the source radiates a monochromatic wave with these properties: (a) the amplitude at any point P is constant and its phase varies linearly with frequency (coherent case); (b) the amplitude and phase at each point vary irregularly (incoherent case).

5-5.a Incoherent Case An object will illuminate incoherently if each of its points radiates independently of all other points. For the incoherent case, we must multiply the integrand of (5.56) by the delta function $\delta(y_0 - y_0')$ to guarantee that the illumination of each point is independent of all other points, no matter how closely they might be packed. Upon introducing this function into (5.56), the result is

$$I(y_1) = \iint_{-\infty}^{\infty} \delta(y_0 - y_0') h(y_1 - y_0) h^*(y_1 - y_0') \Phi_0(y_0) \Phi_0^*(y_0') \, dy_0 \, dy_0'$$

which is

$$I(y_1) = \int_{-\infty}^{\infty} |h(y_1 - y_0)|^2 |\Phi_0(y_0)|^2 \, dy_0 = |h(y_1)|^2 * |\Phi_0(y_1)|^2 \qquad (5.57)$$

This expression shows that for the **incoherent case**, the optical system is **linear in irradiance**.

The Fourier transform of (5.57) is written [see (4.26) and (4.78)]

$$I(\omega_y) = H_{in}(\omega_y) \Phi_{0in}(\omega_y) \qquad \text{(a)} \qquad (5.58)$$

where

$$I(\omega_y) = \int_{-\infty}^{\infty} I(y_1) e^{-j\omega_y y_1} \, dy_1 \qquad \text{(b)}$$

$$H_{in}(\omega_y) = \int_{-\infty}^{\infty} h(y_1) h^*(y_1) e^{-j\omega_y y_1} \, dy_1 \qquad \text{(c)}$$

$$\Phi_{0in}(\omega_y) = \int_{-\infty}^{\infty} \Phi_0(y_0) \Phi_0^*(y_0) e^{-j\omega_y y_0} \, dy_0 \qquad \text{(d)}$$

and where

$$\omega_y = \frac{2\pi y}{\lambda f_l} \tag{e}$$

with f_l being the focal length of the converging lens and λ the wavelength of the optical field. The system function $H(\omega)$ in general systems theory is parallel in the field of optics and optical engineering to

$$H_{in}(\omega_y) = \text{Optical Transfer Function, OTF}$$
$$|H_{in}(\omega_y)| = \text{Modulation Transfer Function, MTF}$$

Employing autocorrelation properties (see Table 4.1), the OTF is written in the form

$$H_{in}(\omega_y) = \int_{-\infty}^{\infty} H(\omega_y')H^*(\omega_y' + \omega_y)\,d\omega_y' \triangleq H(\omega_y) \star H^*(\omega_y) \tag{a}$$
(5.59)

where

$$H(\omega_y) = \mathscr{F}\{h(y)\} \tag{b}$$

so that

$$I(\omega_y) = [H(\omega_y) \star H^*(\omega_y)][\Phi_0(\omega_y) \star \Phi_0^*(\omega_y)] \tag{c}$$

where

$$\Phi_0(\omega_y) = \int_{-\infty}^{\infty} \Phi_0(y_0) e^{-j\omega_y y_0}\,dy_0 \tag{d}$$

A constant $(1/2\pi)^2$ has been dropped because it does not affect the form of the amplitude spectrum.

EXAMPLE 5.10
Find the OTF of an optical system if its $H(\omega_y)$ is a square pulse, as shown in Figure 5.13a.

Solution: Introduce the function $H(\omega_y)$ into (5.59a). We obtain

$$H_{in}(\omega_y) = \begin{cases} \left(1 - \dfrac{|\omega_y|}{2\omega_{y_0}}\right) & -2\omega_{y_0} \leq \omega_y \leq 2\omega_{y_0} \\ 0 & \text{otherwise} \end{cases} \tag{5.60}$$

This function is shown in Figure 5.13b. The inverse transform of this function has been found to be $(4/\pi\omega_{y_0})(\sin^2 \omega_{y_0} y_1)/y_1^2$, which implies from (5.58c) that the impulse response of this system is given by

Figure 5.13
Frequency response of an optical system.

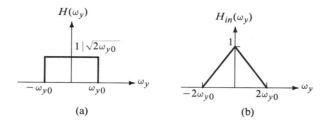

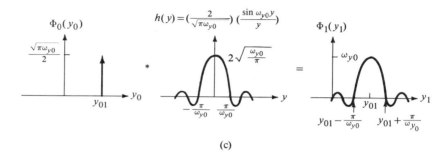

$$h(y) = \frac{2}{\sqrt{\pi\omega_{y_0}}} \frac{\sin \omega_{y_0} y}{y} \qquad (5.61)$$

Therefore, if a point source $\Phi_0(y_0) = (\sqrt{\pi\omega_{y_0}}/2)\delta(y_0 - y_{0_1})$ is located in the object plane, its image is given by the convolution $h(y_1) * \Phi_0(y_1)$, which is equal to

$$\Phi_1(y_1) = \frac{\sin[\omega_{y_0}(y_1 - y_{0_1})]}{y_1 - y_{0_1}} \qquad (5.62)$$

This is shown in Figure 5.13c.

We note also that the relationship [see (4.78)]

$$f_y = \frac{\omega_{y_0}}{2\pi} = \frac{2\pi y_0}{2\pi\lambda f_l} = \frac{\frac{1}{2}}{\frac{\lambda f_l}{2y_0}} = \frac{0.5}{\lambda F^\#} \qquad (5.63)$$

where $F^\#$ is known as the f-number and y_0 is the radius of a lens. From this we can obtain the resolution capabilities of a camera. For example, if a camera has an f-number $F^\# = 1.4$ and we assume an average visual wavelength of $\lambda = 0.5$ μm, the lens of this camera can resolve up to 714 lines/mm. ∎

5-5.b Coherent Case In this case all of the points in the object plane are emitting radiation synchronously in phase. This means that the field at each point on a plane parallel with the object plane is illuminated by monochromatic radiation having the same amplitude and phase. Therefore the phase and amplitude-variation information is contained in the expression for $\Phi_0(y_0)$ and the intensity at the object plane is given by (5.56) without any modifications. Therefore the intensity integral becomes

$$I(y_1) = \int_{-\infty}^{\infty} h(y_1 - y_0)\Phi_0(y_0)\,dy_0 \int_{-\infty}^{\infty} h^*(y_1 - y_0')\Phi_0^*(y_0')\,dy_0'$$

or

$$I(y_1) = \left|\int_{-\infty}^{\infty} h(y_1 - y_0)\Phi_0(y_0)\,dy_0\right|^2 = |h(y_1) * \Phi_0(y_1)|^2 \quad \text{(a)}$$

$$\Phi_{im}(y_1) = \int_{-\infty}^{\infty} h(y_1 - y_0)\Phi_0(y_0)\,dy_0 = h(y_1) * \Phi_0(y_1) \quad \text{(b)}$$

(5.64)

Recalling the convolution and correlation properties, the Fourier transforms of these expressions become

$$I(\omega_y) = [H(\omega_y)\Phi_0(\omega_y)] \star [H(\omega_y)\Phi_0(\omega_y)]^* \quad \text{(a)}$$

$$\Phi_{cim}(\omega_y) = H(\omega_y)\Phi_0(\omega_y) \quad \text{(b)}$$

(5.65)

where

$$\Phi_{cim}(\omega_y) = \int_{-\infty}^{\infty} \Phi_{im}(y_1)e^{-j\omega_y y_1}\,dy_1 \quad \text{(c)}$$

$$H(\omega_y) = \int_{-\infty}^{\infty} h(y_1)e^{-j\omega_y y_1}\,dy_1 \quad \text{(d)}$$

$$\Phi_0(\omega_y) = \int_{-\infty}^{\infty} \Phi_0(y_0)e^{-j\omega_y y_0}\,dy_0 \quad \text{(e)}$$

$$H(\omega_y) = \text{Coherent Transfer Function, CTF} \quad \text{(f)}$$

Observe that no explicit formula has been given for the spread function $h(\cdot)$. It was previously noted in Chapter 2 that the development of $h(\cdot)$ is somewhat involved, but note that its form can be established experimentally in the setup shown in Figure 5.12. The result for a one-dimensional positive lens system is found to be [see (2.88)]

$$h(y_1; y_0) = \int_{-\infty}^{\infty} P(y)e^{-j(k/z_i)(y_1 + My_0)y}\,dy \quad (5.66)$$

where the magnification $M = z_i/z_0$. By introducing the expressions $My_0 = -y_{0_m}$ and $ky/z_i = \omega_y$ into this equation, we obtain

$$h(y_1; y_{0_m}) = h(y_1 - y_{0_m})$$
$$= \int_{-\infty}^{\infty} P\left(\frac{z_i}{k}\omega_y\right) e^{-j(y_1 - y_{0_m})\omega_y} d\omega_y \quad (5.67)$$

where for simplicity we have set the constant factor z_i/k in the integrand to unity. Note that in (5.67) $h(\cdot)$ is now **space-invariant** and equivalent to time-invariant systems, since $h(\cdot)$ is a function only of the coordinate difference $(y_1 - y_{0_m})$.

Since the image of any optical system is the convolution of its object function and its impulse response, we write

$$\Phi_i(y_1) = \int_{-\infty}^{\infty} h(y_1 - y_{0_m})\left[\frac{1}{M}\Phi_0\left(\frac{1}{M}y_{0_m}\right)\right] dy_{0_m} \quad (5.68)$$

where we have changed the y_0 coordinate to $y_{0_m} = My_0$. This equation, in notational form, is given by

$$\Phi_i(y_1) = h(y_1) * \Phi_{01}(y_1) \quad \text{(a)}$$
where
$$\Phi_{01}(y_1) = \frac{1}{M}\Phi_0\left(\frac{1}{M}y_{0_m}\right) \quad \text{(b)} \quad (5.69)$$
$$h(y_1) = \int_{-\infty}^{\infty} P\left(\frac{z_i}{k}\omega_y\right) e^{-jy_1\omega_y} d\omega_y \quad \text{(c)}$$

Equations (5.69a) and (5.69c) indicate that if a system includes diffraction effects, and this is equivalent to saying that the aperture is finite or that the impulse function has nonzero width, the image produced is not an exact replica of the object, but has been smoothed. The smoothing of an image results in a less detailed one, which means that some of the information about the image has been lost. This is the situation when a picture is taken with a camera whose lens is out of focus. An analogous case can be found in electrical and nonelectrical systems when the duration of the impulse response of a system is larger than rapid changes or **fluctuations** of the input signal.

1. *Frequency Reponse of a Coherent System.* From (5.65) we know that the coherent transfer function is the Fourier transform of the space-invariant impulse function. With the help of (5.69c), we can then write

$$H(\omega_y) \triangleq \text{CTF} = \mathscr{F}\{h(y_1)\} = \mathscr{F}\left\{\mathscr{F}^{-1}\left\{P\left(\frac{z_i}{k}\omega_y\right)\right\}\right\} = P\left(-\frac{z_i}{k}\omega_y\right) \quad (5.70)$$

An optical system that has a pupil function of zero or one and creates a converging spherical wave at its exit pupil when a diverging spherical wave is impinging on its entrance pupil is known as the **diffraction-limited** system. When the same system operates under coherent conditions, the CTF (Coherent Transfer Function) of the system is also zero or one. This is equivalent to saying that the system will pass a finite band of spatial frequencies without amplitude or phase distortion. The minus sign that appears in the pupil function arises from the repetition of the Fourier transform rather than from its inverse. That is, in optics positive lenses take the Fourier transforms only of the objects, and this results in image inversion. The minus sign can be ignored by defining the pupil function P in a reflected coordinate system—a procedure that we will follow in our subsequent analysis. Hence we shall write

$$H(\omega_y) \triangleq \text{CTF} = P\left(\frac{z_i}{k}\omega_y\right) \tag{5.71}$$

2. *Frequency Response of an Incoherent Imaging System.* We combine $H(\omega_y)$ from (5.71) with (5.59a) and thereby obtain the OTF in the form

$$H_{in}(\omega_y) = \int_{-\infty}^{\infty} P\left(\frac{z_i}{k}\omega_{y'}\right) P^*\left(\frac{z_i}{k}\omega_{y'} + \frac{z_i}{k}\omega_y\right) d\left(\frac{z_i}{k}\omega_{y'}\right)$$

where we set k/z_i equal to zero without loss of generality. Hence, we have

$$H_{in}(\omega_y) = \int_{-\infty}^{\infty} P(y') P^*\left(y' + \frac{z_i}{k}\omega_y\right) dy' \tag{5.72}$$

This expression can be symmetrized by introducing a change of variable, with $\xi = y' + (z_i/2k)\omega_y$. When this is done, (5.72) becomes

$$H_{in}(\omega_y) \triangleq \text{OTF} = \int_{-\infty}^{\infty} P\left(\xi - \frac{z_i}{2k}\omega_y\right) P^*\left(\xi + \frac{z_i}{2k}\omega_y\right) d\xi \tag{5.73}$$

To normalize this OTF requires dividing this expression by the normalizing factor

$$\int_{-\infty}^{\infty} P(\xi) P^*(\xi) \, d\xi \tag{5.74}$$

For diffraction-limited and aberration-free systems $P(\xi) = 1$ at any point within the lens; in this case, as we find below, the OTF is equal to the distance of overlap of the pupil function.

EXAMPLE 5.11
Find the OTF of an optical system whose pupil function is that shown in Figure 5.14a and is diffraction-limited, aberration-free, and without attenuation.

Figure 5.14
OTF of a diffraction-limited system with slit-type pupil. (a) Geometry of pupil. (b) Two pupil functions, one unshifted. (c) OTF versus $z_i\omega_y/k$. (d) OTF versus ω_y.

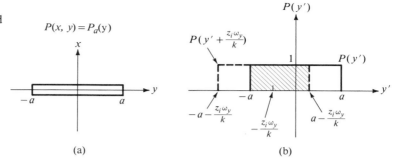

Solution: By using (5.72) together with Figure 5.14b, we find that

$$H_{in}(\omega_y) = \int_{-\infty}^{\infty} P(y')P^*\left(y' + \frac{z_i}{k}\omega_y\right)dy'$$

$$= \int_{-a}^{a-(z_i\omega_y/k)} dy' = 2a - \frac{z_i\omega_y}{k} \tag{5.75}$$

■

EXAMPLE 5.12

Repeat Example 5.11 for a pupil function that has an absorption band, as shown in Figure 5.15a.

Solution: We proceed by employing correlation involving the following steps.

a. For $0 < z_i\omega_y/k < 2a/3$ (see Figure 5.15b)

$$H_{in}(\omega_y) = \int_{-a}^{-a/3-(z_i\omega_y/k)} dy' + \int_{a/3}^{a-(z_i\omega_y/k)} dy' = \frac{4a}{3} - \frac{2z_i\omega_y}{k}$$

b. For $2a/3 < z_i\omega_y/k < 4a/3$ (see Figure 5.15c)

$$H_{in}(\omega_y) = \int_{-a/3-(z_i\omega_y/k)}^{-a/3} dy' = \frac{z_i\omega_y}{k}$$

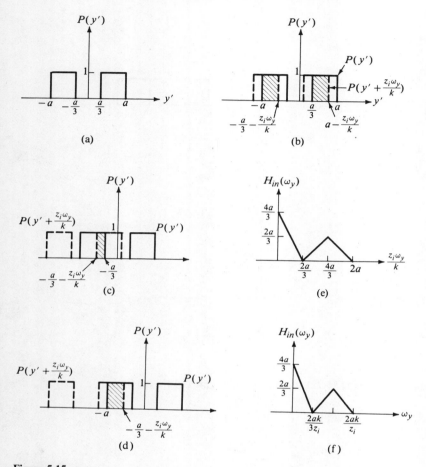

Figure 5.15
OTF of a diffraction-limited system with slit-type pupil and attenuation present. (a) Pupil function where blocking of light at the center is present. (b)–(d) The process in deriving the OTF. (e) OTF versus $z_i\omega_y/k$. (f) OTF versus ω_y.

Thus we have

$$H_{in}(\omega_y) = \frac{z_i\omega_y}{k} - \frac{2}{3}a$$

c. For $4a/3 < \dfrac{z_i\omega_y}{k} < 2a$ (see Figure 5.15d)

$$H_{in}(\omega_y) = \int_{-a}^{-a/3 - (z_i\omega_y/k)} dy' = \frac{2a}{3} - \frac{z_i\omega_y}{k} \quad \text{or} \quad H_{in}(\omega_y) = \frac{6a}{3} - \frac{z_i\omega_y}{k}$$

The OTF of Figures 5.15e and 5.15f show the results of this calculation. Figure 5.15f vividly shows the effect of the pupil function on the frequency response of the optical system. ∎

Figure 5.16
OTF of an optical system with circular pupil. (a) Optical function. (b) Area of overlap. (c) OTF. (d) Three-dimensional OTF.

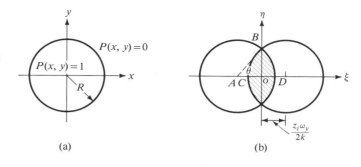

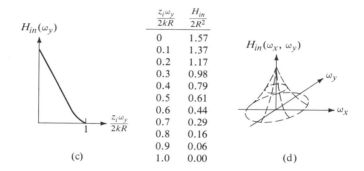

EXAMPLE 5.13

Find the OTF of a diffraction-limited optical system when its pupil function is a circle, as shown in Figure 5.16a.

Solution: This is a more realistic situation than the foregoing examples since lenses of optical systems generally are circular. From the geometry of this system, we may easily conclude that the system is symmetrical—hence we need treat only the one-dimensional case. We can then extrapolate our results to two dimensions. An examination of (5.73) shows that for a diffraction-limited system, the OTF is equal to the distance of overlap of the pupil function. This tells us that in two dimensions the OTF is just the area of overlap of the pupil function. But in our case the area of overlap is 4(area ABD − area ABO), where

$$\text{area } ABD = \frac{1}{2} R^2 \theta = \frac{1}{2} R^2 \cos^{-1}\left(\frac{z_i \omega_y}{2kR}\right)$$

$$\text{area } ABO = \frac{1}{2} \frac{z_i \omega_y}{2k} \left[R^2 - \left(\frac{z_i \omega_y}{2k}\right)^2 \right]^{1/2}$$

It then follows that

$$H_{in}(\omega_y) = 2R^2 \left[\cos^{-1}\left(\frac{z_i \omega_y}{2kR}\right) - \frac{z_i \omega_y}{2kR}\left[1 - \left(\frac{z_i \omega_y}{2kR}\right)^2 \right]^{1/2} \right] \qquad (5.76)$$

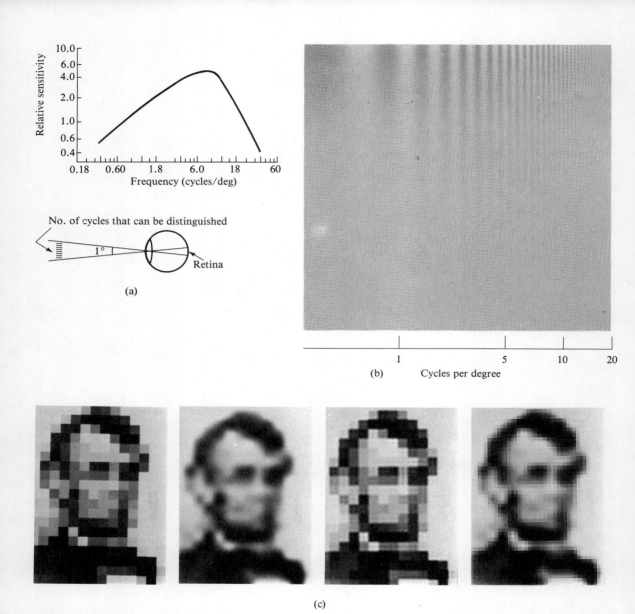

Figure 5.17
(a) Describing the transfer function for the human visual system. (b) A demonstration of the transfer function of your visual system if held at arm's length. (c) Observe the effect of filtering of your visual system by studying the figures when held several feet away. [(a) and (b) reprinted, by permission, from Cornsweet, *Visual Perception*, 341, 343. (c) reprinted, by permission, from Harmon, L. D., "Masking in Visual Recognition: Effects of Two-Dimensional Filtered Noise," 1194.]

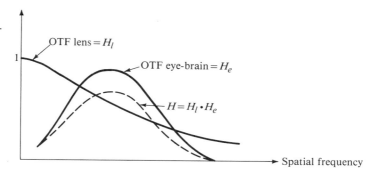

Figure 5.18
The OTF of a lens (magnifying glass) and eye-brain combination.

This has been plotted in Figure 5.16c. Because of symmetry considerations, we see that the three-dimensional OTF is easily obtained by rotating the curve given in Figure 5.16c; this is shown in Figure 5.16d. ∎

In addition to finding the optical transfer function of an optical system, we must also take into consideration the output stage of any optical system—namely, the eye-brain combination. As a case in point, if we are designing an optical system whose output is to be visually examined, it would be pointless to build a high-resolution system since the eye-brain combination has an OTF of the form shown in Figure 5.17a and cannot process the high-resolution data. This assertion is evident by examining Figure 5.17b. An indication of the eye-brain combination filtering of high frequencies is evident by examining Figure 5.17c at varying distances. At large distances the picture loses its sharpness and appears as if it were defocused. This effect exists because the high spatial frequencies responsible for the sharp charges of intensity in the picture are attenuated by the bandpass qualities of our optical system.

If we wish to find the transfer function of a simple optical system, say a magnifying lens plus the eye-brain, we would multiply their transfer functions in the same manner as for other systems consisting of subsystems in cascade. Figure 5.18 shows the procedure for the present case.

■ ■ ■

5-6 PROCESSING OF SIGNALS WITH COHERENT OPTICS—OPTICAL FILTERS

The applications of optical Fourier transforms, optical spatial filtering, holography, incoherent optical signal processing, optical image processing, and many other optical problems have received increased attention since the invention of the laser. One reason that optical processing has been used extensively stems from the fact that parallel processing, in practical terms, can be accomplished instantaneously. We will consider a number of optical applications.

5-6.a Spectrum Analysis Consider the optical system illustrated in Figure 5.19a. The light amplitude field at the (x_1, y_1) plane, which is a complex quantity, is given by [see (4.78)]

$$\Phi(y_1) = \int_{-\infty}^{\infty} \Phi(y_0) e^{-j\omega_{y_1} y_0} dy_0 \quad \text{(a)} \tag{5.77}$$

where

$$\omega_{y_1} = 2\pi \frac{y_1}{\lambda f_l} = 2\pi f_{y_1} \quad \text{(b)}$$

The corresponding irradiance, which is the intensity variation of the film transparency, is

$$I(y_1) = \Phi(y_1)\Phi^*(y_1) = |\mathscr{F}\{\Phi(y_0)\}|^2 \tag{5.78}$$

The intensities of the spectra belonging to a two-dimensional and also a periodic one-dimensional figure are shown in Figure 5.19b. These films were exposed at the plane (x_1, y_1).

5-6.b Convolution If a transparency (filter) with frequency characteristic given by

$$H(\omega_{y_1}) = |H(\omega_{y_1})| e^{j\varphi(\omega_{y_1})} \quad \text{(a)}$$

$$|H(\omega_{y_1})| \leq 1 \quad \text{(b)} \tag{5.79}$$

$$0 \leq \varphi(\omega_{y_1}) \leq 2\pi \quad \text{(c)}$$

is inserted at the plane (x_1, y_1) of Figure 5.19a, then the field distribution (a complex quantity) leaving the plane is

$$\Phi'(\omega_{y_1}) = \Phi(\omega_{y_1}) H(\omega_{y_1}) \quad \text{(a)} \tag{5.80}$$

where

$$\Phi(\omega_{y_1}) = \mathscr{F}\{\Phi(y_0)\} \quad \text{(b)}$$

$$H(\omega_{y_1}) = \mathscr{F}\{h(y_0)\} \quad \text{(c)}$$

In this setup lens L_3 takes the Fourier transform of the field $\Phi(\omega_{y_1})$ and produces a complex field distribution on the (x_2, y_2) plane of the form

$$\Phi(\omega_{y_2}) = \int_{-\infty}^{\infty} \Phi(\omega_{y_1}) H(\omega_{y_1}) e^{-j\omega_{y_2} y_1} dy_1 \quad \text{(a)} \tag{5.81}$$

or equivalently (see Table 4.1)

$$\Phi(\omega_{y_2}) = \int_{-\infty}^{\infty} \Phi(y_0) h(\omega_{y_2} - y_0) dy_0 = \Phi(\omega_{y_2}) * h(\omega_{y_2}) \quad \text{(b)}$$

This expression shows that the field distribution on the (x_2, y_2) plane is the convolution of the object function $\Phi(y_0)$ and the impulse function $h(y_0)$ of the filter.

5-6.c Correlation Suppose that we set the two transparencies $\Phi(y_0)$ and $h(y_0)$ at the plane (x_0, y_0) of Figure 5.19a but with $\Phi(\cdot)$ shifted by an amount τ_y. The

Figure 5.19
(a) A typical one-dimensional spectrum analyzer. (b) Two-dimensional spectra. (From E. Hecht and A. Zajac, *Optics*, © 1974. Addison-Wesley, Reading, MA. Reprinted with permission.)

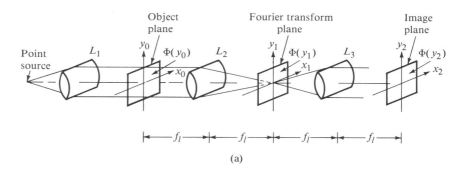

(a)

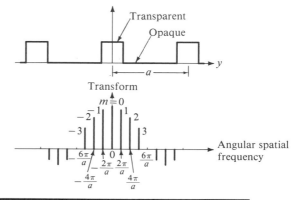

Diffraction pattern

(b)

complex field distribution at the plane (x_1, y_1) becomes

$$\Phi(\omega_{y_1}; \tau_y) = \int_{-\infty}^{\infty} \Phi(y_0 - \tau_y)h(y_0)e^{-j\omega_{y_1}y_0}\,dy_0 \tag{5.82}$$

At the optical center, $y_1 = 0$ so that $\omega_{y_1} = 0$, and the field at the optical center is given by

$$\Phi(\omega_{y_1}; \tau_y) = \int_{-\infty}^{\infty} \Phi(y_0 - \tau_y)h(y_0)\,dy_0 \tag{5.83}$$

By (2.28), this shows that the field is equal to the correlation of the two functions $\Phi(\cdot)$ and $h(\cdot)$. In practice the light at the optical center is detected with a small diode whose current is proportional to the value $|\Phi(\omega_{y_1}; \tau_y)|^2$.

5-6.d Addition-Subtraction Suppose that we place a **filter** in the (x_1, y_1) plane of Figure 5.19a, which has a sinusoidally graded transmittance function of the form

$$H(\omega_{y_1}) = 1 + \cos(\omega_{y_1}y_1 + \varphi) = \frac{1}{2}\left[2 + e^{j(\omega_{y_1}y_1+\varphi)} + e^{-j(\omega_{y_1}y_1+\varphi)}\right] \tag{5.84}$$

If, in addition, we place two objects on the plane (x_0, y_0) symmetrically around the optical axis, the field distribution is given by

$$\Phi(y_0) = \Phi'(y_0 - y_1) + \Phi''(y_0 + y_1)$$

Hence the field distribution of $\Phi(y_0)$ on the plane (x_1, y_1) is the Fourier transform of $\Phi(\cdot)$; it is given by

$$\Phi(\omega_{y_1}) = \Phi'(\omega_{y_1})e^{-j\omega_{y_1}y_1} + \Phi''(\omega_{y_1})e^{j\omega_{y_1}y_1} \tag{5.85}$$

The field leaving the (x_1, y_1) plane is

$$\Phi_1(\omega_{y_1}) = \Phi(\omega_{y_1})H(\omega_{y_1})$$

$$= \frac{1}{2}e^{j\varphi}[\Phi'(\omega_{y_1}) + \Phi''(\omega_{y_1})e^{-2j\varphi}] + 4 \text{ other terms} \tag{5.86}$$

The field at the inverted plane (x_2, y_2) is proportional to

$$\Phi_2(y_2) = \mathscr{F}^{-1}\{\Phi_1(\omega_{y_1})\}$$

$$= \frac{1}{2}e^{j\varphi}[\Phi'(y_2) + \Phi''(y_2)e^{-2j\varphi}] + 4 \text{ terms with centers off axis} \tag{5.87}$$

The extra terms do not appear on the optical axis since they contain frequencies that are diffracted by the grating. If the grating (transparency) is set so that $\varphi = 0$—that is, it has its maximum transmittance on the optical axis—then $\Phi_2 \stackrel{\propto}{=} [\Phi'(y_2) + \Phi(y_2)]$ and addition is accomplished. Correspondingly, if the grating transmittance is shifted by a quarter of a fringe spacing from the optical axis, then $\varphi = 90°$ and $\Phi_2 \stackrel{\propto}{=} [\Phi'(y_2) - \Phi''(y_2)]$, which indicates subtraction.

5-6.e Differentiation If we build a filter whose impulse response is given by

$$h(y) = \delta[y + (a + \epsilon)] - \delta(y + a) \tag{5.88}$$

the differential operation $\partial \Phi(y)/\partial y$ will be performed. The requisite filter can be accomplished by exposing a film on the Fourier plane of two sinusoidal gratings with slightly different frequencies. To see this operation, note that

$$\frac{\partial \Phi(y)}{\partial y} = \lim_{\epsilon \to 0} \frac{1}{\epsilon}\left[\Phi[y + (a + \epsilon)] - \Phi(y + a)\right] = \lim_{\epsilon \to 0} \frac{1}{\epsilon}\left[\Phi(y) * h(y)\right] \quad (5.89)$$

Figure 5.20 shows experimental results of the optical differentiation operation.

5-6.f Contrast Improvement of Images with Amplitude Filter Suppose that an input function exists in the (x_0, y_0) plane (see Figure 5.19a) of the form

$$\Phi(y_0) = A + \Phi_0(y_0) \quad (5.90)$$

where $\Phi_0(y_0)$ is the object field and A is an unwanted uniform illumination. The field distribution on the (x_1, y_1) Fourier plane is

$$\Phi(\omega_{y_1}) = H(\omega_{y_1})[A 2\pi \delta(\omega_{y_1}) + \Phi_0(\omega_{y_1})] \quad (5.91)$$

where $H(\omega_{y_1})$ is the Fourier transform of $h(x, y)$, the impulse response of the optical system. Now suppose that a **high-pass amplitude filter** is inserted at the (x_1, y_1) plane having the characteristics

$$\tau(\omega_{y_1}) = 1 - p_\epsilon(\omega_{y_1}) \quad (5.92)$$

where ϵ is very small. The field leaving the plane is

$$\begin{aligned}\Phi_1(\omega_{y_1}) &= [1 - p_\epsilon(\omega_{y_1})]H(\omega_{y_1})[A 2\pi \delta(\omega_{y_1}) + \Phi_0(\omega_{y_1})] \\ &\doteq H(\omega_{y_1})\Phi_0(\omega_{y_1})\end{aligned} \quad (5.93)$$

Consequently, the field at the (x_2, y_2) plane is the Fourier transform of (5.93), or, equivalently, the inverse Fourier transform at the reflected (x_2, y_2) coordinate system. We therefore have

$$\Phi_2(y_2) = \mathscr{F}^{-1}\{\Phi_1(\omega_{y_1})\} \doteq h(y_2) * \Phi_0(y_2) \quad (5.94)$$

From this, it follows that for a perfect optical system—that is, one for which $h(y_2) = \delta(y_2)$—we will recover the object to very good approximation, but without the background. The physical consequence of this type of filtering is that the image appears sharper, owing to the absence of the diffuse background light. Figure 5.20c shows the results of filtering the high frequencies. It is observed that shades of gray appear and the sharp boundaries vanish. Precisely the same phenomenon occurs if time signals are filtered with low-pass filters.

5-6.g Phase Contrast Improvement with Phase Filter We observe that objects of the form

$$\Phi(y_0) = e^{j\varphi(y_0)} \quad (5.95)$$

cannot be seen or recorded on a photographic film because these detectors respond to $|\Phi(y_0)|^2$. Since, in this case, $|\Phi(y_0)|^2 = 1$, the detected image is just a constant exposure and not a function of $\Phi(y_0)$. For a perfect optical system

Figure 5.20
Optical operations: (a) Subtraction. (b) Optical differentiation: Object pattern and experimental results for $\partial\Phi/\partial y$. (c) Optical filtering. (d) Contrast and edge enhancement. (e) Logic operations with signals image A and image B. [(a) and (c): from E. Hecht and A. Zajac, *Optics*, © 1974. Addison-Wesley, Reading, MA. Reprinted with permission. (b) reprinted, by permission, from Lee, "Review of Coherent Optical Processing," 203. (d) and (e) reprinted, by permission, from Warde and Thackara, "Oblique-cut LiNbO$_3$ Microchannel Spatial Light Modulator," 344, 346.]

THE RESPONSE AND APPLICATIONS OF LINEAR FILTERS

with $\varphi \ll 1$, (5.95) approximates to

$$\Phi(y_0) \doteq 1 + j\varphi(y_0) \tag{5.96}$$

and the light distribution at the Fourier plane will be

$$\Phi(\omega_{y_1}) = 2\pi\, \delta(\omega_{y_1}) + j\varphi(\omega_{y_1}) \tag{5.97}$$

If a **phase filter** (for example, a thin dielectric coating) introducing a $\pi/2$ phase shift is inserted on the focal plane and on the optic axis, the light that emerges from the (x_1, y_1) plane will be

$$\Phi(\omega_{y_1})P(\omega_{y_1}) = 2\pi P(\omega_{y_1})\,\delta(\omega_{y_1}) + P(\omega_{y_1})\varphi(\omega_{y_1}) \quad \text{(a)} \tag{5.98}$$

where

$$P(\omega_{y_1}) = \begin{cases} j & |\omega_{y_1}| < \epsilon \\ 1 & |\omega_{y_1}| > \epsilon \end{cases} \quad \text{(b)}$$

with ϵ being a small number (distance in the frequency plane). The absolute value squared of the inverse Fourier transform of this equation is

$$|1 + \varphi(y_2)|^2 \doteq 1 + 2\varphi(y_2) \tag{5.99}$$

where second- and higher-order terms have been neglected. This expression shows that a spatial phase modulation has been converted to a spatial intensity modulation.

5-6.h The Vander Lugt Filter The Vander Lugt setup allows an impulse function response for the system. To examine this, begin by introducing the impulse function $h(y_0)$ in the object plane, as shown in Figure 5.21a. The field just leaving the (x_0, y_0) plane is

$$\Phi(y_0) = \delta(y_0 - b) + h(y_0) \tag{5.100}$$

The field at the (x_1, y_1) plane, which is the Fourier transform of $\Phi(y_0)$, is

$$\Phi_1(\omega_{y_1}) = e^{-j\omega_{y_1}b} + H(\omega_{y_1}) \tag{5.101}$$

and the detected intensity on the film is

$$|\Phi_1(\omega_{y_1})|^2 = 1 + |H(\omega_{y_1})|^2 + H(\omega_{y_1})e^{j\omega_{y_1}b} + H^*(\omega_{y_1})e^{-j\omega_{y_1}b} \tag{5.102}$$

We make use of the fact that $H(\omega_{y_1})$ is a complex quantity of the form

$$H(\omega_{y_1}) = |H(\omega_{y_1})|e^{j\varphi(\omega_{y_1})} \tag{5.103}$$

When this is combined with (5.102) we obtain

$$|\Phi_1(\omega_{y_1})|^2 = 1 + |H(\omega_{y_1})|^2 + 2|H(\omega_{y_1})|\cos[\omega_{y_1}b + \varphi(\omega_{y_1})] \tag{5.104}$$

Observe that the last term in this equation is similar to a carrier wave that is both amplitude and phase modulated. Hence the amplitude and phase information is recorded, respectively, as amplitude and phase modulations of a high-frequency carrier that has been introduced by the off-axis source. This filter is known as a **hologram**. It was first suggested and demonstrated by Dennis Gabor

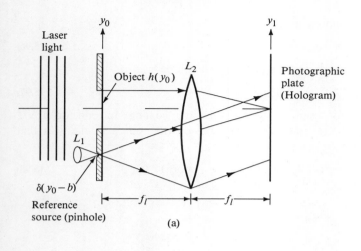

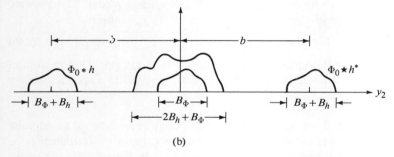

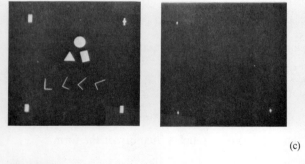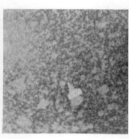

(c)

Figure 5.21
(a) Construction of a hologram (filter). (b) Bandwidth relationship of the signal (object) and filter. (c) Detection of signal using the matched filter technique. The signal in the second case is embedded in noise. (Reprinted, by permission, from Vander Lugt, "Signal Detection by Complex Spatial Filtering," 139.)

in 1948. Figure 5.22 shows the first hologram that was constructed by Gabor together with its input and its output, which is a reconstruction of the input. Holograms have been used in many applications, such as information storage,

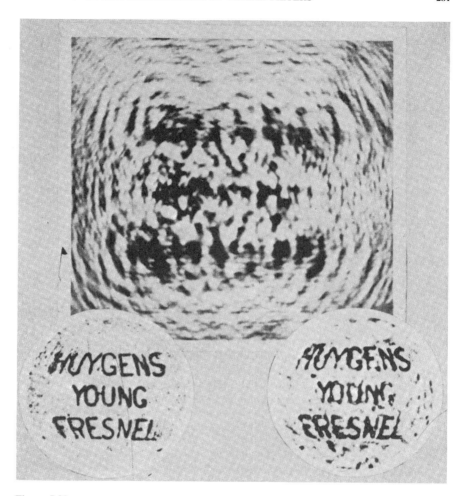

Figure 5.22
First holographic reconstruction (1948). (Reprinted, by permission, from Gabor, "A New Microscopic Principle," 777.)

pattern recognition, vibration analysis, and overcoming effects of turbulence when viewing objects at large distances.

If the filter is introduced at the (x_1, y_1) plane of Figure 5.19a, the field distribution leaving the plane when an object $\Phi_0(y_0)$ is inserted at the (x_0, y_0) plane is

$$\begin{aligned}\Phi'_1(\omega_{y_1}) &\propto \Phi_0(\omega_{y_1})|\Phi_1(\omega_{y_1})|^2 \\ &= \Phi_0(\omega_{y_1}) + \Phi_0(\omega_{y_1})|H(\omega_{y_1})|^2 + \Phi_0(\omega_{y_1})H(\omega_{y_1})e^{j\omega_{y_1}b} \\ &\quad + \Phi_0(\omega_{y_1})H^*(\omega_{y_1})e^{-j\omega_{y_1}b}\end{aligned} \quad (5.105)$$

In this expression we have assumed that the transmittance of the filter is proportional to $|\Phi_1(\omega_{y_1})|^2$. Lens L_3 will Fourier-transform the field $\Phi'_1(\omega_{y_1})$, and the field at the reflected plane $(-x_2, -y_2)$ will be

$$\begin{aligned}\Phi_2(y_2) \propto\ & \Phi_0(y_2) + \Phi_0(y_2) * [h(y_2) \star h(-y_2)] \\ & + \Phi_0(y_2) * h(y_2) * \delta(y_2 + b) \\ & + \Phi_0(y_2) * h^*(y - y_2) * \delta(y_2 - b)\end{aligned} \quad (5.106)$$

Examine the terms in this expansion. The first term is identical with the input function. The second term is the convolution of the input function and the autocorrelation of the impulse response of the filter. Both of these terms are of no particular interest and are centered at (0, 0) in the (x_2, y_2) plane. The third term is of the form

$$\int_{-\infty}^{\infty} \Phi_0(\eta) h(y_2 + b - \eta)\, d\eta \quad (5.107)$$

and is the convolution of $\Phi_0(\cdot)$ and $h(\cdot)$ centered at the point $(0, -b)$ in the (x_2, y_2) plane. The fourth term is of the form

$$\int_{-\infty}^{\infty} \Phi_0(y_2 - b + \eta) h^*(\eta)\, d\eta \quad (5.108)$$

and is the cross-correlation of $\Phi_0(\cdot)$ and $h^*(\cdot)$ centered at $(0, b)$ in the (x_2, y_2) plane.

If we assume a maximum spectral width B_h for h and a maximum width B_Φ for Φ_0, the light distribution along the y_2-axis is that shown in Figure 5.21b.

Of importance in optical pattern and character recognition is an optimum filter, known as a **matched filter**. In 1963 Vander Lugt proposed and synthesized such a filter. We will assume that the matched filter has a transmittance proportional to the complex conjugate of the Fourier transform of the signal. Thus we choose

$$H(\omega_{y_1}) = \Phi_0^*(\omega_{y_1}) \quad (5.109)$$

Introduce the inverse Fourier transform of (5.109) into (5.108), which yields the autocorrelation of the signal Φ_0. Upon introducing (5.109) into (5.105), the fourth term in the expression is equal to

$$|\Phi_0(\omega_{y_1})|^2 e^{-j\omega_{y_1} b} \quad (5.110)$$

This represents a plane wave that is propagating at an angle $\theta = \sin^{-1}(b/f_l)$ with the optic axis. Therefore the complex light field due to the signal from this term will be imaged into a small region on the (x_2, y_2) plane and off the optic axis on the y_2-axis. The effect of the matched filter is to eliminate the phase variation created by the Fourier transform of the object $\Phi_0(x_0, y_0)$. Figure 5.21c shows the results of a detection process by cross-correlation when noise is also included.

■ ■ ■

5-7 INCOHERENT OPTICAL PROCESSING

The basic operation performed by an incoherent optical processor is given by

$$g(u) = \int_{-\infty}^{\infty} f(r)h(r, u)\,dr \tag{5.110}$$

The system shown in Figure 5.23a performs the operation

$$g(0) = \int_{-\infty}^{\infty} f(y)h(y)\,dy \tag{5.111}$$

If this equation is to be used to compute correlations or convolutions, it is necessary that one of the two functions in this integral be shifted by some means. For example, if the input $f(y)$ is mechanically shifted in the y-direction by y_0, a cross-correlation can result, since now

$$g(y_0) = \int_{-\infty}^{\infty} f(y + y_0)h(y)\,dy \tag{5.112}$$

The input function $f(y)$ can be a photographic transparency, the face of a CRT, an erasable electro-optic device, or a temporally modulated input device (that is, a light-emitting diode). The function $h(y)$ could be a photographic transparency, an erasable electro-optic material, or a discretely-coded transparency.

A nonscanning correlator is shown in Figure 5.23b. Observe that all the light rays diverging from a point $(-y_0)$ of the diffuse extended source are collimated by the first lens, pass obliquely through the transparencies $f(y)$ and $h(y)$, and are converged to the point y_0 on the output plane. From simple geometrical considerations, the irradiance on the output is the desired relationship

$$g(y_0) = \int_{-\infty}^{\infty} f\left(y - \frac{d}{f} y_0\right) h(y)\,dy \tag{5.113}$$

Figure 5.23 Incoherent correlation techniques.

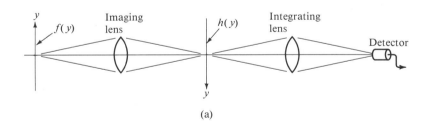

(a)

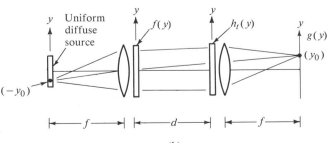

(b)

REFERENCES

1. Basar, E. *Biophysical and Physiological Systems Analysis*. Reading, Mass.: Addison-Wesley, 1976.
2. Cornsweet, T. N. *Visual Perception*. New York: Academic Press, 1971.
3. Gabor, D. "A New Microscopic Principle," *Nature* 161 (1948): 777.
4. Gaskill, J. D. *Linear Systems, Fourier Transforms and Optics*. New York: Wiley, 1978.
5. Goodman, J. W. *Fourier Optics*. New York: McGraw-Hill, 1968.
6. Harmon, L. D. "Masking in Visual Recognition: Effects of Two-Dimensional Filtered Noise," *Science* 180 (1973): 1194.
7. Hecht, E., and A. Zajac. *Optics*. Reading, Mass.: Addison-Wesley, 1974.
8. Lee, S. H. "Review of Coherent Optical Processing," *Applied Physics* 10 (1976): 203.
9. Papoulis, A. *The Fourier Integral and Its Applications*. New York: McGraw-Hill, 1962.
10. Sagawa, K., A. E. Taylor, and A. C. Guyton. "Dynamic Performance and Stability of the Cerebral Ischemic Pressor Response," *American Journal of Physiology* 201 (1961): 1164.
11. Seely, S., and A. D. Poularikas. *Electromagnetics, Classical and Modern Theory and Applications*. New York: Marcel Dekker, 1979.
12. Vander Lugt, A. "Signal Detection by Complex Spatial Filtering," *IEEE Transactions on Information Theory* IT-20 (1964): 139.
13. Warde, C., and J. I. Thackara. "Oblique-cut $LiNbO_3$ Microchannel Spatial Light Modulator," *Optics Letters* 7 (1982): 344, 346.

PROBLEMS

5-1.1 Consider three different systems. The input signals are, respectively,
    a. $u(t)$ **b.** $u(t - t_0) + \delta(t)$ **c.** $f(t) - u(t + 10)$
and the corresponding outputs are
    a. $-2u(t + 2)$ **b.** $3u(t - t_0 - 10) + 3\delta(t - 10)$ **c.** $2f^2(t - t_0) - 2u(t + 10 - t_0)$
Indicate whether any of these filters are distortionless.

5-1.2 If the transfer function of a filter in the frequency domain is $-3\exp[-j2\omega]$, find its output if the input signals are: (a) $u(t - 2)$, (b) $u(t) + \delta(t - 3)$, and (c) $-p_a(t) + \text{sinc}_2(t)$.

5-1.3 Determine the value of the capacitor in the *RC* filter shown in Figure P5-1.3 such that, in the range of frequencies $0 \leq \omega \leq 10{,}000$ rad/s, the amplitude of its transfer function does not fall more than 10%. Note that for frequencies within this range, the filter can be considered to be substantially distortionless.

5-1.4 A network has a transfer function $H(\omega)$; it is initially relaxed and is excited at $t = 0$ by the function $f(t)$, with the spectrum $F(\omega)$. Designate the time response of the network to $f(t)$ by $g(t)$.
    a. Obtain a formula for the network response function such that an excitation $f(t)$ will cause a response $f(t) - g(t)$.

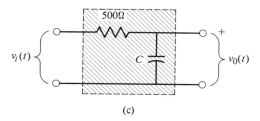

(c)

Figure P5-1.3

b. Apply the result established in (a) to the case where the first network is R and C in series, and $f(t)$ is the unit step voltage $u(t)$. The response is the current.

c. Repeat (b) for the series connection of R and L. Does the proof of the general network of (a) apply rigorously to this case?

5-2.1 Compare the response of the filter shown in Figure P5-2.1 with that of an ideal low-pass filter. Assume a delta function input voltage to the filter.

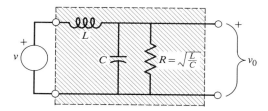

Figure P5-2.1

5-2.2 A periodic square wave shown in Figure P5-2.2 is the input to a low-pass filter that has the frequency characteristic $H(\omega) = p_{\omega_0}(\omega)$. Find the output of this filter if $\omega_0 = \infty$; $\omega_0 = 4$; and $\omega_0 = 1$.

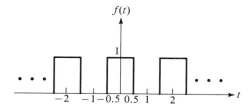

Figure P5-2.2

5-2.3 Show that an ideal high-pass filter can be represented as an all-pass filter plus a low-pass filter.

5-2.4 If the frequency characteristic of an ideal filter is given by

$$H(\omega) = \left[a + b\cos\left(\frac{n}{\omega_0}\right)\omega\right]e^{-j\omega t_0}p_{\omega_0}(\omega)$$

with $b < a$ and both positive constants, show that the filter is equivalent to three ideal low-pass filters in parallel. Plot their frequency characteristics.

5-2.5 The frequency domain transfer functions of a number of systems are shown in Figure P5-2.5. Find their outputs if the inputs are delta functions at $t = 0$. Also, find the time-domain representation of these transfer functions.

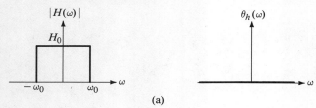

(a)

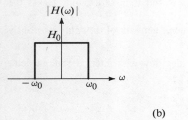 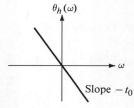

(b)

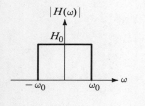 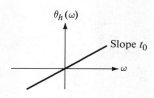

(c)

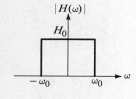 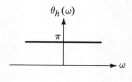

(d)

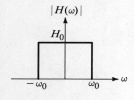 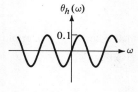

(e)

Figure P5-2.5

5-2.6 Find $H(\omega)$ for the filter shown in Figure P5-2.6. Indicate whether it is a low-pass or high-pass filter. Does this approximate an ideal filter?

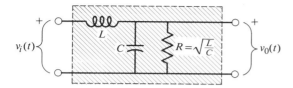

Figure P5-2.6

5-2.7 Sketch the Bode plots for the following frequency response functions:

a. $H(\omega) = 1 + \left(\dfrac{j\omega}{5}\right)$ **b.** $H(\omega) = 1 - \left(\dfrac{j\omega}{5}\right)$ **c.** $H(\omega) = \dfrac{1}{1 + \left(\dfrac{j\omega}{5}\right)}$

d. $H(\omega) = \dfrac{1 + \left(\dfrac{j\omega}{5}\right)}{1 + (j\omega)}$ **e.** $H(\omega) = \dfrac{10}{(1 + j\omega)^2}$ **f.** $H(\omega) = 1 + 2(j\omega) + 4(j\omega)^2$

5-2.8 Find the transfer functions for the systems shown in Figure P5-2.8 and plot their amplitude spectra on a log scale (dB vs. log ω) and their phase spectra on a linear scale (φ vs. log ω). Identify the types of these filters.

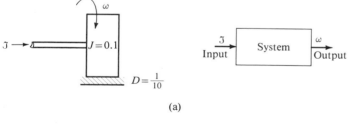

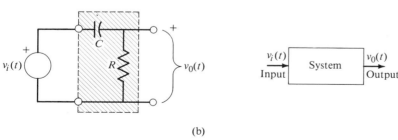

Figure P5-2.8

5-3.1 Find and plot the impulse response for the following phase filters:
  a. $\theta_h(\omega) = 0$ **b.** $\theta_h(\omega) = -\pi$ **c.** $\theta_h(\omega) = -2\pi t_0 \omega$ **d.** $\theta_h(\omega) = -2\pi t_0 \omega - \pi$

5-4.1 Find and plot the impulse response of the bandpass filter shown in Figure P5-4.1.

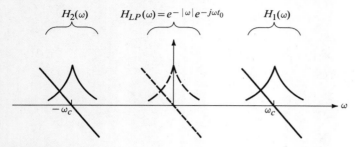

Figure P5-4.1

5-4.2 Use a passband filter, as shown in Figure 5.11 in the text, and show that in the limit when $\omega_0 \to 0$, the output is proportional to the input at $\omega = \omega_c$ times a cosine function with frequency ω_c. This result is used to design spectrum analyzers which comprise a set of narrow bandpass filters having the spectrum $|F(\omega)|_{\omega = \omega_c}$ for each $f(t)$ of the input at $\omega = \omega_c$.

5-4.3 A filter has the transfer function $H(\omega) = H_0(\omega)\exp[-j[\theta_h(\omega) + \omega t_0]]$. Identify the impulse response of this filter in terms of the response of three systems in cascade.

5-4.4 An interfering signal (noise) $n(t)$ is present with a desired signal. Find the frequency characteristics of an ideal filter that will extract the signal.

5-4.5 The quadratic-phase signal $q(t) = \exp(jt^2)$ is employed as shown in Figure P5-4.5. Find the output signal $g(t)$ of the system. If the input to the system is a general signal $f(t)$, what is the output of the system?

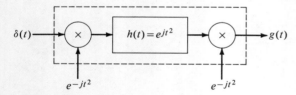

Figure P5-4.5

5-4.6 Find the output from the system shown in Figure P5-4.6. What is the effect of this system on the signal?

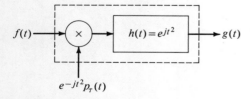

Figure P5-4.6

5-4.7 A matched filter is one that has an impulse response $h(t) = H_0 f^*(-t)$, where $f(t)$ is the input signal. Find the output $g(t)$ if the input is $f(t - t_0)$, with t_0 equal to a constant. This type of filter is often used to detect signals in a noise environment.

5-5.1 Find the OTF if the one-dimensional pupil function is that shown in Figure P5-5.1. Compare these results with those found in Example 5.12.

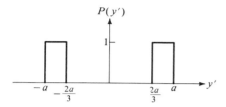

Figure P5-5.1

5-5.2 Find the OTF if the one-dimensional pupil function is that shown in Figure P5-5.2.

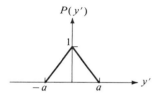

Figure P5-5.2

5-5.3 Verify (5.57), (5.58), and (5.59).

5-6.1 Sketch the amplitude and intensity of the field at the Fourier transform plane versus distance away from the optical axis for an input field given by $\Phi(y_0) = 3 \cos 500 y_0$, $-0.002 \leq x_0 \leq 0.002$. The field is produced by a HeNe laser, $\lambda = 0.63$ μm, and the focal length of the lens is 30 cm.

5-6.2 If a sinusoidal grating placed at the Fourier transform plane has the transmittance $H(\omega_{y_1}) = 1 + \sin(\omega_{y_1} y_1 + \varphi)$, find the conditions for which two objects located at the object plane subtract or add on the image plane.

5-6.3 Verify (5.93).

5-6.4 Specify the characteristics of an amplitude filter that will eliminate the signal $A \cos \omega_{y_0} y_0$ from the signal $\Phi(y_0) = A \cos \omega_{y_0} y_0 + \Phi(y_0)$.

5-6.5 Verify (5.99).

CHAPTER 6

THE LAPLACE TRANSFORM

6-1 INTRODUCTION

The Fourier integral, discussed at some length in Chapter 4, is most useful in theoretical work. It shows that if the steady-state response of a network is prescribed over an infinite band of frequencies, its transient response is uniquely specified. To specify one is to imply the other, and to change one is to change the other. It also gives some general information about the shape of the transient behavior as it is influenced by steady-state characteristics at various parts of the frequency spectrum. In some cases it shows the different effects of the amplitude and phase characteristics of the steady-state response of the network. However, the Fourier integral is not particularly useful in computing actual transient responses of networks. We will discuss the shortcomings of the Fourier integral approach in network analysis as a useful starting point for discussion of the Laplace transform.

To review the situation, suppose that a circuit is excited by a function $v(t)$. Furthermore, suppose that $v(t)$ is a positive time function; that is, it is zero until time $t = 0$. This time function can be represented by the frequency function given by the Fourier transform

$$V(\omega) = \mathscr{F}\{v(t)\} = \int_0^\infty v(t)e^{-j\omega t}\,dt \tag{6.1}$$

The spectrum of the response function of the network is given by the product of $V(\omega)$ and the transfer function $H(\omega)$ of the network

$$I(\omega) = H(\omega)V(\omega) \tag{6.2}$$

The response function is then obtained from the inverse Fourier transform

$$i(t) = \mathscr{F}^{-1}\{I(\omega)\} = \frac{1}{2\pi}\int_{-\infty}^\infty I(\omega)e^{j\omega t}\,d\omega \tag{6.3}$$

This equation gives the current for an initially relaxed system for $t \geq 0$ due to the voltage $v(t)$, which was zero up to $t = 0$. However, it is possible that the circuit was not initially relaxed; that is, initial currents and charges existed in the circuit before $v(t)$ began to act. Then (6.3) is not the complete solution for the current, and it follows that the frequency spectra of the voltage and current are not related by the simple relation $H(\omega)V(\omega)$. Of course, if there are currents and volt-

THE LAPLACE TRANSFORM

ages before zero time, they must be due to some previous excitation that might possibly be incorporated into $v(t)$ to obtain a solution. Usually there is no knowledge of the form of excitations acting prior to $t = 0$, only a knowledge of the initial conditions. Since practical problems are specified in terms of initial currents and charges and a suddenly applied excitation, our desired purpose is a direct solution that automatically takes account of the initial conditions.

In light of the foregoing discussion, note three shortcomings of the Fourier integral for transient analyses:

1. For some functions, such as the unit step, the Fourier integral does not converge.
2. The response function appears as an integral that may be difficult to evaluate.
3. The circuit must be initially relaxed.

These objections are largely overcome by the use of the Laplace transform.

The discussion of the Fourier transform in Chapter 4 shows that we can represent a function $f(t)$ by a continuous sum of exponential functions of the form $\exp[-j\omega t]$. The frequencies of these exponentials are restricted to the $j\omega$-axis in the complex frequency plane. This restriction proves undesirable in many cases and can be removed by representing $f(t)$ by a continuous sum of exponential functions of the form $\exp[-st]$, where $s = \sigma + j\omega$. The result is an integral closely related to the Fourier integral. This integral has the form

$$F(s) = \int_0^\infty f(t) e^{-st}\, dt \tag{6.4}$$

and is called the **Laplace transform** of $f(t)$. This expression reduces to that in (6.1) if $\sigma = 0$. Note that $e^{-st} = e^{-\sigma t} e^{-j\omega t}$; the factor $e^{-\sigma t}$ will make the integral converge for many functions for which the integral in (6.1) will not. If the integral in (6.4) converges for $s = \alpha_1 + j\omega$, it will converge for all s, if $\operatorname{Re}\{s\} \geq \alpha_1$. The parameter α_1 is called the **abscissa of convergence** for the function. It may be positive, negative, or zero, depending on the function.

The extension of the Laplace transform from the $j\omega$-axis to any point in the complex plane results in a number of changes in the properties associated with the Fourier transform. Basically, the Laplace transform pair will now become

$$\mathscr{L}\{f(t)\} \triangleq F(s) = \int_0^\infty f(t) e^{-st}\, dt \qquad \text{(a)}$$

$$\mathscr{L}^{-1}\{F(s)\} \triangleq f(t) = \frac{1}{2\pi j} \int_{\sigma - j\infty}^{\sigma + j\infty} F(s) e^{st}\, ds \qquad \text{(b)} \tag{6.5}$$

In this representation the symbol $\mathscr{L}\{f(t)\}$ is shorthand notation for the integral shown on the right. These equations should be compared with (4.1) for the Fourier transform pair.

■ ■ ■

6-2 THE BILATERAL LAPLACE TRANSFORM

We will proceed in our development of the Laplace transform from considerations of the Fourier transform pair given in (4.1). We begin with the Fourier transform pair

$$F(\omega) = \int_{-\infty}^{\infty} f(t)e^{-j\omega t}\, dt \quad \text{(a)} \tag{6.6}$$

and

$$f(t) = \frac{1}{2\pi} \int_{-\infty}^{+\infty} F(\omega)e^{j\omega t}\, d\omega \quad \text{(b)}$$

Now consider a function $\varphi(t)$ defined as

$$\varphi(t) = f(t)e^{-\sigma t} \tag{6.7}$$

where σ is a real constant. The Fourier transform of this function is

$$\mathscr{F}\{\varphi(t)\} = \int_{-\infty}^{\infty} f(t)e^{-\sigma t}e^{-j\omega t}\, dt = \int_{-\infty}^{\infty} f(t)e^{-(\sigma+j\omega)t}\, dt$$

$$= F(\sigma + j\omega) \tag{6.8}$$

It follows, therefore, that

$$\varphi(t) = \frac{1}{2\pi} \int_{-\infty}^{\infty} F(\sigma + j\omega)e^{j\omega t}\, d\omega$$

Thus we have

$$\varphi(t) \triangleq f(t)e^{-\sigma t} = \frac{1}{2\pi} \int_{-\infty}^{\infty} F(\sigma + j\omega)e^{j\omega t}\, d\omega$$

from which, upon rearrangement,

$$f(t) = \frac{1}{2\pi} \int_{-\infty}^{\infty} F(\sigma + j\omega)e^{(\sigma+j\omega)t}\, d\omega \tag{6.9}$$

The quantity $(\sigma + j\omega)$ is the complex frequency s, and so $d\omega = (1/j)ds$. When we combine this with (6.9), the limits of integration change from $\omega = -\infty$ to $+\infty$ to $s = \sigma - j\infty$ to $\sigma + j\infty$, with the result that

$$f(t) = \frac{1}{2\pi j} \int_{\sigma-j\infty}^{\sigma+j\infty} F(s)e^{st}\, ds \quad \text{(a)} \tag{6.10}$$

Also from (6.8) we have

$$F(s) = \int_{-\infty}^{\infty} f(t)e^{-st}\, dt \quad \text{(b)}$$

The path of integration in (6.10) is the line from $-j\infty$ to $+j\infty$ in the s-plane but displaced to the right of the origin by an amount σ. Observe that the Fourier transform can be considered a special case of the Laplace transform and can be obtained from (6.9) by setting $\sigma = 0$, or writing $s = j\omega$. In the form given, (6.10) defines the bilateral or two-sided Laplace transform pair. More will be said about this later in the chapter.

Because this development proceeded from the Fourier transform pair, it is required that for a function $f(t)$ to be Laplace-transformable, it must also satisfy the Dirichlet conditions (see Section 4-1). Furthermore, since s is a complex variable, (6.10a) involves integration in the complex plane, which will be studied in some detail later in this chapter. Note also that because of the manner of its development, tables of Fourier transforms can be used for Laplace transforms and vice versa, if certain precautions are exercised. We will not pursue this matter.

An advantage of the Laplace transform is that it overcomes the chief difficulties encountered in the use of the Fourier transform. The presence of the convergence factor $\exp[-\sigma t]$ in the integrand of $F(s)$ permits the integral to be evaluated, whereas the Fourier integral may not exist. An example of this is the difficulty in obtaining the Fourier transform of the unit step function and the relative ease in finding its Laplace transform.

■ ■ ■

6-3 THE ONE-SIDED LAPLACE TRANSFORM

It is usually possible to restrict considerations to functions that are zero for negative values of time; these are positive time functions, sometimes called causal functions. The response of physical systems can be determined for all $t > 0$ from a knowledge of the input for $t > 0$ and the energies stored in the system at $t = 0$. That is, the history of the system prior to the reference time $t = 0$ is not necessary. Thus in practical problems the two-sided Laplace transform of (6.10b) is usually replaced by the normal or **one-sided** Laplace transform

$$F(s) = \int_0^\infty f(t)e^{-st}\,dt \triangleq \mathscr{L}\{f(t)\} \tag{6.11}$$

The expression for the inverse transform is the same whether the one-sided or two-sided transform is used.

In our work we will consider piecewise continuous functions such that

$$\lim_{t \to \infty} f(t)e^{-ct} = 0 \qquad c = \text{real constant} \tag{6.12}$$

Functions of this type are known as functions of **exponential order** c. It can also be shown that

$$\int_0^\infty f(t)e^{-st}\,dt$$

converges if

$$\int_0^\infty |f(t)e^{-st}|\,dt = \int_0^\infty |f(t)|e^{-\sigma t}\,dt$$

converges. Therefore, if our function is of exponential order, we write the integral

Figure 6.1
Path of integration for (6.5b).

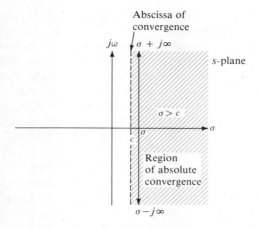

on the right as

$$\int_0^\infty |f(t)|e^{-ct}e^{-(\sigma-c)t}\,dt$$

This shows that for σ in the range

$$\sigma > c \tag{6.13}$$

(σ is the abscissa of convergence), the integral converges and the Laplace transform integral behaves as defined; that is,

$$F(s) = \int_0^\infty |f(t)e^{-st}|\,dt < \infty \qquad \text{Re}\{s\} > c \tag{6.14}$$

The importance of this result is that a finite number of infinite discontinuities are permissible so long as they have finite areas under them. In addition, the convergence is also uniform, which permits us to alter the order of integration in multiple integrals without affecting the results. The restriction in this equation, namely $\text{Re}\{s\} > c$, indicates that when we attempt to find the inverse Laplace transform, we must choose the path of integration as shown in Figure 6.1.

EXAMPLE 6.1
Find the Laplace transform of the unit step function $f(t) = u(t)$.

Solution: By (6.11)

$$\mathcal{L}\{u(t)\} = \int_0^\infty u(t)e^{-st}\,dt = \int_0^\infty e^{-st}\,dt = -\frac{e^{-st}}{s}\bigg|_0^\infty = \frac{1}{s} \tag{6.15}$$

The region of convergence is found from the expression $\int_0^\infty |e^{-st}|\,dt = \int_0^\infty e^{-\sigma t}\,dt < \infty$, which is the entire right half-plane, $\sigma > 0$. ∎

THE LAPLACE TRANSFORM

EXAMPLE 6.2
Find the Laplace transform of the unit impulse function.

Solution: Since $\delta(t)$ is nonzero only at the origin $t = 0$ and there is a finite area equal to unity under this function, then for any function $\psi(t)$ in general [see (1.15)]

$$\int_{-\infty}^{\infty} \psi(t)\delta(t)\,dt = \int_{0-}^{0+} \psi(t)\delta(t)\,dt = \psi(0)\int_{0-}^{0+} \delta(t)\,dt = \psi(0)$$

If we choose $\psi = \exp[-st]$ equal to unity for $t = 0$, then for $t \geq 0$,

$$\int_{-\infty}^{\infty} \delta(t)e^{-st}\,dt = \int_{0-}^{0+} \delta(t)e^{-st}\,dt = e^{-s0} = 1 \qquad (6.16)$$

again using the fact that the area under the unit impulse is unity. Because

$$\int_{-\infty}^{\infty} |\delta(t)e^{-st}|\,dt = \int_{-\infty}^{\infty} \delta(t)e^{-\sigma t}\,dt = 1$$

is independent of the value of σ, this implies that the region of convergence is the entire s-plane. ∎

The application of (6.11) to important common functions of time leads to the results contained in Table 6.1.

Table 6.1 Elementary Laplace Transform Pairs (continued on next page)

Entry No.	$f(t) = \dfrac{1}{2\pi j}\int_{\sigma-j\infty}^{\sigma+j\infty} F(s)e^{st}\,ds$	$F(s) = \int_0^\infty f(t)e^{-st}\,dt$
1	$\delta(t)$	1
2	$u(t)$	$\dfrac{1}{s}$
3	t^n for $n > 0$	$\dfrac{n!}{s^{n+1}}$
4	e^{-at}	$\dfrac{1}{s+a}$
5	te^{-at}	$\dfrac{1}{(s+a)^2}$
6	$\dfrac{t^{n-1}e^{-at}}{(n-1)!}$	$\dfrac{1}{(s+a)^n}$
7	$\dfrac{1}{b-a}(e^{-at} - e^{-bt})\quad a \neq b$	$\dfrac{1}{(s+a)(s+b)}$

Table 6.1 *continued*

Entry No.	$f(t) = \dfrac{1}{2\pi j} \displaystyle\int_{\sigma-j\infty}^{\sigma+j\infty} F(s)e^{st}\,ds$	$F(s) = \displaystyle\int_0^\infty f(t)e^{-st}\,dt$
8	$-\dfrac{1}{b-a}(ae^{-at} - be^{-bt}) \quad a \neq b$	$\dfrac{s}{(s+a)(s+b)}$
9	$\sin \omega t$	$\dfrac{\omega}{s^2 + \omega^2}$
10	$\cos \omega t$	$\dfrac{s}{s^2 + \omega^2}$
11	$e^{-at} \sin \omega t$	$\dfrac{\omega}{(s+a)^2 + \omega^2}$
12	$e^{-at} \cos \omega t$	$\dfrac{s+a}{(s+a)^2 + \omega^2}$
13	$\sinh \omega t$	$\dfrac{\omega}{s^2 - \omega^2}$
14	$\cosh \omega t$	$\dfrac{s}{s^2 - \omega^2}$
15	$\dfrac{\sqrt{a^2 + \omega^2}}{\omega} \sin(\omega t + \varphi), \; \varphi = \tan^{-1} \dfrac{\omega}{a}$	$\dfrac{s+a}{s^2 + \omega^2}$
16	$\dfrac{\omega_n}{\sqrt{1-\zeta^2}} e^{-\zeta\omega_n t} \sin \omega_n\sqrt{1-\zeta^2}\, t \quad \|\zeta\| < 1$	$\dfrac{\omega_n^2}{s^2 + 2\zeta\omega_n s + \omega_n^2}$
17	$\dfrac{1}{a^2 + \omega^2} + \dfrac{1}{\omega\sqrt{a^2+\omega^2}} e^{-at} \sin(\omega t - \varphi),$ $\varphi = \tan^{-1}\left(\dfrac{\omega}{-a}\right)$	$\dfrac{1}{s[(s+a)^2 + \omega^2]}$
18	$1 - \dfrac{1}{\sqrt{1-\zeta^2}} e^{-\zeta\omega_n t} \sin(\omega_n\sqrt{1-\zeta^2}\, t + \varphi),$ $\varphi = \cos^{-1} \zeta \quad \|\zeta\| < 1$	$\dfrac{\omega_n^2}{s(s^2 + 2\zeta\omega_n s + \omega_n^2)}$

■ ■ ■

6-4 PROPERTIES OF THE LAPLACE TRANSFORM

The functions we ordinarily encounter in system theory have discontinuities at $t = 0$, the time at which we close a switch to introduce an excitation. It is usually convenient to define the value of $f(t)$ at $t = 0$ to be the limit of $f(t)$ as t ap-

THE LAPLACE TRANSFORM

proaches zero from the positive direction. This limit is written $f(0+)$, a designation consistent with the choice of system response for $t > 0$. Thus $f(0+)$ denotes the initial condition; correspondingly, $f^{(n)}(0+)$ denotes the value of the nth derivative at time $t = 0+$, and $f^{(-n)}(0+)$ denotes the nth time integral at time $t = 0+$. This means that the definition of the direct Laplace transform can be written

$$F(s) = \lim_{\substack{R \to +\infty \\ a \to 0+}} \int_a^R f(t)e^{-st}\,dt, \qquad R > 0 \quad a > 0 \tag{6.17}$$

The most useful properties of the Laplace transform follow directly from this equation and are contained in Table 6.2 (p. 298). It is interesting to compare these properties with those of the Fourier transform summarized in Table 4.1.

EXAMPLE 6.3

Prove entry 1 in Table 6.2 assuming that the functions $f_1(t)$ and $f_2(t)$ are Laplace-transformable and with K_1 and K_2 being constants.

Solution: From (6.17) we write

$$\mathscr{L}[K_1 f_1(t) + K_2 f_2(t)] = \int_0^\infty [K_1 f_1(t) + K_2 f_2(t)]e^{-st}\,dt$$
$$= \int_0^\infty K_1 f_1(t)e^{-st}\,dt + \int_0^\infty K_2 f_2(t)e^{-st}\,dt$$
$$= K_1 F_1(s) + K_2 F_2(s) \tag{6.18}$$

∎

EXAMPLE 6.4

Prove entry 2 in Table 6.2.

Solution: Begin with the basic definition of the Laplace transform and write

$$\mathscr{L}\left[\frac{df(t)}{dt}\right] = \int_0^\infty \frac{df(t)}{dt} e^{-st}\,dt$$

Integrate by parts by writing

$$u = e^{-st} \qquad du = -se^{-st}\,dt$$

$$dv = \frac{df}{dt}\,dt \qquad v = f$$

Then

$$\mathscr{L}\left[\frac{df(t)}{dt}\right] = f(t)e^{-st}\bigg|_0^\infty + s \int_0^\infty f(t)e^{-st}\,dt$$

Table 6.2 Properties of the Laplace Transform

Property

1. Linearity $\mathscr{L}[K_1 f_1(t) + K_2 f_2(t)] = \mathscr{L}[K_1 f_1(t)] + \mathscr{L}[K_2 f_2(t)] = K_1 F_1(s) + K_2 F_2(s)$

2. Time derivative $\mathscr{L}\left[\dfrac{d}{dt} f(t)\right] = sF(s) - f(0+)$

3. Higher time derivative

$$\mathscr{L}\left[\frac{d^n}{dt^n} f(t)\right] = s^n F(s) - s^{n-1} f(0+) - s^{n-2} f^{(1)}(0+) - \cdots - f^{(n-1)}(0+)$$

where $f^{(i)}(0+)$, $i = 1, 2, \ldots, n - 1$ is the ith derivative of $f(\cdot)$ at $t = 0+$.

4. Integral with zero initial condition $\mathscr{L}\left[\displaystyle\int_0^t f(\xi)\,d\xi\right] = \dfrac{F(s)}{s}$

5. Integral with initial conditions $\mathscr{L}\left[\displaystyle\int_0^t f(\xi)\,d\xi\right] = \dfrac{F(s)}{s} + \dfrac{f^{(-1)}(0+)}{s}$ where

$$f^{(-1)}(0+) = \lim_{t \to 0+} \int_{-\infty}^0 f(\xi)\,d\xi$$

6. Multiplication by exponential $\mathscr{L}[e^{-at} f(t)] = F(s + a)$

7. Multiplication by t $\mathscr{L}[t f(t)] = -\dfrac{d}{ds} F(s)$

8. Time shifting $\mathscr{L}[f(t - \lambda) u(t - \lambda)] = e^{-s\lambda} F(s)$

9. Scaling $\mathscr{L}\left[f\left(\dfrac{t}{a}\right)\right] = aF(as) \qquad a > 0$

10. Time convolution $\mathscr{L}\left[\displaystyle\int_0^t f_1(t - \tau) f_2(\tau)\,d\tau\right] \triangleq \mathscr{L}[f_1(t) * f_2(t)] = F_1(s) F_2(s)$

11. Frequency convolution

$$\mathscr{L}[f_1(t) f_2(t)] = \frac{1}{2\pi j} \int_{x - j\infty}^{x + j\infty} F_1(z) F_2(s - z)\,dz = \frac{1}{2\pi j} [F_1(s) * F_2(s)]$$

where $z = x + jy$, and where x must be greater than the abscissa of absolute convergence for $f_1(t)$ over the path of integration

12. Initial value $\lim\limits_{t \to 0+} f(t) = \lim\limits_{s \to \infty} sF(s)$ provided that this limit exists

13. Final value $\lim\limits_{t \to \infty} f(t) = \lim\limits_{s \to 0} sF(s)$ provided that $sF(s)$ is analytic on the $j\omega$ axis and in the right half of the s-plane

14. Division by t $\mathscr{L}\left\{\dfrac{f(t)}{t}\right\} = \displaystyle\int_s^\infty F(s')\,ds'$

THE LAPLACE TRANSFORM

But $\lim_{t \to \infty} f(t) \exp[-st] = 0$, otherwise the transform would not exist. Thus

$$\mathscr{L}\left[\frac{df(t)}{dt}\right] = s\mathscr{L}[f(t)] - f(0+)$$

so that

$$\mathscr{L}\left[\frac{df(t)}{dt}\right] = sF(s) - f(0+) \qquad (6.19)$$

Note the important fact that this expression contains the term $f(0+)$. By reference to entry 1 in Table 6.1, $f(0+)$ represents an impulse of strength $f(0+)$. This yields an important result in network problems since it shows that initial conditions associated with derivative functions are automatically included as series impulse functions in the network description. For example, if $f(t)$ denotes the current through an inductor, then terms of the form $v = L\,di/dt$ are included in the network equations. In this case, when such a term is Laplace-transformed, there will be terms of the form

$$\mathscr{L}\left[L\frac{di(t)}{dt}\right] = LsI(s) - Li(0+)$$

where $i(0+)$ denotes the initial current through the inductor. This result shows that an inductor with an initial current can be regarded in subsequent calculations as the equivalent of an initially relaxed inductor which leads to the term $sLI(s)$ plus an impulse voltage source $-Li(0+)$ at time $t = 0$ due to the initial current. ∎

EXAMPLE 6.5
Prove entry 3 in Table 6.2.

Solution: For this evaluation we use the result in Example 6.4 by noting that

$$\frac{d^2f(t)}{dt^2} = \frac{d}{dt}\left[\frac{df(t)}{dt}\right]$$

Then

$$\mathscr{L}\left[\frac{d^2f(t)}{dt^2}\right] = \mathscr{L}\left[\frac{d}{dt}\frac{df(t)}{dt}\right] = s\mathscr{L}\left[\frac{df(t)}{dt}\right] - \frac{df(0+)}{dt}$$

Again using (6.19)

$$\mathscr{L}\left[\frac{d^2f(t)}{dt^2}\right] = s^2F(s) - sf(0+) - f^{(1)}(0+) \qquad (6.20)$$

where the bracketed exponent number indicates the degree of the derivative.

The Laplace transform of the nth time derivative follows as a direct extension of the foregoing development. The result is entry 3 in Table 6.2. If all initial values are zero, this expression reduces to

$$\mathscr{L}\left[\frac{d^n f(t)}{dt^n}\right] = s^n F(s) \qquad (6.21)$$

This relationship shows that the differential operator (d^n/dt^n) becomes s^n in the Laplace transforming process. ■

EXAMPLE 6.6
Prove entry 4 in Table 6.2.

Solution: If the function $f(t)$ is Laplace-transformable, then its integral is written

$$\mathscr{L}\left[\int_{-\infty}^{t} f(\xi)\,d\xi\right] = \int_{0}^{\infty}\left[\int_{-\infty}^{t} f(\xi)\,d\xi\right] e^{-st}\,dt$$

This is integrated by parts by writing

$$u = \int_{-\infty}^{t} f(\xi)\,d\xi \qquad du = f(\xi)\,d\xi = f(t)\,dt$$

$$dv = e^{-st}\,dt \qquad v = -\frac{1}{s} e^{-st}$$

Then

$$\mathscr{L}\left[\int_{-\infty}^{t} f(\xi)\,d\xi\right] = \left[-\frac{e^{-st}}{s}\int_{-\infty}^{t} f(\xi)\,d\xi\right]\bigg|_{0}^{\infty} + \frac{1}{s}\int_{0}^{\infty} f(t) e^{-st}\,dt$$

$$= \frac{1}{s}\int_{0}^{\infty} f(t) e^{-st}\,dt + \frac{1}{s}\int_{-\infty}^{0} f(\xi)\,d\xi$$

from which

$$\mathscr{L}\left[\int_{\infty}^{t} f(\xi)\,d\xi\right] = \frac{1}{s} F(s) + \frac{1}{s} f^{(-1)}(0+) \qquad (6.22)$$

where $[f^{(-1)}(0+)/s]$ is the initial value of the integral of $f(t)$ at $t = 0+$ and the negative number in the bracketed exponent indicates integration. Note by entry 2 in Table 6.1 that this term denotes a step function of amplitude $f^{(-1)}(0+)$. This is also an important result in network problems since it shows that initial conditions associated with integral functions are automatically included as step functions in the Laplace transform development. For example, if $f(t)$ denotes a current $i(t)$ through a capacitor, then the voltage across the capacitor is expressed by the relation

$$v(t) = \frac{q(t)}{C} = \frac{1}{C}\int_{-\infty}^{t} i(\xi)\,d\xi$$

The Laplace transform of such a term is then

Figure 6.2
A function $f(t)u(t)$ and the same function delayed by a time $t = \lambda$.

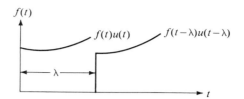

$$\mathscr{L}\left[\frac{q(t)}{C}\right] = \mathscr{L}\left[\frac{1}{C}\int_{-\infty}^{t} i(\xi)\,d\xi\right] = \frac{I(s)}{Cs} + \frac{i^{(-1)}(0+)}{sC}$$

$$= \frac{I(s)}{Cs} + \frac{q(0+)}{sC} = \frac{I(s)}{Cs} + \frac{V(0+)}{s}$$

where $q(0+)$ is the charge on the capacitor at initial time $t = 0+$. This result means that an initially charged capacitor can be regarded, insofar as the subsequent action in a circuit is concerned, as an initially relaxed capacitor plus a series step-function voltage source, or equivalently (because of source transformations), as a shunting delta function current source. ∎

EXAMPLE 6.7
Prove entry 8 in Table 6.2.

Solution: This entry relates to the transform of a function that has been translated along the time axis. This situation is illustrated in Figure 6.2, which shows the translation of a function $f(t)$ to the right by λ units of time, where λ is a positive constant. Upon introducing the translated function into (6.17), we obtain

$$\mathscr{L}[f(t-\lambda)u(t-\lambda)] = \int_{0}^{\infty} f(t-\lambda)u(t-\lambda)e^{-st}\,dt$$

Now introduce a new variable $\tau = t - \lambda$, which converts this equation to the form

$$\mathscr{L}[f(\tau)u(\tau)] = e^{-s\lambda}\int_{-\lambda}^{\infty} f(\tau)u(\tau)e^{-s\tau}\,d\tau$$

$$= e^{-s\lambda}\int_{0}^{\infty} f(\tau)e^{-s\tau}\,d\tau = e^{-s\lambda}F(s) \quad (6.23)$$

since $u(\tau) = 0$ for $-\lambda \leq \tau \leq 0$.

Similarly, we find that

$$f(t+\lambda)u(t+\lambda) = e^{s\lambda}F(s) \quad (6.24)$$

∎

EXAMPLE 6.8
Find the Laplace transform of the pulse function shown in Figure 6.3.

Figure 6.3
Pulse function and its equivalent representation.

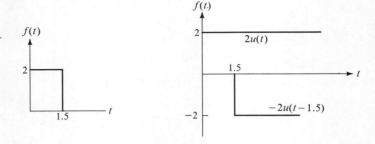

Solution: Since the pulse function can be decomposed into two step functions as shown in the figure, its Laplace transform is given by

$$\mathscr{L}[2[u(t) - u(t - 1.5)]] = 2\left(\frac{1}{s} - \frac{1}{s}e^{-1.5s}\right) = \frac{2}{s}(1 - e^{-1.5s})$$

where the shifting property has been used. ∎

EXAMPLE 6.9
Prove entry 10 in Table 6.2.

Solution: This entry specifies the Laplace transform of the convolution of functions $f_1(t)$ and $f_2(t)$.

$$\mathscr{L}[f_1(t) * f_2(t)] = \mathscr{L}\left[\int_0^\infty f_1(t - \tau) f_2(\tau)\, d\tau\right]$$

$$= \int_0^\infty \left[\int_0^\infty f_1(t - \tau) f_2(\tau)\, d\tau\right] e^{-st}\, dt$$

$$= \int_0^\infty f_2(\tau)\, d\tau \int_0^\infty f_1(t - \tau) e^{-st}\, dt$$

Now by a change of variable, writing $t - \tau = \eta$ and therefore $dt = d\eta$, then

$$= \int_0^\infty f_2(\tau)\, d\tau \int_{-\tau}^\infty f_1(\eta) e^{-s(\eta + \tau)}\, d\eta$$

But for positive time functions $f_1(\eta) = 0$ for $\eta < 0$, which permits changing the lower limit of the second integral, and

$$= \int_0^\infty f_2(\tau) e^{-s\tau}\, d\tau \int_0^\infty f_1(\eta) e^{-s\eta}\, d\eta$$

which is

$$\mathscr{L}[f_1(t) * f_2(t)] = F_1(s) F_2(s) \tag{6.25}$$

∎

Figure 6.4
Illustrating Example 6.10.

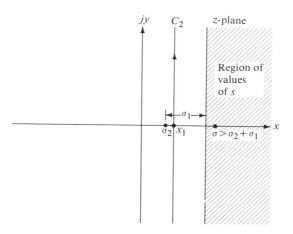

†EXAMPLE 6.10
Prove entry 11 in Table 6.2.

Solution: Begin by considering the following line integral in the z-plane

$$f_2(t) = \frac{1}{2\pi j} \int_{C_2} F_2(z) e^{zt}\, dz \qquad \sigma_2 = \text{abscissa of convergence}$$

It therefore follows that the contour intersects the x-axis at $x_1 > \sigma_2$ (see Figure 6.4). Then we have

$$\int_0^\infty f_1(t) f_2(t) e^{-st}\, dt = \frac{1}{2\pi j} \int_0^\infty f_1(t)\, dt \int_{C_2} F_2(z) e^{(z-s)t}\, dz$$

Assume that the integral of $F_2(z)$ is convergent over the path of integration. Thus we write this equation in the form

$$\int_0^\infty f_1(t) f_2(t) e^{-st}\, dt = \frac{1}{2\pi j} \int_{\sigma_2 - j\infty}^{\sigma_2 + j\infty} F_2(z)\, dz \int_0^\infty f_1(t) e^{-(s-z)t}\, dt$$

$$= \frac{1}{2\pi j} \int_{\sigma_2 - j\infty}^{\sigma_2 + j\infty} F_2(z) F_1(s-z)\, dz \triangleq \mathscr{L}\{f_1(t) f_2(t)\} \quad (6.26)$$

The Laplace transform of $f_1(t)$ converges in the range $\operatorname{Re}\{s - z\} > \sigma_1$, where σ_1 is the abscissa of convergence of $f_1(t)$ (see Section 6-3). In addition, of course, $\operatorname{Re}\{z\} = \sigma_2$ for the z-plane integration involved in (6.26) so that the abscissa of convergence of $f_1(t) f_2(t)$ is specified by

$$\operatorname{Re}\{s\} > \sigma_1 + \sigma_2 \quad (6.27)$$

As far as the integration in the complex plane is concerned, the semicircle can be closed either to the left or to the right as long as $F_1(s)$ and $F_2(s)$ go to zero as $s \to \infty$.

On the basis of the foregoing, we observe the following:

a. Poles* of $F_1(s-z)$ are contained in the region $\text{Re}\{s-z\} < \sigma_1$.
b. Poles of $F_2(z)$ are contained in the region $\text{Re}\{z\} < \sigma_2$.
c. From (a) and (6.27) $\text{Re}\{z\} > \text{Re}\{s - \sigma_1\} > \sigma_2$.
d. Poles of $F_1(s-z)$ are to the right of the path of integration.
e. Poles of $F_2(z)$ are to the left of the path of integration.
f. Poles of $F_1(s-z)$ are functions of s, whereas poles of $F_2(z)$ are fixed in relation to s. ∎

---

†EXAMPLE 6.11
Find the Laplace transform of the function $f(t) = f_1(t)f_2(t) = e^{-t}e^{-2t}u(t)$.

Solution: From Example 6.10 and the absolute convergence region for each function, we have

$$F_1(s) = \frac{1}{s+1} \qquad \sigma_1 > 1$$

$$F_2(s) = \frac{1}{s+2} \qquad \sigma_2 > 2$$

since $f(t) = \exp[-(2+1)t]u(t)$ implies that $\sigma_f = \sigma_1 + \sigma_2 = 3$. We now write

$$F_2(z)F_1(s-z) = \frac{1}{z+2}\frac{1}{s-z+1} = \frac{1}{3+s}\frac{1}{z-(1+s)} - \frac{1}{3+s}\frac{1}{z+2}$$

and we use the contour shown in Figure 6.5, in accordance with our observations relating to the previous example. If we select the contour C_1 and use the residue theorem, we obtain

$$F(s) = \frac{1}{2\pi j}\oint_{C_1} F_2(z)F_1(s-z)\,dz = 2\pi j\,\text{Res}[F_2(z)F_1(s-z)]\bigg|_{z=-2} = \frac{1}{s+3}$$

The inverse transform of this is $\exp[-3t]$. If we select the contour C_2, the residue theorem gives

$$F(s) = \frac{-1}{2\pi j}\oint_{C_2} F_2(z)F_1(s-z)\,dz = -2\pi j\,\text{Res}[F_2(z)F_1(s-z)]\bigg|_{z=(1+s)}$$

$$= -\left[-\frac{1}{s+3}\right] = \frac{1}{s+3}$$

The inverse transform of this is also $\exp[-3t]$, as it should be. ∎

---

* **Poles** of a function are those values of the variable (real or complex) at which a function becomes infinite. Correspondingly, the **zeros** of a function are those values of the variable (real or complex) at which a function becomes zero.

Figure 6.5
Illustrating Example 6.11.

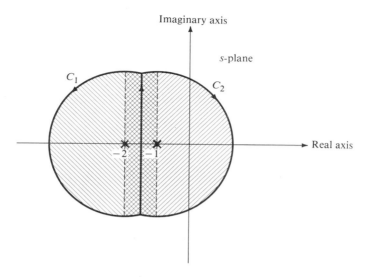

EXAMPLE 6.12
Prove entry 12 in Table 6.2.

Solution: This result is known as the Initial Value Theorem. Recall that for a Laplace-transformable function $f(t)$, we can deduce the corresponding $F(s)$. Further, for a specified $F(s)$, we can find the corresponding $f(t)$ by inversion, and from this calculate the initial value $f(0)$. Entry 12 permits $f(0)$ to be calculated directly from $F(s)$ without the need for inversion.

To establish the result, consider (6.19) as $s \to \infty$. That is, we examine the expression

$$\lim_{s \to \infty} \int_0^\infty \frac{df}{dt} e^{-st}\, dt = \lim_{s \to \infty} [sF(s) - f(0+)]$$

It is assumed here, of course, that $f(t)$ and its first derivative are Laplace-transformable and that the limit of $sF(s)$ as s approaches infinity exists. But the integral vanishes for $s \to \infty$, and $f(0+)$ is independent of s, so that

$$\lim_{s \to \infty} [sF(s) - f(0+)] = 0$$

Furthermore, $f(0+) = \lim_{t \to 0+} f(t)$, so that

$$\lim_{s \to \infty} sF(s) = \lim_{t \to 0+} f(t) \tag{6.28}$$

If $f(t)$ has a discontinuity at the origin, this expression specifies the value of the impulse $f(0+)$. If $f(t)$ contains an impulse term, then the left-hand side does not exist, and the initial value property does not hold. ∎

6-5 SYSTEM ANALYSIS—TRANSFER FUNCTION OF LTI SYSTEMS IN BLOCK DIAGRAM REPRESENTATION

The transfer function $H(s)$ of an LTI system is defined as the ratio of the Laplace transform of its output to the Laplace transform of its input. $H(s)$ describes the properties of the system alone; that is, the system is assumed to be in its quiescent state; hence the initial conditions are assumed to be zero. Thus if the input is $f(t)$ and the output of the system is $g(t)$, we write

$$H(s) = \frac{G(s)}{F(s)} = \text{transfer function or system function} \tag{6.29}$$

The following examples will illustrate how we can find the transfer functions of systems from their time-domain representation. Table 6.3 is provided to help speed the reduction process of complicated block diagrams of systems. The reader should also consult Section 2-11.

Table 6.3 Properties of Block Diagrams

System Diagram	Equivalent Diagram	Observations
$u \rightarrow [a] \rightarrow [b] \rightarrow y$	$u \rightarrow [ab] \rightarrow y$	Two blocks in cascade
$u \xrightarrow{+} \Sigma \xrightarrow{y=u+v}$, v input $+$		Summation point
$u \xrightarrow{+} \Sigma \xrightarrow{y=u-v}$, v input $-$		Subtraction point
$u \rightarrow \bullet$ branching to three u outputs		Pickoff point

THE LAPLACE TRANSFORM

Table 6.3 *continued*

System Diagram	Equivalent Diagram	Observations
	$\dfrac{1}{1 \mp a}$	Feedback loop
	$\dfrac{a}{1 \mp a}$	Special case of unit feedback loop
	$\dfrac{a}{1 \mp ab}$	Complete feedback loop
		Moving a pickoff point ahead
		Moving a pickoff point behind
		Moving a summing point ahead
		Moving a summing point behind

Figure 6.6
(a) An electrical system.
(b) Its block diagram representation.

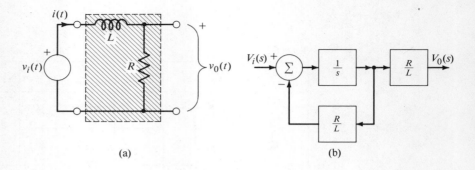

EXAMPLE 6.13
Find the transfer function of the system shown in Figure 6.6a.

Solution: The differential equation describing the system is

$$L\frac{di(t)}{dt} + Ri(t) = v_i(t)$$

or

$$\frac{L}{R}\frac{dv_0(t)}{dt} + v_0(t) = v_i(t) \qquad (6.30)$$

The Laplace transform of both sides of the differential equation is taken, with the linearity and differentiation properties of the Laplace transform and zero initial condition being invoked. We then write

$$\frac{Ls}{R}V_0(s) + V_0(s) = V_i(s)$$

from which we find that

$$H(s) \triangleq \frac{V_0(s)}{V_i(s)} = \frac{\frac{R}{L}}{s + \frac{R}{L}} \qquad (6.31)$$

The block diagram representation of the system is shown in Figure 6.6b (see Section 2-9). ∎

EXAMPLE 6.14
Find the transfer function $H(s) = V_1(s)/F(s)$ for the mechanical system shown in Figure 6.7a.

Figure 6.7
(a) A mechanical system. (b) The network representation. (c) A block diagram representation.

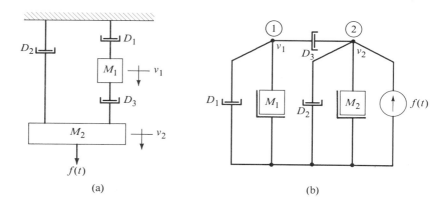

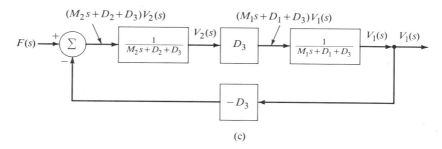

Solution: The circuit representation of the system is shown in Figure 6.7b. The node equations at nodes 1 and 2 are

$$M_1 \frac{dv_1}{dt} + D_1 v_1 + D_3(v_1 - v_2) = 0 \quad \text{(a)}$$

$$M_2 \frac{dv_2}{dt} + D_3(v_2 - v_1) + D_2 v_2 = f \quad \text{(b)}$$

(6.32)

We next Laplace-transform these equations to write

$$(M_1 s + D_1 + D_3)V_1(s) - D_3 V_2(s) = 0 \quad \text{(a)}$$

$$-D_3 V_1(s) + (M_2 s + D_2 + D_3)V_2(s) = F(s) \quad \text{(b)}$$

(6.33)

Solve this set of equations for $V_1(s)$, and then find

$$H(s) \triangleq \frac{V_1(s)}{F(s)} = \frac{D_3}{(M_1 s + D_1 + D_3)(M_2 s + D_2 + D_3) - D_3^2} \quad (6.34)$$

The block diagram representation of the system given in Figure 6.7c was obtained from (6.33a) and (6.33b). ∎

EXAMPLE 6.15

Find the transfer function $H(s) = \Omega(s)/E(s)$ of the rotational electromechanical transducer shown in Figure 6.8a. Mechanical and air friction damping are taken into consideration by the damping constant D. The movement of the cylinder-pointer combination is restrained by a spring, with spring constant K_s. The moment of inertia of the coil assembly is J. There are N turns on the coil. (In this example e's define voltages and v's define velocities.)

Solution: Because there are N turns on the coil, there are $2N$ conductors of length l perpendicular to the magnetic field at a distance a from the center of rotation. Therefore the electrical torque is

$$\mathcal{T}_e = f_t a = (2NBli)a = (2NBla)i = K_e i \tag{6.35}$$

The spring develops an equal and opposite torque that is written

$$\mathcal{T}_s = K_s \theta \qquad K_s = \text{(newton-meter)/degree} \tag{6.36}$$

Due to the movement of the coil in the magnetic field, there is an induced voltage generated in the coil. This is given by

$$e_m = 2NBlv = (2NBla)\frac{d\theta}{dt} = K_m \frac{d\theta}{dt} = K_m \omega \tag{6.37}$$

where K_m is equal to K_e. From the equivalent circuit representation of the system shown in Figure 6.8b, we obtain the equations

$$L\frac{di}{dt} + Ri + K_m \omega = e \qquad \text{Kirchhoff's voltage law} \quad \text{(a)}$$
$$\tag{6.38}$$
$$J\frac{d\omega}{dt} + D\omega + K_s \int \omega \, dt = K_e i \qquad \text{D'Alembert's principle} \quad \text{(b)}$$

The Laplace transform of these equations yields

$$(Ls + R)I(s) + K_m \Omega(s) = E(s) \quad \text{(a)}$$
$$\left(Js + D + \frac{K_s}{s}\right)\Omega(s) - K_e I(s) = 0 \quad \text{(b)} \tag{6.39}$$

Substitute the value of $I(s)$ from the second of these equations into the first and solve for the ratio $\Omega(s)/E(s)$. We obtain

$$H(s) \triangleq \frac{\Omega(s)}{E(s)} = \frac{K_e}{(Ls + R)\left(Js + D + \frac{K_s}{s}\right) + K_e K_m} \tag{6.40}$$

A block diagram representation of this system, which was found using (6.39a) and (6.39b), is shown in Figure 6.8c. ∎

THE LAPLACE TRANSFORM

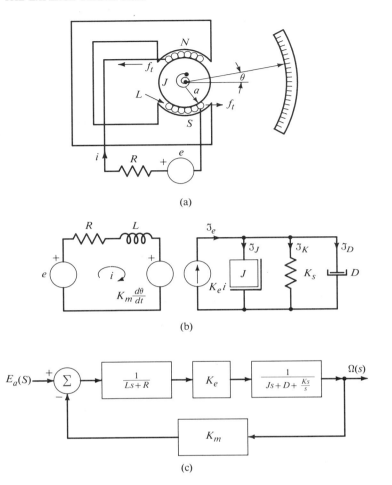

Figure 6.8
A rotational electromechanical transducer. (a) Physical model. (b) Circuit model. (c) Block diagram representation.

EXAMPLE 6.16
Find the transfer function $H(s) = \Omega(s)/E_a(s)$ for a dc motor (Figure 6.9) that operates with constant field current (see also Example 2.28). In this example the symbol e defines voltage and the symbol v defines velocities.

Solution: From Figure 2.45 and the development of Example 2.28, we obtain the following:

Armature circuit equation

$$E_a(s) = (R_a + L_a s)I_a(s) + E_c(s) \qquad \text{[Equation (2.146)]} \qquad (6.41)$$

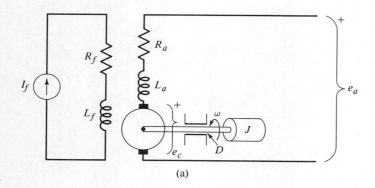

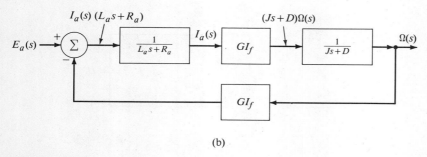

Figure 6.9
The schematic and block diagram representations of a DC motor with load and constant field current.

Electric torque produced by the two currents

$$\mathcal{T}_e(s) = (GI_f)I_a(s) \quad \text{[Equation (2.149)]} \quad (6.42)$$

Counter-electromotive force proportional to the motor speed

$$E_c(s) = (GI_f)\Omega(s) \quad \text{[Equation (2.147)]} \quad (6.43)$$

Load torque

$$\mathcal{T}_L(s) = (Js)\Omega(s) + D\Omega(s) = \mathcal{T}_e(s) \quad \text{[Equation (2.150)]} \quad (6.44)$$

By combining the above equations, we obtain the equation set

$$\begin{aligned} Js\Omega(s) + D\Omega(s) &= (GI_f)I_a(s) &\textbf{(a)}\\ E_a(s) &= (R_a + L_a s)I_a(s) + (GI_f)\Omega(s) &\textbf{(b)} \end{aligned} \quad (6.45)$$

Now eliminate $I_a(s)$ from these equations to obtain the transfer function

$$H(s) \triangleq \frac{\Omega(s)}{E_a(s)} = \frac{GI_f}{(L_a s + R_a)(Js + D) + (GI_f)^2} \quad (6.46)$$

The block diagram representation of the system is shown in Figure 6.9. ∎

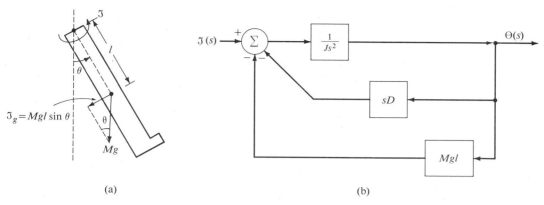

Figure 6.10
The pendulum system and its block diagram representation.

EXAMPLE 6.17

Find the transfer function $T(s) \triangleq \Theta(s)/\mathscr{T}(s)$ for the mechanical system (pendulum) shown in Figure 6.10a. This system represents an idealized model of a stiff human (or animal) limb in order to assess the passive control process of the locomotive action. Assume that friction exists during the movement, which is specified by the friction constant D.

Solution: Apply D'Alembert's principle for rotational systems (see Example 2.12) to write

$$\mathscr{T} = \mathscr{T}_g + \mathscr{T}_D + \mathscr{T}_J \tag{6.47}$$

where:

$\mathscr{T}$ = input torque $\mathscr{T}_g$ = gravity torque = $Mgl \sin \theta$

$\mathscr{T}_D$ = frictional torque = $D\omega = D\dfrac{d\theta}{dt}$

$\mathscr{T}_J$ = inertial torque = $J\dfrac{d\omega}{dt} = J\dfrac{d^2\theta}{dt^2}$

The equation describing the system is

$$J\frac{d^2\theta(t)}{dt^2} + D\frac{d\theta(t)}{dt} + Mgl \sin \theta(t) = \mathscr{T}(t) \tag{6.48}$$

Note that this equation is nonlinear owing to the presence of the $\sin \theta$ term. If we assume small deflections, we can linearize the equation by substituting θ for $\sin \theta$. Now (6.48) is approximated by

$$J\frac{d^2\theta(t)}{dt^2} + D\frac{d\theta(t)}{dt} + Mgl\theta(t) = \mathscr{T}(t) \tag{6.49}$$

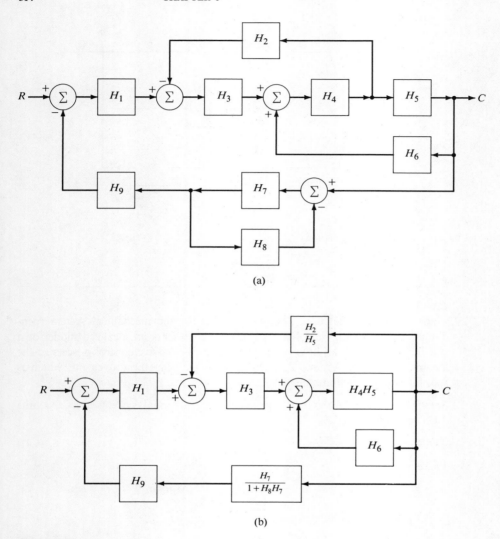

Figure 6.11
Illustrating Example 6.18.

Laplace-transform this equation and solve for the ratio $T(s) = \Theta(s)/\mathcal{T}(s)$ to obtain

$$T(s) \triangleq \frac{\Theta(s)}{\mathcal{T}(s)} = \frac{1}{Js^2 + Ds + Mgl} \qquad (6.50)$$

We note that $T(s)$ is a transfer function but is not $H(s)$ as generally defined. The block diagram representation of this system is shown in Figure 6.10b. ∎

THE LAPLACE TRANSFORM

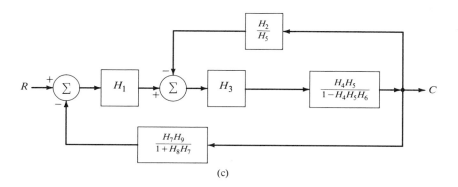

(c)

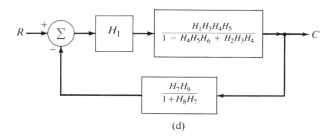

(d)

$$\frac{H_1H_3H_4H_5\,(1+H_7H_8)}{(1-H_4H_5H_6+H_2H_3H_4)(1+H_7H_8)+H_1H_3H_4H_5H_7H_9}$$

(e)

EXAMPLE 6.18

Use Table 6.3 to find the transfer function of the system shown in Figure 6.11a.

Solution: The steps in the reduction process are shown in Figures 6.11b through 6.11e. ∎

EXAMPLE 6.19
Find the transfer function $H(s) \triangleq V_o(s)/V_i(s)$ for the system shown in Figure 6.12a.

Solution: We have seen in the Laplace transform process for zero initial conditions (see the previous examples) that the operator d/dt is replaced by s and the operator $\int dt$ is replaced by $1/s$. Therefore we can write the Kirchhoff voltage law in Laplace form by direct reference to Figure 6.12b. The equations are:

$$(R_1 + L_1 s)I_1(s) - L_1 s I_2(s) = V_i(s) \quad \text{(a)}$$

$$-L_1 s I_1 + (R_2 + L_1 s + L_2 s)I_2(s) = 0 \quad \text{(b)} \quad\quad (6.51)$$

$$L_2 s I_2(s) = V_o(s) \quad \text{(c)}$$

These equations are shown in Figures 6.12c, d, and e, respectively. When the parts are combined, the resulting block diagram is that shown in Figure 6.12f. With the help of entries in Table 6.3, the transfer function is now easily determined and is shown in Figure 6.12g. ∎

■ ■ ■

*6-6 SYSTEM ANALYSIS—SIGNAL FLOW GRAPH (SFG) REPRESENTATION OF LTI SYSTEMS

We will now develop an alternate technique for representing system equations. This technique is known as **signal flow graphs** (SFG). This technique is closely related to the block diagram reduction method discussed in Section 6-5, but there are noticeable differences between the two methods.

The variables in a signal flow graph are represented by points called **nodes.** Connections between nodes are indicated by directed lines specifying the direction of signal flow; these are called **transmittances.** Generally, independent variables (real or assumed) are viewed as system inputs. The dependent variables, representing the unknown quantities, are viewed as system responses. The paths of interaction and the value of the transmittances along them portray the system interconnections and the effect of one quantity on another.

The rules that exist for the SFG allow algebraic transformations to be effected readily, and a given graph can be rapidly transformed into a form that may have some special attribute for subsequent study. Often the SFG makes apparent combinations or transformations that might lead to especially desirable forms. It must be realized, however, that the SFG technique involves the manipulation of the mathematical model of an interconnected physical system. It is less useful in studying the physical system itself.

---

* This section closely follows Chapter 6 of Seely and Poularikas, *Electrical Engineering: Introduction and Concepts*, Matrix Publishers, 1982, by permission.

THE LAPLACE TRANSFORM

Figure 6.12
Illustrating Example 6.19.

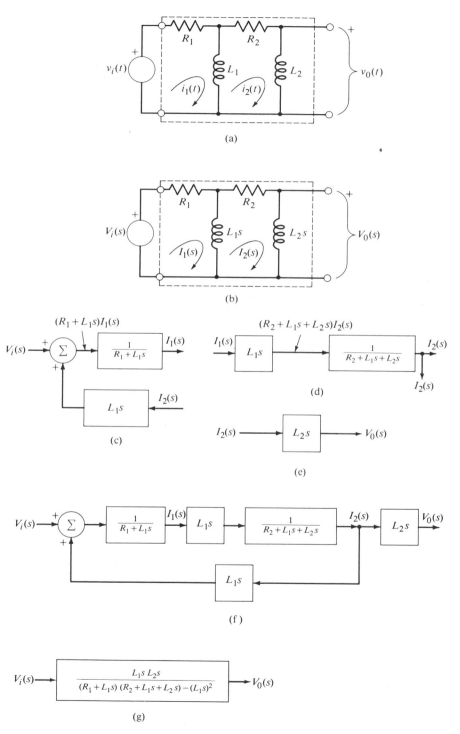

The rules for drawing a signal flow graph are the following:

1. Signals travel along branches only in the direction of the arrows.
2. The value of a node is the value of the nodes from which signals arise multiplied by the transmittances of the branches connecting the nodes.
3. The value of the variable represented by any node is the sum of all signals entering that node.
4. The value of the variable represented by any node is transmitted to all branches leaving that node.
5. The node with outgoing branches only is an input node (or source), and the node with incoming branches only is an output node (or sink).

The following examples illustrate these rules.

EXAMPLE 6.20
Draw the SFG's of the following equations:

$$x_0 = t_{10}x_1 + t_{20}x_2 + t_{30}x_3 \quad \text{(addition rule)} \quad \textbf{(a)}$$

$$x_1 = t_{01}x_0 \quad x_2 = t_{02}x_0 \quad \text{(transmission rule)} \quad \textbf{(b)}$$
$$x_3 = t_{03}x_0$$

$$x_1 = t_{01}x_0 \quad x_2 = t_{12}x_1 \quad \text{(multiplication rule)} \quad \textbf{(c)} \quad (6.52)$$
$$x_3 = t_{23}x_2 = t_{23}t_{12}t_{01}x_0$$

$$x_1 = t_{11}x_1 \quad \text{(loop)} \quad \textbf{(d)}$$

Solution: The graphs are shown in Figure 6.13a–6.13d, respectively. ∎

EXAMPLE 6.21
Refer to the electrical network shown in Figure 6.14a. Draw an SFG in a form that shows the signal flow characteristics through the network from input to output.

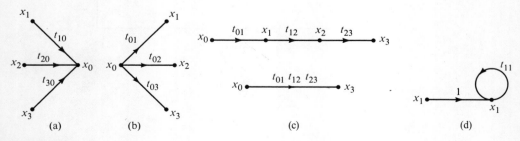

Figure 6.13
The SFG's of the equations given in Example 6.20.

Figure 6.14
A circuit and its signal flow representations.

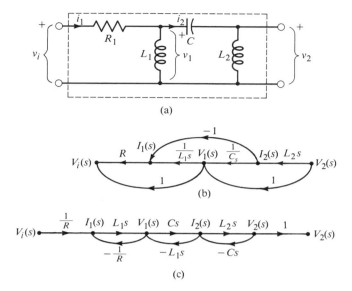

Solution: From an inspection of the given circuit we write

$$v_i = Ri_1 + v_1 \qquad v_1 = \frac{1}{C}\int i_2\, dt + v_2$$

$$v_1 = L_1 \frac{d}{dt}(i_1 - i_2) \qquad v_2 = L_2 \frac{di_2}{dt} \tag{6.53}$$

The corresponding Laplace-transformed equations are

$$V_i(s) = RI_1(s) + V_1(s) \qquad V_1(s) = \frac{1}{Cs} I_2(s) + V_2(s)$$

$$V_1(s) = L_1 s[I_1(s) - I_2(s)] \qquad V_2(s) = L_2 s I_2(s) \tag{6.54}$$

An SFG is drawn directly from these equations and is given in Figure 6.14b. While this is a valid SFG, it does not meet the requirement that signal flow through the network be from input to output. Rather, it shows the flow from output to input. Let us rearrange the above equations into the form

$$I_1(s) = \frac{1}{R}[V_i(s) - V_1(s)] \qquad I_2(s) = Cs[V_1(s) - V_2(s)]$$

$$V_1(s) = L_1 s[I_1(s) - I_2(s)] \qquad V_2(s) = L_2 s I_2(s) \tag{6.55}$$

These equations are graphed in Figure 6.14c. Observe that this SFG does satisfy the specified requirement of input-output signal flow. ∎

It is clear from this last example that the form of the SFG will depend upon how the equations relating selected variables are chosen and arranged. This lack

Figure 6.15
An electrical system and its SFG representation.

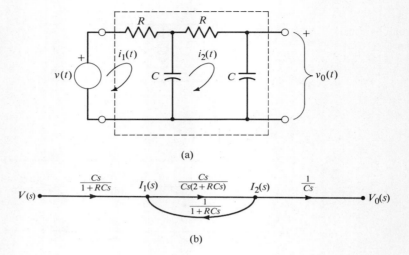

(a)

(b)

of a single unique SFG for a given set of equations is both the strength and the weakness of the SFG portrayal. Its strength lies in the fact that the system interconnection can be selected from a number of alternatives, allowing selection of the form that will best meet the needs of a given situation. This versatility is often a real advantage. Its weakness lies in the fact that it may be necessary to examine a number of alternative SFG's to ascertain the one best suited to the specific need.

Note that nothing has yet been said concerning the inclusion of initial conditions in an SFG. This matter will be addressed in Section 6-8; the details of the inclusion of initial conditions are referred to Section 12-10.

EXAMPLE 6.22
Find the SFG for the circuit shown in Figure 6.15a.

Solution: The equations describing the system are:

$$Ri_1 + \frac{1}{C}\int(i_1 - i_2)\,dt = v(t) \quad \text{(a)}$$

$$Ri_2 + \frac{1}{C}\int i_2\,dt + \frac{1}{C}\int(i_2 - i_1)\,dt = 0 \quad \text{(b)} \qquad (6.56)$$

$$\frac{1}{C}\int i_2\,dt = v_0(t) \quad \text{(c)}$$

Rearrange the Laplace transform of these equations to the form

$$I_1(s) = \frac{1}{R + \dfrac{1}{Cs}}V(s) + \frac{\dfrac{1}{Cs}}{R + \dfrac{1}{Cs}}I_2(s) \quad \text{(a)}$$

THE LAPLACE TRANSFORM

$$I_2(s) = \frac{1}{Cs\left(R + \dfrac{2}{Cs}\right)} I_1(s) \qquad \text{(b)} \qquad (6.57)$$

$$V_0(s) = \frac{1}{Cs} I_2(s) \qquad \text{(c)}$$

To draw the SFG, we arrange the four nodes in a convenient array and then use branches appropriate to (6.57). The resulting SFG is shown in Figure 6.15b. ∎

To facilitate the reduction of SFG's, we include Table 6.4. By using the rules introduced in Example 6.20, it is easy to prove the entries contained in this table. Proofs have been left as exercises.

Table 6.4 Signal Flow Graph Properties

System Diagram	Equivalent Diagram	Observations
$u_1 \xrightarrow{T_{12}} u_2$		$u_2 = T_{12} u_1$
$u_1, u_2 \to u_3$ with T_{13}, T_{23}		$u_3 = T_{13} u_1 + T_{23} u_2$
$u_1 \rightleftarrows u_2$ with T_{12} and T_{12}	$u_1 \xrightarrow{T_{12} + T_{12}'} u_2$	Superposition $u_2 = (T_{12} + T_{12}') u_1$
$u_1 \xrightarrow{T_{12}} u_2 \xrightarrow{T_{23}} u_3$	$u_1 \xrightarrow{T_{12} T_{23}} u_3$	Cascade nodes $u_3 = T_{12} T_{23} u_1$
$u_1, u_2 \to u_3 \xrightarrow{T_{34}} u_4$ with T_{13}, T_{23}	$u_1, u_2 \to u_4$ with $T_{13} T_{34}, T_{23} T_{34}$	$u_3 = u_1 T_{13} + u_2 T_{23}$ $u_4 = T_{13} T_{34} u_1 + T_{23} T_{34} u_2$

Table 6.4 *continued*

System Diagram	Equivalent Diagram	Observations
		Feedback loop $u_3 = \left(\dfrac{T_{12}T_{23}}{1 - T_{23}T_{32}} \right) u_3$
		$u_3 = \dfrac{T_{12}T_{23}}{1 - T_{22}} u_1$
		Absorption of a node
		$u_3 = \dfrac{T_{12}T_{23}}{1 - T_{22}} u_1$

In the case of more complicated graphs, the results are dependent on whether feedback loops are touching other loops. We will show that a loop that shares nodes with a path modifies the transmittance of that path. We find it convenient to express the results in the following notation for the overall transmittance T

$$T = \frac{P}{1 - L} \qquad (6.58)$$

Figure 6.16
(a) Single isolated loop.
(b) Two isolated loops.

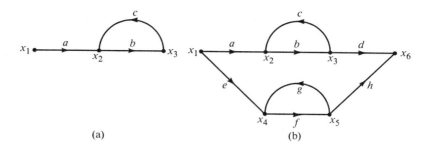

where P denotes the **path transmittance** and L denotes the **loop transmittance**. Specifically, refer to Figure 6.16, which shows two different graphs. For Figure 6.16a the transmittance is written (see also Table 6.4)

$$T = \frac{x_3}{x_1} = \frac{ab}{1-bc} = \frac{P}{1-L} \tag{6.59}$$

For the two isolated loops shown in Figure 6.16b, the result is readily shown to be

$$T = \frac{x_6}{x_1} = \frac{abd}{1-bc} + \frac{efh}{1-fg} \tag{6.60}$$

which can be written as

$$T = \frac{x_6}{x_1} = \frac{P_1}{1-L_1} + \frac{P_2}{1-L_2} \tag{6.61}$$

where the path transmittances P_1 and P_2 and the loop transmittances L_1 and L_2 are directly identifiable.

Consider now Figure 6.17, which shows two nontouching loops on a single transmission path. This graph is conveniently divided into two cascaded parts, as shown. The total transmittance is the product of the transmittance of each subgraph, or

$$T = \frac{x_6}{x_1} = \frac{P_1}{1-L_1} \frac{P_2}{1-L_2} = \frac{P_1 P_2}{1 - L_1 - L_2 + L_1 L_2}$$

$$= \frac{P}{1 - L_1 - L_2 + L_1 L_2} \tag{6.62}$$

Figure 6.17
Two nontouching loops on a single transmission path.

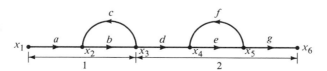

Figure 6.18
An SFG with interacting loops.

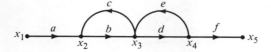

which is

$$T = \frac{x_6}{x_1} = \frac{abdeg}{1 - bc - ef + bcef} \tag{6.63}$$

Refer now to Figure 6.18, which is similar to Figure 6.17 but which shows touching or interacting loops, with two loops sharing a single node. The isolated loop procedure used in conjunction with Figure 6.17 cannot be used in this case, and we must give detailed attention to the SFG. By inspection the following set of equations is obtained:

$$\begin{aligned} x_2 &= ax_1 + cx_3 &\text{(a)} \\ x_3 &= bx_2 + ex_4 &\text{(b)} \\ x_4 &= dx_3 &\text{(c)} \\ x_5 &= fx_4 &\text{(d)} \end{aligned} \tag{6.64}$$

A systematic elimination of x_2, x_3, and x_4 from this set of equations yields, for the transmittance from x_1 to x_5,

$$T = \frac{x_5}{x_1} = \frac{abdf}{1 - bc - de} \tag{6.65}$$

In terms of our general notation, this is

$$T = \frac{P}{1 - L_1 - L_2} \tag{6.66}$$

By comparing this expression with (6.62), we see, because of the **interaction** of the loops, that the product term $L_1 L_2$ is missing from the equation.

To generalize these results, refer to Figure 6.19, which shows a more complicated situation. The total transmittance of this graph can be written directly

Figure 6.19
An SFG of two sets of interacting loops.

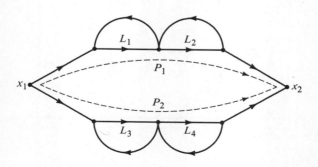

THE LAPLACE TRANSFORM

on the basis of our prior discussion. The result is

$$T = \frac{x_2}{x_1} = \frac{P_1}{1 - L_1 - L_2} + \frac{P_2}{1 - L_3 - L_4} \tag{6.67}$$

Expand this expression and write the result in the form

$$T = \frac{P_1(1 - L_3 - L_4) + P_2(1 - L_1 - L_2)}{1 - (L_1 + L_2 + L_3 + L_4) + (L_1 L_3 + L_1 L_4 + L_2 L_3 + L_2 L_4)} \tag{6.68}$$

This expression is now written in the form

$$T = \frac{P_1 \Delta_1 + P_2 \Delta_2}{\Delta} \tag{6.69}$$

This shows that the total transmittance is the sum of the individual path transmittances P_k weighted by path factors Δ_k, the sum being divided by a quantity Δ that involves all of the loops. Note, however, that Δ reflects the manner of the interconnection.

In its general form, the transmittance of an SFG can be written as

$$\boxed{T = \sum_k \frac{P_k \Delta_k}{\Delta}} \tag{6.70}$$

where P_k denotes the path transmittance from input to output for every possible direct path through the network. In selecting the direct paths P_k, no node should be encountered more than once along path k. The quantity Δ, which is called the graph determinant, involves only the closed loops of the graph and their interconnections, if any [refer to (6.69)]. The rule for evaluating Δ in terms of the loop transmittances $L_1, L_2, \ldots$ is

$$\Delta = 1 - \begin{pmatrix} \text{sum of all} \\ \text{separate loop} \\ \text{transmittances} \end{pmatrix} + \begin{pmatrix} \text{sum of all transmittance} \\ \text{products of all possible} \\ \text{pairs of nontouching loops} \end{pmatrix}$$

$$- \begin{pmatrix} \text{sum of all transmittance} \\ \text{products of all possible} \\ \text{triples of nontouching loops} \end{pmatrix} + \cdots$$

The path factor Δ_k, which is a weighting factor for each path transmittance, involves all of the loops in the graph that are isolated from path k. When a path touches all of the loops of a graph, the path factor is 1. Also, the path factor is 1 when the path contains no loops. In general, the path factor Δ_k is the value of the graph determinant with path k removed from the network. This equation is known as the **Mason rule**. A general proof of this theorem is rather involved and will not be given here.

Figure 6.20
An interconnected SFG.

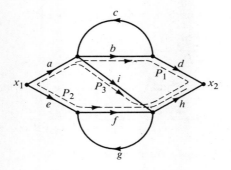

EXAMPLE 6.23
Specify the path factors P_k, Δ_k, and Δ for the SFG shown in Figure 6.20.

Solution: The required quantities are obtained by an inspection of the graph in accordance with the Mason rule. These quantities are:

$$P_1 = abd \qquad P_2 = efh \qquad P_3 = aih$$
$$\Delta_1 = 1 - fg \qquad \Delta_2 = 1 - bc \qquad \Delta_3 = 1$$
$$\Delta = 1 - (bc + fg) + bcfg$$

EXAMPLE 6.24
Specify the path factors P_k, Δ_k, and Δ for the SFG shown in Figure 6.21.

Solution: From an inspection of the graph, we write the following:

$$P_1 = t_{02}t_{21} \qquad P_2 = t_{01}$$
$$\Delta_1 = 1 \qquad \Delta_2 = 1 - t_{22}$$
$$\Delta = 1 - (t_{11} + t_{22} + t_{12} + t_{21}) + t_{11}t_{22}$$

Figure 6.21
An interconnected SFG.

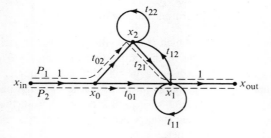

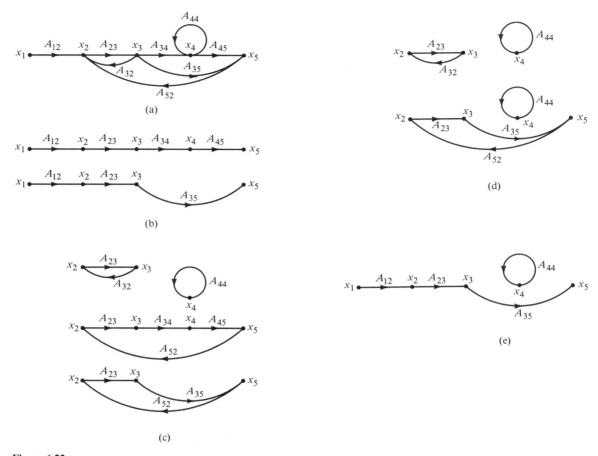

Figure 6.22
An application of the Mason formula. (a) The system SFG. (b) Two forward loops: $P_1 = A_{12}A_{23}A_{34}A_{45}$, $P_2 = A_{12}A_{23}A_{35}$. (c) Four individual loops with gains: $P_{11} = A_{23}A_{32}$, $P_{22} = A_{44}$, $P_{31} = A_{23}A_{34}A_{45}A_{52}$, $P_{41} = A_{23}A_{35}A_{52}$. (d) Two possible combinations of two non-touching loops with loop gain products: $P_{12} = A_{23}A_{32}A_{44}$, $P_{22} = A_{23}A_{35}A_{52}A_{44}$. (e) One forward path which is not in touch with one loop.

EXAMPLE 6.25
Apply the Mason rule to find the transfer function of the SFG shown in Figure 6.22.

Solution: For the SFG of Figure 6.22a, the Mason formula takes the form

$$T = \frac{x_5}{x_1} = \frac{P_1\Delta_1 + P_2\Delta_2}{\Delta}$$

where:

$$P_1 = A_{12}A_{23}A_{34}A_{45}$$

$$P_2 = A_{12}A_{23}A_{35}$$

$$\Delta = 1 - (A_{23}A_{32} + A_{44} + A_{23}A_{34}A_{45}A_{52} + A_{23}A_{35}A_{52}) + (A_{23}A_{32}A_{44} + A_{23}A_{35}A_{52}A_{44})$$

Δ_1 = first forward path touches all loops = 1

Δ_2 = second forward path does not touch one loop = $1 - A_{44}$ ∎

EXAMPLE 6.26

Find the transfer function $H(s) \triangleq V_1(s)/F(s)$ for the system shown in Figure 6.23a.

Solution: The differential equations describing the system are obtained by inspection of Figure 6.23b.

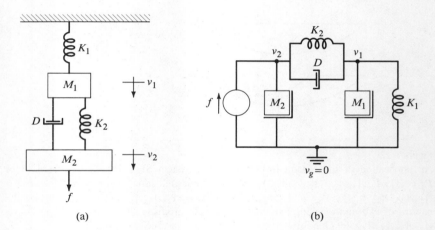

Figure 6.23
(a) A mechanical system;
(b) its network model;
(c) its SFG.

$$K_2 \int (v_2 - v_1)\,dt + D(v_2 - v_1) + M_2 \frac{dv_2}{dt} = f$$

$$K_2 \int (v_1 - v_2)\,dt + D(v_1 - v_2) + M_1 \frac{dv_1}{dt} + K \int v_1\,dt = 0$$

Laplace-transform both equations and arrange the resulting equations in the form

$$V_2(s) = \frac{1}{\frac{K_2}{s} + D + M_2 s} F(s) + \frac{\frac{K_2}{s} + D}{\frac{K_2}{s} + D + M_2 s} V_1(s) \triangleq aF + bV_1$$

$$V_1(s) = \frac{D + \frac{K_2}{s}}{\frac{(K_1 + K_2)}{s} + D + M_1 s} V_2(s) \triangleq cV_2$$

The SFG is shown in Figure 6.23c from which we find, using Table 6.4,

$$H(s) \triangleq \frac{V_1(s)}{F(s)} = \frac{ac}{1 - cb}$$

■ ■ ■

6-7 THE INVERSE LAPLACE TRANSFORM

As already discussed, the inverse Laplace transform is the integral that changes $F(s)$ back to the equivalent $f(t)$. This was specified by (6.5b)

$$\mathscr{L}^{-1}\{F(s)\} \triangleq f(t) = \frac{1}{2\pi j} \int_{\sigma - j\infty}^{\sigma + j\infty} F(s) e^{st}\,ds \qquad (6.71)$$

This requires that the integration be performed in the s-plane along a line $\text{Re}\{s\} = \sigma$ where σ is a constant factor chosen to ensure convergence of the integral

$$\int_0^\infty |f(t)| e^{-\sigma t}\,dt < \infty \qquad (6.72)$$

This is actually a necessary condition for $f(t)\exp[-\sigma t]$ to be Fourier-integral transformable, as already discussed.

Integration in the s-plane is not generally required for linear problems of the type with which we are concerned. This follows from the uniqueness property of the Laplace transform; namely, corresponding to a function $f(t)$, the transform $F(s)$ is unique, and vice versa. This means there is a one-to-one correspondence between the direct and the inverse transforms, as expressed by the pair

$$F(s) = \mathscr{L}\{f(t)\} \qquad f(t) = \mathscr{L}^{-1}\{F(s)\} \tag{6.73}$$

Consequently, we need merely refer to Table 6.1 to write the $f(t)$ appropriate to a given $\mathscr{L}^{-1}\{F(s)\}$. In the event that the available tables do not include a given $F(s)$ and the $F(s)$ cannot be reduced to available forms, recourse to the inversion integral will be necessary. The following examples will explain some of the common methods used in finding the inverse Laplace transform.

EXAMPLE 6.27
Find the inverse Laplace transform of the function

$$F(s) = \frac{s-3}{s^2 + 5s + 6} \tag{6.74}$$

Solution: Observe that the denominator can be factored into the form $(s+2) \times (s+3)$. Thus $F(s)$ can be written in partial-fraction form as

$$F(s) = \frac{s-3}{(s+2)(s+3)} = \frac{A}{s+2} + \frac{B}{s+3} \tag{6.75}$$

where A and B are constants that must be determined.

To evaluate A, multiply both sides of (6.75) by $(s+2)$ and then set $s = -2$. This gives

$$A = F(s)(s+2)\big|_{s=-2} = \frac{s-3}{s+3}\bigg|_{s=-2} = -5$$

where $B(s+2)/(s+3)\big|_{s=-2}$ is identically zero. In the same manner, for the constant B we have, multiplying both sides by $(s+3)$,

$$B = F(s)(s+3)\big|_{s=-3} = \frac{s-3}{s+2}\bigg|_{s=-3} = 6$$

The inverse transform is given by $\qquad t > 0$

$$\mathscr{L}^{-1}\{F(s)\} = -5\mathscr{L}^{-1}\left[\frac{1}{s+2}\right] + 6\mathscr{L}^{-1}\left[\frac{1}{s+3}\right] = -5e^{-2t} + 6e^{-3t}$$

where entry 4 from Table 6.1 is used. ■

EXAMPLE 6.28
Find the inverse Laplace transform of the function

$$F(s) = \frac{s+1}{[(s+2)^2 + 1](s+3)}$$

Solution: This function is written in the form

$$F(s) = \frac{A}{s+3} + \frac{Bs+C}{[(s+2)^2+1]} = \frac{s+1}{[(s+2)^2+1](s+3)}$$

The value of A is evaluated by multiplying both sides of this equation by $(s+3)$ and then setting $s = -3$. This gives

$$A = (s+3)F(s)\big|_{s=-3} = \frac{-3+1}{(-3+2)^2+1} = -1$$

To evaluate B and C, combine the two fractions and equate the coefficients of like powers of s in the numerators. This yields

$$\frac{-1[(s+2)^2+1] + (s+3)(Bs+C)}{[(s+2)^2+1](s+3)} = \frac{s+1}{[(s+2)^2+1](s+3)}$$

from which it follows that

$$-(s^2+4s+5) + Bs^2 + (C+3B)s + 3C = s+1$$

Combine like-powered terms to write

$$(-1+B)s^2 + (-4+C+3B)s + (-5+3C) = s+1$$

Therefore

$$-1+B = 0 \qquad -4+C+3B = 1 \qquad -5+3C = 1$$

From these equations we obtain

$$B = 1 \qquad C = 2$$

The function $F(s)$ is written in the equivalent form

$$F(s) = \frac{-1}{s+3} + \frac{s+2}{(s+2)^2+1}$$

Now using Table 6.1, the result is

$$f(t) = -e^{-3t} + e^{-2t}\cos t \qquad t > 0 \qquad \blacksquare$$

In many cases $F(s)$ is the quotient of two polynomials with real coefficients. If the numerator polynomial is of the same or higher degree than the denominator polynomial, first divide the numerator polynomial by the denominator polynomial; the division is carried forward until the numerator polynomial of the remainder is of one degree less than the denominator polynomial. This results in a polynomial in s plus a proper fraction. The proper fraction can be expanded into a partial-fraction expansion. The result of such an expansion is an expression of the form

$$F'(s) = B_0 + B_1 s + \cdots + \frac{A_1}{s - s_1} + \frac{A_2}{s - s_2} + \cdots$$

$$+ \frac{A_{p1}}{s - s_p} + \frac{A_{p2}}{(s - s_p)^2} + \cdots + \frac{A_{pr}}{(s - s_p)^r} \qquad (6.76)$$

CHAPTER 6

This expression has been written in a form to show three types of terms: polynomial, simple partial fraction including all terms with distinct roots, and partial fraction appropriate to multiple roots.

To find the constants $A_1, A_2, \ldots$, the polynomial terms are removed, leaving the proper fraction

$$F'(s) - (B_0 + B_1 s + \cdots) = F(s) \tag{6.77}$$

where

$$F(s) = \frac{A_1}{s - s_1} + \frac{A_2}{s - s_2} + \cdots + \frac{A_k}{s - s_k} + \frac{A_{p1}}{s - s_p} + \frac{A_{p2}}{(s - s_p)^2}$$

$$+ \cdots + \frac{A_{pr}}{(s - s_p)^r}$$

To find the constants A_k, which in complex variable terminology are the residues of the function $F(s)$ at the simple poles s_k, it is only necessary to note that as $s \to s_k$, the term $A_k/(s - s_k)$ will become large compared with all other terms. In the limit

$$\boxed{A_k = \lim_{s \to s_k} \left[(s - s_k) F(s)\right]} \tag{6.78}$$

Upon taking the inverse transform for each simple pole, the result will be a simple exponential of the form

$$\mathscr{L}^{-1}\left[\frac{A_k}{s - s_k}\right] = A_k e^{s_k t} \tag{6.79}$$

Note also that since $F(s)$ contains only real coefficients, if s_k is a complex pole with residue A_k, there will also be a conjugate pole s_k^* with residue A_k^*. For such complex poles

$$\mathscr{L}^{-1}\left[\frac{A_k}{s - s_k} + \frac{A_k^*}{s - s_k^*}\right] = A_k e^{s_k t} + A_k^* e^{s_k^* t}$$

These may be combined in the following way

$$\begin{aligned}
\text{response} &= (a_k + jb_k)e^{(\sigma_k + j\omega_k)t} + (a_k - jb_k)e^{(\sigma_k - j\omega_k)t} \\
&= e^{\sigma_k t}(a_k + jb_k)(\cos \omega_k t + j \sin \omega_k t) \\
&\quad + (a_k - jb_k)(\cos \omega_k t - j \sin \omega_k t) \\
&= 2e^{\sigma_k t}(a_k \cos \omega_k t - b_k \sin \omega_k t) \\
&= 2A_k e^{\sigma_k t} \cos(\omega_k t + \theta_k)
\end{aligned} \tag{6.80}$$

where $\theta_k = \tan^{-1}(b_k/a_k)$ and $A_k = a_k/\cos \theta_k$.

When the proper fraction contains a multiple pole of order r, the coefficients in the partial-fraction expansion $A_{p1}, A_{p2}, \ldots A_{pr}$, which are involved in the terms

THE LAPLACE TRANSFORM

$$\frac{A_{p1}}{(s-s_p)} + \frac{A_{p2}}{(s-s_p)^2} + \cdots + \frac{A_{pr}}{(s-s_p)^r}$$

must be evaluated. A simple application of (6.78) is not adequate. Now the procedure is to multiply both sides of (6.77) by $(s - s_p)^r$, which gives

$$(s-s_p)^r F(s) = (s-s_p)^r \left[\frac{A_1}{s-s_1} + \frac{A_2}{s-s_2} + \cdots + \frac{A_k}{s-s_k} \right]$$
$$+ A_{p1}(s-s_p)^{r-1} + \cdots + A_{p(r-1)}(s-s_p) + A_{pr} \qquad (6.81)$$

In the limit as $s = s_p$, all terms on the right vanish with the exception of A_{pr}. Suppose now that this equation is differentiated once with respect to s. The constant A_{pr} will vanish in the differentiation, but $A_{p(r-1)}$ will be determined by setting $s = s_p$. This procedure will be continued to find each of the coefficients A_{pk}. Specifically, this procedure is specified by

$$A_{pk} = \frac{1}{(r-k)!} \left\{ \frac{d^{r-k}}{ds^{r-k}} \left[F(s)(s-s_p)^r \right] \right\}_{s=s_p} \qquad k = 1, 2, \ldots r \qquad (6.82)$$

EXAMPLE 6.29
Find the inverse transform of the following function

$$F'(s) = \frac{s^3 + 2s^2 + 3s + 1}{s^2(s+1)}$$

Solution: This is not a proper fraction. The numerator polynomial is divided by the denominator polynomial by simple long division. The result is

$$F'(s) = 1 + \frac{s^2 + 3s + 1}{s^2(s+1)}$$

The proper fraction is expanded into partial-fraction form

$$F(s) = \frac{s^2 + 3s + 1}{s^2(s+1)} = \frac{A_{11}}{s} + \frac{A_{12}}{s^2} + \frac{A_2}{s+1}$$

The value of A_2 is deduced by using (6.78)

$$A_2 = [(s+1)F(s)]_{s=-1} = \frac{s^2 + 3s + 1}{s^2} \bigg|_{s=-1} = -1$$

To find A_{11} and A_{12}, we proceed as specified by (6.82)

$$A_{12} = [s^2 F(s)]_{s=0} = \frac{s^2 + 3s + 1}{s+1} \bigg|_{s=0} = 1$$

$$A_{11} = \frac{1}{1!} \left\{ \frac{d}{ds} [s^2 F(s)] \right\}_{s=0} = \frac{d}{ds} \left[\frac{s^2 + 3s + 1}{s+1} \right] \bigg|_{s=0}$$

$$= -\frac{s^2 + 3s + 1}{(s+1)^2} + \frac{2s+3}{s+1}\bigg|_{s=0} = 2$$

Therefore

$$F'(s) = 1 + \frac{2}{s} + \frac{1}{s^2} - \frac{1}{s+1}$$

From Table 6.1 the inverse transform is

$$f(t) = \delta(t) + 2 + t - e^{-t} \qquad \text{for } t \geq 0$$

∎

6-8 INITIAL CONDITIONS

We will find a need to establish initial conditions when solving systems problems using Laplace transform methods. This stems from the fact that systems problems are described in terms of differential equations, and in their solution the Laplace transform of derivatives and integrals automatically introduces such initial conditions, as discussed in connection with Examples 6.5 and 6.6. These initial conditions are present and must be considered at time $t = 0$ when some switching operation occurs that actuates the circuit whose response is desired. These initial conditions will be the known charges on all capacitors (usually given as initial voltages across the capacitors) and the known currents through all inductors. If these initial voltages and currents are zero, the system is **initially relaxed.** In the more general case, one or all may be different from zero.

A variety of switching operations are possible, and each must be accounted for separately. These include switching excitation sources into or out of the circuit, instantaneously changing the system capacitance, and instantaneously changing the system inductance. We will discuss each of these.

1. The most common switching operation involves the introduction of an excitation source into the circuit. The more common input waveforms are given in Figure 6.24. If the circuit is initially relaxed, then the excitation function

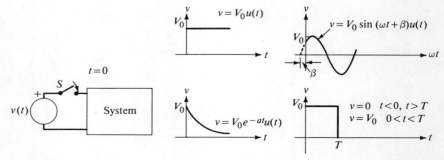

Figure 6.24
Common waveforms introduced at time $t = 0$.

establishes the resulting response function. If nonzero initial conditions exist in the circuit, they will also contribute to the resulting response function, as already noted.

2. If initial voltages across capacitors are not zero and if during the switching operation the total system capacitance remains unchanged, the voltage across each capacitor will be the same before and after the instant of switching. This condition assumes the absence of switching impulses to the capacitors. This result follows from the fact that for the capacitor $i = C\,dv/dt$, with C constant over the switching operation from $t = 0-$ to $t = 0+$,

$$\int_{v(0-)}^{v(0+)} C\,dv = \int_{0-}^{0+} i\,dt \tag{6.83}$$

The value of the integral on the right is zero unless i is an impulse function. This is a statement of conservation of charge (conservation of momentum in mechanics), which states that if no impulse is applied during the switching interval, then

$$Cv(0+) - Cv(0-) = 0 \quad \text{(a)}$$
or
$$\text{across-variable}\,(0+) = \text{across-variable}\,(0-) \quad \text{(b)} \tag{6.84}$$

3. If initial currents exist in inductors and if during the switching operation the total system inductance remains unchanged, the current through each inductor will remain unchanged over the switching instant. This condition is the dual of that for the voltage across the capacitor, and assumes the absence of switching voltage impulses. This result follows from the fact that the terminal relation for an inductor is $v = L\,di/dt$, and, with L constant over the switching interval,

$$\int_{i(0-)}^{i(0+)} L\,di = \int_{0-}^{0+} v\,dt \tag{6.85}$$

The right-hand side will be zero in the absence of voltage impulses. This is a statement of conservation of flux linkages, which states that in the absence of voltage impulses during the switching operation

$$Li(0+) - Li(0-) = 0 \quad \text{(a)}$$
or
$$\text{through-variable}\,(0+) = \text{through-variable}\,(0-) \quad \text{(b)} \tag{6.86}$$

4. In the event that L or C or both are changed instantaneously during a switching operation and in the absence of switching impulses, (6.84) and (6.86)

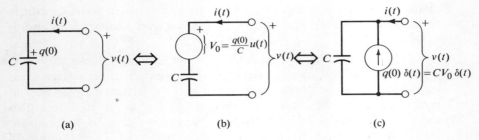

Figure 6.25
Equivalent circuits of initially charged capacitor.

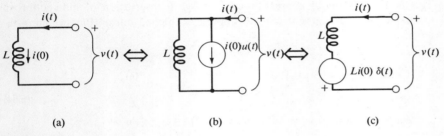

Figure 6.26
Equivalent circuits of initially current-carrying inductor.

must be modified to the forms:

$$\begin{array}{l} \text{for capacitors, conservation of charge} \\ q(0+) = q(0-) \quad \text{or} \quad C(0+)v(0+) = C(0-)v(0-) \\ \text{for inductors, conservation of flux linkages} \\ \psi(0+) = \psi(0-) \quad \text{or} \quad L(0+)i(0+) = L(0-)i(0-) \end{array} \quad (6.87)$$

As a practical matter, circuits with switched L or C are easily constructed by placing switches across all or a part of the L or C of the circuit.

When initial charges (voltages) across capacitors and currents through inductors are present, we can represent them in the alternative forms shown in Figures 6.25 and 6.26, respectively (see Problem 6-8.1).

■ ■ ■

6-9 PROBLEM SOLVING BY LAPLACE TRANSFORMS

We will consider several examples that show the use of the Laplace transform in the solution of network problems.

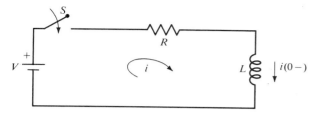

Figure 6.27
The series RL circuit.

EXAMPLE 6.30
Assume that an initial current $i(0-)$ exists in the inductor (due to a circuit not shown) when a dc source is switched into the simple RL circuit shown in Figure 6.27.

Solution: By application of the Kirchhoff voltage law, the controlling differential equation for the circuit shown is

$$L\frac{di}{dt} + Ri = v(t) = Vu(t)$$

By taking the Laplace transform of each term in this differential equation

$$L[sI(s) - i(0+)] + RI(s) = \frac{V}{s}$$

This is rearranged to

$$(Ls + R)I(s) = \frac{V}{s} + Li(0+)$$

Solve for $I(s)$.

$$I(s) = \frac{V}{L}\frac{1}{s\left(s + \frac{R}{L}\right)} + \frac{i(0+)}{s + \frac{R}{L}}$$

This is written

$$I(s) = \frac{V}{R}\left(\frac{1}{s} - \frac{1}{s + \frac{R}{L}}\right) + \frac{1}{s + \frac{R}{L}}i(0+)$$

Invoking (6.86), since the system L does not change, then $i(0+) = i(0-)$. Using the appropriate entries in Table 6.1, the inverse transform of this equation is

$$i(t) = \frac{V}{R}(1 - e^{-Rt/L}) + i(0+)e^{-Rt/L} \quad \text{for } t \geq 0$$

EXAMPLE 6.31

A force $f(t) = au(t)$ is applied to the mechanical system of negligible mass shown in Figure 6.28. Find the velocity of the system.

Solution: From Figure 6.28b we write

$$Dv + K \int v\, dt = au(t)$$

Now define a new variable $y(t) = \int_0^t v\, dt$; this equation takes the form

$$D \frac{dy}{dt} + Ky = au(t)$$

This equation, when Laplace-transformed and rearranged, yields

$$Y(s) = \frac{a}{D} \frac{1}{\left(s + \frac{K}{D}\right)s}$$

since $\int_{0-}^{0+} v\, dt = 0$, (no initial displacement), so that $y(0+) = 0$. The inverse transform of this expression is (see Example 6.30)

$$y(t) = \frac{a}{K}(1 - e^{-Kt/D}) \quad \text{for } t \geq 0$$

and therefore

$$v(t) \triangleq \frac{dy}{dt} = \frac{a}{D} e^{-Kt/D} \quad \text{for } t > 0 \qquad \blacksquare$$

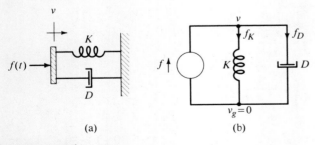

Figure 6.28
(a) A simple mechanical system (for example, car shock absorber).
(b) Circuit equivalent form.

EXAMPLE 6.32

Refer to Figure 6.29a, which shows the switching of L in a circuit with initial current. Prior to switching the inductance is L_1; after switching the total circuit inductance is $L_1 + L_2$. The switching occurs at $t = 0$. Find the current in the circuit.

Solution: The current at $t = 0-$ is

$$i(0-) = \frac{V}{R}$$

To find the current after switching at $t = 0+$, we employ the law of conservation of flux linkages [see (6.87)]. We can write, over the switching period,

$$L_1 i(0-) = (L_1 + L_2) i(0+)$$

from which

$$i(0+) = \frac{L_1}{L_1 + L_2} i(0-) = \frac{L_1}{L_1 + L_2} \frac{V}{R}$$

The differential equation that governs the circuit response just after the switch is closed is

$$(L_1 + L_2) \frac{di}{dt} + Ri = Vu(t)$$

The Laplace transform of this equation yields

$$I(s) = \frac{V}{R} \left(\frac{1}{s} - \frac{1}{s + \dfrac{R}{L_1 + L_2}} \right) + \frac{1}{s + \dfrac{R}{L_1 + L_2}} i(0+)$$

Include the value of $i(0+)$ in this expression and then take the inverse Laplace transform. The result is

$$i(t) = \frac{V}{R} \left(1 - \frac{L_2}{L_1 + L_2} e^{-Rt/(L_1 + L_2)} \right) \qquad \text{for } t \geq 0 \qquad \blacksquare$$

Figure 6.29
(a) Switching L in a circuit with initial current. (b) Response of the RL circuit when L is switched.

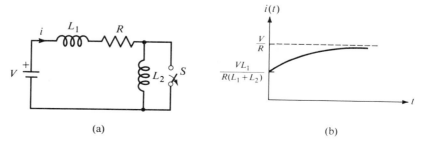

EXAMPLE 6.33

Find an expression for the response $v_2(t)$ for $t > 0$ in the circuit of Figure 6.30. The source $v_1(t)$, the current $i_L(0-)$ through $L = 2H$, and the voltage $v_c(0-)$ across the capacitor $C = 1F$ at the switching instant are all assumed to be known.

Solution: After the switch is closed, the circuit is described by the two loop equations

$$\left(3i_1 + 2\frac{di_1}{dt}\right) - \left(1i_2 + 2\frac{di_2}{dt}\right) = v_1(t)$$

$$-\left(1i_1 + 2\frac{di_1}{dt}\right) + \left(3i_2 + 2\frac{di_2}{dt} + \int i_2\,dt\right) = 0$$

$$v_2(t) = 2i_2(t)$$

All terms in these equations are Laplace-transformed. The result is the set of transformed equations

$$(3 + 2s)I_1(s) - (1 + 2s)I_2(s) = V_1(s) + 2[i_1(0+) - i_2(0+)]$$

$$-(1 + 2s)I_1(s) + \left(3 + 2s + \frac{1}{s}\right)I_2(s) = 2[-i_1(0+) + i_2(0+)] - \frac{q_2(0+)}{s}$$

$$V_2(s) = 2I_2(s)$$

Since the current through the inductor is

$$i_L(t) = i_1(t) - i_2(t)$$

then we can choose $t = 0+$ and write

$$i_L(0+) = i_1(0+) - i_2(0+)$$

Also,

$$\frac{1}{C}q_2(t) = \frac{1}{C}\int_{-\infty}^{t} i_2(t)\,dt$$

$$= \frac{1}{C}\lim_{t=0+}\int_0^t i_2(t)\,dt + \frac{1}{C}\int_{-\infty}^0 i_2(t)\,dt = 0 + v_c(0-)$$

Figure 6.30
Circuit for Example 6.33.

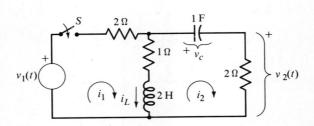

THE LAPLACE TRANSFORM

so that

$$\frac{q_2(0+)}{C} \triangleq v_c(0+) = v_c(0-) = i_2^{(-1)}(0)$$

The equation set is solved for $I_2(s)$, which is written by Cramer's rule

$$I_2(s) = \frac{\begin{vmatrix} 3 + 2s & V_1(s) + 2i_L(0+) \\ -(1+2s) & -2i_L(0+) - \dfrac{v_c(0+)}{s} \end{vmatrix}}{\begin{vmatrix} 3 + 2s & -(1+2s) \\ -(1+2s) & 3 + 2s + \dfrac{1}{s} \end{vmatrix}}$$

$$= \frac{(3+2s)\left[-2i_L(0+) - \dfrac{v_c(0+)}{s}\right] + (1+2s)[V_1(s) + 2i_L(0+)]}{(3+2s)\left(\dfrac{2s^2 + 3s + 1}{s}\right) - (1+2s)^2}$$

$$= \frac{-(2s+3)v_c(0+) - 4si_L(0+) + (2s^2+s)V_1(s)}{8s^2 + 10s + 3}$$

Hence, with

$$V_2(s) = 2I_2(s)$$

then, upon taking the inverse transform,

$$v_2(t) = 2\mathscr{L}^{-1}[I_2(s)]$$

If the circuit contains no stored energy at $t = 0$, then $i_L(0+) = v_c(0+) = 0$ and now

$$v_2(t) = 2\mathscr{L}^{-1}\left[\frac{(2s^2+s)V_1(s)}{8s^2+10s+3}\right]$$

The details of the evaluation follow the procedure already discussed. ∎

EXAMPLE 6.34

The input to the *RL* circuit shown in Figure 6.31a is the recurrent series of impulse functions shown in Figure 6.31b. Find the output current.

Solution: The differential equation that characterizes the system is

$$\frac{di(t)}{dt} + i(t) = v(t)$$

For zero initial current through the inductor, the Laplace transform of this equation is

$$(s+1)I(s) = V(s)$$

By entry 1 of Table 6.1 and the shifting property (entry 8 of Table 6.2), we can write the explicit form for $V(s)$, which is

$$V(s) = 2 + e^{-s} + 2e^{-2s} + e^{-3s} + 2e^{-4s} + \cdots$$
$$= (2 + e^{-s})(1 + e^{-2s} + e^{-4s} + \cdots)$$
$$= \frac{2 + e^{-s}}{1 - e^{-2s}}$$

Thus we must evaluate $i(t)$ from

$$I(s) = \frac{2 + e^{-s}}{1 - e^{-2s}} \frac{1}{s + 1} = \frac{2}{(1 - e^{-2s})(s + 1)} + \frac{e^{-s}}{(1 - e^{-2s})(s + 1)}$$

Expand these expressions into

$$I(s) = \frac{2}{(s + 1)}(1 + e^{-2s} + e^{-4s} + e^{-6s} + \cdots)$$
$$+ \frac{1}{(s + 1)}(e^{-s} + e^{-3s} + e^{-5s} + e^{-7s} + \cdots)$$

The inverse transform of these expressions yields

$$i(t) = 2e^{-t}u(t) + 2e^{-(t-2)}u(t - 2) + 2e^{-(t-4)}u(t - 4) + \cdots$$
$$+ e^{-(t-1)}u(t - 1) + e^{-(t-3)}u(t - 3) + e^{-(t-5)}u(t - 5) + \cdots$$

This result has been plotted in Figure 6.31c. ∎

Figure 6.31
Illustrating the steady-state response to periodic signals.

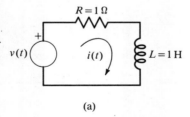

(a)

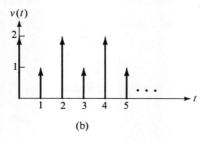

(b)

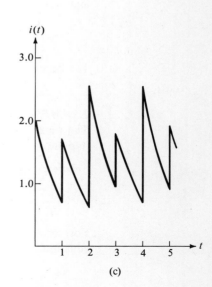

(c)

THE LAPLACE TRANSFORM

EXAMPLE 6.35
Find the velocity of the system shown in Figure 6.32a when the applied force is $f(t) = e^{-t}u(t)$. Use the Laplace transform method and assume zero initial conditions. Solve the same problem by means of the convolution technique. The input is the force and the output is the velocity.

Solution: From Figure 6.32b write the controlling equation

$$\frac{dv}{dt} + 5v + 4 \int_0^t v \, dt = e^{-t}u(t)$$

Laplace-transform this equation and then solve for $V(s)$. We obtain

$$V(s) = \frac{s}{(s+1)(s^2+5s+4)} = \frac{s}{(s+1)^2(s+4)}$$

Write this in the form

$$V(s) = \frac{A}{s+4} + \frac{B}{s+1} + \frac{C}{(s+1)^2}$$

where

$$A = \frac{s}{(s+1)^2}\bigg|_{s=-4} = -\frac{4}{9}$$

$$B = \frac{1}{1!}\frac{d}{ds}\left(\frac{s}{s+4}\right)\bigg|_{s=-1} = \frac{4}{9}$$

$$C = \frac{s}{s+4}\bigg|_{s=-1} = -\frac{1}{3}$$

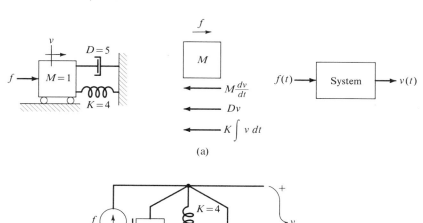

Figure 6.32
Illustrating Example 6.35. (a) Mechanical system. (b) Network representation.

The inverse transform of $V(s)$ is given by

$$v(t) = -\frac{4}{9}e^{-4t} + \frac{4}{9}e^{-t} - \frac{1}{3}te^{-t}$$

To find $v(t)$ by the use of the convolution integral, first find the impulse response of the system. This is specified by

$$\frac{dh}{dt} + 5h + 4\int_0^t h\,dt = \delta(t)$$

The Laplace transform of this equation yields

$$H(s) = \frac{s}{s^2 + 5s + 4} = \frac{s}{(s+4)(s+1)} = \frac{4}{3}\frac{1}{s+4} - \frac{1}{3}\frac{1}{s+1}$$

The inverse transform of this expression is easily found to be

$$h(t) = \frac{4}{3}e^{-4t} - \frac{1}{3}e^{-t} \qquad t > 0$$

The output of the system to the input $\exp(-t)u(t)$ is written

$$v(t) = \int_{-\infty}^{\infty} h(\tau)f(t-\tau)\,d\tau = \int_0^t e^{-(t-\tau)}\left[\frac{4}{3}e^{-4\tau} - \frac{1}{3}e^{-\tau}\right]d\tau$$

$$= e^{-t}\left[\frac{4}{3}\int_0^t e^{-3\tau}\,d\tau - \frac{1}{3}\int_0^t d\tau\right] = e^{-t}\left[\frac{4}{3}\left(\frac{1}{-3}\right)e^{-3\tau}\Big|_0^t - \frac{1}{3}t\right]$$

$$= -\frac{4}{9}e^{-4t} + \frac{4}{9}e^{-t} - \frac{1}{3}te^{-t} \qquad t > 0$$

This result is identical with that found using the Laplace transform technique. ∎

■ ■ ■

6-10 STABILITY OF LTI SYSTEMS

There is a direct relationship between the input $v(t)$ and the output $y(t)$ for a linear time-invariant system in the absence of initial stored energies. The system function or transfer function was defined in Section 6-5 as the ratio

$$H(s) = \frac{Y(s)}{V(s)} = \frac{\mathscr{L}\{\text{output}\}}{\mathscr{L}\{\text{input}\}} \tag{6.88}$$

The output may be a voltage or a current anywhere in the system, and the $H(s)$ is then appropriate to the selected output for a specified input. That is, $H(s)$ may be an impedance, an admittance, or a transfer entity in any given case. The corresponding output time function is then

THE LAPLACE TRANSFORM

$$y(t) = \mathscr{L}^{-1}[H(s)V(s)] \qquad (6.89)$$

Recall, however, that for a given system the output-input relationship is given by the differential equation that describes the system. Thus $H(s)$ can be written from an inspection of the system's differential equation, but with the time-differential Heaviside operator p being replaced by the Laplace variable s.

Another interpretation of the system function follows when (6.89) is used for the special case when $v(t) = \delta(t)$. Now, since $\mathscr{L}\{\delta(t)\} = 1$, then

$$h(t) = \mathscr{L}^{-1}[H(s)] \qquad (6.90)$$

This shows that **the system function $H(s)$ is the transform of the impulse response of the system.**

We note specifically that (6.89) is, from entry 10 of Table 6.2,

$$y(t) = \mathscr{L}^{-1}[V(s)H(s)] = \int_0^t v(\tau)h(t-\tau)\,d\tau = \int_0^t v(t-\tau)h(\tau)\,d\tau \qquad (6.91)$$

This is the convolution integral, which has already been discussed and was shown to represent the response of a given system with impulse response $h(t)$ to any excitation function $v(t)$, starting at $t = 0$.

Reference to Examples 6.30 and 6.33 will show that the system function, as noted, is just the transformed form of the differential equation for zero initial conditions and appears as an essential term in the complete solution of the problem, both steady state and transient. However, in simple system analysis by Laplace transform methods, it is not necessary to isolate and identify the system function, as this automatically appears through the transform of the differential equation and is included in the mathematical operations. The situation changes considerably in those cases where the system is identified as a number of subsystems interconnected to form the completed system. In that case one defines each subsystem by its system function. Specifically, consider two subsystems identified as $H_1(s)$ and $H_2(s)$, as in Figure 6.33a. These might be connected in cascade, as in Figure 6.33b, or in the feedback configuration, as in Figure 6.33c.

Figure 6.33
(a) Two subsystems $H_1(s)$ and $H_2(s)$. (b) Cascade interconnection. (c) Feedback interconnection.

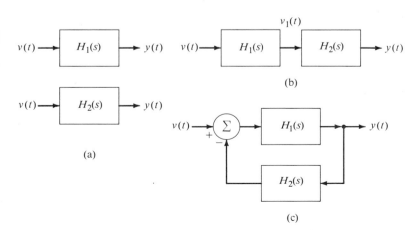

CHAPTER 6

One can now proceed directly to find for the cascade circuit of Figure 6.33b the total system function

$$H(s) = H_1(s)H_2(s) \qquad (6.92)$$

and for the feedback circuit of Figure 6.33c (see also Section 2-11)

$$H(s) = \frac{H_1(s)}{1 + H_1(s)H_2(s)} \qquad (6.93)$$

The total output response would then use the resultant $H(s)$ in conjunction with (6.89).

EXAMPLE 6.36
Two identical amplifiers are connected in cascade, as shown in Figure 6.34. Assume that isolation exists between them so that neither circuit loads the other, but the signals do transfer from the output of the first to the input of the second amplifier. The impulse response of each amplifier is $h(t) = te^{-t}$ for $t \geq 0$. Find the unit step response of the combined circuit.

Solution: Since by definition $H(s)$ is the Laplace transform of the impulse response of a system, then each amplifier is defined by

$$\mathscr{L}\{te^{-t}\} \triangleq H_1(s) = \frac{1}{(s+1)^2}$$

and the total system function is

$$H(s) = \frac{1}{(s+1)^4}$$

The total unit step response is then, since $V(s) = 1/s$

$$y(t) = \mathscr{L}^{-1}\left[\frac{1}{s(s+1)^4}\right]$$

$$= \mathscr{L}^{-1}\left[\frac{A_{11}}{(s+1)} + \frac{A_{12}}{(s+1)^2} + \frac{A_{13}}{(s+1)^3} + \frac{A_{14}}{(s+1)^4} + \frac{A}{s}\right]$$

By (6.82)

$$A_{14} = \frac{1}{0!}\frac{d^0}{ds^0}\left[(s+1)^4 H(s)\right]_{s=-1} = 1\left[\frac{1}{s}\right]_{s=-1} = -1$$

$$A_{13} = \frac{1}{1!}\frac{d}{ds}\left[\frac{1}{s}\right]_{s=-1} = -1 \times \frac{1}{s^2}\bigg|_{s=-1} = -1$$

$$A_{12} = \frac{1}{2!}\frac{d^2}{ds^2}\left[\frac{1}{s}\right]_{s=-1} = \frac{1}{2}\left(\frac{+2}{s^3}\right) = \frac{1}{s^3}\bigg|_{s=-1} = -1$$

$$A_{11} = \frac{1}{3!}\frac{d^3}{ds^3}\left[\frac{1}{s}\right]_{s=-1} = \frac{1}{6}\left(\frac{-6}{s^4}\right)\bigg|_{s=-1} = -1$$

Figure 6.34
Two amplifiers in cascade.

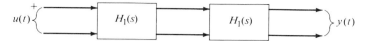

$$A = \left[s \frac{1}{s(s+1)^4} \right]_{s=0} = \frac{1}{(s+1)^4} \bigg|_{s=0} = 1$$

Then

$$y(t) = \mathscr{L}^{-1} \left[\frac{1}{s} - \frac{1}{s+1} - \frac{1}{(s+1)^2} - \frac{1}{(s+1)^3} - \frac{1}{(s+1)^4} \right]$$

$$= u(t) - e^{-t} - \frac{te^{-t}}{1!} - \frac{t^2 e^{-t}}{2!} - \frac{t^3 e^{-t}}{3!}$$

■

The physical idea of stability is closely related to the response of a bounded system to a sudden disturbance or input. If the system is disturbed and is displaced slightly from its equilibrium state, several different behaviors are possible. If the system remains near the equilibrium state, the system is said to be **stable**. If the system tends to return to the equilibrium state or tends to a bounded or limited state, it is said to be **asymptotically stable**. The matter of stability will be considered in some detail from a state point of view in Chapter 12. Here we note that stability can be examined by studying a system either through its impulse response $h(t)$ or its Laplace-transformed system function $H(s)$. We will examine these approaches to a study of system stability.

Assume that we are able to expand the transfer function in terms of its roots, as follows.

$$H(s) = \frac{A_1}{s - s_1} + \frac{A_2}{s - s_2} + \cdots + \frac{A_k}{s - s_k} + \cdots + \frac{A_n}{s - s_n} \quad (6.94)$$

Further, because the controlling differential equation that describes the system has real coefficients, the roots either are real or, if complex, will occur only in complex conjugate pairs. Further, the time response due to the kth pole will be of the form $A_k \exp(s_k t)$, and so the nature of the response will depend on the location of the root s_k in the s-plane.

Three general cases exist that depend intimately on the location of s_k in the s-plane. These are:

1. The point representing s_k lies to the left of the imaginary axis in the s-plane.
2. The point representing s_k lies on the $j\omega$-axis.
3. The point representing s_k lies to the right of the imaginary axis in the s-plane.

We examine each of these possibilities.

Case 1. The root is a real number $s_k = \sigma_k$, and it is located on the negative real σ-axis. The response due to this root will be of the form

$$\text{response} = A_k e^{\sigma_k t} \qquad \sigma_k < 0 \tag{6.95}$$

This indicates that after a lapse of time the response will become vanishingly small.

For the case when a pair of complex conjugate roots exist, the response due to these is given by

$$\text{response} = A_k e^{s_k t} + A_k^* e^{s_k^* t} \tag{6.96}$$

where A_k and A_k^* specify the appropriate amplitude factors. The response terms can be combined, noting that $A_k = a + jb$ and $s_k = \sigma_k + j\omega_k$ (see 6.80),

$$\text{response} = (a + jb)e^{(\sigma_k + j\omega_k)t} + (a - jb)e^{(\sigma_k - j\omega_k)t}$$

or

$$\text{response} = 2\sqrt{a^2 + b^2}\, e^{\sigma_k t} \cos(\omega_k t + \beta_k) \qquad \sigma_k < 0 \quad \text{(a)} \tag{6.97}$$

where

$$\beta_k = \tan^{-1}\left(\frac{b}{a}\right) \tag{b}$$

This response is a damped sinusoid and it ultimately decays to zero.

As a general conclusion, we find that systems with only simple poles in the left half-plane are stable.

Case 2. s_k lies on the imaginary axis. This is a special case under Case 1, but now $\sigma_k = 0$. The response for complex conjugate poles is, from (6.97),

$$\text{response} = 2\sqrt{a^2 + b^2}\cos(\omega_k t + \beta_k) \tag{6.98}$$

Observe that there is no damping, and the response is thus a sustained oscillatory function. Such a system has a bounded response to a bounded input, and the system is defined as **stable** even though it is oscillatory.

Case 3. s_k lies in the right half-plane. The response function will be of the form

$$\text{response} = A_k e^{s_k t} \tag{6.99}$$

for real roots, and will be of the form

$$\text{response} = 2\sqrt{a^2 + b^2} e^{\sigma_k t} \cos(\omega_k t + \beta_k) \qquad \sigma_k > 0 \tag{6.100}$$

for conjugate complex roots. Because both functions increase with time without limit even for bounded inputs, the system for which these are roots is said to be **unstable**.

The conclusions that follow from these three cases are that **a system with simple poles is unstable if one or more of the poles of its transfer function appear in the right half plane. Conversely, a system whose transfer function has simple poles is stable when all of the poles are in the left half plane or on its boundary.** In fact, the distance of the poles from the imaginary axis gives a measure of the decay rate of the response with time.

Figure 6.35
Location of poles and corresponding impulse response of the system.

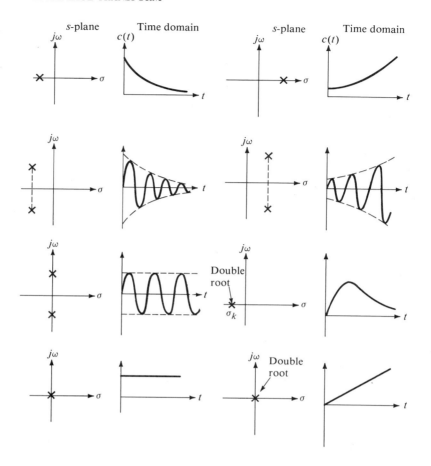

The impulse response of a system and the location of its simple poles are shown in Figure 6.35.

We now wish to reexamine the situation when multiple-order poles exist. We again consider the three cases.

Case 1—Multiple real poles in the left half plane. As previously discussed (see Table 6.1), a second-order real pole (two repeated roots) gives rise to the response function

$$\text{response} = (A_{k1} + A_{k2}t)e^{\sigma_k t} \tag{6.101}$$

For negative values of σ_k, the exponential time function decreases faster than the linearly increasing time factor. The response ultimately dies out, the rapidity of decay depending on the value of σ_k. A system with such poles is stable.

Case 2—Multiple poles on the imaginary axis. The response function is made up of the responses due to each pair of poles, and is

$$\text{response} = (A_{k1} + A_{k2}t)e^{j\omega_k t} + (A_{k1}^* + A_{k2}^*t)e^{-j\omega_k t} \tag{6.102}$$

This result can be written, following the procedure discussed above for the simple complex poles

$$\text{response} = 2\sqrt{a^2 + b^2}\cos(\omega_k t + \beta_k) + 2\sqrt{c^2 + d^2}\,t\cos(\omega_k t + \gamma_k) \quad (6.103)$$

The first term on the right is a sustained oscillatory function. The second term is a time-modulated oscillatory function that increases with time. Clearly, the system in this case is unstable.

Case 3—Multiple roots in the right half plane. The solutions in this case will be, for real roots,

$$\text{response} = (A_{k1} + A_{k2}t)e^{\sigma_k t} \quad (6.104)$$

and for complex roots,

$$\begin{aligned}\text{response} = e^{\sigma_k t}[&2\sqrt{a^2 + b^2}\cos(\omega_k t + \beta_k)\\ &+ 2\sqrt{c^2 + d^2}\cos(\omega_k t + \gamma_k)]\end{aligned} \quad (6.105)$$

In both cases, owing to the factor $e^{\sigma_k t}$, the response increases with time and the system is unstable.

The foregoing considerations can be summarized as follows: **A system with multiple poles is unstable if one or more of its poles appear on the $j\omega$-axis or in the right half plane. Conversely, when all the poles of the system are confined to the left half plane, the system is stable.**

■ ■ ■

†6-11 THE INVERSION INTEGRAL

When the Laplace transform $F(s)$ is known, the function of time can be found by (6.56), which is rewritten

$$f(t) = \mathscr{L}^{-1}\{F(s)\} = \frac{1}{2\pi j}\int_{\sigma - j\infty}^{\sigma + j\infty} F(s)e^{st}\,ds \quad (6.106)$$

This equation applies equally well to both the two-sided and the one-sided transforms.

It was pointed out in Section 6-2 that the path of integration in (6.106) is restricted to values of σ for which the direct transform formula converges. In fact, for the two-sided Laplace transform, the region of convergence must be specified in order to determine uniquely the inverse transform. That is, for the two-sided transform, the regions of convergence for functions of time that are zero for $t > 0$, zero for $t < 0$, or in neither category must be distinguished. For the one-sided transform, the region of convergence is given by σ, where σ is the abscissa of absolute convergence.

The path of integration in (6.106) is usually taken as shown in Figure 6.36, and consists of the straight line ABC displaced to the right of the origin by σ and extending in the limit from $-j\infty$ to $+j\infty$ with connecting semicircles. The evaluation of the integral usually proceeds by using the Cauchy integral theorem

THE LAPLACE TRANSFORM

(see Appendix 1 and Section 4-6), which specifies that

$$f(t) = \frac{1}{2\pi j} \lim_{R \to \infty} \oint_{\Gamma_1} F(s)e^{st}\, ds$$

$$= \sum [\text{residues of } F(s)e^{st} \text{ at the singularities to the left of } ABC]$$
for $t > 0$ **(a)**   (6.107)

As we shall find, the contribution to the integral around the circular path with $R \to \infty$ is zero, leaving the desired integral along path ABC, and

$$f(t) = \frac{1}{2\pi j} \lim_{R \to \infty} \oint_{\Gamma_2} F(s)e^{st}\, ds$$

$$= -\sum [\text{residues of } F(s)e^{st} \text{ at the singularities to the right of } ABC]$$
for $t < 0$ **(b)**

We will present a number of examples involving these equations.

EXAMPLE 6.37
Use the inversion integral to find $f(t)$ for the function

$$F(s) = \frac{1}{s^2 + \omega^2}$$

Note that by entry 9 of Table 6.1, this is $\sin \omega t/\omega$.

Solution: The inversion integral is written in a form that shows the poles of the integrand

$$f(t) = \frac{1}{2\pi j} \oint \frac{e^{st}}{(s + j\omega)(s - j\omega)}\, ds$$

The path chosen is Γ_1 in Figure 6.36. Evaluate the residues

$$\text{Res}\left[(s - j\omega)\frac{e^{st}}{s^2 + \omega^2}\right]_{s=j\omega} = \frac{e^{st}}{s + j\omega}\bigg|_{s=j\omega} = \frac{e^{j\omega t}}{2j\omega}$$

Figure 6.36
The path of integration in the s-plane.

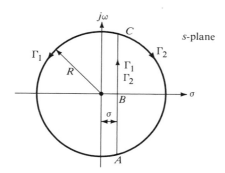

$$\text{Res}\left[(s+j\omega)\frac{e^{st}}{s^2+\omega^2}\right]_{s=-j\omega} = \frac{e^{st}}{s-j\omega}\bigg|_{s=-j\omega} = \frac{e^{-j\omega t}}{-2j\omega}$$

Therefore

$$f(t) = \sum \text{Res} = \frac{e^{j\omega t} - e^{-j\omega t}}{2j\omega} = \frac{\sin \omega t}{\omega} \qquad \blacksquare$$

EXAMPLE 6.38
Find $\mathscr{L}^{-1}[1/\sqrt{s}]$.

Solution: The function $F(s) = 1/\sqrt{s}$ is a double-valued function because of the square root operation. That is, if s is represented in polar form by $re^{j\theta}$, then $re^{j(\theta+2\pi)}$ is a second acceptable representation, and $\sqrt{s} = \sqrt{re^{j(\theta+2\pi)}} = -\sqrt{re^{j\theta}}$, thus showing two different values for $\sqrt{s}$. But a double-valued function is not analytic and requires a special procedure in its solution.

The procedure is to make the function analytic by restricting the angle of s to the range $-\pi < \theta < \pi$ and by excluding the point $s = 0$. This is done by constructing a **branch cut** along the negative real axis, as shown in Figure 6.37. The end of the branch cut, which is the origin in this case, is called a **branch point**. Since a branch cut can never be crossed, this essentially ensures that $F(s)$ is single-valued. Now, however, the inversion integral (6.107) becomes, for $t > 0$,

$$f(t) = \lim_{R \to \infty} \frac{1}{2\pi j} \int_{GAB} F(s)e^{st}\, ds = \frac{1}{2\pi j} \int_{\sigma-j\infty}^{\sigma+j\infty} F(s)e^{st}\, ds$$

$$= -\frac{1}{2\pi j}\left[\int_{BC} + \int_{\Gamma_2} + \int_{l-} + \int_{\gamma} + \int_{l+} + \int_{\Gamma_3} + \int_{FG}\right] \qquad (6.108)$$

which does not include any singularity.

First we will show that for $t > 0$ the integrals over the contours BC and CD vanish as $R \to \infty$, from which $\int_{\Gamma_2} = \int_{\Gamma_3} = \int_{BC} = \int_{FG} = 0$. Note from Figure 6.37 that $\beta = \cos^{-1}(\sigma/R)$ so that the integral over the arc BC is, since $|e^{j\theta}| = 1$,

$$|I| \le \int_{BC} \left|\frac{e^{\sigma t}e^{j\omega t}}{R^{1/2}e^{j\theta/2}} jRe^{j\theta}\, d\theta\right| = e^{\sigma t}R^{1/2}\int_{\beta}^{\pi/2} d\theta = e^{\sigma t}R^{1/2}\left(\frac{\pi}{2} - \cos^{-1}\frac{\sigma}{R}\right)$$

$$= e^{\sigma t}R^{1/2}\sin^{-1}\frac{\sigma}{R}$$

But for small arguments $\sin^{-1}(\sigma/R) = \sigma/R$, and in the limit as $R \to \infty$, $I \to 0$. By a similar approach, we find that the integral over CD is zero. Thus the integrals over the contours Γ_2 and Γ_3 are also zero as $R \to \infty$.

For evaluating the integral over γ, let $s = re^{j\theta} = r(\cos\theta + j\sin\theta)$ and

$$\int_{\gamma} F(s)e^{st}\, ds = \int_{\pi}^{-\pi} \frac{e^{r(\cos\theta + j\sin\theta)}}{\sqrt{r}e^{j\theta/2}} jre^{j\theta}\, d\theta$$

$$= 0 \quad \text{as} \quad r \to 0$$

Figure 6.37
The integration contour for $\mathscr{L}^{-1}[1/\sqrt{s}]$.

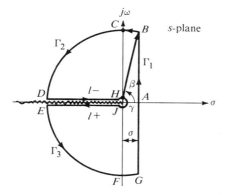

The remaining integrals in (6.108) are written

$$f(t) = -\frac{1}{2\pi j}\left[\int_{l-} F(s)e^{st}\,ds + \int_{l+} F(s)e^{st}\,ds\right] \tag{6.109}$$

Along path $l-$, let $s = -u$; $\sqrt{s} = j\sqrt{u}$, and $ds = -du$, where u and $\sqrt{u}$ are real positive quantities. Then

$$\int_{l-} F(s)e^{st}\,ds = -\int_\infty^0 \frac{e^{-ut}}{j\sqrt{u}}\,du = \frac{1}{j}\int_0^\infty \frac{e^{-ut}}{\sqrt{u}}\,du$$

Along path $l+$, $s = -u$, $\sqrt{s} = -j\sqrt{u}$ (not $+j\sqrt{u}$), and $ds = -du$. Then

$$\int_{l+} F(s)e^{st}\,ds = -\int_0^\infty \frac{e^{-ut}}{-j\sqrt{u}}\,du = \frac{1}{j}\int_0^\infty \frac{e^{-ut}}{\sqrt{u}}\,du$$

Combine these results to find

$$f(t) = -\frac{1}{2\pi j}\left[\frac{2}{j}\int_0^\infty u^{-1/2}e^{-ut}\,du\right] = \frac{1}{\pi}\int_0^\infty u^{-1/2}e^{-ut}\,du$$

which is a standard form integral listed in most handbooks of mathematical tables, with the result

$$f(t) = \frac{1}{\pi}\sqrt{\frac{\pi}{t}} = \frac{1}{\sqrt{\pi t}} \qquad t > 0 \tag{6.110}$$

EXAMPLE 6.39
Find the inverse Laplace transform of the given function with an infinite number of poles.

$$F(s) = \frac{1}{s(1 + e^{-s})}$$

Figure 6.38
Illustrating Example 6.39, the Laplace inversion for the case of infinitely many poles.

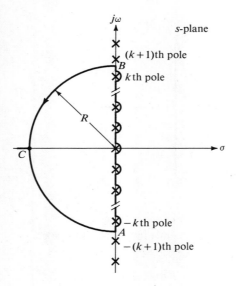

Solution: The integrand in the inversion integral $e^{st}/s(1 + e^{-s})$ possesses simple poles at

$$s = 0 \quad \text{and} \quad s = jn\pi \quad n = \pm 1, \pm 3, +\ldots \quad \text{(odd values)} \tag{6.111}$$

These are illustrated in Figure 6.38. This means that the function $e^{st}/s(1 + e^{-s})$ is analytic in the s-plane except at the simple poles at $s = 0$ and $s = jn\pi$. Hence the integral is specified in terms of the residues in the various poles. We thus have:

For $s = 0$

$$\text{Res}\left\{\frac{se^{st}}{s(1 + e^{-s})}\right\}\bigg|_{s=0} = \frac{1}{2}$$

For $s = jn\pi$

$$\text{Res}\left\{\frac{(s - jn\pi)e^{st}}{s(1 + e^{-s})}\right\}\bigg|_{s=jn\pi} = \frac{0}{0} \tag{6.112}$$

The problem we now face in this evaluation is that

$$\text{Res}\left\{(s - a)\frac{n(s)}{d(s)}\right\}\bigg|_{s=a} = \frac{0}{0}$$

where the roots of $d(s)$ are such that $s = a$ cannot be factored. However, we have discussed such a situation in Appendix 1 for complex variables and we have the following result

$$\frac{d[d(s)]}{ds}\bigg|_{s=a} = \lim_{s \to a}\frac{d(s) - d(a)}{s - a} = \lim_{s \to a}\frac{d(s)}{s - a}$$

THE LAPLACE TRANSFORM

since $d(a) = 0$. Combine this expression with the above equation to obtain

$$\text{Res}\left\{(s-a)\frac{n(s)}{d(s)}\right\}\Bigg|_{s=a} = \frac{n(s)}{\frac{d}{ds}[d(s)]}\Bigg|_{s=a} \quad (6.113)$$

We use (6.113) in evaluating (6.112) to find

$$\text{Res}\left\{\frac{e^{st}}{s\frac{d}{ds}(1+e^{-s})}\right\}\Bigg|_{s=jn\pi} = \frac{e^{jn\pi t}}{jn\pi} \quad (n \text{ odd})$$

We obtain, by adding all of the residues,

$$f(t) = \frac{1}{2} + \sum_{\substack{n=-\infty \\ n \text{ odd}}}^{\infty} \frac{e^{jn\pi t}}{jn\pi}$$

This can be rewritten as follows

$$f(t) = \frac{1}{2} + \left[\cdots + \frac{e^{-j3\pi t}}{-j3\pi} + \frac{e^{-j\pi t}}{-j\pi} + \frac{e^{j\pi t}}{j\pi} + \frac{e^{j3\pi t}}{j3\pi} + \cdots\right]$$

$$= \frac{1}{2} + \sum_{\substack{n=1 \\ n \text{ odd}}}^{\infty} \frac{2j \sin n\pi t}{jn\pi}$$

which we write, finally

$$f(t) = \frac{1}{2} + \frac{2}{\pi}\sum_{k=1}^{\infty} \frac{\sin(2k-1)\pi t}{2k-1} \quad (6.114)$$

As a second approach to a solution to this problem, we will show the details in carrying out the contour integration for this problem. We choose the path shown in Figure 6.38 that includes semicircular hooks around each pole, the vertical connecting line from hook to hook, and the semicircular path at $R \to \infty$. Thus we have

$$f(t) = \frac{1}{2\pi j}\oint \frac{e^{st}\,ds}{s(1+e^{-s})}$$

$$= \frac{1}{2\pi j}\left[\underbrace{\int_{BCA}}_{I_1} + \underbrace{\int_{\text{Vertical connecting lines}}}_{I_2} + \underbrace{\sum\int_{\text{Hooks}}}_{I_3} - \sum \text{Res}\right] \quad (6.115)$$

We consider the several integrals:

Integral I_1. By setting $s = re^{j\theta}$ and taking into consideration that $\cos\theta = -\cos\theta$ for $\theta > \pi/2$, the integral $I_1 \to 0$ as $r \to \infty$.

Integral I_2. Along the Y-axis, $s = jy$ and

$$I_2 = j\int_{\substack{-\infty \\ r\to 0}}^{\infty} \frac{e^{jyt}}{jy(1+e^{-jy})}\,dy$$

Note that the integrand is an odd function, whence $I_2 = 0$.

Integral I_3. Consider a typical hook at $s = jn\pi$. Since

$$\lim_{\substack{r \to 0 \\ s \to jn\pi}} \left[\frac{(s - jn\pi)e^{st}}{s(1 + e^{-s})} \right] = \frac{0}{0}$$

this expression is evaluated (as considered in 6.113 above) and yields $e^{jn\pi t}/jn\pi$. Thus for all poles

$$I_3 = \frac{1}{2\pi j} \int_{\substack{-\pi/2 \\ r \to 0 \\ s \to jn\pi}}^{\pi/2} \frac{e^{st}}{s(1 + e^{-s})} ds$$

$$= \frac{j\pi}{2\pi j} \left[\sum_{\substack{n = -\infty \\ n \text{ odd}}}^{\infty} \frac{e^{jn\pi t}}{jn\pi} + \frac{1}{2} \right] = \frac{1}{2} \left[\frac{1}{2} + \frac{2}{\pi} \sum_{\substack{n=1 \\ n \text{ odd}}}^{\infty} \frac{\sin n\pi t}{n} \right]$$

Finally, the residues enclosed within the contour are

$$\text{Res} \frac{e^{st}}{s(1 + e^{-s})} = \frac{1}{2} + \sum_{\substack{n=-\infty \\ n \text{ odd}}}^{\infty} \frac{e^{jn\pi t}}{jn\pi} = \frac{1}{2} + \frac{2}{\pi} \sum_{\substack{n=1 \\ n \text{ odd}}}^{\infty} \frac{\sin n\pi t}{n}$$

which is seen to be twice the value around the hooks. Then when all terms are included in (6.115)

$$f(t) = \frac{1}{2} + \frac{2}{\pi} \sum_{\substack{n=1 \\ n \text{ odd}}}^{\infty} \frac{\sin n\pi t}{n} = \frac{1}{2} + \frac{2}{\pi} \sum_{k=1}^{\infty} \frac{\sin(2k-1)\pi t}{2k-1} \qquad \blacksquare$$

■ ■ ■

6-12 COMPLEX INTEGRATION AND THE BILATERAL LAPLACE TRANSFORM

We have discussed the fact that the region of absolute convergence of the unilateral Laplace transform is the region to the left of the abscissa of convergence. This is not true for the bilateral Laplace transform: the region of convergence must be specified to invert a function $F(s)$ obtained using the bilateral Laplace transform. This requirement is necessary because different time signals might have the same Laplace transform but different regions of absolute convergence.

To establish the region of convergence, write (6.10b) in the form

$$F_2(s) = \int_0^\infty e^{-st} f(t) \, dt + \int_{-\infty}^0 e^{-st} f(t) \, dt \qquad (6.116)$$

If the function $f(t)$ is of exponential order ($e^{\sigma_1 t}$), then the region of convergence for $t > 0$ is $\text{Re}\{s\} > \sigma_1$. If the function $f(t)$ for $t < 0$ is of exponential order $\exp(\sigma_2 t)$, then the region of convergence is $\text{Re}\{s\} < \sigma_2$. Hence the function $F_2(s)$ exists and is analytic in the vertical strip defined by

$$\sigma_1 < \text{Re}\{s\} < \sigma_2 \qquad (6.117)$$

THE LAPLACE TRANSFORM

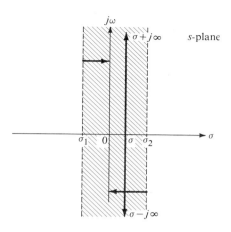

Figure 6.39
Region of convergence for the bilateral Laplace transform.

provided, of course, that $\sigma_1 < \sigma_2$. If $\sigma_1 > \sigma_2$, no region of convergence would exist and the inversion process could not be performed. This region of convergence is shown in Figure 6.39.

EXAMPLE 6.40
Find the bilateral Laplace transform of the signals $f(t) = e^{-at}u(t)$ and $f(t) = -e^{-at}u(-t)$ and specify their regions of convergence.

Solution: Using the basic definition of the transform, we obtain

a. $F_2(s) = \int_{-\infty}^{\infty} e^{-at}u(t)e^{-st}\,dt = \int_0^{\infty} e^{-(s+a)t}\,dt = \dfrac{1}{s+a}$

and its region of convergence is

$\operatorname{Re}\{s\} > -a$

b. For the second signal

$F_2(s) = \int_{-\infty}^{\infty} -e^{-at}u(-t)e^{-st}\,dt = -\int_{-\infty}^{0} e^{-(s+a)t}\,dt = \dfrac{1}{s+a}$

and its region of convergence is

$\operatorname{Re}\{s\} < -a$ ∎

EXAMPLE 6.41
Find the function, if its Laplace transform is given by

$$F_2(s) = \dfrac{3}{(s-4)(s+1)(s+2)} \qquad -2 < \operatorname{Re}\{s\} < -1$$

Figure 6.40
Illustrating Example 6.41.

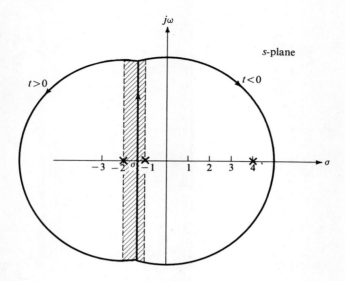

Solution: The region of convergence and the paths of integration are shown in Figure 6.40.

For $t > 0$, we close the contour to the left, and we obtain

$$f(t) = \left.\frac{3e^{st}}{(s-4)(s+1)}\right|_{s=-2} = \frac{1}{2}e^{-2t} \qquad t > 0$$

For $t < 0$, the contour closes to the right, and now

$$f(t) = \left.\frac{3e^{st}}{(s-4)(s+2)}\right|_{s=-1} + \left.\frac{3e^{st}}{(s+1)(s+2)}\right|_{s=4} = -\frac{3}{5}e^{-t} + \frac{e^{4t}}{10} \qquad t < 0$$

These examples confirm that we must know the region of convergence to find the inverse transform.

■ ■ ■

REFERENCES

1. Carslaw, H. S., and J. C. Jaeger. *Operational Methods in Applied Mathematics.* New York: Dover, 1963.
2. DeRusso, P. M., R. J. Roy, and C. M. Close. *State Variables for Engineers.* New York: Wiley, 1965.
3. LePage, W. R. *Complex Variables and the Laplace Transform for Engineers.* New York: McGraw-Hill, 1961.
4. Papoulis, A. *The Fourier Integral and its Applications.* New York: McGraw-Hill, 1962.

THE LAPLACE TRANSFORM

5. Seely, S., and A. D. Poularikas. *Electrical Engineering—Introduction and Concepts*. Beaverton, Ore.: Matrix Publishers, 1982.
6. Sneddon, I. N. *The Use of Integral Transforms*. New York: McGraw-Hill, 1972.
7. Widder, D. V. *The Laplace Transform*. Princeton, N.J.: Princeton University Press, 1946.

PROBLEMS

6-3.1 Deduce the Laplace transforms of the following:
   a. Entries 5, 6, 9, 12 in Table 6.1.
   b. $t^2 + 2t + 1$ $\quad \dfrac{1 + \cos 4t}{2} \quad$ $t \sin \omega t \quad$ $\dfrac{\sin \omega t}{\omega t} \quad$ for $t \geq 0$

6-3.2 Find the Laplace transform and the region of convergence for the following functions, $t \geq 0$:
   a. $2 + 3t$ **b.** $4e^{-0.1t}$ **c.** $\sinh 2t$ **d.** $1 + \sin t$ **e.** e^{2t}

6-3.3 Find the Laplace transform of the following functions, $t \geq 0$:
   a. $2 - 8t^3$ **b.** $t \cos 2t$ **c.** $e^{-t} \cosh t$ **d.** $e^{-t} \cos t$ **e.** $\cos t - \sin t$ **f.** $\cos 2t - e^t \sin 3t$

6-4.1 Prove entries 6, 7, 9, 13, 14 in Table 6.2. Find the Laplace transform of $\int_0^t (t - \lambda) \cos \lambda \, d\lambda$.

6-4.2 Find the Laplace transform of the functions illustrated in Figure P6-4.2.

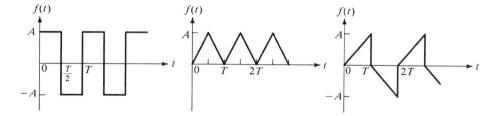

Figure P6-4.2

6-4.3 Apply the initial value and final value theorems to the functions given below:
   a. $F_1(s) = \dfrac{s}{s^2 + 3}$ **b.** $F_2(s) = \dfrac{s^2 + s + 3}{s^2 + 3}$ **c.** $F_3(s) = \dfrac{s^2}{s^2 + b^2}$ **d.** $F_4(s) = \dfrac{s + a}{(s + a)^2 + b^2}$

6-4.4 Find the Laplace transform of the following expressions:
   a. $f(t) = e^{-t} u(t - 2) + \dfrac{dg(t)}{dt}$ **b.** $f(t) = e^{-2t} g(t) + \int_0^t e^{-|\xi|} d\xi$

6-4.5 Show that if a signal $f(t)$ is of finite duration and if there is one value of s for which the Laplace transform converges, then it converges for all values of s.

6-4.6 If $f(t)$ is an even function, find the relationship between $F(s)$ and $F(-s)$. Repeat for the case when $f(t)$ is an odd function.

6-5.1 Draw the equivalent block diagram representation of the systems shown in Figure P6-5.1 and find the transfer function, as indicated, for each case.

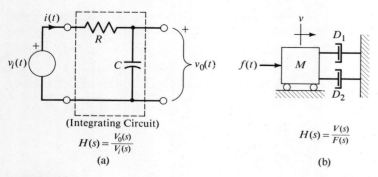

Figure P6-5.1

6-5.2 Draw the equivalent block diagram representation of the system (a microphone) shown in Figure P6-5.2 and find the transfer function $H(s) = E_0(s)/F_a(s)$. In this example the voltage is represented by e and the velocity by v.

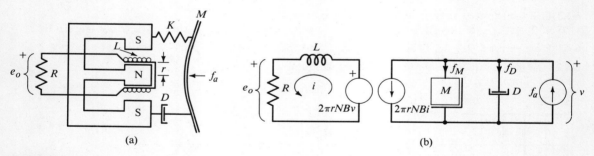

Figure P6-5.2

6-5.3 Find the equivalent block diagram representation of   in Figure P6-5.3 and determine the transfer function, as indicated, in each case.

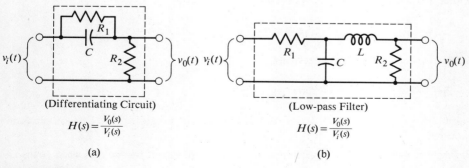

Figure P6-5.3

6-5.4 Determine the transfer functions $H(s) = C(s)/R(s)$ for the systems shown in Figure P6-5.4.

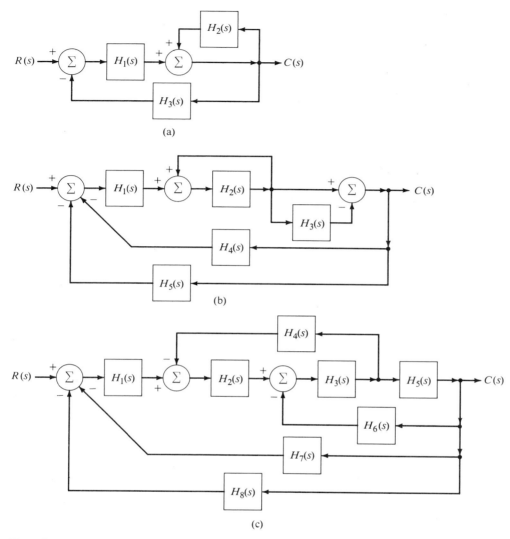

Figure P6-5.4

6-6.1 Draw the SFG's for the systems shown in Figure P6-6.1 and find their transfer functions, as indicated in each case.

6-6.2 Determine the transfer functions for the SFG's shown in Figure P6-6.2.

6-7.1 Find the inverse Laplace transforms of the following functions by means of partial-fraction expansions:

a. $F_1(s) = \dfrac{1}{s} - \dfrac{1}{s} e^{-2s}$ **b.** $F_2(s) = \dfrac{2s + 1}{s^2 + 4s + 7}$ **c.** $F_3(s) = \dfrac{1}{(s + 2)^4}$ **d.** $F_4(s) = \dfrac{s + 1}{(s + 2)(s + 3)^2}$

e. $F_5(s) = \dfrac{s^2 + 2s + 1}{s^2(s + 1)}$ **f.** $F_6(s) = \dfrac{s^2 + 2s + 1}{s^2 + 3s + 5}$ **g.** $F_7(s) = \dfrac{e^{-2s}}{(s + 1)^2(s^2 + 3)}$

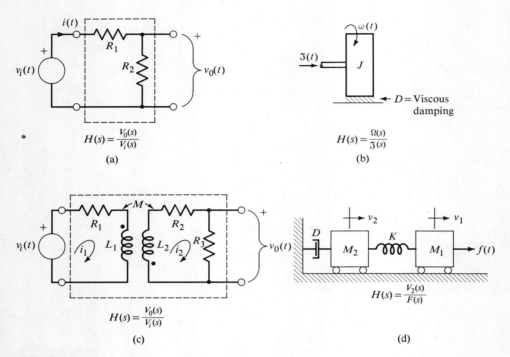

Figure P6-6.1

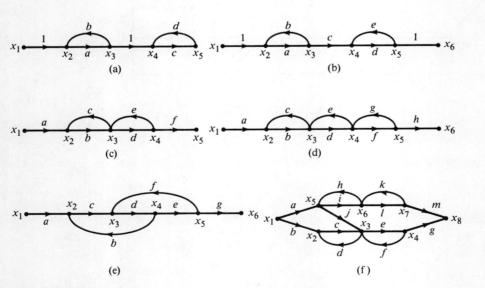

Figure P6-6.2

6-7.2 Find the inverse Laplace transform of each of the following:

a. $\dfrac{a}{s^2 + b^2}$ b. $\dfrac{2}{s^2 - 5s + 4}$ c. $\dfrac{2}{(s^2 + 1)^2}$ d. $\dfrac{2s - 3}{(s+1)^2 + 25}$ e. $\dfrac{4}{(s-1)(s^2+1)}$ f. $\dfrac{s}{s^2 - a^2}$

6-8.1 Verify the equivalence of the circuit representations shown in Figures 6.25 and 6.26.

6-8.2 Find the equivalent circuits of Figures 6.25 and 6.26 in the Laplace format.

6-9.1 Solve the following differential equations by Laplace transform methods ($t \geq 0$):

a. $\dfrac{d^2y}{dt^2} + 3\dfrac{dy}{dt} + 2y = 0$ $y(0+) = 5$ $\dfrac{dy(0+)}{dt} = 0$

b. $\dfrac{d^2y}{dt^2} + 3\dfrac{dy}{dt} + 2y = \delta(t)$ Initially relaxed

c. $\dfrac{d^2y}{dt^2} + 5y = \sin 2t + e^{-3t}$ Initially relaxed

d. $\dfrac{d^2y}{dt^2} + 3\dfrac{dy}{dt} + 2y = t^2 + 3t$ $y(0+) = 2$ $\dfrac{dy(0+)}{dt} = -8$

6-9.2 Determine the driving point current in the circuits of Figure P6-9.2. Assume that these circuits are initially relaxed.

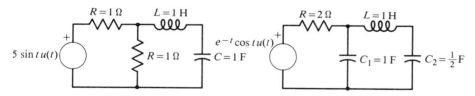

Figure P6-9.2

6-9.3 Determine the currents $i_1(t)$ and $i_2(t)$ in the network shown in Figure P6-9.3 subject to the following switching sequence: S_1 closed at $t = 0$, S_2 then closed at $t = 3$ s.

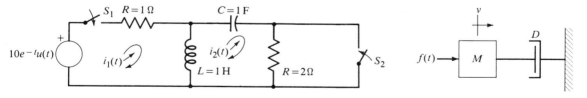

Figure P6-9.3   **Figure P6-9.4**

6-9.4 Find the velocity v for the system shown in Figure P6-9.4 if $f(0) = V_0$ and $f(t) = \sin(\omega t + \varphi)u(t)$.

6-9.5 Find the Laplace transform of the output of the system shown in Figure P6-9.5 when the input is $v(t) = te^{-2t}u(t)$. Use the convolution property.

6-9.6 Refer to Figure P6-9.6a. Prove that the portion external to the rectangle can be replaced by the circuit of Figure P6-9.6b for initially relaxed conditions. Hint: Consider the similarity of this with the Thévenin theorem for the steady state.

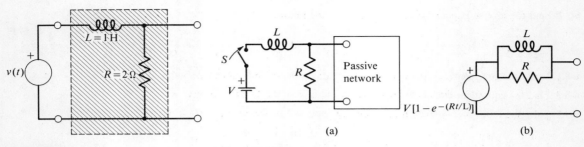

Figure P6-9.5   **Figure P6-9.6**

6-9.7 Determine the impulse response of a series RLC circuit for the following values of the circuit constants:
  a. $R = 4$, $C = 1$, $L = 1$ **b.** $R = 1$, $C = 4$, $L = 1$ **c.** $R = 2$, $C = 0.1$, $L = 1$.

6-9.8 Determine the impulse response for the systems shown in Figure P6-9.8.

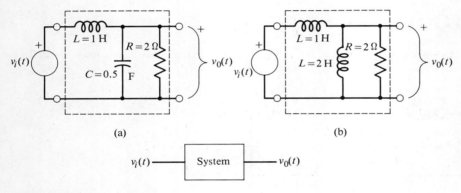

Figure P6-9.8

6-9.9 Find the impulse response of the system shown in Figure P6-9.9. The input is $f(t)$ and the output is $v_2(t)$.

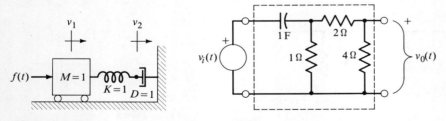

Figure P6-9.9   **Figure P6-9.10**

6-9.10 Use Laplace transform techniques to find the output voltage of the relaxed circuit shown in Figure P6-9.10 for an input voltage $v_i(t) = e^{-t}u(t)$. Verify your results using the convolution integral method.

6-10.1 Systems are described by the following differential equations:
  a. $(p^2 + 3p + 7)y(t) = (2p + 1)v(t)$ **b.** $(p^3 + 2p^2 + 5p + 2)y(t) = (p^3 + 3p + 2)v(t)$
  Determine the system functions. From these determine the impulse response $h(t)$.

6-10.2 Determine the values of the constant k that cause the system shown in Figure P6-10.2 to be unstable and to be stable.

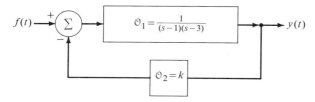

Figure P6-10.2

6-10.3 The output of a system is given by the relation $V_o(s) = H(s)V_i(s)$. If $v_i(t) = \sin \omega t$, show that the steady-state output is given by $|H(j\omega)| \sin[\omega t + \theta(j\omega)]$ provided that $H(s)$ has singularities only on the left-hand plane. What conclusion can you draw about the steady-state response of a linear time-invariant system?

6-10.4 Sketch the surface $|H(s)|$ over the s-plane for the transfer functions given below, and draw conclusions by observing the intersection between the surface $|H(s)|$ and the plane $\sigma = 0$.

  a. $H(s) = \dfrac{1}{s - 0.5}$ **b.** $H(s) = \dfrac{1}{s - 4}$

6-11.1 Find the function $u(t)$ by means of the Laplace inversion integral, given $F(s) = 1/s$.

6-11.2 Solve Problem 6-7.1 by the method of residues.

6-11.3 Find the inverse Laplace transform of the function $F(s) = e^{-\sqrt{s}}/\sqrt{s}$.

6-11.4 Find the Laplace transform of the periodic function shown in Figure P6-11.4 and compare these results with those in Example 6.39.

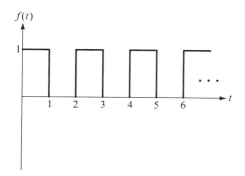

Figure P6-11.4

6-12.1 Find the function $f(t)$ if its two-sided Laplace transform is $F(s) = 2/(s + 2)$ and the region of absolute convergence is $\text{Re}\{s\} > -2$.

6-12.2 If the region of convergence of $f(t)$ is $\sigma_1 < \text{Re}\{s\} < \sigma_2$, find the region of convergence of the function $f(at)$, with a a positive constant.

CHAPTER 7
SAMPLING OF SIGNALS

The sampling of continuous signals at periodic intervals has become an important practical, as well as mathematical, operation. Also important to the sampling operation is the fact that most engineering systems have frequency response limitations; that is, they can respond only to some upper frequency limit. As a result the output signal of these systems is band-limited; for example, ordinary house telephones have an upper frequency limit of about 4000 Hz and television signals have an upper frequency limit of about 4 MHz. An important consequence of a signal having a finite bandwidth is that it can be accurately represented by short time-duration sample sequences taken at discrete and periodic instants.

The representation of continuous signals with their narrow time-limited samples taken at equal intervals in time has important practical utility. When transmitting signals through a transmitting channel, the time space between the samples of one signal can be used to accommodate without interference the samples of a different signal. This process is known as time-division multiplexing. Often the samples are digitized, a process readily accomplished with an analog-to-digital (A/D) converter, and the output is amplitude information in digital form. In this form sampled and digitized signals enter a computer for further processing.

Of concern when signals are sampled is how accurately the sampled values represent the sampled functions and what sampling interval must be used in order for an optimum recovery of the original signal from the sampled values. We wish to address these matters in some detail.

■ ■ ■

7-1 INTRODUCTION

The values of the function at the sampling points are called **sampled values,** the time that separates the sampling points is called the **sampling interval,** and the reciprocal of the sampling interval is the **sampling frequency** or the **sampling rate.** The value of any continuous function $f(t)$ at the point nT_s, where T_s is the sampling interval, is specified by

$$f(t)\,\delta(t - nT_s) = f(nT_s)\,\delta(t - nT_s) \tag{7.1}$$

SAMPLING OF SIGNALS

The sampling interval T_s is chosen here to be constant, and $n = 0, \pm 1, \pm 2, \ldots$. The sampled signal is (see also Figure 1.17a)

$$f_s(t) \triangleq f(t) \sum_{n=-\infty}^{\infty} \delta(t - nT_s) = f(t) \, \mathrm{comb}_{T_s}(t)$$

$$= \sum_{n=-\infty}^{\infty} f(nT_s) \, \delta(t - nT_s) \tag{7.2}$$

The Fourier transform of this quantity is

$$F_s(\omega) \triangleq \mathscr{F}\{f_s(t)\} = \sum_{n=-\infty}^{\infty} f(nT_s) \mathscr{F}\{\delta(t - nT_s)\}$$

$$= \sum_{n=-\infty}^{\infty} f(nT_s) e^{-jn\omega T_s} \tag{7.3}$$

since the Fourier transform of the shifted delta function is $\exp(-jn\omega T_s)$. By (4.60) and the symmetry property of Fourier transforms, (7.3) becomes

$$\sum_{n=-\infty}^{\infty} f(nT_s) e^{-jn\omega T_s} = \frac{1}{T_s} \sum_{n=-\infty}^{\infty} F(\omega + n\omega_s) \qquad \omega_s = \frac{2\pi}{T_s} \tag{7.4}$$

Therefore it follows that

$$F_s(\omega) = \frac{1}{T_s} \sum_{n=-\infty}^{\infty} F(\omega + n\omega_s) \tag{7.5}$$

The factor $1/T_s$ ensures conservation of the integrated area between the function $F(\omega)$ and its sampled representation.

Equation (7.5) can also be obtained if we use the relation

$$f_s(t) = f(t) \, \mathrm{comb}_{T_s}(t)$$

The Fourier transform of this expression which yields the convolution of the Fourier transforms of the two terms on the right becomes

$$F_s(\omega) = \mathscr{F}\{f(t) \, \mathrm{comb}_{T_s}(t)\} = \frac{1}{2\pi} F(\omega) * \mathscr{F}\{\mathrm{comb}_{T_s}(t)\}$$

$$= \frac{1}{2\pi} F(\omega) * \left[\frac{2\pi}{T_s} \sum_{n=-\infty}^{\infty} \delta(\omega - n\omega_s) \right]$$

$$= \frac{1}{T_s} \sum_{n=-\infty}^{\infty} \int_{-\infty}^{\infty} F(\tau) \delta(\omega - n\omega_s - \tau) d\tau$$

$$= \frac{1}{T_s} \sum_{n=-\infty}^{\infty} F(\omega - n\omega_s) = \frac{1}{T_s} \sum_{n=-\infty}^{\infty} F(\omega + n\omega_s)$$

Figure 7.1
The Fourier spectrum of a sampled signal.

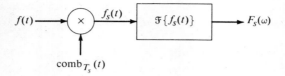

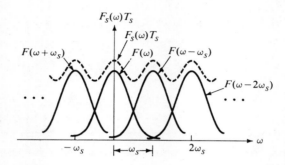

In the foregoing development, we used (4.58) and the frequency convolution property (see Table 4.1).

This discussion shows that if we know the Fourier transform of $f(t)$, its sampled version $f_s(t)$ is uniquely determined. Moreover, if we set $\omega = \omega + m\omega_s$ in (7.5) we obtain

$$F_s(\omega + m\omega_s) = \frac{1}{T_s} \sum_{n=-\infty}^{\infty} F[\omega + (n+m)\omega_s]$$

$$= \frac{1}{T_s} \sum_{k=-\infty}^{\infty} F(\omega + k\omega_s) = F_s(\omega) \quad (7.6)$$

This shows that $F_s(\omega)$ is periodic in the frequency domain, as shown in Figure 7.1. We will say more about this figure in Section 7-2.

When the function $f(t)$ is causal (positive time), $f(t) = 0$ for $t < 0$, then

$$f_s(t) = \sum_{n=0}^{\infty} f(nT_s)\,\delta(t - nT_s) \quad \textbf{(a)} \quad (7.7)$$

and

$$F_s(\omega) = \sum_{n=0}^{\infty} f(nT_s) e^{-jn\omega T_s} = \frac{f(0+)}{2} + \frac{1}{T_s} \sum_{n=-\infty}^{\infty} F(\omega + n\omega_s) \quad \textbf{(b)}$$

This last equality was shown to be true in (4.62) Example 4.18.

EXAMPLE 7.1
Find the Fourier transform of the sampled functions

    a. $f_s(t) = e^{-|t|} \text{comb}_{T_s}(t)$ **b.** $f_s(t) = e^{-t} u(t) \text{comb}_{T_s}(t)$

SAMPLING OF SIGNALS

Solution: By (7.5), (7.7), and Table 4.2, we obtain, respectively,

$$\mathcal{F}\{e^{-|t|} \operatorname{comb}_{T_s}(t)\} \triangleq F_s(\omega)$$

$$= \frac{1}{T_s} \sum_{n=-\infty}^{\infty} \frac{2}{1 + (\omega + n\omega_s)^2}$$

$$\omega_s = \frac{2\pi}{T_s} \qquad \text{(a)} \qquad (7.8)$$

$$\mathcal{F}\{e^{-t} u(t) \operatorname{comb}_{T_s}(t)\} \triangleq F_s(\omega)$$

$$= \frac{1}{2} + \frac{1}{T_s} \sum_{n=-\infty}^{\infty} \frac{1}{1 + j(\omega + n\omega_s)}$$

$$\omega_s = \frac{2\pi}{T_s} \qquad \text{(b)} \qquad \blacksquare$$

EXAMPLE 7.2

Consider three functions: $f_1(t)$, $f_2(t)$, and $f_3(t)$ with respective frequency characteristics $F_1(\omega)$, $F_2(\omega)$, and $F_3(\omega)$, as shown in Figure 7.2b. Find the maximum sampling interval T_s in order that the function $f(t) = f_1(t) + f_2(t)f_3(t)$ shown in Figure 7.2a can be recovered from its sampled version $f_s(t)$ using a low-pass filter.

Solution: The Fourier transform of the sampled function is given by

$$F_s(\omega) = \mathcal{F}\{f_1(t) \operatorname{comb}_{T_s}(t) + f_2(t)f_3(t) \operatorname{comb}_{T_s}(t)\}$$

$$F_s(\omega) = \frac{1}{2\pi} F_1(\omega) * \frac{2\pi}{T_s} COMB_{\omega_s}(\omega)$$

$$+ \frac{1}{(2\pi)^2} F_2(\omega) * F_3(\omega) * \frac{2\pi}{T_s} COMB_{\omega_s}(\omega)$$

The convolution of $F_1(\omega)$ and $COMB_{\omega_s}(\omega)$ gives us a periodic repetition of the spectrum $F_1(\omega)$. The convolution of $F_2(\omega) * F_3(\omega)$ with $COMB_{\omega_s}(\omega)$ gives us a periodic repetition of the spectrum of $F_2(\omega) * F_3(\omega)$. But the spectral width of $F_2(\omega) * F_3(\omega)$ is equal to the sum of the spectral widths of $F_2(\omega)$ and $F_3(\omega)$; hence in the present case $\omega_N = \omega_{N_2} + \omega_{N_3} = 2\pi \times 10^4 + 2\pi \times 10^4 = 4\pi \times 10^4$. The spectrum of $F_s(\omega)$ is shown in Figure 7.2c. Observe that the minimum ω_s in order that the spectra of $F_2(\omega) * F_3(\omega)$ do not overlap is $8\pi \times 10^4$, or equivalently, the maximum $T_s = 2\pi/\omega_s = 2\pi/8\pi \times 10^4 = 0.25 \times 10^{-4}$s. Because the spectral width of $F_2(\omega) * F_3(\omega)$ is greater than the spectral width of $F_1(\omega)$, the value of T_s is determined by the spectral width of $F_2(\omega) * F_3(\omega)$. However, if the spectral width of $F_1(\omega)$ were greater than the spectral width of $F_2(\omega) * F_3(\omega)$, the value of T_s would be determined from the spectral width of $F_1(\omega)$. The spectra in Figure 7.2c have been normalized to unity.

■ ■ ■

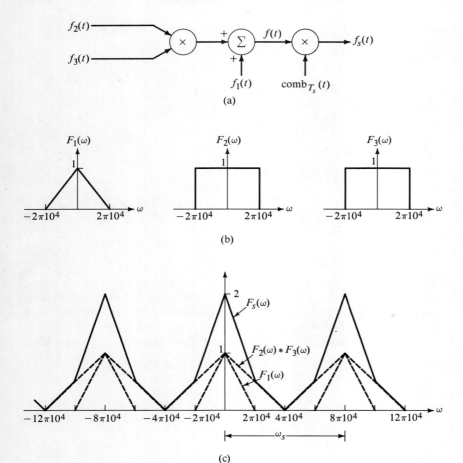

Figure 7.2
Illustrating Example 7.2.

7-2 THE SAMPLING THEOREM

We will now show that it is possible for a band-limited signal $f(t)$ to be exactly specified by its sampled values provided that the time distance between sampled values does not exceed a critical sampling interval. A limited view of sampling was discussed in Theorem 3.1 for periodic band-limited signals. For a more general view of sampling, we state the following theorem:

Theorem 7.1 A finite energy function $f(t)$ having a band-limited Fourier transform (that is, $F(\omega) = 0$ for $|\omega| \geq \omega_N$) can be completely reconstructed from its sampled values $f(nT_s)$ (see Figure 7.3), with

SAMPLING OF SIGNALS

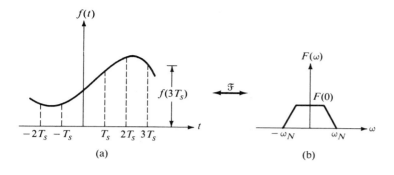

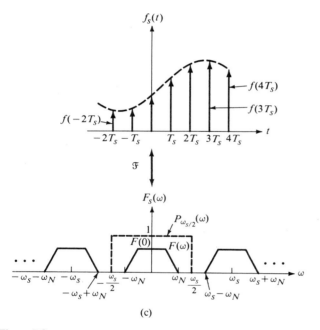

Figure 7.3
Illustrations of the sampling theorem.

$$f(t) = \sum_{n=-\infty}^{\infty} T_s f(nT_s) \frac{\sin\left[\frac{\omega_s(t-nT_s)}{2}\right]}{\pi(t-nT_s)} \qquad \omega_s = \frac{2\pi}{T_s} \qquad \text{(a)} \quad (7.9)$$

provided that

$$\frac{2\pi}{\omega_s} \triangleq T_s = \frac{\pi}{\omega_N} = \frac{\pi}{2\pi f_N} = \frac{1}{2f_N} \triangleq \frac{T_N}{2} \qquad \text{(b)}$$

The function within the braces, which is a sinc function, is often called the **interpolation function** to indicate that it allows an interpolation between the sampled values to find $f(t)$ for all t.

Proof: Employ (7.5) and Figure 7.3c to write, $(n = 0)$,

$$F(\omega) = p_{\omega_s/2}(\omega) T_s F_s(\omega) \tag{7.10}$$

By (7.3) this equation becomes

$$F(\omega) = p_{\omega_s/2}(\omega) T_s \sum_{n=-\infty}^{\infty} f(nT_s) e^{-jn\omega T_s} \tag{7.11}$$

From this we have

$$f(t) \triangleq \mathscr{F}^{-1}\{F(\omega)\} = \mathscr{F}^{-1}\left\{ p_{\omega_s/2}(\omega) T_s \sum_{n=-\infty}^{\infty} f(nT_s) e^{-j\omega nT_s} \right\}$$

$$= T_s \sum_{n=-\infty}^{\infty} f(nT_s) \mathscr{F}^{-1}\{p_{\omega_s/2}(\omega) e^{-j\omega nT_s}\} \tag{7.12}$$

By an application of the frequency shift property of Fourier transforms, this equation proves the theorem.

Equation (7.9) demonstrates that the generation of $f(t)$ from the sequence $\{f(nT)\}$ is a complex task. This equation is reserved mainly for theoretical considerations.

■ ■ ■

To avoid frequencies beyond ω_N, a low-pass filter is included. If the frequency width of the filter ω_1 is other than $\omega_s/2$, (7.9) must be appropriately modified. This involves substituting this frequency value for $\omega_s/2$, with the result that the sine term within the braces of (7.9a) becomes $\sin[\omega_1(t - nT_s)]$ and $p_{\omega_s/2}(\omega)$ becomes $p_{\omega_1/2}(\omega)$.

For the case when $\omega_s = 2\omega_N$, (7.9a) becomes (see Problem 7-2.2)

$$f(t) = \sum_{n=-\infty}^{\infty} f(nT_s) \frac{\sin[\omega_N(t - nT_s)]}{\omega_N(t - nT_s)} \tag{7.13}$$

and the spectrum of $F_s(\omega)$ is that shown in Figure 7.5d, which shows that the spectrum for $F_s(\omega)$ just touches the successive replicas of $F(\omega)$.

The sampling time

$$\boxed{T_s = \frac{T_N}{2} = \frac{1}{2f_N}} \tag{7.14}$$

is called the **Nyquist interval.** It is the longest time interval that can be used for sampling a band-limited signal and still allow recovery of the signal without distortion. Figure 7.4 shows how a signal can be reconstructed from its samples

Figure 7.4 Reconstruction of a signal from its samples.

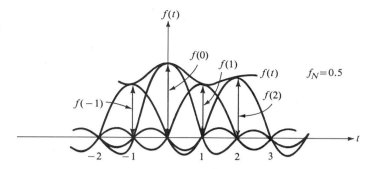

using (7.9a). Observe that the sinc functions tend to cancel between the sampling times and to reinforce at the sampling points. In the limit as $n \to \infty$, the recovery of the signal is complete.

Figure 7.5a shows the overall procedure of delta function sampling and transform representation of a band-limited signal. Note, however, that if the sampling time interval T_s is too large, the product $F(\omega) * COMB_{\omega_s}(\omega)$ would look like Figure 7.5b. The overlapping of spectra is known as **aliasing**. Aliasing disappears if the sampling time diminishes at least to the value $T_s = 1/2f_N$, where f_N is the highest frequency component belonging to the signal. It is clear from Figure 7.5b that there is no available filter capable of extracting the frequency content of the signal without including additional frequencies not contained in the signal itself. The recovery of the signal with aliasing present results in a signal with artifacts, as shown in Figure 7.5c for a two-dimensional periodic signal (spokewheel) that has been under-sampled.

The effect of aliasing can be used to our advantage in some cases. Suppose that in a visual observation we want to stop a repetitive action—for example, the wing undulation of a bee or the turning of a wheel; to accomplish this we flash the object with a strobe light. If we adjust the repetition of the strobe flashes to equal the wing repetition rate or the wheel rotation rate, these events will appear to be stationary. If the strobe frequency is much higher than twice that of the periodic phenomenon, the speed of the phenomenon does not appear to change. However, if the frequency of the strobe flashes is less than twice the frequency of the phenomenon under observation, the repetition slows down; thus we observe a slow-flying bee or a slowly rotating wheel. This phenomenon is commonly observed in movies when the wheels of a moving stagecoach appear to be stationary or turning slowly (sometimes backwards). In movies the sampling rate is about 1/20 sec since the frame rate of the film is about 20 frames per sec.

Figure 7.5d shows the product $F(\omega) * COMB_{\omega_s}(\omega)$ when the sampling time $T_s = 1/2f_N$ is just at the Nyquist rate. Figure 7.5e shows the procedure to recover the signal. In this case we write

$$\mathscr{F}^{-1}\{F(\omega)\} \triangleq f(t) = \mathscr{F}^{-1}\left\{\left[\frac{1}{2\pi}F(\omega) * COMB_{\omega_s}(\omega)\right] p_{\omega_s/2}(\omega)\right\}$$

$$= \mathscr{F}^{-1}\left\{\frac{1}{2\pi}F(\omega) * COMB_{\omega_s}(\omega)\right\} * \mathscr{F}^{-1}\{p_{\omega_s/2}(\omega)\}$$

$$= \left[T_s f(t)\, \text{comb}_{T_s}(t)\right] * \frac{\sin\left(\frac{\omega_s t}{2}\right)}{\pi t}$$

$$= \frac{T_s}{\pi}\left[\sum_{n=-\infty}^{\infty} f(nT_s)\,\delta(t - nT_s)\right] * \frac{\sin\left(\frac{\omega_s t}{2}\right)}{t}$$

$$= T_s \sum_{n=-\infty}^{\infty} f(nT_s)\frac{\sin\left[\frac{\omega_s(t - nT_s)}{2}\right]}{\pi(t - nT_s)}$$

which is identical with (7.9a).

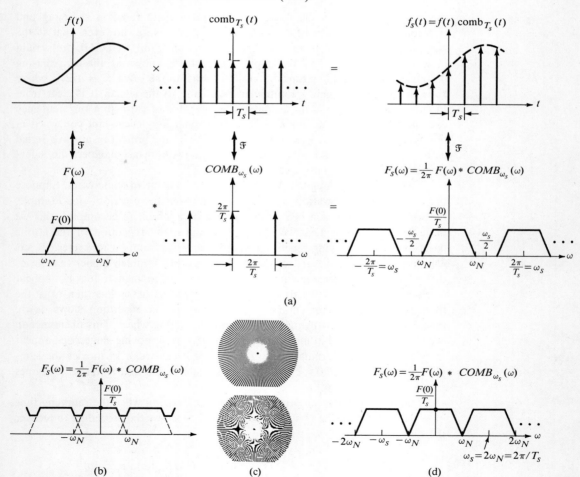

Figure 7.5 *(continued on page 375)*

SAMPLING OF SIGNALS

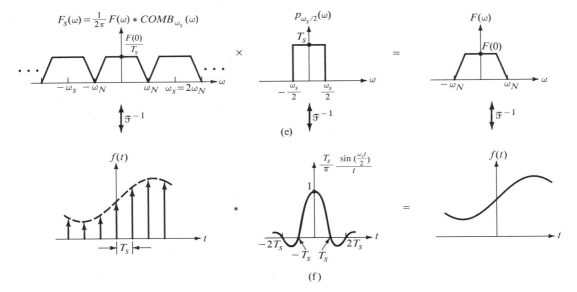

Figure 7.5
Delta sampling, representation, and recovery of signals. [(c) Original image of spoke-wheel and its under-sampled image. The effect of aliasing appears clearly. Reprinted by permission from Leger and Lee, *Signal Processing Using Hybrid Systems*.]

EXAMPLE 7.3
Show the aliasing phenomenon by decreasing ω_s or, equivalently, by increasing the sampling time T_s associated with a pure cosine function, $f(t) = \cos \omega_0 t$. Use a low-pass filter of bandwidth ω_s in the output.

Solution: The Fourier transform of $f(t)$ is $F(\omega) = \pi\delta(\omega - \omega_0) + \pi\delta(\omega + \omega_0)$. Thus the Fourier transform of the sampled function $f_s(t)$ is

$$F_s(\omega) = \frac{1}{T_s} F(\omega) * COMB_{\omega_s}(\omega)$$

$$= \frac{\pi}{T_s} [\delta(\omega - \omega_0) * COMB_{\omega_s}(\omega) + \delta(\omega + \omega_0) * COMB_{\omega_s}(\omega)]$$

Figure 7.6a shows the spectrum of $F_s(\omega)$ when $\omega_s \gg \omega_0$ or, equivalently, when $T_s \ll T_0 = 2\pi/\omega_0$. Figure 7.6b is the result of the convolution of $\delta(\omega - \omega_0)$ and $COMB_{\omega_s}(\omega)$. Figure 7.6d shows the convolution of $\delta(\omega + \omega_0)$ and $COMB_{\omega_s}(\omega)$. By adding the spectra of Figures 7.6b and 7.6d, we obtain the total spectrum

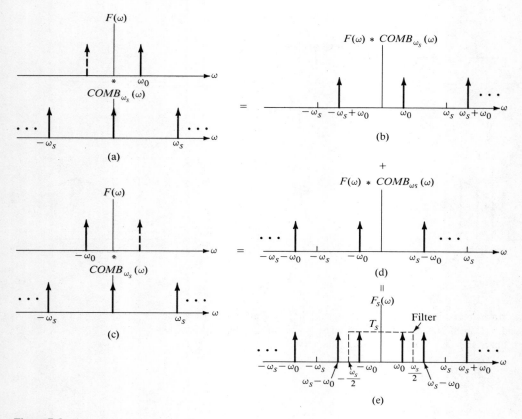

Figure 7.6
Illustrating the aliasing problem using a pure sinusoid function (*continued on page 377*).

of $F_s(\omega)$ as specified by the above equation. If we incorporate a filter with a frequency bandwidth of $\omega_s/2$, we regain our signal since

$$\frac{\pi}{2\pi} \frac{T_s}{T_s} \int_{-\infty}^{\infty} p_{\omega_s/2}(\omega) F_s(\omega) e^{j\omega t} \, d\omega$$

$$= \frac{1}{2} \int_{-\omega_s/2}^{\omega_s/2} [\delta(\omega + \omega_0) + \delta(\omega - \omega_0)] e^{j\omega t} \, d\omega$$

$$= \frac{1}{2} [e^{j\omega_0 t} + e^{-j\omega_0 t}] = \cos \omega_0 t$$

We follow the same procedure for the steps shown in Figures 7.6f–j and develop the function $\cos(\omega_s - \omega_0)t$, a function that is quite different from $\cos \omega_0 t$. Hence when aliasing occurs, the original frequency ω_0 assumes another value, which is the "alias" of a lower frequency $\omega_s - \omega_0$. ■

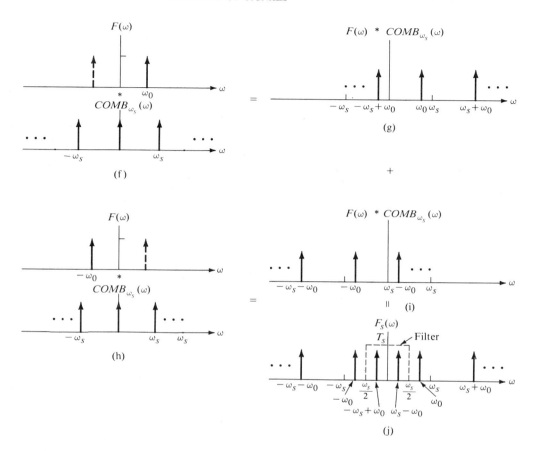

Frequency Sampling Analogous to the time sampling theorem, there also exists a parallel **frequency sampling** theorem. This can be stated as follows:

Theorem 7.2 A time function $f(t)$ that is time-limited so that

$$f(t) = 0 \qquad |t| > T_N \tag{7.15}$$

possesses a Fourier transform that can be uniquely determined from its samples at distances $n\pi/T_N$, and is given by

$$F(\omega) = \sum_{n=-\infty}^{\infty} F\left(n\frac{\pi}{T_N}\right) \frac{\sin(\omega T_N - n\pi)}{\omega T_N - n\pi} \tag{7.16}$$

when the sampling is at the Nyquist rate (see Problem 7-2.3).

■ ■ ■

Figure 7.7
Illustrating the time-domain aliasing problem.

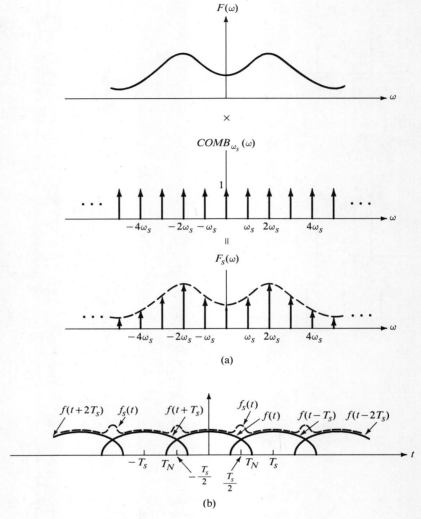

(a)

(b)

EXAMPLE 7.4
Find the time-limited function $f(t)$ from the sampled frequency function when ω_s, the distance of the samples in the frequency domain, is larger than $\omega_N/2 = 2\pi/2T_N$ or, equivalently, $T_N > T_s/2$, where $2T_N$ is the time bandwidth of the signal.

Solution: The sampled function $F_s(\omega) = F(\omega)\,COMB_{\omega_s}(\omega)$ is shown in Figure 7.7a. Its time-domain representation is determined by the Fourier inversion of $F_s(\omega)$. Hence we obtain (see Table 4.1, entry 10 and Problem 7-2.3)

SAMPLING OF SIGNALS

$$f_s(t) = \mathscr{F}^{-1}\{F(\omega)\ COMB_{\omega_s}(\omega)\} = f(t) * \left\{\frac{T_s}{2\pi}\ \text{comb}_{T_s}t\right\}$$

$$f_s(t) = \frac{T_s}{2\pi} \sum_{n=-\infty}^{\infty} f(t - nT_s) \tag{7.17}$$

A plot of $f_s(t)$ is shown in Figure 7.7b that indicates **time-domain aliasing.** ∎

Sampling of Shifted Functions If the function $f(t)$ is band-limited (that is, $F(\omega) = 0$ for $|\omega| \geq \omega_N$) then $f(t - a)$ is also band-limited, since $\mathscr{F}\{f(t - a)\} = \exp(-ja\omega)\ F(\omega)$, and (7.9) applies equally well for the shifted function. We find that

$$f(t - a) = \sum_{n=-\infty}^{\infty} T_s f(nT_s - a) \frac{\sin\left[\dfrac{\omega_s(t - nT_s)}{2}\right]}{\pi(t - nT_s)} \tag{7.18}$$

Upon shifting the time coordinates by a, the resulting expression is

$$f(t) = \sum_{n=-\infty}^{\infty} T_s f(nT_s - a) \frac{\sin\left[\dfrac{\omega_s(t + a - nT_s)}{2}\right]}{\pi(t + a - nT_s)} \tag{7.19}$$

Sampling with a Train of Rectangular Pulses As a practical matter, one cannot produce infinitely narrow time pulses for sampling needs. For this reason sampling involves a train of rectangular pulses rather than comb(t). To examine the consequences of rectangular pulse sampling, both the sampling pulses and $f(t)$ are specified by their Fourier spectra, as shown in Figure 7.8. The Fourier transform of the pulse train is [see (4.64)]

$$P(\omega) = \sum_{n=-\infty}^{\infty} 2\pi \frac{\sin\left(\dfrac{n\omega_s\tau}{2}\right)}{\dfrac{n\omega_s\tau}{2}} \delta(\omega - n\omega_s) \tag{7.20}$$

where τ is the width of the sampling pulses. Now employ the frequency convolution property (see Table 4.1)

$$F_s(\omega) = \mathscr{F}\{f(t)p_{\tau/2}(t)\} = \frac{1}{2\pi} F(\omega) * P(\omega)$$

$$= \sum_{n=-\infty}^{\infty} \frac{\sin\left(\dfrac{n\omega_s\tau}{2}\right)}{\dfrac{n\omega_s\tau}{2}} \int_{-\infty}^{\infty} \delta(\xi - n\omega_s) F(\omega - \xi)\, d\xi$$

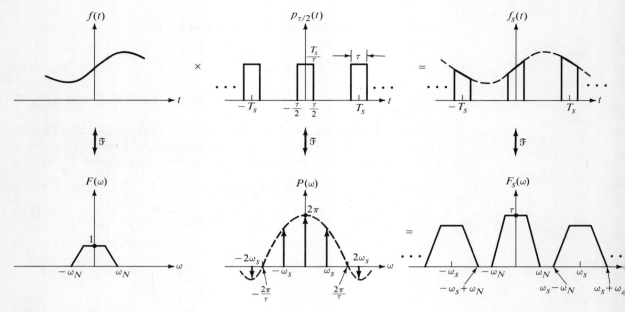

Figure 7.8
Sampling with rectangular pulse train.

or

$$F_s(\omega) = \sum_{n=-\infty}^{\infty} \frac{\sin\left(\frac{n\omega_s \tau}{2}\right)}{\frac{n\omega_s \tau}{2}} F(\omega - n\omega_s) \quad (7.21)$$

This expression indicates that as long as $\omega_s > 2\omega_N$, the spectrum of the sampled signal contains no overlapping spectra of $f(t)$ and can be recovered using a low-pass filter.

Finite Pulse Train Sampling—Samples with Flat Tops Suppose that sampling is limited in time, a situation that occurs if sampling proceeds for a finite time and is then stopped. The sampled signal is made up of pulses with flat tops and is given by (see Figure 7.9a)

$$f_s(t) = [f(t) \, \text{comb}_{T_s}(t)] * p_{\tau/2}(t) \quad (7.22)$$

The resulting frequency spectrum is

SAMPLING OF SIGNALS

$$F_s(\omega) = [F(\omega) * COMB_{\omega_s}(\omega)]P(\omega)$$

$$= \frac{4\pi}{T_s} \frac{\sin\left(\frac{\omega\tau}{2}\right)}{\omega} \sum_{n=-\infty}^{\infty} F(\omega - n\omega_s) \qquad (7.23)$$

Note that because the frequency content in the band $-\omega_N < \omega < \omega_N$ with $\omega_s > 2\omega_N$ is not just $F(\omega)$ but is $[\sin(\omega\tau/2)/\omega]F(\omega)$, the use of a low-pass filter will not result in the recovery of the signal without distortion, as it did previously. To effect distortionless signal recovery requires the addition of an equalizing filter of the form $[\omega/\sin(\omega\tau/2)]$ within the band $|\omega| < \omega_N$. Characteristics of the equalizing filter are shown in Figure 7.9b. A block diagram representation of the process is shown in Figure 7.9c.

If, instead of the time limiter $p_{\tau/2}(t)$, a general time-limiting function $s(t)$ with Fourier transform $S(\omega)$ is used, the result given in (7.23) is replaced by the expression

$$F_s(\omega) = \frac{2\pi}{T_s} S(\omega) \sum_{n=-\infty}^{\infty} F(\omega - n\omega_s) \quad \omega_s = \frac{2\pi}{T_s} \qquad (7.24)$$

Reconstruction of Sampled Signals We have found that if a signal is band-limited, we can always sample it at such a rate that its periodic Fourier spectrum is without overlap. Under these conditions a low-pass filter will capture all the harmonics contained in the signal and will permit the signal to be reproduced

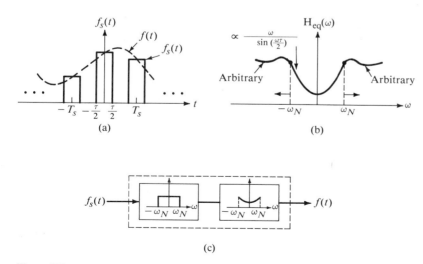

Figure 7.9
Sampled signal made up of pulses with flat tops (compare Figure 7.8) and its recovery process.

Figure 7.10
Estimating signal bandwidth.

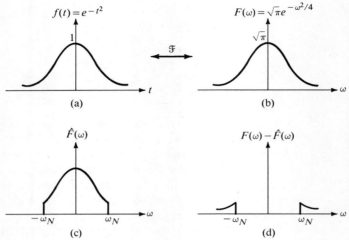

exactly. However, as a practical matter, all signals that are of interest possess spectra of infinite extent and are impossible to reconstruct without some aliasing. One possible solution to this problem is first to pass the signal through a low-pass filter to band-limit it, and then to sample it appropriately, as we have discussed above. This will distort the signal, of course, and we must impose some kind of quality factor to ensure acceptable results. One such factor is the amount of energy rejected during the band-limiting process.

To clarify the foregoing discussion, consider the Gaussian signal shown in Figure 7.10a and its Fourier spectrum shown in Figure 7.10b (see Table 4.2). Our purpose is to establish an appropriate bandwidth ω_N of the signal such that the energy error

$$\Delta E = \frac{1}{2\pi} \int_{-\infty}^{\infty} |F(\omega) - \hat{F}(\omega)|^2 \, d\omega \tag{7.25}$$

is, say, 2 percent of the energy of the signal $f(t)$. In this expression $\hat{F}(\omega)$ is the truncated Fourier spectrum of $F(\omega)$ shown in Figure 7.10c. The rejected spectrum $F(\omega) - \hat{F}(\omega)$ is shown in Figure 7.10d.

Because of symmetry (7.25) becomes

$$\Delta E = \frac{1}{\pi} \int_0^\infty |F(\omega) - \hat{F}(\omega)|^2 \, d\omega = \frac{1}{\pi} \int_{\omega_N}^\infty (\sqrt{\pi} e^{-\omega^2/4})^2 \, d\omega = \int_{\omega_N}^\infty e^{-\omega^2/2} \, d\omega$$

$$= \int_0^\infty e^{-\omega^2/2} \, d\omega - \int_0^{\omega_N} e^{-\omega^2/2} \, d\omega = \sqrt{\frac{\pi}{2}} - \sqrt{\frac{\pi}{2}} \, \text{erf}\left(\frac{\omega_N}{\sqrt{2}}\right) \tag{7.26}$$

where $\text{erf}(\cdot)$ is the **error function.** The error function is defined by the relation (see Appendix 7-1)

$$\text{erf}(t) = \frac{2}{\sqrt{\pi}} \int_0^t e^{-\tau^2} \, d\tau \tag{7.27}$$

SAMPLING OF SIGNALS

The total energy of the function $f(t)$ is given by $\lim_{\omega_N \to 0} \Delta E = \sqrt{(\pi/2)} = E_f$. Therefore

$$\frac{\Delta E}{E_f} = 1 - \text{erf}\left(\frac{\omega_N}{\sqrt{2}}\right) \qquad (7.28)$$

From the error function table in Appendix 7-1, we see that when $\omega_N/\sqrt{2} = 1.65$, $\text{erf}(1.65) = 0.98037$. This gives an energy error of $\Delta E/E_f = 1 - 0.98037 = 0.01963$, which is approximately equal to 2 percent. Hence for $\omega_N \geq 2.33$, the error will be maintained within $\Delta E \leq 0.02 \, E_f$.

■ ■ ■

7-3 TELEPHONE TRANSMISSION

We have already mentioned that the time space between the samples of a given signal can be used for a second signal if a properly conditioned sample of another signal is available. This interlacing of pulses representing different signals for transmission over the same channel is known as time-division multiplexing (TDM), and this practice is used extensively in communication systems and computers. Figure 7.11 shows the elements of a system for multiplexing two signals.

A major problem in a TDM system is that of synchronizing the commutators to ensure that the transmitted and received signals belong to the same message. The synchronization problem is more difficult when many signals are transmitted per unit time and when the transmitter and receiver are far apart. Synchronization is achieved in such systems by using very stable **clock** oscillators at both ends of the system. They are synchronized by occasionally transmitting special synchronizing signals.

A much more sophisticated TDM system is used in the Bell T-1 system. In this system, which accommodates 24 voice channels, each signal is passed through a low-pass filter and then through synchronized sampling gates. These are combined and passed through a compressor and amplifier, then through a pulse code modulation (PCM) (see below) encoder, and then onto the transmission line. Each voice channel is encoded into an 8-bit group; the 8th bit of this group provides supervisory and signal information for effecting synchronization between transmitting and receiving demodulators. The 24-signal group requires a 193-bit frame, with a repetition of 125 μs. On reception the signal group is passed through a PCM decoder, an expander and amplifier, sampling gates synchronized with the transmitter sampling gates, and then through low-pass filters.

The radio transmission of multiplexed signals is often accomplished by pulse amplitude modulation (PAM). The signals to be transmitted are sampled and the resulting pulses have amplitudes specified by the continuous signals at the points of sampling. That is, the pulse amplitudes are dictated by the continuous waves being sampled, but the pulse duration is fixed by electronic circuits contained within the pulse generators. Each pulse modulates a carrier wave, which involves the multiplication of the sinusoidal carrier frequency by the pulses, to

Figure 7.11
Time-division multiplexing.

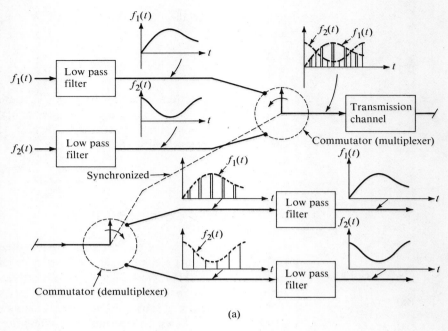

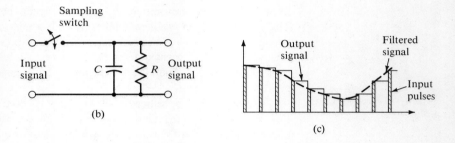

give blocks of radio frequency waves. These radio frequency waves are transmitted over a given channel, usually free space.

Because we are interested in recovering the sampled signals, we must pay attention to the channel bandwidth if distortion is to be avoided. For example, if we send N telephone signals, each having a bandwidth $B = 3.5$ kHz, then as already found we must sample each signal at a rate of at least $2 \times 3.5 = 7$ kHz (or 7000 samples/s). Therefore the length of each sampling interval cannot exceed $1/(2 \times 3.5 \times 10^3 \times N)$ seconds. For simplicity let the samples be square pulses of width $2a$ and define the bandwidth B_s as the distance to the first zero crossing of the sinc function in the frequency domain. Hence the bandwidth of each pulse is

$$B_p = \frac{1}{2a} \quad \text{Hz} \tag{7.29}$$

SAMPLING OF SIGNALS

The maximum width of each pulse is $2a = 1/(2B_s N)$, and so the total bandwidth of the channel must be equal to the pulse bandwidth or

$$B_c = B_p = \frac{1}{\frac{1}{2B_s N}} = 2B_s N \quad \text{Hz} \tag{7.30}$$

In general, the sampling rate is somewhat larger than $2B_s$ by a small amount, to provide a **guard band** between pulses.

When the PAM pulses arrive at the receiving terminal, a demodulation process must take place to recover the individual signals from the carrier frequency. This downshifting in frequency from the carrier level to the zero frequency level is accomplished in the circuit shown in Figure 7.11b, which is known as a **sample-and-hold** circuit. The sampling switch and a parallel RC circuit comprise the demodulator. The switch is closed at the times when the particular assigned channel is to be recovered. The values of the resistor and the capacitor are so related that the time constant $RC > T$, where T is the time between pulses to be detected. When the switch is closed, the capacitor is charged to the value of the pulse height and essentially retains the charge until the next pulse arrives, when the new pulse height is imposed on the capacitor. In this way the output consists of steps, the heights of which are dictated by the heights of the incoming pulses. A low-pass filter is added to the output of the sample-and-hold circuit in order to smooth the output signal. The input and output signals to the circuit are shown in Figure 7.11c.

Interface devices are available that convert time-division multiplexed (TDM) signals into frequency-division multiplexed (FDM) signals. The use of these transmultiplexors allows accommodating many more signals over the circuit path than is possible with TDM. Such transmultiplexors are used in telephone service.

■ ■ ■

7-4 ADDITIONAL PULSE MODULATION TECHNIQUES

A number of other pulse modulation schemes are of commercial importance, in addition to PAM. The simplest and most basic modulation techniques used in practice are:

- PDM: pulse duration modulation
- PWM: pulse width modulation
- PPM: pulse position modulation
- PCM: pulse code modulation
- DM: delta modulation

Figure 7.12 graphically shows the three types of analog pulse modulation: PAM, PDM, and PPM.

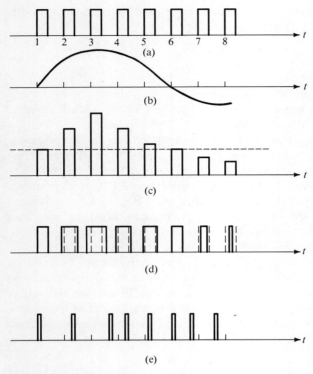

Figure 7.12
Pulse modulation methods. (a) Pulsing circuit (switch). (b) Signal. (c) Unipolar PAM. (d) PDM output. (e) PPM output.

To achieve PDM signals, the input continuous signal is sampled at a rate that satisfies the Nyquist criterion. At each sample time, a pulse is generated that has a width proportional to the signal amplitude, but the amplitude of all pulses so generated is a constant. What is accomplished by this method is a conversion of amplitude information into pulse-width or pulse-time information.

In the case of PPM, pulses of common width and constant height are generated, but the position of the pulses relative to a fixed time reference is adjusted in accordance with the height of the sampled signal.

The advantage of PDM and PPM over PAM arises because the pulse heights in PDM and PPM are of constant amplitude, and as a result noise that is an amplitude variation occurring in the transmission process has a lesser impact on PDM and PPM. Furthermore, since these pulses might be characterized by the rising and falling edges of the pulses, noise variations will not affect these features appreciably.

To achieve PCM we must first sample the signal at a minimum of twice its highest frequency, quantize it, and then encode it. A typical telephone signal,

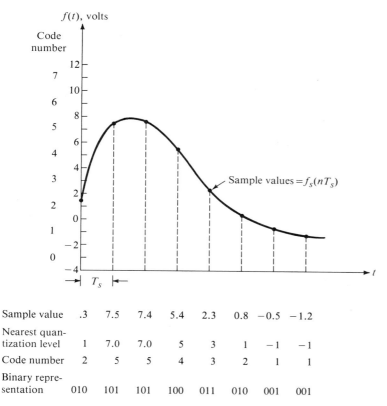

Figure 7.13
Sampling, quantization, and binary encoding for PCM system.

for example, has a bandwidth of approximately 4 kHz. The sampling rate is thus 8 kHz or, equivalently, the distances between samples is $T_s = 1/8 \times 10^3 = 125$ μs. Figure 7.13 shows the features of quantization and coding. Here, for simplicity, we have used only $2^3 = 8$ distinct levels, although for better results more levels should be employed. In this method of transmission, the signal level is converted into binary form and a string of binary digits is transmitted. At the receiving point, decoding and reconstruction takes place using a digital-to-analog (D/A) converter.

Delta modulation is another technique for converting sampled signals into a string of binary digits, but in this case the circuitry is less complicated than that of an A/D converter. Figure 7.14a shows a block diagram representation of a delta modulating system, and Figures 7.14b and 7.14c show the modulation process and the string of digits that are transmitted through the communication channel to the receiver. Observe that the transmitted pulses correspond to the rising and falling edges of the sampled signal.

Figure 7.14
Delta modulation technique.
(a) System. (b) The signal $f(t)$ and its approximation $\hat{f}(t)$. (c) The transmitted pulse train.

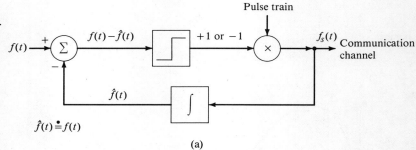

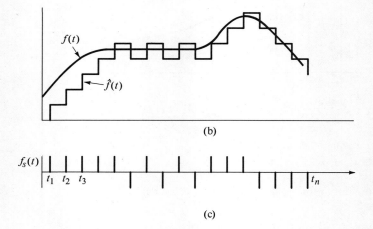

REFERENCES

1. Bracewell, R. *The Fourier Transform and Its Applications.* New York: McGraw-Hill, 1985.
2. Leger and Lee. "Signal Processing Using Hybrid Systems." *Applications of Optical Fourier Transforms,* Ed. H. Stark, Academic Press, 1980.
3. Linden, D. A. "A Discussion of Sampling Theorems." *Proceedings of the Institute of Radio Engineers* 47 (1959): 1219.
4. Papoulis, A. *The Fourier Integral and Its Applications.* New York: McGraw-Hill, 1968.
5. Shannon, C. E. "Communication in the Presence of Noise." *Proceedings of the Institute of Radio Engineers* 37 (1949): 10.
6. Stark, H. *Applications of Optical Fourier Transforms.* New York: Academic Press, 1982.
7. Stark, H., and F. B. Tuteur. *Modern Electrical Communication Theory and Systems.* Englewood Cliffs, N.J.: Prentice-Hall, 1979.

8. Stremler, F. G. *Introduction to Communication Systems*. Reading, Mass.: Addison-Wesley, 1977.

9. Whittaker, E. T. "On the Functions Which Are Represented by the Expansions of the Interpolation Theory." *Proc. Roy. Soc.*, Edinburgh, sec. A, 35 (1915): 181.

PROBLEMS

7-1.1 Prove the following identities:

a. $\sum_{n=-\infty}^{\infty} \delta(\omega + n\omega_s) = \frac{1}{\omega_s} \sum_{n=-\infty}^{\infty} e^{-jnT_s\omega} \qquad \omega_s = \frac{2\pi}{T_s}$

b. $F_s(\omega) = \frac{1}{T_s} \sum_{n=-\infty}^{\infty} F(\omega + n\omega_s)$ (Equation 7.5)

7-1.2 Sketch the Fourier transform of the sampled function $f_s(t) = e^{-|t|} \operatorname{comb}_{T_s}(t)$ for $T_s = 0.1$ and $T_s = 10$. What do you conclude from these results?

7-1.3 Sketch the Fourier spectra of $f_s(t)$ for the systems shown in Figure P7-1.3 (p. 390). Assume that ω_s is sufficiently large that no overlap of the repeated spectra occurs.

7-1.4 Carry out the reverse of Example 7.3, for the signal $f(t) = \cos \omega_0 t$; that is, hold the sampling frequency constant but change the frequency ω_0. Show that aliasing will occur at $T_s > T_0/2$.

7-2.1 Verify (7.9a) using (7.12).

7-2.2 Expand the band-limited $F(\omega)$, with $|\omega| \leq \omega_N$, in a Fourier series, and prove (7.13).

7-2.3 Verify (7.16) and (7.17).

7-2.4 Show the following relation, if $F(\omega) = 0$ for $|\omega| > \omega_N$,

$$f^2(t) = \sum_{n=-\infty}^{\infty} T_s f^2(nT_s) \frac{\sin\left[\omega_s \frac{t - nT_s}{2}\right]}{\pi(t - nT_s)}$$

7-2.5 Find a sampling time for the signal $f(t) = 2(\sin 2t/t)$ such that 90% of the energy is contained within the bandwidth $-\omega_N \leq \omega \leq \omega_N$.

7-2.6 Sketch two superposed sinusoidal signals of the same amplitude but with frequencies in the ratio of 10:1, approximately. Consider the points of intersection of the two waves, one per cycle of the higher frequency wave. Ascertain sampling times such that only the lower frequency wave can be extracted from the sampled points.

7-2.7 Verify (7.18) and (7.20).

7-2.8 Let a sampled signal be represented by $f_s(t) = f(t)s(t)$, where

$$s(t) = \sum_{n=-\infty}^{\infty} p_{\tau/2}(t - nT_s)$$

Use the Fourier representation of $s(t)$ to find the Fourier spectrum of $f_s(t)$. Sketch the Fourier spectrum of $f_s(t)$ if $f(t)$ is a band-limited function.

7-2.9 Determine the truncated spectrum $\hat{F}(\omega)$ from the spectrum $F(\omega)$ of the time function $f(t) = 2/(t^2 + 1)$ with the energy error between the two functions $f(t)$ and $\hat{f}(t) = \mathscr{F}^{-1}\{\hat{F}(\omega)\}$ less than 0.05.

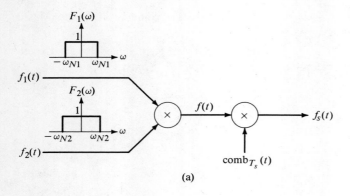

(a)

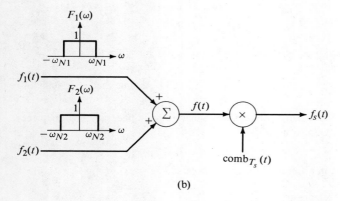

(b)

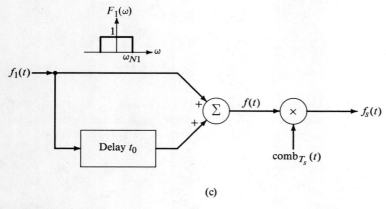

(c)

Figure P7-1.3

7-2.10 A signal has the spectrum shown in Figure P7-2.10. Find the maximum sampling time T_s, the filter type, and its bandwidth so that the signal is completely recovered from its sampled version $f_s(t)$.

7-2.11 The Fourier spectrum of $f(t)$ is shown in Figure P7-2.11. Determine the maximum sampling time T_s, the filter type, and its bandwidth so that the signal is completely recovered from its sampled version $f_s(t)$.

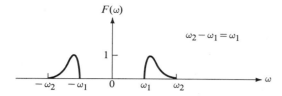

Figure P7-2.10

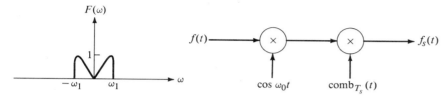

Figure P7-2.11

7-3.1 Graphically show the steps in multiplexing two sinusoidal signals that have frequencies in the ratio of 2:1. Show also their output forms obtained from sample-and-hold circuits.

7-3.2 Four input signals are sampled and time-multiplexed using the pulse amplitude modulation technique. Three of the input signals are band-limited with peak frequencies to 4.5 kHz; the fourth has a 9 kHz bandwidth. What is the minimum sampling rate if all signals are sampled equally?

7-4.1 Apply the delta modulation technique to the two signals shown in Figure P7-4.1 and state your observations.

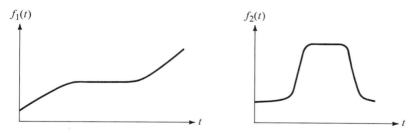

Figure P7-4.1

CHAPTER 8

DISCRETE SIGNALS, DIFFERENCE EQUATIONS, AND SYSTEM DESCRIPTION

8-1 INTRODUCTION

The general features of discrete time signals have received considerable attention in Chapter 7. A discrete time system is one that transforms an input sequence of numbers, $\{x(k)\}$ into an output sequence of numbers $\{y(k)\}$ according to some recursion formula that represents the solution to a **difference equation** describing the system. For a physical system, the difference equation will express the features of what is often termed a digital filter or a signal processor.

We can represent the discrete system in a form similar to the representation of the continuous system discussed in Chapter 2. Some simple discrete systems are shown in Figure 8.1. The system in Figure 8.1b, a divider, has an output that is independent of the previous inputs; therefore it has no memory. Accordingly, it is called a **memoryless** system. The discrete system shown in Figure 8.1c does have **memory**. The system in Figure 8.1d is a differentiator since $y(k) = x(k) - x(k-1)$. The unit delay element illustrated in these figures represents a device with output delayed by one time unit relative to the input. Thus if the input signal to the unit delay is $x(k)$, the output is $x(k-1)$.

A discrete system is **causal** if its output at any time depends on the present and past inputs. For example, the system defined by $y(k) = y(k-1) + x(k)$ is causal, but a system defined by $y(k) = x(k-1) + y(k+1)$ is not causal.

Linear discrete systems are characterized by the superposition property. For example, if the input $x_1(k)$ to a discrete system produces an output $y_1(k)$ and the input $x_2(k)$ produces an output $y_2(k)$, then for a linear system the input $x_1(k) + x_2(k)$ produces the output $y_1(k) + y_2(k)$. A system is called **homogeneous** if the input $ax_1(k) + bx_2(k)$ produces the output $ay_1(k) + by_2(k)$.

Stability is another important property of systems. Stable discrete systems are characterized by decaying outputs as time progresses. If their output increases without bounds in time, the systems are unstable. For example, the system with output $y(k) = a^k$, $k \geq 0$, is unstable if $a > 1$ and is stable if $a < 1$.

Another important property of discrete systems is the time-invariance property. A system is **time invariant** when the inputs $x(k)$ and $x(k - k_0)$ produce outputs $y(k)$ and $y(k - k_0)$, respectively. The system $y(k) = 2kx(k)$ is not time invariant because the output to $x(k - k_0)$ is $y(k) = 2kx(k - k_0)$. This is quite different from introducing a time shift of $k - k_0$ into the system which results in an output of $y(k - k_0) = 2(k - k_0)x(k - k_0)$.

DISCRETE SIGNALS, DIFFERENCE EQUATIONS, AND SYSTEM DESCRIPTION

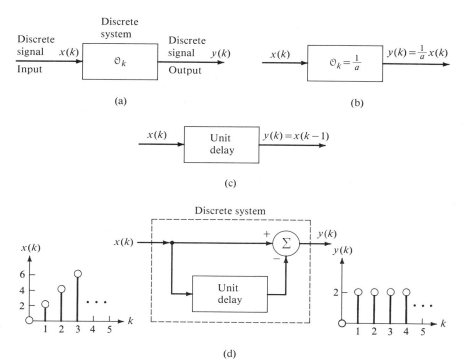

Figure 8.1 Simple discrete systems.

The description of discrete time systems and the techniques for writing and solving difference equations describing them will receive considerable attention in this chapter. Consideration will be given to methods that are designated as classical methods. More important to us is the use of the Z-transform in the solution of difference equations, which will receive detailed attention in Chapter 9.

■ ■ ■

8-2 LINEAR DIFFERENCE EQUATIONS

Suppose that the discrete time system or signal processor that transforms an input sequence of numbers $\{x(k)\}$ into an output sequence of numbers $\{y(k)\}$ is expressed by the difference equation

$$y(k) = x(k) + 3x(k-1) + x(k-2) \tag{8.1}$$

This expression specifies that the kth member of the output sequence is formed by adding the following terms: the present (or kth) input $x(k)$, three times the value of the previous or once delayed input $x(k-1)$, and the second delayed value of the input. In particular, if the input sequence to this system beginning with $k = 0$ is the set of numbers $\{1, 2, 0, 3, 1, 5, 0, \ldots\}$, the corresponding output will be the sequence of numbers $\{1, 5, 7, 5, 10, 11, 16, \ldots\}$.

A discrete time process familiar to most people is that associated with a savings account in a bank. Consider a savings account that pays interest at the rate of $r\%$ per year compounded n times per year ($n = 4$ would correspond to quarterly compounding). The interest is computed at the rate of $r/n\%$ for each compounding period. We will assume that deposits during any period earn no interest until the next compounding period. We denote the following:

$y(k)$ = total bank account balance at the end of the kth compounding period

$x(k)$ = total of deposits during the kth compounding period.

Clearly, at the conclusion of any compounding period, the total bank account balance is equal to the sum of the following: the bank account balance at the beginning of the compounding period, the interest accrued on this balance, and the deposits made during the period. This result can be expressed mathematically by the expression

$$y(k) = y(k-1) + \frac{r}{n} y(k-1) + x(k) = \left(1 + \frac{r}{n}\right) y(k-1) + x(k) \qquad (8.2)$$

This is a linear first-order difference equation since k and $k-1$ in the unknown (output) variable differ by one unit. The solution to such an equation can be accomplished numerically by hand or by machine, it can be accomplished analytically, and it can be accomplished using Z-transform methods, as we will see in the next chapter.

To solve this equation by hand or by machine requires that the initial value $y(0)$ and the sequence of inputs $x(k)$, $k = 1, 2, 3, \ldots$, be recorded, say in the storage registers in a computer. The quantity $(1 + r/n)$ must be evaluated, and then the multiplication with $y(k-1)$ must be carried out. This result is stored in the accumulator; it is combined with the appropriate $x(k)$ and the updated $y(k)$ is deduced. During the next period, the present $y(k)$ will become the $y(k-1)$ and the process can be repeated. Thus for a known interest rate r, an initial deposit $y(0)$, and a specified sequence of deposits, the difference equation is readily solved to establish all successive values.

Note that a notational distinction is made between systems that generate difference equations of the form given in (8.1) in which $y(k)$ depends only on $x(k)$ and past values of the input and systems that generate difference equations such as those in (8.2), which specify the value of $y(k)$ in terms of past values of $y(k)$, present values of $x(k)$, and perhaps past values of $x(k)$. The first type is called **nonrecursive, transversal,** or **finite duration impulse response** (FIR). The second type is called a **recursive** or **infinite impulse response** (IIR) system. Specifically, an IIR system is described mathematically by an Nth-order difference equation of the form

$$\begin{aligned}y(k) = a_1 y(k-1) + \cdots + a_N y(k-N) + b_0 x(k) + b_1 x(k-1) \\ + \cdots + b_M x(k-M)\end{aligned} \qquad (8.3)$$

which can be written in compact form as

DISCRETE SIGNALS, DIFFERENCE EQUATIONS, AND SYSTEM DESCRIPTION   395

$$y(k) = \sum_{n=1}^{N} a_n y(k-n) + \sum_{n=0}^{M} b_n x(k-n) \tag{8.4}$$

If all values $a_n = 0$, the system description is given by

$$y(k) = \sum_{n=0}^{M} b_n x(k-n) \tag{8.5}$$

This is the general description of an FIR system. These general forms will be discussed later in this chapter and also in Chapter 11 as forms in digital filter design.

■ ■ ■

8-3 FIRST-ORDER LINEAR DIFFERENCE EQUATIONS WITH CONSTANT COEFFICIENTS

We proceed by considering a general first-order linear discrete time system of the IIR type characterized by the difference equation

$$y(k) - \alpha_1 y(k-1) = \beta_0 x(k) + \beta_1 x(k-1) \tag{8.6}$$

where α_1, β_0, and β_1 are constants whose values are specified by the system under survey. This system is shown graphically in Figure 8.2. We will assume that the input signal is applied at time $k = 0$, which requires that $x(k) = 0$ for k negative. Specifically, suppose that the input is the sequence $\{x(0), x(1), x(2), \ldots\}$. We proceed to build up the solution directly from (8.6). We write the first equation of the set

$$y(0) = \beta_0 x(0) + \beta_1 x(-1) + \alpha_1 y(-1)$$

Figure 8.2
Graphic representation of a first-order IIR system.

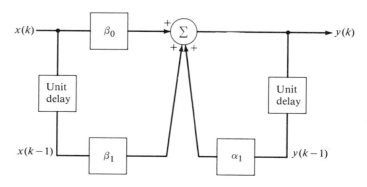

Figure 8.3
Inductor excited by a discrete current source.

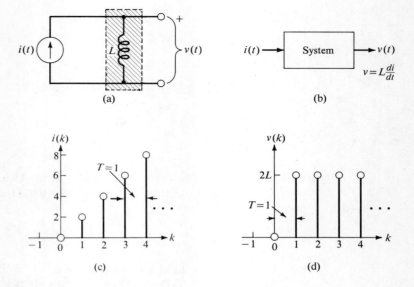

But since $x(-1) = 0$, this equation becomes

$$y(0) = \beta_0 x(0) + \beta_1 y(-1)$$

This requires knowing the value of the output signal $y(-1)$ just prior to the application of the input signal in order to calculate $y(0)$. The next element in the output sequence $y(1)$ becomes, by writing $k = 1$ in (8.6),

$$y(1) = \beta_0 x(1) + \beta_1 x(0) + \alpha_1 y(0)$$

where $y(0)$ is the value determined one iteration previously. This procedure is continued to deduce all successive values of $y(k)$.

The value $y(-1)$ is the system's **initial condition** and specifies the state of the system just prior to the application of the input signal. Clearly, input signals can be first applied at times other than $k = 0$.

EXAMPLE 8.1
Determine the voltage across the inductor shown in Figure 8.3 if a current source $i(t) = 2t$ ($t \geq 0$), which is impulse-sampled at intervals $t = T$, is applied.

Solution: Replace the differential relationship given in Figure 8.3 by an equivalent difference relationship. We accomplish this transformation by replacing the derivative by the approximate form

$$\boxed{\frac{di(t)}{dt} \doteq \frac{i(kT) - i[(k-1)T]}{T}} \quad \text{(a)} \tag{8.7}$$

Figure 8.4
Block diagram of the system process described by (8.7c).

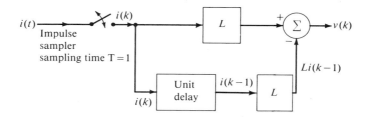

or

$$\frac{di}{dt} \doteq i(k) - i(k-1) \quad \text{for } T = 1 \quad \textbf{(b)}$$

Multiply by L, and knowing that our equation is an approximation, we obtain

$$v(k) = L[i(k) - i(k-1)] \quad \textbf{(c)}$$

Choose an input function and initial condition of the form

$$i(k) = 2k \quad k \geq 0 \quad v(-1) = 0 \quad \textbf{(d)}$$

We now build up the solution as follows:

$$v(0) = L[i(0) - i(-1)] = L(0 - 0) = 0$$
$$v(1) = L[i(1) - i(0)] = L(2 - 0) = 2L$$
$$v(2) = L[i(2) - i(1)] = L(4 - 2) = 2L$$
$$v(3) = L[i(3) - i(2)] = L(6 - 4) = 2L$$
$$\vdots$$

This result is shown in Figure 8.3d.

The numerical process described by (8.7c) is shown in Figure 8.4. Observe that the instantaneous or impulse sampling is shown as a switch that is closed momentarily at the sampling time intervals. ∎

EXAMPLE 8.2

Determine the voltage across an initially uncharged capacitor when a sampled current source $i(t) = 2u(t)$, $t \geq 0$, is applied as shown in Figure 8.5.

Solution: We must replace the integral relationship between the voltage and the current for the simple capacitor by an equivalent discrete form. This is done by considering a time interval T, and then writing the integral relationship

$$v(t) = \frac{1}{C} \int_0^t i(t') \, dt'$$

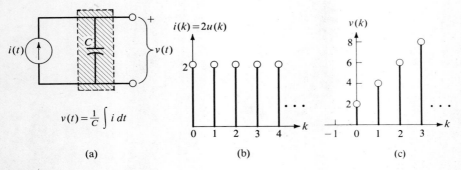

Figure 8.5
A capacitor with pulse excitation.

by the expression

$$v(kT) = \frac{1}{C} \int_0^{kT} i(t')\,dt' \qquad k = 0, 1, 2, \ldots \tag{8.8}$$

Write this in the form

$$v(kT) = \frac{1}{C} \int_0^{kT-T} i(t')\,dt' + \frac{1}{C} \int_{kT-T}^{kT} i(t')\,dt'$$

which, by (8.8), can be written

$$v(kT) = v(kT - T) + \frac{1}{C} \int_{kT-T}^{kT} i(t')\,dt' \tag{8.9}$$

But the integral represents the area under the curve of $i(t)$ in the interval $kT - T \le t \le kT$, and this is approximately equal to $Ti(kT)$. Equation (8.9) thus becomes

$$\boxed{v(kT) = v(kT - T) + \frac{1}{C} Ti(kT) \qquad k = 0, 1, 2, \ldots} \tag{8.10}$$

Note that this relationship would have resulted directly if we had written $i = C\,dv/dt$ and then used the approximation for the derivative in (8.7).

If we choose $C = T = 1$, then the calculation proceeds as follows:

$$v(0) = v(-1) + i(0) = 0 + 2 = 2$$
$$v(1) = v(0) + i(1) = 2 + 2 = 4$$
$$v(2) = v(1) + i(2) = 4 + 2 = 6$$
$$\vdots$$

This result is plotted in Figure 8.5c.

Figure 8.6
Numerical integration algorithm for (8.10).

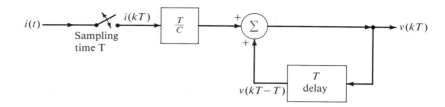

Figure 8.7
A series RL circuit.

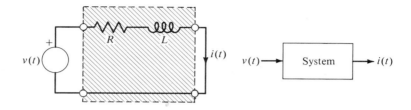

The numerical integration algorithm specified by (8.10) is shown in the block diagram in Figure 8.6. ∎

For a more detailed study, consider the RL circuit shown in Figure 8.7. The circuit equation is written by a simple application of the Kirchhoff voltage law, from which

$$\frac{di(t)}{dt} + \frac{R}{L} i(t) = \frac{1}{L} v(t) \tag{8.11}$$

The difference equation approximation of this equation is

$$\frac{i(kT) - i(kT - T)}{T} = \frac{1}{L} v(kT) - \frac{R}{L} i(kT) \tag{8.12}$$

Rearrange this expression to the form

$$i(kT) = \frac{\frac{T}{L}}{1 + \frac{R}{L} T} v(kT) + \frac{1}{1 + \frac{R}{L} T} i(kT - T)$$

Set $T = 1$ for simplicity, and this equation becomes

$$i(k) = \frac{\frac{1}{L}}{1 + \frac{R}{L}} v(k) + \frac{1}{1 + \frac{R}{L}} i(k - 1) \tag{8.13}$$

This equation can be written in the form of (8.6)

$$y(k) = \beta_0 x(k) + \alpha_1 y(k - 1) \tag{8.14}$$

where the constants are

$$\beta_0 = \frac{\dfrac{1}{L}}{1 + \dfrac{R}{L}} \qquad \alpha_1 = \frac{1}{1 + \dfrac{R}{L}}$$

We wish to study the response of this first-order system to a unit step sequence

$$v(k) = u(k) = \begin{cases} 1 & k = 0, 1, 2, \ldots \\ 0 & k \text{ negative} \end{cases}$$

Further, the system is assumed to be initially relaxed so that $y(-1) = 0$. Now we proceed systematically as follows:

$k = 0 \qquad y(0) = \beta_0 x(0) + \alpha_1 y(-1) = \beta_0 + 0 = \beta_0$

$k = 1 \qquad y(1) = \beta_0 x(1) + \alpha_1 y(0) = \beta_0 \cdot 1 + \alpha_1 \beta_0 = (1 + \alpha_1)\beta_0$

$k = 2 \qquad y(2) = \beta_0 x(2) + \alpha_1 y(1)$
$\qquad\qquad = \beta_0 \cdot 1 + \alpha_1(1 + \alpha_1)\beta_0 = (1 + \alpha_1 + \alpha_1^2)\beta_0$

$k = 3 \qquad y(3) = \beta_0 x(3) + \alpha_1 y(2)$
$\qquad\qquad = \beta_0 \cdot 1 + \alpha_1(1 + \alpha_1 + \alpha_1^2)\beta_0 = (1 + \alpha_1 + \alpha_1^2 + \alpha_1^3)\beta_0$

$\vdots \qquad\qquad \vdots$

By induction, we write

$$y(k) = (1 + \alpha_1 + \alpha_1^2 + \cdots + \alpha_1^k)\beta_0 \qquad k = 0, 1, 2, \ldots$$

But the finite series can be written as

$$1 + \alpha_1 + \alpha_1^2 + \cdots + \alpha_1^k = \frac{1 - \alpha_1^{k+1}}{1 - \alpha_1} \qquad \text{for } \alpha_1 \neq 1$$

so that finally

$$y(k) = \frac{1 - \alpha_1^{k+1}}{1 - \alpha_1} \beta_0 \tag{8.15}$$

For values of $\alpha_1 > 1$, the factor $(1 - \alpha_1^{k+1})$ becomes arbitrarily large as k increases; this indicates a condition of **instability**. For $|\alpha_1| < 1$, $(1 - \alpha_1^{k+1})$ approaches 1 as k increases, and the unit step response approaches the value

$$y(k) = \frac{\beta_0}{1 - \alpha_1} \qquad k \text{ large} \tag{8.16}$$

Note that for our particular circuit specified by (8.14), we obtain

$$y(k) \triangleq i(k) = \frac{1}{R} \qquad k \text{ large}$$

which is the steady-state current for this physical example.

EXAMPLE 8.3

Deduce the solution to the difference equation

$$y(k) + 2y(k-1) = 3.5u(k)$$

with $y(-1) = 0$.

Solution: This equation is precisely of the form of (8.14) with

$$\beta_0 = 3.5 \quad x(k) \triangleq u(k) = 1 \quad \alpha_1 = -2$$

The solution is given by (8.15), and for the parameters used here

$$y(k) = 3.5 \frac{[1-(-2)^{k+1}]}{1-(-2)} = \frac{3.5}{3}[1-(-2)^{k+1}]$$

This is written

$$y(k) = 1.167[1-(-2)^k(-2)^1]$$
$$= 1.167 + 2.33(-2)^k \quad \text{for } k = 0, 1, 2, \ldots$$

Observe that this system is unstable. ∎

We now investigate the solution to the more general first-order difference equations with constant coefficients when the input functions are of a form particularly appropriate to engineering applications. Such equations are of the form

$$y(k) = \alpha_1 y(k-1) + x(k) \quad \textbf{(a)} \tag{8.17}$$

with their homogeneous counterparts

$$y(k) = \alpha_1 y(k-1) \quad \textbf{(b)}$$

Any constant multiplier of the input is absorbed in $x(k)$. The special class of excitation functions $x(k)$ now under survey is defined by

$$\{x(k)\} = \psi_p^k \tag{8.18}$$

Important forms for ψ_p are 1, $\exp[\pm aT]$, and $\exp[\pm j\omega T]$. Note that excitations of these forms bear much the same place in the solution of difference equations as the solution $x(t) = \exp[s_p t]$ does in finding the solution of differential equations.

Specifically, assume that the input signal to (8.17a) is given by

$$x(k) = \begin{cases} X\psi_p^k & k \geq 0 \\ 0 & k < 0 \end{cases} \tag{8.19}$$

where X is a constant. Furthermore, assume that the output to such inputs will be of the form

$$y_p(k) = Y\psi_p^k \tag{8.20}$$

where Y is an unknown constant. To determine constant Y, substitute this trial solution into (8.17a). This yields

$$Y\psi_p^k - \alpha_1 Y\psi_p^{k-1} = X\psi_p^k$$

This expression must be true for any k; thus we consider that $k = 1$. Therefore

$$Y(\psi_p - \alpha_1) = X\psi_p$$

from which

$$Y = \frac{\psi_p}{\psi_p - \alpha_1} X \tag{8.21}$$

The particular solution is

$$y_p(k) = \frac{\psi_p}{\psi_p - \alpha_1} X\psi_p^k \tag{8.22}$$

To find the solution to the homogeneous equation (8.17b), we assume a solution of the form

$$y_h(k) = A\lambda^k \tag{8.23}$$

where A is another unknown constant. Note that this function λ is not related to ψ_p^k. Substitute this trial solution in (8.17b), which yields

$$A\lambda^k - \alpha_1 A\lambda^{k-1} = 0$$

from which it follows that (set $k = 0$)

$$\lambda = \alpha_1 \tag{8.24}$$

and the solution is

$$y_h(k) = A\alpha_1^k \tag{8.25}$$

The complete solution of (8.17a) is

$$y(k) = y_p(k) + y_h(k) = \frac{\psi_p}{\psi_p - \alpha_1} X\psi_p^k + A\alpha_1^k \qquad k \geq 0 \tag{8.26}$$

To evaluate the constant A, use is made of the difference equation beginning with $k = 0$, the initial time, so that

$$y(0) = x(0) + \alpha_1 y(-1) = X\psi_p^0 + \alpha_1 y(-1) = X + \alpha_1 y(-1)$$

If the system is at rest prior to the application of the excitation, then

$$y(-1) = 0 \quad \text{whence} \quad y(0) = X$$

This result is combined with (8.26) for $k = 0$, from which we have

$$X = \frac{\psi_p}{\psi_p - \alpha_1} X + A$$

so that

… DISCRETE SIGNALS, DIFFERENCE EQUATIONS, AND SYSTEM DESCRIPTION

$$A = -\frac{\alpha_1}{\psi_p - \alpha_1} X \tag{8.27}$$

The complete solution is thus

$$y(k) = \frac{\psi_p}{\psi_p - \alpha_1} X\psi_p^k - \frac{\alpha_1}{\psi_p - \alpha_1} X\alpha_1^k \qquad k \geq 0$$

or

$$y(k) = \frac{X}{\psi_p - \alpha_1} [\psi_p^{k+1} - \alpha_1^{k+1}] \qquad k \geq 0 \quad y(-1) = 0 \tag{8.28}$$

For the special case when $\psi_p = 1$ and $X = 1$, the excitation is the unit step function $u(k)$, and the solution is given by

$$y(k) = \frac{1}{1 - \alpha_1}[1 - \alpha_1^{k+1}] = \frac{1 - \alpha_1^{k+1}}{1 - \alpha_1} \qquad k \geq 0 \tag{8.29}$$

This is the same result as deduced in (8.15) for the same problem ($\beta_0 = 1$).

Observe from the general solution (8.28) (for both ψ_p and α_1 real) that as k assumes large values, the particular term ψ_p^{k+1} will be the dominant one if $|\psi_p| > |\alpha_1|$. Moreover, for the solution to remain bounded, it is necessary that $|\psi_p| < 1$ and $|\alpha_1| < 1$.

If the system has the initial condition $y(-1) = C$, then

$$X = y(0) - \alpha_1 y(-1) = y(0) - \alpha_1 C$$

This result is combined with (8.26) for $k = 0$, from which we have

$$A = \alpha_1 C - \frac{\alpha_1}{\psi_p - \alpha_1} X \tag{8.30}$$

The complete solution is

$$y(k) = \frac{X}{\psi_p - \alpha_1}[\psi_p^{k+1} - \alpha_1^{k+1}] + \alpha_1^{k+1} C$$

$$k \geq 0 \quad y(-1) = C \tag{8.31}$$

EXAMPLE 8.4

Use Laplace transform techniques to find the output of the system shown in Figure 8.8a. Now replace the system by an equivalent discrete system and find the output of this discrete system. Compare the results. Assume zero initial conditions.

Figure 8.8
Illustrating Example 8.4.

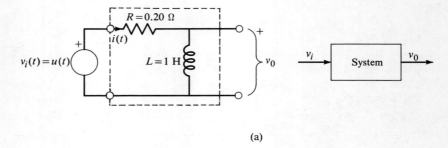

(a)

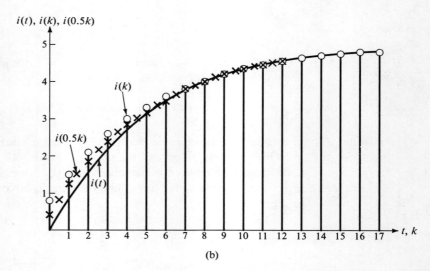

(b)

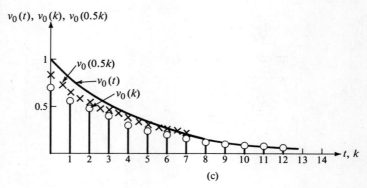

(c)

Solution: The controlling differential equation for the system of Figure 8.8a is

$$\frac{di(t)}{dt} + 0.2i(t) = u(t) \qquad (8.32)$$

The Laplace transform of this equation leads to the expression for $I(s)$

$$I(s) = \frac{1}{s(s+0.2)} = 5\left[\frac{1}{s} - \frac{1}{s+0.2}\right]$$

The inverse Laplace transform of this expression yields the relationship

$$i(t) = 5(1 - e^{-0.2t}) \tag{8.33}$$

The output voltage is given by

$$v_o(t) = L\frac{di(t)}{dt} = e^{-0.2t} \tag{8.34}$$

Now write (8.32) in its approximate discrete form

$$\frac{i(kT) - i(kT-T)}{T} + 0.2i(kT) = u(kT)$$

or

$$(1 + 0.2T)i(kT) - i(kT-T) = Tu(kT) \tag{8.35}$$

with the initial condition $y(-T) = 0$. Select $T = 1$ and this equation becomes

$$i(k) = \frac{1}{1.2}i(k-1) + \frac{1}{1.2}u(k) \tag{8.36}$$

Proceed to build up the solution to this difference equation, as shown:

$$i(0) = \frac{1}{1.2}0 + \frac{1}{1.2}1 = \frac{1}{1.2} = 0.833$$

$$i(1) = \frac{1}{1.2}\frac{1}{1.2} + \frac{1}{1.2}1 = \frac{2.2}{(1.2)^2} = 1.528$$

$$i(2) = \frac{1}{1.2}\frac{2.2}{(1.2)^2} + \frac{1}{1.2}1 = 2.11$$

$$i(3) = \frac{1}{1.2}\frac{3.64}{(1.2)^3} + \frac{1}{1.2}1 = 2.59$$

$$i(4) = \frac{1}{1.2}\frac{5.368}{(1.2)^4} + \frac{1}{1.2}1 = 2.99$$

$$\vdots$$

$$\tag{8.37}$$

The data given in (8.37) and the results specified by (8.33) are plotted in Figure 8.8b. The output voltage for $T = 1$ is given by

$$L\frac{di(t)}{dt} \doteq i(k+1) - i(k) \tag{8.38}$$

This equation together with successive values of $i(k)$ contained in (8.37) yields the curve shown in Figure 8.8c. For comparison we have also plotted the variation specified by (8.34).

If we choose $T = 0.5$ s, (8.35) becomes

$$i(0.5k) = \frac{i[(k-1)0.5]}{1.1} + \frac{0.5u(0.5k)}{1.1}$$

The solution to this equation is shown in Figure 8.8b by crosses. Similarly, for the output voltage, the difference equation is

$$v_0(k) = \frac{i[0.5(k+1)] - i(0.5k)}{0.5}$$

A plot of this equation is shown by crosses in Figure 8.8c.

Observe the large improvement in the results of the difference equation approximation by decreasing the sampling time T by only one-half. ∎

■ ■ ■

8-4 DELAY OPERATIONS AND SIGNALS

We have already introduced the idea of unit delay. Clearly, the concept can be extended to unit delay operations in cascade and, as shown in Figure 8.9, this becomes an essential part of the graphical display of difference equations. Observe that delayers in cascade are additive, a property quite different from scalars or system elements in cascade, which are multiplicative. This suggests that delayers should be written in an exponential representation. It is customary to write z^{-1} to denote a delay of one unit, and z^{-T} to denote a delay of T units. Thus corresponding to Figure 8.9, we can show the delay process in either block diagram or flow graph representation as shown in Figure 8.10.

Figure 8.9
Unit delayers in cascade.

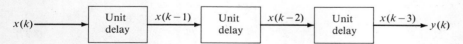

(a)

(b)

Figure 8.10
Representation of delayers. (a) Block diagram representation. (b) Flow graph representation.

DISCRETE SIGNALS, DIFFERENCE EQUATIONS, AND SYSTEM DESCRIPTION

EXAMPLE 8.5
A signal generator produces a signal with the properties

$$x(k) = \begin{cases} 0 & k < 0 \\ k & 0 \le k \le 2 \\ 0 & k > 2 \end{cases} \qquad (8.39)$$

With this signal source as a basic unit, draw block diagrams and flow graphs of systems that will produce the signals shown in Figures 8.11a–c.

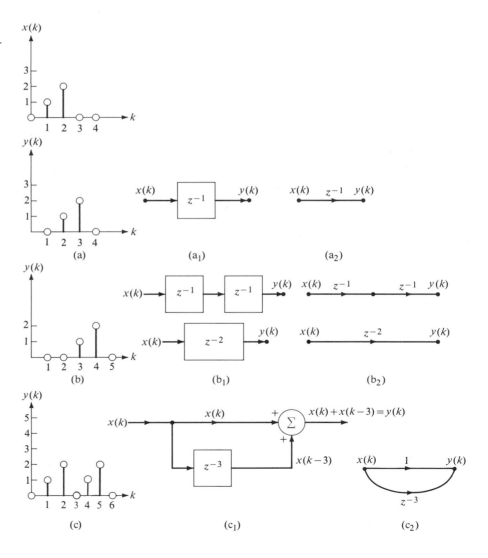

Figure 8.11 Diagrammatic representation of delayed signals.

Solution: The first signal is the original signal delayed by one unit. The block diagram and flow graph representations are shown in Figures 8.11a_1 and a_2. The second signal is the original signal delayed by two time units, thus requiring two delayers, as shown in Figures 8.11b_1 and b_2. The third signal is the sum of the original signal plus a replica of itself that has been delayed by three time units. Figures 8.11c_1 and c_2 show the graphical representation of this. ∎

EXAMPLE 8.6

A signal generator produces an impulse signal specified by

$$\delta(k) = \begin{cases} 0 & k < 0 \\ 1 & k = 0 \\ 0 & k > 0 \end{cases} \tag{8.40}$$

Using this signal source, draw block diagrams and flow graphs of systems that will produce the outputs shown in Figures 8.12a and b.

Solution: Notice that the signal of Figure 8.12a is the sum of an unshifted signal with an amplitude of one, another signal that has been shifted by one unit and has an amplitude of two, and a third signal that has been shifted by two time units and has an amplitude of three. The graphical representation is shown in Figures 8.12a_1 and a_2.

The second signal shown is the unit step function $u(k)$, a repetition of the input pulse with constant amplitude and successive time shifts of one time unit. Since the signal components have the same amplitudes, the resulting signal is given by the equation

$$y(k) = \delta(k) + \delta(k)z^{-1} + \delta(k)z^{-2} + \cdots$$

$$= \delta(k)(1 + z^{-1} + z^{-2} + \cdots) = \delta(k)\frac{1}{1-z^{-1}} \tag{8.41}$$

The two equivalent graphical representations are shown in Figures 8.12b_1 and b_2. Observe that the result is that of a simple feedback system.

It follows from the block diagram that

$$\delta(k) + y(k)z^{-1} = y(k) \tag{8.42}$$

or

$$y(k) = \delta(k)\frac{1}{1-z^{-1}} \tag{8.43}$$

Refer to (8.42), which we write

$$y(k) = \delta(k) + y(k-1)$$

This difference equation can be used to generate the signal of Figure 8.12b. From

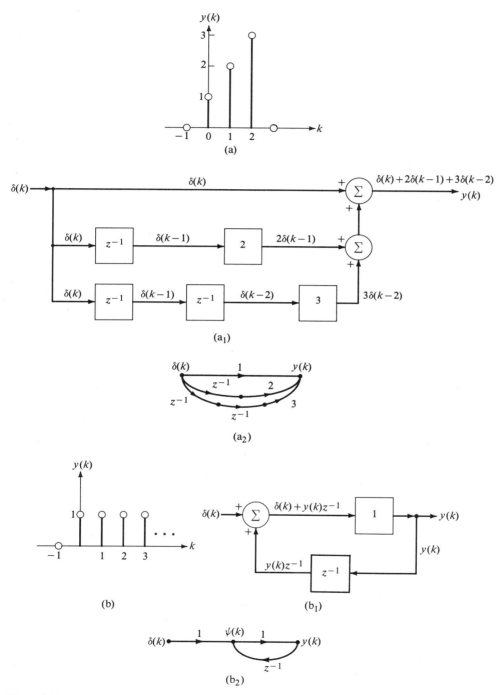

Figure 8.12
Diagrammatic representation of delayed signals.

the fact that $\delta(k) = 0$ for all $k < 0$, we see that

$\delta(0) = y(0) = 1$

$\delta(1) = 0 \qquad y(1) = y(0) = 1$

$\vdots$

which creates the signal shown in Figure 8.12b. ∎

A question of some interest relates to the stability of the system described by (8.43). In particular, let us examine a system described by

$$y(k) = x(k)\frac{1}{1 - az^{-1}} = x(k) + x(k)az^{-1} + x(k)a^2z^{-2} + \cdots \quad \text{(a)} \qquad (8.44)$$

or

$$y(k) = x(k) + ax(k-1) + a^2x(k-2) + \cdots \quad \text{(b)}$$

This system, for $a > 1$, is unstable since $y(k) \to \infty$ as $k \to \infty$; the system is stable if $|a| < 1$, since $y(k) \to 0$ as $k \to \infty$. These results are shown graphically in Figure 8.13, in one case with $a = 2$ and in the other with $a = 0.5$. The system description is shown in block diagram and signal flow graph form in Figures 8.13a$_1$ and a$_2$, and the system output in Figure 8.13b. The corresponding output for $a = 0.5$ is shown in Figure 8.13c, which indicates a stable output.

∎ ∎ ∎

8-5 HIGHER-ORDER LINEAR DIFFERENCE EQUATIONS WITH CONSTANT COEFFICIENTS

The class of linear discrete systems now under survey is described by the difference equation

$$a_n y(k + n) + a_{n-1} y(k + n - 1) + \cdots + a_1 y(k + 1) + a_0 y(k) = r(k) \quad (8.45)$$

where n is a positive integer, the coefficients a_i are constants, and $r(k)$ is a function of k defined for $k = 0, 1, 2, \ldots$. Such an equation can arise when a differential equation is transformed into an equivalent difference equation using the approximate conversions:

$$\frac{dy}{dt} \doteq \frac{y[(k+1)T] - y(kT)}{T} \qquad \text{(a)}$$

$$\frac{d^2y}{dt^2} = \frac{d}{dt}\left(\frac{dy}{dt}\right)$$

$$\doteq \frac{1}{T}\left[\frac{y[(k+2)T] - y[(k+1)T]}{T} - \frac{y[(k+1)T] - y(kT)}{T}\right] \qquad (8.46)$$

$$= \frac{y[(k+2)T] - 2y[(k+1)T] + y(kT)}{T^2} \qquad \text{(b)}$$

Figure 8.13
Description of a stable and an unstable system.

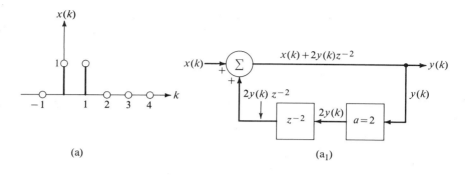

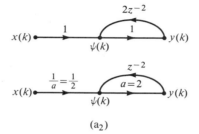

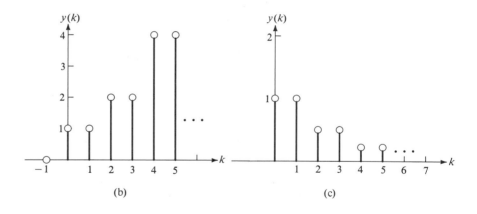

Similarly, for the third-order derivative we will find that

$$\frac{d^3 y}{dt^3} \doteq \frac{y[(k+3)T] - 3y[(k+2)T] + 3y[(k+1)T] - y(kT)}{T^3} \quad \text{(c)}$$

and so on.

If we substitute $k-1, k-2, k-3, \ldots$ in (8.46), we find another equivalent system of identities. When the coefficients a_i are independent of k, the equation describes a **time-invariant** system; otherwise the system is **time varying**. In the following, we assume that the coefficients a_n and a_0 are nonzero for all $k = 0, 1, 2, \ldots$, in which case (8.45) is of **order** n. We will proceed to a detailed study of an equation of order $n = 2$. The particular difference equation now under survey has the form

$$a_2 y(k+2) + a_1 y(k+1) + a_0 y(k) = r(k) \tag{8.47}$$

a form for which $T = 1$. We write (8.47) in the equivalent form

$$\boxed{y(k+2) + \alpha_1 y(k+1) + \alpha_0 y(k) = x(k)} \quad \text{(a)} \tag{8.48}$$

where

$$\alpha_1 = \frac{a_1}{a_2} \qquad \alpha_0 = \frac{a_0}{a_2} \qquad x(k) = \frac{r(k)}{a_2} \quad \text{(b)}$$

where $a_2 \neq 0$. The homogeneous equivalent of this equation is

$$\boxed{y(k+2) + \alpha_1 y(k+1) + \alpha_0 y(k) = 0} \tag{8.49}$$

We assert that a complete and unique solution to (8.48a) can be found if the initial conditions are known and

$$y(0) = Y(0) \qquad y(1) = Y(1) \tag{8.50}$$

where $Y(0)$ and $Y(1)$ are constants. In this connection we state certain theorems without proof (see the Finizio and Ladas text for proofs).

Definition 8.1 If $\{a(k)\}$ and $\{b(k)\}$ denote two sequences, the determinant

$$C[a(k), b(k)] = \begin{vmatrix} a(k) & b(k) \\ a(k+1) & b(k+1) \end{vmatrix} \tag{8.51}$$

is known as their **Casoratian** or their **Wronskian**.

Theorem 8.1 Two solutions $y_1(k)$ and $y_2(k)$ of the linear homogeneous difference equation (8.49) are linearly independent if and only if their Casoratian

$$C[y_1(k), y_2(k)] = \begin{vmatrix} y_1(k) & y_2(k) \\ y_1(k+1) & y_2(k+1) \end{vmatrix} \tag{8.52}$$

is different from zero for all values of $k = 0, 1, 2, \ldots$.

Theorem 8.2 If $y_1(k)$ and $y_2(k)$ are two linearly independent solutions of the homogeneous equation (8.49) and if $y_p(k)$ is the particular solution to the nonhomogeneous equation (8.48a), then the general solution to (8.48a) is

DISCRETE SIGNALS, DIFFERENCE EQUATIONS, AND SYSTEM DESCRIPTION

$$y(k) = y_h(k) + y_p(k) = C_1 y_1(k) + C_2 y_2(k) + y_p(k)$$

where C_1 and C_2 are arbitrary constants determined from appropriate initial conditions.

We call attention to the fact that we also used this result in Section 8-3.

Theorem 8.2 specifies that the general solution to a nonhomogeneous difference equation consists of two parts: (a) the solution to the homogeneous equation and (b) the particular solution to the nonhomogeneous one. For the homogeneous solution of the second-order difference equation, we state the following theorem:

Theorem 8.3 The difference equation

$$a_2 y(k+2) + a_1 y(k+1) + a_0 y(k) = 0 \tag{8.53}$$

with constant and real coefficients, with $a_2, a_0 \neq 0$, and with λ_1 and λ_2 as the roots of its characteristic equation

$$a_2 \lambda^2 + a_1 \lambda + a_0 = 0 \tag{8.54}$$

has the possible solutions shown in Table 8.1.

Proof: Assume that $y(k) = c\lambda^k$ as in (8.23). This trial solution in (8.53) leads to

$$a_2 \lambda^{k+2} + a_1 \lambda^{k+1} + a_0 \lambda^k = 0$$

from which the result follows.

EXAMPLE 8.7

Find the discrete time solution for the distance $x(t)$ for the system shown in Figure 8.14a.

Solution: The equation governing the displacement of this system is given by

$$M \frac{d^2 x(t)}{dt^2} + D \frac{dx(t)}{dt} = 0 \tag{8.55}$$

This equation is converted into the approximate difference form using (8.46). The result is:

$$\frac{M}{T^2} x[(k+2)T] - \left(\frac{2M}{T^2} - \frac{D}{T}\right) x[(k+1)T] + \left(\frac{M}{T^2} - \frac{D}{T}\right) x(kT) = 0$$

A block diagram representation of this equation is shown in Figure 8.14b. For the particular choice of parameters $T = 1$, $M = 2$, and $D = 5$, this equation

Table 8.1 Solutions to Homogeneous Difference Equations.

Difference Equation $a_2 y(k+2) + a_1 y(k+1) + a_0 y(k) = 0$			Characteristic Equation $a_2 \lambda^2 + a_1 \lambda + a_0 = 0$
Characteristic roots	$\lambda_1 \neq \lambda_2$	Solutions	$y(k) = c_1 \lambda_1^k + c_2 \lambda_2^k$
	$\lambda_1 = \lambda_2 = \lambda$		$y(k) = c_1 \lambda^k + c_2 k \lambda^k$
	$\lambda_1 = a + jb$ $\lambda_2 = a - jb$		$y(k) = c_1 r^k \cos k\theta + c_2 r^k \sin k\theta$ $r = [a^2 + b^2]^{1/2} \quad \cos \theta = \dfrac{a}{r} \quad \sin \theta = \dfrac{b}{r}$ $-\pi < \theta \leq \pi$
$a_n y(k+n) + a_{n-1} y(k+n-1)$ $\cdots + a_0 y(k) = 0$			$a_n \lambda^n + a_{n-1} \lambda^{n-1} + \cdots + a_0 = 0$
Characteristic roots	$\lambda_1 \neq \lambda_2 \neq \cdots \neq \lambda_n$	Solutions	$y(k) = \sum_{i=1}^{n} c_i \lambda_i^k$
	$\lambda_j \equiv$ root of multiplicity m		$y(k) = \sum_{i=1}^{n-m} c_i \lambda_i^k + c_j \lambda_j^k + c_{j+1} k \lambda_j^k$ $+ \cdots + c_{j+m} k^{m-1} \lambda_j^k$
	pair of roots $a \pm jb$ of multiplicity m		$y(k) = \sum_{i=1}^{n-2m} c_i \lambda_i^k + d_1 r^k \cos k\theta + d_2 k r^k \cos k\theta$ $+ \cdots + d_m k^{m-1} r^k \cos k\theta + d_{m+1} r^k \sin k\theta$ $+ d_{m+2} k r^k \sin k\theta + \cdots + d_{2m} k^{m-1} r^k \sin k\theta$

becomes

$$x(k+2) + \frac{1}{2} x(k+1) - \frac{3}{2} x(k) = 0 \tag{8.56}$$

The characteristic equation is

$$\lambda^2 + \frac{1}{2} \lambda - \frac{3}{2} = 0 \tag{8.57}$$

The roots of this equation are $\lambda_1 = -1.5$, $\lambda_2 = +1$, and the two solutions are

$$x_1(k) = (-1.5)^k \quad \text{and} \quad x_2(k) = (1)^k$$

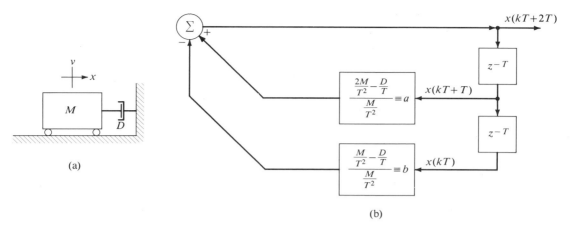

Figure 8.14
A mechanical unforced linear system and its discrete block diagram representation.

To show that these solutions are independent, we consider their Casoratian as given by (8.52); namely,

$$\begin{vmatrix} (-1.5)^k & (1)^k \\ (-1.5)^{k+1} & (1)^{k+1} \end{vmatrix} = (-1.5)^k - (-1.5)^{k+1} \neq 0$$

for any $k = 0, 1, 2, \ldots$. Hence by Theorem 8.3 the solution to the homogeneous equation is

$$x_h(k) = C_1(-1.5)^k + C_2(1)^k \tag{8.58}$$

∎

EXAMPLE 8.8

Find the ray trajectories of a system of thin lenses with identical focal lengths, as shown in Figure 8.15a.

Solution: First refer to the two lenses shown in Figure 8.15b. In the figure $r(k)$ denotes the height of the ray entering the thin lens L_k, and $\dot{r}(k)$ denotes the slope of this ray. Also the slope of the ray between the two lenses is

$$\dot{r}(k+1) = \frac{\dot{r}(k)f - r(k)}{f} = -\frac{1}{f}r(k) + \dot{r}(k) \tag{8.59}$$

where f is the focal length of the lens.

By examination of the similar triangles ABC and BDE, we find the height

$$r(k+1) = r(k) - \frac{d}{f}r(k) + \dot{r}(k)d \tag{8.60}$$

Figure 8.15
A system of thin lenses.

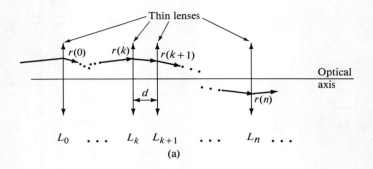

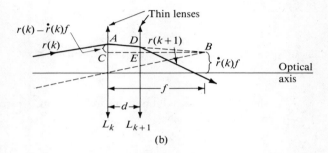

Increase the index in this equation by writing $k+1$ for k and combine the resulting equation with (8.59) to eliminate $\dot{r}(k)$. From these we obtain

$$r(k+2) - \left(2 - \frac{d}{f}\right) r(k+1) + r(k) = 0 \tag{8.61}$$

The auxiliary equation of this difference equation is given by

$$\lambda^2 - \left(2 - \frac{d}{f}\right)\lambda + 1 = 0 \tag{8.62}$$

with the roots

$$\lambda_1 = \frac{2 - \dfrac{d}{f} + \sqrt{\dfrac{d^2}{f^2} - 4\dfrac{d}{f}}}{2} \qquad \lambda_2 = \frac{2 - \dfrac{d}{f} - \sqrt{\dfrac{d^2}{f^2} - 4\dfrac{d}{f}}}{2} \tag{8.63}$$

Let us consider several important special cases.

Case 1. If $d = 4f$, then the roots become $\lambda_1 = \lambda_2 = -1$, and the general solution is

$$r(k) = (C_1 + C_2 k)(-1)^k \tag{8.64}$$

The value of $r(k)$ becomes unbounded with increasing k, thus denoting an unstable system. Physically this means that the rays will diverge from the guiding system.

DISCRETE SIGNALS, DIFFERENCE EQUATIONS, AND SYSTEM DESCRIPTION

Case 2. If $d > 4f$, then λ_1 and λ_2 are real and different. For the specific case $d = 8f$, the general solution is readily found to be

$$r(k) = C_1 \left(\frac{-6 + \sqrt{32}}{2}\right)^k + C_2 \left(\frac{-6 - \sqrt{32}}{2}\right)^k \tag{8.65}$$

Case 3. If $d < 4f$, λ_1 and λ_2 are complex conjugate roots $\lambda_1 = a + jb$, $\lambda_2 = a - jb$. The general solution will be of the form

$$\begin{aligned} r(k) &= Ce^{j\varphi}(a + jb)^k + Ce^{-j\varphi}(a - jb)^k \\ &= Ce^{j\varphi} R^k e^{jk\theta} + Ce^{-j\varphi} R^k e^{-jk\theta} \\ &= 2CR^k \cos(k\theta + \varphi) \end{aligned} \tag{8.66}$$

where C and φ are arbitrary constants, $R = \sqrt{a^2 + b^2}$, and $\theta = \tan^{-1}(b/a)$. This form of the solution shows that $r(k)$ is a real quantity. The same result can be found by direct reference to Table 8.1. ∎

Theorem 8.4 The particular solution to the nonhomogeneous difference equation

$$y(k + 2) + \alpha_1 y(k + 1) + \alpha_0 y(k) = x(k) \tag{8.67}$$

where α_1 and α_0 are constants, $x(k)$ is a given sequence, $\alpha_0 \neq 0$, and $y_1(k)$ and $y_2(k)$ are two linear independent solutions to the corresponding homogeneous equation, is

$$y_p(k) = \sum_{n=0}^{k-1} x(n) \frac{y_1(n+1)y_2(k) - y_2(n+1)y_1(k)}{y_1(n+1)y_2(n+2) - y_2(n+1)y_1(n+2)} \tag{8.68}$$

EXAMPLE 8.9

Consider the discrete time system shown in Figure 8.16. Find the general solution if the input sequence is $x(k) = 3^k$.

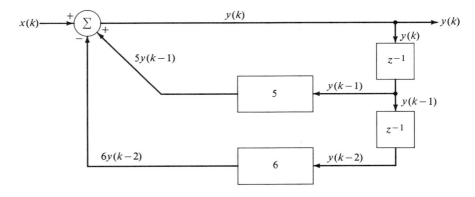

Figure 8.16 A discrete time system.

Solution: From Figure 8.16 we deduce the controlling difference equation

$$y(k) - 5y(k-1) + 6y(k-2) = 3^k \tag{8.69}$$

The characteristic equation obtained from the corresponding homogeneous equation is

$$\lambda^2 - 5\lambda + 6 = 0 \tag{8.70}$$

The roots are $\lambda_1 = 2$ and $\lambda_2 = 3$, thus yielding as solutions

$$y_1(k) = 2^k \quad \text{and} \quad y_2(k) = 3^k \tag{8.71}$$

We employ these solutions in (8.68) to determine the particular solution. However, (8.67) is shifted by 2 time units. Also, we shift (8.69) by 2 time units to write

$$y(k+2) - 5y(k+1) + 6y(k) = 3^{k+2} = 9 \times 3^k$$

Hence we can write

$$y_p(k) = 9 \sum_{n=0}^{k-1} 3^n \frac{2^{n+1} 3^k - 3^{n+1} 2^k}{2^{n+1} 3^{n+2} - 3^{n+1} 2^{n+2}} = 9 \sum_{n=0}^{k-1} 3^n \frac{2^{n+1} 3^k - 3^{n+1} 2^k}{2^n 3^n (2 \times 3^2 - 3 \times 2^2)}$$

$$= \frac{9}{6} \left[\sum_{n=0}^{k-1} 2 \times 3^k - \sum_{n=0}^{k-1} \frac{3^{n+1}}{2^n} 2^k \right] = 9k \times 3^{k-1} - 9 \times 2^{k-1} \sum_{n=0}^{k-1} \left(\frac{3}{2}\right)^n$$

$$= 3k \times 3^k - 9 \times 2^{k-1} \left[\frac{1 - \left(\frac{3}{2}\right)^k}{1 - \left(\frac{3}{2}\right)} \right] = 3k \times 3^k - 9(3^k - 2^k) \tag{8.72}$$

However, the second term of the particular solution is a linear combination of the homogeneous solution, and so can be eliminated. Therefore the general solution is given by

$$y(k) = C_1 2^k + C_2 3^k + 3k \times 3^k \tag{8.73}$$

If the starting conditions are assumed to be $y(0) = 1$ and $y(1) = 0$, we have $C_1 + C_2 = 1$ and $2C_1 + 3C_2 = -9$, from which we find that $C_1 = 12$ and $C_2 = -11$. ∎

Another method for finding the particular solution of a nonhomogeneous equation is the method of **undetermined coefficients.** This method proves to be particularly efficient for input functions that are linear combinations of the following:

1. k^n, where n is a positive integer or zero.
2. β^k, where β is a nonzero constant.
3. $\cos \gamma k$, where γ is a nonzero constant.
4. $\sin \gamma k$, where γ is a nonzero constant.
5. A product of two or more sequences of these four types.

DISCRETE SIGNALS, DIFFERENCE EQUATIONS, AND SYSTEM DESCRIPTION   419

This method works because any derivative of the input function $x(k)$ is also possible as a linear combination of functions of the five types listed. For example, the function $2k^2$ or any derivative of $2k^2$ is a linear combination of the sequences k^2, k, and 1, all of which are of type 1. Hence what is required in any case is to seek the appropriate sequences for which any derivative of the input function $x(k)$ can be constructed by a linear combination of these sequences. Clearly, if $x(k) = \cos 3k$, the appropriate sequences are $\cos 3k$, and $\sin 3k$. The following examples will clarify these ideas.

EXAMPLE 8.10
Find the particular solution for the problem of Example 8.9.

Solution: Since 3^k is a solution of the homogeneous equation, we assume a solution of the form $y_p(k) = Ak \times 3^k$. Substitute this assumed solution into (8.69), which yields

$$Ak \times 3^k - 5A(k-1)3^{k-1} + 6A(k-2)3^{k-2} = 3^k$$

from which

$$\frac{3}{9} A3^k = 3^k$$

Hence $A = 3$, and the particular solution is $y_p(k) = 3k \times 3^k$, as we have already found. ∎

EXAMPLE 8.11
Find the particular solution for the system shown in Figure 8.17.

Figure 8.17
A second-order discrete system.

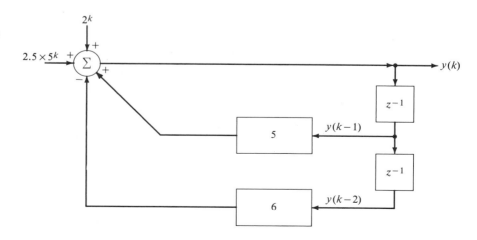

Solution: Observe that this is the system of Figure 8.16 with a driving function $2.5 \times 5^k + 2^k$. We wish to find the particular solution to the difference equation

$$y(k) - 5y(k-1) + 6y(k-2) = 2.5 \times 5^k + 2^k \tag{8.74}$$

As found in Example 8.9, the roots of the characteristic equation are $\lambda_1 = 2$, $\lambda_2 = 3$, and the general solution to the homogeneous equation is

$$y_h(k) = C_1 2^k + C_2 3^k \tag{8.75}$$

Observe that $x(k)$ is a linear combination of two sequences of type 2— namely, 2^k and 5^k. However, 2^k is already a solution to the homogeneous equation, which suggests that we multiply 2^k by k for one of the sequences. Thus we try as a particular solution

$$y_p(k) = A5^k + Bk2^k \tag{8.76}$$

where the constants A and B are **undetermined coefficients.** These can be found by substituting (8.76) into (8.74), from which we find

$$[A5^k + Bk2^k] - 5[A5^{k-1} + B(k-1)2^{k-1}] + 6[A5^{k-2} + B(k-2)2^{k-2}]$$
$$= 2.5 \times 5^k + 2^k$$

Rearranging terms we have

$$\left(A - 5A \times \frac{1}{5} + 6A \times \frac{1}{25}\right)5^k + \left(Bk + \frac{5B - 5Bk}{2} + \frac{6Bk - 12B}{4}\right)2^k$$
$$= 2.5 \times 5^k + 2^k$$

By equating coefficients of similar terms we find that

$$A = \frac{62.5}{6} \qquad B = -2$$

The particular solution is then

$$y_p(k) = \frac{62.5}{6} \times 5^k - 2k2^k \tag{8.77}$$

∎

EXAMPLE 8.12

Find the particular solution of the nonhomogeneous equation

$$y(k) - y(k-2) = 5k^2 \tag{8.78}$$

Solution: The roots of the characteristic equation are readily found to be $\lambda_1 = 1$, $\lambda_2 = -1$, and the solution to the homogeneous equation is

$$y_h(k) = C_1(1)^k + C_2(-1)^k$$

We observe that the function $5k^2$ and its derivatives can be found by the linear combination of sequences k^2, k, and 1. But 1 is a solution of the homogeneous

DISCRETE SIGNALS, DIFFERENCE EQUATIONS, AND SYSTEM DESCRIPTION

Table 8.2 Method of Undetermined Coefficients.

$x(k)$	$y_p(k)$
k^n	$A_1 k^n + A_2 k^{n-1} + \cdots + A_n k + A_{n+1}$
β^k	$A\beta^k$
$\cos \gamma k$ or $\sin \gamma k$	$A_1 \sin \gamma k + A_2 \cos \gamma k$
$k^n \beta^k$	$\beta^k(A_1 k^n + A_2 k^{n-1} + \cdots + A_n k + A_{n+1})$
$\beta^k \sin \gamma k$ or $\beta^k \cos \gamma k$	$\beta^k(A_1 \sin \gamma k + A_2 \cos \gamma k)$

equation, so we choose as a trial solution k times the sequences k^2, k, and 1; namely

$$y_p(k) = Ak^3 + Bk^2 + Ck \tag{8.79}$$

By substituting this function into (8.78) and equating coefficients of like-power terms, the coefficients are found to be $A = 5/6$, $B = 5/2$, $C = 5/3$. Table 8.2 gives the corresponding particular solutions for specified $x(k)$. ∎

EXAMPLE 8.13

A set of atoms in a one-dimensional lattice is spaced at a distance d apart. Refer to Figure 8.18a to develop the difference equation for the system, and solve the equation if the displacements of the first $k = 0$ and the last $k = K + 1$ atoms are zero. A constant force is applied to all the atoms.

Solution: From the figure it is evident that the kth atom will experience a force towards the axis

$$f(k) = -\frac{T}{d}[y(k) - y(k-1) + y(k) - y(k+1)]$$

or

$$y(k+1) - 2y(k) + y(k-1) = f(k)\frac{d}{T} \tag{8.80}$$

where T is the tension due to atomic forces holding the atoms to their equilibrium points. In the continuous case, the force is equal to $m(d^2y/dt^2)$, which is Newton's law.

For a constant force $f(k) = f_0$ applied to each atom, (8.80) becomes

$$y(k+1) - 2y(k) + y(k-1) = \frac{f_0 d}{T} \tag{8.81}$$

A block diagram representation of this equation is shown in Figure 8.18b. The auxiliary second-order equation is readily found to have a double root,

$$\lambda_1 = \lambda_2 = 1$$

Figure 8.18
Three adjacent atoms in a lattice.

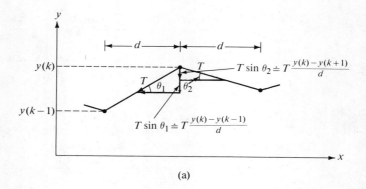

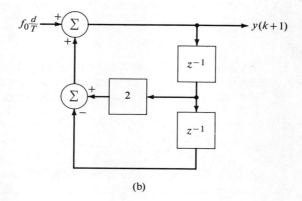

The homogeneous solution is

$$y_h(k) = C_1 + C_2 k \tag{8.82}$$

Since 1 and k are solutions of the homogeneous equation, we guess at the particular solution

$$y_p(k) = Ak^2 \tag{8.83}$$

Substituting this into (8.81), we find that it is the appropriate form with the specific value

$$A = \frac{f_0 d}{2T} \tag{8.84}$$

Therefore the general solution of (8.81) is

$$y(k) = C_1 + C_2 k + \frac{f_0 d}{2T} k^2 \tag{8.85}$$

Upon applying the boundary conditions, we obtain $C_1 = 0$ and $C_2 =$

DISCRETE SIGNALS, DIFFERENCE EQUATIONS, AND SYSTEM DESCRIPTION

$-(f_0 d/2T)(K+1)$. The final general solution is then

$$y(k) = \frac{f_0 d}{2T} k^2 - \frac{f_0 d}{2T}(K+1)k$$

$$= -\frac{f_0 d}{2T} k(K+1-k) \qquad 0 \le k \le K+1 \tag{8.86}$$

∎ ∎ ∎

8-6 FREQUENCY RESPONSE OF DISCRETE TIME SYSTEMS

It was found in Section 2-10g that if an input signal of the form $e^{j\omega t}$ is applied to a relaxed linear continuous time-invariant system, the output is of the form $H(\omega)e^{j\omega t}$, where $H(\omega)$ is the system function and is a complex function, in general. Similarly in discrete time-invariant systems, if the input is of the form $e^{jk\omega}$, the output is $H(e^{j\omega})e^{jk\omega}$. $H(e^{j\omega})$ is a function of ω and specifies the frequency response of the system. A schematic representation of these input-output relationships is shown in Figure 8.19. $H(e^{j\omega})$ is a complex function, in general, and can be represented by either

$$H(e^{j\omega}) = H_r(e^{j\omega}) + jH_i(e^{j\omega}) \qquad \textbf{(a)} \tag{8.87}$$

or

$$H(e^{j\omega}) = |H(e^{j\omega})|e^{j\varphi} \qquad \varphi = \tan^{-1} \frac{H_i(e^{j\omega})}{H_r(e^{j\omega})} \qquad \textbf{(b)}$$

where $H_r(\cdot)$ and $H_i(\cdot)$ are real functions. The input-output relation is thus

$$y(k) = H(e^{j\omega})e^{jk\omega} \qquad \textbf{(c)}$$

and for $T \ne 1$

$$\boxed{y(kT) = H(e^{j\omega T})e^{jk\omega T}} \qquad \textbf{(d)}$$

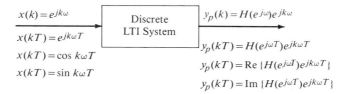

Figure 8.19 Schematic representation of a discrete time system with an $e^{jk\omega}$ input.

EXAMPLE 8.14

Find the output of a discrete LTI system for which the input is

$$x(k) = A \cos \omega_0 k \tag{8.88}$$

Solution: We can write the input in the form

$$x(k) = \frac{A}{2} e^{jk\omega_0} + \frac{A}{2} e^{-jk\omega_0}$$

and expect that the output is (preserving the even symmetry of $|H|$)

$$\begin{aligned} y(k) &= \frac{A}{2} H(e^{j\omega_0}) e^{jk\omega_0} + \frac{A}{2} H(e^{-j\omega_0}) e^{-jk\omega_0} \\ &= \frac{A}{2} H(e^{j\omega_0}) e^{jk\omega_0} + \frac{A}{2} [H(e^{j\omega_0}) e^{jk\omega_0}]^* \\ &= A |H(e^{j\omega_0})| \cos(k\omega_0 + \varphi) \quad \text{(a)} \end{aligned} \tag{8.89}$$

where

$$\varphi = \tan^{-1} \frac{H_i(e^{j\omega_0})}{H_r(e^{j\omega_0})} \quad \text{(b)}$$

We observe from this equation that:

- $H(e^{j\omega})$ is a continuous function of ω.
- $H(e^{j\omega})$ is a periodic function of ω with period 2π.
- For a sinusoidal input to a linear discrete and time-invariant system, the output is a sinusoidal signal modified in amplitude and with a phase shift, as given in (8.89). ■

EXAMPLE 8.15

Evaluate the frequency response of the system shown in Figure 8.20a for an input $v(k) = e^{jk\omega}$.

Solution: This circuit is described in discrete form by [see (8.14)]

$$2i(k) - i(k-1) = e^{jk\omega} \tag{8.90}$$

To use the method of undetermined coefficients, we assume a solution of the form

$$i(k) = Ae^{jk\omega}$$

where A is to be found. Substitute the assumed solution into (8.90) and equate the coefficients of similar terms. This yields

$$A = \frac{1}{2 - e^{-j\omega}}$$

Figure 8.20
The system function of an *RL* circuit and its frequency response.

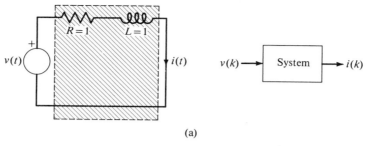

(a)

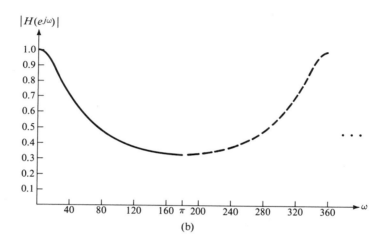

(b)

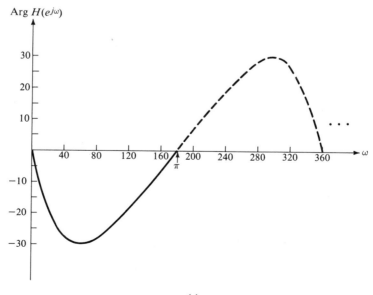

(c)

so that we can write

$$i_p(k) = \frac{1}{2 - e^{-j\omega}} e^{j\omega k} \tag{8.91}$$

Since $i_p(k) = H(e^{j\omega})v(k)$, it then follows that the system function is

$$H(e^{j\omega}) = \frac{1}{2 - e^{-j\omega}} \tag{8.92}$$

From this we can write

$$|H(e^{j\omega})| = \frac{1}{\sqrt{(2 - \cos \omega)^2 + \sin^2 \omega}} \quad \text{(a)}$$

$$\text{Arg } H(e^{j\omega}) = -\tan^{-1}\left(\frac{\sin \omega}{2 - \cos \omega}\right) \quad \text{(b)} \tag{8.93}$$

The magnitude function $|H(e^{j\omega})|$ is shown in Figure 8.20b, which is seen to be a **periodic** function with period 2π (360°) and a function that is **symmetric;** that is, $|H(e^{j\omega})| = |H(e^{-j\omega})|$. The phase response of the system function is given in Figure 8.20c, which shows that Arg $H(e^{j\omega})$ is **periodic** with period 2π, and that it possesses **odd symmetry** with respect to the origin; that is, Arg $H(e^{j\omega}) = -\text{Arg } H(e^{-j\omega})$. The points $\pm \pi$ are called the **fold-over** frequencies of the frequency response.

Note that had we used the difference equation with delay T explicitly included and with sampling taken at T seconds, our ω would have been taken to be $\omega T = 2\pi f T$ and the normalized interval $-\pi \le \omega \le \pi$ would have been $-\pi \le 2\pi f T \le \pi$ or $-1/2T < f < 1/2T$. The fold-over frequency is $1/2T$ Hz, which is the **Nyquist frequency;** this shows that to increase the bandwidth of the filter, we must make the sampling time small.

The foregoing discussion of the periodicity of the system function of a discrete time filter indicates that we cannot build a low-pass filter (for example) in the same sense as can be done for the continuous case using circuit elements. This requires that care be exercised in analyzing the response of digital filters. ∎

Chapter 3 demonstrates that any periodic function can be expressed in a Fourier series representation. Since $H(e^{j\omega})$ is a periodic function, it can be expressed in the form

$$H(e^{j\omega}) = \sum_{k=-\infty}^{\infty} h(k) e^{j\omega k} \qquad \omega_0 = \frac{2\pi}{2\pi} = 1 \quad \text{(a)} \tag{8.94}$$

where

$$h(k) = \frac{1}{2\pi} \int_{-\pi}^{\pi} H(e^{j\omega}) e^{-jk\omega} d\omega \quad \text{(b)}$$

provided that the series given by (8.94a) converges. This suggests that for any input sequence $x(k)$, we can define the Fourier representation

DISCRETE SIGNALS, DIFFERENCE EQUATIONS, AND SYSTEM DESCRIPTION

$$X(e^{j\omega}) = \sum_{k=-\infty}^{\infty} x(k)e^{j\omega k} \qquad \text{(a)} \qquad (8.95)$$

and its inverse

$$x(k) = \frac{1}{2\pi} \int_{-\pi}^{\pi} X(e^{j\omega}) e^{-j\omega k} d\omega \qquad \text{(b)}$$

This latter expression for $x(k)$ can be viewed as a summation of exponentials having amplitudes equal to $X(e^{j\omega})\,d\omega/2\pi$; the response of an LTI discrete system to $x(k)$ is just the superposition of the effects of these exponentials, which make up the input signal $x(k)$. In light of this, the output can be written

$$y(k) = \frac{1}{2\pi} \int_{-\pi}^{\pi} H(e^{j\omega}) X(e^{j\omega}) e^{-j\omega k} d\omega \qquad (8.96)$$

We conclude from this that the Fourier transform of the output is equal to

$$Y(e^{j\omega}) = H(e^{j\omega}) X(e^{j\omega}) \qquad (8.97)$$

and that

$$\boxed{y(k) = \sum_{n=-\infty}^{\infty} h(k-n) x(n)} \qquad \text{(a)} \qquad (8.98)$$

and for $T \neq 1$

$$\boxed{y(kT) = \sum_{n=-\infty}^{\infty} h[(k-n)T] x(nT)} \qquad \text{(b)}$$

This is the **convolution** relation for discrete, linear, and time-invariant systems where both $h(k)$ and $x(k)$ have values for both positive and negative k's. More will be said about convolution in the next chapter.

EXAMPLE 8.16

Find the impulse response of the discrete time filter that has a bandpass frequency response $H(e^{j\omega})$ as shown in Figure 8.21a.

Solution: The impulse response is found from (8.94b)

$$h(k) = \frac{1}{2\pi} \int_{-\pi}^{\pi} H(e^{j\omega}) e^{-j\omega k} d\omega = \frac{1}{2\pi} \left[\int_{-\omega_2}^{-\omega_1} e^{-j\omega k} d\omega + \int_{\omega_1}^{\omega_2} e^{-j\omega k} d\omega \right]$$

which becomes

$$h(k) = \frac{\sin k\omega_2}{\pi k} - \frac{\sin k\omega_1}{\pi k}$$

Figure 8.21
A passband digital filter and its impulse response.

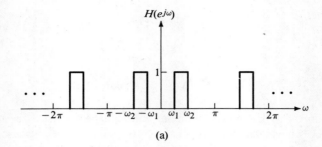

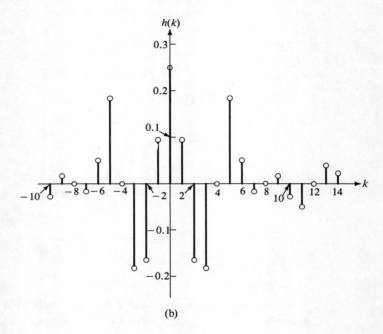

The impulse response for $\omega_2 = \pi/2$ and $\omega_1 = \pi/4$ is shown in Figure 8.21b. ∎

As already noted in (8.3), digital systems of order n are characterized by difference equations of the form

$$y(k) + a_1 y(k-1) + a_2 y(k-2) + \cdots + a_n y(k-n)$$
$$= b_0 x(k) + b_1 x(k-1) + \cdots + b_m x(k-m) \qquad (8.99)$$

where the coefficients a_i and b_i are assumed constant. Since we know that the particular solution to any linear discrete and time-invariant system is $y_p(k) = H(e^{j\omega})e^{j\omega k}$ when the input is $e^{j\omega k}$, it is apparent that for the input $e^{j\omega k}$, (8.99) becomes

Figure 8.22
Block diagram representation of a discrete system.

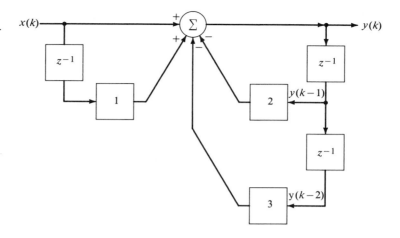

$$H(e^{j\omega})e^{j\omega k} + a_1 H(e^{j\omega})e^{j\omega(k-1)} + \cdots + a_n H(e^{j\omega})e^{j\omega(k-n)}$$
$$= b_0 e^{j\omega k} + b_1 e^{j\omega(k-1)} + \cdots + b_m e^{j\omega(k-m)}$$

from which it follows that

$$H(e^{j\omega}) = \frac{b_0 + b_1 e^{-j\omega} + \cdots + b_m e^{-j\omega m}}{1 + a_1 e^{-j\omega} + \cdots + a_n e^{-j\omega n}} \tag{8.100}$$

This important formula shows that we can find the transfer function of a given system without finding its particular solution. In this respect the result parallels that of finding the system function of a continuous system from the differential equation without having to find a particular solution to the differential equation.

EXAMPLE 8.17
Use (8.100) to find the system function for the system shown in Figure 8.22.

Solution: The controlling difference equation for the system is, from an inspection of Figure 8.22,

$$y(k) = -2y(k-1) - 3y(k-2) + x(k) + x(k-1)$$

or

$$y(k) + 2y(k-1) + 3y(k-2) = x(k) + x(k-1)$$

This is precisely in the form of (8.99), where $a_1 = 2$, $a_2 = 3$, $b_0 = 1$, $b_1 = 1$. It follows from (8.100) that the system function is

$$H(e^{j\omega}) = \frac{1 + e^{-j\omega}}{1 + 2e^{-j\omega} + 3e^{-j2\omega}}$$

∎

REFERENCES

1. Cadzow, J. A. *Discrete Time Systems*. Englewood Cliffs, N.J.: Prentice-Hall, 1973.
2. Finizio, N., and G. Ladas, *An Introduction to Differential Equations with Difference Equations, Fourier Series, and Partial Differential Equations*. Belmont, Calif.: Wadsworth, 1982.
3. Gabel, R. A., and R. A. Roberts. *Signals and Linear Systems*. 2d ed. New York: Wiley, 1980.
4. Seely, S. *An Introduction to Engineering Systems*. New York: Pergamon Press, 1972.
5. Seely, S., and A. D. Poularikas. *Electrical Engineering—Introduction and Concepts*. Beaverton, Ore.: Matrix, 1982.

PROBLEMS

8-1.1 Specify whether the systems given below are memoryless, time-invariant, linear, causal, or stable. The input to each system is $x(k)$ and the output is $y(k)$.

   a. $y(k) = -x(-k)$ **b.** $y(k) = x(k-1) - x(k-4)$ **c.** $y(k) = \sum_{k=0}^{k+2} x(k)$ **d.** $y(k) = (k+1)x(k)$

8-2.1 Determine the savings account balance specified by (8.2) after 12 months, for the following conditions:

$x(0) = 100$, $x(1) = -200$, $x(5) = 50$, $x(6) = 40$, $x(8) = 50$, $x(11) = 50$

$y(-1) = 300$; 6% annual interest rate, compounded monthly

8-3.1 Select $\beta_0 = 1 - \alpha_1$ in (8.15) so that the step response is $y(k) = 1 - \alpha_1^{k+1}$. Sketch the response functions for the following values of α_1: 0.2, 0.5, 0.8.

8-3.2 Determine the difference equations relating $y(k)$ to $x(k)$ for the systems shown in block diagram form in Figure P8-3.2.

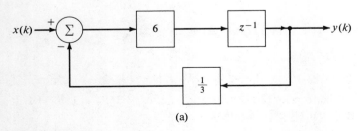

(a)

Figure P8-3.2

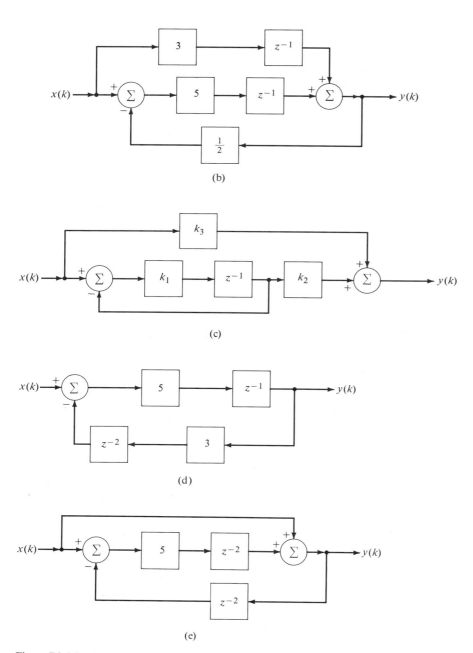

Figure P8-3.2 (*continued*)

8-3.3 Find the velocity of the systems shown in Figure P8-3.3 if $f(k) = au(k)$. Assume zero starting conditions, $v(-1) = 0$.

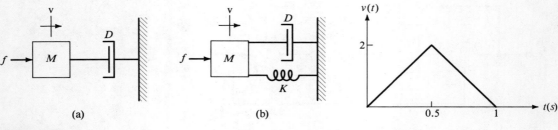

Figure P8-3.3

Figure P8-3.4

8-3.4 Find the approximate area under the curve shown in Figure P8-3.4 and compare it with the exact value. Assume $T = 0.2$ and use the appropriate equation similar to (8.10). Repeat for $T = 0.1$, and compare your results.

8-3.5 a. Show that the general solution to the first-order difference equation

$$y(k) - \alpha_1(k)y(k-1) = x(k)$$

is given by

$$y(k) = \left(\prod_{i=0}^{k} \alpha_1(i)\right) y(-1) + \sum_{n=0}^{k} \left(\prod_{i=n+1}^{k} \alpha_1(i)\right) x(n)$$

where $y(-1)$ is the initial condition.
b. Use this equation to verify (8.29).

8-3.6 Find the general solutions to the following difference equations:
  a. $y(k) - y(k-1) = 2k + 1$ **b.** $y(k) - y(k-1) = k$
  c. $y(k+1) - 3y(k) = 2$ **d.** $y(k) - (k+1)y(k-1) = 0$

8-3.7 Determine the voltage across the resistor in the block diagram representation of the system shown in Figure P8-3.7. The input is $v(k) = u(k) - u(k-4)$ and the system is assumed to be initially relaxed.

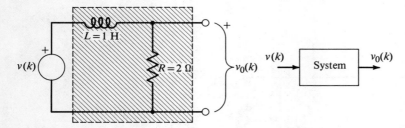

Figure P8-3.7

8-3.8 Find and sketch the response of the systems shown in Figure P8-3.8. Assume zero initial conditions.

8-3.9 Find and sketch the output of the systems shown in Figure P8-3.9. The initial condition for each system is $y(-1) = 1$.

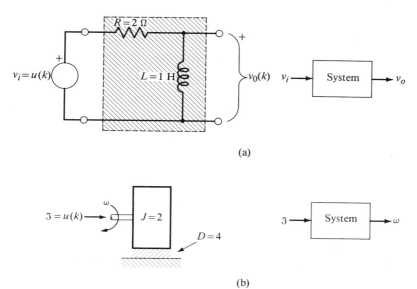

Figure P8-3.8

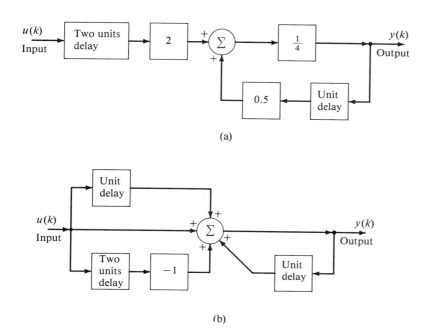

Figure P8-3.9

8-4.1 Plot the signals for the systems shown in Figure P8-4.1.

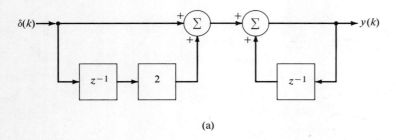

(a)

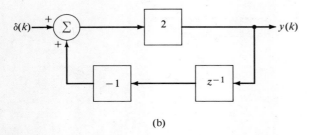

(b)

Figure P8-4.1

8-5.1 Find the complete solution to the following difference equations:

a. $y(k) - 4y(k-2) = 2\cos\dfrac{k\pi}{2}$
b. $y(k) - 7y(k-1) + 10y(k-2) = 2 \times 3^k$

c. $y(k) - 7y(k-1) + 10y(k-2) = 3 \times 2^k$
d. $y(k) + 5y(k-1) - 6y(k-2) = \cos\dfrac{k\pi}{2}$

8-5.2 Solve the difference equation $y(k) - 7y(k-1) + 10y(k-2) = 2 \times 3^k$ with the starting conditions $y(0) = -4$, $y(-1) = -20$.

8-5.3 Find and plot the voltage across the capacitor in the circuit shown in Figure P8-5.3 if the input voltage is the function 0.9^k for $k = 0, 1, 2, \ldots$. Assume that the capacitor had an initial charge of $Q = 1$ C and that the initial current in the circuit was zero at $t = 0$.

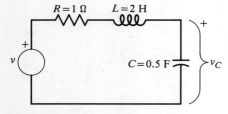

Figure P8-5.3

8-5.4 The **natural modes** of free lattice vibration in an atomic structure (see Example 8.13) are dictated by the equation

$$y(k+1) - 2y(k) + y(k-1) = \frac{md}{T}\frac{d^2y(k)}{dt^2}$$

Assume a solution of the form $y(k) = A(k)\cos(\omega t + \varphi)$, where $A(k)$ is the displacement associated with the kth atom and φ is a constant phase. Find ω. Additional assumptions are boundary conditions $y(0) = 0$, $y(K+1) = 0$, and $(md/T)\omega^2 < 2$.

8-5.5 Use the method of undetermined coefficients to find the form of a particular solution to the difference equations:
  a. $y(k) - 6y(k-1) + 8y(k-2) = 3 \times 2^k$ **b.** $y(k) - 2y(k-1) + y(k-2) = 3(k^2 - 1)$
  c. $y(k) - 4y(k-1) + 4y(k-2) = 3k\,2^k + 3(-2)^k$

8-5.6 Find the particular solution to the equation

$$6y(k) - y(k-2) = \cos\frac{k\pi}{2}$$

8-5.7 Design the digital system block diagrams described by the following difference equations. $x(k)$ is the input and $y(k)$ is the output:
  a. $y(k) - 2y(k-1) + 4y(k-2) + 2y(k-3) = x(k)$ **b.** $y(k) - ay(k-2) = x(k) + 2x(k-1)$
  c. $y(k) = a_1 x(k) + a_2 x(k-1) + a_3 x(k-2)$

8-5.8 Describe the type of operations being performed in the discrete systems shown in Figure P8-5.8.

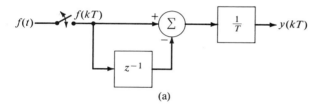

(a)

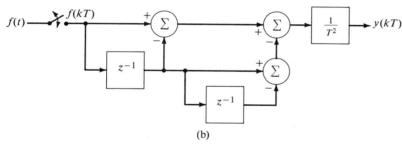

(b)

Figure P8-5.8

8-5.9 The output of a discrete system is given by

$$y(k) = \sum_{j=0}^{K} h_j \delta(k - jT)$$

when the input is an impulse function. Draw its block diagram representation. This type of diagram is known as a **tapped delay line**.

8-6.1 Find the amplitude response of the system shown in Figure 8.7 for the following sets of values:
  a. $L = 1$, $R = 4$ **b.** $L = 1$, $R = 0.2$.
  Discuss the different results as a function of frequency. Choose $i(k) = Ae^{jk\omega}$.

8-6.2 Find the amplitude response of the systems described by the following input-output relationships for an excitation $x(k) = e^{jk\omega}$:
a. $y(k) = 2x(k) + 2x(k-1)$ **b.** $y(k) = 2x(k) - 2x(k-1)$
Identify the type of filter that each of these represents.

8-6.3 Evaluate the frequency response of the system shown in Figure P8-6.3 for an input $i(k) = e^{jk\omega}$.

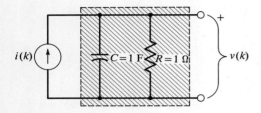

Figure P8-6.3

8-6.4 Find the impulse response of the discrete time filter that has the frequency response $H(e^{j\omega})$ as shown in Figure P8-6.4.

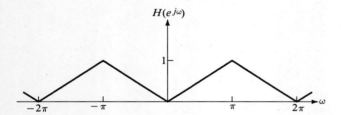

Figure P8-6.4

8-6.5 Find the impulse response sequence of the FIR system specified by the block diagram in Figure P8-6.5.

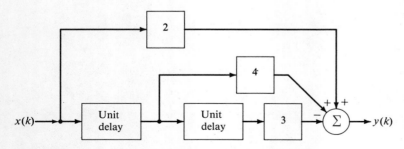

Figure P8-6.5

8-6.6 Evaluate and sketch the frequency response of the system shown in Figure P8-6.6 for an input $f(0.5k) = e^{j0.5k\omega}$. What is the Nyquist frequency for this system?

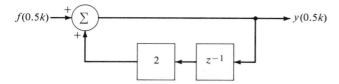

Figure P8-6.6

8-6.7 Evaluate and sketch the frequency response of the system shown in Figure P8-6.7 for an input $f(0.01k) = \cos(0.01k\omega)$. What is the Nyquist frequency for this system?

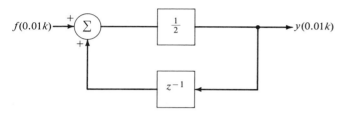

Figure P8-6.7

8-6.8 Develop Equations (8.94) through (8.98) when $T \neq 1$.

CHAPTER 9

THE Z-TRANSFORM

The classical techniques for the solution of difference equations discussed in Chapter 8 do not lend themselves well to a systematic study of linear discrete time systems of orders higher than one. The Z-transform method introduced here is a powerful method for solving difference equations. Although application of the Z-transform is relatively new, the essential features of this mathematical technique date back to the early 1730s when DeMoivre introduced the concept of a generating function, which is identical with that for the Z-transform.

The Z-transform method provides a technique for transforming a difference equation into an algebraic equation. Specifically, the Z-transform converts a sequence of numbers $\{y(k)\}$ into the function $Y(z)$ of a complex variable, thereby allowing algebraic processes and well-developed mathematical procedures to become part of the solution procedure. In this sense the Z-transform plays the same general role in the solution of difference equations that the Laplace transform plays in the solution of differential equations. Inversion procedures that parallel one another also exist. We will study the Z-transform method in considerable detail in this chapter.

■ ■ ■

9-1 THE Z-TRANSFORM

To understand the essential features of the Z-transform, consider a **one-sided sequence** of numbers $\{y(k)\}$ taken at uniform time intervals,

$$\{y(k)\} = y(0), y(1), y(2), \ldots, y(k), \ldots \tag{9.1}$$

We associate with this sequence the function

$$Y(z) = \frac{y(0)}{z^0} + \frac{y(1)}{z^1} + \frac{y(2)}{z^2} + \cdots \tag{9.2}$$

We can interpret z^{-1} as the unit delay operator introduced in Chapter 8. This operation allows a ready interpretation of each element of $\{y(k)\}$ and its position

THE Z-TRANSFORM

in the sequence. We will often make use of this interpretation to relate terms of $Y(z)$ with the corresponding terms in the sequence $\{y(k)\}$.

A second and more formal mathematical interpretation compatible with the above interpretation exists for (9.2). Now we interpret z as a general complex variable, and $Y(z)$ denotes the Z-transform of the sequence $\{y(k)\}$. In its more general form, the one-sided Z-transform of a function $y(k)$ is

$$Y(z) = \mathscr{Z}\{y(k)\} \triangleq \sum_{k=0}^{\infty} y(k) z^{-k} \tag{9.3}$$

This equation can be taken as the definition of the Z-transform operation.

EXAMPLE 9.1
Find the Z-transform of the discrete function

$$y(k) = \begin{cases} 0 & k \leq 0 \\ 1 & k = 1 \\ 2 & k = 2 \\ 0 & k \geq 3 \end{cases}$$

Solution: From the defining equation (9.3), we write

$$Y(z) = \mathscr{Z}\{y(k)\} = \frac{1}{z^1} + \frac{2}{z^2} = \frac{z+2}{z^2}$$

Observe that this function possesses a second-order pole at the origin and a zero at -2. ∎

EXAMPLE 9.2
Write the Z-transform of the function

$$f(t) = Ae^{-at} \qquad t \geq 0 \tag{9.4}$$

which is sampled every T seconds; that is, $t = kT$.

Solution: The sampled values are written

$$\{f(kT)\} = A, Ae^{-aT}, Ae^{-2aT}, \ldots$$

The Z-transform of this sequence is written

$$F(z) = A\left[1 + \left(\frac{e^{-aT}}{z}\right) + \left(\frac{e^{-aT}}{z}\right)^2 + \cdots\right]$$

This series can be written in closed form by recalling that the expression

$$\frac{1}{1-x} = 1 + x + x^2 + x^3 + \cdots \qquad x < 1$$

Thus we have

$$F(z) = \frac{A}{1 - \left(\dfrac{e^{-aT}}{z}\right)} = \frac{Az}{z - e^{-aT}} \qquad (9.5)$$

∎

EXAMPLE 9.3
Find the Z-transform of the function

$$f(t) = e^{-t} + 2e^{-2t} \qquad t \geq 0$$

which is sampled at time intervals $T = 0.1$ seconds.

Solution: Use the results of Example 9.2 to write

$$F(z) = \frac{z}{z - e^{-0.1}} + \frac{2z}{z - e^{-0.2}}$$

This is

$$F(z) = \frac{z}{z - 0.905} + \frac{2z}{z - 0.819}$$

which, written in this form, is the partial-fraction expansion of the function

$$F(z) = \frac{z(z - 0.819) + 2z(z - 0.905)}{(z - 0.905)(z - 0.819)} = \frac{3z^2 - 2.629z}{z^2 - 1.724z + 0.741}$$

∎

9-2 CONVERGENCE OF THE Z-TRANSFORM

The function $F(z)$ for a specific value of z may be either finite or infinite. The set of z in the complex z-plane for which the magnitude of $F(z)$ is finite is the **region of convergence** for $F(z)$, whereas the set of z for which the magnitude of $F(z)$ is infinite is the **region of divergence** (see also Appendix 1). We examine the region of convergence by considering the defining expression (9.3) and examining the complex values of z for which $\sum_{k=0}^{\infty} |f(k)z^{-k}|$ has a finite value. If we write z in polar form $z = re^{j\theta}$, we have

$$\sum_{k=0}^{\infty} |f(k)z^{-k}| = \sum_{k=0}^{\infty} |f(k)(re^{j\theta})^{-k}|$$

$$= \sum_{k=0}^{\infty} |f(k)r^{-k}e^{-jk\theta}| = \sum_{k=0}^{\infty} |f(k)r^{-k}| \qquad (9.6)$$

For this sum to be finite, we find numbers M and R such that $|f(k)| \leq MR^k$ for $k \geq 0$. Thus

THE Z-TRANSFORM

$$\sum_{k=0}^{\infty} |f(k)z^{-k}| \le M \sum_{k=0}^{\infty} R^k r^{-k} = M \sum_{k=0}^{\infty} \left(\frac{R}{r}\right)^k \qquad (9.7)$$

For the sum to be finite, it is required that $R/r < 1$. That is, $F(z)$ is absolutely **convergent** for all z in the region outside the circle of radius R. Conversely, the region $|z| < R$ is the **region of divergence**. A separate test is required to establish whether the boundary belongs to the region of convergence or the region of divergence. The following example demonstrates this.

EXAMPLE 9.4

Find the Z-transform of the signal given, and discuss its properties.

$$f(k) = \begin{cases} c^k & k = 0, 1, 2, \ldots \\ 0 & k = -1, -2, \ldots \end{cases} \qquad (9.8)$$

The constant c takes the following values: (a) $0 < c < 1$, (b) $c > 1$.

Solution: The time sequences for the two cases are shown in Figures 9.1a and b. The Z-transform is given by [see (9.3)]

$$F(z) = \sum_{k=0}^{\infty} c^k z^{-k} = \sum_{k=0}^{\infty} (c^{-1}z)^{-k}$$
$$= 1 + cz^{-1} + c^2 z^{-2} + \cdots + c^n z^{-n} + \cdots \qquad (9.9)$$

Figure 9.1
The discrete signal c^k and the regions of convergence and divergence in the z-plane.

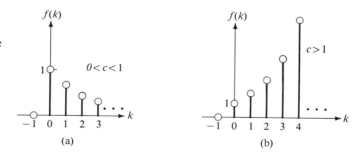

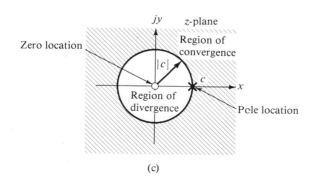

Initially, consider the sum of the first n terms of this geometric series. This is given by

$$F_n(z) = \frac{1 - (cz^{-1})^{n+1}}{1 - cz^{-1}} \qquad (9.10)$$

We set $cz^{-1} = |cz^{-1}|e^{j\theta}$, where θ is the argument of the complex number z^{-1}; hence we can write that

$$(cz^{-1})^{n+1} = |cz^{-1}|^{n+1} e^{jn\theta}$$

We now observe that for values of z for which $|cz^{-1}| < 1$, the magnitude of the complex number $(cz^{-1})^{n+1}$ approaches zero as $n \to \infty$. As a consequence

$$F(z) = \lim_{n \to \infty} F_n(z) = \frac{1}{1 - cz^{-1}} = \frac{z}{z - c} \qquad |cz^{-1}| < 1 \qquad (9.11)$$

For the general case where c is a complex number, the inequality $|cz^{-1}| < 1$ leads to $|c| < |z|$, which implies that the series converges when the magnitude of $|z| > |c|$ and diverges for $|z| < |c|$. Thus we see that the regions of convergence and divergence in the complex plane for $F(z)$ are those shown in Figure 9.1c.

To establish whether the boundary of the circle in Figure 9.1c belongs to the region of convergence or the region of divergence, we apply L'Hospital's rule to (9.10), so that

$$\lim_{z \to c} F_n(z) = \lim_{z \to c} \frac{\frac{d}{d(cz^{-1})}[1 - (cz^{-1})^{n+1}]}{\frac{d}{d(cz^{-1})}[1 - (cz^{-1})]} = \lim_{z \to c} \frac{-(n+1)(cz^{-1})^n}{-1} = n + 1$$

and therefore

$$\lim_{n \to \infty} F_n(z) \to \infty$$

Clearly, the boundary belongs to the region of divergence. ■

EXAMPLE 9.5
Find the Z-transform and discuss the properties of the impulse functions

$$y(k) = \delta(k) = \begin{cases} 1 & k = 0 \\ 0 & k \neq 0 \end{cases} \quad \text{(a)}$$

$$y(k) = \delta(k - N) = \begin{cases} 1 & k = N \\ 0 & k \neq N \end{cases} \quad \text{(b)} \qquad (9.12)$$

which are shown in Figure 9.2.

Solution: **a.** From the basic definition, we can write that

$$Y(z) = y(0)z^{-0} = 1$$

THE Z-TRANSFORM

Figure 9.2
The discrete delta function and its Z-transform.

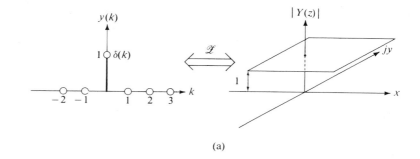

(a)

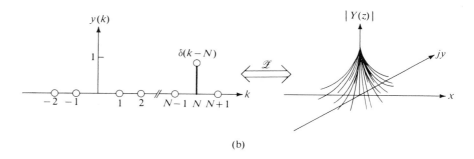

(b)

Since $Y(z)$ is independent of z, the region of convergence is the entire z-plane.

b. An application of the definition of the Z-transform to the function $\delta(k - N)$ leads to

$$\mathscr{Z}\{\delta(k - N)\} \triangleq Y(z) = \sum_{k=0}^{\infty} \delta(k - N) z^{-k}$$
$$= 0 \times z^{-0} + 0 \times z^{-1} + \cdots + 1 \times z^{-N} + 0 \times z^{-(N+1)} + \cdots$$
$$= z^{-N}$$

Since $Y(z) \to \infty$ only for $z = 0$, the region of convergence is the entire z-plane except for an infinitesimal region around the origin. ∎

EXAMPLE 9.6
Deduce the Z-transform of the function

$$y(k) = \begin{cases} a^k \sin(k\omega) & k \geq 0 \quad a > 0 \\ 0 & k < 0 \end{cases} \qquad (9.13)$$

Indicate the region of divergence, the region of convergence, and the poles and zeros in the z-plane.

Solution: The given function is shown in Figures 9.3a and b for two different values of a. Clearly, the function is a sinusoidal discrete signal for $a = 1$. The Z-transform is given by

Figure 9.3
The discrete signal $a^k \sin(k\omega)$, its poles and zeros, and regions of convergence.

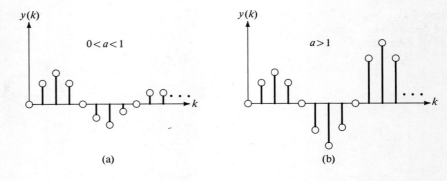

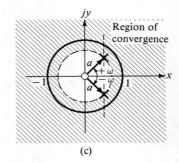

$$Y(z) = \sum_{k=0}^{\infty} a^k \sin(k\omega) z^{-k} = \sum_{k=0}^{\infty} \frac{a^k(e^{jk\omega} - e^{-jk\omega}) z^{-k}}{2j}$$

$$= \frac{1}{2j} \sum_{k=0}^{\infty} (ae^{j\omega} z^{-1})^k - \frac{1}{2j} \sum_{k=0}^{\infty} (ae^{-j\omega} z^{-1})^k \tag{9.14}$$

Sum the geometric series of (9.14) in the manner of (9.9) to write

$$Y(z) = \frac{1}{2j} \left[\frac{1}{1 - ae^{j\omega} z^{-1}} - \frac{1}{1 - ae^{-j\omega} z^{-1}} \right]$$

$$= \frac{z^{-1} a \sin \omega}{1 - 2az^{-1} \cos \omega + a^2 z^{-2}} \qquad |z| > a \tag{9.15}$$

Multiply the numerator and denominator of this expression by z^2 to find

$$Y(z) = \frac{za \sin \omega}{z^2 - 2a(\cos \omega)z + a^2}$$

$$= \frac{za \sin \omega}{[z - a(\cos \omega + j \sin \omega)][z - a(\cos \omega - j \sin \omega)]}$$

$$= \frac{za \sin \omega}{(z - ae^{j\omega})(z - ae^{-j\omega})} \tag{9.16}$$

THE Z-TRANSFORM

The zeros and poles are shown in Figure 9.3c for the case $a < 1$.

To examine the region of convergence, consider (9.14). Notice that each of the series converges if $|ae^{j\omega}z^{-1}| = |az^{-1}||e^{j\omega}| = |az^{-1}| < 1$ or $|z| > a$. This region is shown in Figure 9.3c. ∎

When the sequence $\{y(k)\}$ has values for both positive and negative k, the region of convergence of $Y(z)$ becomes an annular ring around the origin. To see this, consider the specific sequence

$$y(k) = \begin{cases} 3^k & \text{for } k \geq 0 \\ 4^k & \text{for } k < 0 \end{cases}$$

This is a bilateral function, and the definition for the bilateral Z-transform equivalent to (9.3) is used

$$\mathscr{Z}\{y(k)\} = \sum_{k=-\infty}^{\infty} y(k) z^{-k} \qquad (9.17)$$

For our specific function

$$Y(z) = \sum_{k=0}^{\infty} 3^k z^{-k} + \sum_{k=-\infty}^{-1} 4^k z^{-k} = \sum_{k=0}^{\infty} 3^k z^{-k} + \sum_{k=1}^{\infty} 4^{-k} z^{k}$$

The first summation converges as $k \to \infty$, provided that $|3z^{-1}| < 1$ or $|z| > 3$. If we set $R^+ = 3$ for positive k's, we see that the region of convergence for positive k's is $|z| > R^+$. Similarly, the second summation converges if $|4^{-1}z| < 1$ or $|z| < 4$, and the region of convergence for the negative k's is $|z| < R^-$ with $R^- = 4$. The sequence $y(k)$ and the region of convergence (depicted as the double-lined region) are shown in Figure 9.4.

Following steps parallel to the above, the reader can easily see that the sequence

$$y(k) = \begin{cases} 4^k & \text{for } k \geq 0 \\ 3^k & \text{for } k < 0 \end{cases}$$

has no region of convergence.

From the foregoing discussion we conclude the following:

- The region of convergence of a two-sided sequence is a ring in the z-plane centered at the origin.
- The region of convergence of a sequence of finite duration is the entire z-plane, except possibly the points $z = 0$ and/or $z = \infty$.
- If the sequence is right-handed (that is, $k \geq 0$), then the region of convergence is beyond a circle of finite radius.
- If the sequence is left-handed, (that is, $k < 0$), then the region of convergence is within a circle of finite radius.

Figure 9.4
Region of convergence for a two-sided sequence.

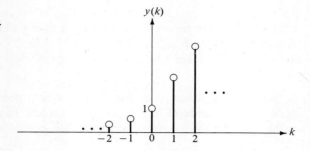

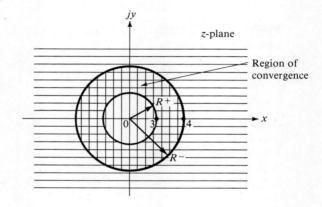

The following examples show why it is important to specify the region of convergence.

EXAMPLE 9.7
Specify the region of convergence for the two sequences

$$y_1(k) = a^k u(k) \quad \textbf{(a)}$$
$$y_2(k) = -a^k u(-k-1) \quad \textbf{(b)}$$ (9.18)

Solution: The Z-transform of the first sequence is

$$Y_1(z) = \sum_{k=0}^{\infty} (az^{-1})^k = \frac{1}{1 - az^{-1}} = \frac{z}{z-a}$$ (9.19)

For convergence we must have $|az^{-1}| < 1$; this implies that the region of convergence is $|z| > |a|$.

The Z-transform of the second sequence is

$$Y_2(z) = -\sum_{k=-\infty}^{-1} a^k z^{-k} = -\sum_{k=1}^{\infty} a^{-k} z^k = 1 - \sum_{k=0}^{\infty} (a^{-1}z)^k = 1 + \frac{a}{z-a}$$

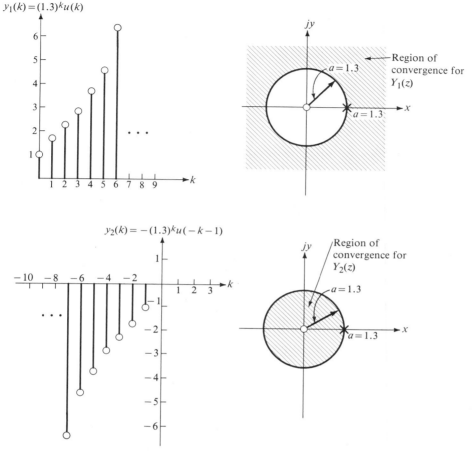

Figure 9.5
Illustrating Example 9.7.

$$= \frac{z}{z-a} = \frac{1}{1-az^{-1}} \qquad (9.20)$$

The region of convergence is found from the relation $|a^{-1}z| < 1$, or $|z| < |a|$. The sequences for $a > 1.3$ with their zero-pole configuration and their regions of convergence are shown in Figure 9.5.

We observe from (9.19), (9.20), and Figure 9.5 that two completely different sequences can have the same analytical form in the z-domain and the same zero-pole configuration. Clearly, to obtain a unique sequence in the time domain if we are given the Z-transform of a sequence, we must also know the region of convergence.

As we discussed in Chapter 7, whenever we use the digital computer to analyze a time function $f(t)$, we must first sample it every T sec using an A/D converter and then use the sample values $f(kT)$ for $k = 0, 1, 2, \ldots$. Hence for sampled functions, the one-sided Z-transform is given by

$$\boxed{\mathscr{Z}\{f(kT)\} = \sum_{k=0}^{\infty} f(kT)z^{-k}} \tag{9.21}$$

EXAMPLE 9.8
Find the Z-transform of the functions given below when sampled every T sec:

a. $f(t) = u(t)$ **b.** $f(t) = tu(t)$ **c.** $f(t) = e^{-\beta t}u(t)$ **d.** $f(t) = \sin \omega t \, u(t)$

Solution:

a. $\mathscr{Z}\{f(kT)\} = \mathscr{Z}\{u(kT)\} = \sum_{k=0}^{\infty} u(kT)z^{-k}$

$= (1 + z^{-1} + z^{-2} + \cdots)$

$= \dfrac{1}{1 - z^{-1}} = \dfrac{z}{z - 1} \qquad |z| > 1 \tag{9.22}$

b. $\mathscr{Z}\{f(kT)\} = \mathscr{Z}\{kTu(kT)\} = \sum_{k=0}^{\infty} kTu(kT)z^{-k}$

$= Tz^{-1} + 2Tz^{-2} + 3Tz^{-3} + \cdots$

$= -Tz\dfrac{d}{dz}(z^{-1} + z^{-2} + z^{-3} + \cdots)$

$= -Tz\dfrac{d}{dz}[z^{-1}(1 + z^{-1} + z^{-2} + \cdots)]$

$= -Tz\dfrac{d}{dz}\left[z^{-1}\dfrac{z}{z-1}\right] = \dfrac{Tz}{(z-1)^2} \qquad |z| > 1 \tag{9.23}$

c. $\mathscr{Z}\{f(kT)\} = \mathscr{Z}\{u(kT)e^{-\beta kT}\}$

$= \mathscr{Z}\{u(kT)c^{-k}\}; \; (e^{-\beta kT} = c^{-k} \text{ for convenience})$

$= \sum_{k=0}^{\infty} u(kT)c^{-k}z^{-k} = 1 + c^{-1}z^{-1} + c^{-2}z^{-2} + \cdots$

$= \dfrac{1}{1 - \dfrac{1}{cz}} = \dfrac{cz}{cz - 1}$

$= \dfrac{ze^{\beta T}}{ze^{\beta T} - 1} = \dfrac{z}{z - e^{-\beta T}} \qquad |z| > e^{-\beta T} \tag{9.24}$

d. $\mathscr{Z}\{f(kT)\} = \mathscr{Z}\{u(kT)\sin \omega kT\} = \mathscr{Z}\left\{u(kT)\dfrac{e^{j\omega kT} - e^{-j\omega kT}}{2j}\right\}$

$$= \sum_{k=0}^{\infty} \frac{u(kT)}{2j} c_1^{-k} z^{-k} - \sum_{k=0}^{\infty} \frac{u(kT)}{2j} c_2^{-k} z^{-k}$$

$$= \frac{1}{2j} \left[\frac{c_1 z}{c_1 z - 1} - \frac{c_2 z}{c_2 z - 1} \right]$$

$$= \frac{z}{2j} \left[\frac{e^{-j\omega T}}{e^{-j\omega T} z - 1} - \frac{e^{j\omega T}}{e^{j\omega T} z - 1} \right]$$

$$= \frac{z \sin \omega T}{z^2 - 2z \cos \omega T + 1} \qquad |z| > 1 \qquad (9.25)$$

∎

■ ■ ■

9-3 PROPERTIES OF THE Z-TRANSFORM

We wish to develop the more important basic properties of the Z-transform for one-sided sequences—that is, those with zero elements for $k < 0$. The one-sided sequence is of great importance because all detected signals are of finite extent and can always be referenced so that their starting-point is at $t = 0$ ($k = 0$). From this, several properties follow.

1. Linearity.

$$\mathscr{Z}\{ay_1(k) + by_2(k)\} = a\mathscr{Z}\{y_1(k)\} + b\mathscr{Z}\{y_2(k)\} \qquad (9.26)$$

where a and b are any constants. The region of convergence is $|z| > \max(R_1, R_2)$, where $|z| > R_1$ and $|z| > R_2$ are the regions of convergence of $y_1(k)$ and $y_2(k)$, respectively.

2. Right-Shifting Property.

$$\mathscr{Z}\{y(k-n)\} = z^{-n}\mathscr{Z}\{y(k)\} = z^{-n}Y(z) \qquad (9.27)$$

Proof: From the definition of the Z-transform, we have

$$Y(z) = \sum_{k=0}^{\infty} y(k) z^{-k}$$

Multiply through by z^{-n} and then substitute $-m$ for $-n-k$. The result is

$$z^{-n} Y(z) = \sum_{m=n}^{\infty} y(m-n) z^{-m} = \sum_{m=0}^{\infty} y(m-n) z^{-m} = \sum_{k=0}^{\infty} y(k-n) z^{-k}$$

where, since m is a dummy index, we changed it to letter k. The third term in this expression was obtained by invoking the one-sided character of $y(k)$, with y (negative number) $= 0$. For the case when $y(-k)$ has values, we must

Figure 9.6
Illustration of a shifted discrete signal.

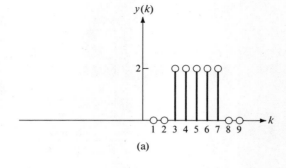
(a)

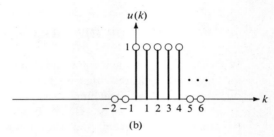
(b)

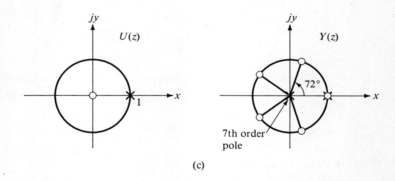
(c)

add the quantity $y(-n) + y(1-n)z^{-1} + \cdots + y(-1)z^{-(n-1)}$ to the right-hand member of (9.27); thus m runs from 0 to $n-1$. The multiplication by z^{-n} $(n > 0)$ creates a pole at $z = 0$ and deletes a pole at infinity. Therefore the region of convergence is the same as the region of convergence of $y(k)$, with the possible exclusion of the origin.

EXAMPLE 9.9
Find the Z-transform of the function shown in Figure 9.6a using the shifting property.

Solution: The Z-transform of the unit step function $u(k)$ shown in Figure 9.6b

THE Z-TRANSFORM

is given by

$$\mathscr{L}\{u(k)\} = \sum_{k=0}^{\infty} u(k)z^{-k} = 1 + z^{-1} + z^{-2} + \cdots = \frac{1}{1-z^{-1}} = \frac{z}{z-1}$$

The discrete time function in Figure 9.6a is

$$y(k) = 2u(k-3) - 2u(k-8)$$

It follows therefore that

$$\mathscr{L}\{y(k)\} = 2\mathscr{L}\{u(k-3)\} - 2\mathscr{L}\{u(k-8)\} = 2z^{-3}\mathscr{L}\{u(k)\} - 2z^{-8}\mathscr{L}\{u(k)\}$$

$$= \frac{2z}{z-1}\left[\frac{1}{z^3} - \frac{1}{z^8}\right] = \frac{2(z^5-1)}{z^7(z-1)}$$

The pole-zero configurations for $U(z)$ and $Y(z)$ are shown in Figure 9.6c. From the figure we see that while $U(z)$ does not have poles at zero, the linear combination of its shifted versions $y(k)$ does possess poles at the origin. ∎

3. Left-Shifting Property.

$$\mathscr{L}\{y(k+n)\} = z^n Y(z) - \sum_{k=0}^{n-1} y(k)z^{n-k} \tag{9.28}$$

Proof: From the basic definition

$$\mathscr{L}\{y(k+1)\} = \sum_{k=0}^{\infty} y(k+1)z^{-k}$$

Now set $k+1 = m$ to find

$$\mathscr{L}\{y(m)\} = \sum_{m=1}^{\infty} y(m)z^{-m+1} = z\sum_{m=1}^{\infty} y(m)z^{-m} = z\sum_{m=0}^{\infty} y(m)z^{-m} - zy(0)$$

$$= zY(z) - zy(0)$$

By a similar procedure, we can show that

$$\mathscr{L}\{y(k+n)\} = z^n Y(z) - z^n y(0) - z^{n-1}y(1) - \cdots - zy(n-1)$$

$$= z^n Y(z) - \sum_{k=0}^{n-1} y(k)z^{n-k}$$

Because of the factor z^n, zeros are introduced at $z=0$ and at infinity. Observe that if the function is defined by

$$y(k+n) = \begin{cases} 0 & \text{for } k = -n, -n-1, -n-2, \ldots \\ f(k+n) & \text{for } k = 0, 1, 2, \ldots \end{cases}$$

then the shifting property gives

$$\mathscr{L}\{y(k+n)\} = z^n Y(z) \tag{9.29}$$

Figure 9.7
A first-order discrete system.

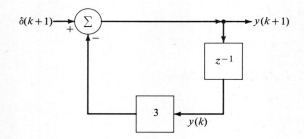

This result indicates that if n is a positive integer and $f(k) = 0$ for $k < n$, then the Z-transform is that of (9.29).

EXAMPLE 9.10
Find the Z-transform of the output of the system shown in Figure 9.7.

Solution: The difference equation that describes the system is derived from the figure to be

$$y(k + 1) + 3y(k) = \delta(k + 1)$$

Take the Z-transform of both sides of this expression, recalling the linearity property, with the result

$$\mathscr{L}\{y(k + 1)\} + 3\mathscr{L}\{y(k)\} = \mathscr{L}\{\delta(k + 1)\}$$

which, by applying the left-shifting property, yields

$$zY(z) - zy(0) + 3Y(z) = z$$

Solve for $Y(z)$

$$Y(z) = \frac{z(1 + y(0))}{z + 3}$$

Note that by setting $k = 0$ in the difference equation, we obtain $y(1) + 3y(0) = \delta(1)$, which indicates that $y(0) = 0$ because the input is applied one time unit later than $T = 0$. Therefore $Y(z)$ becomes

$$Y(z) = \frac{z}{z + 3}$$

Suppose that we shift the time axis to the right by one time unit. The difference equation now assumes the form

$$y(k) + 3y(k - 1) = \delta(k)$$

Invoking the right-shifting property, the Z-transform of all terms in this equation yields

THE Z-TRANSFORM

$$Y(z) + 3z^{-1}Y(z) = 1$$

or

$$Y(z) = \frac{z}{z+3}$$

which is identical with our previous result. ∎

4. Time Scaling.

$$\mathscr{L}\{a^k y(k)\} = Y(a^{-1}z) = \sum_{k=0}^{\infty} (a^{-1}z)^{-k} y(k) \tag{9.30}$$

Proof: From the definition of the Z-transform

$$\mathscr{L}\{a^k y(k)\} = \sum_{k=0}^{\infty} a^k y(k) z^{-k} = \sum_{k=0}^{\infty} y(k)(a^{-1}z)^{-k}$$

Recall that we found in Example 9.8 that the Z-transform of $y(k) = \sin k\omega$, $k = 0, 1, 2, \ldots$, is equal to $z \sin \omega/(z^2 - 2z \cos \omega + 1)$. By an application of (9.30), we can write the Z-transform of the function $a^k y(k) = a^k \sin k\omega$ by inserting the value $a^{-1}z$ for z. This leads to the result

$$\mathscr{L}\{a^k \sin k\} = \frac{a^{-1}z \sin \omega}{a^{-2}z^2 - 2a^{-1}z \cos \omega + 1}$$

This result is the same as (9.15), which was deduced by a different approach.

5. Periodic Sequences.

$$\mathscr{L}\{y(k)\} = \frac{z^N}{z^N - 1} \mathscr{L}\{y_{(1)}(k)\} \tag{9.31}$$

where N indicates the number of time units in a period, $y_{(1)}(k)$ is the first period of the periodic sequence, and $y(k) = y(k + N)$.

Proof: The Z-transform of the first period is

$$\mathscr{L}\{y_{(1)}(k)\} = \sum_{k=0}^{N-1} y_{(1)}(k) z^{-k} = Y_{(1)}(z)$$

Because the period is repeated every N discrete time units, we can use Property 2 (Right-Shifting Property) to write

$$\mathscr{L}\{y(k)\} = \mathscr{L}\{y_{(1)}(k)\} + \mathscr{L}\{y_{(1)}(k - N)\} + \mathscr{L}\{y_{(1)}(k - 2N)\} + \cdots$$
$$= Y_1(z) + z^{-N} Y_1(z) + z^{-2N} Y_1(z) + \cdots$$

Figure 9.8
Periodic discrete function.

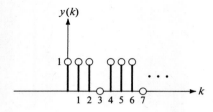

$$= Y_1(z)(1 + z^{-N} + z^{-2N} + \cdots) = \frac{z^N}{z^N - 1} Y_1(z)$$

EXAMPLE 9.11
Find the Z-transform of the signal shown in Figure 9.8.

Solution: Use (9.31) with $N = 4$ to find

$$Y(z) = \frac{z^4}{z^4 - 1} \mathscr{L}\{y_{(1)}(k)\} = \frac{z^4}{z^4 - 1}(1 + z^{-1} + z^{-2})$$

$$= \frac{z^2(z^2 + z + 1)}{z^4 - 1}$$ ∎

6. Multiplication by k.

$$\boxed{\mathscr{L}\{ky(k)\} = -z\frac{d}{dz} Y(z)} \qquad (9.32)$$

Proof: From the basic definition

$$\mathscr{L}\{ky(k)\} = \sum_{k=0}^{\infty} ky(k)z^{-k} = z \sum_{k=0}^{\infty} y(k)(kz^{-k-1})$$

$$= z \sum_{k=0}^{\infty} y(k)\left[-\frac{d}{dz} z^{-k}\right] = -z\frac{d}{dz} \sum_{k=0}^{\infty} y(k)z^{-k} = -z\frac{d}{dz} Y(z)$$

As an example of this procedure, since $\mathscr{L}\{u(k)\} = z/(z - 1)$, by (9.32) we have

$$\mathscr{L}\{ku(k)\} = -z\frac{d}{dz}\left(\frac{z}{z-1}\right) = \frac{z}{(z-1)^2}$$

This result was shown in (9.23). We can continue this procedure to find

$$\mathscr{L}\{k^2 u(k)\} = \mathscr{L}\{k[ku(k)]\} = -z\frac{d}{dz}\left(\frac{z}{(z-1)^2}\right) = \frac{z(z+1)}{(z-1)^3}$$

THE Z-TRANSFORM

This result was already shown in (9.23). But $k(k + 1) = k^2 + k$, and we thus have

$$\mathscr{L}\{k(k + 1)u(k)\} = \frac{z(z + 1)}{(z - 1)^3} + \frac{z}{(z - 1)^2} = \frac{2z^2}{(z - 1)^3}$$

Similarly we have

$$\mathscr{L}\{k(k - 1)u(k)\} = \frac{z(z + 1)}{(z - 1)^3} - \frac{z}{(z - 1)^2} = \frac{2z}{(z - 1)^3}$$

7. Division by $k + a$ (a is any real number).

$$\mathscr{L}\left\{\frac{y(k)}{k + a}\right\} = -z^a \int_0^z \frac{Y(\tilde{z})}{\tilde{z}^{a+1}} d\tilde{z} \tag{9.33}$$

Proof: In a direct manner, where $\tilde{z}$ is a dummy integration variable.

$$\mathscr{L}\left\{\frac{y(k)}{k + a}\right\} = \sum_{k=0}^{\infty} \frac{y(k)}{k + a} z^{-k} = \sum_{k=0}^{\infty} y(k) z^a \left(-\int_0^z \tilde{z}^{-k-a-1} d\tilde{z}\right)$$

This is written

$$\mathscr{L} = -z^a \int_0^z \frac{1}{\tilde{z}^{a+1}} \sum_{k=0}^{\infty} y(k) \tilde{z}^{-k} d\tilde{z} = -z^a \int_0^z \frac{Y(\tilde{z})}{\tilde{z}^{a+1}} d\tilde{z}$$

8. Initial value. Given a sequence that is zero for $k < k_0$, the value of the point $y(k_0)$ is

$$y(k_0) = z^{k_0} Y(z)\big|_{z \to \infty} \tag{9.34}$$

Proof: Consider the series

$$Y(z) = \sum_{k=k_0}^{\infty} y(k) z^{-k} = y(k_0) z^{-k_0} + y(k_0 + 1) z^{-(k_0 + 1)} + \cdots$$

Multiply both sides by z^{k_0} to obtain

$$y(k_0) + y(k_0 + 1) z^{-1} + y(k_0 + 2) z^{-2} + \cdots = z^{k_0} Y(z)$$

All terms except the first approach zero as $z \to \infty$.

For example, consider the function given by (9.15). For this function we have

$$y(0) = z^0 Y(z)\big|_{z \to \infty} = \frac{z^{-1} a \sin \omega}{1 - 2az^{-1} \cos \omega + a^2 z^{-2}}\bigg|_{z \to \infty} = 0$$

which is the correct answer, as can be easily found from (9.13).

9. Final Value.

$$\lim_{k \to \infty} y(k) = \lim_{z \to 1} (1 - z^{-1})Y(z), \qquad \text{if } y(\infty) \text{ exists} \qquad (9.35)$$

Proof: Begin with $y(k) - y(k-1)$, and consider

$$\mathscr{L}\{y(k) - y(k-1)\} = Y(z) - z^{-1}Y(z) = \sum_{k=0}^{\infty} [y(k) - y(k-1)]z^{-k}$$

by Property 2 (The Right-Shifting Property). This is written

$$(1 - z^{-1})Y(z) = \lim_{N \to \infty} \sum_{k=0}^{N} [y(k) - y(k-1)]z^{-k}$$

Take the limit as $z \to 1$, or

$$\lim_{z \to 1} (1 - z^{-1})Y(z) = \lim_{z \to 1} \lim_{N \to \infty} \sum_{k=0}^{N} [y(k) - y(k-1)]z^{-k}$$

Interchange the summations on the right

$$= \lim_{N \to \infty} \lim_{z \to 1} \sum_{k=0}^{N} [y(k) - y(k-1)]z^{-k}$$

$$= \lim_{N \to \infty} \sum_{k=0}^{N} [y(k) - y(k-1)]$$

$$= \lim_{N \to \infty} [y(0) - y(-1) + y(1) - y(0) + y(2) - y(1) + \cdots]$$

$$= \lim_{N \to \infty} y(N)$$

since $y(-1) = 0$. The limit $z \to 1$ will give meaningful results only when the point $z = 1$ is located within the region of convergence of $Y(z)$.

10. Convolution.

$$\mathscr{L}\{y(k)\} = \mathscr{L}\{h(k) * x(k)\} = \mathscr{L}\left\{\sum_{n=0}^{\infty} h(k-n)x(n)\right\} = H(z)X(z) \qquad (9.36)$$

where $h(k - n) = 0$ for $n > k$.

Proof: The Z-transform of the convolution summation is

$$\mathscr{L}\{y(k)\} \triangleq Y(z) = \sum_{k=0}^{\infty} z^{-k} \sum_{n=0}^{k} h(k-n)x(n)$$

Since $h(\text{negative})$ is zero, we can replace k by ∞ in the summation in n so that

THE Z-TRANSFORM

$$\mathscr{Z}\{y(k)\} = \sum_{k=0}^{\infty} z^{-k} \sum_{n=0}^{\infty} h(k-n)x(n)$$

Write $k - n = m$, and invert the order of summation

$$Y(z) = \sum_{n=0}^{\infty} x(n) \sum_{k=0}^{\infty} h(k-n)z^{-k} = \sum_{n=0}^{\infty} x(n)z^{-n} \sum_{m=-n}^{\infty} h(m)z^{-m}$$

$$= \sum_{n=0}^{\infty} x(n)z^{-n} \sum_{m=0}^{\infty} h(m)z^{-m} = X(z)H(z)$$

since $h(m) = 0$ for $m < 0$.

EXAMPLE 9.12

The input signal sequence $x(m)$ and the impulse response $v(m)$ of a system are shown in Figures 9.9a and b. Deduce the output of the system $w(k)$.

Solution: Figure 9.9c shows the folded signal $x(-m)$, and Figures 9.9d–h show steps in carrying out the convolution process. Note that when the shift $v(4 - m)$ is introduced, the two functions do not coincide and their product is zero. The output signal $w(k)$ of this system is shown in Figure 9.9h. ∎

EXAMPLE 9.13

Find the Z-transform of the convolution of the following three functions:

$$y_1(k) = a_0 \delta(k) + a_1 \delta(k-1)$$

$$y_2(k) = b_0 \delta(k) + b_1 \delta(k-1)$$

$$y_3(k) = c_0 \delta(k) + c_1 \delta(k-1)$$

Solution: The Z-transforms of these three functions are:

$$Y_1(z) = a_0 + a_1 z^{-1} \qquad Y_2(z) = b_0 + b_1 z^{-1} \qquad Y_3(z) = c_0 + c_1 z^{-1}$$

We observe that

$$[Y_1(z)Y_2(z)]Y_3(z) = Y_1(z)[Y_2(z)Y_3(z)]$$

from which it follows that

$$[y_1(k) * y_2(k)] * y_3(k) = y_1(k) * [y_2(k) * y_3(k)]$$

This shows that convolution is **associative**; it is also **commutative** (see Problem 9-3.11). Thus we can write

$$\mathscr{Z}\{y_1(k) * y_2(k) * y_3(k)\} = Y_1(z)Y_2(z)Y_3(z)$$
$$= (a_0 b_0 + (a_1 b_0 + a_0 b_1)z^{-1} + a_1 b_1 z^{-2})(c_0 + c_1 z^{-1})$$
$$= a_0 b_0 c_0 + (a_1 b_0 c_0 + a_0 b_1 c_0 + a_0 b_0 c_1)z^{-1}$$
$$+ (a_1 b_0 c_1 + a_0 b_1 c_1 + a_1 b_1 c_0)z^{-2} + a_1 b_1 c_1 z^{-3}$$

Figure 9.9
The convolution of two discrete functions.

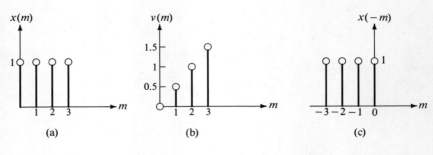

(a)   (b)   (c)

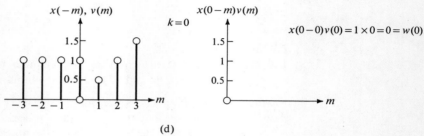

(d)

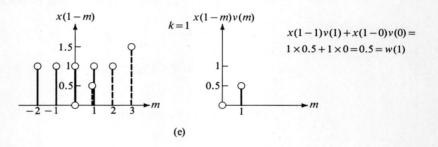

(e)

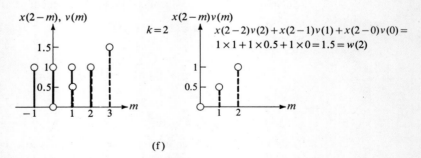

(f)

Now apply the definition of the Z-transform to see that this is

$$y_1(k) * y_2(k) * y_3(k) = a_0 b_0 c_0 \, \delta(k) + (a_1 b_0 c_0 + a_0 b_1 c_0 + a_0 b_0 c_1) \, \delta(k-1)$$
$$+ (a_1 b_0 c_1 + a_0 b_1 c_1 + a_1 b_1 c_0) \, \delta(k-2)$$
$$+ a_1 b_1 c_1 \, \delta(k-3)$$

■

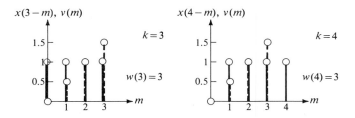
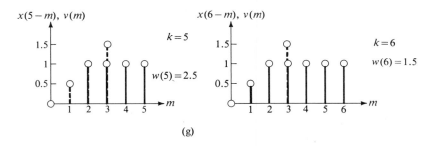

(g)

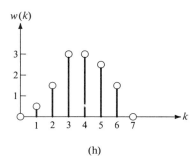

(h)

EXAMPLE 9.14

Find the output of the system shown in Figure 9.10a if the input is that shown in Figure 9.10b. Express the system in its discrete form.

Solution: A direction application of the Kirchhoff voltage law yields the equation

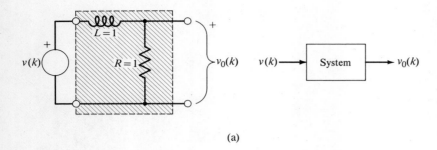

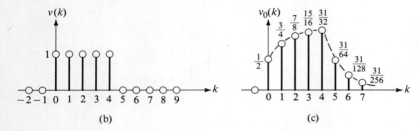

Figure 9.10
The impulse response of an *RL* circuit in discrete form.

$$\frac{di(t)}{dt} + i(t) = v(t) \quad \text{or} \quad \frac{dv_0(t)}{dt} + v_0(t) = v(t)$$

The second equation follows from the first since here $v_0(t) = i(t) \times 1$ with $R = 1$ ohm. A direct conversion to approximate discrete form using (8.7)

$$\frac{dv}{dt} = v(k) - v(k-1) \text{ for } T = 1$$

leads to

$$v_0(k) = \frac{1}{2} v(k) + \frac{1}{2} v_0(k-1)$$

We now proceed by determining the impulse response of the system, which will then be used in convolution with the actual input to obtain the complete solution. For a delta function input $v(k) = \delta(k)$, the Z-transform of the given equation gives the impulse reponse $H(z)$. This is

$$H(z) = \frac{1}{2} \frac{1}{1 - \frac{1}{2} z^{-1}} = \frac{1}{2}\left(1 + \frac{1}{2} z^{-1} + \frac{1}{4} z^{-2} + \cdots\right)$$

THE Z-TRANSFORM

It follows from this that

$$h(k) = \frac{1}{2}\left(\frac{1}{2}\right)^k \qquad k \geq 0$$

This result is used in the expression

$$v_0(k) = \sum_{n=0}^{k} h(k-n)v(n)$$

We determine the output at successive time steps; namely,

$$v_0(0) = h(0)v(0) = \frac{1}{2} \times 1 = \frac{1}{2}$$

$$v_0(1) = h(1)v(0) + h(0)v(1) = \frac{1}{4} \times 1 + \frac{1}{2} \times 1 = \frac{3}{4}$$

$$v_0(2) = h(2)v(0) + h(1)v(1) + h(0)v(2) = \frac{1}{8} \times 1 + \frac{1}{4} \times 1 + \frac{1}{2} \times 1 = \frac{7}{8}$$

$$\vdots$$

The resulting output is shown in Figure 9.10c. The resulting shape is identical with that of the corresponding continuous system found in Problem 2-8.2. ∎

11. Bilateral Convolution. When the index k takes on both positive and negative values, the bilateral convolution is specified as

$$y(k) = \sum_{n=-\infty}^{\infty} h(k-n)x(n) = \sum_{n=-\infty}^{\infty} h(n)x(k-n) \qquad (9.37)$$

The Z-transform is given by the product

$$\boxed{\mathscr{L}\{y(k)\} = H(z)X(z)} \qquad (9.38)$$

For convenience we tabulate the Z-transform properties (Table 9.1).

Table 9.2 presents the properties of the Z-transform for sampled continuous signals at time intervals T. The verification of this table has been left as exercises in solving selected problems.

■ ■ ■

9-4 Z-TRANSFORM PAIRS

Implicit in the foregoing discussion is the idea that we can proceed in two directions. Given an $f(k)$, the Z-transform $F(z)$ can be deduced according to Section 9-1. We wish to show that given an $F(z)$, a reverse process leads to $f(R)$. These processes are shown functionally as

CHAPTER 9

Table 9.1 Properties of the Z-transform ($k \geq 0$).

1.	$\mathscr{L}\{ay_1(k) + by_2(k)\}$	$aY_1(z) + bY_2(z)$
2.	$\mathscr{L}\{y(k-n)\}$	$z^{-n}Y(z) + \sum_{m=1}^{n} y(-m)z^{-(n-m)}$
3.	$\mathscr{L}\{y(k+n)\}$	$z^n Y(z) - \sum_{k=0}^{n-1} y(k)z^{n-k}$
4.	$\mathscr{L}\{a^k y(k)\}$	$Y\left(\dfrac{z}{a}\right)$
5.	$\mathscr{L}\{y_{(1)}(k)\}$	$\dfrac{z^N}{z^N - 1} Y_1(z)$ where $y_{(1)}(k)$ is the first period of a periodic sequence $y(k) = y(k+N)$
6.	$\mathscr{L}\{h(k) * x(k)\}$	$H(z)X(z)$
7.	$\mathscr{L}\{ky(k)\}$	$-z\dfrac{d}{dz}Y(z)$
8.	$\mathscr{L}\left\{\dfrac{y(k)}{k+a}\right\}$	$-z^a \int_0^z \dfrac{Y(\tilde{z})}{\tilde{z}^{a+1}} d\tilde{z}$
9.	$y(k_0)$	$z^{k_0} Y(z)\big|_{z \to \infty}$ where k_0 is the initial value of the sequence and $Y(z) = \sum_{k=0}^{\infty} y(k_0 + k)z^{-k}$
10.	$\lim_{k \to \infty} y(k)$	$\lim_{z \to 1} (1 - z^{-1})Y(z)$

Table 9.2 Properties of the Z-transform of Sampled Functions ($k \geq 0$).

1.	$\mathscr{L}\{ay_1(kT) + by_2(kT)\}$	$aY_1(z) + bY_2(z)$
2.	$\mathscr{L}\{y(kT - mT)\}$	$z^{-m} \sum_{k=0}^{\infty} y(kT)z^{-k} = z^{-m}Y(z)$
3.	$\mathscr{L}\{y(kT + mT)\}$	$z^m \left[Y(z) - \sum_{k=0}^{m-1} y(kT)z^{-k} \right]$
4.	$\mathscr{L}\{a^{kT} y(kT)\}$	$Y(a^{-T}z)$
5.	$\mathscr{L}\{kTy(kT)\}$	$-Tz\dfrac{dY(z)}{dz}$
6.	$y(k_0 T)$	$z^{k_0} Y(z)\big|_{z \to \infty}$ $Y(z) = \sum_{k=0}^{\infty} y[(k_0 + k)T]z^{-k}$
7.	$\lim_{k \to \infty} y(kT)$	$\lim_{z \to 1} (1 - z^{-1})Y(z)$
8.	$\mathscr{L}\{h(kT) * x(kT)\}$	$H(z)X(z)$

THE Z-TRANSFORM

$$\mathscr{Z}\{f(k)\} \triangleq F(z) = \sum_{k=0}^{\infty} f(k)z^{-k} \quad \text{direct transform} \quad \text{(a)}$$

$$\mathscr{Z}^{-1}[F(z)] \triangleq f(k) = \frac{1}{2\pi j} \oint_C F(z)z^{k-1}\, dz \quad \text{(9.39)}$$
$$\text{inverse transform} \quad \text{(b)}$$

where $\mathscr{Z}^{-1}$ denotes the inverse Z-transform.

As is evident from (9.39b), determining the inverse transform involves contour integration in the complex plane, where C is a circle of radius r that encloses all of the singularities of $F(z)z^{k-1}$.

A proof of (9.39b) follows from (9.39a): multiply both sides of (9.39a) by $z^{n-1}dz$ and integrate over a path within the regions of convergence. Thus

$$\oint F(z)z^{n-1}\, dz = \sum_{k=0}^{\infty} f(k) \oint z^{n-k-1}\, dz$$

By the Cauchy integral theorem (see Appendix 1)

$$\oint z^{n-k-1}\, dz = \begin{cases} 0 & \text{for } k \neq n \\ 2\pi j & \text{for } k = n \end{cases}$$

Thus

$$\oint F(z)z^{n-1}\, dz = f(n) \oint z^{-1}\, dz = 2\pi j f(n)$$

from which

$$f(n) = \frac{1}{2\pi j} \oint F(z)z^{n-1}\, dz$$

For the case when a time function $f(t)$ is sampled at time units T, (9.39) takes the forms

$$\mathscr{Z}\{f(kT)\} \triangleq F(z) = \sum_{k=0}^{\infty} f(kT)z^{-k} \quad \text{(a)}$$

$$\mathscr{Z}^{-1}\{F(z)\} \triangleq f(kT) = \frac{1}{2\pi j} \oint F(z)z^{k-1}\, dz \quad \text{(b)} \quad \text{(9.40)}$$

Carrying out the mathematical details for the Z-transform inversion integral is not usually necessary. By its very nature, the transform pair obtained according to (9.39) is valid either for a specified $\mathscr{Z}\{f(k)\}$ resulting in $F(z)$ or for the inverse transform of $F(z)$, $\mathscr{Z}^{-1}[F(z)]$, which results in the original $f(k)$. The situation here is seen to parallel that of the Laplace transform pair. Conse-

Table 9.3 Common Z-transform Pairs.

Entry Number	$f(k)$ for $k \geq 0$	$F(z) = \sum_{k=0}^{\infty} f(k) z^{-k}$	Radius of Convergence $|z| > R$
1.	$\delta(k)$	1	0
2.	$\delta(k-m)$	z^{-m}	0
3.	1	$\dfrac{z}{z-1}$	1
4.	k	$\dfrac{z}{(z-1)^2}$	1
5.	k^2	$\dfrac{z(z+1)}{(z-1)^3}$	1
6.	k^3	$\dfrac{z(z^2+4z+1)}{(z-1)^4}$	1
7.	a^k	$\dfrac{z}{z-a}$	$|a|$
8.	ka^k	$\dfrac{az}{(z-a)^2}$	$|a|$
9.	$k^2 a^k$	$\dfrac{az(z+a)}{(z-a)^3}$	$|a|$
10.	$\dfrac{a^k}{k!}$	$e^{a/z}$	0
11.	$(k+1)a^k$	$\dfrac{z^2}{(z-a)^2}$	$|a|$
12.	$\dfrac{(k+1)(k+2)a^k}{2!}$	$\dfrac{z^3}{(z-a)^3}$	$|a|$
13.	$\dfrac{(k+1)(k+2)\cdots(k+m)a^k}{m!}$	$\dfrac{z^{m+1}}{(z-a)^{m+1}}$	$|a|$

quently, what is needed is a table of Z-transform pairs. A table of common Z-transform pairs is given in Table 9.3.

A number of the entries in this table follow directly from the results in Section 9-1. For example, by the proper selection of the exponent a in Example 9.2, we deduce the following: for $a = 0$, we have entry 3; for $a = \pm j\omega T$, entries 14 and 15 follow. The other entries can be found by direct application of (9.39).

■ ■ ■

Table 9.3 (*continued*)

Entry Number	$f(k)$ for $k \geq 0$	$F(z) = \sum_{k=0}^{\infty} f(k) z^{-k}$	Radius of Convergence $\|z\| > R$
14.	$\sin k\omega T$	$\dfrac{z \sin \omega T}{z^2 - 2z \cos \omega T + 1}$	1
15.	$\cos k\omega T$	$\dfrac{z(z - \cos \omega T)}{z^2 - 2z \cos \omega T + 1}$	1
16.	$a^k \sin k\omega T$	$\dfrac{az \sin \omega T}{z^2 - 2az \cos \omega T + a^2}$	$\|a\|^{-1}$
17.	$a^{kT} \sin k\omega T$	$\dfrac{a^T z \sin \omega T}{z^2 - 2a^T z \cos \omega T + a^{2T}}$	$\|a\|^{-T}$
18.	$a^k \cos k\omega T$	$\dfrac{z(z - a \cos \omega T)}{z^2 - 2az \cos \omega T + a^2}$	$\|a\|^{-1}$
19.	$e^{-\alpha kT} \sin k\omega T$	$\dfrac{ze^{-\alpha T} \sin \omega T}{z^2 - 2e^{-\alpha T} z \cos \omega T + e^{-2\alpha T}}$	$\|z\| > \|e^{-\alpha T}\|$
20.	$e^{-\alpha kT} \cos k\omega T$	$\dfrac{z(z - e^{-\alpha T} \cos \omega T)}{z^2 - 2e^{-\alpha T} z \cos \omega T + e^{-2\alpha T}}$	$\|z\| > \|e^{-\alpha T}\|$
21.	$\dfrac{k(k-1)}{2!}$	$\dfrac{z}{(z-1)^3}$	1
22.	$\dfrac{k(k-1)(k-2)}{3!}$	$\dfrac{z}{(z-1)^4}$	1
23.	$\dfrac{k(k-1)(k-2)\cdots(k-m+1)}{m!} a^{k-m}$	$\dfrac{z}{(z-a)^{m+1}}$	1
24.	$e^{-\alpha kT}$	$\dfrac{z}{z - e^{-\alpha T}}$	$\|e^{-\alpha T}\|$
25.	$ke^{-\alpha kT}$	$\dfrac{ze^{-\alpha T}}{(z - e^{-\alpha T})^2}$	$\|e^{-\alpha T}\|$

9-5 THE INVERSE Z-TRANSFORM

In our studies we will assume that $F(z)$ corresponds to a sequence $\{f(k)\}$ that is bounded as $k \to +\infty$. To find the inverse Z-transform, we will cast the transformed functions into forms amenable to the use of Table 9.3. The functions with which we will be concerned are rational functions of z; that is, they are the ratio of two polynomials. Ordinarily these are **proper fractions** since the degree of the numerator polynomial is less than the degree of the denominator polynomial. If these are not proper fractions, the numerator polynomial is divided by the denominator polynomial and the long-division process continued until the numerator polynomial of the remainder is of one degree less than the denominator polynomial. This results in power terms plus a proper fraction. The following examples will illustrate the most commonly used procedures. Where existing tables are inadequate, recourse can be made to the inversion integral, with the solution carried out using details of complex function theory.

EXAMPLE 9.15
Determine the inverse Z-transform of the function

$$F(z) = \frac{1}{1 - 0.1z^{-1}} \tag{9.41}$$

Solution: The function possesses a simple pole at $z = 0.1$, with the function converging outside the circle of radius $\frac{1}{10}$ in the complex plane. Proceed by carrying out long division, which results in an infinite series in powers of z^{-1}

$$\begin{array}{r} 1 + 0.1z^{-1} + (0.1)^2 z^{-2} + \cdots \\ 1 - 0.1z^{-1} \overline{\smash{\big)}\ 1} \\ \underline{1 - 0.1z^{-1}} \\ 0.1z^{-1} \\ \underline{0.1z^{-1} - (0.1z^{-1})^2} \\ (0.1z^{-1})^2 - \vdots \end{array}$$

Thus we have

$$F(z) = 1 + 0.1z^{-1} + (0.1)^2 z^{-2} + (0.1)^3 z^{-3} + \cdots$$

But from the basic definition, (9.3), we see that the sequence is

$$\{f(k)\} = \begin{cases} 1, 0.1, (0.1)^2, (0.1)^3, \ldots & k \geq 0 \\ 0 & k < 0 \end{cases} \quad \textbf{(a)} \tag{9.42}$$

which is the sequence

$$\{f(k)\} = 0.1^k \qquad k \geq 0 \qquad \textbf{(b)} \qquad \blacksquare$$

EXAMPLE 9.16
Find the inverse Z-transform of the function

$$F(z) = \frac{1}{(1 - 0.2z^{-1})(1 + 0.2z^{-1})} = \frac{1}{1 - 0.04z^{-2}} \tag{9.43}$$

Solution: a. One approach is to proceed as in the foregoing example. By long division the following polynomial will result:

THE Z-TRANSFORM

$$F(z) = 1 + 0.04z^{-2} + (0.04)^2 z^{-4} + \cdots = (0.2)^{2k}(z^{-1})^{2k}$$

with region of convergence $|z| > 0.2$. This corresponds to the sequence

$$f(k') = \begin{cases} (0.2)^{k'} & k' = 2k \quad k \geq 0 \\ 0 & k < 0 \end{cases} \tag{9.44}$$

b. A separate approach calls for separating the function $F(z)$ into partial-fraction form as follows:

$$\frac{1}{1 - 0.04z^{-2}} = \frac{A}{1 - 0.2z^{-1}} + \frac{B}{1 + 0.2z^{-1}} = \frac{(A + B) + 0.2z^{-1}(A - B)}{1 - (0.2z^{-1})^2}$$
$$\tag{9.45}$$

The first and last fractions must be the same, which requires that $A + B = 1$, $A - B = 0$. These conditions lead to $A = B = \frac{1}{2}$, and (9.45) becomes

$$F(z) = \frac{1}{2}\left[\frac{1}{1 - 0.2z^{-1}} + \frac{1}{1 - (-0.2)z^{-1}}\right]$$

From appropriate entries in Table 9.3, the inverse transform is

$$f(k) = \begin{cases} \frac{1}{2}[(0.2)^k + (-0.2)^k] & k \geq 0 \\ 0 & k < 0 \end{cases} \tag{9.46}$$

The reader can easily verify that (9.44) and (9.46) yield identical results. ■

EXAMPLE 9.17
Find the inverse Z-transform of the function

$$F(z) = \frac{1}{(1 - 0.2z^{-1})z^{-2}} \tag{9.47}$$

Solution: This function is written as

$$F(z) = \frac{z^3}{z - 0.2} = Az^2 + Bz + \frac{Cz}{z - 0.2}$$

$$= \frac{Az^3 - 0.2Az^2 + Bz^2 - 0.2Bz + Cz}{z - 0.2}$$

Equate terms having the same powers of z. This yields: $A = 1, B = 0.2, C = (0.2)^2$
Equation (9.47) is then

$$F(z) = z^2 + 0.2z + (0.2)^2 \frac{z}{z - 0.2}$$

From Table 9.3 the inverse transform is

$$f(k) = \delta(k + 2) + 0.2\,\delta(k + 1) + (0.2)^2(0.2)^k$$

where the last term is applicable for $k \geq 0$. Therefore this equation is equivalent to

$$f(k) = \begin{cases} 0.2^{k+2} & k \geq -2 \\ 0 & k < -2 \end{cases} \tag{9.48}$$

Recall that (9.47) could be expanded into the form

$$F(z) = z^2 + 0.2z + (0.2)^2 z^0 + (0.2)^3 z^{-1} + (0.2)^4 z^{-2} + \cdots$$

$$= z^2 \left(1 + \frac{0.2}{z} + \frac{(0.2)^2}{z^2} + \cdots \right)$$

The inverse transform of the bracketed term is 0.2^k, and the factor z^2 indicates a shift to the left of two sample periods. Thus (9.48) is realized.

Note the following. To find the inverse Z-transform, we must (a) initially ignore any factor of the form z^k, where k is an integer, (b) expand the remaining part into a partial-fraction expansion, (c) use the Z-transform table or Z-transform properties to obtain the inverse Z-transform of each term in the expansion, and (d) combine the results and perform the necessary shifting due to z^k being omitted in step (a). ∎

EXAMPLE 9.18
Find the inverse Z-transform of the function

$$F(z) = \frac{z^2 - 3z + 8}{(z-2)(z+2)(z+3)}$$

Solution: Expand this in partial-fraction form to

$$F(z) = \frac{z^2 - 3z + 8}{(z-2)(z+2)(z+3)} = A + \frac{Bz}{z-2} + \frac{Cz}{z+2} + \frac{Dz}{z+3}$$

where

$$A = F(z)\big|_{z=0} = -\frac{2}{3}$$

$$B = \frac{(z-2)F(z)}{z}\bigg|_{z=2} = \frac{z^2 - 3z + 8}{(z+2)(z+3)}\bigg|_{z=2} = \frac{3}{20}$$

$$C = \frac{(z+2)F(z)}{z}\bigg|_{z=-2} = \frac{z^2 - 3z + 8}{(z-2)(z+3)}\bigg|_{z=-2} = \frac{9}{4}$$

$$D = \frac{(z+3)F(z)}{z}\bigg|_{z=-3} = \frac{z^2 - 3z + 8}{(z-2)(z+2)}\bigg|_{z=-3} = -\frac{26}{15}$$

Therefore

$$F(z) = -\frac{2}{3} + \frac{3}{20}\frac{z}{z-2} + \frac{9}{4}\frac{z}{z+2} - \frac{26}{15}\frac{z}{z+3}$$

THE Z-TRANSFORM

This leads to the following value for $f(k)$ using Table 9.3:

$$f(k) = -\frac{2}{3}\delta(k) + \frac{3}{20}2^k + \frac{9}{4}(-2)^k - \frac{26}{15}(-3)^k$$

∎

EXAMPLE 9.19
Find the inverse Z-transform of the function

$$F(z) = \frac{z^2 - 9}{(z-1)(z-2)^3} \tag{9.49}$$

Observe that this function has a single and a multiple-order pole.

Solution: This function is expanded in partial fraction form as follows

$$F(z) = \frac{z^2 - 9}{(z-1)(z-2)^3} = A + \frac{Bz}{z-1} + \frac{Cz}{z-2} + \frac{Dz^2}{(z-2)^2} + \frac{Ez^3}{(z-2)^3} \tag{9.50}$$

We can find three of these unknown constants using the relations (see Section 6-5)

$$A = F(z)\big|_{z=0} = -\frac{9}{8}$$

$$B = \frac{F(z)(z-1)}{z}\bigg|_{z=1} = \frac{z^2 - 9}{(z-2)^3}\bigg|_{z=1} = 8$$

$$E = \frac{F(z)(z-2)^3}{z}\bigg|_{z=2} = \frac{z^2 - 9}{z-1}\bigg|_{z=2} = -\frac{5}{8}$$

These constants can be combined with the above equations, leaving a relation involving the two remaining constants C and D. One procedure for finding these is to select any two appropriate values for z, thereby creating a set of two equations with two unknowns C and D. In particular, if we choose $z = 3$ and $z = 4$, we obtain the following expressions

$$C + 3D = 2$$

$$C + 2D = -\frac{51}{24}$$

from which we find that

$$C = -\frac{83}{8} \qquad D = \frac{99}{24}$$

The solution is obtained using Table 9.3. The result is

$$f(k) = -\frac{9}{8}\delta(k) + 8u(k) - \frac{83}{8} \times 2^k + \frac{99}{24}(k+1)2^k$$

$$-\frac{5}{8}\frac{(k+1)(k+2)}{2!}2^k$$

$$= -\frac{9}{8}\delta(k) + 8u(k) + \left[-\frac{83}{8} + \frac{99}{24}(k+1) - \frac{5}{8}\frac{(k+1)(k+2)}{2}\right]2^k \quad \blacksquare$$

A more formal method for the evaluation of the constants in the partial-fraction expansion of a function with multiple roots is essentially that discussed for Laplace transform expansions, which is given in (6.52). This method considers the expansion for one of these multiple-order roots to be of the form

$$F(z) = \frac{F_1(z)}{(z-p)^n} = \frac{A_1}{z-p} + \frac{A_2}{(z-p)^2} + \frac{A_3}{(z-p)^3} + \cdots + \frac{A_n}{(z-p)^n} \quad (9.51)$$

The constants A_i are found using the relations

$$A_n = (z-p)^n F(z)\big|_{z=p}$$

$$A_{n-1} = \frac{d}{dz}\left[(z-p)^n F(z)\right]\bigg|_{z=p}$$

$$\vdots \quad \vdots$$

$$A_{n-k} = \frac{1}{k!}\frac{d^k}{dz^k}\left[(z-p)^n F(z)\right]\bigg|_{z=p} \quad (9.52)$$

$$\vdots \quad \vdots$$

$$A_1 = \frac{1}{(n-1)!}\frac{d^{n-1}}{dz^{n-1}}\left[(z-p)^n F(z)\right]\bigg|_{z=p}$$

Table 9.4 is particularly useful in carrying out the details of (9.52).

■ ■ ■

9-6 THE TRANSFER FUNCTION

As has already been noted and will receive additional attention later, the use of the Z-transform provides an important technique in the solution of difference equations. The transfer function plays an important part in the analysis and representation of linear discrete time and time-invariant systems. Consider the circuit shown in Figure 9.11, which was discussed in Chapter 8 (see Figure 8.7). The current was specified by (8.14); that is $y(k) = \beta_0 x(k) + \alpha_1 y(k-1)$. The Z-transform of (8.14) is, using the appropriate entries in Table 9.1,

$$I(z) = \beta_0 V(z) + \alpha_1 z^{-1} I(z) \quad \text{(a)} \quad (9.53)$$

from which

$$I(z) = \frac{\beta_0}{1 - \alpha_1 z^{-1}} V(z) = H(z)V(z) \quad \text{(b)}$$

This equation relates the input-output relation explicitly in the transformed domain of a discrete system. The proportionality function $H(z) = I(z)/V(z)$ is the **system function or transfer function** for the discrete system.

THE Z-TRANSFORM

Table 9.4 Z-transforms for the Expansion (9.51).

Entry Number	$F(z)$	$f(k-1)$		Radius of Convergence
1.	$\dfrac{1}{z-a}$	a^{k-1}	$k \geq 1$	$|z| > a$
2.	$\dfrac{1}{(z-a)^2}$	$(k-1)a^{k-2}$	$k \geq 1$	$|z| > a$
3.	$\dfrac{1}{(z-a)^3}$	$\dfrac{1}{2!}(k-1)(k-2)a^{k-3}$	$k \geq 1$	$|z| > a$
	$\vdots$			
4.	$\dfrac{1}{z-a}$	$-a^{k-1}$	$k \leq 0$	$|z| < a$
5.	$\dfrac{1}{(z-a)^2}$	$-(k-1)a^{k-2}$	$k \leq 0$	$|z| < a$
6.	$\dfrac{1}{(z-a)^3}$	$-\dfrac{1}{2!}(k-1)(k-2)a^{k-3}$	$k \leq 0$	$|z| < a$

Figure 9.11 A series RL circuit.

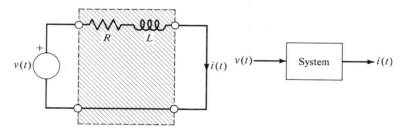

When the input is a unit impulse $\delta(k)$ so that $\mathscr{L}\{\delta(k)\} = V(z) = 1$ (see Example 9.5), the output of the system is equal to its system function $H(z)$. The inverse transform of the system function $H(z)$ is the **impulse response** $h(k)$ of the system. Thus for our circuit with a δ-function excitation

$$I(z) = \frac{\beta_0}{1 - \alpha_1 z^{-1}} \triangleq H(z) \tag{9.54}$$

the inverse transform is

$$i(k) \triangleq h(k) = (\alpha_1)^k \beta_0 \qquad k = 0, 1, 2, \ldots \tag{9.55}$$

EXAMPLE 9.20
Determine the response of the first-order system specified by (9.53) to a unit step sequence.

Solution: The unit step sequence, which is written

$$u(k) = \begin{cases} 1 & \text{for } k = 0, 1, 2, \ldots \\ 0 & \text{for } k < 0 \end{cases}$$

has the Z-transformed value

$$U(z) = \frac{z}{z-1}$$

The response is given by (writing y for i)

$$Y(z) = \frac{\beta_0}{1 - \alpha_1 z^{-1}} \frac{1}{1 - z^{-1}} = \beta_0 \left[\frac{A}{1 - \alpha_1 z^{-1}} + \frac{B}{1 - z^{-1}} \right]$$

where

$$A = \left. \frac{1}{1 - z^{-1}} \right|_{z^{-1} = 1/\alpha_1} = \frac{\alpha_1}{\alpha_1 - 1}$$

$$B = \left. \frac{1}{1 - \alpha_1 z^{-1}} \right|_{z^{-1} = 1} = \frac{1}{1 - \alpha_1}$$

Thus

$$Y(z) = \frac{\beta_0}{1 - \alpha_1} \left(\frac{-\alpha_1}{1 - \alpha_1 z^{-1}} + \frac{1}{1 - z^{-1}} \right)$$

The inverse transform is

$$y(k) = \frac{\beta_0}{1 - \alpha_1} [-\alpha_1(\alpha_1)^k + (1)^k] = \frac{\beta_0}{1 - \alpha_1} (1 - \alpha_1^{k+1})$$

This result is identical with (8.15), which was obtained directly from the difference equation. ∎

We could also find the output in the foregoing example by using the convolution equation. Here we write

$$y(k) = \sum_{n=0}^{k} h(k - n)u(n) = \sum_{n=0}^{k} \beta_0 \alpha_1^{k-n} u(n)$$

The output at successive time steps is

$$y(0) = \beta_0 \alpha_1^0 = \beta_0$$

$$y(1) = \beta_0(\alpha_1 + 1)$$

$$y(2) = \beta_0(\alpha_1^2 + \alpha_1 + 1)$$

$$\vdots \qquad \vdots$$

$$y(k) = \beta_0(\alpha_1^k + \alpha_1^{k-1} + \cdots + 1) = \beta_0 \left(\frac{1 - \alpha_1^{k+1}}{1 - \alpha_1} \right)$$

which is identical with the expression above using the Z-transform method.

Further, since $H(z)$ is the Z-transform of $h(k)$, we can write

$$H(z) = \sum_{k=0}^{\infty} h(k) z^{-k} \tag{9.56}$$

THE Z-TRANSFORM

For the first-order system described above, as already found

$$H(z) = \frac{\beta_0}{1 - \alpha_1 z^{-1}} = \beta_0(1 + \alpha_1 z^{-1} + \alpha_1^2 z^{-2} + \cdots)$$

$$= \sum_{k=0}^{\infty} \beta_0 \alpha_1^k z^{-k} \qquad (9.57)$$

By comparing (9.57) and (9.56), we see that $h(k) = \beta_0(\alpha_1)^k$, the result given by (9.55), as expected.

When systems are interconnected, the rules that apply to continuous systems are also applicable for discrete systems. For example, if two systems with impulse response functions $h_1(k)$ and $h_2(k)$, respectively, are connected in cascade, the combined impulse response of the total system is

$$\boxed{h(k) = h_1(k) * h_2(k)} \qquad \text{(a)} \qquad (9.58)$$

In the Z-domain this specifies that

$$\boxed{H(z) = H_1(z)H_2(z)} \qquad \text{(b)}$$

EXAMPLE 9.21
Find the transfer function for the system shown in Figure 9.12a.

Solution: Consider initially the portion of the system between $x(k)$ and $y_1(k)$. This is described by the difference equation

$$y_1(k) = \beta x(k) + \alpha_1 y_1(k-1)$$

The Z-transform of this expression is

$$Y_1(z) = \beta X(z) + \alpha_1 z^{-1} Y_1(z)$$

from which

$$Y_1(z) = \frac{\beta}{1 - \alpha_1 z^{-1}} X(z) = H_1(z)X(z)$$

The portion of the system between $y_1(k)$ and $y(k)$ is described by a similar equation whose Z-transform is

$$Y(z) = \frac{\beta}{1 - \alpha_1 z^{-1}} Y_1(z) = H_2(z)Y_1(z)$$

We substitute the known expression for $Y_1(z)$ into this final expression to obtain

$$Y(z) = \left(\frac{\beta}{1 - \alpha_1 z^{-1}}\right)\left(\frac{\beta}{1 - \alpha_1 z^{-1}}\right) X(z) = H_1(z)H_2(z)X(z) = H(z)X(z)$$

where

$$H(z) = H_1(z)H_2(z) = \left(\frac{\beta}{1 - \alpha_1 z^{-1}}\right)^2$$

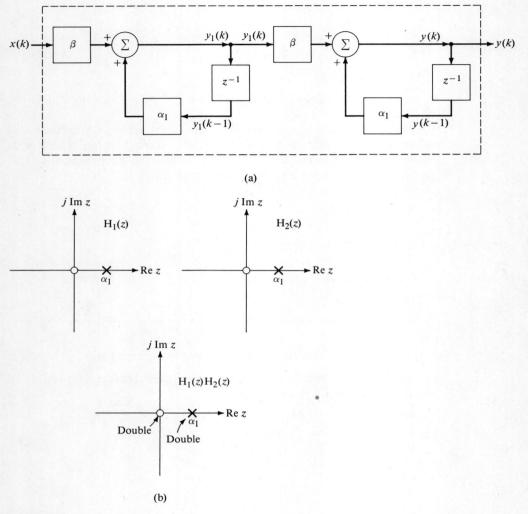

Figure 9.12
Two first-order discrete systems in cascade.

The zero-pole configurations for each system and for the combined system are shown in Figure 9.12b. ∎

In the general case of an nth-order discrete system for which the difference equation is of the form

$$y(k) + a_1 y(k-1) + \cdots + a_n y(k-n)$$
$$= b_0 x(k) + b_1 x(k-1) + \cdots + b_m x(k-m) \tag{9.59}$$

Figure 9.13
A second-order discrete system.

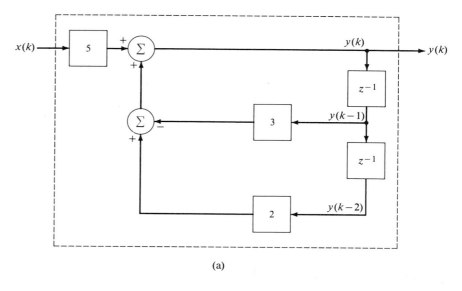

(a)

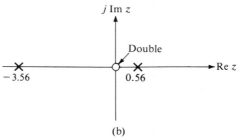

(b)

we obtain, by taking the Z-transform of all terms in this equation and solving for the ratio $Y(z)/X(z)$,

$$H(z) = \frac{Y(z)}{X(z)} = \frac{b_0 + b_1 z^{-1} + \cdots + b_m z^{-m}}{1 + a_1 z^{-1} + \cdots + a_n z^{-n}} = \frac{\sum_{k=0}^{m} b_k z^{-k}}{1 + \sum_{k=1}^{n} a_k z^{-k}} \quad (9.60)$$

This relation shows that if we know $H(z)$ of a system, then the output to any input $X(z)$ [or equivalently $x(k)$] can be determined.

EXAMPLE 9.22
Find the transfer function for the second-order system shown in Figure 9.13a.

Solution: The difference equation describing the system is

$$y(k) + 3y(k-1) - 2y(k-2) = 5x(k)$$

Figure 9.14
One possible way of representing three types of filters.

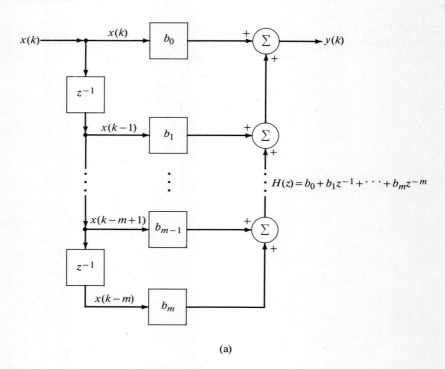

(a)

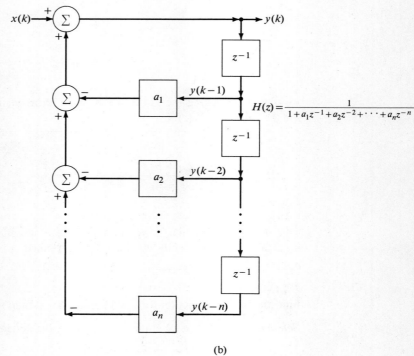

(b)

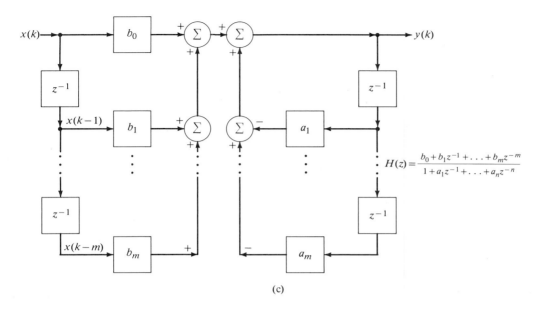

(c)

The transfer function is written directly by using (9.60) and is

$$H(z) = \frac{5}{1 + 3z^{-1} - 2z^{-2}}$$

The zero-pole configuration is shown in Figure 9.13b. ∎

If we set $a_1, a_2, \ldots a_n$ equal to zero, (9.59) becomes

$$y(k) = b_0 x(k) + b_1 x(k-1) + \cdots + b_m x(k-m) \qquad (9.61)$$

This defines an **FIR filter,** as already noted in (8.5). The block diagram of such a system is given in Figure 9.14a.

If we set $b_0 = 1$ and $b_1, b_2, b_3 \ldots b_m$ equal to zero, the difference equation becomes

$$y(k) + a_1 y(k-1) + \cdots + a_n y(k-n) = x(k) \qquad (9.62)$$

This equation, as noted in (8.4), describes an nth-order **IIR filter.** A block diagram representation of this equation is shown in Figure 9.14b.

Finally, if none of the constants in (9.59) is zero, the block diagram representation is that shown in Figure 9.14c.

■ ■ ■

9-7 TRANSFER FUNCTION AND POLE LOCATIONS

It is important to know the effect on the output of the location of the poles belonging to the transfer function of the system and those belonging to the input signal. To study this we begin with the general function given by (9.60), but we

CHAPTER 9

assume for simplicity that all poles belonging to the system are distinct. We write (9.60), in the form

$$Y(z) = H(z)X(z)$$
$$= \frac{A(z)}{(z - p_{h1})(z - p_{h2}) \cdots (z - p_{hn})} \cdot \frac{B(z)}{(z - p_{x1})(z - p_{x2}) \cdots (z - p_{xm})} \quad (9.63)$$

where the poles p_{hi} belong to the transfer function and the poles p_{xi} belong to the input function. Now express this equation as

$$Y(z) = a_0 + \left[\frac{a_1 z}{z - p_{x1}} + \frac{a_2 z}{z - p_{x2}} + \cdots + \frac{a_m z}{z - p_{xm}}\right]$$
$$+ \left[\frac{b_1 z}{z - p_{h1}} + \frac{b_2 z}{z - p_{h2}} + \cdots + \frac{b_n z}{z - p_{hn}}\right] \quad (9.64)$$

This shows that the response function depends on the poles of the transfer function and those due to the input function.

Take the inverse Z-transform of this last equation, with the result that

$$y(k) = a_0 \delta(k) + a_1 (p_{x1})^k + a_2 (p_{x2})^k + \cdots + a_m (p_{xm})^k$$
$$+ b_1 (p_{h1})^k + b_2 (p_{h2})^k + \cdots + b_n (p_{hn})^k \quad (9.65)$$

This expression shows that the location of the poles is very important in system consideration. If we would like to reject the effect of any particular pole or, equivalently, to cause any particular term, say $a_i(p_{xi})^k$, to be zero, we must add a zero to the transfer function at the point p_{xi}. This requirement is obvious from an inspection of (9.63), since the added zero will cancel the pole.

We can infer from (9.65) that if the poles p_{hi} are close to the origin (for stable systems they must be less than unity), then the output signal will closely resemble the input and will do so in a short time. On the other hand, if the poles p_{hi} are located close to the unit circle, the system response will not resemble the input for a long time. This means that the poles located close to the unit circle are the **dominant poles** and that the time response of a linear time-invariant discrete system is directly dependent on the location of these poles. As the poles approach the unit circle, the system response depends less and less on the form of the input. If the poles extend beyond the unit circle, the system is unstable.

EXAMPLE 9.23
Ascertain the constants a and b in order that the discrete system characterized by the second-order difference equation with a unit step input

$$y(k) = u(k) + ay(k-1) + by(k-2) \quad (9.66)$$

has a transfer function $H(z)$, which can be written in the form

$$H(z) = H_1(z) + H_2(z) \quad (9.67)$$

where $H_1(z)$ contains the dominant mode and $H_2(z)$ contains the nondominant mode.

Solution: The Z-transform of (9.66) yields the relationship

$$Y(z) = H(z)U(z) = \frac{1}{1 - az^{-1} - bz^{-2}} \cdot \frac{1}{1 - z^{-1}}$$

From this we write

$$H(z) = \frac{z^2}{z^2 - az - b} = \frac{z^2}{\left[z - \left(\frac{a}{2} + \frac{\sqrt{a^2 + 4b}}{2}\right)\right]\left[z - \left(\frac{a}{2} - \frac{\sqrt{a^2 + 4b}}{2}\right)\right]}$$

An examination of this equation shows that to establish dominance we should choose a to be unity and choose b to be small. We will select $a = 1$ and $b = -0.1$, and this equation becomes

$$H(z) = \frac{z^2}{(z - 0.887)(z - 0.112)} = A + \frac{Bz}{z - 0.887} + \frac{Cz}{z - 0.112}$$

The constants B and C are found to be

$$B = H(z)\frac{(z - 0.887)}{z}\bigg|_{z=0.887} = 1.144$$

$$C = H(z)\frac{(z - 0.112)}{z}\bigg|_{z=0.112} = -0.144$$

Also, setting $z = 0$ in the expression for $H(z)$ yields for A, $A = 0$. Thus

$$H(z) = \frac{1.144z}{z - 0.887} - \frac{0.144z}{z - 0.112} = H_1(z) + H_2(z) \tag{9.68}$$

$H_1(z)$ contains the dominant pole, and $H_2(z)$ contains the nondominant pole.

Of course, once $H(z)$ in the form of (9.68) is known, we can deduce the corresponding impulse response function

$$h(k) = h_1(k) + h_2(k) = 1.144(0.887)^k - 0.144(0.112)^k$$

The response to a delta function input for the dominant and nondominant modes is shown in Figure 9.15. It is apparent that the response due to the dominant pole approaches its steady-state value (zero) at a slower rate than the response due to the nondominant pole. The system response to any excitation $x(k)$ would then be written in the form

$$y(k) = \sum_{n=0}^{\infty} h(n)x(k - n) = \sum_{n=0}^{\infty} h_1(n)x(k - n) + \sum_{n=0}^{\infty} h_2(n)x(k - n) \tag{9.69}$$

∎

We sometimes wish to suppress a sinusoidal frequency component of ω_0, a component that might be contained in the input signal. Such a sinusoidal component produces a Z-transformed configuration having a denominator of the form $1 - 2z^{-1}\cos\omega_0 + z^{-2}$ (see Table 9.3). Hence if we can create a transfer

Figure 9.15
The behavior of the response due to dominant and non-dominant poles.

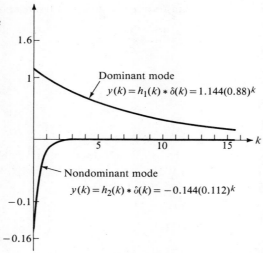

function of the form $H(z) = H_1(z)(1 - 2z^{-1} \cos \omega_0 + z^{-2})$, the poles created by this sinusoidal component in the input function will be cancelled.

The discussion of the system function in this section can be summarized in the following observations:

1. First- and multiple-order poles inside the unit circle represent stable terms.
2. First- and multiple-order poles outside the unit circle represent unstable terms.
3. First-order poles on the unit circle represent marginally stable terms, but multiple-order poles on the unit circle represent unstable terms.
4. A system is no more stable than its least stable part.
5. Zeros can appear on any part of the plane.

■ ■ ■

9-8 FREQUENCY RESPONSE OF SYSTEMS

Suppose that the input to the system is a function of z^k. Then, using the convolution property of system response, the output is given by

$$y(k) = z^k * h(k) = \sum_{n=0}^{\infty} h(n) z^{k-n} = z^k \sum_{n=0}^{\infty} h(n) z^{-n}$$
$$\triangleq z^k H(z) \tag{9.70}$$

This is a more general relationship than that discussed in Section 8-5, where the input to the system was $\exp(j\omega k)$. If we set $z = \exp(j\omega)$ in this expression we have

$$y(k) = e^{j\omega k} H(e^{j\omega}) \tag{9.71}$$

THE Z-TRANSFORM

With this chosen form for z, (9.60) becomes

$$H(e^{j\omega}) = \frac{b_0 + b_1 e^{-j\omega} + b_2 e^{-j2\omega} + \cdots + b_m e^{-jm\omega}}{1 + a_1 e^{-j\omega} + a_2 e^{-j2\omega} + \cdots + a_n e^{-jn\omega}} \triangleq H(z)\Big|_{z=e^{j\omega}} \quad (9.72)$$

which is the **frequency response function**.

Recall that our discussion in Section 8-6 showed that $H(e^{j\omega})$ is an even function of ω, whereas the phase function arg $H(e^{j\omega})$ is an odd function of ω. We may write, therefore,

$$|H(e^{j\omega})|^2 = H(e^{j\omega})H(e^{j\omega})^* = H(e^{j\omega})H(e^{-j\omega}) = H(z)H(z^{-1})\Big|_{z=e^{j\omega}} \quad (9.73)$$

The second equality is valid because the coefficients of $H(e^{j\omega})$ are real. Of course, $H(z^{-1})$ possesses the property that its poles and zero are the reciprocals of those of $H(z)$. It follows, therefore, that the poles and zeros of $H(z)H(z^{-1})$ will occur in pairs that are arranged in an inverse manner with respect to the unit circle. If, for example, we have the transfer function

$$H(z) = \frac{b_0 + b_1 z^{-1} + b_2 z^{-2}}{a_0 + a_1 z^{-1} + a_2 z^{-2}} \quad (9.74)$$

the amplitude squared is given by

$$\begin{aligned}|H(z)|^2\Big|_{z=e^{j\omega}} &= H(z)H(z^{-1})\Big|_{z=e^{j\omega}} \\ &= \frac{d_2 z^2 + d_1 z + d_0 + d_1 z^{-1} + d_2 z^{-2}}{c_2 z^2 + c_1 z + c_0 + c_1 z^{-1} + c_2 z^{-2}}\Big|_{z=e^{j\omega}} \\ &= \frac{d_0 + \sum_{k=1}^{2} 2d_k \cos k\omega}{c_0 + \sum_{k=1}^{2} 2c_k \cos k\omega}\end{aligned} \quad (9.75)$$

where

$$c_k = \sum_{n=0}^{n-k} a_n a_{k+n} \qquad d_k = \sum_{m=0}^{m-k} b_m b_{k+m} \quad (9.76)$$

For example, $c_0 = \sum_{n=0}^{2-0} a_n a_{0+n} = a_0 a_0 + a_1 a_1 + a_2 a_2$

$$c_1 = \sum_{n=0}^{2-1} a_n a_{1+n} = a_0 a_1 + a_1 a_2$$

$$c_2 = \sum_{n=0}^{2-2} a_n a_{2+n} = a_0 a_2$$

EXAMPLE 9.24

Find the frequency response of the systems shown in Figures 9.16a, b, and c. Note that a single basic unit is used in different configurations.

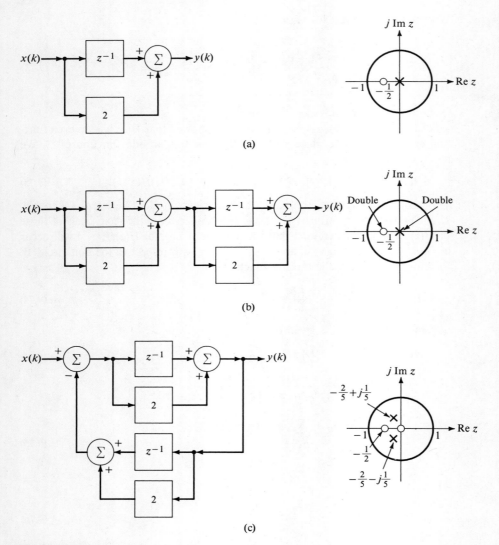

Figure 9.16
Three types of interconnected systems.

Solution: The difference equation that describes the system shown in Figure 9.16a is

$$y(k) = 2x(k) + x(k-1)$$

The system function for this system follows from the Z-transform of this equation

$$H(z) = \frac{Y(z)}{X(z)} = 2 + z^{-1}$$

THE Z-TRANSFORM

The frequency response is obtained from (9.75) and is

$$[H(z)H(z^{-1})|_{z=e^{j\omega}}]^{1/2} = (5 + 4\cos\omega)^{1/2}$$

This expression is plotted in Figure 9.17a. The values of z as ω varies, which are substituted into the expression for $|H(z)|$, are also shown.

Observe that Figure 9.16b, shows two system units connected in cascade. In this configuration the frequency response is given by

$$[|H^2(z)H^2(z^{-1})|_{z=e^{j\omega}}]^{1/2} = (33 + 40\cos\omega + 8\cos 2\omega)^{1/2}$$

This function is plotted in Figure 9.17b.

Figure 9.16c shows two basic units connected in a feedback configuration (refer to Section 2-9). The frequency response is given by

$$\left[\left|\frac{H(z)}{1+H^2(z)} \frac{H(z^{-1})}{1+H^2(z^{-1})}\right|_{z=e^{j\omega}}\right]^{1/2} = \left(\frac{5 + 4\cos\omega}{42 + 48\cos\omega + 10\cos 2\omega}\right)^{1/2}$$

A plot of this expression is given in Figure 9.17c. ∎

Note that if we set $z = e^{j\omega T}$ in (9.70), we obtain the expression

$$y(k) = e^{jk\omega T} H(e^{j\omega T}) \tag{9.77}$$

and (9.72) has the form

$$H(e^{j\omega T}) = \frac{b_0 + b_1 e^{-j\omega T} + \cdots + b_m e^{-jm\omega T}}{1 + a_1 e^{-j\omega T} + \cdots + a_n e^{-jn\omega T}} \triangleq H(z)|_{z=e^{j\omega T}} \tag{9.78}$$

When the sampling time T is different from unity, we observe that the frequency response function is a function of T and the folding frequency is equal to $\omega_s/2 = \pi/T$.

EXAMPLE 9.25

(a) Find the frequency response function of the system shown in Figure 9.16a if the input is sampled every T seconds. (b) Find the output of the system if the input is $x(kT) = \cos k\omega T$.

Solution: From the figure, taking into consideration the sampling interval T, we obtain the difference equation

$$y(kT) = 2x(kT) + x[(k-1)T]$$

The system function for this system is

$$H(z) = \frac{Y(z)}{X(z)} = 2 + z^{-1}$$

and the frequency response function is then given by

$$[H(z)H(z^{-1})|_{z=e^{j\omega T}}]^{1/2} = (5 + 4\cos\omega T)^{1/2}$$

Figure 9.17
Amplitude frequency responses of the systems shown in Figure 9.16.

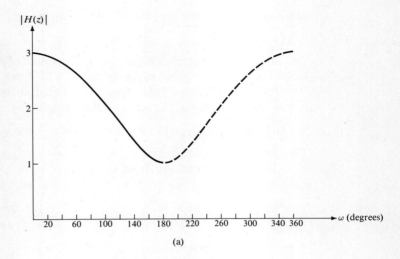

(a)

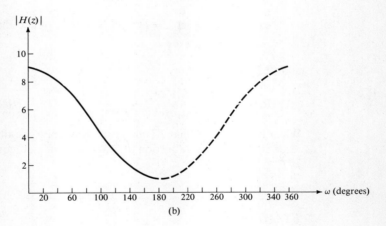

(b)

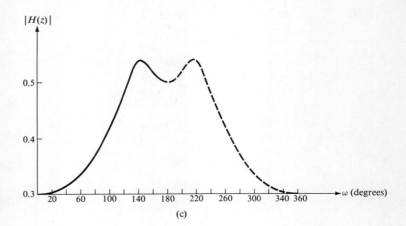

(c)

THE Z-TRANSFORM

Also we find that

$$\theta \triangleq \arg H(z) = \arg\left[\frac{2z+1}{z}\right]\bigg|_{z=e^{j\omega T}} = \arg\left[\frac{2\cos\omega T + 1 + j2\sin\omega T}{e^{j\omega T}}\right]$$

$$= \tan^{-1}\left(\frac{2\sin\omega T}{1+2\cos\omega T}\right) - \omega T$$

The output is the function

$$y(kT) = |H(e^{j\omega T})|\cos(k\omega T + \theta)$$

∎

■ ■ ■

9-9 SOLUTION OF DIFFERENCE EQUATIONS

We will study here the use of Z-transform methods in the solution of linear difference equations with constant coefficients. Although some problems have already been completed, we wish to review this in a more systematic manner by presenting several examples.

EXAMPLE 9.26
Solve the discrete time problem defined by the equation

$$y(k) + 2y(k-1) = 3.5u(k) \tag{9.79}$$

with $y(-1) = 0$, and

$$u(k) = \begin{cases} 1 & k = 0, 1, \ldots \\ 0 & k = -1, -2, \ldots \end{cases}$$

Solution: Begin by taking the Z-transform of both sides of (9.79). This is

$$\mathscr{Z}\{y(k)\} + 2\mathscr{Z}\{y(k-1)\} = 3.5\mathscr{Z}\{u(k)\}$$

By Tables 9.1 and 9.3, we write

$$Y(z) + 2z^{-1}Y(z) = 3.5\frac{z}{z-1}$$

Solve for $Y(z)$ to obtain

$$Y(z) = 3.5\frac{z}{z-1}\frac{z}{z+2} = \frac{7}{6}\frac{z}{z-1} + \frac{7}{3}\frac{z}{z-(-2)}$$

The inverse Z-transform of this equation is (see Example 8.3 also)

$$y(k) = 1.167u(k) + 2.33(-2)^k$$

∎

EXAMPLE 9.27
Solve the difference equation

$$ay(k-2) - by(k-1) + y(k) = u(k-1) \tag{9.80}$$

CHAPTER 9

The values of the constants a and b are to be selected so that the second-order system is: (a) critically damped, (b) underdamped, and (c) overdamped. The starting conditions are assumed to be zero: $y(-2) = y(-1) = 0$.

Solution: The Z-transform of both sides of (9.80) yields for $Y(z)$

$$Y(z) = \frac{z}{z^2 - bz + a} \times \frac{z}{z - 1} \tag{9.81}$$

The denominator of the first factor has two roots, which are specified by

$$z_{1,2} = \frac{b \pm \sqrt{b^2 - 4a}}{2} \tag{9.82}$$

a. Critically Damped Case (two equal real roots). We set $b^2 = 4a$, and select $b = 0.8$ in (9.81). We thus obtain

$$Y(z) = \frac{z}{(z - 0.4)^2} \frac{z}{(z - 1)}$$

This is expanded into the form

$$Y(z) = \frac{Az}{z - 1} + \frac{Bz}{z - 0.4} + \frac{Cz^2}{(z - 0.4)^2} \tag{9.83}$$

By straightforward methods the constants in this expansion are found to be: $A = 1/(0.6)^2$; $B = -(0.4)/(0.6)^2$; $C = -1/0.6$. The inverse transform of this equation is the expression

$$y(k) = \frac{1}{(0.6)^2} [u(k) - (1 + 0.6k)0.4^k] \tag{9.84}$$

The quantity in the bracket, which is the normalized value of $y(k)$, is shown in Figure 9.18. Had we selected the value of b larger than 0.8, the curve would have taken longer to reach the value unity, the steady-state response of the system.

b. Underdamped Case. For $b^2 < 4a$, two complex conjugate roots of (9.82) exist, and these are poles of $Y(z)$. To proceed, write the denominator of $Y(z)$ in the form

$$z^2 - bz + a = (z - ce^{j\theta})(z - ce^{-j\theta}) \tag{9.85}$$

By expanding the right-hand side and equating like powers of z, we find that

$$a = c^2 \quad \text{and} \quad b = 2c \cos \theta$$

This indicates that if a and b are known, c and θ are readily obtained. By combining (9.85) with (9.81), we write

$$Y(z) = \frac{z}{(z - ce^{j\theta})(z - ce^{-j\theta})} \times \frac{z}{z - 1}$$

Figure 9.18
Step response of a second-order system.

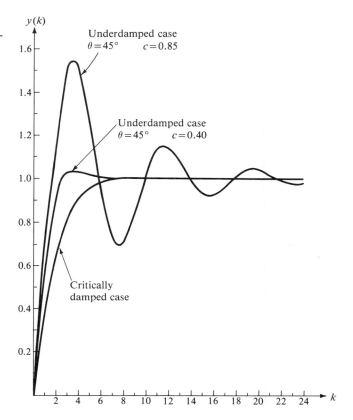

This is now expanded into fractional form, which is

$$Y(z) = \frac{1}{1 - 2c\cos\theta + c^2} \frac{z}{z-1} - \frac{e^{j\theta}}{2j\sin\theta(1 - ce^{j\theta})} \frac{z}{z - ce^{j\theta}}$$

$$+ \frac{e^{-j\theta}}{2j\sin\theta(1 - ce^{-j\theta})} \frac{z}{z - ce^{-j\theta}}$$

The inverse Z-transform of this equation is

$$y(k) = \frac{1}{1 - 2c\cos\theta + c^2} u(k) - \frac{e^{j\theta}}{2j\sin\theta(1 - ce^{j\theta})} c^k e^{jk\theta}$$

$$+ \frac{e^{-j\theta}}{2j\sin\theta(1 - ce^{-j\theta})} c^k e^{-jk\theta}$$

This can be written in more convenient form by writing $1 - ce^{j\theta} = re^{-j\varphi}$, then $1 - ce^{-j\theta} = re^{j\varphi}$, and the equation for $y(k)$ takes the form

$$y(k) = \frac{1}{r^2}\left[u(k) - \frac{rc^k}{\sin\theta}\sin[(k+1)\theta + \varphi]\right] \tag{9.86}$$

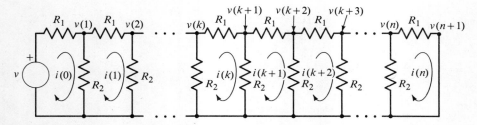

Figure 9.19
Resistor ladder network.

If we choose $c = 0.85$ and $\theta = 45°$, we find that $r = 0.72$ and $\varphi = 56.4°$. The normalized value of $y(k)$ under these conditions is plotted in Figure 9.18. If we had chosen $c = 0.4$ and $\theta = 45°$, then $r = 0.77$ and $\varphi = 21.5°$. The curve resulting from this choice of r, θ is also contained in Figure 9.18. From these two curves we can conclude that as c decreases, the speed with which $y(k)$ approaches its steady-state value increases.

c. Overdamped Case. The requirement in this case is that $b^2 > 4a$, and the two roots given by (9.82) are real and different. The response of the system is easily found by proceeding in the manner discussed above (see Problem 9-9.1). ∎

EXAMPLE 9.28
Determine the voltage $v(k)$ for the resistor ladder circuit shown in Figure 9.19.

Solution: Write the Kirchhoff voltage law for the $k + 1$ loop. This gives

$$R_2 i(k + 2) - (R_1 + 2R_2)i(k + 1) + R_2 i(k) = 0 \tag{9.87}$$

This equation applies for any loop except the input and output loops. The input and output loops are sufficient to determine the two unknowns that arise in the solution of the second-order difference equation.

The Z-transform of this equation results in the expression (see Table 9.1)

$$I(z) = \frac{R_2 i(0) z^2 + R_2 i(1) z - (R_1 + 2R_2) i(0) z}{R_2 z^2 - (R_1 + 2R_2) z + R_2} \tag{9.88}$$

For simplicity, let $R_1 = 1$ and $R_2 = 2$. The equation for $I(z)$ becomes

$$I(z) = \frac{z[i(0)z + i(1) - 2.5 i(0)]}{z^2 - 2.5z + 1} \tag{9.89}$$

The voltage law in the input or first loop gives us the relationship (for $R_1 = 1$ and $R_2 = 2$)

$$i(1) = 1.5 i(0) - 0.5 v \tag{9.90}$$

THE Z-TRANSFORM

Combine this equation with (9.89) and we find

$$I(z) = i(0) \frac{z\left(z - \frac{3}{2}\right) + z\left(\frac{1}{2} - \frac{1}{2}\frac{v}{i(0)}\right)}{z^2 - 2.5z + 1}$$

which can be written

$$I(z) = i(0)\left[\frac{z\left(z - \frac{3}{2}\right)}{(z - 2)(z - 0.5)} + \frac{z\left(0.5 - \frac{0.5v}{i(0)}\right)}{(z - 2)(z - 0.5)}\right] \quad (9.91)$$

The inverse transform of this equation yields an expression for $i(k)$. In detail, the expression is first expanded into partial-fraction form and $i(k)$ is deduced from this. The unknown $i(0)$, which appears in the final expression, can be found from the last or output loop by noting that the solution must be satisfied there also. ∎

■ ■ ■

†9-10 THE TWO-SIDED Z-TRANSFORM. REGION OF CONVERGENCE

The formal definition of the two-sided Z-transform is

$$F_{\text{II}}(z) = \mathscr{Z}_{\text{II}}\{f(k)\} = \sum_{k=-\infty}^{\infty} f(k)z^{-k} \quad (9.92)$$

Note, however, that this equation defines $F(z)$ only for those values of z for which the infinite series converges.

From this definition we find that the sequence

$$f(k) = \begin{cases} \left(\frac{1}{4}\right)^k & k \geq 0 \\ \left(\frac{1}{2}\right)^{-k} & k < 0 \end{cases} \quad (9.93)$$

has the following two-sided Z-transform

$$\begin{aligned} F_{\text{II}}(z) &= \cdots + \left(\tfrac{1}{2}\right)^2 z^2 + \left(\tfrac{1}{2}\right)z + 1 + \tfrac{1}{4}z^{-1} + \left(\tfrac{1}{4}\right)^2 z^{-2} + \cdots \\ &= 1 + \tfrac{1}{2}z + \left(\tfrac{1}{2}\right)^2 z^2 + \cdots + 1 + \tfrac{1}{4}z^{-1} + \left(\tfrac{1}{4}\right)^2 z^{-2} + \cdots - 1 \\ &= \frac{1}{1 - \tfrac{1}{2}z} + \frac{1}{1 - \tfrac{1}{4}z^{-1}} - 1 = \frac{7z}{(2 - z)(4z - 1)} \end{aligned} \quad (9.94)$$

The first summation converges for $|z| < 2$, and the second converges for $|z| > 0.25$. Thus the infinite series of (9.94) converges in the ring $0.25 < |z| < 2$.

The one-sided Z-transform of the sequence $f(k) = \left(\frac{1}{4}\right)^k - 2^k$, $k \geq 0$, is algebraically identical with the two-sided Z-transform of (9.94). However, the region of absolute convergence for this sequence is $|z| > 2$, which is quite different from that for the sequence of (9.93). This observation tells us that to define a sequence uniquely, the region of convergence must be specified or, alternatively, we must know if the sequence is one-sided or two-sided. This requirement was previously pointed out in Section 9-2.

Inverse Transformation (the series method). Equation (9.92) shows that a sequence $f(k)$ can be found if we know $F(z)$ in expanded form involving positive and negative powers of z. The expansion requires that the series converge absolutely within the region of convergence. For example, the transform

$$F(z) = \frac{1}{z-a} \qquad (9.95)$$

can be expanded into the two series

$$F(z) = z^{-1}(1 + az^{-1} + a^2z^{-2} + \cdots) \quad \text{(a)} \qquad (9.96)$$

and

$$F(z) = -\frac{1}{a}\left(1 + \frac{z}{a} + \left(\frac{z}{a}\right)^2 + \cdots\right) \quad \text{(b)}$$

The first expansion has a region of convergence $|z| > a$, with its equivalent sequence (see Table 9.4)

$$f(k) = \begin{cases} a^{k-1} & k \geq 1 \\ 0 & k < 0 \end{cases} \qquad (9.97)$$

The second series has a region of convergence $|z| < a$, with its equivalent sequence

$$f(k) = \begin{cases} -a^{k-1} & k \leq 0 \\ 0 & k > 0 \end{cases} \qquad (9.98)$$

EXAMPLE 9.29
Find the sequence whose Z-transform is given by

$$F(z) = \frac{3}{(1-z)(2z-1)} \qquad (9.99)$$

with the region of convergence $\frac{1}{2} < |z| < 1$.

Solution: Apply the partial-fraction expansion method to this function to write

$$F(z) = \frac{3}{1-z} + \frac{6}{2z-1} = 3(1 + z + z^2 + \cdots)$$

$$+ \frac{6}{2z}(1 + 2^{-1}z^{-1} + 2^{-2}z^{-2} + \cdots) \qquad (9.100)$$

THE Z-TRANSFORM

The region of convergence is $|z| < 1$ for the first summation and $|z| > \frac{1}{2}$ for the second summation. Hence the sequence corresponding to the given transform is

$$f(k) = \begin{cases} 3(\frac{1}{2})^{k-1} & k \geq 1 \\ 3 & k \leq 0 \end{cases} = \begin{cases} -3\delta(k) + 3(\frac{1}{2})^{k-1} & k \geq 0 \\ 3 & k < 0 \end{cases} \qquad (9.101)$$

■

EXAMPLE 9.30

Find the sequence corresponding to the Z-transform

$$F(z) = \frac{-z^2 + 0.4z - 0.4}{(z - 2)(z - 0.2)} \qquad (9.102)$$

with the region of convergence $0.2 < |z| < 2$.

Solution: From the data we recognize that the factor $(z - 2)$ results from the signal with negative k's, the other factor being due to the signal with positive k's. Therefore, we split $F(z)$ into the form

$$F(z) = \frac{A}{z - 2} + F_+(z) \qquad (9.103)$$

We readily find A

$$A = F(z)(z - 2)\big|_{z=2} - F + (z)(z - 2)\big|_{z=2} = \frac{-z^2 + 0.4z - 0.4}{z - 0.2}\bigg|_{z=2}$$

$$A = \frac{-2^2 + 0.4 \times 2 - 0.4}{2 - 0.2} = -2$$

We observe that

$$\frac{-2}{z - 2} = \frac{2}{2 - z} = \frac{1}{1 - \frac{1}{2}z} = 1 + (\tfrac{1}{2})z + (\tfrac{1}{2})^2 z^2 + \cdots$$

which arises from

$$f_-(k) = (\tfrac{1}{2})^{-k} \qquad k \leq 0$$

Next we subtract the quantity $-2/(z - 2)$ from $F(z)$ to find $F_+(z)$, which corresponds to the sequence for $k \geq 0$. By performing this operation, we find

$$F_+(z) = \frac{-z + 0.4}{z - 0.2} = -2 + \frac{z}{z - 0.2}$$

The inverse transform of this function is

$$f_+(k) = -2\delta(k) + (0.2)^k \qquad k \geq 0 \qquad (9.104)$$

CHAPTER 9

Note that for $k = 0$, $f_-(0) = 1$ and $f_+(0) = -1$ and the sequence $f(0) = f_-(0) + f_+(0) = 0$, so that the result is

$$f(k) = \begin{cases} -\delta(k) + (0.2)^k & k \geq 0 \\ \left(\frac{1}{2}\right)^{-k} & k < 0 \end{cases} \qquad (9.105)$$

■

EXAMPLE 9.31
Find the sequences corresponding to the transform

$$F(z) = \frac{2z+1}{3z^2 - 7z + 2} = \frac{1}{2} - \frac{z}{z - \frac{1}{3}} + \frac{1}{2}\frac{z}{z-2} \qquad (9.106)$$

Solution: We observe that the function has two poles: one at $z = 1/3$ and the other at $z = 2$. If we accept the region $|z| > 2$ as the region of absolute convergence, we can expand the given function in the form

$$F(z) = \frac{1}{2} - \left(1 + \frac{1}{3}z^{-1} + \left(\frac{1}{3}\right)^2 z^{-2} + \cdots\right)$$

$$+ \frac{1}{2}(1 + 2z^{-1} + 2^2 z^{-2} + \cdots) \qquad (9.107)$$

The region of convergence for the first series is $|z| > 1/3$; that for the second series is $|z| > 2$. Therefore both series converge in the region $|z| > 2$. The corresponding sequence is

$$f(k) = \frac{1}{2}\delta(k) - \left(\frac{1}{3}\right)^k + \frac{1}{2} \times 2^k \qquad k \geq 0 \qquad (9.108)$$

If the region of convergence is taken as $1/3 < |z| < 2$, we would expand (9.106) in the form

$$F(z) = \frac{1}{2} - \frac{z}{z - \frac{1}{3}} + \frac{1}{2}\frac{z}{z-2} = \frac{1}{2} - \frac{1}{1 - \frac{1}{3z}} + \frac{1}{2}\left(1 - \frac{1}{1 - \frac{z}{2}}\right)$$

$$= \frac{1}{2} - \left[1 + \frac{1}{3}z^{-1} + \left(\frac{1}{3}\right)^2 z^{-2} + \cdots\right]$$

$$- \frac{1}{2}\left[\frac{1}{2}z + \left(\frac{1}{2}\right)^2 z^2 + \cdots\right] \qquad (9.109)$$

The first series converges in the region $|z| > 1/3$ and the second converges in the region $|z| < 2$. Hence the corresponding sequence is

$$f(k) = \begin{cases} \frac{1}{2}\delta(k) - \left(\frac{1}{3}\right)^k & k \geq 0 \\ -\frac{1}{2}\left(\frac{1}{2}\right)^{-k} & k \leq -1 \end{cases} \qquad (9.110)$$

THE Z-TRANSFORM

Finally, if the region of convergence is $|z| < 1/3$, then $F(z)$ has the expansion

$$F(z) = \frac{1}{2} - \left(1 + \frac{1}{3z - 1}\right) + \frac{1}{2}\left(1 - \frac{1}{1 - \frac{z}{2}}\right)$$

$$= \frac{1}{2} - 1 + (1 + 3z + 3^2 z^2 + \cdots) - \frac{1}{2}\left(\frac{1}{2}z + \left(\frac{1}{2}\right)^2 z^2 + \cdots\right) \quad (9.111)$$

The corresponding sequence is

$$f(k) = \begin{cases} \frac{1}{2}\delta(k) & k = 0 \\ (3)^{-k} - \frac{1}{2}(\frac{1}{2})^{-k} & k \leq -1 \end{cases} \quad (9.112)$$

From the above results, we observe the following (see Figure 9.20):

- When the region of convergence is $|z| > 2$, all of the poles are **interior** to contour C and the sequence is valid for $k \geq 0$ **only.**
- When the region of convergence is $1/3 < |z| < 2$, one pole is **interior** to contour C and one is **exterior** to contour C. The sequence is valid for both $k \geq 0$ and $k < 0$.
- When the region of convergence is $|z| < 1/3$, both poles are **exterior** to contour C and the sequence is valid only for $k \leq 0$. ∎

Inverse Transformation (The integral method). The inverse Z-transform is evaluated by using the inversion integral

$$f(k) = \frac{1}{2\pi j} \oint_C F(z) z^{k-1} \, dz \quad (9.113)$$

This expression is valid for either the one-sided or the two-sided Z-transform. To apply this integral, we must select a contour C that lies wholly within the region of convergence of $F(z)$, as shown in Figure 9.20.

An application of the method of residues to the evaluation of this integral leads to

$$f(k) = \begin{cases} \Sigma \text{ (residues of } F(z)z^{k-1}) \text{ at poles inside contour } C, \ k \geq 0 \\ -\Sigma \text{ (residues of } F(z)z^{k-1}) \text{ at poles outside contour } C, \ k < 0 \end{cases} \quad (9.114)$$

As we discussed in connection with the Laplace transform (also in Appendix 1), the residue for a simple pole at $z = z_0$ is

$$(z - z_0) F(z) z^{k-1} \big|_{z = z_0} \quad (9.115)$$

Figure 9.20
The contours for three different convergence regions.

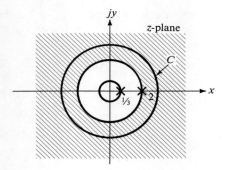

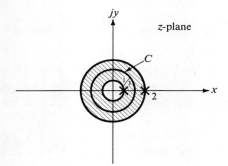

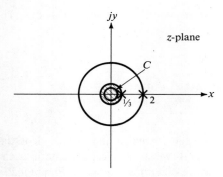

and for an nth-order pole at $z = z_0$, the residue is

$$\left. \frac{1}{(n-1)!} \frac{d^{n-1}}{dz^{n-1}} \left[(z - z_0)^n F(z) z^{k-1} \right] \right|_{z = z_0} \tag{9.116}$$

Let us apply the method of residues to the function of Example 9.30

$$F(z) = \frac{-z^2 + 0.4z - 0.4}{(z - 2)(z - 0.2)} \qquad 0.2 < |z| < 2 \quad \text{(region of convergence)}$$

(9.117)

THE Z-TRANSFORM

The contour C lies between 0.2 and 2; hence the pole at $z = 2$ is exterior to C. Thus for $k < 0$, we obtain

$$f(k) = -F(z)z^{k-1}(z-2)\big|_{z=2}$$

$$= -\left[\frac{(-2)^2 + 0.4 \times 2 - 0.4}{(2 - 0.2)}\right]2^{k-1} = 2^k \qquad k < 0 \qquad (9.118)$$

The poles at $z = 0.2$ and $z = 0$ are interior to C, and so

$$f(k) = \begin{cases} F(z)z^{k-1}(z-0.2)\big|_{z=0.2} = \left[\dfrac{-(0.2)^2 + 0.4 \times 0.2 - 0.4}{(0.2 - 2)}\right](0.2)^{k-1} \\ \qquad\qquad\qquad\qquad\qquad = (0.2)^k \qquad k > 0 \\ F(z)z^{-1}\big|_{z=1} = \left[\dfrac{-0^2 + 0.4 \times 0 - 0.4}{(0-2)(0-0.2)}\right] = -1 \qquad k = 0 \end{cases}$$

or equivalently

$$f(k) = -\delta(k) + (0.2)^k \qquad k \geq 0 \qquad (9.119)$$

Observe that the pole at $z = 0$ appears only for the value $k = 0$ and thus has to be treated separately. The final expression is obtained by combining (9.118) and (9.119). It is

$$f(k) = \begin{cases} -\delta(k) + (0.2)^k & k \geq 0 \\ 2^k & k < 0 \end{cases} \qquad (9.120)$$

which is identical to (9.105), as it must be.

EXAMPLE 9.32
Find the sequence $f(k)$, given the function

$$F(z) = \frac{z(2z-1)}{(z-1)^2(z-2)} \qquad 1 < |z| < 2 \quad \text{region of convergence}$$

Solution: We consider the integrand of the inversion integral

$$F(z)z^{k-1} = \frac{z^k(2z-1)}{(z-1)^2(z-2)}$$

For $k \geq 0$, only the pole at $z = 1$ is inside the contour of integration. The residue at this pole is

$$\frac{d}{dz}\left(\frac{z^k(2z-1)}{z-2}\right)\bigg|_{z=1} = -3 - k$$

From this we have

$$f(k) = -3 - k \qquad k \geq 0$$

The pole outside of contour C is at $z = 2$, with the result

$$\text{Res} \left. \frac{z^k(2z - 1)}{(z - 1)^2} \right|_{z=2} = 3(2)^k \qquad k < 0$$

From this

$$f(k) = 3(2)^k \qquad k < 0$$

These results are combined to specify the desired sequence

$$f(k) = \begin{cases} -3 - k & k \geq 0 \\ 3(2)^k & k < 0 \end{cases}$$

∎

EXAMPLE 9.33
Suppose that the function of Example 9.32 is a one-sided function. Find $f(k)$.

Solution: There are two poles, a double pole at $z = 1$ and a single pole at $z = 2$. The residues are as follows:

at $z = 1$

$$\text{Res} = \frac{d}{dz} \left[\frac{z^k(2z - 1)}{z - 2} \right]_{z=1} = -3 - k$$

at $z = 2$

$$\text{Res} = \left. \frac{z^k(2z - 1)}{(z - 1)^2} \right|_{z=2} = 3(2)^k$$

The final result is

$$f(k) = -3 - k + 3(2)^k \qquad \text{for } k \geq 0$$

In this particular case, we could have proceeded in the manner of Section 9-5 by expressing $F(z)$ in partial-fraction form

$$\frac{F(z)}{z} = \frac{A}{z - 1} + \frac{B}{(z - 1)^2} + \frac{C}{z - 2}$$

where

$$A = \frac{d}{dz} \left. \frac{2z - 1}{z - 2} \right|_{z=1} = -3$$

$$B = (z - 1)^2 F(z) \big|_{z=1} = -1$$

$$C = (z - 2) F(z) \big|_{z=2} = 3$$

Therefore

$$F(z) = -3 \frac{z}{z - 1} - \frac{z}{(z - 1)^2} + 3 \frac{z}{z - 2}$$

Using Table 9.3 we then have

$$f(kT) = -3 - k + 3(2)^k$$

which agrees with the results above.

■ ■ ■

†9-11 RELATIONSHIP OF THE Z-TRANSFORM TO THE FOURIER AND LAPLACE TRANSFORMS

The relationship between the Laplace transform of a continuous function and the Z-transform of a sequence of samples of a function at time instants ..., $-T$, 0, T, $2T$, ... is readily developed. Let $f(t)$ be a continuous function being sampled at time instants ..., $-T$, 0, T, $2T$, ... The sampled function is [see (7.2)]

$$f_s(t) = f(t) \operatorname{comb}_T(t) = \sum_{n=-\infty}^{\infty} f(nT) \delta(t - nT) \tag{9.121}$$

The Laplace transform of this expression is

$$F_s(s) = \mathscr{L}\{f_s(t)\} = \mathscr{L}\left\{\sum_{n=-\infty}^{\infty} f(nT)\delta(t - nT)\right\}$$

$$= \sum_{n=-\infty}^{\infty} f(nT)e^{-nsT} \tag{9.122}$$

If we make the substitution $z = \exp(sT)$, then

$$F_s(s)\big|_{z = e^{sT}} = \sum_{n=-\infty}^{\infty} f(nT)z^{-n} \triangleq F(z) \tag{9.123}$$

where $F(z)$ is the Z-transform of the sequence of samples of $f(t)$—namely, $f(nT)$ with $n = 0, \pm 1, \pm 2, +\ldots$.

Several observations follow from this development. It shows that the Z-transform may be viewed as the Laplace transform of the sampled time function $f(t)$, with an appropriate change of variable. Also, as independently developed, $F(z)$ is the transform of a sequence of values $\{f(n)\}$, for $n = \ldots, -2, -1, 0, 1, 2, \ldots$. Another aspect of the development above is that with the transformation $z = \exp[sT]$, the complex s-plane maps into the complex z-plane, with the imaginary axis $\operatorname{Re}(s) = 0$ mapping onto the unit circle $|z| = 1$ in the z-plane. Further, the left-hand s-plane $\operatorname{Re}(s) < 0$ corresponds to the interior of the unit circle $|z| = 1$ in the z-plane.

If we restrict s to the $j\omega$-axis in the s-plane, then $F_s(s)$ becomes $F_s(j\omega)$, which is the Fourier transform of the sample function $f_s(t)$. In this case (9.123) becomes

$$F_s(j\omega) = \sum_{n=-\infty}^{\infty} f(nT)e^{-jn\omega T} = F(z)\big|_{z = e^{j\omega T}} \tag{9.124}$$

However, $F_s(j\omega)$ is periodic with period $2\pi/T$ (see also Section 7-1); that is, $F_s(j\omega) = F_s[j(\omega + 2\pi/T)]$, and in evaluating $F(z)$, z is restricted to the unit circle. One period of $F_s(j\omega)$ can be obtained by evaluating $F(z)$ once around the unit circle.

Correspondingly, the inversion integral in the Z-domain can be developed from the inversion integral for the two-sided Laplace transform. This development is in addition to the development of the inversion integral using the Cauchy Integral theorem given in Section 9-4. We begin here with the Laplace inversion integral

$$f(t) = \frac{1}{2\pi j} \oint_C F(s)e^{st}\,ds \qquad (9.125)$$

where the contour C is any path from $\sigma - j\infty$ to $\sigma + j\infty$ within the region of convergence. Further, to be consistent with the above discussion of the periodicity of $F_s(j\omega)$, we will break the contour into individual sections $\ldots$, $-3\pi/T < \omega < -\pi/T$; $-\pi/T < \omega < \pi/T$; $\ldots$. Now write $f(nT)$ instead of $f(t)$ since we are interested in $F(z)$, and (9.125) is written

$$f(nT) = \frac{1}{2\pi j} \sum_{k=-\infty}^{\infty} \int_{\sigma+j(2k-1)\pi/T}^{\sigma+j(2k+1)\pi/T} F(s)e^{nsT}\,ds \qquad (9.126)$$

Now replace the dummy variable s by $s + j(2\pi k/T)$, noting that $\exp[j2\pi nk] = 1$. Then

$$f(nT) = \frac{1}{2\pi j} \sum_{k=-\infty}^{\infty} \int_{\sigma-j\pi/T}^{\sigma+j\pi/T} F\!\left(s + j\frac{2\pi k}{T}\right) e^{nsT}\,ds$$

Upon interchanging the order of the summation and integration

$$f(nT) = \frac{T}{2\pi j} \int_{\sigma-j\pi/T}^{\sigma+j\pi/T} \frac{1}{T} \sum_{k=-\infty}^{\infty} F\!\left(s + j\frac{2\pi k}{T}\right) e^{nsT}\,ds$$

$$= \frac{T}{2\pi j} \int_{\sigma-j\pi/T}^{\sigma+j\pi/T} F^*(s)e^{nsT}\,ds \qquad (9.127)$$

since $F^*(s + j(2\pi k/T)) = (1/T)\sum_{k=-\infty}^{\infty} F(s + j(2\pi k/T))$. When $z = \exp[sT]$, then $dz = Tz\,ds$, and since the vertical line from $-\pi/T < \omega < \pi/T$ corresponds to a circle of radius $\exp[\sigma T]$ in the z-plane, and $F^*(s)$ is equivalently $F(z)$, then

$$f(nT) = \frac{1}{2\pi j} \oint F(z)z^{n-1}\,dz \qquad (9.128)$$

■ ■ ■

†9-12 THE Z-TRANSFORM, TAYLOR SERIES, AND FOURIER SERIES

An important characteristic of the Z-transform is that it provides a bridge between the Taylor series and the Fourier series. If we consider an analytic function $Y(z)$ of the complex variable z, we can expand this function in a Taylor series about the point $z^{-1} = 0$ to obtain the quantity

$$Y(z) = \sum_{n=0}^{\infty} a_n z^{-n}$$

The radius of convergence of this series extends from $z^{-1} = 0$ to the first singular point, say z_0^{-1}. That is, the region of convergence of the Taylor series of $Y(z)$ is the region in the z-plane outside the circle of radius $|z_0|$; hence the region of convergence for all points z is such that $|z^{-1}| < |z_0^{-1}|$ or equivalently $|z| > |z_0|$.

If we write the Taylor series expansion for points on the unit circle $z = \exp[-j\theta]$, we have

$$Y(e^{-j\theta}) = \sum_{n=0}^{\infty} a_n e^{jn\theta} = \sum_{n=0}^{\infty} a_n (\cos n\theta + j \sin n\theta)$$

which is in the form of a complete Fourier series expansion in angle θ. Three cases can occur. In the first case, the singular point z_0 is inside the unit circle in the z-plane. In this case the function is analytic on the unit circle, and the Fourier series is an analytic representation of this analytic function. In the second case, the singular point z_0 is outside the unit circle. Now the Taylor series does not represent the function and need not be considered. In the third case, the singular point lies on the unit circle, and the Taylor series will not converge at some or all points on the unit circle. Thus the Taylor series defines an analytic function that is differentiable to any order outside the unit circle but becomes nonanalytic at some or all points on the unit circle. The Fourier series in θ is the Taylor series for z on the unit circle and represents a function in the variable θ, which is nonanalytic at some or all points in its range $-\pi \leq \theta \leq \pi$. A small modification of the Fourier coefficients that would move the singular point z_0 from a point just on the unit circle to a point just inside the unit circle would change a nonanalytic Fourier representation to an analytic one. When this is done, it changes the given nondifferentiable function θ to one that can be differentiated any number of times.

■ ■ ■

†9-13 STABILITY OF TIME-INVARIANT DISCRETE SYSTEMS

Several different methods are available for testing the stability of a discrete system. We will consider only the technique that involves the use of the bilinear transformation and the Routh-Hurwitz test.

As we know, if the poles of the transfer function of a discrete system are contained inside the unit circle, the system is stable. Therefore, if we can find a transformation from the variable z to a variable w such that the inside of the unit circle in the z-domain is transformed into the left half of the w-plane, we can then apply the Routh-Hurwitz criterion as a test for stability. The bilinear transformation is such a transformation function where

$$z = \frac{1+w}{1-w} \qquad w = \frac{z-1}{z+1} \qquad (9.129)$$

To examine the properties of this transformation, we select certain values of z and then find the corresponding values of w. These values are contained in the following tabulation.

z	w
1	0
j	$+j$
$-j$	$-j$
-1	$\pm j\infty$

These results show that some points on the unit circle in the z-plane transform to some points on the imaginary axis on the w-plane. Further, if we set $z = 2$, $w = 1/3$ (in fact, for any value of z greater than 1), we will find these will map onto the right half of the w-plane. Correspondingly, any value of $|z|$ less than 1, which corresponds to the inside of the unit circle, will map onto the left half of the w-plane. This proves that the bilinear transformation possesses the requisite properties for our needs.

In the general case, therefore, given the transfer function of a time-invariant system that has the general form

$$H(z) = \frac{b_0 + b_1 z + \cdots + b_m z^m}{a_0 + a_1 z + \cdots + a_n z^n} \qquad (9.130)$$

then, upon applying the bilinear transformation to the characteristic equation

$$C(z) = a_0 + a_1 z + a_2 z^2 + \cdots + a_n z^n = 0 \qquad (9.131)$$

we will obtain a function $C(w) = 0$ to which the Routh-Hurwitz test can be applied.

EXAMPLE 9.34
Determine the stability of a discrete system whose characteristic equation is given by

$$C(z) = 4z^2 + 5z + 3$$

Solution: Introduce the bilinear transformation of (9.129) into $C(z)$. This yields

$$C(w) = 4\left(\frac{1+w}{1-w}\right)^2 + 5\left(\frac{1+w}{1-w}\right) + 3$$

Expand this expression and focus attention on the equation

$$C(w) = 2w^2 + 2w + 12 = 0$$

The Routh-Hurwitz schedule for this equation is

1	6
1	0
6	0
0	

Since all terms in the first column have the same sign (that is, there are no changes in sign), all of the roots of $C(w)$ are in the left half of the w-plane. This implies that all the roots of $C(z)$ are inside the unit circle; hence the system is stable. ∎

EXAMPLE 9.35
Determine the stability of a discrete system whose characteristic equation is

$$C(z) = 4z^3 - 4z^2 - 7z - 3 = 0$$

Solution: Apply the bilinear transformation to $C(z)$ to obtain

$$C(w) = 4\left(\frac{1+w}{1-w}\right)^3 - 4\left(\frac{1+w}{1-w}\right)^2 - 7\left(\frac{1+w}{1-w}\right) - 3 = 0$$

Expanding this expression leads to

$$C(w) = \frac{4w^3 + 14w^2 + 24w - 10}{(1-w)^3} = 0$$

Because there is a negative sign in the numerator, there is at least one root of $C(w) = 0$ in the right half of the w-plane, rendering this an unstable system.

REFERENCES

1. Ahmed, N., and T. Natarajan, *Discrete-Time Signals and Systems*, Reston, Va.: Reston, 1983.
2. Cadzow, J. A. *Discrete Time Systems*. Englewood Cliffs, N.J.: Prentice-Hall, 1973.
3. DeRusso, P. M., R. J. Roy, and C. M. Close. *State Variables for Engineers*. New York: Wiley, 1965.
4. Gabel, R. A., and R. A. Roberts. *Signals and Linear Systems*, 2d. ed. New York: Wiley, 1980.
5. Jury, E. I. *Theory and Applications of the Z-Transform Method*. Huntington, N.Y.: Krieger, 1973.
6. Seely, S. *An Introduction to Engineering Systems*. New York: Pergamon Press 1972.
7. Seely, S., and A. D. Poularikas. *Electrical Engineering—Introduction and Concepts*. Beaverton, Ore.: Matrix, 1982.

PROBLEMS

9-1.1 Write $X(z)$ for the sequences $\{x(k)\}$ shown in Figure P9-1.1.

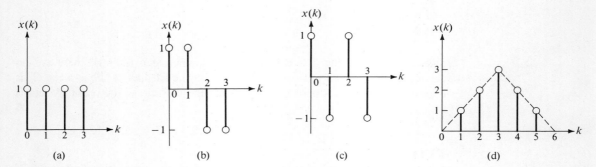

Figure P9-1.1

9-1.2 Determine the Z-transforms of the following sequences:

a. $f(k) = \begin{cases} (\frac{1}{2})^k & k = 0, 1, 2, \ldots \\ 0 & k \text{ negative} \end{cases}$
b. $f(k) = \begin{cases} 0 & k \leq 0 \\ -1 & k = 1 \\ a^k & k = 2, 3, \ldots \end{cases}$

c. $f(t) = \begin{cases} t = 0 & t < 0 \\ t^2 & t \geq 0, \text{ sampled every } T \text{ seconds} \end{cases}$

9-1.3 Find the time sequences $f(k)$ corresponding to the following z-domain functions:

a. $\dfrac{z^2}{(z-1)(z-3)}$ b. $\dfrac{z+4}{(z-2)^2}$ c. $\dfrac{3z^4 - 3z^3 + 6z^2 + 4z + 7}{z(z^2 - 6z + 9)}$

9-2.1 Find the Z-transforms and their region of convergence, and plot the pole-zero configuration of the sequences, for $k \geq 0$,

a. $\{y(k)\} = 1 + k$ b. $\{y(k)\} = k^2$ c. $\{y(k)\} = a^k + a^{-k}$ d. $\{y(k)\} = e^{jk\theta}$

9-2.2 Determine the Z-transform of the function

$$y(k) = \begin{cases} a^k \cos kb & k \geq 0 \quad a > 0 \\ 0 & k < 0 \end{cases}$$

Indicate the regions of convergence and divergence, and show the zeros and poles on the z-plane.

9-2.3 Determine the Z-transforms and their regions of convergence, and plot the pole configuration of the sequences

a. $y(k) = \begin{cases} 2 & \text{for } k = 0, 1, 2, 3, 4 \\ 3^k & \text{for } k = 5, 6, \ldots \end{cases}$ b. $y(k) = \left(\dfrac{1}{3}\right)^k + \left(\dfrac{1}{4}\right)^k$ for $k = 0, 1, 2, \ldots$

9-2.4 Determine the Z-transform and the region of convergence, and plot the pole-zero configuration of the sequence

$$y(k) = \begin{cases} 2^k & 0 \leq k \leq 5 \\ 0 & \text{otherwise} \end{cases}$$

9-2.5 Show that the region of convergence of the sequence $y(k) = a(b)^k$ for $k = 0, 1, 2, \ldots$ depends only on b.

9-2.6 Determine the Z-transform and the region of convergence of the sequence

$y(k) = e^{j\omega k}$ for $k = 0, 1, 2, \ldots$

9-2.7 Find the Z-transforms of the functions noted, when sampled every T seconds.
  a. $f(t) = \cos \omega t\, u(t)$ **b.** $f(t) = a^t \sin \omega t\, u(t)$ **c.** $f(t) = a^t \cos \omega t\, u(t)$
  d. $f(t) = e^{-at} \sin \omega t\, u(t)$ **e.** $f(t) = e^{-at} \cos \omega t\, u(t)$ **f.** $f(t) = a^k u(t)$

9-3.1 Generalize (9.32) and show that $\mathscr{Z}\{k^n y(k)\} = [-z(d/dz)]^n Y(z)$.

9-3.2 Two sequences are given. Find their Z-transforms, their regions of convergence, and their pole-zero configurations. Compare the two results and state your observations.
  a. $y(k) = 2e^{-k}\quad k = 0, 1, 2, \ldots$ **b.** $y(k) = 2e^{-(k-2)}u(k-2)\quad k = 0, 1, 2, \ldots$

9-3.3 Find the Z-transforms and their regions of convergence for the following sequences:
  a. $y(k) = e^{\pm j\omega_0 k} x(k)$ **b.** $y(k) = z_0^k x(k)$
  Compare your results with (9.30).

9-3.4 Find the Z-transform of the sequence given in Figure P9-3.4 and compare the zero-pole configurations between $y(k)$ and $y_{(1)}(k)$.

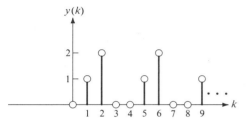

Figure P9-3.4

9-3.5 Given the relationship $\mathscr{Z}\{u(k)\} = 1/(1 - z^{-1})$, use (9.32) to obtain the inverse transform of
$$Y(z) = z^{-1}/(1 - z^{-1})^2.$$

9-3.6 Prove that the operation of convolution is **commutative**; hence we have
$$y(k) = \sum_{n=0}^{k} h(k-n)x(n) = \sum_{n=0}^{k} h(n)x(k-n)$$

9-3.7 Find the Z-transform of the output $y(k)$ of the system shown in Figure P9-3.7 if the input is a unit step function $u(k)$ and $h_3(k) = 0.5^k$.

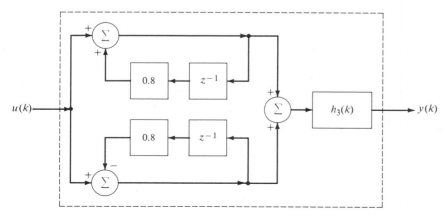

Figure P9-3.7

9-3.8 Use the convolution approach to find the voltage output of the system shown in Figure P9-3.8, if the input current is $i(k) = e^{-0.2k}$, $k = 0, 1, 2, \ldots$. Compare your results with the results obtained for the continuous time case.

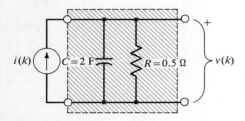

Figure P9-3.8

9-3.9 Show that the Z-transform of the function $y(t)u(t)$ sampled every T sec and shifted to the right by mT sec is

$$\mathscr{Z}\{y(kT - mT)u(kT - mT)\} = z^{-m} \sum_{k=0}^{\infty} y(kT)z^{-k} = z^{-m}Y(z)$$

9-3.10 Find the Z-transform of the shifted function $(kT - 2T)[u(kT - 2T) - u(kT - 5T)]$.

9-3.11 Show that the Z-transform of the function $y(t)u(t)$ sampled every T sec and shifted to the left by mT sec is

$$\mathscr{Z}\{y(kT + mT)u(kT + mT)\} = z^m \left[Y(z) - \sum_{k=0}^{m-1} y(kT)z^{-k} \right]$$

9-3.12 Find the Z-transform of the output of the system shown in Figure P9-3.12.

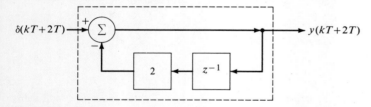

Figure P9-3.12

9-3.13 Find the Z-transforms of the following functions using the time-scaling property:
    a. $f(k) = a^k e^{-\alpha k} u(k)$ **b.** $f(k) = a^k \cos k\omega \, u(k)$ **c.** $f(k) = ka^k u(k)$ **d.** $f(k) = a^k e^{-\alpha k} \sin k\omega \, u(k)$

9-3.14 Show that the Z-transform of the function $y(t) = a^t y(t)u(t)$ sampled at T sec (time-scaling property of sampled signals) is

$$\mathscr{Z}\{a^{kT} y(kT) u(kT)\} = Y(a^{-T} z)$$

9-3.15 Find the Z-transform of the sequences shown in Fig. P9-3.15.

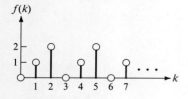 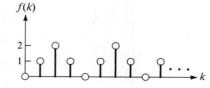

Figure P9-3.15

9-3.16 Show that the Z-transform of the function $y(t) = ty(t)u(t)$ sampled every T seconds is

$$\mathscr{Z}\{kTy(kT)u(kT)\} = -Tz\frac{dY(z)}{dz} \qquad Y(z) = \sum_{k=0}^{\infty} y(kT)z^{-k}$$

9-3.17 Use the result of Problem 9-3.16 to find the Z-transforms of the following functions:
  a. $y(kT) = kTu(kT)$
  b. $y(kT) = (kT)^2 u(kT)$
  c. $y(kT) = kT(kT+1)u(kT)$
  d. $y(kT) = kT(kT-1)u(kT)$

9-3.18 Show that the initial and final values of a sampled function $y(t)$ are given by

  a. $y(k_0 T) = z^{k_0} Y(z)|_{z \to \infty} \qquad Y(z) = \sum_{k=0}^{\infty} y[(k_0 + k)T]z^{-k}$

  b. $\lim_{k \to \infty} y(kT) = \lim_{z \to 1}(z-1)Y(z) \qquad Y(z) = \sum_{k=0}^{\infty} y(kT)z^{-k}$

9-3.19 Show that the Z-transform of the convolution of two sampled functions is given by

$$\mathscr{Z}\{y(kT)\} \triangleq \mathscr{Z}\{h(kT)*x(kT)\} = H(z)X(z)$$

9-3.20 Find the response of the system shown in Figure P9-3.20 in discrete form. Assume $T = 0.5$. Comment on the similarities and differences between the results of this problem and those of Example 9.14. Use the convolution property of the input-output relationship of LTI systems.

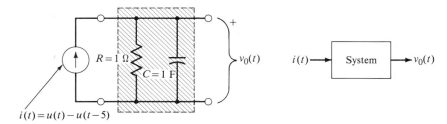

Figure P9-3.20

9-4.1 Find the functions $f(k)$ specified by the following one-sided Z-transforms. Employ fraction expansions.

  a. $\dfrac{1}{(1 - z^{-1})(1 - 2z^{-1})}$
  b. $\dfrac{z^{-1}}{(2 - z^{-1})^2}$
  c. $\dfrac{z^2(z+2)}{z^2 + 4z + 3}$

9-5.1 Determine the inverse functions for the following Z-transforms:

  a. $\dfrac{z}{(z-1)(z-2)}$
  b. $\dfrac{z^2}{(z-1)^2(z-2)}$
  c. $\dfrac{z^3}{(z-1)(z-2)^2}$

9-5.2 Find the inverse Z-transform of the function

$$F(z) = \frac{z^2 - 3z + 8}{(z-2)(z+2)(z+3)}$$

using the expansion form given by (9.51).

9-5.3 Determine the inverse Z-transform of the function

$$F(z) = \frac{z \sin \alpha}{z^2 - (2 \cos \alpha)z + 1} \qquad |z| > 1$$

9-5.4 Find the inverse transforms of the following functions:

a. $F(z) = \dfrac{z+2}{z^2(z-1)}$ **b.** $F(z) = \dfrac{z^3 + 2z^2 + z + 1}{z^3 + z^2 - 5z + 3}$

9-5.5 Find the Z-transforms of the sequences shown in Figure P9-5.5 and then find the inverse Z-transforms.

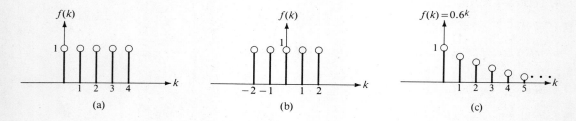

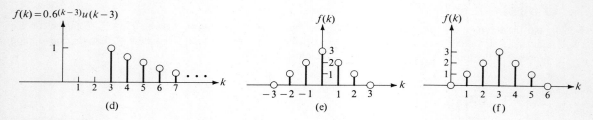

Figure P9-5.5

9-6.1 A discrete time system function is

$$H(z) = \dfrac{2(3 - z^{-1})}{1 - 2z^{-1} + z^{-2}}$$

The input consists of the signal: $x(0) = 0$, $x(1) = 1$, $x(2) = 2$, $x(3) = 0$. Determine $y(k)$.

9-6.2 A discrete time system function is

$$H(z) = \dfrac{2 + z^{-1}}{1 + 3z^{-1} + z^{-2}}$$

The input is the unit step sequence $u(k) = \begin{cases} 1 & k \geq 0 \\ 0 & k < 0 \end{cases}$

Determine the system response.

9-6.3 Write the difference equations for the systems specified by the SFG's given in Figure P9-6.3. From these determine $H(z)$ by finding the impulse response, assuming initially relaxed conditions.

9-6.4 A system is described by its system function

$$H(z) = \dfrac{\alpha_1 z}{z - \beta_1} + \dfrac{\alpha_2 z}{z - \beta_2}$$

β_1 and β_2 are real and have magnitudes less than unity. Show three block diagram configurations that have this $H(z)$.

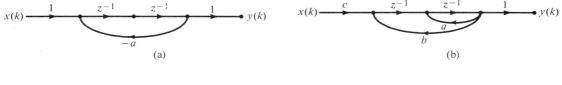

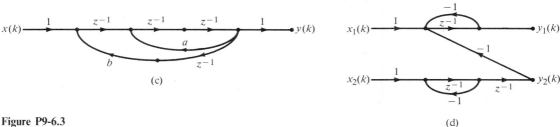

Figure P9-6.3

9-6.5 Draw two SFG's or block diagrams showing the implementation of

$$H(z) = \frac{z(z^2 - 1)}{(z - \frac{1}{2})(z^2 - 1.6z + 0.8)}$$

One realization should be a parallel connection of a first- and second-order section. The second realization should be a cascade of first- and second-order sections.

9-6.6 A system is described by the difference equation

$$y(k) + 3y(k - 1) + 5y(k - 2) = x(k - 2) - 2x(k - 1)$$

a. Draw a signal flow graph that displays this system.
b. The input sequence is $\{x(k)\} = 3^k$. Write the Z-transform of the system.
c. Deduce the system function.
d. If the initial conditions are zero, find the output response.

9-6.7 A system is described by the difference equation

$$y(k + 3) - 5y(k + 2) + 2y(k + 1) + 7y(k) = x(k + 2)$$

With $x(k) = 0$ for $k \geqslant 0$, $y(0) = 0$, $y(1) = 1$, and $y(2) = 5$, determine the output for $k \geq 1$.

9-7.1 Find the transfer functions and locate the poles of the systems shown in Figure P9-7.1. The real constants a and b are less than one.

9-7.2 The transfer function of a system is given by

$$H(z) = \frac{z(z - 0.6)}{(z - 0.8)(z - 0.1)}$$

a. Decompose the transfer function into its dominant and nondominant poles. Find the response of each of these modes to a unit step function, and plot your results.
b. Find the difference equation from the given transfer function for the system.

9-8.1 Sketch the general shape of $H(e^{j\omega})$ as a function of ω, given

$$H(z) = \frac{(z^2 + 1)(z^2 - 1)}{z^4 + 0.8}$$

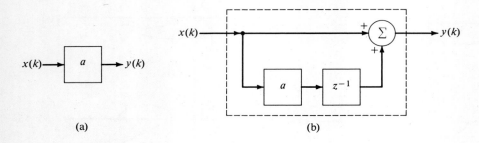
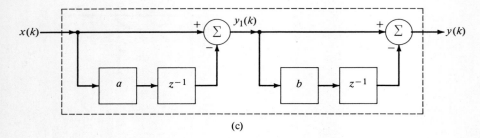
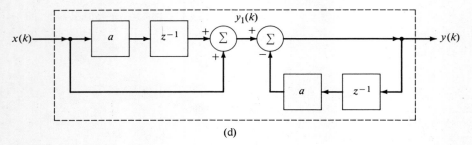

Figure P9-7.1

9-8.2 Find the frequency response of the systems shown in Figure P9-8.2.

9-9.1 A system is defined by the difference equation

$$y(k) = \alpha y(k-1) + (1-\alpha)x(k)$$

Determine the response of the system to the input signal

$$x(k) = \begin{cases} \sin k\omega T & k = 0, 1, 2, \ldots \\ 0 & k < 0 \end{cases}$$

9-9.2 a. Find the voltage equation for the ladder network shown in Figure P9-9.2.
  b. Assume that the resistors all have resistances equal to 1. Find the voltage v_k for $k = 0, 1, 2, \ldots$.

9-9.3 Use the Z-transform technique to find the solution to the following difference equations, and state your observations:
  a. $3y(k) + 2y(k-1) = 5u(k-1)$ $y(-1) = 0$
  b. $3y(k+1) + 2y(k) = 5u(k)$ $y(0) = 0$

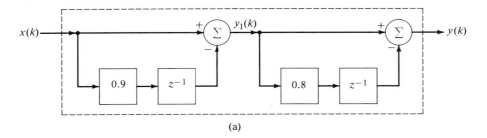

(a)

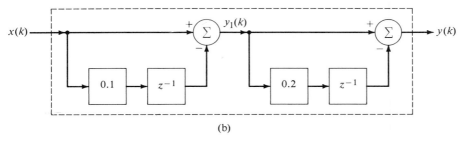

(b)

Figure P9-8.2

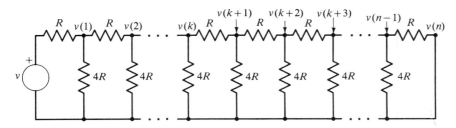

Figure P9-9.2

9-9.4 Use Z-transform methods to find the solution to the following difference equations:

  a. $y(k + 1) + y(k) = -2$

  b. $y(k + 1) + y(k) = 2 \sin \dfrac{k\pi}{4}$

  c. $y(k + 2) - y(k) = 9$

  d. $y(k + 2) - 7y(k + 1) + 10y(k) = 2^k$

9-10.1 Solve Problem 9-5.1 by the use of residues. Use the regions $1 < |z| < 2$ in all three cases.

9-12.1 Use the relation $1 = |z| = |(1 + w)/(1 - w)|$ to show that the real part of w is zero, and that the unit circle on the z-plane is mapped onto the imaginary axis of the w-plane.

9-12.2 Prove that the inside of the unit circle in the z-plane is mapped on the left-hand side of the w-plane by using the transformation (9.129).

CHAPTER 10
DISCRETE TRANSFORMS

10-1 INTRODUCTION

The previous chapters have included discussion of the use of Fourier transform techniques in continuous signal processing studies. However, in many problems the signals may be sampled or the signals may be experimentally obtained, and analytic functions are not available for the integrations involved. There are two general approaches we might pursue in such cases. One method calls for approximating the functions and carrying out the integrations by numerical means. A second method, and one that is used extensively, calls for replacing the continuous Fourier transform by an equivalent **discrete Fourier transform** (DFT) and then evaluating the DFT using the discrete data. However, direct solution of the DFT requires, for each sample, N complex multiplications, N complex additions, and the access to N coefficients $\exp(-j\Omega Tnk)$ that appear in the DFT; hence with 10^4 samples (a small number in some cases), more than 10^8 mathematical operations are required in the solution. It was development of the **Fast Fourier Transform** (FFT), a computational technique that reduces the number of mathematical operations in the evaluation of the DFT to $N \log_2 N$, that made the DFT a useful form (in digital filtering design, for example).

The discrete Fourier transform is particularly suitable for describing phenomena related to a discrete time series. It can be developed from the Fourier integral transform of the continuous waveform from which samples have been taken to form the time series. However, we will proceed by defining the DFT and will later show its relationship to the continuous Fourier transform. Therefore the mathematical properties of the DFT are analogous to those of the Fourier integral transform.

We established in Section 7-1 that if a time function is sampled uniformly in time, its Fourier spectrum is a periodic function. Therefore, corresponding to any sampled function in the frequency domain, a periodic function exists in the time domain. As a result we anticipate that we can relate the sampled signal values in both domains.

As a practical matter, we are only able to manipulate a certain length of signal. Suppose, as is often the case, that the data sequence is available from only a finite time window from $n = 0$ to $n = N - 1$. The transform is discretized for N values by taking samples at the frequencies $2\pi/NT$, where T is the time

DISCRETE TRANSFORMS

interval between sample points. Hence we define the discrete Fourier transform (DFT, written $\mathscr{D}$) of a sequence of N samples $\{f(kT)\}$ for $0 \leq k \leq N - 1$ by the relation

$$F(n\Omega) \triangleq \mathscr{D}\{f(kT)\} = \sum_{k=0}^{N-1} f(kT)e^{-j2\pi nkT/NT}$$

$$= \sum_{k=0}^{N-1} f(kT)e^{-j\Omega Tnk} \qquad n = 0, 1, \ldots, N-1 \quad (10.1)$$

where

N = number of sample values (even number)

T = sampling time interval

$(N - 1)T$ = signal length

$$\Omega = \frac{\omega_s}{N} = \frac{2\pi}{NT}$$

$e^{-j\Omega T}$ = Nth principal root of unity

Observe that with this specification of Ω there are only N distinct values computable by (10.1). We may view the discrete Fourier transform as an evaluation of the Z-transform of the finite sequence $\{f(kT)\}$ at N points in the z-plane equally spaced along a unit circle at angles $k\Omega$ radians.

The Inverse Discrete Fourier Transform (IDFT, written $\mathscr{D}^{-1}$) is related to the direct DFT in much the same way that the direct Fourier transform is related to the inverse Fourier integral. We will show that the IDFT is given by

$$f(kT) \triangleq \mathscr{D}^{-1}\{F(n\Omega)\}$$

$$= \frac{1}{N} \sum_{n=0}^{N-1} F(n\Omega)e^{j\Omega Tnk} \qquad k = 0, 1, \ldots, N-1 \quad (10.2)$$

The relationship between (10.1) and (10.2) is proved as follows:

$$\frac{1}{N} \sum_{n=0}^{N-1} F(n\Omega)e^{j2\pi nkT/NT} = \frac{1}{N} \sum_{n=0}^{N-1} \left[\sum_{m=0}^{N-1} f(mT)e^{-j2\pi nmT/NT} \right] e^{j2\pi nkT/NT}$$

$$= \frac{1}{N} \sum_{m=0}^{N-1} f(mT) \sum_{n=0}^{N-1} e^{-j2\pi(m-k)nT/NT} \quad (10.3)$$

Since we have already established that

$$\sum_{n=0}^{N-1} e^{-j2\pi(m-k)n/N} = \begin{cases} N & m = k \\ 0 & m \neq k \end{cases}$$

(10.3) yields

$$\frac{1}{N} \sum_{n=0}^{N-1} F(n\Omega)e^{j2\pi nkT/NT} = \frac{1}{N} \sum_{n=0}^{N-1} F(n\Omega)e^{j\Omega Tnk} = f(kT)$$

Hence we have the DFT pair

$$
\begin{aligned}
\mathscr{D}\{f(kT)\} &\triangleq F(n\Omega) \\
&= \sum_{k=0}^{N-1} f(kT)e^{-j2\pi nkT/NT} \\
&= \sum_{k=0}^{N-1} f(kT)e^{-jn\Omega kT} \quad k = 0, 1, \ldots, N-1 \quad \text{(a)} \\
\mathscr{D}^{-1}\{F(n\Omega)\} &\triangleq f(kT) \\
&= \frac{1}{N} \sum_{n=0}^{N-1} F(n\Omega)e^{j2\pi nkT/NT} \\
&= \frac{1}{N} \sum_{n=0}^{N-1} F(n\Omega)e^{jn\Omega kT} \quad n = 0, 1, \ldots, N-1 \quad \text{(b)}
\end{aligned}
\quad (10.4)
$$

with both the time and the frequency domain sequences being **periodic**. The periodicity of these sequences stems from the fact that the functions $\exp(-j2\pi nk/N)$ are periodic; that is,

$$e^{-j\pi nk/N} = e^{-j\pi(n+N)k/N} \quad k, n = 0, \pm 1, \pm 2, \ldots \quad (10.5)$$

In general, $F(n\Omega)$ is complex and can be written in the form

$$F(n\Omega) = |F(n\Omega)|e^{j\varphi(n\Omega)}$$

where $|F(n\Omega)|$ and $\varphi(n\Omega) = \text{Arg } F(n\Omega)$ are discrete frequency functions. The plots of $|F(n\Omega)|$ and $\varphi(n\Omega)$ versus $n\Omega$ are referred to as the **amplitude** and **phase** spectra of the sequence $f(kT)$.

Figure 10.1 presents the sine and cosine basis functions for $N = 8$, which are the real and imaginary parts of the complex exponential $\exp(j2\pi nk/N)$.

■ ■ ■

10-2 PROPERTIES OF THE DFT

Because the DFT is related to the Fourier transform, we can anticipate that many of the properties of the DFT will parallel those of the Fourier transform. The important properties will be discussed. For convenience of notation, we replace kT by k and $n\Omega$ by n in the functional forms that follow.

1. Linearity

$$
\begin{aligned}
\mathscr{D}\{af_1(k) + bf_2(k)\} &= a\mathscr{D}\{f_1(k)\} + b\mathscr{D}\{f_2(k)\} \\
&= aF_1(n) + bF_2(n)
\end{aligned}
\quad (10.6)
$$

DISCRETE TRANSFORMS

Figure 10.1
(a) The function $\cos(n2\pi k/8)$.
(b) The function $\sin(n2\pi k/8)$.

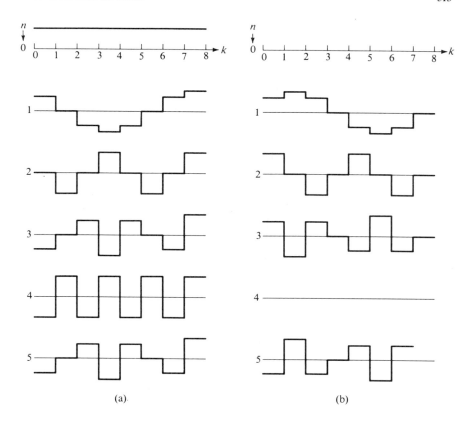

The property is the direct result of (10.4).

2. Symmetry. If $f(k)$ and $F(n)$ are a DFT pair, then

$$\mathscr{D}\left\{\frac{1}{N}F(k)\right\} = f(-n) \tag{10.7}$$

Proof: Rewrite (10.4b) in the form

$$f(-k) = \frac{1}{N}\sum_{n=0}^{N-1} F(n)e^{jn(-k)}$$

Now interchange the parameters k and n to yield

$$f(-n) = \frac{1}{N}\sum_{k=0}^{N-1} F(k)e^{-jnk} \triangleq \mathscr{D}\left\{\frac{1}{N}F(k)\right\}$$

3. Time Shifting. For any real integer

$$\mathscr{D}\{f(k-i)\} = F(n)e^{-jni} \tag{10.8}$$

Proof: Substitute $m = k - i$ into (10.4b) so that

$$f(m) = \frac{1}{N} \sum_{n=0}^{N-1} F(n)e^{jnm}$$

$$f(k-i) = \frac{1}{N} \sum_{n=0}^{N-1} F(n)e^{jn(k-i)} = \frac{1}{N} \sum_{n=0}^{N-1} [F(n)e^{-jni}]e^{jnk} \triangleq \mathscr{D}^{-1}\{F(n)^{-jni}\}$$

EXAMPLE 10.1
Find the DFT of the two sequences shown in Figure 10.2a.

Figure 10.2 Illustrating Example 10.1.

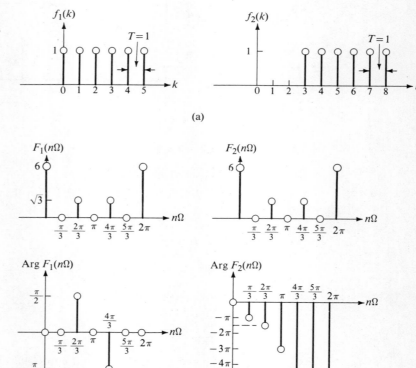

DISCRETE TRANSFORMS

Solution: From (10.4a) we obtain, respectively, where $\Omega = 2\pi/NT = \pi/3$,

$$F_1\left(\frac{n\pi}{3}\right) = \sum_{k=0}^{5} f_1(k)e^{-jn\pi k/3} \qquad F_2\left(\frac{n\pi}{3}\right) = \sum_{k=3}^{8} f_2(k)e^{-jn\pi k/3}$$

The specific numerical values are:

$$F_1\left(\frac{0\pi}{3}\right) = f_1(0) + f_1(1) + f_1(2) + f_1(3) + f_1(4) + f_1(5) = 6$$

$$F_1\left(\frac{1\pi}{3}\right) = f_1(0) + f_1(1)\left[\cos\frac{\pi}{3} - j\sin\frac{\pi}{3}\right] + f_1(2)\left[\cos\frac{2\pi}{3} - j\sin\frac{2\pi}{3}\right]$$

$$+ f_1(3)[\cos\pi - j\sin\pi] + f_1(4)\left[\cos\frac{4\pi}{3} - j\sin\frac{4\pi}{3}\right]$$

$$+ f_1(5)\left[\cos\frac{5\pi}{3} - j\sin\frac{5\pi}{3}\right]$$

$$= 1 + 0.5 - j\frac{\sqrt{3}}{2} - 0.5 - j\frac{\sqrt{3}}{2} - 1 - 0.5$$

$$+ j\frac{\sqrt{3}}{2} + 0.5 + j\frac{\sqrt{3}}{2} = 0$$

By continuing with this procedure, we find the following:

$$F_1\left(\frac{2\pi}{3}\right) = j\sqrt{3} \qquad F_1\left(\frac{3\pi}{3}\right) = 0 \qquad F_1\left(\frac{4\pi}{3}\right) = -j\sqrt{3}$$

$$F_1\left(\frac{5\pi}{3}\right) = 0 \qquad F_1\left(\frac{6\pi}{3}\right) = 6$$

Performing a similar procedure for the second sequence, we have

$$F_2\left(\frac{0\pi}{3}\right) = f_2(3) + f_2(4) + f_2(5) + f_2(6) + f_2(7) + f_2(8) = 6$$

$$F_2\left(\frac{1\pi}{3}\right) = f_2(3)\left[\cos\frac{3\pi}{3} - j\sin\frac{3\pi}{3}\right] + f_2(4)\left[\cos\frac{4\pi}{3} - j\sin\frac{4\pi}{3}\right]$$

$$+ f_2(5)\left[\cos\frac{5\pi}{3} - j\sin\frac{5\pi}{3}\right] + f_2(6)\left[\cos\frac{6\pi}{3} - j\sin\frac{6\pi}{3}\right]$$

$$+ f_2(7)\left[\cos\frac{7\pi}{3} - j\sin\frac{7\pi}{3}\right] + f_2(8)\left[\cos\frac{8\pi}{3} - j\sin\frac{8\pi}{3}\right]$$

$$= -1 - 0.5 + j\frac{\sqrt{3}}{2} + 0.5 + j\frac{\sqrt{3}}{2}$$

$$+ 1 + 0.5 - j\frac{\sqrt{3}}{2} - 0.5 - j\frac{\sqrt{3}}{2}$$

$$= -\left[1 + 0.5 - j\frac{\sqrt{3}}{2} - 0.5 - j\frac{\sqrt{3}}{2} - 1 - 0.5\right.$$
$$\left. + j\frac{\sqrt{3}}{2} + 0.5 + j\frac{\sqrt{3}}{2}\right]$$
$$= e^{-j\pi}F_1\left(\frac{1\pi}{3}\right)$$

By following the same procedure, we find that

$$F_2\left(\frac{2\pi}{3}\right) = j\sqrt{3} = e^{-j2\pi}F_1\left(\frac{2\pi}{3}\right)$$

$$F_2\left(\frac{3\pi}{3}\right) = -F_1\left(\frac{3\pi}{3}\right) = e^{-j3\pi}F_1\left(\frac{3\pi}{3}\right)$$

$$F_2\left(\frac{4\pi}{3}\right) = -j\sqrt{3} = e^{-j4\pi}F_1\left(\frac{4\pi}{3}\right)$$

$$F_2\left(\frac{5\pi}{3}\right) = -F_1\left(\frac{5\pi}{3}\right) = e^{-j5\pi}F_1\left(\frac{5\pi}{3}\right)$$

$$F_2\left(\frac{6\pi}{3}\right) = e^{-j6\pi}F_1\left(\frac{6\pi}{3}\right)$$

These results are in accord with (10.8). The amplitude and phase spectra are shown in Figure 10.2b. ∎

4. Frequency Shifting

$$\boxed{f(k)e^{jki} = \mathcal{D}^{-1}\{F(n-i)\}} \qquad (10.9)$$

Proof: Write (10.4a) as

$$F(m) = \sum_{k=0}^{N-1} f(k)e^{-jmk}$$

Now write $m = n - i$ in this expression

$$F(n-i) = \sum_{k=0}^{N-1} f(k)e^{-j(n-i)k} = \sum_{k=0}^{N-1} [f(k)e^{jik}]e^{-jnk} \triangleq \mathcal{D}\{f(k)e^{jki}\}$$

5. Alternative Inversion Formula. The inversion formula of (10.4b) can also be written in the form

DISCRETE TRANSFORMS

$$f(k) = \frac{1}{N} \left[\sum_{k=0}^{N-1} F^*(n) e^{-jnk} \right]^*$$
$$= \frac{1}{N} \left[\mathscr{D}\{F^*(n)\} \right]^* = \mathscr{D}^{-1}\{F(n)\} \quad (10.10)$$

Proof: The proof proceeds by writing $F(n) = R(n) + jI(n)$ in terms of its real and imaginary parts, and so $F^*(n) = R(n) - jI(n)$ in (10.10). By carrying out the successive steps, it will lead to

$$f(k) = \frac{1}{N} \sum_{k=0}^{N-1} F(n) e^{jnk}$$

The usefulness of (10.10) is that it shows we can obtain the IDFT using a DFT algorithm, scale the results by $1/N$, and take their conjugate value.

6. Even Functions. If $f(k)$ is an even function $f_e(k) = f_e(-k)$, then

$$\mathscr{D}\{f_e(k)\} \triangleq F_e(n) = \sum_{k=0}^{N-1} f_e(k) \cos nk \quad (10.11)$$

Proof: This follows from the defining equation, with

$$F_e(n) = R_e(n) + jI_e(n)$$
$$= \sum_{k=0}^{N-1} f_e(k) e^{-jnk} = \sum_{k=0}^{N-1} f_e(k) \cos nk - j \sum_{k=0}^{N-1} f_e(k) \sin nk$$
$$= \sum_{k=0}^{N-1} f_e(k) \cos nk = R_e(n)$$

The summation of the j terms is zero because the summation is taken over an even number of cycles of an odd function. Furthermore, $f_e(k) \cos nk$ is an even function and the summation is even. Additionally $F_e(n)$ is an even function, so that $F_e(n) = F_e(-n)$.

7. Odd Functions. If $f(k)$ is an odd function so that $f_0(k) = -f_0(-k)$, its discrete Fourier transform is an odd and imaginary function

$$\mathscr{D}\{f_0(k)\} \triangleq F_0(n) = -j \sum_{k=0}^{N-1} f_0(k) \sin nk \quad (10.12)$$

Proof: Write

$$\mathscr{D}\{f_o(k)\} \triangleq F_o(n) = \sum_{k=0}^{N-1} f_o(k)e^{-jnk}$$

Expand this to

$$F_o(n) \triangleq R_o(n) + jI_o(n) = \sum_{k=0}^{N-1} f_o(k)\cos nk - j\sum_{k=0}^{N-1} f_o(k)\sin nk$$

$$= -j\sum_{k=0}^{N-1} f_o(k)\sin nk = jI_o(n)$$

This latter step follows because the summation of an odd function over an even number of cycles is zero.

8. Time Convolution. The discrete convolution is defined by the expression

$$y(k) \triangleq f(k) * g(k) = \sum_{i=0}^{N-1} f(i)g(k-i) \qquad \textbf{(a)} \qquad (10.13)$$

where $f(k)$ and $g(k)$ are periodic and of the same period N

$$f(k) = f(k + pN) \qquad p = 0, \pm 1, \pm 2, \ldots \qquad \textbf{(b)}$$
$$g(k) = g(k + pN) \qquad p = 0, \pm 1, \pm 2, \ldots \qquad \textbf{(c)}$$

This type of convolution is known as **circular** or **cyclic** convolution. The DFT of this expression yields

$$\boxed{Y(n) \triangleq \mathscr{D}\{f(k) * g(k)\} = F(n)G(n) \qquad \textbf{(d)}}$$

Proof: Begin with the function

$$\sum_{i=0}^{N-1} f(i)g(k-i) = \sum_{i=0}^{N-1} \frac{1}{N}\sum_{n=0}^{N-1} F(n)e^{jin} \times \frac{1}{N}\sum_{m=0}^{N-1} G(m)e^{jm(k-i)}$$

This is rearranged to

$$= \frac{1}{N}\sum_{n=0}^{N-1}\sum_{m=0}^{N-1} F(n)G(m)e^{jmk}\left[\frac{1}{N}\sum_{i=0}^{N-1} e^{jin}e^{-jim}\right]$$

But as already discussed in connection with (10.3),

$$\frac{1}{N}\sum_{i=0}^{N-1} e^{jin}e^{-jim} = \begin{cases} 1 & \text{for } n = m \\ 0 & \text{for } n \neq m \end{cases}$$

Hence for $n = m$ in the second sum, we find finally

$$y(k) = \sum_{i=0}^{N-1} f(i)g(k-i) = \frac{1}{N}\sum_{n=0}^{N-1} F(n)G(n)e^{jnk} \triangleq \mathscr{D}^{-1}\{F(n)G(n)\}$$

which implies (10.13).

Figure 10.3
Convolution of two periodic sequences.

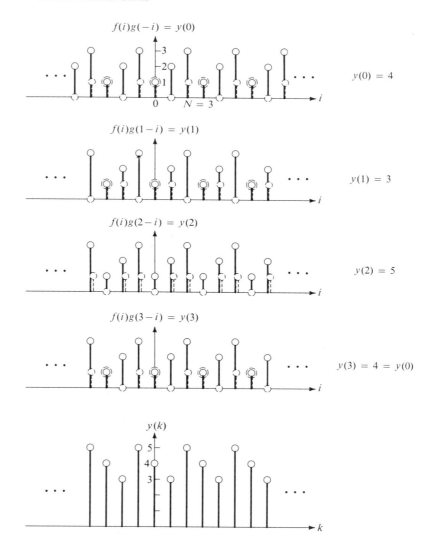

Figure 10.3 shows graphically the convolution of two periodic sequences, $f(k) = \{1, 2, 3\}$ and $g(k) = \{1, 1, 0\}$. Observe that $y(k)$ is periodic and, since $F(n)$ and $G(n)$ are periodic, $Y(n)$ is also periodic. Figure 10.4 shows how we can arrange the two sequences and perform cyclic convolution.

EXAMPLE 10.2
Consider the two periodic sequences

$$f(k) = \{1, -1, 4\} \qquad g(k) = \{0, 1, 3\}$$

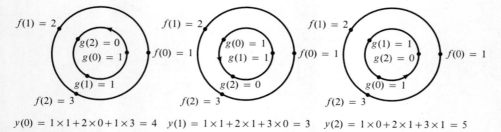

Figure 10.4
Arrangement of sequences for cyclic convolution.

Verify (10.13) by showing that, for $T = 1$, where $\Omega = 2\pi/NT = 2\pi/3$,

$$y(2) = \sum_{i=0}^{N-1} f(i)g(2-i) \triangleq \mathscr{D}^{-1}\{F(n)G(n)\} = \frac{1}{N}\sum_{n=0}^{N-1} F(n)G(n)e^{j2\pi n2/N}$$

Solution: First we find the summation

$$\sum_{i=0}^{2} f(i)g(2-i) = f(0)g(2) + f(1)g(1) + f(2)g(0)$$

$$= 1 \times 3 + (-1) \times 1 + 4 \times 0 = 2$$

Next, we obtain $F(n)$ and $G(n)$ using (10.4a)

$$F(0) = \sum_{k=0}^{2} f(k)e^{-j0k2\pi/3} = f(0) + f(1) + f(2) = 1 - 1 + 4 = 4$$

$$F(1) = \sum_{k=0}^{2} f(k)e^{-jk2\pi/3}$$

$$= f(0) + f(1)e^{-j2\pi/3} + f(2)e^{-j4\pi/3} = -0.5 + j5 \times 0.866$$

$$F(2) = \sum_{k=0}^{2} f(k)e^{-j2k2\pi/3}$$

$$= f(0) + f(1)e^{-j4\pi/3} + f(2)e^{-j8\pi/3} = -0.5 - j5 \times 0.866$$

Similarly, we obtain

$$G(0) = 4 \quad G(1) = -2 + j2 \times 0.866 \quad G(2) = -2 - j2 \times 0.866$$

The second summation given above becomes

$$\frac{1}{3}\sum_{n=0}^{2} F(n)G(n)e^{jn4\pi/3} = \frac{1}{3}\left[16 + 11.55e^{j115.7} + 11.55e^{j604.3}\right] = 2$$

This shows the validity of (10.13). ∎

Because we are primarily interested in linear rather than circular convolution, we can modify the two sequences in such a manner that circular convolution

DISCRETE TRANSFORMS

gives identical results with those of linear convolution. The advantage in using circular convolution lies in the fact that we can use DFT techniques in the convolution process. If we have two finite and unequal sequences $f(k)$ and $g(k)$, we select the period for each function according to the relation

$$N = F + G - 1$$

where F denotes the number of samples in the $f(k)$ sequence and G denotes the number of samples in the $g(k)$ sequence. Under this modification both the linear and circular convolution give identical results. Since F and G are each less than N, we pad the sequences with zeros. However, if one of the signals has infinite extent, end effects will appear. Note that it is impossible to convolve two discrete sequences of infinite extent. Refer to Section 10-7 for further discussion.

9. Frequency Convolution. Consider the frequency convolution

$$Y(n) = \sum_{i=0}^{N-1} F(i)G(n - i) \triangleq F(n) * G(n)$$

The inverse DFT of this expression yields

$$\mathscr{D}^{-1}\{Y(n)\} = \mathscr{D}^{-1}\left[\frac{1}{N} \sum_{i=0}^{N-1} F(i)G(n - i)\right] = f(k)g(k) \qquad (10.14)$$

Proof: Substitute known forms into

$$\sum_{i=0}^{N-1} F(i)G(n - i) = \sum_{i=0}^{N-1} \left[\sum_{m=0}^{N-1} f(m)e^{-jmi}\right]\left[\sum_{k=0}^{N-1} g(k)e^{-jk(n-i)}\right]$$

$$= \sum_{m=0}^{N-1} \sum_{k=0}^{N-1} f(m)g(k)e^{-jkn}\left[\sum_{i=0}^{N-1} e^{-jmi}e^{jki}\right]$$

The bracketed term is the orthogonality relationship and is equal to N if $m = k$. Therefore

$$\sum_{i=0}^{N-1} F(i)G(n - i) = N \sum_{k=0}^{N-1} f(k)g(k)e^{-jnk} \triangleq N\mathscr{D}\{f(k)g(k)\}$$

from which (10.14) follows by taking the inverse DFT of both sides of the equation. Because $F(n)$ and $G(n)$ are periodic, (10.14) indicates a circular convolution in the frequency plane.

EXAMPLE 10.3
Use the results of Example 10.2 to verify the frequency convolution property.

Solution: Figure 10.5 shows the circular convolution of $F(n)$ and $G(n)$, with the values obtained from Example 10.2. From (10.14) we obtain the periodic sequence $y(k)$ from the relation

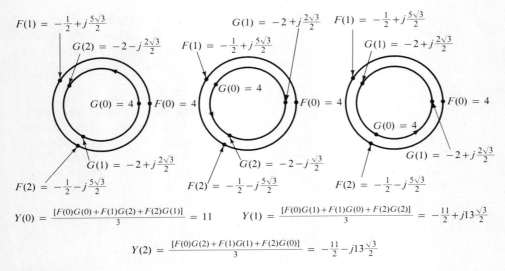

Figure 10.5
Circular convolution in the frequency domain.

$$y(k) = \mathscr{D}^{-1}\{Y(n)\} \quad \text{or} \quad \{y(k)\} = \{f(k)g(k)\} = \{0, -1, 12\}$$

Thus from the results of Figure 10.5 we obtain

$$y(0) = \mathscr{D}^{-1}\{Y(n)\}$$
$$= \frac{1}{3}\sum_{n=0}^{N-1} Y(n)e^{j0n2\pi/3} = \frac{1}{3}\left[11 - \frac{11}{2} - \frac{11}{2} + j13\frac{\sqrt{3}}{2} - j13\frac{\sqrt{3}}{2}\right] = 0$$

$$y(1) = \mathscr{D}^{-1}\{Y(n)\}$$
$$= \frac{1}{3}\left[11 + \left(-\frac{11}{2} + j13\frac{\sqrt{3}}{2}\right)\left(\cos\frac{2\pi}{3} + j\sin\frac{2\pi}{3}\right) \right.$$
$$\left. + \left(-\frac{11}{2} - j13\frac{\sqrt{3}}{2}\right)\left(\cos\frac{4\pi}{3} + j\sin\frac{4\pi}{3}\right)\right] = -1$$

$$y(2) = \mathscr{D}^{-1}\{Y(n)\}$$
$$= \frac{1}{3}\left[11 + \left(-\frac{11}{2} + j13\frac{\sqrt{3}}{2}\right)\left(\cos\frac{4\pi}{3} + j\sin\frac{4\pi}{3}\right) \right.$$
$$\left. + \left(-\frac{11}{2} - j13\frac{\sqrt{3}}{2}\right)\left(\cos\frac{8\pi}{3} + j\sin\frac{8\pi}{3}\right)\right] = 12$$

The results obtained thus verify (10.14). ∎

10. Parseval's Theorem. For discrete functions the relationship between power computed in the time domain and in the frequency domain is

DISCRETE TRANSFORMS

$$\boxed{\sum_{k=0}^{N-1} f^2(k) = \frac{1}{N} \sum_{n=0}^{N-1} |F(n)|^2}$$ (10.15)

Proof: To prove this relationship, let $y(k) = f(k)f(k)$. By (10.14) this is given by

$$\sum_{k=0}^{N-1} f^2(k) e^{-jnk} = \frac{1}{N} \sum_{i=0}^{N-1} F(i) F(n-i)$$

Now set $n = 0$ in this expression. Therefore

$$\sum_{k=0}^{N-1} f^2(k) = \frac{1}{N} \sum_{i=0}^{N-1} F(i) F(-i) = \frac{1}{N} \sum_{i=0}^{N-1} |F(i)|^2$$

From this we may define a discrete energy spectral density or periodogram spectral estimate

$$S(n) = |F(n)|^2 \qquad 0 \leq n \leq N - 1$$ (10.16)

Note, however, that $S(n)$ and $F_s(\omega)$ in (7.3) when evaluated at $\omega = 2\pi n/NT$ for $n = 0, 1, \ldots, (N-1)$ do not yield identical values. $S(n)$ is, in effect, a sampled version of a spectrum determined from the convolution of $F(\omega)$ with the transform of the rectangular window containing the sampled data. Thus the discrete spectrum $S(n)$ based on a finite data set is a distorted version of the continuous spectrum $S_s(\omega)$ based on an infinite data set.

EXAMPLE 10.4
Verify Parseval's theorem using the sequence $f(k) = \{1, -1, 4\}$.

Solution: We have directly that

$$\sum_{k=0}^{2} f^2(k) = 1 + 1 + 16 = 18$$

The values of $F(n)$ for this sequence are given in Example 10.2, so that

$$\frac{1}{3} \sum_{n=0}^{2} |F(n)|^2 = \frac{1}{3}[16 + |-0.5 + j5 \times 0.866|^2 + |-0.5 - j5 \times 0.866|^2]$$

$$= \frac{1}{3}[16 + 19 + 19] = 18 \qquad \blacksquare$$

11. Time Reversal

$$\boxed{\mathscr{D}\{f(-k)\} = F(-n)}$$ (10.17)

Proof: From (10.4b) we have

$$f(-k) = \frac{1}{N} \sum_{n=0}^{N-1} F(n) e^{-jnk}$$

Now set $n = -m$ on the right. Then

$$f(-k) = \frac{1}{N} \sum_{m=0}^{-(N-1)} F(-m) e^{jmk}$$

Because of the periodic nature of $F(-m)$ and $\exp(jmk)$, the sum over $(-N+1, 0)$ and $(0, N-1)$ is the same. Thus

$$f(-k) = \frac{1}{N} \sum_{m=0}^{N-1} F(-m) e^{jmk} \triangleq \mathscr{D}^{-1}\{F(-m)\}$$

which proves (10.17).

12. Conjugate Functions

$$\boxed{\begin{aligned} \mathscr{D}\{f^*(k)\} &= F^*(-n) \quad &\text{(a)} \\ \mathscr{D}\{f^*(-k)\} &= F^*(n) \quad &\text{(b)} \end{aligned}} \qquad (10.18)$$

Proof: From (10.4b) we obtain

$$f^*(k) = \frac{1}{N} \sum_{n=0}^{N-1} F^*(n) e^{-jnk}$$

Set $n = -m$ and use the periodicity property, as was done in Property 11, to find

$$f^*(k) = \frac{1}{N} \sum_{m=0}^{-(N-1)} F^*(-m) e^{jmk} = \frac{1}{N} \sum_{m=0}^{N-1} F^*(-m) e^{jmk} = \mathscr{D}^{-1}\{F^*(-m)\}$$

This proves the first identity. The second identity is easily proved using (10.17).

13. Delta Function

$$\boxed{\mathscr{D}\{\delta(k)\} = 1} \qquad (10.19)$$

Proof: This relationship is deduced directly from (10.4).

14. Amplitude Relations

DISCRETE TRANSFORMS

$$f(0) = \frac{1}{N} \sum_{n=0}^{N-1} F(n)$$

$$F(0) = \sum_{k=0}^{N-1} f(k)$$

(10.20)

Proof: These relationships are the direct consequence of (10.4).

EXAMPLE 10.5
Find the DFT of the function shown in Figure 10.6a discretized with $T = 0.5$ and 1.

Solution: For the case $T = 0.5$ and $NT = 16$, we obtain $N = 32$. This yields 8 discrete values from the triangular function. The value of Ω is $2\pi/NT = 2\pi/16 = \pi/8$ rad/s, and the values of $F(n\Omega)$ for different values of n are (0.5 is the scale factor in 10.4a):

$$F(0\Omega) = (0 + 1 + 2 + 3 + 4 + 3 + 2 + 1) \times 0.5 = 8$$

$$F(1\Omega) = \left[0 + 1\cos\frac{\pi}{16} + 2\cos\frac{2\pi}{16} + 3\cos\frac{3\pi}{16} + 4\cos\frac{4\pi}{16}\right.$$
$$\left. + 3\cos\frac{5\pi}{16} + 2\cos\frac{6\pi}{16} + 1\cos\frac{7\pi}{16}\right] \times 0.5$$

$$-j\left[0 + 1\sin\frac{\pi}{16} + 2\sin\frac{2\pi}{16} + 3\sin\frac{3\pi}{16} + 4\sin\frac{4\pi}{16}\right.$$
$$\left. + 3\sin\frac{5\pi}{16} + 2\sin\frac{6\pi}{16} + 1\sin\frac{7\pi}{16}\right] \times 0.5$$

$$= 5.39 - j5.39$$

$$F(2\Omega) = \left[0 + 1\cos\frac{2\pi}{16} + 2\cos\frac{4\pi}{16} + 3\cos\frac{6\pi}{16} + 4\cos\frac{8\pi}{16} + 3\cos\frac{10\pi}{16}\right.$$
$$\left. + 2\cos\frac{12\pi}{16} + 1\cos\frac{14\pi}{16}\right] \times 0.5$$

$$-j\left[0 + 1\sin\frac{2\pi}{16} + 2\sin\frac{4\pi}{16} + 3\sin\frac{6\pi}{16} + 4\sin\frac{8\pi}{16} + 3\sin\frac{10\pi}{16}\right.$$
$$\left. + 2\sin\frac{12\pi}{16} + 1\sin\frac{14\pi}{16}\right] \times 0.5$$

$$= 0 - j6.568$$

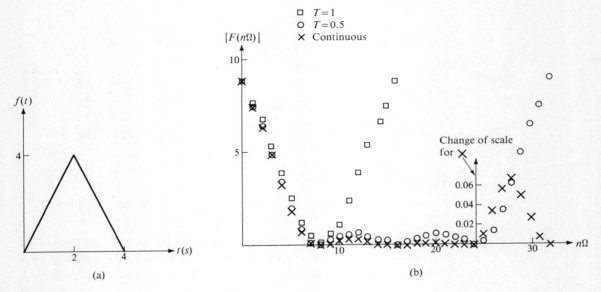

Figure 10.6
(a) The signal and (b) its discrete Fourier transform, for Example 10.5.

and so on for subsequent values of n. The values of $F(n\Omega)$ are summarized in Table 10.1. Figure 10.6b shows the discrete Fourier transform for $T = 1$ and $T = 0.5$, and it also shows a comparison with the values obtained from a continuous Fourier transform at the same values of ω. ∎

15. Relationship between the DFT and the Fourier Transform. A real-valued continuous signal of duration $(N - 1)T$ and its sampled version are shown in Figure 1.16. Also, assume that the function is band-limited, with $F(\omega) = 0$ for $|\omega| \geq \omega_s/2 = \pi/T$. The impulse at $t = kT$ has a strength $f(kT)$, and the sampled function assumes the form [see (7.2)]

$$f_s(t) = \sum_{k=0}^{N-1} f(kT)\delta(t - kT)$$

The Fourier transform of this equation is [see (7.3)]

$$F_s(\omega) = \sum_{k=0}^{N-1} f(kT) \int_{-\infty}^{\infty} \delta(t - kT)e^{-j\omega t}\, dt = \sum_{k=0}^{N-1} f(kT)e^{-j\omega kT} \quad (10.21)$$

If we are only interested in the set of discrete values of $F_s(\omega)$ evaluated for $\omega = n\Omega$, this equation takes the form

$$F_s(n\Omega) = \sum_{k=0}^{N-1} f(kT)e^{-jn\Omega kT} \quad \textbf{(a)} \quad (10.22)$$

DISCRETE TRANSFORMS

Table 10.1 The values of $F(n\Omega)$ in Example 10.5

n	$\omega = n\pi/8$	$\mathrm{Re}\{F(n\Omega)\}$	$\mathrm{Im}\{F(n\Omega)\}$	$|F(n\Omega)|$
0	0	8.0000	0.0000	8.0000
1	$\pi/8$	5.3892	−5.3892	7.6215
2	$2\pi/8$	0.0000	−6.5685	6.5685
3	$3\pi/8$	−3.5812	−3.5812	5.0646
4	$4\pi/8$	−3.4142	0.0000	3.4142
5	$5\pi/8$	−1.3580	1.3580	1.9205
6	$6\pi/8$	0.0000	0.8099	0.8099
7	$7\pi/8$	0.1286	0.1286	0.1819
8	$8\pi/8$	0.0000	0.0000	0.0000
9	$9\pi/8$	0.0866	−0.0866	0.1225
10	$10\pi/8$	0.0000	−0.3616	0.3616
11	$11\pi/8$	−0.3879	−0.3880	0.5487
12	$12\pi/8$	−0.5857	0.0000	0.5857
13	$13\pi/8$	−0.3295	0.3295	0.4660
14	$14\pi/8$	0.0000	0.2598	0.2598
15	$15\pi/8$	0.0522	0.0522	0.0739
16	$16\pi/8$	0.0000	0.0000	0.0000
17	$17\pi/8$	0.0522	−0.0522	0.0739
18	$18\pi/8$	0.0000	−0.2598	0.2598
19	$19\pi/8$	−0.3295	−0.3295	0.4660
20	$20\pi/8$	−0.5857	0.0000	0.5857
21	$21\pi/8$	−0.3880	0.3879	0.5487
22	$22\pi/8$	0.0000	0.3616	0.3616
23	$23\pi/8$	0.0866	0.0866	0.1225
24	$24\pi/8$	0.0000	0.0000	0.0000
25	$25\pi/8$	0.1286	−0.1286	0.1819
26	$26\pi/8$	0.0000	−0.8099	0.8099
27	$27\pi/8$	−1.3579	−1.3580	1.9205
28	$28\pi/8$	−3.4141	−0.0001	3.4141
29	$29\pi/8$	−3.5813	3.5811	5.0646
30	$30\pi/8$	−0.0002	6.5684	6.5684
31	$31\pi/8$	5.3890	5.3894	7.6215
32	$32\pi/8$	8.0000	0.0002	8.0000

which shows that

$$F_s(\omega)|_{\omega = n\Omega} = F(n\Omega). \qquad \text{(b)}$$

We have already discussed the procedure for obtaining $F(\omega)$ from $F_s(\omega)$. The important question remains of deducing $F(\omega)$ or $F_s(\omega)$ from $F(n\Omega)$. This involves combining (10.4b) with (10.21), where $\Omega = 2\pi/NT$,

$$F_s(\omega) = \sum_{k=0}^{N-1} f(kT)e^{-j\omega kT} = \sum_{k=0}^{N-1} \frac{1}{N} \sum_{n=0}^{N-1} F(n\Omega)e^{jn\Omega kT} e^{-j\omega kT}$$

$$= \sum_{n=0}^{N-1} \frac{1}{N} F(n\Omega) \sum_{k=0}^{N-1} e^{j(n\Omega - \omega)kT} \tag{10.23}$$

But

$$\sum_{k=0}^{N-1} e^{j(n\Omega - \omega)kT} = e^{j[(N-1)/2](n\Omega - \omega)T} \frac{\sin\left(\dfrac{n\Omega - \omega}{2}\right)NT}{\sin\left(\dfrac{n\Omega - \omega}{2}\right)T}$$

Further, $n\Omega T/2 = (\pi/N)n$; hence (10.23) becomes

$$\boxed{F_s(\omega) = \frac{1}{N} \sin\frac{\omega NT}{2} e^{-j\omega NT/2} \sum_{n=0}^{N-1} \frac{F(n\Omega)\, e^{j[\omega T/2 - \pi n/N]}}{\sin\left(\dfrac{\omega T}{2} - \dfrac{\pi n}{N}\right)}} \tag{10.24}$$

16. Relationship between the DFT and the Z-transform. If the sequence $f(kT)$ is zero for $k < 0$, it has a Z-transform

$$F(z) = \sum_{k=0}^{\infty} f(kT)z^{-k}$$

By comparing this equation with (10.21), we see that

$$F_s(\omega) = F(z = e^{j\omega T}) \tag{10.25}$$

as already shown in (9.124).

17. Discrete Fourier Series (DFS). We have already noted that a discrete signal is periodic if for some integer N the following relation holds:

$$f(k) = f(k + rN) \qquad r = \pm 1, \pm 2, \ldots \tag{10.26}$$

One such periodic function is the discrete exponential signal

$$v(k) = e^{(j2\pi k/N)} = e^{j(2\pi/N)(k+rN)} \qquad r = \pm 1, \pm 2, \ldots \tag{10.27}$$

In a manner analogous to our handling of the continuous case, we may use these periodic exponential signals to represent a general periodic discrete signal $f(k)$ in the form

$$f(k) = \sum_{n=0}^{N-1} \alpha_n v_n(k) = \sum_{n=0}^{N-1} \alpha_n e^{jk 2\pi n/N} \tag{10.28}$$

The value of n can start at any integer such as $n = 4, 5, \ldots, N + 3$ or $n = -3, -2, \ldots, N - 2$. This equation is the discrete Fourier series (DFS), with α_n being the unknown coefficients.

The Fourier coefficients of the periodic discrete function shown in Figure 10.7 are found by using (10.28) with $N = 3$. We can write

DISCRETE TRANSFORMS

Figure 10.7
A periodic discrete signal.

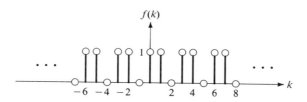

$$f(0) = 1 = \alpha_0 + \alpha_1 + \alpha_2$$
$$f(1) = 1 = \alpha_0 + \alpha_1 e^{j120°} + \alpha_2 e^{j240°} \quad (10.29)$$
$$f(2) = 0 = \alpha_0 + \alpha_1 e^{j240°} + \alpha_2 e^{j480°}$$

The unknown coefficients in this set of equations are found to be

$$\alpha_0 = \frac{2}{3} \quad \alpha_1 = \frac{1}{6} - j\frac{\sqrt{3}}{6} \quad \alpha_2 = \frac{1}{6} + j\frac{\sqrt{3}}{6} \quad (10.30)$$

We can also evaluate the Fourier coefficients in a rather different way by following the procedure developed in Chapter 3 for the continuous case. This involves multiplying both sides of (10.28) by $\exp(-jr(2\pi/N)k)$ and then summing over N terms. Therefore we obtain

$$\sum_{k=0}^{N-1} f(k) e^{-jr(2\pi/N)k} = \sum_{k=0}^{N-1} \sum_{n=0}^{N-1} \alpha_n e^{jk(2\pi/N)n} e^{-jr(2\pi/N)k}$$
$$= \sum_{n=0}^{N-1} \alpha_n \sum_{k=0}^{N-1} e^{j(2\pi/N)(n-r)k} \quad (10.31)$$

As already discussed, the second summation on the right is zero unless $n - r = 0$ or an integral multiple of N. Therefore for $n = r$, where the values of r are taken from the integers $n = 0$ to $N - 1$, (10.31) becomes

$$\alpha_r = \frac{1}{N} \sum_{k=0}^{N-1} f(k) e^{-jr(2\pi/N)k} \quad (10.32)$$

Because r in this equation is a dummy variable, we can replace it by n. Thus we have the discrete Fourier series pair

$$f(k) = \sum_{n=0}^{N-1} \alpha_n e^{jk(2\pi/N)n} \quad \text{(a)}$$
$$\alpha_n = \frac{1}{N} \sum_{k=0}^{N-1} f(k) e^{-jn(2\pi/N)k} \quad \text{(b)} \quad (10.33)$$

The running index in these equations k and n can start at any integer provided they span N consecutive integers.

Figure 10.8
A periodic discrete signal and its amplitude spectrum.

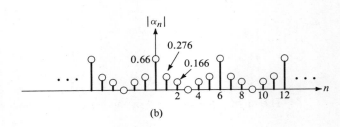

If we use the values of the periodic sequence shown in Figure 10.7, we find, for example, that

$$\alpha_1 = \frac{1}{3}\sum_{k=0}^{2} f(k)e^{-j1(2\pi/3)k} = \frac{1}{3}[1 + e^{-j120°}] = \frac{1}{6} - j\frac{\sqrt{3}}{6}$$

which agrees with the value found above.

EXAMPLE 10.6
Find the Fourier series coefficients for the periodic discrete function shown in Figure 10.8a. Plot the amplitude spectrum $|\alpha_n|$.

Solution: An application of (10.33b) yields the following values ($N = 6$)

$$\alpha_0 = \frac{1}{6}(1 + 1) = \frac{2}{3} \qquad \alpha_1 = \frac{1}{6}(1 + e^{-j60°}) = \frac{3}{12} - j\frac{\sqrt{3}}{12}$$

$$\alpha_2 = \frac{1}{6}(1 + e^{-j120°}) = \frac{1}{12} - j\frac{\sqrt{3}}{12} \qquad \alpha_3 = \frac{1}{6}(1 - 1) = 0$$

$$\alpha_4 = \frac{1}{6}(1 + e^{-j240°}) = \frac{1}{12} + j\frac{\sqrt{3}}{12} \qquad \alpha_5 = \frac{3}{12} + j\frac{\sqrt{3}}{12}$$

The spectrum $|\alpha_n|$ is shown in Figure 10.8b. ∎

The foregoing important properties of the DFT are contained in Table 10.2. Table 10.3 gives the properties for odd and even functions in both time and frequency domains.

DISCRETE TRANSFORMS

Table 10.2 Properties of Discrete Fourier Transforms

	Property	Function	Transform
1	Linearity	$af_1(kT) + bf_2(kT)$	$aF_1(n\Omega) + bF_2(n\Omega)$
2	Symmetry	$(1/N)F(kT)$	$f(-n\Omega)$
3	Time Shifting	$f(kT - iT)$	$F(n\Omega)e^{-jn\Omega iT}$
4	Frequency Shifting	$f(kT)e^{\pm jkTi\Omega}$	$F(n\Omega \mp i\Omega)$
5	Even Functions	$f_e(kT)$	$\sum_{k=0}^{N-1} f_e(kT)\cos(kTn\Omega)$
6	Odd Functions	$f_o(kT)$	$-j\sum_{k=0}^{N-1} f_o(kT)\sin(kTn\Omega)$
7	Time Convolution	$f(kT)*g(kT)$	$F(n\Omega)G(n\Omega)$
8	Frequency Convolution	$f(kT)g(kT)$	$F(n\Omega)*G(n\Omega)/N$
9	Time Reversal	$f(-kT)$	$F(-n\Omega)$
10	Conjugate Functions	$f^*(kT)$	$F^*(-n\Omega)$
		$f^*(-kT)$	$F^*(n\Omega)$
11	Delta Function	$\delta(kT)$	1
12	Amplitude Relations	$f(0) = \dfrac{1}{N}\sum_{n=0}^{N-1} F(n\Omega)$	
		$F(0) = \sum_{n=0}^{N-1} f(kT)$	
13	Parseval's Theorem	$\sum_{k=0}^{N-1} f^2(kT) = \dfrac{1}{N}\sum_{n=0}^{N-1} |F(n\Omega)|^2$	
14	Relation to continuous FT	$F_s(\omega)\big|_{\omega=n\Omega} = F(n\Omega)$	
		$= FT$ of the sampled function $f_s(t)$	
15	Relation to Fourier series	$\alpha_k = F(n\Omega)/N$	

■ ■ ■

10-3 THE FAST FOURIER TRANSFORM—DECIMATION IN TIME

As noted in Section 10-1, the Fast Fourier Transform (FFT) is a computational algorithm that reduces the number of multiplications and additions required for determining the coefficients of the DFT. The FFT method, first developed by Cooley and Tukey, has produced major changes in the computational techniques used in digital spectral analysis, filter simulation, and related fields. The ideas

Table 10.3 Even and Odd Functions

$f(kT)$	$F(n\Omega)$
real	$F(n\Omega) = F^*(-n\Omega) = F^*[(N-n)\Omega]$
$f(kT) = f^*(-kT) = f[(N-k)T]$	real
imaginary	$F(n\Omega) = -F^*(-n\Omega) = -F^*[(N-n)\Omega]$
$f(kT) = -f^*(-kT) = -f^*[(N-k)T]$	imaginary
real and even	real and even
real and odd	imaginary and odd
imaginary and even	imaginary and even
imaginary and odd	real and odd

behind the FFT will be considered in detail. Note that a number of FFT algorithms have been programmed.

For purposes of computation, it is convenient to define the quantity

$$W = e^{-j\Omega T} = e^{-j2\pi/N} \qquad (10.34)$$

which denotes a unit distance at the angle $-2\pi/N$. With this notation the DFT pair given in (10.4) is written

$$F(n) = \sum_{k=0}^{N-1} f(k)W^{nk} \qquad \text{(a)}$$

$$f(k) = \frac{1}{N}\sum_{n=0}^{N-1} F(n)W^{-nk} \qquad \text{(b)} \qquad (10.35)$$

Observe that W^{nk} denotes N equally spaced points on the unit circle.

To appreciate the ideas behind an FFT, we examine the process known as **decimation in time**. This name arises because the algorithm is structured to suggest sample rate reduction. Suppose that the number of samples N is divisible by 2. In this case it is advantageous to consider the DFT of two shorter sequences, one containing the even-numbered position samples $f(2l)$ and the second containing the odd-numbered position samples $f(2l + 1)$, where $l = 0, 1, \ldots, (N/2 - 1)$. That is, we begin with

$$F(k) = \sum_{l=0}^{N-1} f(l)W^{lk} \qquad l = 0, 1, \ldots, (N-1) \qquad (10.36)$$

which, in expanded form, is

DISCRETE TRANSFORMS

$$F(k) = f(0)W^0 + f(1)(W^k)^1 + f(2)(W^k)^2$$
$$+ \cdots + f(N-1)(W^k)^{(N-1)} \tag{10.37}$$

In the **divide-by-2** process, the terms are rearranged into two sums

$$F(k) = [f(0)W^0 + f(2)(W^k)^2 + \cdots]$$
$$+ [f(1)(W^k)^1 + f(3)(W^k)^3 + \cdots]$$
$$= \sum_{l=0}^{N/2-1} f(2l)(W^k)^{2l} + \sum_{l=0}^{N/2-1} f(2l+1)(W^k)^{(2l+1)} \tag{10.38}$$

Observe that these two sums can be combined into a single expression

$$F(k) = \sum_{m=0}^{1} \sum_{l=0}^{N/2-1} f(2l+m)(W^k)^{(2l+m)}$$
$$= \sum_{m=0}^{1} \sum_{l=0}^{N/2-1} f(2l+m)(W^k)^{2l}(W^k)^m \tag{10.39}$$

Now carry out a second divide-by-2 process on $\sum_{l=0}^{N/2-1} f(2l+m)(W^k)^{2l}$. This operation leads to

$$\sum_{l=0}^{N/2-1} f(2l+m)(W^k)^{2l} = \sum_{l=0}^{N/4-1} f(4l+m)(W^k)^{4l}$$
$$+ \sum_{l=0}^{N/4-1} f(4l+2+m)(W^k)^{(4l+2)}$$

But this result can be combined into the single expression

$$= \sum_{r=0}^{1} \sum_{l=0}^{N/4-1} f(4l+2r+m)(W^k)^{(4l+2r)} \tag{10.40}$$

The result for $F(k)$ after two such divide-by-2 processes is

$$F(k) = \sum_{m=0}^{1} \sum_{r=0}^{1} \sum_{l=0}^{N/4-1} f(4l+2r+m)(W^k)^{4l}(W^k)^{2r}(W^k)^m \tag{10.41}$$

This decimation process is continued until the limits on l are 0 and 1. Thus for $N = 8$, (10.41) would denote the complete decimation process, with the result that

$$F(k) = F(l, m, r) = f_3(4, 2, 1) \quad \text{(a)} \tag{10.42}$$

where we would write

$$F(k) = \sum_{m=0}^{1} \sum_{r=0}^{1} \sum_{l=0}^{1} f(4l+2r+m)(W^k)^{4l}(W^k)^{2r}(W^k)^m \quad \text{(b)}$$

$$\underbrace{\qquad f_1(4, 0, 0) \qquad}$$
$$\underbrace{\qquad\qquad f_2(4, 2, 0) \qquad\qquad}$$
$$\underbrace{\qquad\qquad\qquad f_3(4, 2, 1) \qquad\qquad\qquad}$$

To calculate $F(k)$ using this expression in the order shown requires many fewer mathematical operations than proceeding directly from (10.36).

■ ■ ■

10-4 THE FFT–MATRIX APPROACH

We observe that (10.35) can be written in matrix form. The matrix form for $F(n)$ is

$$\boxed{\mathbf{F}(n) = \mathbf{W}^{nk} \times \mathbf{f}(k)} \qquad (10.43)$$

where $\mathbf{F}(n)$ and $\mathbf{f}(k)$ are $N \times 1$ column matrices and where $\mathbf{W}^{nk}$ is an $N \times N$ square matrix. We consider the special case for $N = 4$ sample points. Equation (10.43) is shown explicitly as

$$\begin{bmatrix} F(0) \\ F(1) \\ F(2) \\ F(3) \end{bmatrix} = \begin{bmatrix} W^0 & W^0 & W^0 & W^0 \\ W^0 & W^1 & W^2 & W^3 \\ W^0 & W^2 & W^4 & W^6 \\ W^0 & W^3 & W^6 & W^9 \end{bmatrix} \times \begin{bmatrix} f_0(0) \\ f_0(1) \\ f_0(2) \\ f_0(3) \end{bmatrix} \qquad (10.44)$$

Note that we have changed the notation from $f(k)$ to $f_0(k)$ in order to provide for subsequent steps in the development. Equation (10.44) constitutes a set of simultaneous equations and will require 4^2 complex multiplications and 4×3 additions in its solution. In the general case for an $N \times N$ matrix, N^2 multiplications and $N(N - 1)$ additions are required. This can make the details of carrying out a problem time-consuming, even when a large digital computer is used.

To proceed, suppose that $N = 2^i$, where i is an integer; in the form given in (10.44) for $N = 4$, $i = 2$. It is convenient to write W^{nk} in the form

$$\boxed{W^{nk} = W^{nk \bmod N}} \qquad (10.45)$$

where $nk \bmod N$ denotes the remainder, upon division of nk by N. For example, take $N = 4$, which means that the unit circle has been equally divided along its circumference to specify 4 phasors spaced $90°$. When $N = 4$ with $n = 2$, $k = 3$ so that $nk = 6$, then $nk \bmod N = 2$, which means that $W^6 = W^2$. Based on these considerations, (10.44) is written

$$\mathbf{F}(n) \triangleq \begin{bmatrix} F(0) \\ F(1) \\ F(2) \\ F(3) \end{bmatrix} = \begin{bmatrix} 1 & 1 & 1 & 1 \\ 1 & W^1 & W^2 & W^3 \\ 1 & W^2 & W^0 & W^2 \\ 1 & W^3 & W^2 & W^1 \end{bmatrix} \begin{bmatrix} f_0(0) \\ f_0(1) \\ f_0(2) \\ f_0(3) \end{bmatrix} \qquad (10.46)$$

The next step, and this is a critical one, is to factor the square matrix into the form

DISCRETE TRANSFORMS

$$\mathbf{F}(n) \triangleq \begin{bmatrix} F(0) \\ F(2) \\ F(1) \\ F(3) \end{bmatrix} = \begin{bmatrix} 1 & W^0 & 0 & 0 \\ 1 & W^2 & 0 & 0 \\ 0 & 0 & 1 & W^1 \\ 0 & 0 & 1 & W^3 \end{bmatrix} \begin{bmatrix} 1 & 0 & W^0 & 0 \\ 0 & 1 & 0 & W^0 \\ 1 & 0 & W^2 & 0 \\ 0 & 1 & 0 & W^2 \end{bmatrix} \begin{bmatrix} f_0(0) \\ f_0(1) \\ f_0(2) \\ f_0(3) \end{bmatrix} \quad (10.47)$$

Observe that even though W^0 is equivalent to unity, both W^0 and 1 are used in the factored form. Observe also that rows 1 and 2 have been interchanged in (10.47). Clearly, of course, the method of factorization constitutes an essential step in the FFT, and the means for factorization constitutes the essence of the FFT.

The basis for the operation in going from (10.46) to (10.47) is best examined graphically. First we consider the two matrices on the right, which we write

$$\begin{bmatrix} f_1(0) \\ f_1(1) \\ f_1(2) \\ f_1(3) \end{bmatrix} = \begin{bmatrix} 1 & 0 & W^0 & 0 \\ 0 & 1 & 0 & W^0 \\ 1 & 0 & W^2 & 0 \\ 0 & 1 & 0 & W^2 \end{bmatrix} \begin{bmatrix} f_0(0) \\ f_0(1) \\ f_0(2) \\ f_0(3) \end{bmatrix} \quad (10.48)$$

The steps in the expansion of this matrix equation are shown graphically in Figure 10.9. The next step in the evaluation is the matrix

$$\begin{bmatrix} F(0) \\ F(2) \\ F(1) \\ F(3) \end{bmatrix} = \begin{bmatrix} 1 & W^0 & 0 & 0 \\ 1 & W^2 & 0 & 0 \\ 0 & 0 & 1 & W^1 \\ 0 & 0 & 1 & W^3 \end{bmatrix} \begin{bmatrix} f_1(0) \\ f_1(1) \\ f_1(2) \\ f_1(3) \end{bmatrix} = \begin{bmatrix} f_2(0) \\ f_2(1) \\ f_2(2) \\ f_2(3) \end{bmatrix} \quad (10.49)$$

The steps in the expansion of this matrix equation are shown in Figure 10.10. The total process of this expansion is shown in Figure 10.11. The total process, which shows that the 4-point DFT is completely reduced to complex multiplica-

Figure 10.9 A step in the FFT algorithm.

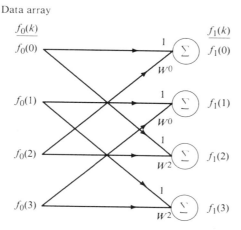

Figure 10.10
A second step in the FFT algorithm.

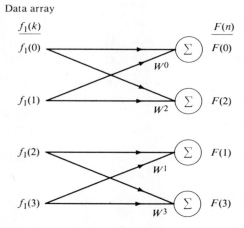

Figure 10.11
The FFT algorithm for evaluating
$\mathbf{F}(n) = \mathbf{W}^{nk}\mathbf{f}(k)$.

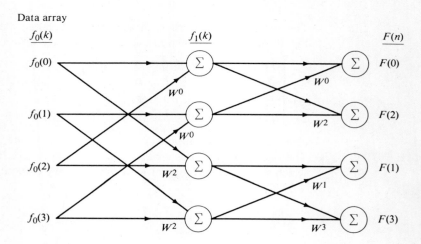

tions and additions by repeated **decimation**, has been reduced to data flow graph form.

It is of interest to determine the number of multiplications and additions involved in the expansion of (10.48). Note, therefore, that $f_1(0)$ is determined by one complex multiplication and one addition since

$$f_1(0) = f_0(0) + W^0 f_0(2) \tag{10.50}$$

To find $f_1(1)$ also requires one complex multiplication and one addition. To find $f_1(2)$, we examine the expansion

$$\begin{aligned} f_1(2) &= f_0(0) + W^2 f_0(2) \\ &= f_0(0) - W^0 f_0(2) \end{aligned} \tag{10.51}$$

DISCRETE TRANSFORMS

since $W^2 = -W^0$. But $W^0 f_0(2)$ has already been evaluated in carrying out (10.50), so carrying out the expansion in (10.51) requires one additional complex addition. By the same reasoning, $f_1(3)$ requires one complex addition. Thus the evaluation of the vector $f_1(k)$ requires a total of two multiplications and four additions. In the next step, we determine the number of computations involved in carrying out the requirements of (10.49). Consider the expansion for $F(0)$ which is

$$F(0) = f_1(0) + W^0 f_1(1) \tag{10.52}$$

This involves one multiplication and one addition. The determination of $F(2)$ requires one addition, because $W^2 = -W^0$. Continuing with this procedure, we find that $F(1)$ requires one multiplication and one addition, and $F(3)$ requires one addition. In total, therefore, the vector $\mathbf{F}(n)$ has been computed by a total of $Ni/2 = 4$ multiplications, and $Ni = 8$ additions. This shows a total of 12 complex multiplications and additions, as compared with $N^2 = 16$ operations for finding $F(n)$ in (10.35). For large N, the savings in time and effort are huge. For example, a 1024 sequence requires 1,048,576 multiplications and 1,047,552 additions when a DFT is applied and only 5120 multiplications and 10,240 additions when an FFT is used.

In its more general form, the Cooley-Tukey FFT algorithm can be considered a method for factoring an $N \times N$ matrix into $iN \times N$ matrices ($N = 2^i$) so that each of the factored matrices has the special property of minimizing the number of complex multiplications. This optimum matrix factoring process rearranges the $\mathbf{F}(n)$ of (10.46) to the $\mathbf{F}(n)$ of (10.47). We will find that such a scheme is not difficult; it calls for replacing the argument n in the $\mathbf{F}(n)$ of (10.46) by its binary equivalent and then effecting a bit-reversal.

■ ■ ■

10-5 THE BASE 2 FFT ALGORITHM

We now wish to consider a theoretical basis for the matrix factorization introduced in the previous section. To do this we again consider the quantity W^{nk} for $N = 2^\gamma$ so that both n and k are γ-bit binary numbers and can be written in binary form. Specifically, attention is given to the respective weights of the binary digits making up the number. Thus if $N = 12$, k would range from 0 to 11; in binary notation, k would take on the values 0000, 0001, ..., 1011, where the least significant bit has a weight of 1 and the most significant bit has a weight of 8. If k is written in binary form

$$k = k_3 k_2 k_1 k_0$$

then the effective weights would allow k to be written

$$k = 8k_3 + 4k_2 + 2k_1 + 1k_0 \tag{10.53}$$

to denote the number k in binary form. A similar writing exists for n.

To carry out the calculations for $N = 4$ (corresponding to the discussion in Section 10-4), we examine

$$F(n) = \sum_{k=0}^{N-1} f_0(k) W^{nk} = \sum_{k_0=0}^{1} \sum_{k_1=0}^{1} f_0(k_1, k_0) W^{(2n_1+n_0)(2k_1+k_0)} \quad (10.54)$$

We focus our attention on the factor W and write this

$$W^{(2n_1+n_0)(2k_1+k_0)} = W^{(2n_1+n_0)2k_1} \cdot W^{(2n_1+n_0)k_0}$$
$$= W^{4n_1k_1} \cdot W^{2n_0k_1} \cdot W^{(2n_1+n_0)k_0}$$

Note that the term [Equation (10.34)]

$$W^{4n_1k_1} = W^{4(n_1k_1)} = 1^{n_1}k1 = 1$$

because $W^4 = [e^{-j2\pi/4}]^4 = e^{-j2\pi} = 1$. Equation (10.54) is thus written

$$F(n_1, n_0) = \sum_{k_0=0}^{1} \left[\sum_{k_1=0}^{1} f_0(k_1, k_0) W^{2n_0k_1} \right] W^{(2n_1+n_0)k_0} \quad (10.55)$$

This equation is the keystone for the FFT algorithm. We apply it to the specific case already considered.

Consider the inner summation included in the brackets as

$$f_1(n_0, k_0) = \sum_{k_1=0}^{1} f_0(k_1, k_0) W^{2n_0k_1} \quad (10.56)$$

Expand this summation to yield

$$\begin{aligned} f_1(0, 0) &= f_0(0, 0) + f_0(1, 0)W^0 \\ f_1(0, 1) &= f_0(0, 1) + f_0(1, 1)W^0 \\ f_1(1, 0) &= f_0(0, 0) + f_0(1, 0)W^2 \\ f_1(1, 1) &= f_0(0, 1) + f_0(1, 1)W^2 \end{aligned} \quad (10.57)$$

Observe that these constitute (10.48) with the index k being written in binary form.

Now consider the outer summation of (10.55), which is written

$$f_2(n_0, n_1) = \sum_{k_0=0}^{1} f_1(n_0, k_0) W^{(2n_1+n_0)k_0} \quad (10.58)$$

The expansion of this expression, in matrix form, is

$$\begin{bmatrix} f_2(0, 0) \\ f_2(0, 1) \\ f_2(1, 0) \\ f_2(1, 1) \end{bmatrix} = \begin{bmatrix} 1 & W^0 & 0 & 0 \\ 1 & W^2 & 0 & 0 \\ 0 & 0 & 1 & W^1 \\ 0 & 0 & 1 & W^3 \end{bmatrix} \begin{bmatrix} f_1(0, 0) \\ f_1(0, 1) \\ f_1(1, 0) \\ f_1(1, 1) \end{bmatrix} \quad (10.59)$$

This is (10.49), which is the second of the factored matrices. From (10.55) and (10.58), it follows that

$$F(n_1, n_0) = f_2(n_0, n_1) \quad (10.60)$$

DISCRETE TRANSFORMS

This shows that the final result $f_2(n_0, n_1)$ obtained from the outer sum is the bit-reversed form of the desired $F(n_1, n_0)$.

The key result of this development is the set of equations

$$f_1(n_0, k_0) = \sum_{k_1=0}^{1} f_0(k_1, k_0) W^{2n_0 k_1}$$

$$f_2(n_0, n_1) = \sum_{k_0=0}^{1} f_1(n_0, k_0) W^{(2n_1 + n_0)k_0} \tag{10.61}$$

$$F(n_1, n_0) = f_2(n_0, n_1)$$

This represents the Cooley-Tukey formulation for $N = 4$.

Suppose we extend the results for the case $N = 8 = 2^3$. For this case

$$n = 4n_2 + 2n_1 + n_0$$
$$k = 4k_2 + 2k_1 + k_0 \tag{10.62}$$

Now (10.55) becomes

$$F(n_2, n_1, n_0) = \sum_{k_0=0}^{1} \sum_{k_1=0}^{1} \sum_{k_2=0}^{1} f_0(k_2, k_1, k_0) W^{(4n_2 + 2n_1 + n_0)(4k_2 + 2k_1 + k_0)}$$

$$\tag{10.63}$$

Expand the W function

$$W^{(4n_2 + 2n_1 + n_0)(4k_2 + 2k_1 + k_0)} = W^{(4n_2 + 2n_1 + n_0)4k_2} W^{(4n_2 + 2n_1 + n_0)2k_1}$$
$$\times W^{(4n_2 + 2n_1 + n_0)k_0}$$

But since $W^8 = [e^{-j2\pi/8}]^8 = 1$, then

$$W^{(4n_2 + 2n_1 + n_0)4k_2} = W^{8(2n_2 k_2)} W^{8(n_1 k_2)} W^{4n_0 k_2} = W^{4(n_0 k_2)}$$

$$W^{(4n_2 + 2n_1 + n_0)2k_1} = W^{8(n_2 k_1)} W^{(2n_1 + n_0)2k_1} = W^{(2n_1 + n_0)2k_1}$$

and (10.63) can be written

$$F(n_2, n_1, n_0) = \sum_{k_0=0}^{1} \sum_{k_1=0}^{1} \sum_{k_2=0}^{1} f_0(k_2, k_1, k_0)$$
$$\times W^{4n_0 k_2} W^{(2n_1 + n_0)2k_1} W^{(4n_2 + 2n_1 + n_0)k_0} \tag{10.64}$$

This can be reduced to the set of equations

$$f_1(n_0, k_1, k_0) = \sum_{k_2=0}^{1} f_0(k_2, k_1, k_0) W^{4n_0 k_2} \quad \text{(a)}$$

$$f_2(n_0, n_1, k_0) = \sum_{k_1=0}^{1} f_1(n_0, k_1, k_0) W^{(2n_1 + n_0)2k_1} \quad \text{(b)}$$

$$f_3(n_0, n_1, n_2) = \sum_{k_0=0}^{1} f_2(n_0, n_1, k_0) W^{(4n_2 + 2n_1 + n_0)k_0} \quad \text{(c)}$$

$$F(n_2, n_1, n_0) = f_3(n_0, n_1, n_2) \quad \text{(d)}$$

$$\tag{10.65}$$

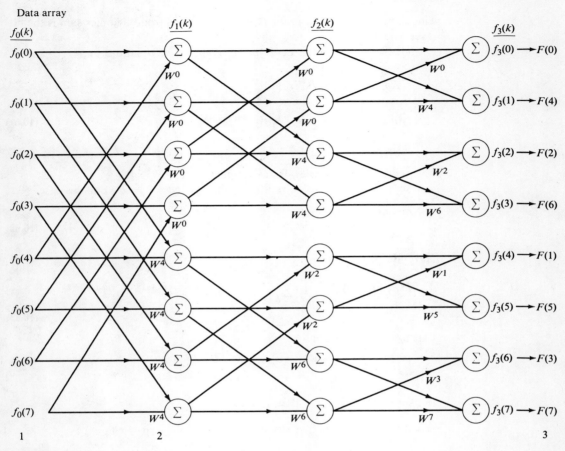

Figure 10.12
FFT data flow graph for $N = 8$.

The factorization given by these equations is shown graphically in Figure 10.12. This Cooley-Tukey procedure can be extended to the formulation of the FFT for $N = 2^\gamma$.

■ ■ ■

10-6 THE SANDE-TUKEY FFT

Another distinct form of the FFT was developed by Sande. To understand the features of the Sande approach, we will study the case for $N = 4$. The procedure again begins with (10.54) but the factor of W is separated into the components of n instead of the components of k. Thus we write

DISCRETE TRANSFORMS

Figure 10.13
The Sande-Tukey FFT algorithm.

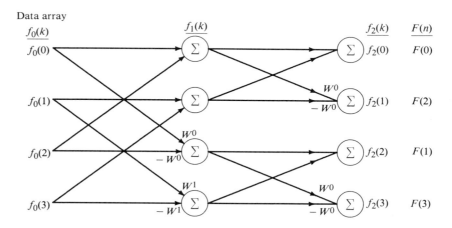

$$W^{(2n_1 + n_0)(2k_1 + k_0)} = W^{2n_1(2k_1 + k_0)} W^{n_0(2k_1 + k_2)}$$
$$= [W^{4n_1 k_1}] W^{2n_1 k_0} W^{n_0(2k_1 + k_0)}$$
$$= W^{2n_1 k_0} W^{n_0(2k_1 + k_0)}$$

because $W^4 = 1$.

Equation (10.54) can be written

$$F(n_1, n_0) = \sum_{k_0=0}^{1} \left[\sum_{k_1=0}^{1} f_0(k_1, k_0) W^{2n_0 k_1} W^{n_0 k_0} \right] W^{2n_1 k_0} \quad (10.66)$$

This expression can be written in the form

$$f_1(n_0, k_0) = \sum_{k_1=0}^{1} f_0(k_1, k_0) W^{2n_0 k_1} W^{n_0 k_0} \quad \text{(a)}$$

$$f_2(n_0, n_1) = \sum_{k_0=0}^{1} f_1(n_0, k_0) W^{2n_1 k_0} \quad \text{(b)} \quad (10.67)$$

$$F(n_1, n_0) = f_2(n_0, n_1) \quad \text{(c)}$$

The data flow graph describing these equations is given in Figure 10.13. The details indicated in this figure should be compared with those in Figure 10.11. Note that if the data array is displayed in inverse order, the form of the FFT data flow graphs in Figure 10.11 and in Figure 10.13 will be altered. This is illustrated in Figure 10.14.

The Cooley-Tukey algorithm is often referred to as decimation in time. Correspondingly, the Sande-Tukey algorithm is often referred to as decimation in frequency.

As a final word note that FFT algorithms can be developed for N = arbitrary factors. We considered the special case for $N = 2^\gamma$, where γ is an integer. The procedure for the case $N = r_1 r_2$ parallels the cases considered, but the FFT data flow graphs will vary with the base considered.

■ ■ ■

Figure 10.14
(a) The Cooley-Tukey and (b) Sande-Tukey FFT algorithms with data in inverse order.

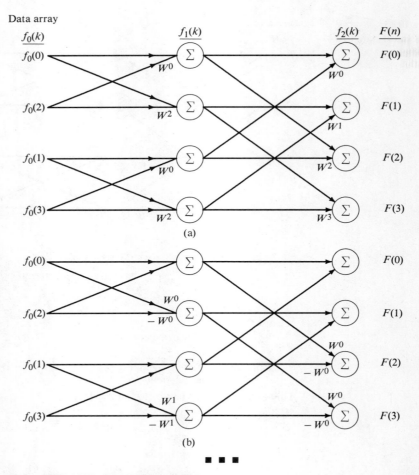

10-7 FFT AND CONVOLUTION

The discrete convolution relationship given in entry 8 of Section 10-2 is

$$y(n) = \sum_{k=0}^{N-1} f(k)g(n-k) \qquad (10.68)$$

where both $f(k)$ and $g(k)$ are periodic functions with period N. The use of the FFT algorithm greatly reduces computational time in this evaluation. Essentially what is required is to compute the DFT's of $f(k)$ and $g(k)$, which are given by

$$F(n) = \sum_{k=0}^{N-1} f(k)e^{-j2\pi nk/N} = \sum_{k=0}^{N-1} f(k)W^{nk} \quad \text{(a)} \qquad (10.69)$$

$$G(n) = \sum_{k=0}^{N-1} g(k)e^{-j2\pi nk/N} = \sum_{k=0}^{N-1} g(k)W^{nk} \quad \text{(b)}$$

DISCRETE TRANSFORMS

and then form the product

$$Y(n) = F(n)G(n) \qquad (c)$$

Finally, we deduce

$$y(k) = \frac{1}{N}\sum_{n=0}^{N-1} Y(n)e^{j2\pi nk/N} = \frac{1}{N}\sum_{n=0}^{N-1} Y(n)W^{-nk} \qquad (d)$$

Even though the sequence involves four steps, the total computing time is often substantially reduced.

To calculate correlation sequences, we use the same procedure as used for convolution sequences. We find in this case (refer to Problem 10-2.2)

$$y(n) = \sum_{k=0}^{N-1} f(k)g(n+k) \qquad (a)$$

$$Y(n) = F^*(n)G(n) \qquad (b) \qquad (10.70)$$

$$y(n) = \mathscr{D}^{-1}\{F^*(n)G(n)\} \qquad (c)$$

It is important to carefully consider the results of the DFT (or FFT) in studying real signals. The FFT applies only to the discrete Fourier transform. Further, the multiplication of DFT's corresponds to the convolution of periodic functions, not aperiodic ones. Thus the convolution of periodic functions can be regarded as a circular convolution in which the values of the functions are shifted from one end of a period and circulated into the other end. One may force a periodic convolution to yield results numerically identical to those of an aperiodic one by augmenting functions with zero values. The extent of the augmentation dictates how much of the periodic convolution is rendered into an aperiodic equivalent. A second method employs a window to reduce the effects of the truncation resulting from the selection of finite samples before taking the DFT. Specifically, if the signal $f(t)$ extends beyond the total sampling period NT, the resulting frequency spectrum is an approximation of the exact one. If, for example, we take the DFT of a truncated sinusoidal signal, we will find that the Fourier spectrum consists of additional lines that are the result of the truncation process. Therefore, if N is small and the sampling covers neither a large number nor an integral number of cycles of the signal, a large error in spectral representation may occur. This phenomenon is known as **leakage** and is the direct result of truncation. Since the truncated portion of the signal is equal to $f(t)p_a(t)$, the leakage is the result of the rectangular window $p_a(t)$. For this reason many types of windows have been suggested; one type with considerable appeal has zero values and derivatives at $t = 0$ and $t = NT$.

EXAMPLE 10.7
Find the DFT of the exponential function shown in Figure 10.15a for durations $t = 1$ s and $t = 1.5$ s.

Figure 10.15
Illustration of the truncation effect.

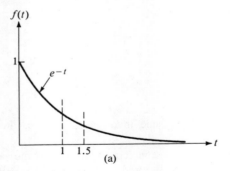

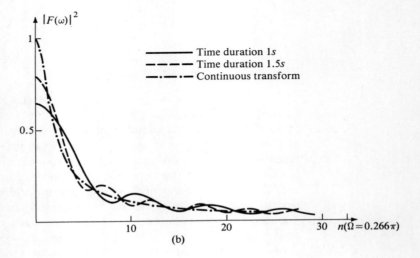

Solution: The results are shown in Figure 10.15b and were obtained by a direct application of the DFT expression. Observe that as NT increases (that is, as we incorporate more and more of the function in our calculations), the variations (or noise) on its spectrum decreases. This phenomenon is caused by the truncation, which is equivalent to multiplying the function by a pulse of adjustable width. ∎

To avoid leakage we must multiply the length of the data record by a rounded window before taking the DFT. That is, we determine $\mathscr{D}\{f(k)w(k)\}$ rather than $\mathscr{D}\{f(k)\}$, which is equivalent to $\mathscr{D}\{f(k)p_a(k)\}$. A number of important window functions exist. These include:

Hann function: This is given by

$$w(t) = \frac{1}{2} - \frac{1}{2}\cos\frac{2\pi t}{NT} \qquad 0 \leq t \leq NT \qquad \textbf{(a)} \qquad (10.71)$$

Figure 10.16
Different truncating windows: (a) Hann. (b) Hamming. (c) Blackman. (d) Hann-Poisson with $\alpha = 3$. (e) Hann-Poisson with $\alpha = 0.5$.

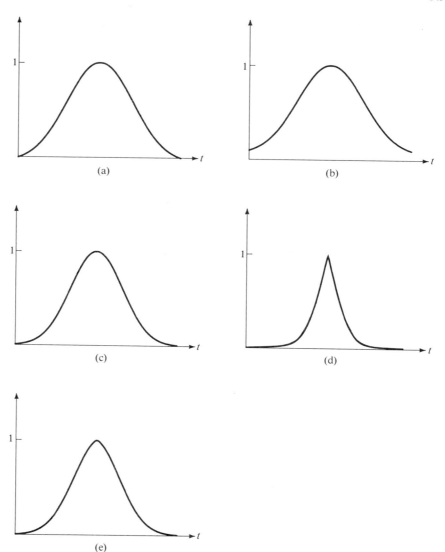

or in discrete form

$$w(k) = \frac{1}{2} - \frac{1}{2} \cos \frac{2\pi k}{N-1} \qquad 0 \le k \le N-1 \quad \textbf{(b)}$$

Figure 10.16a shows the Hann function, which has zero first-order derivatives at its two endpoints. In addition to the Hann window, we also show some additional ones that are commonly used in practice (see Figures 10.16b–10.16e).

Hamming window:

$$w(k) = 0.54 - 0.46 \cos \frac{2\pi k}{N-1} \qquad k = 0, 1, 2, \ldots, N-1 \tag{10.72}$$

Blackman window:

$$w(k) = 0.42 - 0.5 \cos \frac{2\pi k}{N-1} + 0.08 \cos \frac{4\pi k}{N-1} \tag{10.73}$$

$$k = 0, 1, \ldots, N-1$$

Hann-Poisson:

$$w(k) = 0.5 \left[1 - \cos\left(\frac{2\pi k}{N-1}\right)\right] e^{-\alpha 2|k-(N/2)|/N} \tag{10.74}$$

$$k = 0, 1, \ldots, N-1$$

Triangle

$$w(k) = \begin{cases} \dfrac{k}{N/2} & k = 0, 1, \ldots, \dfrac{N}{2} \\ w(N-k) & k = \dfrac{N}{2}, \ldots, N-1 \end{cases} \tag{10.75}$$

$\sin^\alpha(k)$ type:

$$w(k) = \sin^\alpha\left(\frac{k\pi}{N}\right) \qquad k = 0, 1, \ldots, N-1 \tag{10.76}$$

$$w(k) = \sin^2\left(\frac{k\pi}{N}\right) = 0.5\left[1 - \cos\frac{2\pi k}{N}\right] \qquad k = 0, 1, \ldots N-1 \tag{10.77}$$

■ ■ ■

10-8 WALSH TRANSFORMS

A number of linear transformations exist that have the form of the one-dimensional discrete Fourier transform relations. Such transformations can be written in the form

$$F(n) = \sum_{k=0}^{N-1} f(k) a(k, n) \quad \text{(a)}$$

$$f(k) = \sum_{n=0}^{N-1} F(n) b(k, n) \quad \text{(b)} \tag{10.78}$$

where $a(k, n)$ is the **forward transformation kernel** and $b(k, n)$ is the **inverse transformation kernel**. In the case of the DFT, the functions $a(k, n)$ and $b(k, n)$ are

Figure 10.17
Walsh-ordered functions of sequency $N = 8$.

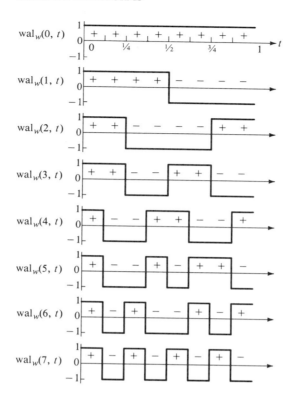

$\exp[-j2\pi kn/N]$ and $[\exp[j2\pi kn/N]/N]$, respectively. Another transformation pair of interest that is of this general type is the Walsh transformation and its related Hadamard-ordered Walsh transformation.

The Walsh functions form an orthogonal set defined over the interval [0, 1]. These functions were discussed in Section 1-4 and in Appendix 1-2, but the approach here is different. Because these functions form an orthogonal set, they are useful in creating additional transforms in much the same way that sine and cosine functions are useful in a number of transforms.

If we sample the Walsh functions shown in Figure 10.17 at the designated $8 = 2^3$ points (more generally $2^r = N$ at N equidistant points), we will obtain an 8×8 matrix (or equivalently, an $N \times N$ matrix for N points). The corresponding matrix is given in Figure 10.18, which shows only the signs (all amplitudes are of magnitude one). Observe that the rows (or columns) are orthogonal to each other. The one-dimensional kernel for this function is written

$$a_w(k, n) = (-1)^{\sum_{r=0}^{r-1} p_r(k) q_r(n)} \quad \text{(a)} \tag{10.79}$$

where
$$q_0(n) = p_{r-1}(n)$$
$$q_1(n) = p_{r-1}(n) + p_{r-2}(n)$$
$$q_2(n) = p_{r-2}(n) + p_{r-3}(n) \quad \text{(b)}$$
$$\vdots$$
$$q_{r-1}(n) = p_1(n) + p_0(n)$$

and where the arguments of $p_r(k)$ and $q_r(n)$ are written in binary form, with the value of $p_r(n)$ equal to the rth bit in the binary expansion of n. Specifically, the Walsh-ordered transformation functions are written explicitly

$$F(n) = \sum_{k=0}^{N-1} f(k)(-1)^{\sum_{r=0}^{r-1} p_r(k)q_r(n)} \quad \text{(a)}$$

$$f(k) = \frac{1}{N} \sum_{n=0}^{N-1} F(n)(-1)^{\sum_{r=0}^{r-1} p_r(k)q_r(n)} \quad \text{(b)}$$

(10.80)

Observe from these that the inverse kernel is equal to the forward kernel divided by N.

EXAMPLE 10.8
Find the element $a_w(4, 3)$ of the matrix shown in Figure 10.18.

Solution: We apply (10.79) and have

$$a_{w43}(r = 3) = (-1)^{\sum_{r=0}^{2} p_r(4)q_r(3)} = (-1)^{[p_0(4)q_0(3) + p_1(4)q_1(3) + p_2(4)q_2(3)]}$$

$$= (-1)^{[p_0(4)p_2(3) + p_1(4)[p_2(3) + p_1(3)] + p_2(4)[p_1(3) + p_0(3)]]}$$

Figure 10.18 Values of the Walsh-ordered transformation kernel for $r = 3$. The plus and minus signs substitute for $+1$ and -1, respectively, for convenience in writing.

$$\mathbf{a}_w(r=3) = \begin{array}{c} \\ n \downarrow \end{array} \begin{array}{c} k \longrightarrow \\ \begin{array}{c} 0 \\ 1 \\ 2 \\ 3 \\ 4 \\ 5 \\ 6 \\ 7 \end{array} \left[\begin{array}{cccccccc} 0 & 1 & 2 & 3 & 4 & 5 & 6 & 7 \\ + & + & + & + & + & + & + & + \\ + & + & + & + & - & - & - & - \\ + & + & - & - & - & - & + & + \\ + & + & - & - & + & + & - & - \\ + & - & - & + & + & - & - & + \\ + & - & - & + & - & + & + & - \\ + & - & + & - & - & + & - & + \\ + & - & + & - & + & - & + & - \end{array} \right] \end{array}$$

DISCRETE TRANSFORMS

But we have

$$p_0(4) \triangleq p_0(100) = 0 \quad p_1(4) \triangleq p_1(100) = 0 \quad p_2(4) \triangleq p_2(100) = 1$$

$$p_0(3) \triangleq p_0(011) = 1 \quad p_1(3) \triangleq p_1(011) = 1 \quad p_2(3) \triangleq p_2(011) = 0$$

Therefore we have

$$a_{w43}(r = 3) = (-1)^{[0 \times 0 + 0 \times [0+1] + 1 \times [1+1]]} = (-1)^2 = +1 \triangleq +$$

which is the value of this element shown in Figure 10.18.

Note that (10.80) can also be written in matrix form; namely,

$$\mathbf{F}(r) = \mathbf{a}_w(r)\mathbf{f}(r) \qquad N = 2^r \quad \text{(a)}$$

$$\mathbf{f}(r) = \frac{1}{N} \mathbf{a}_w^{-1}(r)\mathbf{F}(r) \qquad \text{(b)} \qquad (10.81)$$

where $\mathbf{a}_w(r)$ is the Walsh-ordered matrix that is shown in Figure 10.18 for the case $r = 3$.

EXAMPLE 10.9

Find the Walsh-ordered transform of the sequence $f(k) = \{0, 1, 2, 3\}$.

Solution: From a reduced form of Figure 10.17 for $N = 4$ $[a_w(r = 2)]$, we write

$$\mathbf{F}(2) \triangleq \begin{bmatrix} F(0) \\ F(1) \\ F(2) \\ F(3) \end{bmatrix} = \begin{bmatrix} 1 & 1 & 1 & 1 \\ 1 & 1 & -1 & -1 \\ 1 & -1 & -1 & 1 \\ 1 & -1 & 1 & -1 \end{bmatrix} \begin{bmatrix} 0 \\ 1 \\ 2 \\ 3 \end{bmatrix} = \begin{bmatrix} 6 \\ -4 \\ 0 \\ -2 \end{bmatrix}$$

from which we have

$$F(n) = \{6, -4, 0, -2\}$$

A review of the number of additions and subtractions to compute this transform shows there to be N^2 operations. ■

By analogy with the power spectrum for a DFT, we can write the power spectrum for the Walsh-ordered transform. This is defined as follows:

$$P_w(0) = F^2(0)$$

$$P_w(n) = F^2(2n - 1) + F^2(2n) \qquad n = 1, 2, \ldots, \left(\frac{N}{2} - 1\right) \qquad (10.82)$$

$$P_w\left(\frac{N}{2}\right) = F^2(N - 1)$$

The power spectrum of Example 10.9 is readily found to be:

$$P_w(0) = 36 \qquad P_w(1) = F^2(1) + F^2(2) = (-4)^2 + 0^2 = 16$$
$$P_w(2) = (-2)^2 = 4$$

Associated with the amplitude function $a_w(r)$ is a phase spectrum defined as

$$\varphi_w(0) = 0, \pi$$
$$\varphi_w\left(\frac{N}{2}\right) = 2k\pi \pm \frac{\pi}{2} \qquad k = 0, 1, 2, \ldots \tag{10.83}$$
$$\varphi_w(n) = \tan^{-1}\left\{\frac{F(2n-1)}{F(2n)}\right\} \qquad n = 1, 2, \ldots, \left(\frac{N}{2} - 1\right)$$

■ ■ ■

10-9 HADAMARD TRANSFORM

The set of Hadamard-ordered Walsh functions is illustrated in Figure 10.19 (also Figure 1.25b). A comparison of these functions with those shown in Figure 10.17 shows them to be of the same form, although there is a change in the numbering system. In fact, the Hadamard-ordered Walsh function transfor-

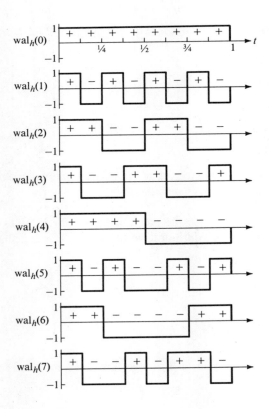

Figure 10.19
The Hadamard-ordered Walsh functions, $N = 8$.

DISCRETE TRANSFORMS

mations are precisely the same as the Walsh function transformations given in Equations 10.80, which we repeat here for convenience,

$$F(n) = \sum_{k=0}^{N-1} f(k)(-1)^{\sum_{r=0}^{r-1} p_r(k)q_r(n)} \quad (a)$$

$$f(k) = \frac{1}{N} \sum_{n=0}^{N-1} F(n)(-1)^{\sum_{r=0}^{r-1} p_r(k)q_r(n)} \quad (b)$$

(10.84)

except for the manner in which one writes the values for k and n in the values of $p_r(k)$ and $q_r(n)$. For each k or n, the binary equivalent is now written first, as for the Walsh-ordered functions. Then the binary forms are bit-reversed and then the Gray code conversion of this bit-reversed number is used. The Gray-code-to-binary-code transformations are contained in Appendix 1-2 and the Walsh- to Hadamard-ordering is contained in Table 1.2.

If we sample the Hadamard-ordered Walsh functions (see Figure 10.19) in 8 equal samples, we create an 8×8 matrix, which is shown in Figure 10.20a. Observe the presence in this matrix of an array of 2nd-order matrices (some with simple sign changes). This second-order Hadamard matrix, shown in Fig-

$$\mathbf{a}_h(r=3) = \begin{bmatrix} + & + & + & + & + & + & + & + \\ + & - & + & - & + & - & + & - \\ + & + & - & - & + & + & - & - \\ + & - & - & + & + & - & - & + \\ + & + & + & + & - & - & - & - \\ + & - & + & - & - & + & - & + \\ + & + & - & - & - & - & + & + \\ + & - & - & + & - & + & + & - \end{bmatrix}$$

(a)

$$\mathbf{a}_{h2} = \begin{bmatrix} + & + \\ + & - \end{bmatrix}$$

(b)

$$\mathbf{a}_{h2 \cdot 2} \triangleq \mathbf{a}_{h4} = \begin{bmatrix} \mathbf{a}_{h2} & \mathbf{a}_{h2} \\ \mathbf{a}_{h2} & -\mathbf{a}_{h2} \end{bmatrix}$$

(c)

$$\mathbf{a}_{h2 \cdot 4} \triangleq \mathbf{a}_{h8} = \begin{bmatrix} \mathbf{a}_{h4} & \mathbf{a}_{h4} \\ \mathbf{a}_{h4} & -\mathbf{a}_{h4} \end{bmatrix} = \begin{bmatrix} \mathbf{a}_{h2} & \mathbf{a}_{h2} & \mathbf{a}_{h2} & \mathbf{a}_{h2} \\ \mathbf{a}_{h2} & -\mathbf{a}_{h2} & \mathbf{a}_{h2} & -\mathbf{a}_{h2} \\ \mathbf{a}_{h2} & \mathbf{a}_{h2} & -\mathbf{a}_{h2} & -\mathbf{a}_{h2} \\ \mathbf{a}_{h2} & -\mathbf{a}_{h2} & -\mathbf{a}_{h2} & \mathbf{a}_{h2} \end{bmatrix}$$

(d)

Figure 10.20
The Hadamard-ordered matrix.

ure 10.20b, can be used to construct, recursively, any higher order ($2N$) matrices, as shown in Figures 10.20c and 10.20d. If we substitute the values for the $\mathbf{a}_{h2}$ matrices in Figure 10.20d, we obtain Figure 10.20a, as we should.

As for the Walsh-ordered transformation in matrix form, (10.84) is given by

$$\begin{aligned} \mathbf{F}(r) &= \mathbf{a}_h(r)\mathbf{f}(r) \qquad N = 2^r \quad &\text{(a)} \\ \mathbf{f}(r) &= \frac{1}{N}\mathbf{a}_h^{-1}(r)\mathbf{F}(r) \quad &\text{(b)} \end{aligned} \qquad (10.85)$$

where $\mathbf{a}_h(r)$ is the Hadamard-ordered matrix shown in Figure 10.20a for the case $r = 3$.

■ ■ ■

10-10. FAST HADAMARD AND WALSH TRANSFORMATIONS

We wish to show that the Walsh and Hadamard transformations can be cast in forms quite like the FFT as a routine in calculations. Because of the binary nature of the Walsh (and related) transforms, these transforms have an advantage over the Fourier transform because of their computational simplicity. That is, the transform matrices are made up only of ± 1's and the required computations in any digital signal processing application are only additions and subtractions. The resulting hardware implementation of such transformations is relatively simple, and these transformations are being applied in such diverse fields as image processing, signal coding, imaging in astronomy, filter design, imaging spectrometry, and photoacoustic spectroscopy and imaging.

We begin with (10.85) and choose $r = 3$ ($N = 8$) for the Hadamard transform. Using the results contained in Figure 10.20, we can write explicitly

$$\begin{bmatrix} F(0) \\ F(1) \\ F(2) \\ F(3) \\ \hline F(4) \\ F(5) \\ F(6) \\ F(7) \end{bmatrix} = \begin{bmatrix} \mathbf{a}_{h4} & \mathbf{a}_{h4} \\ \mathbf{a}_{h4} & \mathbf{a}_{h4} \end{bmatrix} \begin{bmatrix} f(0) \\ f(1) \\ f(2) \\ f(3) \\ \hline f(4) \\ f(5) \\ f(6) \\ f(7) \end{bmatrix} \qquad (10.86)$$

We now employ the matrix equality property (see Appendix 2) to split the matrix expression into two matrices, as follows

$$\begin{bmatrix} F(0) \\ F(1) \\ \hline F(2) \\ F(3) \end{bmatrix} = \mathbf{a}_{h4} \begin{bmatrix} f(0) + f(4) \\ f(1) + f(5) \\ \hline f(2) + f(6) \\ f(3) + f(7) \end{bmatrix} \triangleq \mathbf{a}_{h4} \begin{bmatrix} f_1(0) \\ f_1(1) \\ f_1(2) \\ f_1(3) \end{bmatrix} = \begin{bmatrix} \mathbf{a}_{h2} & \mathbf{a}_{h2} \\ \mathbf{a}_{h2} & \mathbf{a}_{h2} \end{bmatrix} \begin{bmatrix} f_1(0) \\ f_1(1) \\ f_1(2) \\ f_1(3) \end{bmatrix} \quad \text{(a)}$$

$$(10.87)$$

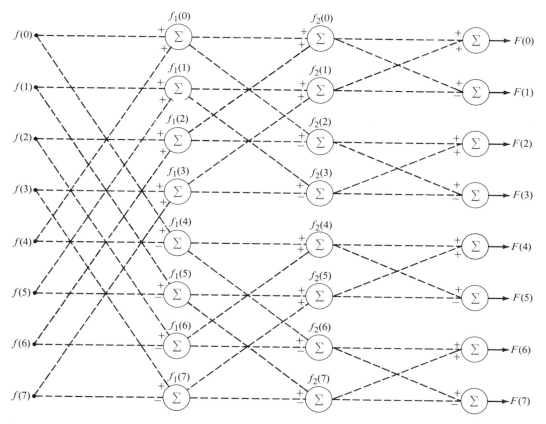

Figure 10.21
Data flow graph of the fast Hadamard transform.

$$\begin{bmatrix} F(4) \\ F(5) \\ \hline F(6) \\ F(7) \end{bmatrix} = \mathbf{a}_{h4} \begin{bmatrix} f(0)-f(4) \\ f(1)-f(5) \\ \hline f(2)-f(6) \\ f(3)-f(7) \end{bmatrix} \triangleq \mathbf{a}_{h4} \begin{bmatrix} f_1(4) \\ f_1(5) \\ \hline f_1(6) \\ f_1(7) \end{bmatrix} = \begin{bmatrix} \mathbf{a}_{h2} & \mathbf{a}_{h2} \\ \mathbf{a}_{h2} & \mathbf{a}_{h2} \end{bmatrix} \begin{bmatrix} f_1(4) \\ f_1(5) \\ f_1(6) \\ f_1(7) \end{bmatrix} \quad \text{(b)}$$

These two equations can similarly be split into the following matrices:

$$\begin{bmatrix} F(0) \\ F(1) \end{bmatrix} = \mathbf{a}_{h2} \begin{bmatrix} f_1(0)+f_1(2) \\ f_1(1)+f_1(3) \end{bmatrix} \triangleq \mathbf{a}_{h2} \begin{bmatrix} f_2(0) \\ f_2(1) \end{bmatrix} \quad \text{(a)}$$

$$\begin{bmatrix} F(2) \\ F(3) \end{bmatrix} = \mathbf{a}_{h2} \begin{bmatrix} f_1(0)-f_1(2) \\ f_1(0)-f_1(3) \end{bmatrix} \triangleq \mathbf{a}_{h2} \begin{bmatrix} f_2(2) \\ f_2(3) \end{bmatrix} \quad \text{(b)}$$

$$\begin{bmatrix} F(4) \\ F(5) \end{bmatrix} = \mathbf{a}_{h2} \begin{bmatrix} f_1(4)+f_1(6) \\ f_1(5)+f_1(7) \end{bmatrix} \triangleq \mathbf{a}_{h2} \begin{bmatrix} f_2(4) \\ f_2(5) \end{bmatrix} \quad \text{(c)} \qquad (10.88)$$

$$\begin{bmatrix} F(6) \\ F(7) \end{bmatrix} = \mathbf{a}_{h2} \begin{bmatrix} f_1(4)-f_1(6) \\ f_1(5)-f_1(7) \end{bmatrix} \triangleq \mathbf{a}_{h2} \begin{bmatrix} f_2(6) \\ f_2(7) \end{bmatrix} \quad \text{(d)}$$

The data flow graph of these operations is shown in Figure 10.21. The number

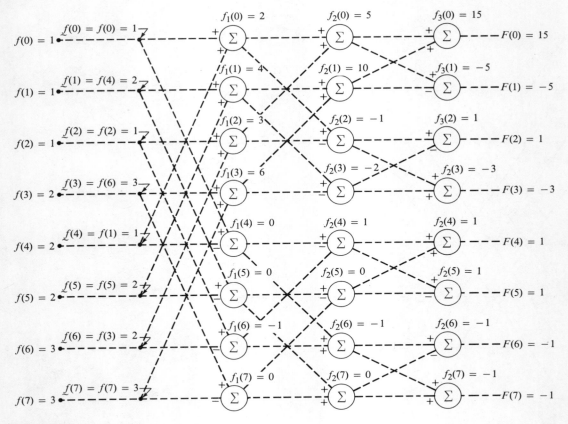

Figure 10.22
Data flow graph of the fast Walsh transform.

of operations (additions and subtractions) required in these calculations equals $8 \log_2 8 = 24$ ($N \log_2 N$ in the general case).

The fast Walsh-ordered transform algorithm, which uses a Cooley-Tukey type data flow graph, was developed by Manz. It is illustrated in Figure 10.22. In this figure the second column specifies a set of functions in an ascending index order, functions that are created from the first column by bit-reversal. Note that the first iteration is identical with the first iteration of the Hadamard data flow graph shown in Figure 10.21. The second iteration (column 4) is deduced from two prior blocks. One block contains rows 0–3 and the second block contains rows 4–7 [see (10.87)] for the general pattern. By comparing columns 3 and 4, we observe that the second block of iterations has additions and subtractions with reversed signs. The same situation exists in the third iteration for block 2 (rows 3 and 4) and block 4 (rows 5 and 7).

REFERENCES

1. Ahmed, N., and T. Natarajan. *Discrete-Time Signals and Systems*, Reston, Va.: Reston, 1983.
2. Ahmed, N., and K. R. Rao. *Orthogonal Transforms for Digital Signal Processing*. New York: Springer-Verlag, 1973.
3. Andrews, H. C., and K. L. Caspari. "A Generalized Technique for Spectral Analysis." *IEEE Transactions on Computers* C-19 (1970): 16.
4. Bergland, G. D. "Guided Tour of the Fast Fourier Transform." *IEEE Spectrum* 6 (1969): 41.
5. Brigham, E. O. *The Fast Fourier Transform*. Englewood Cliffs, N.J.: Prentice-Hall, 1974.
6. Cooley, J. W., P. A. W. Lewis, and P. D. Welch. "The Finite Fourier Transform." *IEEE Transactions on Audio and Electroacoustics* AU-17 (1969): 77.
7. Harmuth, H. F. *Transmission of Information by Orthogonal Functions*. 2d ed. New York: Springer-Verlag, 1972
8. Manz, J. W. "A Sequency-Ordered Fast Walsh Transform." *IEEE Transactions on Audio and Electroacoustics* AU-20 (1972): 204.
9. Openheim, A. V., and R. W. Schafer, *Digital Signal Processing*. Englewood Cliffs, N.J.: Prentice-Hall, 1975.
10. Papoulis, A. *Signal Analysis*. New York: McGraw-Hill, 1977.
11. Pratt, W. K., H. C. Andrews, and J. Kane. "Hadamard Transform Image Coding," *Proceedings of the IEEE* 57 (1969): 58.
12. Rabiner, L. R., and B. Gold. *Theory and Application of Digital Signal Processing*, Englewood Cliffs, N.J.: Prentice-Hall, 1975.

PROBLEMS

10-1.1 Prepare both time and frequency sketches showing the steps in obtaining the DFT of the waveform given in Figure P10-1.1.

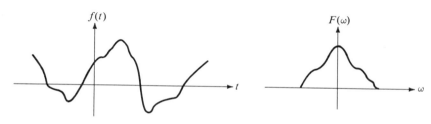

Figure P10-1.1

10-1.2 Write out several terms to establish the orthogonality relationship

$$\sum_{n=0}^{N-1} e^{2\pi j(m-n)k/N} = \begin{cases} N & \text{if } m = n \\ 0 & \text{if } m \neq n \end{cases} \quad m \text{ and } n \text{ integers}$$

10-1.3 Suppose that $f(t) = e^{-t}u(t)$. Sample this function at $T = 0.25$ s and compute the DFT for $N = 16$, $N = 32$. Compare these results and discuss the differences.

10-1.4 Complete all the required steps to prove (10.2).

10-1.5 Show that both $F(n\Omega)$ and $f(kT)$ are periodic and determine their periods.

10-1.6 Find the DFT of the exponential function $f(kT) = \exp(j\omega_0 kT)$ and compare it with the Fourier transform of $f(t) = \exp(j\omega_0 t)$. Next set $\omega_0 = m\omega_s/N = m2\pi/NT$ and find the final form of the DFT.

10-1.7 Compute the DFT of the signals shown in Figure P10-1.7.

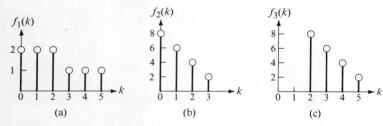

Figure P10-1.7

10-2.1 Show that the DFT of an arbitrary periodic function $f(k)$ can be written in terms of even and odd parts

$$F(n) = R(n) + jI(n) = F_e(n) + F_o(n)$$

10-2.2 Discrete correlation is defined by

$$y(k) = \sum_{i=0}^{N-1} f(i)u(k+i)$$

Show that $Y(n) = F^*(n)U(n)$.

10-2.3 Evaluate $\mathscr{D}\{f(k-6)\}$. Compare the results with (10.8).

10-2.4 Evaluate $\mathscr{D}^{-1}\{F(n-2)\}$ using (10.4). Compare the results with (10.9).

10-2.5 Compute $\mathscr{D}\{f(k-2)\}$. What can be said about the even-odd relationship of $f(k-2)$ given by (10.11) and (10.12)?

10-2.6 Suppose that $y(k) = f(k) + g(k-2)$, where $f(k)$ and $g(k)$ are even functions. Compute $\mathscr{D}\{y(k)\}$ and compare the results with (10.11) and (10.12).

10-2.7 Suppose that we have a sequence $f(n)$ with period N diluted with zeros between each member of the sequence to give a "stretched" sequence $\hat{f}(n)$ of period $2N$, where

$$\hat{f}(n) = f\left(\frac{n}{2}\right) \quad n \text{ even}$$
$$= 0 \quad n \text{ odd}$$

Show that $\hat{F}(k)$ is $F(k)$ with twice the number of terms, thereby providing interpolation midway between $F(k)$ samples. Sketch a sequence $f(n)$ and $\hat{f}(n)$ for $N = 4$.

10-2.8 Consider the time sequence $x(k)$, where

$$X(k) = \mathscr{D}\{x(k)\} \qquad k = 0, 1, 2, \ldots, (N-1)$$

Define a new sequence $w(k)$, where

$$w(k) = x(k) \qquad k = 0, 1, \ldots, (N-1)$$
$$ = 0 \qquad k = N, (N+1), \ldots, (2N-1)$$

Determine $\mathscr{D}\{w(k)\}$ and relate the results to $X(k)$. Show that odd values of k give interpolation between the even values.

10-2.9 Given two 3-period sequences $\{f(k)\} = \{1, 0, 2\}$ and $\{g(k)\} = \{-1, 1, 4\}$. Verify that

$$\sum_{i=0}^{2} f(i)g(2-i) = \frac{1}{N} \sum_{n=0}^{2} F(n)G(n)W^{-n2}$$

by considering each side independently.

10-2.10 Find the DFT of $f(kT)$, if

$$f(kT) = \begin{cases} 1 & \text{for } 1 \leq k \leq 3 \\ 0 & \text{for } k = 0, 4, 5. \end{cases}$$

Assume $N = 5$.

10-2.11 Verify the alternative inversion formula given in (10.10).

10-2.12 Find the DFT of a complex time function: (a) having both real and imaginary parts, (b) having only an imaginary part.

10-2.13 Use circular convolution to find the linear convolution of the two signals shown in Figure P10-2.13.

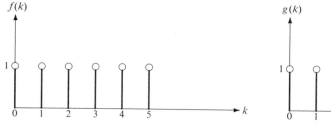
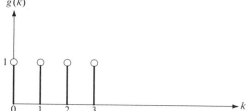

Figure P10-2.13

10-2.14 Show that the convolution of two periodic sequences with $N = 3$ is given by the matrix relationship

$$\begin{bmatrix} y(0) \\ y(1) \\ y(2) \end{bmatrix} = \begin{bmatrix} f(0) & f(1) & f(2) \\ f(1) & f(2) & f(0) \\ f(2) & f(0) & f(1) \end{bmatrix} \begin{bmatrix} g(0) \\ g(2) \\ g(1) \end{bmatrix}$$

By observing the relationships among the elements in the above matrix, write the matrix expression for $N = 4$ and then generalize it for any N.

10-2.15 Find the circular convolution of the two sequences shown in Figure P10-2.15 so that it is identical with their linear convolution.

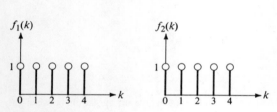
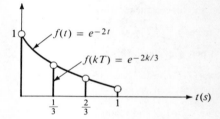

Figure P10-2.15

Figure P10-2.17

10-2.16 Show that the following relation applies for the discrete correlation of two functions:

$$\mathscr{D}\left\{\sum_{i=0}^{N-1} f(i)g(k+i)\right\} \triangleq \mathscr{D}\{f(k) \star g(k)\} = F^*(n)G(n)$$

10-2.17 Show that Parseval's relation holds for the discretized function shown in Figure P10-2.17. Show that these results also verify (10.20).

10-2.18 Plot the complex discrete signals $\exp[jk(2\pi/8)n]$ in the complex plane over the range $n = 0, 1, 2, \ldots, 7$ for different values of k.

10-2.19 Find the Fourier coefficients of the periodic sequences shown in Figure P10-2.19 and plot their spectra.

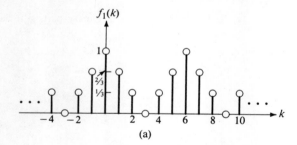

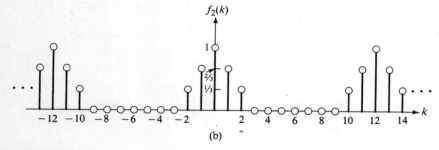

Figure P10-2.19

10-3.1 Develop an FFT for the computation of a 9-point DFT.

10-4.1 Let $f_0(k) = k + 1$ for $k = 0, 1, 2, 3$. Find $\mathscr{D}\{f_0(k)\}$ using (10.4) and note the number of multiplications and additions. Repeat the calculation using the FFT discussed in Section 10-4. Compare the number of additions and multiplications by each method.

10-4.2 Show the several terms that equal unity in the factorization of W^{nk} for $N = 8$ and $N = 16$.

10-5.1 Deduce the set of Cooley-Tukey equations for $N = 2^4 = 16$.

10-8.1 Find the element a_{w71} of the matrix shown in Figure 10.17.

10-8.2 Find the Walsh-ordered transforms and the spectra of the sequences: $\{f_1(k)\} = \{0, 1, 2, 3\}$ and $\{f_2(k)\} = \{0, \frac{1}{2}, 1, \frac{3}{2}, 2, \frac{5}{2}, 3, \frac{7}{2}\}$. Also, find the power spectrum.

10-9.1 Find the Hadamard transform of the sequence $\{f(k)\} = \{0, 1, 2, 3\}$.

10-9.2 Find the element a_{h31} of the matrix shown in Figure 10.20a.

10-9.3 The Paley-ordered Walsh functions differ from the Walsh order in that the binary digits appropriate to the values of k and n are replaced by their Gray code equivalents (see Figure 1.25a) (unlike the Hadamard transform, which considers the Gray code of the bits in reverse order). The Paley matrix is shown:

$$\mathbf{a}_p(r=3) = \begin{array}{c} n \\ \downarrow \end{array} \begin{array}{c} k \to \\ 0 \\ 1 \\ 2 \\ 3 \\ 4 \\ 5 \\ 6 \\ 7 \end{array} \begin{bmatrix} 0 & 1 & 2 & 3 & 4 & 5 & 6 & 7 \\ + & + & + & + & + & + & + & + \\ + & + & + & + & - & - & - & - \\ + & + & - & - & + & + & - & - \\ + & + & - & - & - & - & + & + \\ + & - & + & - & + & - & + & - \\ + & - & + & - & - & + & - & + \\ + & - & - & + & + & - & - & + \\ + & - & - & + & - & + & + & - \end{bmatrix}$$

Show that each entry of the matrix can be found from the one-dimensional kernel of the transform, which is given by

$$a_p(k, n) = \prod_{r=0}^{r=1}(-1)^{p_r(k)q_{[(r-1)-r]}(n)}$$

$$= (-1)^{p_0(k)q_{r-1}(n)}(-1)^{p_1(k)q_{[(r-1)-1]}(n)} \cdots (-1)^{p_{r-1}(k)q_0(n)}$$

10-10.1 Find the data flow diagram for the fast Hadamard transform, with $N = 16$.

10-10.2 Find the Walsh transform of the sequence $\{f(k)\} = \{1, 1, 1, 2, 2, 2, 3, 3\}$.

CHAPTER 11
ELEMENTS OF DIGITAL FILTER DESIGN

11-1 INTRODUCTION

A filter is a network or system that operates on an input signal in a specified way to produce a desired output. The signals may be continuous-time entities and may be stated in time or frequency terms. On the other hand, the signals may be discrete time and may also be stated in time or frequency terms. Because digital filter design draws heavily upon the methods of continuous-frequency filter design, we will first review the basic features of the latter.

Depending upon the service to which the filter is to be applied, the design process may vary considerably. The simplest type of filters might be referred to as "brute force." This would include the simple shunt C or series L, RC, or LC type filters used for filtering the ac ripple produced when a rectifier is used to convert ac into dc. Here the requirement is to permit the dc to pass unimpeded but to attenuate the ac ripple component. Hence such filters are essentially low-pass devices with cutoff at 0 Hz. Actually, these filters are not very frequency selective, but multiple sections can be used in cascade if improved ac attenuation is desired. In addition to such passive filters, one may use an electronic regulator, a device that not only serves to maintain the output voltage constant but also greatly reduces the ripple. From this point of view, the electronic regulator may be considered an active filter.

Perhaps more typical are filters that are frequency-selective assemblies designed to pass signals of certain frequencies and block signals of other frequencies. There are many classes of analog domain (continuous) filters categorized according to their behavior in the frequency domain and specified in terms of their magnitude or phase characteristics. Based on their magnitude or transfer response, filters are classified as low-pass, high-pass, bandpass, or bandstop. In the ideal cases these response characteristics are as shown in Figure 11.1 (see also Chapter 5). However, these ideal characteristics are not physically realizable, and a number of different approaches to filter design have been developed over the years to achieve acceptable approximations to the ideal responses. This has led to the formulation of constant-K filter design, m-derived filter design, and variants of these approaches. These filter designs, which are usually called classical filter design, have resulted in very acceptable filters.

ELEMENTS OF DIGITAL FILTER DESIGN

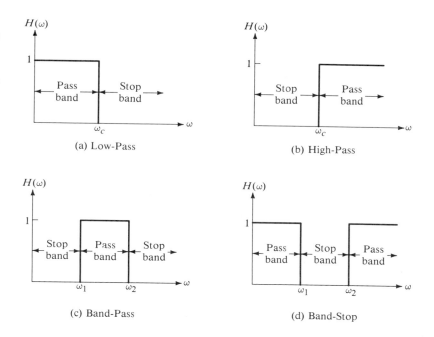

Figure 11.1 Ideal frequency response characteristics of analog filters: (a) Low-Pass. (b) High-Pass. (c) Bandpass. (d) Bandstop.

In modern filter design, which stems from the 1930s, the design problem is approached in a different manner. Essentially, the modern technique is first to find an analytic approximation to the specified filter characteristics as a transfer function, and then to develop a network that realizes this desired transfer function. For the filters shown in Figure 11.1 certain well-developed procedures exist, and these lead to such functions as Butterworth, Chebyshev, and elliptic. The use of these functions has the advantage that the formulas for them are well established and design tables are available. The second step in the design is the realization of the transfer function by passive or active networks. An extensive literature has been developed on active networks for use in such filter design problems.

An important feature of modern filter design is that the problems of approximation and of realization are solved separately to achieve optimum results. For our purposes we are mainly interested in the approximation problem. We will study the means for converting from the s-plane, in which the $H(s)$ approximation exists, to the z-plane and to a corresponding $H(z)$. $H(z)$ can then be realized by discrete systems either by transformation to difference equation form, which can then be adapted for computer calculations, or by direct hardware implementation. That is, the resulting difference equation can be considered to denote a digital filter approximation to the analog filter.

We will develop this approach to digital filter design. However, it is important to know that greater flexibility in filter specifications can be obtained using optimization techniques than by using classical analog-to-digital conversion. A number of different techniques have been developed, but we will not pursue them here.

From a practical point of view, the implementation of the filter on a general-purpose digital computer places accuracy constraints on the realization of the transfer function because the registers can only contain a finite number of binary bits at any one instant. Therefore, the filter coefficients, which can be defined with infinite precision over the real number field, can be represented only by a finite number of binary digits in a hardware register. As a result filter coefficients as well as the arithmetic performed within a digital filter are subject to approximation errors and uncertainty. These limitations can cause the frequency response of the filter to differ measurably from that of the design model. This is true because the filter that is specified is very sensitive to the polynomial coefficients of $H(s)$. These sources of error within a digital filter are referred to as **roundoff noise** and require attention in practical implementation practices.

■ ■ ■

11-2 THE BUTTERWORTH FILTER

We will examine the use of the Butterworth function to approximate the low-pass filter shown in Figure 11.1a. The features of this function are illustrated graphically in Figure 11.2. Note that attention in this case is being given only to the amplitude function.

The amplitude response of the nth-order normalized Butterworth filter is given by

$$|H_n(j\omega)| = \frac{1}{\sqrt{1 + \omega^{2n}}} \qquad n = 1, 2, 3, \ldots \qquad (11.1)$$

Figure 11.2
Butterworth amplitude response.

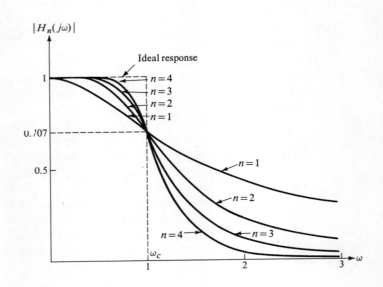

ELEMENTS OF DIGITAL FILTER DESIGN

As shown in Figure 11.2, the response is monotonically decreasing, having its maximum value $|H_n(j\omega)|_{max} = 1$ at $\omega = 0$. Further, the cutoff point at the normalized $\omega_c = 1$ is

$$|H_n(j1)| = \frac{1}{\sqrt{2}} = 0.707 |H_n(j\omega)|_{max}$$

for all orders. Because the functions all approach the value $H_n(j0) = 1$ smoothly, this function is called **maximally flat**. It is observed that the approximation to the square wave improves as n increases.

To obtain the transfer function form of the Butterworth function, we make use of the fact that $H_n(j\omega) = H_n(s)|_{s=j\omega}$ and rewrite (11.1) in squared form

$$H_n(s)H_n(-s) = \frac{1}{1 + \left(\frac{(j\omega)^2}{j^2}\right)^n} = \frac{1}{1 + (-s^2)^n} \tag{11.2}$$

Let us write the denominator polynomial in the form

$$D(s)D(-s) = 1 + (-s^2)^n \tag{11.3}$$

The roots of this function are obtained from

$$1 + (-s^2)^n = 0$$

Therefore

$$(-1)^n s^{2n} = -1 = e^{j(2k-1)\pi} \qquad k = 1, 2, \ldots, 2n \tag{11.4}$$

from which

$$s^{2n} = e^{j(2k-1)\pi} e^{j\pi n}$$

The kth root is

$$s_k \triangleq \sigma_k + j\omega_k = e^{j(2k+n-1)\pi/2n} = je^{j(2k-1)\pi/2n} \tag{11.5}$$

From this we write

$$s_k = -\sin\left[(2k-1)\frac{\pi}{2n}\right] + j\cos\left[(2k-1)\frac{\pi}{2n}\right] \qquad 1 \leq k \leq 2n \tag{11.6}$$

It is clear from (11.5) that the roots of s_k are on a unit circle and are spaced π/n radians apart. Moreover, no s_k can occur on the $j\omega$-axis since $(2k-1)$ cannot be an even integer. We thus see that there are n left half plane roots and n right half plane roots. The left half plane roots are associated with $H(s)$ since σ_k is negative for these. These results are shown in Figure 11.3 for the case $n = 4$:

$$s_1 = e^{j5\pi/8} \qquad s_2 = e^{j7\pi/8} \qquad s_3 = e^{j9\pi/8} \qquad s_4 = e^{j11\pi/8}$$

so that

$$D(s) = (s - s_1)(s - s_2)(s - s_3)(s - s_4)$$
$$= 1 + 2.6131s + 3.4142s^2 + 2.6131s^3 + s^4$$

Figure 11.3
Location of the zeros of the function (11.5), shown for $n = 4$.

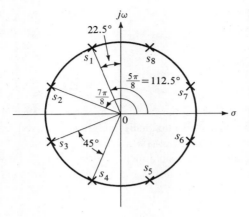

If we write $D(s)$ of the normalized Butterworth function of order n in the general form

$$D(s) = 1 + a_1 s + a_2 s^2 + \cdots + a_{n-1} s^{n-1} + s^n \tag{11.7}$$

then the coefficients can be computed in the manner shown above. Note that a_0 and a_n are always unity because the poles are all on the unit circle. Table 11.1 gives the functions for n to 8.

Table 11.2 gives the Butterworth polynomials [of (11.7)] in factored form.

EXAMPLE 11.1
Find the transfer function of a Butterworth filter that has an attenuation of at least 10 dB at twice the cutoff frequency $\omega_c = 2.5 \times 10^3$ rad/s.

Solution: We initially find the normalized Butterworth filter. At $\omega = 2\omega_c = 2 \times 1$ ($\omega_c = 1$, normalized)

$$|H_n(j2)|^2 = \frac{1}{1 + 2^{2n}} \tag{11.8}$$

The dB attenuation is given by

$$-10 \log_{10} |H_n(j2)|^2 = 10 \log_{10}(1 + 2^{2n}) \geq 10$$

from which $\mathrm{antilog}_{10} 1 \leq 1 + 2^{2n}$. Hence we find that

$$1 + 2^{2n} \geq 10 \quad \text{or} \quad 2^{2n} \geq 9$$

The order of the filter must then be $n = (\ln_2 9)/2 = 1.584$; hence a second-order Butterworth filter will satisfy this requirement. The corresponding transfer function of the normalized filter is (see Table 11.2)

$$H_n(s) = \frac{1}{s^2 + 1.4142s + 1} \tag{11.9}$$

Table 11.1 Coefficients of Butterworth Polynomials

n	a_1	a_2	a_3	a_4	a_5	a_6	a_7
2	1.4142						
3	2.0000	2.0000					
4	2.6131	3.4142	2.6131				
5	3.2361	5.2361	5.2361	3.2361			
6	3.8637	7.4641	9.1416	7.4641	3.8637		
7	4.4940	10.0978	14.5918	14.5918	10.0978	4.4940	
8	5.1528	13.1371	21.8462	25.6884	21.8462	13.1371	5.1258

Table 11.2 Factors of Butterworth Polynomials

n	Factored Polynomial
1	$s + 1$
2	$s^2 + 1.4142s + 1$
3	$(s + 1)(s^2 + s + 1)$
4	$(s^2 + 0.7654s + 1)(s^2 + 1.8478s + 1)$
5	$(s + 1)(s^2 + 0.6180s + 1)(s^2 + 1.6180s + 1)$
6	$(s^2 + 0.5176s + 1)(s^2 + 1.4142s + 1)(s^2 + 1.9319s + 1)$
7	$(s + 1)(s^2 + 0.4450s + 1)(s^2 + 1.2470s + 1)(s^2 + 1.8019s + 1)$
8	$(s^2 + 0.3902s + 1)(s^2 + 1.1111s + 1)(s^2 + 1.6639s + 1)(s^2 + 1.9616s + 1)$
9	$(s + 1)(s^2 + 0.3473s + 1)(s^2 + s + 1)(s^2 + 1.5321s + 1)(s^2 + 1.8794s + 1)$

The amplitude and phase of this filter are shown in Figure 11.4. For a cutoff frequency $\omega_c = 2.5 \times 10^3$, the transfer function is

$$H(s) = H_n\left(\frac{s}{\omega_c}\right) = \frac{1}{1.6 \times 10^{-7}s^2 + 5.6569 \times 10^{-4}s + 1} \tag{11.10}$$

The denominator polynomial has conjugate complex roots with negative real parts. If we split the function $D(s)$ of (11.10) into two polynomials including, respectively, the even and odd powers of s

$$D(s) = e(s) + o(s) = (1.6 \times 10^{-7}s^2 + 1) + (5.6569 \times 10^{-4}s) \tag{11.11}$$

their roots are purely imaginary (zero included) and, additionally, alternate $-2.5 \times 10^3, 0, 2.5 \times 10^3$. Polynomials that possess these properties are known as **Hurwitz polynomials**.

Observe also that (11.10) is of the general form

$$H(s) = \frac{k}{a_n s^n + a_{n-1} s^{n-1} + \cdots + a_0} = \frac{k}{D(s)} \tag{11.12}$$

Figure 11.4
The phase and amplitude characteristics of the second-order Butterworth filter.

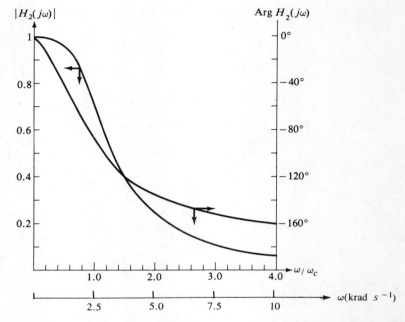

It can be shown that a transfer function of this form can be realized with passive elements if and only if $D(s)$ is a Hurwitz polynomial.

Suppose that we wish to find a two-port network of the terminated form shown in Figure 11.5a appropriate to (11.10). Since $D(s)$ is of second order, the lossless 2-port network shown in Figure 11.5b appears to be an appropriate form. The transfer function of this circuit is

$$\frac{V_o(s)}{V_i(s)} = H(s) = \frac{1}{CLs^2 + (C + L)s + 2} \tag{11.13}$$

This form differs from (11.10) by the factor 2 in the denominator. If we multiply the numerator and denominator of (11.10) by 2, we obtain an equivalent transfer function

$$\frac{H(s)}{2} = \frac{1}{3.2 \times 10^{-7}s^2 + 1.1314 \times 10^{-3}s + 2} \tag{11.14}$$

By comparing these two equations, we obtain the equations

$$CL = 3.2 \times 10^{-7}$$

$$C + L = 1.1314 \times 10^{-3}$$

These can be solved to yield $C = 5.7 \times 10^{-4}$ F; $L = 5.6 \times 10^{-4}$ H. ∎

EXAMPLE 11.2
A Butterworth filter must have an attenuation of at least 20 dB at twice the cut-

Figure 11.5
Second-order Butterworth filter.

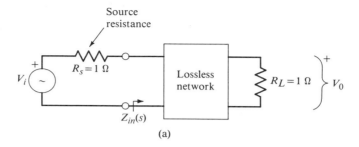

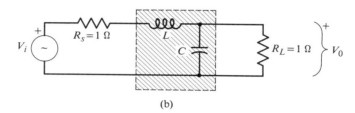

off frequency of 3 kHz. Find the transfer function of the appropriate low-pass filter.

Solution: First determine the normalized Butterworth filter. This requires that at twice the cutoff frequency

$$|H_n(j2)|^2 = \frac{1}{1 + 2^{2n}} \qquad (11.15)$$

The dB attenuation is then given by

$$-10 \log_{10}|H_n(j2)|^2 = 10 \log_{10}(1 + 2^{2n}) \geq 20$$

From this antilog$_{10}$ $2 \leq 1 + 2^{2n}$ or

$$1 + 2^{2n} \geq 10^2 = 100$$

and the order of the filter must be $n = 3.3$. Hence a fourth-order Butterworth will satisfy this requirement. The transfer function of the normalized filter is

$$H_n(s) = \frac{1}{1 + 2.6131s + 3.4142s^2 + 2.6131s^3 + s^4} \qquad (11.16)$$

The amplitude and phase characteristics of this filter are shown in Figure 11.6. With a cutoff frequency of 3 kHz, the denormalizing factor is

$$\omega_c = 2\pi \times 3000 = 18{,}849.54$$

Hence the desired transfer function is

Figure 11.6
Amplitude and phase characteristics of the fourth-order Butterworth filter.

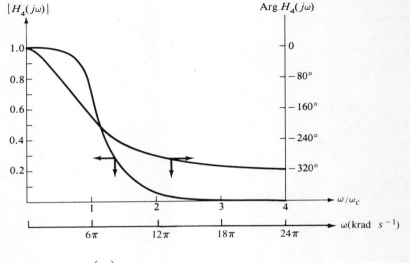

$$H(s) = H_n\left(\frac{s}{\omega_c}\right)$$

$$= \frac{1}{1 + 1.3863 \times 10^{-4}s + 9.609 \times 10^{-9}s^2 + 3.9017 \times 10^{-13}s^3 + 7.9213 \times 10^{-18}s^4} \quad (11.17)$$

∎

■ ■ ■

11-3 THE CHEBYSHEV LOW-PASS FILTER

An examination of Figure 11.2 shows that the Butterworth low-pass amplitude response is very good in the region of small ω and also in the region of large ω but is not very good in the neighborhood of the cutoff frequency ($\omega = 1$). The Chebyshev low-pass filter possesses sharper cutoff response, but it does possess amplitude variations within the passband. The features of the Chebyshev response are shown in Figure 11.7 for n even ($= 4$) and n odd ($= 5$). Several general features are contained in these figures; the oscillations in the passband have equal amplitudes for a given value of ϵ; the curves for n even always start from the trough of the ripple whereas the curves for n odd always start from the peak; and at the normalized cutoff frequency of 1, all curves pass through the same point, shown in the figures as $1/(1 + \epsilon^2)$.

The amplitude response of the Chebyshev low-pass filter is defined by

$$\boxed{|H_n(j\omega)| = \frac{1}{\sqrt{1 + \epsilon^2 C_n^2(\omega)}}} \qquad n = 1, 2, 3, \ldots \quad (11.18)$$

ELEMENTS OF DIGITAL FILTER DESIGN

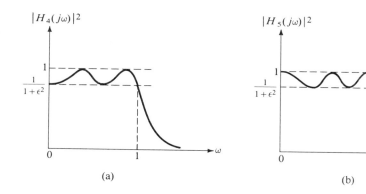

Figure 11.7 General shape of Chebyshev approximations. (a) n even ($= 4$). (b) n odd ($= 5$).

Table 11.3 Chebyshev Polynomials

n	$C_n(\omega)$
0	1
1	ω
2	$2\omega^2 - 1$
3	$4\omega^3 - 3\omega$
4	$8\omega^4 - 8\omega^2 + 1$
5	$16\omega^5 - 20\omega^3 + 5\omega$
6	$32\omega^6 - 48\omega^4 + 18\omega^2 - 1$
7	$64\omega^7 - 112\omega^5 + 56\omega^3 - 7\omega$
8	$128\omega^8 - 256\omega^6 + 160\omega^4 - 32\omega^2 + 1$
9	$256\omega^9 - 576\omega^7 + 432\omega^5 - 120\omega^3 + 9\omega$
10	$512\omega^{10} - 1280\omega^8 + 1120\omega^6 - 400\omega^4 + 50\omega^2 - 1$

where ϵ is a constant, and where $C_n(\omega)$ is the Chebyshev polynomial given by the equation

$$\begin{aligned} C_n(\omega) &= \cos(n \cos^{-1} \omega) & \text{for } |\omega| \leq 1 & \quad \text{(a)} \\ &= \cosh(n \cosh^{-1} \omega) & \text{for } |\omega| > 1 & \quad \text{(b)} \end{aligned} \quad (11.19)$$

as already discussed in Section 1-6. The Chebyshev polynomials of orders up to 5 are shown in Figure 1.29 for the range $|\omega| \leq 1$. The analytic form of the Chebyshev polynomials from orders 0 to 10 are tabulated in Table 11.3. Figure 11.8 shows the Chebyshev polynomials $C_n(\omega)$ of (11.19) for $\omega \geq 0$.

By taking into consideration (11.19) and the recurrence relationship [see (1.56)]

$$C_{n+1}(\omega) = 2\omega C_n(\omega) - C_{n-1}(\omega) \qquad n = 1, 2, \ldots$$

The nth-order Chebyshev polynomial has the following properties:

a. For any n $\quad 0 \leq |C_n(\omega)| \leq 1 \quad$ for $0 \leq |\omega| \leq 1$
$\quad\quad\quad\quad\quad\quad |C_n(\omega)| > 1 \quad$ for $|\omega| \geq 1$

b. $C_n(\omega)$ is monotonically increasing for $\omega \geq 1$ for all n

c. $C_n(\omega)$ is an odd polynomial, if n is odd
$\quad C_n(\omega)$ is an even polynomial, if n is even

d. $|C_n(0)| = 0 \quad$ for n odd
$\quad |C_n(0)| = 1 \quad$ for $n =$ even

The curves shown in Figure 11.8 together with (11.18) show that $|H_n(j\omega)|$ attains its maximum value of 1 at the zeros of $C_n(\omega)$ and for $|\omega| \leq 1$ attains its minimum value of $1/\sqrt{1 + \epsilon^2}$ at the points where $C_n(\omega)$ attains its maximum

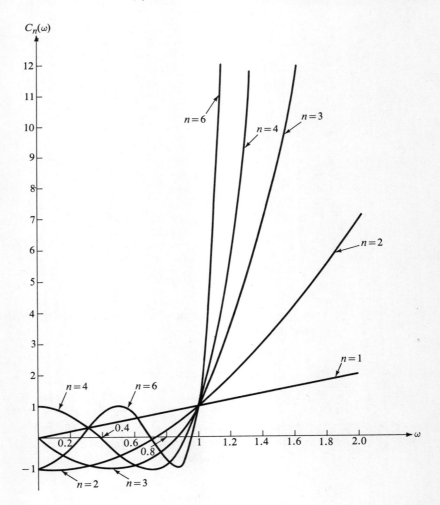

Figure 11.8
Chebyshev polynomials $C_n(\omega)$ for $\omega \geq 0$.

ELEMENTS OF DIGITAL FILTER DESIGN

values of 1. Thus the ripples in the passband $0 \le \omega \le 1$ have a peak-to-peak amplitude of

$$r = 1 - \frac{1}{\sqrt{1 + \epsilon^2}} \tag{11.20}$$

The ripple in decibels is given by

$$r_{dB} = -20 \log_{10} \frac{1}{\sqrt{1 + \epsilon^2}} = 10 \log_{10}(1 + \epsilon^2) \tag{11.21}$$

Outside of the passband $\omega > 1$, $|H_n(j\omega)|$ is monotonically decreasing.

To find the pole locations of $H_n(s)$ where $s = j\omega$, we consider the denominator of the function

$$H_n(s)H_n(-s) = \frac{1}{1 + \epsilon^2 C_n^2\left(\frac{s}{j}\right)} \tag{11.22}$$

More specifically, the poles of interest occur when $C_n(s/j) = \pm\sqrt{-1/\epsilon^2}$ or when

$$\cos\left[n \cos^{-1}\left(\frac{s}{j}\right)\right] = \pm\frac{j}{\epsilon} \tag{11.23}$$

We proceed by defining

$$\cos^{-1}\left(\frac{s}{j}\right) = \alpha - j\beta \tag{11.24}$$

Combine this with (11.23) from which

$$\cos n\alpha \cosh n\beta + j \sin n\alpha \sinh n\beta = \pm\frac{j}{\epsilon} \tag{11.25}$$

Real and imaginary parts must be equal on both sides of the equation so that

$$\cos n\alpha \cosh n\beta = 0 \quad \text{(a)}$$
$$\sin n\alpha \sinh n\beta = \pm\frac{1}{\epsilon} \quad \text{(b)} \tag{11.26}$$

From (11.26a) it is found, since $\cosh n\beta \ne 0$, that

$$\alpha = (2k - 1)\frac{\pi}{2n} \qquad k = 1, 2, 3, \ldots, 2n \tag{11.27}$$

From the second equation (11.26b) together with (11.27), β is found to be, because $\sin n\alpha = \pm 1$,

$$\beta = \pm\frac{1}{n}\sinh^{-1}\left(\frac{1}{\epsilon}\right) \tag{11.28}$$

CHAPTER 11

Equation (11.24) can be used to give the poles

$$s_k = j\cos(\alpha - j\beta)$$
$$= -\sin\left[(2k-1)\frac{\pi}{2n}\right]\sinh\left[\frac{1}{n}\sinh^{-1}\left(\frac{1}{\epsilon}\right)\right] \qquad (11.29)$$
$$+ j\cos\left[(2k-1)\frac{\pi}{2n}\right]\cosh\left[\frac{1}{n}\sinh^{-1}\left(\frac{1}{\epsilon}\right)\right]$$

These points are located on an ellipse in the s-plane, as illustrated in Figure 11.9 for $n = 4$. To prove that the locus is an ellipse, let

$$s_k = \sigma_k + j\omega_k \qquad \text{(a)} \qquad (11.30)$$

so that

$$\sigma_k = -\sin\left[(2k-1)\frac{\pi}{2n}\right]\sinh\left[\frac{1}{n}\sinh^{-1}\left(\frac{1}{\epsilon}\right)\right] \qquad \text{(b)}$$

$$\omega_k = \cos\left[(2k-1)\frac{\pi}{2n}\right]\cosh\left[\frac{1}{n}\sinh^{-1}\left(\frac{1}{\epsilon}\right)\right] \qquad \text{(c)}$$

It follows from (11.30b) and (11.30c) that

$$\frac{\sigma_k^2}{\left\{\sinh\left[\frac{1}{n}\sinh^{-1}\left(\frac{1}{\epsilon}\right)\right]\right\}^2} + \frac{\omega_k^2}{\left\{\cosh\left[\frac{1}{n}\sinh^{-1}\left(\frac{1}{\epsilon}\right)\right]\right\}^2} = 1 \qquad (11.31)$$

This is the equation of the ellipse shown in Figure 11.9. Equation 11.30c shows that the imaginary parts are the same as if the zeros had been uniformly spaced

Figure 11.9
Location of zeros of function (11.23) on the s-plane shown for $n = 4$ and $\epsilon = 0.458$.

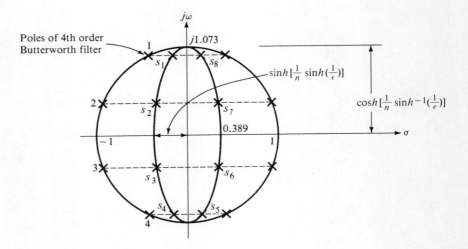

Table 11.4 Coefficients of the Polynomial in (11.32)

n	a_0	a_1	a_2	a_3	a_4	a_5	a_6	a_7
			$r = 0.5$dB		($\epsilon = 0.3493$)			
1	2.8628							
2	1.5162	1.4256						
3	0.7157	1.5349	1.2529					
4	0.3791	1.0255	1.7169	1.1974				
5	0.1789	0.7525	1.3096	1.9374	1.1725			
6	0.0948	0.4324	1.1719	1.5898	2.1718	1.1592		
7	0.0447	0.2821	0.7557	1.6479	1.8694	2.4127	1.1512	
8	0.0237	0.1525	0.5736	1.1486	2.1840	2.1492	2.6567	1.1461
			$r = 1.0$dB		($\epsilon = 0.5088$)			
1	1.9652							
2	1.1025	1.0977						
3	0.4913	1.2384	0.9883					
4	0.2756	0.7426	1.4539	0.9528				
5	0.1228	0.5805	0.9744	1.6888	0.9368			
6	0.0689	0.3071	0.9393	1.2021	1.9308	0.9283		
7	0.0307	0.2137	0.5486	1.3575	1.4288	2.1761	0.9231	
8	0.0172	0.1073	0.4478	0.8468	1.8369	1.6552	2.4230	0.9198

on a circle of radius $\cosh[(1/n)\sinh^{-1}(1/\epsilon)]$. Therefore, the graphical construction indicated can be used to locate the roots.

We can use the same reasoning as in Section 11-2 to show that the desired response can be obtained by limiting consideration to the zeros on the left half plane. The transfer function is thus

$$H(s) = \frac{K}{a_0 + a_1 s + a_2 s^2 + \cdots + a_{n-1} s^{n-1} + s^n} \qquad (11.32)$$

For a specified dB ripple, ϵ can be found from (11.21), and the poles of an nth-order Chebyshev filter can then be found from (11.30). The constant K must be selected to meet the specified dc gain level. Table 11.4 gives the coefficients in the denominator of (11.32) for two values of r.

EXAMPLE 11.3
Repeat Example 11.2 using a Chebyshev filter with a 1 dB ripple in the passband.

Table 11.5 Factors of the Polynomial in (11.32)

	Factored Polynomial
n	$r = 0.5\text{dB}$ $\quad \epsilon = 0.3493$
1	$s + 2.8628$
2	$s^2 + 1.4256s + 1.5162$
3	$(s + 0.6265)(s^2 + 0.6265s + 1.1424)$
4	$(s^2 + 0.3507s + 1.0635)(s^2 + 0.8467s + 0.3564)$
5	$(s + 0.3623)(s^2 + 0.2239s + 1.0358)(s^2 + 0.5862s + 0.4768)$
6	$(s^2 + 0.1553s + 1.0230)(s^2 + 0.4243s + 0.5900)(s^2 + 0.5796s + 0.1570)$

n	$r = 1 \text{ dB}$ $\quad \epsilon = 0.5088$
1	$s + 1.9652$
2	$s^2 + 1.0978s + 1.1025$
3	$(s + 0.4942)(s^2 + 0.4941s + 0.9942)$
4	$(s^2 + 0.2791s + 0.9865)(s^2 + 0.6737s + 0.2794)$
5	$(s + 0.2895)(s^2 + 0.1789s + 0.9883)(s^2 + 0.4684s + 0.4293)$
6	$(s^2 + 0.1244s + 0.9907)(s^2 + 0.3398s + 0.5577)(s^2 + 0.4641s + 0.1247)$

Solution: By (11.21) for the 1 dB ripple

$$\epsilon^2 = 10^{1/10} - 1 = 0.2589$$

from which

$$\epsilon = 0.5088$$

The attenuation can be found from (11.18), beginning with

$$|H_n(j\omega)|^2 = \frac{1}{1 + \epsilon^2 C_n^2(\omega)}$$

The dB attenuation for high frequencies is given by $[\epsilon^2 C_n^2(\omega) \gg 1]$

$$-10 \log_{10}|H_n(j\omega)|^2 = 10 \log_{10} \epsilon^2 + 10 \log_{10} C_n^2(\omega) \geq 20$$

Therefore

$$\log_{10} \epsilon + \log_{10}[\cosh(n \cosh^{-1} 2)] \geq 1$$

so that

ELEMENTS OF DIGITAL FILTER DESIGN

$$\log_{10}[\cosh(n \cosh^{-1} 2)] \geq 1 - \log_{10} \epsilon = 1 - (-.2935) = 1.2935$$
$$\cosh[(n \cosh^{-1} 2)] \geq 19.65$$
$$n \cosh^{-1} 2 \geq 3.670$$

from which we have

$$n \geq \frac{3.670}{1.317} = 2.8$$

Hence the conditions of the problem can be met with a third-order Chebyshev filter, and the normalized transfer function is

$$H_n(s) = \frac{0.4913}{0.4913 + 1.2384s + 0.9883s^2 + s^3} \tag{11.33}$$

With the denormalizing factor specified by

$$\omega_c = 2\pi \times 3000 = 18,849.54$$

the desired transfer function is

$$H(s) = H_n\left(\frac{s}{\omega_c}\right) \tag{11.34}$$

$$= \frac{0.4913}{0.4913 + 6.5699 \times 10^{-5}s + 2.7816 \times 10^{-9}s^2 + 1.4931 \times 10^{-13}s^3}$$

We could also proceed in the following manner. Using the data $\epsilon = 0.5088$ and $n = 3$, we obtain

$$\frac{1}{3}\sinh^{-1}\left(\frac{1}{\epsilon}\right) = \frac{1.4280}{3} = 0.4760$$

Therefore,

$$\sinh\left[\frac{1}{3}\sinh^{-1}\left(\frac{1}{0.5088}\right)\right] = 0.4942$$

$$\cosh\left[\frac{1}{3}\sinh^{-1}\left(\frac{1}{0.5088}\right)\right] = 1.1154$$

and by (11.30b) and (11.30c) we obtain the following poles

$$s_1 = \sigma_1 + j\omega_1 = -0.2471 + j0.9660$$
$$s_2 = \sigma_2 + j\omega_2 = -0.4942$$
$$s_3 = \sigma_3 + j\omega_3 = -0.2471 - j0.9660$$

When these pole positions are substituted into the transfer function expression

$$H_n(s) = \frac{0.4913}{(s - s_1)(s - s_2)(s - s_3)} \tag{11.35}$$

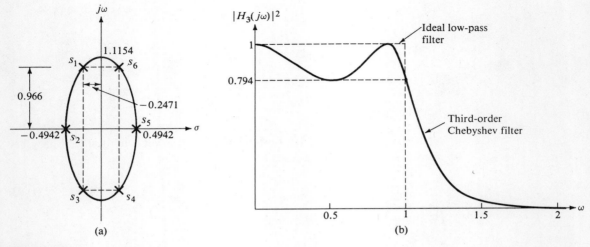

Figure 11.10
Illustrating Example 11.3.

the resulting form for $H_n(s)$ is precisely that given by (11.33). Figure 11.10a indicates graphically the location of the three poles, and Figure 11.10b shows a plot of the normalized third-order Chebyshev filter of (11.33).

$$|H_3(j\omega)|^2 = \frac{1}{1 + 0.2589(4\omega^3 - 3\omega)^2} \qquad (11.36)$$

■ ■ ■

11-4 ELLIPTIC FILTERS

We have found that the Butterworth approximation possesses a monotonic characteristic in both the passband and the stopband, while the Chebyshev approximation has a magnitude response that varies between equal maximum and equal minimum values in the passband and decreases monotonically in the stopband. Moreover, because of the willingness to accept a ripple in the passband, the Chebyshev filter possesses sharper cutoff characteristics in the stopband.

Another type of approximation is characterized by a magnitude response that is equiripple in both the passband and the stopband, as shown in Figure 11.11. This approximation is given by the amplitude function

$$|H(j\omega)| = \frac{1}{\sqrt{1 + \epsilon^2 R_n^2(\omega)}} \qquad (11.37)$$

where $R_n(\omega)$ is a Chebyshev rational function. The roots of the rational function ω_k are related to the Jacobi elliptic sine functions and the resulting filter is called the **elliptic filter**.

Figure 11.11
Magnitude-squared response of an elliptic filter.

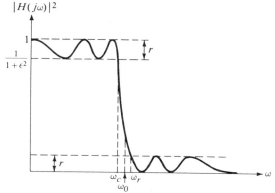

We will not undertake a discussion of the elliptic function filter, but note that the response curve is an improvement over the Chebyshev and Butterworth filters.

■ ■ ■

11-5 PHASE CHARACTERISTICS

Our prior discussion has focused on the amplitude response of low-pass filters. These results show that in most respects the Chebyshev filter is superior to the Butterworth and in some cases the elliptic filter is superior to both. However, in these discussions, we have totally ignored the phase characteristics, which become progressively worse (less linear) as the amplitude response is improved. Often the phase characteristic is an important factor, since a linear phase response is necessary if one wishes to transmit a pulse through a network without distortion (although a time delay will ensue). We will find that FIR digital filters possess linear phase characteristics.

Instead of beginning with considerations of the amplitude function, it is possible to consider obtaining a realizable approximation to an ideal constant delay function e^{-s}. One method for obtaining a realizable approximation leads to an $H(s)$ specified in terms of Bessel polynomials. Pursuing this matter is beyond our present concerns.

■ ■ ■

11-6 LOW-PASS TO HIGH-PASS TRANSFORMATION

By an appropriate frequency transformation, the $H_n(s)$ for a normalized low-pass filter can be used to obtain $H_n(s)$ for a normalized high-pass filter. If we use p to denote the low-pass case and s to denote the high-pass case, then $H(s)$ can be obtained from $H(p)$ through the frequency transformation

$$p = \frac{\omega_{oh}}{s}$$

(11.38)

where ω_{oh} is a constant chosen to meet the specifications of the high-pass filter. Evidently $p = j0$ maps into $s = \pm j\infty$, and $p = \pm j1$ maps into $s = \mp \omega_{oh}$. For a Butterworth or Chebyshev filter, ω_{oh} is chosen as the passband cutoff frequency and will correspond to the cutoff frequency of 1 rad/s associated with these low-pass filters. For the elliptic filter, the normalized frequency ω_o is the geometric mean of the two edge frequencies shown in Figure 11.11.

■ ■ ■

11-7 LOW-PASS TO BANDPASS TRANSFORMATION

In parallel with the discussion in Section 11-6, $H(s)$ for a normalized low-pass filter can be used to obtain the $H(s)$ for a normalized bandpass filter when a proper low-pass to bandpass transformation is effected. The required transformation is

$$p = \frac{\omega_{ob}}{B}\left(\frac{s}{\omega_{ob}} + \frac{\omega_{ob}}{s}\right) \qquad (11.39)$$

where ω_{ob} and B are constants to be chosen to satisfy certain frequency specifications of the bandpass filter. Suppose that we let $s = j\omega$ be the point that corresponds to $p = j1$, the cutoff point of the low-pass filter. Under this condition, (11.39) yields the two values corresponding to the cutoff values of the bandpass filter. These are, from solving (11.39) in the standard quadratic form,

$$\omega_1 = \frac{B}{2} - \sqrt{\left(\frac{B}{2}\right)^2 + \omega_{ob}^2} \quad \text{(a)}$$

$$\omega_2 = \frac{B}{2} + \sqrt{\left(\frac{B}{2}\right)^2 + \omega_{ob}^2} \quad \text{(b)}$$

(11.40)

The negative sign that appears in (11.40a) is ignored. Taking account of this negative value, we find from (11.40) that

$$B = \omega_2 - \omega_1 \qquad (11.41)$$

which specifies the bandwidth of the filter. Also we find

$$\omega_{ob} = \sqrt{\omega_1 \omega_2} \qquad (11.42)$$

which shows that ω_{ob} is the geometric mean of ω_1 and ω_2, the two cutoff frequencies.

It is important to realize that frequency transformations do not necessarily preserve the stability of filters. Unstable filters can be stabilized, but a better approach is to seek frequency transformations that will preserve the stability in the first place. We will not pursue this matter.

■ ■ ■

ELEMENTS OF DIGITAL FILTER DESIGN

11-8 DIGITAL FILTERS

We have already noted that any device or process that will transform an input sequence of numbers into an output sequence of numbers might be called a digital filter. If we consider the digital filter to be a computational algorithm for carrying out this transformation process according to some prescribed rule, this rule is either a difference equation (as discussed in Chapter 8) or the convolution summation (discussed in Chapter 9). Digital filter design is concerned with the selection of the coefficients of the difference equation or with the unit sample response $h(k)$ used in the convolution summation.

As already noted, digital filter design often stems from analog filters of the low-pass and high-pass class by the use of a transformation that yields an equivalent z-plane expression for a given analog description in the s-plane or in the time domain. In essence this means that we establish a roughly equivalent sampled form for the given analog function. We will discuss the impulse invariant response method and also the use of a bilinear transformation for effecting this analog-digital transformation.

We discussed in Section 8-2 that the difference equation description of a discrete time system (filter) was of two types—FIR and IIR. The general form of the difference equation for the IIR system, given by (8.4), relates the present output value with immediate past values of the output and the present and past values of the input. This involves a recursive process using these present and past values to update the output. Similarly, if the output depends only upon the present and past values of the input, the system is nonrecursive or finite duration impulse response (FIR). More precisely, however, FIR and IIR describe digital filters relative to the length of their unit sample response sequence; it is possible to implement an FIR digital filter in a recursive fashion, and an IIR digital filter can be implemented in a nonrecursive fashion.

■ ■ ■

11-9 THE IMPULSE INVARIANT RESPONSE METHOD

Suppose that the system function of an analog filter that has certain desired properties is specified by

$$H(s) = \sum_{i=1}^{m} \frac{A_i}{(s + s_i)} \quad \text{(a)} \qquad (11.43)$$

Assume that all poles are distinct, with the impulse response function being

$$h(t) = \mathscr{L}^{-1}\{H(s)\} = \sum_{i=1}^{m} A_i e^{-s_i t} \quad \text{(b)}$$

If $h(kT)$ is the corresponding sampled version of $h(t)$, then we can write

$$H(z) = \sum_{k=0}^{\infty} h(kT) z^{-k}$$

CHAPTER 11

Therefore,

$$H(z) = \sum_{k=0}^{\infty} z^{-k} \sum_{i=1}^{m} A_i e^{-s_i kT} = \sum_{i=1}^{m} A_i \sum_{k=0}^{\infty} z^{-k} e^{-s_i kT}$$

which is, using the well-known formula of geometric series,

$$H(z) = \sum_{i=1}^{m} \frac{A_i}{1 - e^{-s_i T} z^{-1}} \qquad (11.44)$$

A comparison of (11.43) and (11.44) shows that a continuous-time filter specified by the system function $H(s)$ transforms, via **impulse invariant techniques,** by setting

$$s + s_i = 1 - e^{-s_i T} z^{-1} \qquad (11.45)$$

into a digital filter specified by $H(z)$. As already noted, some degree of approximation exists in this transformation because the digital filter is necessarily band-limited whereas $H(s)$, being a rational function of s, is not band-limited.

Let us examine how well the frequency response of the digital filter corresponds to the original analog filter frequency response. We use the fact that [see (4.57)]

$$\sum_{n=-\infty}^{\infty} \delta(t - nT) = \frac{1}{T} \sum_{n=-\infty}^{\infty} e^{jn\omega_s t} \qquad \omega_s = \frac{2\pi}{T}$$

The Laplace transform is

$$F_s(s) = \mathscr{L}\{f(nT)\} = \mathscr{L}\left[\sum_{n=-\infty}^{\infty} f(t)\,\delta(t-nT)\right]$$

$$= \frac{1}{T} \sum_{n=-\infty}^{\infty} \int_0^{\infty} f(t) e^{-(s-jn\omega_s)t}\, dt = \frac{1}{T} \sum_{n=-\infty}^{\infty} F(s - jn\omega_s) \qquad (11.46)$$

where ω_s is the radian sampling frequency. Also, the Z-transform of $f(nT)$ is

$$\mathscr{Z}\{f(nT)\} = F(z)$$

But we know that $z = e^{sT}$ from Section 9-11, and thus

$$F(z) = F_s\left(s = \frac{1}{T} \ln z\right) \qquad z = e^{j\omega_d}, \quad \ln z = j\omega_d \qquad (11.47)$$

where ω_d is the frequency in the discrete domain. We apply these results to (11.44) to write

$$H(z) = \sum_{i=1}^{m} \frac{A_i}{1 - e^{-s_i T} z^{-1}} = H_s\left(s = \frac{1}{T} \ln z\right) \triangleq \mathscr{L}\left\{\sum_{n=-\infty}^{\infty} h(t)\,\delta(t-nT)\right\}$$

$$= \frac{1}{T} \sum_{k=-\infty}^{\infty} H_s\left(s = \frac{1}{T} \ln z - jk\omega_s\right)$$

ELEMENTS OF DIGITAL FILTER DESIGN

$$= \frac{1}{T} \sum_{k=-\infty}^{\infty} H_s\left(\frac{j\omega_d}{T} - jk\omega_s\right) \tag{11.48}$$

From this equation it follows that in the base band

$$\frac{-\omega_s}{2} \leq \omega_d \leq \frac{\omega_s}{2} \qquad k = 0$$

the frequency response characteristic of the digital filter $H(z)$ will differ from that of the analog filter $H(s)$, the difference being the amount "added" or "folded in" from the additional terms of the form

$$H_s\left(\left[s = \frac{1}{T}\ln z\right] - jk\omega_s\right)$$

which make up the summation in (11.48). As the discussion of the sampling theorem in Chapter 7 demonstrates, no folding error exists if

$$|H_s(s)| \triangleq |H_s(j\omega)| = 0 \qquad |\omega| > \frac{\omega_s}{2}$$

and in this case the frequency response of the digital filter is identical with that of the continuous filter, when

$$H(e^{j\omega_d}) = \frac{1}{T} H_s(s) = \frac{1}{T} H_s\left(j\frac{\omega_d}{T}\right) \qquad |\omega_d| \leq \pi \tag{11.49}$$

If $H_s(s)$ includes a repeated root, then, in addition to terms such as those in (11.43), we would have such terms as (see Table 6.1)

$$h(t) = \mathscr{L}^{-1}\left\{\frac{A}{(s + s_i)^r}\right\} = \frac{A}{(r-1)!} t^{r-1} e^{-s_i t} u(t) \tag{11.50}$$

The sampled version of this is

$$h(kT) = \begin{cases} \dfrac{A}{(r-1)!}(kT)^{r-1} e^{-s_i kT} & k \geq 0 \\ 0 & k < 0 \end{cases} \tag{11.51}$$

The corresponding Z-transform is

$$H(z) = \frac{A}{(r-1)!} T^{r-1} \sum_{k=0}^{\infty} k^{r-1}(e^{-s_i T} z^{-1})^k \tag{11.52}$$

In accordance with our discussion above, we can follow the following steps in carrying out a digital filter design:

1. A set of filter specifications are given.

2. Create an analog transfer function $H(s)$ that meets the specifications of Step 1.

3. Determine the impulse response of the analog filter by means of the Laplace inversion technique, $h(t) = \mathscr{L}^{-1}\{H(s)\}$.

4. Sample $h(t)$ at T second intervals, thus creating a sequence $\{h(kT)\}$
5. Deduce $H(z)$ of the resulting digital filter by taking the Z-transform of the discrete function $h(kT)$, $H(z) = \sum_{k=0}^{\infty} h(kT)z^{-k}$.

EXAMPLE 11.4
Determine the digital equivalent of the first-order Butterworth filter. The cutoff frequency is $20/2\pi$ Hz.

Solution: The normalized analog transfer function is found in Table 11.2 for the designated filter

$$H_n(s) = \frac{1}{s+1}$$

The system transfer function is given by

$$H\left(\frac{s}{\omega_c}\right) = H\left(\frac{s}{2\pi \times \frac{20}{2\pi}}\right) = \frac{1}{\frac{s}{20}+1} = \frac{20}{s+20}$$

The impulse response of this filter is given by

$$h(t) = \mathscr{L}^{-1}\left\{H\left(\frac{s}{20}\right)\right\} = \mathscr{L}^{-1}\left\{\frac{20}{s+20}\right\} = 20e^{-20t} \quad t \geq 0$$

The Z-transform of the discrete function $h(kT)$ is [see (9.3) and (11.44)]

$$H(z) = 20\sum_{k=0}^{\infty} e^{-20kT}z^{-k} = 20\sum_{k=0}^{\infty}(e^{-20T}z^{-1})^k = \frac{20}{1-e^{-20T}z^{-1}}$$

To proceed, for the specific sampling time $T = 0.005$ s we deduce the absolute value of the analog transfer function

$$\left|H\left(\frac{j\omega}{20}\right)\right| = \frac{1}{\left|\frac{j\omega}{20}+1\right|} = \frac{1}{\sqrt{1+\left(\frac{j\omega}{20}\right)^2}}$$

It is convenient to plot this function versus ω_d, where $\omega = \omega_d/T$. The equivalent expression becomes

$$|H(j\omega_d)| = \frac{1}{\sqrt{1+(10\omega_d)^2}}$$

This is shown plotted as the lower curve in Figure 11.12.

The equivalent expression for the discrete function is [see (11.44)]

$$|H(e^{j\omega_d})| = \frac{20}{|1-0.9048e^{-j\omega_d}|}$$

Note, however, that $H(e^{j0}) = 210.084$ is the value of $H(e^{j\omega_d})$ for $\omega_d = 0$ (or

Figure 11.12
Comparison between the analog and its corresponding digital filter for two different sampling times.

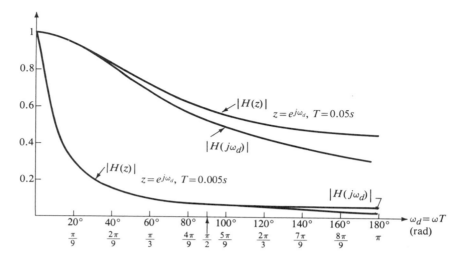

$z = 1$). We thus consider the normalized function

$$\frac{|H(e^{j\omega_d})|}{210.084} = \frac{20}{210.084} \frac{1}{[(1 - 0.9048 \cos \omega_d)^2 + (0.9048 \sin \omega_d^2)]^{1/2}}$$

which is also plotted in Figure 11.12 (lower curve).
For the case when $T = .05$ s

$$H(e^{j\omega_d}) = \frac{20}{1 - 0.3679 e^{-j\omega_d}}$$

The value of $H(e^{j0}) = 31.641$ at $\omega_d = 0$, and the normalized relation becomes

$$\frac{|H(e^{j\omega_d})|}{31.641} = \frac{20}{31.641} \frac{1}{[(1 - 0.3679 \cos \omega_d)^2 + (0.3679 \sin \omega_d)^2]^{1/2}}$$

This relation is plotted in Figure 11.12 (upper curve) together with $|H(j\omega_d)| = 1/\sqrt{1 + \omega_d^2}$ since $\omega = \omega_d/0.05$.

Figure 11.12 clearly shows the effect of sampling times T. The upper curves start to separate at about $\omega_d = 30°$, which is equal to $30/57.2958 = 0.5236$ radians; hence at $T = 0.05$ s the error begins to be pronounced at a frequency roughly equal to $f = \omega_d/2\pi T = 0.5236/(2\pi \times .05) = 1.666$ Hz. Furthermore, the error for the lower curves begins at about 90° or 1.571 radians; hence for $T = .005$ s the error starts at the frequency $f = 1.571/(2\pi \times .005) = 50$ Hz, which is above the cutoff frequency. Thus an aliasing error is present in one case but not in the other. This indicates that good accuracy is achieved with our digital filter if we choose the sampling time $T \ll 0.05$ s.

Had we continued the plots of the above curves from $\omega_d = 180° = \pi$ to $\omega_d = 2\pi$ the curves of $H(z)$ would have repeated themselves, being symmetric about $\omega_d = \pi$. However, the curves for $|H(j\omega)|$ would have continued to decrease, as expected from the form of $H(j\omega)$. ∎

EXAMPLE 11.5
Determine the digital equivalent of a Butterworth third-order low-pass filter, and find T if the sampling frequency is 15 times the cutoff frequency.

Solution: The normalized system function of this filter is deduced from Table 11.1 or Table 11.2. It is

$$H_n(s) = \frac{1}{1 + 2s + 2s^2 + s^3} = \frac{1}{(s+1)(s^2+s+1)} = \frac{A_1}{s+s_1} + \frac{A_2}{s+s_2} + \frac{A_3}{s+s_3}$$

where

$$s_1 = 1 \quad s_2 = \frac{1}{2}(1 - j\sqrt{3}) \quad s_3 = \frac{1}{2}(1 + j\sqrt{3})$$

Evaluating the constants A in this expression, we find that

$$A_1 = 1 \quad A_2 = \frac{2}{-3 + j\sqrt{3}} \quad A_3 = A_2^* = \frac{2}{-3 - j\sqrt{3}}$$

The impulse response of this system is given by

$$h(t) = A_1 e^{-t} + A_2 e^{-s_2 t} + A_3 e^{-s_3 t}$$

Further, to find the sampling time $\omega_s = 2\pi/T = 15 \cdot \omega_c = 15$, from which the sampling time is $T = 0.419$. The sampled impulse response function is

$$h(0.419k) = A_1 e^{-0.419k} + A_2 e^{-0.419 s_2 k} + A_3 e^{-0.419 s_3 k}$$

The Z-transform of this function is

$$H(z) = \frac{z}{z - e^{-0.419}} + A_2 \frac{z}{z - e^{-0.419 s_2}} + A_3 \frac{z}{z - e^{-0.419 s_3}}$$

$$= \frac{z}{z - 0.658} + \frac{z(-z + 0.924)}{z^2 - 1.516z + 0.658}$$

$$= \frac{0.066 z^{-1} + 0.0497 z^{-2}}{1 - 2.174 z^{-1} + 1.656 z^{-2} - 0.433 z^{-3}}$$

A realization of this filter is shown in Figure 11.13. ■

In general, a digital filter designed using the invariant impulse method results in a transfer function in the form of the ratio of two polynomials. The difference equation written from this sytem function is a recursive expression and the filter so realized is an infinite impulse response (IIR) filter. But note that the invariant impulse response method is equivalent to analog filtering of an impulse-sampled input signal. We have already discussed the fact that for a sampled signal to approximate the continuous signal, the sampling rate (Nyquist rate) must be at least twice the highest frequency component contained in the signal. However,

Figure 11.13
Digital filter realization of a third-order Butterworth low-pass filter.

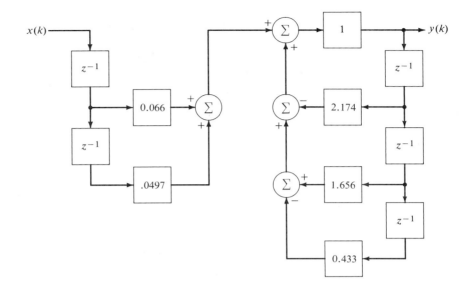

a practical analog filter $H(s)$ is never strictly band-limited. Therefore, an aliasing error will occur when this design method is used. As a practical matter, if the sampling frequency is 5 or more times the cutoff frequency of the low-pass analog filter, the aliasing effect on the frequency response is extremely small.

If $H(z)$ is used to obtain the appropriate $h(k)$, and if this is used in a convolution summation expansion, the result will be a finite-duration impulse response (FIR) filter, if $h(k)$ is truncated to N terms. The FIR filter obtained this way suffers from the fact that a large number of samples will be required to approximate $h(k)$. This same limitation exists if $h(k)$ is to approximate the $h(t)$ of an analog filter directly. The disadvantage of this procedure is that the computation time required to generate an output sample is longer than with a recursive filter.

Another problem with an FIR realization developed this way results from the abrupt truncation of the unit response function, which introduces the problem of the Gibbs phenomenon. In such cases one may use a window function to alleviate the problem by smoothing the sampled data in the neighborhood of the truncation region. An advantage of the FIR approach is that any errors in the computations are not recycled into subsequent calculations.

EXAMPLE 11.6
Find the digital filter equivalent of the *RC* network in Figure 11.14 where the 3 dB cutoff frequency is 1/5 of the sampling frequency.

Solution: By a simple determination, the transfer function is

Figure 11.14
The *RC* network under survey.

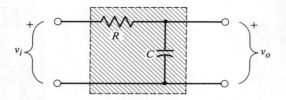

$$H(s) = \frac{\dfrac{1}{RC}}{s + \dfrac{1}{RC}} = \frac{\omega_c}{s + \omega_c}$$

where

$$\omega_c = \frac{1}{RC} = \frac{1}{5}\frac{2\pi}{T}$$

We then write $H(s)$ in the form

$$H(s) = \frac{\dfrac{0.4\pi}{T}}{s + \dfrac{0.4\pi}{T}}$$

By the invariant impulse method, the digital equivalent $H(z)$ is (for numerical convenience it was multiplied by T)

$$\frac{Y(z)}{X(z)} \triangleq H(z) = \frac{T \times \dfrac{0.4\pi}{T}}{1 - e^{-0.4\pi}z^{-1}} = \frac{1.2566}{1 - 0.285z^{-1}}$$

The equivalent difference equation is then found from

$$(1 - 0.285z^{-1})Y(z) = 1.2566X(z)$$

from which we can write

$$y(k) = 1.2566x(k) + 0.285y(k-1) \qquad \blacksquare$$

EXAMPLE 11.7
The normalized transfer function (11.33) of the third-order Chebyshev filter of Example 11.3 is

$$H_n(s) = \frac{0.4913}{0.4913 + 1.2384s + 0.9883s^2 + s^3} \qquad (11.53)$$

Determine the following:

a. The corresponding impulse-invariant digital filter.

ELEMENTS OF DIGITAL FILTER DESIGN

  b. The amplitude characteristics of the digital filter.
  c. The impulse response $h(t)$ of $H_n(s)$.
  d. The unit sample response $h(kT)$ of the digital filter for $T = 1$.

Solution: a. Begin with the Chebyshev function in factored form

$$H_n(s) = \frac{0.4913}{(s + 0.2471 - j0.9660)(s + 0.2471 + j0.9660)(s + 0.4942)}$$

This is written in partial fraction form

$$H_n(s) = \frac{A_1}{s + 0.2471 - j0.9660} + \frac{A_2}{s + 0.2471 + j0.9660} + \frac{A_3}{s + 0.4942}$$

(11.54)

where

$$A_1 = \frac{0.4913}{2j \times 0.9660(0.2471 + j0.9660)}$$

$$A_2 = \frac{0.4913}{-2j \times 0.9660(0.2471 - j0.9660)}$$

$$A_3 = \frac{0.4913}{(0.2471 + j0.9660)(0.2471 - j0.9660)} = 0.4942$$

Now set $s + s_i = 1 - e^{-s_i T} z^{-1}$ in (11.54), which gives the impulse invariant digital filter representation,

$$H_n(z) = \frac{A_1}{1 - e^{-0.2471T + j0.9660T} z^{-1}} + \frac{A_2}{1 - e^{-0.2471T - j0.9660T} z^{-1}}$$

$$+ \frac{A_3}{1 - e^{-0.4942T} z^{-1}} \qquad (11.55)$$

  b. The corresponding frequency response is obtained by substituting $z = \exp(j\omega_d) = \exp(j\omega T)$ in (11.55). The resulting expression is

$$H_n(e^{j\omega T}) = \frac{A_1}{1 - e^{-0.2471T + j0.9660T} e^{-j\omega T}} + \frac{A_2}{1 - e^{-0.2471T - j0.9660T} e^{-j\omega T}}$$

$$+ \frac{A_3}{1 - e^{-0.4942T} e^{-j\omega T}} \qquad (11.56)$$

This expression is plotted in Figure 11.15 for $T = 1$.

  c. Apply the inverse Laplace transform to (11.54) to find the impulse response. This is given by

$$h(t) = A_1 e^{-0.2471t} e^{j0.9660t} + A_2 e^{-0.2471t} e^{-j0.9660t} + A_3 e^{-0.4942t}$$
$$= e^{-0.2471t} \times [A_1 e^{j0.9660t} + A_1^*(e^{j0.9660t})^*] + 0.4942 e^{-0.4942t}$$
$$= e^{-0.2471t} \times 0.51 \cos(0.966t - 165.65°) + 0.4942 e^{-0.4942t} \qquad (11.57)$$

Figure 11.15
The frequency response of the Chebyshev filter under review.

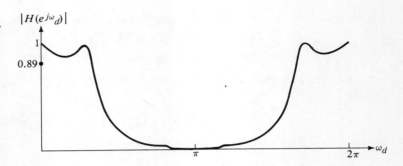

Figure 11.16
The impulse responses $h(t)$ and $h(kT)$, $T = 1$.

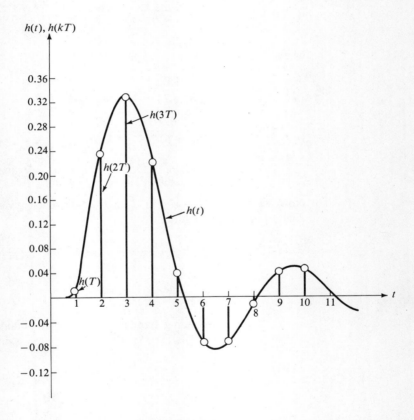

Figure 11.16 shows the impulse response corresponding to (11.57). ∎

■ ■ ■

11-10 THE BILINEAR TRANSFORMATION

To circumvent the "folding" problem of the impulse invariant response transformation noted in the previous section, a transformation from the s-plane to the p-plane can be employed that will map the entire s-plane into a horizontal strip in the p-plane bounded by the lines $p = -j\omega_s/2$ and $p = +j\omega_s/2$. Moreover, since $H(z)$ is also periodic in ω (with period ω_s), this transformation may also cause $H(s)$ to be mapped identically into each of the other horizontal strips bounded by the lines $p = j(n - \tfrac{1}{2})\omega_s$ and $p = j(n + \tfrac{1}{2})\omega_s$, where n is an integer (see Figure 11.17). A transform having the requisite properties is the bilinear transformation, which is defined by

$$s = \frac{2}{T}\tanh\frac{pT}{2} \tag{11.58}$$

Using the identity

$$\tanh x = \frac{e^x - e^{-x}}{e^x + e^{-x}}$$

we have

$$s = \frac{2}{T}\frac{e^{pT/2} - e^{-pT/2}}{e^{pT/2} + e^{-pT/2}}$$

Upon substituting the quantity $z = e^{pT}$ in this expression, we have

$$\boxed{s = \frac{2}{T}\left(\frac{z-1}{z+1}\right)} \quad \text{(a)} \tag{11.59}$$

and

$$\boxed{z = \frac{1 + Ts/2}{1 - Ts/2}} \quad \text{(b)}$$

In terms of the z-plane, this algebraic transformation uniquely maps the left half of the s-plane into the interior of the unit-circle in the z-plane, as shown in Figure 11.17. Because no folding occurs, no folding errors will arise. However, a shortcoming of this transformation is that the frequency response is nonlinear (that is, warped) in the digital domain.

If we insert $z = \exp(j\omega_d)$ into (11.59a), we obtain a relationship between the frequency ω of the analog filter and ω_d of the digital filter. We find that

$$s \triangleq \sigma + j\omega = \frac{2(e^{j\omega_d} - 1)}{T(e^{j\omega_d} + 1)} = \frac{2e^{j\omega_d/2}(e^{j\omega_d/2} - e^{-j\omega_d/2})}{Te^{j\omega_d/2}(e^{j\omega_d/2} + e^{-j\omega_d/2})}$$

$$= j\frac{2}{T}\tan\frac{\omega_d}{2} \tag{11.60}$$

Figure 11.17
(a) A step in finding the inverse-Z transform. (b) Mapping of the s-plane onto the z-plane.

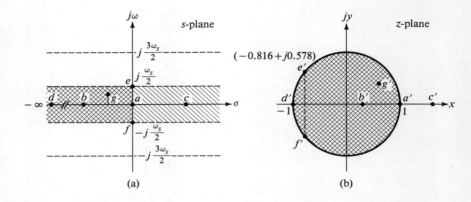

From this equation, by separately equating real and imaginary parts, we obtain

$$\sigma = 0 \quad \text{(a)}$$

$$\boxed{\omega = \frac{2}{T} \tan \frac{\omega_d}{2}} \quad \text{(b)}$$

(11.61)

Observe that the relationship between the two frequencies ω and ω_d is a nonlinear one. This is the **warping** effect. Figure 11.18a shows the warping relationship graphically. It is evident from the plot that the sampling time T changes

Figure 11.18
Graphic illustration of ω and ω_d in bilinear transformation.

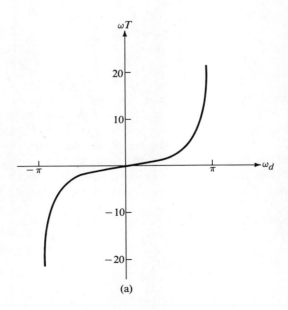

(a)

ELEMENTS OF DIGITAL FILTER DESIGN

the ωT axis by stretching or compressing it. Figure 11.18b below clearly shows the warping effect. Note that the shapes are similar, but the higher frequency bands are reduced disproportionately.

EXAMPLE 11.8

Determine the characteristics of a digital filter if the corresponding analog filter has the transfer function

$$H(s) = \frac{H_0}{s - s_1} \qquad (11.62)$$

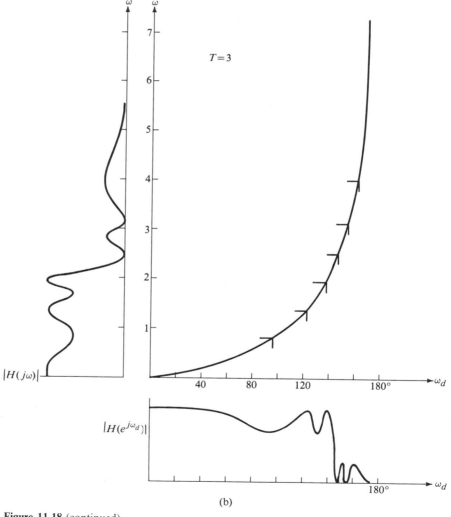

Figure 11.18 (continued)

Solution: By use of (11.59) we obtain

$$H(z) = H(s)\Big|_{s=2(z-1)/T(z+1)} = \frac{H_0}{\dfrac{2}{T}\dfrac{z-1}{z+1} - s_1}$$

$$= H_0 T \frac{z+1}{(2-s_1 T)z - (2+s_1 T)} \tag{11.63}$$

To illustrate the variation graphically, we set $H_0 = 0.8$, $s_1 = 0.8$, and $T = 1$. Figures 11.19a and 11.19b give the frequency response of the analog filter and the digital filter, respectively. The two transfer functions are given by

$$|H(s)| = \frac{0.8}{\sqrt{\omega^2 + (0.8)^2}} \quad \text{(a)}$$

$$|H(z)| = 0.8 \frac{\sqrt{(\cos \omega_d + 1)^2 + \sin^2 \omega_d}}{\sqrt{(1.2 \cos \omega_d - 2.8)^2 + (1.2 \sin \omega_d)^2}} \quad \text{(b)} \tag{11.64}$$

The figures show the corresponding frequencies of the two filters, including the cutoff value of the digital filter. The nonlinear relation between the two curves is clearly evident. ∎

EXAMPLE 11.9
We wish to design a digital Butterworth filter that will meet the following conditions:

a. The 3 dB cutoff point ω_d is to occur at 0.4π rad/s.

b. $T = 50$ μs.

c. At $2\omega_d$ the attenuation is to be 15 dB.

Solution: First we find the analog equivalent criteria for the requisite digital filter. Thus we have, using (11.61)

$$\omega_c = \frac{2}{T} \tan \frac{\omega_d}{2} = \frac{2}{50 \times 10^{-6}} \tan \frac{0.4\pi}{2} = 29.1 \times 10^3 \text{ rad/s}$$

$$\omega\big|_{15\text{ dB}} = \frac{2}{T} \tan \frac{2\omega_d}{2} = \frac{2}{50 \times 10^{-6}} \tan \frac{2 \times 0.4\pi}{2} = 123.1 \times 10^3 \text{ rad/s}$$

Now using (11.1) with $\omega = \omega/\omega_c$, we obtain

$$-10 \log |H(j123.1 \times 10^3)|^2 = -10 \log \left| \frac{1}{1 + \left(\dfrac{123.1 \times 10^3}{29.1 \times 10^3}\right)^{2n}} \right| \geq 15$$

Figure 11.19
Frequency response of a first-order analog filter and its corresponding digital filter.

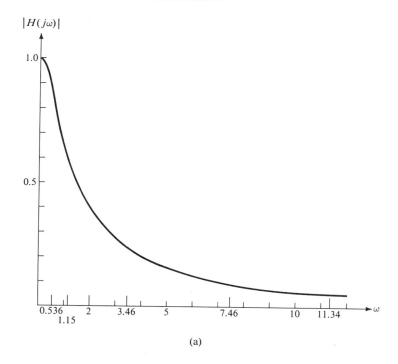

(a)

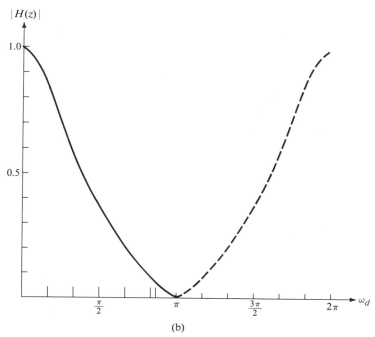

(b)

from which

$$1 + \left(\frac{123.1 \times 10^3}{29.1 \times 10^3}\right)^{2n} \geq 31.62$$

From this we find that

$n \geq 1.18$

hence the minimal order of the Butterworth filter to meet the specifications is $n = 2$. The normalized form of the filter is, from Table 11.1,

$$H_n(s) = \frac{1}{s^2 + 1.4142s + 1}$$

The analog filter satisfying the specifications is

$$H(s) = H_n\left(\frac{s}{\omega_c}\right) = H_n\left(\frac{s}{29.1 \times 10^3}\right) = \frac{1}{\left(\frac{s}{29.1 \times 10^3}\right)^2 + \left(\frac{1.4142}{29.1 \times 10^3}\right)s + 1}$$

$$H(s) = \frac{(29.1 \times 10^3)^2}{s^2 + 41.1532 \times 10^3 s + (29.1 \times 10^3)^2}$$

Introduce (11.59) into this equation, which leads to

$$H(z) = H(s)|_{s = 2(z-1)/T(z+1)}$$

$$= \frac{29.1^2 \times 10^6}{\left(\frac{2}{T}\right)^2 \frac{(z-1)^2}{(z+1)^2} + 41.1532 \times 10^3 \left(\frac{2}{T}\right)\frac{(z-1)}{(z+1)} + 29.1^2 \times 10^6}$$

$$= \frac{2.117 + 4.234z^{-1} + 2.117z^{-2}}{10.232 - 3.766z^{-1} + 2.002z^{-2}}$$

This is the final digital Butterworth design. ∎

EXAMPLE 11.10

Determine the characteristics of a digital Butterworth filter to meet the specifications of Example 11.2. The sampling frequency is 10 times the cutoff frequency.

Solution: We find that we must choose a filter with $n = 4$ to meet the conditions of the problem. The normalized function is then known to be

$$H(s) = \frac{1}{s^4 + 2.6131s^3 + 3.4142s^2 + 2.6131s + 1}$$

which can also be written

$$H(s) = \frac{1}{(s^2 + 0.7654s + 1)(s^2 + 1.8478s + 1)}$$

ELEMENTS OF DIGITAL FILTER DESIGN

Now apply the bilinear z-transformation for the normalized function

$$s = \frac{2}{10}\left[\frac{(z-1)}{(z+1)}\right] = A\frac{z-1}{z+1} \qquad A = 0.2$$

so that

$$H(z) = \frac{1}{\left(A^2\left[\frac{z-1}{z+1}\right]^2 + 0.7654A\left[\frac{z-1}{z+1}\right] + 1\right)}$$

$$\times \frac{1}{\left(A^2\left[\frac{z-1}{z+1}\right]^2 + 1.8478A\left[\frac{z-1}{z+1}\right] + 1\right)}$$

Expand this expression to find

$$H(z) = \frac{0.5946(z+1)^4}{(z^2 + 0.9943z + 0.7434)(z^2 + 1.3621z + 0.4756)}$$

■

■ ■ ■

†11-11 PREWARPING

Because of frequency warping, as discussed earlier, the bilinear z-transformation is most useful in obtaining filter approximations for continuous filters whose magnitude characteristics can be divided along the frequency scale into successive stop and passbands, where the loss or gain is essentially constant in the band. Compensation can be made for the effect of warping by prewarping the continuous filter design in such a way that upon applying the bilinear transformation, the critical frequencies will be shifted back to the desired values.

To examine the prewarping process in general terms (see Example 11.9) suppose that the system function $H(s)$ is expressed in partial-fraction form, with a typical term being $H_1/(s - p_1)$. By applying the bilinear z-transformation (11.59) to such a term, we obtain

$$\frac{H_1}{s - p_1} \rightarrow \frac{H_1 T(z + 1)}{(2 - p_1 T)z - 2 - p_1 T} \qquad (11.65)$$

A partial-fraction expansion of the right-hand expression gives

$$H_b(z) = \frac{H_1 T}{2 - p_1 T} + \frac{\left[\dfrac{4H_1 T}{4 - p_1^2 T^2}\right]}{\left[z - \dfrac{2 + p_1 T}{2 - p_1 T}\right]} \qquad (11.66)$$

This shows that the bilinear z-transformation results in the pole in the z-plane being at $(2 + p_1 T)/(2 - p_1 T)$ rather than at $e^{p_1 T}$, as in the impulse invariant response method. To prewarp the function, we wish to apply a transformation that moves the pole location of the digital filter to $e^{p_1 T}$, the pole of the original

CHAPTER 11

filter. The transformation sequence is the following: first the transfer function is modified by moving the pole [see (11.58)]

$$\frac{H_1}{s - p_1} \rightarrow \frac{H_1}{s - \frac{2}{T}\tanh\frac{p_1 T}{2}} \qquad (11.67)$$

and then applying the bilinear z-transformation to this modified transfer function. This leads, using (11.59), to

$$\frac{H_1}{s - \frac{2}{T}\tanh\frac{p_1 T}{2}} \rightarrow \frac{H_1 T(z + 1)}{2\left[z - 1 - (z + 1)\tanh\frac{p_1 T}{2}\right]} \qquad (11.68)$$

By simple arithmetic manipulation it can be shown that the transfer function becomes

$$H_{pb}(z) = \frac{H_1 T(1 + e^{p_1 T})}{4}\left[\frac{z + 1}{z - e^{+p_1 T}}\right] \qquad (11.69)$$

We will consider applying the bilinear z-transformation to the digitization of the Butterworth low-pass filter. The magnitude-squared characteristic of these filters is written [see (11.2)]

$$H(s)H(-s) = \frac{1}{1 + (-1)^n \left(\frac{s}{\omega_c}\right)^{2n}} \qquad (11.70)$$

where ω_c is the cutoff frequency. From (11.61) replace ω_c by its prewarped value

$$\omega_c = \frac{2}{T}\tan\frac{(\omega_c T)}{2} \qquad (11.71)$$

Then the magnitude squared characteristic of the filter is

$$H(s) \times H(-s) = \frac{1}{1 + (-1)^n \left[\frac{s}{\frac{2}{T}\tan\frac{\omega_c T}{2}}\right]^{2n}} \quad (a) \qquad (11.72)$$

which becomes, by (11.59),

$$|H(z)|^2 = \frac{1}{1 + (-1)^n \left[\frac{\frac{z-1}{z+1}}{\tan\frac{\omega_c T}{2}}\right]^{2n}} \quad (b)$$

To find the system function $H(z)$ from this, we write the expression

ELEMENTS OF DIGITAL FILTER DESIGN

$$|H(z)|^2 = \frac{\tan^{2n}\left(\frac{\omega_c T}{2}\right)}{\tan^{2n}\left(\frac{\omega_c T}{2}\right) + (-1)^n \left[\frac{z-1}{z+1}\right]^{2n}} \tag{11.73}$$

or

$$|H(z)|^2 = \frac{\tan^{2n}\left(\frac{\omega_c T}{2}\right)[z+1]^{2n}}{\tan^{2n}\left(\frac{\omega_c T}{2}\right)[z+1]^{2n} + (-1)^n[z-1]^{2n}}$$

The denominator polynomial in this expression can be expanded to the form

$$\left[\tan^{2n}\left(\frac{\omega_c T}{2}\right) + (-1)^n\right]\left[z^{2n} + \beta\binom{2n}{1}z^{2n-1} + \binom{2n}{2}z^{2n-2}\right.$$

$$\left. + \beta\binom{2n}{3}z^{2n-3} + \binom{2n}{4}z^{2n-4} + \cdots + \beta\binom{2n}{2n-1}z + 1\right] \tag{11.74}$$

where

$$\beta = \frac{\tan^{2n}\left(\frac{\omega_c T}{2}\right) - (-1)^n}{\tan^{2n}\left(\frac{\omega_c T}{2}\right) + (-1)^n} \qquad \binom{n}{k} = \frac{n(n-1)\cdots(n-k+1)}{k!} = \frac{n!}{(n-k)!k!}$$

A calculation shows that the $2n$ roots of (11.74) are given by

$$p_i = \frac{1 - \tan^2\left(\frac{\omega_c T}{2}\right) + j2\tan\left(\frac{\omega_c T}{2}\right)\sin\theta_i}{1 - 2\tan\left(\frac{\omega_c T}{2}\right)\cos\theta_i + \tan^2\left(\frac{\omega_c T}{2}\right)} \qquad i = 1, 2, \ldots, 2n \tag{11.75}$$

where

$$\theta_i = \begin{cases} (i-1)\pi/n & n \text{ odd} \\ (2i-1)\pi/2n & n \text{ even} \end{cases}$$

The squared gain factor is then

$$|H(z)|^2 = \frac{\tan^{2n}\left(\frac{\omega_c T}{2}\right)}{\tan^{2n}\left(\frac{\omega_c T}{2}\right) + (-1)^n}\left[\frac{(z+1)^{2n}}{(z-p_1)(z-p_2)\cdots(z-p_{2n})}\right] \tag{11.76}$$

Now, if $p = re^{j\theta}$ is a root of the denominator polynomial, then so are $re^{-j\theta}$, $(1/r)e^{j\theta}$ and $(1/r)e^{-j\theta}$. Hence $|H(z)|^2$ will have n poles inside and n poles outside the unit circle ($r = 1$). If $p_1, \ldots, p_n$ denote the poles inside the unit circle, then

$$H(z) = \frac{b(z+1)^n}{(z-p_1)(z-p_2)\cdots(z-p_n)} \quad \text{(a)} \tag{11.77}$$

where b is selected so that $H(1) = 1$; therefore

$$b = \frac{(1-p_1)(1-p_2)\cdots(1-p_n)}{2^n} \quad \textbf{(b)}$$

The remaining poles, $p_{n+1}, \ldots, p_{2n}$ associated with $|H(z)|^2$ are outside the unit circle. By selecting $H(z)$ according to (11.77) for bounded input signal, the output signal goes to zero for $n \to \infty$; this establishes the bounded-input, bounded-output (bibo) stability property, a matter to be considered in Section 12-17.

EXAMPLE 11.11

Determine the characteristics of a digital Butterworth filter that has 20 dB attenuation at a frequency of 2.6 times its cutoff frequency. The sampling frequency is 10 times the cutoff frequency. Assume that prewarping has been employed.

Solution: The product

$$\omega_c T = 2\pi \times 3000 \frac{1}{30{,}000} = \frac{2\pi}{10} = \frac{\pi}{5}$$

From (11.72) we write

$$|H(e^{j\omega T})|^2 = \frac{1}{1 + (-1)^n \left[\dfrac{e^{j\omega T} - 1}{e^{j\omega T} + 1} \bigg/ \tan\left(\dfrac{\omega_c T}{2}\right) \right]^{2n}}$$

$$= \frac{1}{1 + (-1)^{2n} \left[\dfrac{\tan^{2n}\left(\dfrac{\omega T}{2}\right)}{\tan^{2n}\left(\dfrac{\pi}{10}\right)} \right]}$$

From the requirement that the attenuation is 20 dB down ($= 0.01$) when $\omega T = 2.6\pi/5$, we have

$$|H(e^{j\omega T})|^2 = \frac{1}{1 + \dfrac{\tan^{2n} 0.26\pi}{\tan^{2n} 0.1\pi}} = |0.01|^2$$

from which

$$10^4 \leq 1 + (3.28)^{2n}$$

From this we find that $n \geq 4$. Also, from (11.75) we find the four roots contained within the unit circle to be:

$$z_1, z_2 = 0.6604 \pm j0.4432$$

$$z_3, z_4 = 0.5243 \pm j0.1458$$

ELEMENTS OF DIGITAL FILTER DESIGN

Thus the Butterworth function is, from (11.77),

$$H(z) = \frac{b(z+1)^4}{(z^2 - 1.3208z + 0.6325)(z^2 - 1.0486z + 0.2972)}$$

where b is the value specified by (11.77b).

∎

∎ ∎ ∎

11-12 FINITE IMPULSE RESPONSE (FIR) FILTERS

The design considerations in the previous two sections were based largely on the impulse response of analog filters, and these led to transfer functions of the IIR (recursive filter) type. We now wish to present two methods for designing FIR filters (nonrecursive), whose present output is computed by using only the present and past inputs, but none of its previous outputs. Because no feedback is present, this type of filter is stable. Furthermore, such filters are associated with linear phase characteristics, and so phase distortion in the output may be eliminated.

Two related methods will be discussed: (a) Fourier series method, and (b) the DFT method. Our discussion will be confined to a low-pass filter. Transformations for other types are discussed below. Refer to Fig. 11.20.

In the Fourier series method, the specified $H(\omega)$ is expanded into a Fourier expansion, assuming zero axis symmetry, thereby including only cosine terms. The procedure now continues as follows:

1. Truncate the series expansion to N terms and evaluate the coefficients.
2. Write the cosine terms in the expansion in exponential form and then write the exponential terms as functions of z through the transformation $z = \exp(jn\theta)$. This yields an expression for $H'(z)$.
3. Multiply the expression for $H'(z)$ by z^{-N} to yield $H(z)$, the system function for the FIR filter. Recall that z^{-N} is just a phase factor which will not alter the amplitude expression for $H(z)$.

Figure 11.20
Ideal low-pass filter characteristics.

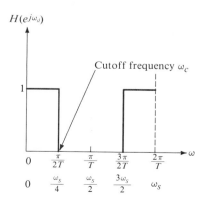

CHAPTER 11

In the DFT method, we sample the frequency response function at distances $1/NT$. This provides a sequence $\{H(n\Omega)\}$ which will be the DFT of an impulse sequence $\{h(kT)\}$. From the results obtained in Chapter 10, we write (10.4)

$$H(n\Omega) = \sum_{k=0}^{N-1} h(kT)e^{-j(2\pi kn/N)} = \sum_{k=0}^{N-1} h(kT)e^{-jn\Omega kT}$$

$$k = 0, 1, \ldots, N-1 \qquad \Omega = \frac{\omega_s}{N} = \frac{2\pi}{NT} \qquad \omega_s = \frac{2\pi}{T} \qquad (11.78)$$

$$h(kT) = \frac{1}{N}\sum_{n=0}^{N-1} H(n\Omega)e^{jn\Omega kT} \qquad n = 0, 1, \ldots, N-1 \qquad (11.79)$$

The steps required to design the appropriate discrete filter are:

1. From the given amplitude frequency characteristics, obtain the desired sampled values $H(n\Omega)$ for a specified N.
2. Incorporate into (11.79) the values found in Step 1, and find the sequence $\{h(kT)\}$.
3. Use the values $\{h(kT)\}$ in the Z-transform relationship given in (9.3) to find $H(z)$.
4. Plot $|H(e^{j\omega T})|$ versus ω to obtain the amplitude characteristics.

EXAMPLE 11.12
Find the nonrecursive filter corresponding to the ideal filter shown in Figure 11.21a for $N = 16$.

Solution: From Figure 11.21a the sampled values of $H(e^{j\omega T})$ are:

$\{1, 1, 0.5, 0, 0, 0, 0, 0, 0, 0, 0, 0, 0, 0, 0.5, 1\}$

Observe that we have used the average value of the function at the point of discontinuity. Next apply (11.79) to obtain

$$h(0T) = \frac{1}{16}(1 + 1 \times e^{j1(\pi/8)0} + 0.5 \times e^{j2(\pi/8)0} + 0 \times e^{j3(\pi/8)0} + \cdots$$

$$+ 0 \times e^{j13(\pi/8)0} + 0.5 \times e^{j14(\pi/8)0} + 1 \times e^{j15(\pi/8)0}$$

$$= \frac{4}{16} = 0.25$$

$$h(1T) = \frac{1}{16}(1 + 1e^{j1(\pi/8)1} + 0.5e^{j2(\pi/8)1} + 0 + \cdots + 0.5e^{j14(\pi/8)1}$$

$$+ 1e^{j15(\pi/8)1})$$

$$= \frac{1}{16}(1 + 0.924 + j0.383 + 0.354 + j0.354 + \cdots + 0.354 - j0.354$$

$$+ 0.924 - j0.383) = 0.222$$

Figure 11.21
The frequency response of the FIR filter under review.

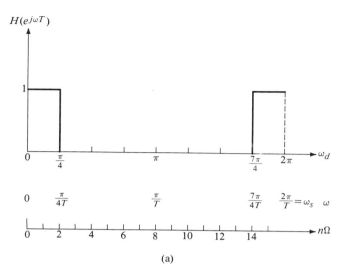

(a)

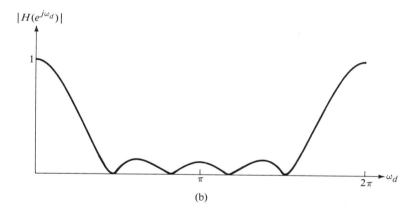

(b)

$$h(2T) = \frac{1}{16}(1 + 1e^{j1(\pi/8)2} + 0.5e^{j2(\pi/8)2} + 0 + \cdots + 0.5e^{j14(\pi/8)2}$$
$$+ 1e^{j15(\pi/8)2})$$
$$= 0.151$$

From Step 3, we obtain

$$H'(z) = 0.151z^2 + 0.222z^1 + 0.25z^0 + 0.222z^{-1} + 0.151z^{-2}$$

The positive power on z indicates a time advance, which in turn requires that the filter must have input data for $t < 0$. To obtain a causal filter, we multiply $H'(z)$ by z^{-2}, which yields, for the desired filter,

$$H(z) = 0.151 + 0.222z^{-1} + 0.25z^{-2} + 0.222z^{-3} + 0.151z^{-4}$$

The amplitude characteristics $|H(e^{j\omega T})|$ are plotted in Figure 11.21b.

The time shift does not alter the amplitude characteristics of the filter but only its phase, which here is linear; this feature is often desired. ∎

11-13 USE OF WINDOW FUNCTIONS FOR FIR FILTERS

In this section we will develop the Fourier series method of filter design. Also, we will discuss the use of window functions to produce a smoother response function than would be possible without the use of an appropriate window.

We know that any periodic function can be expanded into a Fourier series. Thus $H(e^{j\omega T})$, which is periodic, can be written in the form

$$H(e^{j\omega T}) = \sum_{k=-\infty}^{\infty} h(kT)e^{jk\omega T} \tag{11.80}$$

where

$$h(kT) = \frac{1}{\omega_s}\int_{-\omega_s/2}^{\omega_s/2} H(e^{j\omega T})e^{jk\omega T}\,d\omega \tag{11.81}$$

If we set $z = e^{j\omega T}$ [see (11.47)], then (11.80) becomes

$$H(z) = \sum_{k=-\infty}^{\infty} h(kT)z^{-k} \tag{11.82}$$

In practice, however, we use a finite number of samples, and we therefore set

$$h(kT) = 0 \quad \text{for } k > \frac{N-1}{2} \quad \text{and} \quad k < -\frac{N-1}{2} \tag{11.83}$$

For a finite number of samples, (11.82) becomes

$$H'(z) = h(0) + \sum_{k=1}^{(N-1)/2} [h(-kT)z^k + h(kT)z^{-k}] \tag{11.84}$$

To create a causal filter, we multiply $H'(z)$ by $z^{-(N-1)/2}$, which then gives as a final configuration the expression

$$H(z) = z^{-(N-1)/2}H'(z) \tag{11.85}$$

The truncation of the infinite series specified in (11.80) which implies the use of a rectangular window results in a modified FIR filter that closely resembles the exact one. The truncation operation creates Gibbs phenomenon overshoots at the points of discontinuities of $H(e^{j\omega T})$ with the ripples existing in the neighborhood of the discontinuities. Here, as in several other points of our study, **weighting sequences** (windows) are used to modify the truncated function in order to decrease the overshoot. Thus the modified truncated sequence of the Fourier coefficients will be specified by

$$h_w(kT) = \begin{cases} h(kT)w(kT) & 0 \le k \le N-1 \\ 0 & \text{otherwise} \end{cases} \tag{11.86}$$

where

ELEMENTS OF DIGITAL FILTER DESIGN

$$w(kT) = \begin{cases} w(kT) & 0 \le k \le N-1 \\ 0 & \text{otherwise} \end{cases} \quad (11.87)$$

To design an FIR filter using the DFT approach, the following steps are required:

1. From the given amplitude characteristics, obtain the desired sampled values $H(n\Omega)$ for a specified N.
2. Incorporate in (11.81) the values found in Step 1 and find the sequence $\{h(kT)\}$.
3. Apply an appropriate window function to the finite sequence found in Step 2.
4. Use (11.84) to obtain $H'(z)$.
5. Multiply $H'(z)$ by $z^{-(N-1)/2}$.
6. Plot $|H(e^{j\omega T})|$ versus ω to obtain the amplitude characteristic of the resulting filter.

EXAMPLE 11.12

Apply the Hamming window [see (10.72)] to the impulse response sequence $h(kT)$ that corresponds to the desired frequency characteristics of the filter shown in Figure 11.20. Choose the values: $0 \le k \le 4 = N - 1$ and $T = 1$.

Solution: From (11.81) we obtain

$$h(kT) = \frac{T}{2\pi} \int_{-\pi/2T}^{\pi/2T} e^{jk\omega T} d\omega = \frac{1}{\pi k} \sin \frac{\pi k}{2}$$

$$= \frac{1}{\pi k} \sin\left(T\frac{1}{4} \times \frac{2\pi}{T} k\right) = \frac{1}{\pi k} \sin(\omega_c kT)$$

where ω_c is the cutoff frequency of the filter and $\omega_s = 2\pi/T$ is the sampling frequency. If we set $T = 1$, a given condition, the impulse sequence is

$$\{h(k)\} = \frac{1}{2}, \frac{1}{\pi}, 0, -\frac{1}{3\pi}, 0$$

For the Hamming window in a form that is symmetric around the origin, we must write (10.72) in the form

$$w_{\text{hm}}(k) = 0.54 + 0.46 \cos \frac{k\pi}{K}$$

where the constant K is equal to the number of terms to be included in the range on each side of $h(0T)$. Thus we find, (where $k = 4$)

$w_{\text{hm}}(0) = 1 \qquad w_{\text{hm}}(1) = 0.865 \qquad w_{\text{hm}}(2) = 0.541$
$w_{\text{hm}}(3) = 0.215 \qquad w_{\text{hm}}(4) = 0.081$

The resulting windowed sequence is

Figure 11.22
(a) Unsmoothed amplitude versus frequency characteristic of the filter under consideration. (b) Smoothed (windowed) amplitude versus frequency characteristic with the Hamming window.

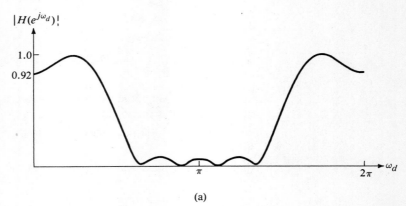

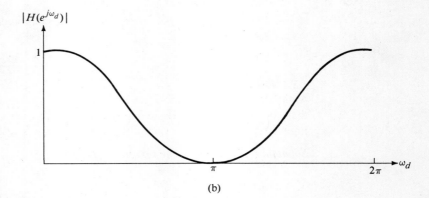

$$\{h_w(k)\} = \{h(k)w_{\text{hm}}(k)\} = \frac{1}{2}, \frac{0.865}{\pi}, 0, -\frac{0.215}{3\pi}, 0$$

This sequence is used in (11.84) to obtain

$$H'(z) = \frac{1}{2}z^{-0} + \frac{0.865}{\pi}z^{-1} + 0 \times z^{-2} - \frac{0.215}{3\pi}z^{-3} + 0 \times z^{-4}$$

$$+ \frac{0.865}{\pi}z^1 + 0 \times z^2 - \frac{0.215}{3\pi}z^3 + 0 \times z^4$$

Since our largest positive exponent is 3, we multiply $H'(z)$ by z^{-3} to find

$$H(z) = z^{-3}H'(z)$$

$$= -\frac{0.215}{3\pi} + \frac{0.865}{\pi}z^{-2} + \frac{1}{2}z^{-3} + \frac{0.865}{\pi}z^{-4} - \frac{0.215}{3\pi}z^{-6}$$

A plot of $|H(e^{j\omega T})|$ with $T = 1$ is given in Figure 11.22 for both the unmodified and the windowed cases. It is apparent from the plots that the effect of the window tends to smooth the ripples. ∎

Figure 11.23
(a) Approximation to ideal low-pass filter, exhibiting Gibbs phenomenon. (b) The low-pass filter designed with the use of a window function.

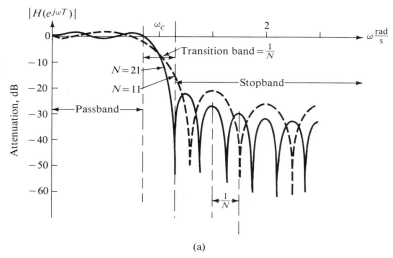

(a)

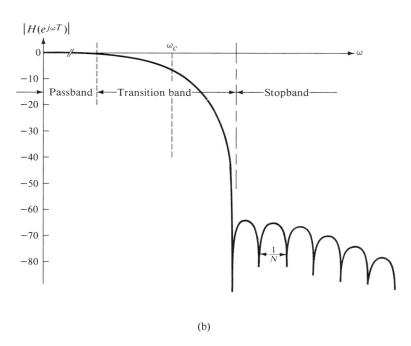

(b)

The data contained in Figure 11.22 are often presented by plotting the magnitude function $|H(e^{j\omega T})|$ on a dB scale. Figure 11.23a shows such a plot for a low-pass filter with $N = 11$ and $N = 21$. We see that the points for $|H(e^{j\omega T})| = 0$ are placed in sharp perspective since the magnitude function becomes $-\infty$ at these points. Figure 11.23b shows by comparison the effect obtainable using an appropriate window for the filter (with $N = 11$) shown in Figure 11.23a. The enlarged transition band and the increased attenuation in the stopband are clearly evident. As shown here, the transition bandwidth is approximately $5/N$.

The three most commonly used window functions for modifying nonrecursive digital filters are:

1. Hamming window

$$w_{\text{hm}} = 0.54 + 0.46 \cos(k\pi/K) \tag{11.88}$$

2. Blackman window

$$w_b = 0.42 + 0.5 \cos(k\pi/K) + 0.08 \cos(2k\pi/K) \tag{11.89}$$

3. Hann window

$$w_h = 0.5 + 0.5 \cos(k\pi/K) \tag{11.90}$$

The constant K is equal to the number of terms to be included in the expansion for $h(kT)$.

■ ■ ■

11-14 THE DFT AS A FILTER

The discrete Fourier transform (DFT) can be used in digital filter design. This requires that the discrete signal to be filtered is transformed to the frequency domain via an FFT algorithm, and the frequency samples of the signal are then multiplied by the desired frequency response characteristic of the filter. The filtered frequency samples are then transformed to the time domain through an inverse FFT procedure. Since the filtering operation is accomplished in the frequency domain, there is no need for determining the coefficients of the DFT. We first assume that the desired frequency response is specified at discrete frequency points, and the filter coefficients are found from the discrete frequency samples through the inverse DFT. Suppose therefore that

$$H(n) \triangleq H\left(\frac{2\pi n}{NT}\right) = H(n\Omega) \qquad n = 0, 1, 2, \ldots, N-1$$

denotes the samples of the frequency response at equally spaced frequency points. The unit sample response sequence is

$$h(k) = \frac{1}{N} \sum_{n=0}^{N-1} H(n) e^{j2\pi kn/N}$$

$$\triangleq \frac{1}{N} \sum_{n=0}^{N-1} H(n\Omega) e^{jn\Omega kT} \qquad k = 0, 1, \ldots, N-1 \tag{11.91}$$

The transfer function $H(z)$ is then

$$H(z) = \sum_{k=0}^{N-1} h(k) z^{-k} = \sum_{k=0}^{N-1} \left(\frac{1}{N} \sum_{n=0}^{N-1} H(n) e^{j2\pi kn/N} \right) z^{-k}$$

$$= \frac{1}{N} \sum_{n=0}^{N-1} H(n) \sum_{k=0}^{N-1} (e^{j2\pi kn/N}) z^{-k} \tag{11.92}$$

ELEMENTS OF DIGITAL FILTER DESIGN

But the summation

$$\sum_{k=0}^{N-1} e^{j2\pi nk/N} z^{-k} = 1 + e^{j(2\pi n)/N} z^{-1} + \cdots + e^{j(2\pi n)(N-1)/N} z^{-(N-1)}$$

$$= \frac{1 - (e^{j2\pi n/N} z^{-1})^N}{1 - e^{j2\pi n/N} z^{-1}} = \frac{1 - e^{j2\pi n} z^{-N}}{1 - e^{j2\pi n/N} z^{-1}}$$

$$= \frac{1 - z^{-N}}{1 - e^{j2\pi n/N} z^{-1}} = \frac{z^N - 1}{z^{N-1}(z - e^{j2\pi n/N})}$$

and (11.92) becomes

$$H(z) = \frac{1}{N} \sum_{n=0}^{N-1} H(n) \frac{1 - z^{-N}}{1 - e^{j2\pi n/N} z^{-1}} \tag{11.93}$$

This shows that the transfer function can be found, given the samples of the frequency response of the FIR filter. This equation has N poles and N zeros. The N zeros are located at the principal roots of 1. There are $N - 1$ poles located at $z = 0$, and there is one pole located at $z = \exp[j2\pi n/N]$. This pole will cancel the zero for the value $n = N$. Thus the resulting sequence has $N - 1$ poles, all located at $z = 0$ and $N - 1$ zeros located on the unit circle. This means, of course, that $H(z)$ is given as the ratio of two polynomials. Therefore the FIR filter that is realized from this transfer function is a recursive one.

EXAMPLE 11.13
Show that each term in the summation of (11.93) specifies a filter. Choose $N = 3$, and consider the function for $n = 8$.

Solution: We examine the specific function

$$H_n(z) = \frac{1 - z^{-N}}{1 - e^{j2\pi n/N} z^{-1}} = \frac{z^N - 1}{z^{N-1}(z - e^{j2\pi n/N})}$$

For $T = 1/f_s$,

$$|H_n(e^{j\omega T})| = \left| \frac{e^{j\omega NT} - 1}{e^{j\omega T} - e^{j2\pi n/N}} \right| = \left| \frac{e^{j2\pi Nf/f_s} - 1}{e^{j2\pi f/f_s} - e^{j2\pi n/N}} \right|$$

$$= \frac{|e^{j\pi Nf/f_s} - e^{-j\pi Nf/f_s}|}{|e^{j\pi(f/f_s + n/N)} - e^{-j\pi(f/f_s + n/N)}|} = \frac{\sin \pi Nf/f_s}{\sin \pi \left(\dfrac{f}{f_s} + \dfrac{n}{N} \right)}$$

We observe that $H_n(z)$ includes $N - 1$ poles at $z = 0$ and N zeros on the unit circle. However, one of the poles and one of the zeros cancel each other. The situations for the zero-poles constellation and the response function are illustrated in Figure 11.24. ∎

The frequency response of this filter is obtained from (11.93) by setting $z = \exp[j\omega T]$. We then have

Figure 11.24
The DFT filter function.
(a) Pole-zero configuration.
(b) Frequency response characteristics.

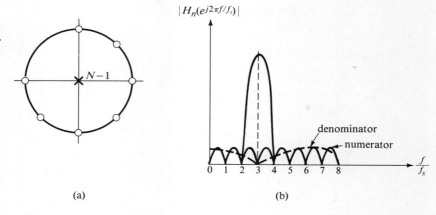

(a)　　　　　　　　　　　　(b)

$$H(e^{j\omega T}) = \frac{1}{N} \sum_{n=0}^{N-1} H(n) \frac{1 - e^{-jN\omega T}}{1 - e^{j2\pi n/N} e^{-j\omega T}}$$

$$= \frac{1}{N} \sum_{n=0}^{N-1} H(n) \frac{1 - e^{-jN(\omega T - 2\pi n/N)}}{1 - e^{-j(\omega T - 2\pi n/N)}}$$

This can be expanded to

$$H(e^{j\omega T}) = \frac{1}{N} \sum_{n=0}^{N-1} H(n) \frac{e^{-jN(\omega T - 2\pi n/N)/2}}{e^{-j(\omega T - 2\pi n/N)/2}} \frac{\sin\left[N \frac{\omega T - \frac{2\pi n}{N}}{2}\right]}{\sin\left[\frac{\omega T - \frac{2\pi n}{N}}{2}\right]}$$

$$= e^{-j(N-1)\omega T/2} \sum_{n=0}^{N-1} H(n) e^{-j\pi n/N} \frac{\sin\left(\frac{N\omega T}{2}\right)}{N \sin\left[\frac{\left(\omega T - \frac{2\pi n}{N}\right)}{2}\right]} \quad (11.94)$$

As we know from DFT properties, for the unit sample response sequence $h(k)$ of the FIR filter to be a real valued quantity, the frequency samples $H(n)$ must satisfy the symmetry property; that is, the real part must be an even function and the imaginary part must be an odd function. Therefore the frequency samples must be so selected that $H(N - n) = H(n)$. This requires that the amplitudes and phase be selected so that

$$A(N - n) = A(n) \qquad \theta(N - n) = 2m\pi - \theta(n)$$
$$\text{for} \quad n = 1, 2, \ldots, N - 1 \quad (11.95)$$

where m is an integer. Therefore we must have

ELEMENTS OF DIGITAL FILTER DESIGN

$$H(0) = \text{real} \quad \text{or} \quad \theta(0) = m\pi \quad \text{(a)}$$
$$H\left(\frac{N}{2}\right) = A\left(\frac{N}{2}\right) \geq 0 \quad \theta\left(\frac{N}{2}\right) = 2m\pi \quad \text{(b)} \quad (11.96)$$

■ ■ ■

11-15 FREQUENCY TRANSFORMATION OF IIR FILTERS

In parallel with the availability of special transformations for analog filters for converting low-pass to high-pass, we can find special transformations for digital filters. These will permit transformation of a low-pass filter to another low-pass filter, a high-pass filter, a bandpass filter, or a bandstop digital filter. We consider these transformations:

1. Low-Pass to Low-Pass Transformation. This transformation is

$$z^{-1} \rightarrow \frac{z^{-1} - \alpha}{1 - \alpha z^{-1}} \quad (11.97)$$

where

$$\alpha = \frac{\sin\left[\left(\frac{\omega_{dc} - \omega'_{dc}}{2}\right)\right]}{\sin\left[\left(\frac{\omega_{dc} + \omega'_{dc}}{2}\right)\right]}$$

ω_{dc} = cutoff frequency of the given filter

ω'_{dc} = cutoff frequency of the desired filter

EXAMPLE 11.14

Design a low-pass filter with $\omega'_{dc} = 0.5$ if the given filter is that specified in Example 11.9.

Solution: We first determine the value of

$$\alpha = \frac{\sin\left[\left(\frac{0.4\pi - 0.5\pi}{2}\right)\right]}{\sin\left[\left(\frac{0.4\pi + 0.5\pi}{2}\right)\right]} = -0.158$$

Next we combine (11.97) with the results of Example 11.9. We find

$$H(z) = \frac{2.117 + 4.234\left[\dfrac{z^{-1} + 0.158}{1 + 0.158z^{-1}}\right] + 2.117\left[\dfrac{z^{-1} + 0.158}{1 + 0.158z^{-1}}\right]^2}{10.232 - 3.766\left[\dfrac{z^{-1} + 0.158}{1 + 0.158z^{-1}}\right] + 2.002\left[\dfrac{z^{-1} + 0.158}{1 + 0.158z^{-1}}\right]^2}$$

$$= \frac{2.839 + 5.683z^{-1} + 2.839z^{-2}}{3.963 + 9.682z^{-2}}$$

2. Low-Pass to High-Pass Transformation.

$$z^{-1} \to -\frac{z^{-1} + \alpha}{1 + \alpha z^{-1}} \tag{11.98}$$

where

$$\alpha = -\frac{\cos\left[\left(\frac{\omega_{dc} + \omega'_{dc}}{2}\right)\right]}{\cos\left[\left(\frac{\omega_{dc} - \omega'_{dc}}{2}\right)\right]}$$

3. Low-Pass to Bandpass Transformation.

$$z^{-1} \to -\frac{z^{-2} + \frac{2\alpha\beta}{\beta+1}z^{-1} + \frac{\beta-1}{\beta+1}}{\frac{\beta-1}{\beta+1}z^{-2} + \frac{2\alpha\beta}{\beta+1}z^{-1} + 1} \tag{11.99}$$

where

$$\alpha = \frac{\cos\left[\left(\frac{\omega'_{du} + \omega'_{dl}}{2}\right)\right]}{\cos\left[\left(\frac{\omega'_{du} - \omega'_{dl}}{2}\right)\right]}$$

$$\beta = \cot\left(\frac{\omega'_{du} - \omega'_{dl}}{2}\right)\tan\frac{\omega_{dc}}{2}$$

ω'_{du} = desired upper cutoff frequency

ω'_{dl} = desired lower cutoff frequency

4. Low-Pass to Bandstop Transformation.

$$z^{-1} \to \frac{z^{-2} - \frac{2\alpha}{1+\beta}z^{-1} + \frac{1-\beta}{1+\beta}}{\frac{1-\beta}{1+\beta}z^{-2} - \frac{2\alpha}{1+\beta}z^{-1} + 1} \tag{11.100}$$

where

$$\alpha = \frac{\cos\left[\left(\frac{\omega'_{du} + \omega'_{dl}}{2}\right)\right]}{\cos\left[\left(\frac{\omega'_{du} - \omega'_{dl}}{2}\right)\right]}$$

$$\beta = \tan\left(\frac{\omega'_{du} - \omega'_{dl}}{2}\right) \tan\frac{\omega_{dc}}{2}$$

ω'_{du} = desired upper cutoff frequency

ω'_{dl} = desired lower cutoff frequency

■ ■ ■

11-16 RECURSIVE VERSUS NONRECURSIVE DESIGNS: GENERAL REMARKS

A comparison of the important features of different filter designs is helpful. In recursive filters the poles of the transfer function can be placed anywhere within the unit circle. As a result, high selectivity can be achieved using low-order transfer functions. With nonrecursive filters, on the other hand, the poles are fixed at the origin and high selectivity can be achieved only by using a relatively high order for the transfer function. For the same filter specifications, the order of the nonrecursive filter might be as much as 5 to 10 times that of the recursive structure, with the consequent need for more electronic parts. Often, however, a recursive filter might not meet the specifications, and in such cases the nonrecursive filter can be used by the designer.

An important advantage of the nonrecursive filter is that it can be implemented using the FFT method, in the manner discussed in Section 11-14.

Hardware filter implementation requires that storage of input and output data and also arithmetic operations are implemented by using finite word-length registers (for example, 8, 12, or 16 bits). As a result, certain errors will occur; these are categorized as follows:

1. Quantization errors due to arithmetic operations such as rounding off and truncation.

2. Quantization errors due to representing the input signal by a set of discrete values.

3. Quantization errors when the filter coefficients are represented by a finite number of bits.

It is left to the filter designer to decide on the various trade-offs between cost and precision in trying to reach a specified goal.

REFERENCES

1. Ahmed, N., and T. Natarajan. *Discrete-Time Signals and Systems*. Reston, Va.: Reston, 1983.

2. Antoniou, A. *Digital Filters; Analysis and Design*. New York: McGraw-Hill, 1979.

3. Gold, B., and C. M. Rader. *Digital Processing of Signals*. New York: McGraw-Hill, 1969.

PROBLEMS

11-2.1 A low-pass filter is to be designed to have a nominal cutoff of 5 kHz. It is to be maximally flat (Butterworth) and is to be down 1 dB at the edge of the passband. The response function is to have 3 poles.
a. Locate the poles in the s-plane.
b. What is the rate of attenuation remote from cutoff?

11-2.2 Derive the transfer function for a third-order Butterworth low-pass filter and locate its poles.

11-2.3 The squared normalized amplitude of a Butterworth filter is $|H_n(j\omega)|^2 = 1/(1 + \omega^6)$. Find the normalized transfer function $H_n(j\omega)$.

11-2.4 Show that the high frequency roll-off of an nth-order Butterworth filter is $20n$ dB/decade. Also, show that the first $(2n - 1)$ derivatives of an nth-order Butterworth filter are zero at $\omega = 0$.

11-3.1 Repeat Problem 11-2.1 for a Chebyshev filter.

11-3.2 Find the value of n for a Butterworth and a Chebyshev filter that will satisfy the conditions specified in Figure P11-3.2.

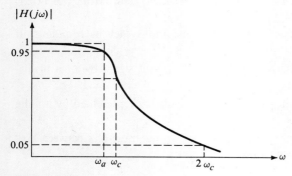

Figure P11-3.2

11-3.3 A Chebyshev low-pass filter is designed to have a passband ripple ≤ 2 dB with a cutoff frequency at 1000 rad/s. The attenuation is to be at least 50 dB at 5000 rad/s. Specify ϵ, n, and $H(j\omega)$.

11-3.4 Compare the attenuation at high frequencies of Butterworth and Chebyshev low-pass filters of the same order when the 3 dB cutoff frequencies are the same for both filters. Sketch the results on a graph of dB versus ω.

11-3.5 The characteristics to be met by a low-pass filter are:

passband: 0 to 4 kHz

attenuation: amplitude to be down 60 dB at 6 kHz

ELEMENTS OF DIGITAL FILTER DESIGN

Determine the value of n for:
a. Butterworth.
b. Chebyshev with 1/2 dB ripple in the passband.
c. Chebyshev with 1 dB ripple in the passband.

11-7.1 A bandpass filter is to have nominal cutoff points at frequencies 300 and 600 kHz. It is to have a maximally flat shape (Butterworth) and is to be down 3 dB at the edge of the band. The response function is to have 3 poles.
a. Locate the poles of the response function.
b. What is the ultimate rate of attenuation on each side of the band?

11-9.1 Suppose that two RC sections, as discussed in Example 11.6, are cascaded, with a buffer amplifier between the sections to avoid loading. Find the digital filter equivalent to this combination. How does it relate to the results of Example 11.6?

11-9.2 Design a low-pass digital filter consisting of a cascade of identical sections, each of which has $H(z) = \dfrac{(1-\alpha)z}{z-\alpha}$. The nominal cutoff is 1 kHz with a gain at $\omega = 2\omega_c$ that is less than 0.35. The sampling frequency is $10 f_c$. Specify the $H(e^{j\omega T})$ of this filter.

11-9.3 Repeat Example 11.4 with $T = 0.1$ s and $T = .0025$ s.

11-9.4 Find the second-order Butterworth digital filter with cutoff frequency $f_c = 100$ Hz and sampling rate of 1250 Hz.

11-10.1 Design a Butterworth low-pass filter, given the following data: half-power point at $\omega_1 = 250\pi$, sampling period $T = .0005$ s, $\omega_2 = 500\pi$, gain ≤ 0.2. Refer to Figure 11.2.

11-10.2 Locate the points $s = 1$, $s = 1 + j\omega_s/2$, $s = 1 - j\omega_s/2$, $s = 10$, and $s = \infty$ on the s-plane when mapped by a bilinear transformation onto the z-plane. Assume that $T = 0.1$.

11-10.3 Show that warping also affects the phase curves.

11-11.1 Determine analytically the frequency response of the nonrecursive filter specified by $H(z) = \sum_{n=0}^{20} z^{-n}$.

11-11.2 Show that if $p = re^{j\theta}$ is a root of the polynomial in z given in (11.67), then $re^{-j\theta}$, $(1/r)e^{j\theta}$, and $(1/r)e^{-j\theta}$ also are roots.

11-11.3 Consider the Butterworth low-pass filter given by (11.73). Show that by rotating the poles and zeros, the resulting function has the squared gain

$$|H(e^{j\omega T})|^2 = \dfrac{1}{1 + \dfrac{\cot^{2n}\left(\dfrac{\omega T}{2}\right)}{\cot^{2n}\left(\dfrac{\omega_2 T}{2}\right)}}$$

where $\omega_2 = (\pi/T) - \omega_1$, and where $H(-1) = 1$. This is a high-pass filter.

11-12.1 The low-pass filter of Example 11.11 is modified by the use of a Hann window. Find the resulting frequency response.

11-12.2 Refer to Figure P11-12.2.
a. Determine a nonrecursive filter that will approximate the frequency response shown using 10 terms from the Fourier series design method.
b. Using the Hamming window function, determine a nonrecursive filter that will approximate the given frequency response using 10 terms from the Fourier series.
c. Plot the frequency response of the filters developed in a. and b. Comment on the results.

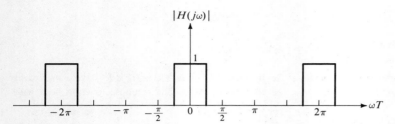

Figure P11-12.2

11-14.1 Determine the frequency response of a DFT filter that has a linear phase characteristic, with the constant delay equal to $(N - 1)/2$ units of the sampling interval.

11-14.2 Show that a nonrecursive transfer function obtained by the DFT method can be realized by means of a set of parallel second-order recursive sections in cascade with an elementary Nth-order nonrecursive section.

CHAPTER 12
STATE VARIABLES AND STATE EQUATIONS

A. CONTINUOUS-TIME (ANALOG) SYSTEMS

12-1 INTRODUCTION

We introduced the notion of state variables and state equations in Section 2-12 as one means for describing the dynamics of a system. In this representation, instead of developing system descriptions in terms of input-output relationships, the systems are represented by sets of equations that describe unique relations among the input, output, and state of the system.

Among the important features of a state description of a system are the following:

a. This type of description has been extensively investigated and used for both analysis and synthesis in controlled systems.

b. The techniques can be extended to time-varying and nonlinear systems.

c. The state variables are part of the solution and are often important to know.

d. This form is compact in its representation and is suitable for analog and digital computer solution.

e. The form of solution is common to all systems.

f. General system characteristics can be discussed and developed from the state equations.

A feature of a state description of any system is its formulation in matrix form. This allows complicated systems to be studied using relatively simple mathematical constructs. The purpose of this chapter is to explore the state approach in some detail.

We discussed in Section 2-12 how to write the appropriate set of equations for a given analog time system. These ideas will be extended here; we begin with the continuous-time systems, which were found to be expressible in state equation form by the set of equations (see 2.134)

$$\dot{\mathbf{x}}(t) = \mathbf{A}\mathbf{x}(t) + \mathbf{B}\mathbf{w}(t) \quad \text{(a)}$$
$$\mathbf{y}(t) = \mathbf{C}\mathbf{x}(t) + \mathbf{D}\mathbf{w}(t) \quad \text{(b)}$$

(12.1)

Equation (12.1a) denotes a set of n first-order differential equations. In these expressions, $\mathbf{x}(t)$ is the state vector that describes the system at any time t; $\mathbf{A}$, $\mathbf{B}$, $\mathbf{C}$, and $\mathbf{D}$ are the matrices determined from the constants of the system; $\mathbf{y}(t)$ is the output vector; $\mathbf{w}(t)$ is the input vector; and the dot above $\dot{\mathbf{x}}(t)$ denotes time differentiation. If $\mathbf{y}(t)$ depends only on $\mathbf{w}(t)$, the system is said to be **memoryless**. If the coefficient matrices have time-varying elements, the system is **time-varying**. We will restrict our study to systems with constant elements; that is, we assume that in the electrical systems to be treated the values of resistors, capacitors, and inductors do not change with time.

We will find that under certain conditions (12.1a) may be of the form

$$\dot{\mathbf{x}}(t) = \mathbf{A}\mathbf{x}(t) + \mathbf{B}\mathbf{w}(t) + \mathbf{G}\dot{\mathbf{w}}(t) \tag{12.2}$$

However, if we define a new state vector

$$\hat{\mathbf{x}}(t) = \mathbf{x}(t) - \mathbf{G}\mathbf{w}(t) \tag{12.3}$$

then (12.1a) can be written

$$\dot{\hat{\mathbf{x}}}(t) = \mathbf{A}\hat{\mathbf{x}}(t) + (\mathbf{B} + \mathbf{A}\mathbf{G})\mathbf{w}(t) \tag{12.4}$$

which is of the same general form as (12.1a). Because of this, we assume in subsequent developments that the state equations for the general system are given by (12.1).   ∎ ∎ ∎

12-2 STATE EQUATIONS FOR LINEAR SYSTEMS

The state vector $\mathbf{x}(t)$ specifies the minimal set of state variables that uniquely determines the future state of a dynamic system, if the present values and the input $\mathbf{w}(t)$ are known. This means that given $\mathbf{x}(t_0)$ (the state of the system at time $t = t_0$), then the state vector at any future time $\mathbf{x}(t)$ for $t > t_0$ is uniquely determined by (12.1) We will show that many acceptable state variables exist for any given system. Actually, any minimum set of variables that completely describes the system will be a suitable state vector. However, these must be chosen to allow formulation in terms of a set of first-order differential equations.

A reasonably systematic procedure for the selection of the state variables of a system is possible. Clearly, as we mentioned in Chapter 2, if the voltage across a capacitor and the current through an inductor are known at some initial time t_0, then the circuit equations will allow a description of the system behavior for all subsequent times. This suggests the following guidelines for the selection of acceptable state variables:

1. Currents (through-variables) associated with inductor-type elements (inductors, springs).
2. Voltages (across-variables) associated with capacitor-type elements (capacitors, mass elements).

STATE VARIABLES AND STATE EQUATIONS

3. Dissipative-type elements do not specify independent state variables.
4. When closed loops of capacitors or junctions of inductors exist, not all state variables chosen according to Rules 1 and 2 are independent.

Note that when the output-input description is given in terms of a differential equation, a systematic method exists for converting an nth-order differential equation into n first-order equations, each associated with a state variable. However, this state formulation will involve states different, in general, from the states obtained using the rules given above. We will find that a linear transformation exists among the states.

EXAMPLE 12.1

Deduce the state equations for the circuit shown in Figure 12.1a.

Solution: As suggested by the rules above, the state variables are first selected to be: i_1, i_2, v_2, v_3. Next, the mesh equations and the relationships between mesh currents and state variables are written (these equations represent inductor and capacitor currents in terms of state variables and their derivatives). In our case we write by inspection

$$v = R_1 i_1 + L_1 \frac{di_1}{dt} + R_3(i_1 - i_2) + v_3$$

$$0 = R_2 i_2 + v_2 + L_2 \frac{di_2}{dt} - v_3 - R_3(i_1 - i_2)$$

$$C_3 \frac{dv_3}{dt} = i_1 - i_2$$

$$C_2 \frac{dv_2}{dt} = i_2$$

These equations are rearranged and then combined in matrix form:

$$\frac{d}{dt}\begin{bmatrix} v_2 \\ v_3 \\ i_1 \\ i_2 \end{bmatrix} = \begin{bmatrix} 0 & 0 & 0 & \frac{1}{C_2} \\ 0 & 0 & \frac{1}{C_3} & -\frac{1}{C_3} \\ 0 & -\frac{1}{L_1} & -\frac{R_1+R_3}{L_1} & \frac{R_3}{L_1} \\ -\frac{1}{L_2} & \frac{1}{L_2} & \frac{R_3}{L_2} & -\frac{R_2+R_3}{L_2} \end{bmatrix} \begin{bmatrix} v_2 \\ v_3 \\ i_1 \\ i_2 \end{bmatrix} + \begin{bmatrix} 0 \\ 0 \\ \frac{v}{L_1} \\ 0 \end{bmatrix}$$

A block diagram representation of this system with two desired outputs v_3 and i_1 is given in Figure 12.1b. The dots on the i's and v's indicate time differentiation. ∎

CHAPTER 12

Figure 12.1
An electrical system and its block diagram representation.

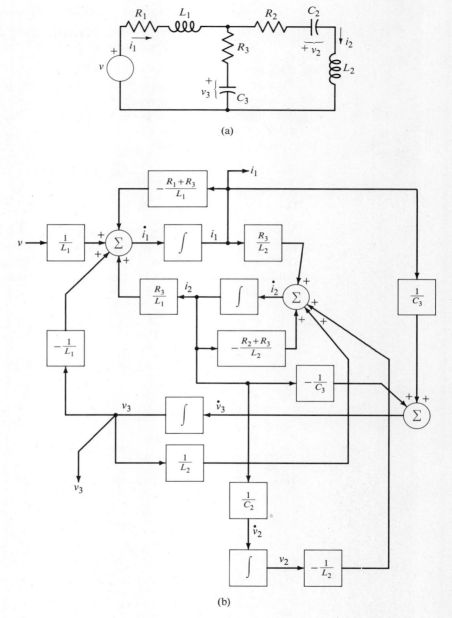

EXAMPLE 12.2
Write a state equation description of the circuit shown in Figure 12.2. Observe that a dependent source is included in this circuit.

STATE VARIABLES AND STATE EQUATIONS

Figure 12.2
An electrical system with dependent source.

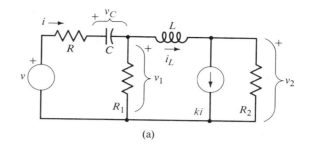

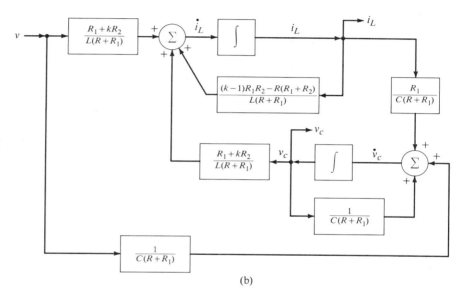

Solution: Using our guide to the selection of state variables, these are chosen as i_L and v_c. We write the following equations by inspection:

$$v = iR + v_c + v_1$$

$$i_L = i - \frac{v_1}{R_1} = ki + \frac{v_2}{R_2}$$

$$v_2 = v_1 - L\frac{di_L}{dt}$$

Combine the first and second equations to write

$$v = iR + v_c + R_1(i - i_L)$$

from which

$$i = C\frac{dv_c}{dt} = \frac{v - v_c + R_1 i_L}{R + R_1}$$

Now consider

$$L\frac{di_L}{dt} = v_1 - v_2 = (i - i_L)R_1 - (i_L - ki)R_2$$

$$= -(R_1 + R_2)i_L + (R_1 + kR_2)i$$

$$= -(R_1 + R_2)i_L + (R_1 + kR_2)\frac{(v - v_c + R_1 i_L)}{R + R_1}$$

$$= \left[\frac{(R_1 + kR_2)R_1}{R + R_1} - (R_1 + R_2)\right]i_L + \frac{(R_1 + kR_2)}{R + R_1}(v - v_c)$$

These two equations are expressed in matrix form

$$\frac{d}{dt}\begin{bmatrix} i_L \\ v_c \end{bmatrix} = \begin{bmatrix} \dfrac{(k-1)R_1 R_2 - R(R_1 + R_2)}{L(R + R_1)} & -\dfrac{R_1 + kR_2}{L(R + R_1)} \\ \dfrac{R_1}{C(R + R_1)} & -\dfrac{1}{C(R + R_1)} \end{bmatrix}\begin{bmatrix} i_L \\ v_c \end{bmatrix}$$

$$+ \begin{bmatrix} \dfrac{R_1 + kR_2}{L(R + R_1)} \\ \dfrac{1}{C(R + R_1)} \end{bmatrix} v$$

Figure 12.2 also includes a block diagram representation showing the two desired outputs. The dots on the i's and v's indicate time differentiation. ∎

■ ■ ■

12-3 DIFFERENTIAL EQUATIONS IN NORMAL FORM

If the input-output description of a system is given in terms of an nth-order differential equation or in terms of a set of second-order differential equations deduced by application of the Kirchhoff current and voltage laws, it is readily possible to convert these higher-order differential equations into sets of first-order differential equations. In this case the higher-order differential equation is transformed into **normal form**. We will find that the normal form set is expressible in state matrix form.

EXAMPLE 12.3
Find the normal form of the differential equation

$$4\frac{d^2 y}{dt^2} + 3\frac{dy}{dt} - 2y = 2\frac{dw}{dt} + 5w \tag{12.5}$$

where y is the output and w is the input to the system.

Solution: We first focus our attention on the reduced equation

STATE VARIABLES AND STATE EQUATIONS

$$4\frac{d^2\eta}{dt^2} + 3\frac{d\eta}{dt} - 2\eta = w \tag{12.6}$$

Observe that this reduced equation has the same form as (12.5) on the left, but has a simple function w on the right. We observe again the important property of linear differential equations, which relate the response of a system to an rth-order derivative input with the rth derivative of the input excitation function. Specifically, if $\eta(t)$ is the response of (12.6) to the excitation $w(t)$, then the response to the input $(d^r w/dt^r)$ will be $(d^r\eta/dt^r)$. This conclusion follows by straightforward differentiation of (12.6), with the result

$$4\frac{d^2}{dt^2}\left(\frac{d^r\eta}{dt^r}\right) + 3\frac{d}{dt}\left(\frac{d^r\eta}{dt^r}\right) - 2\left(\frac{d^r\eta}{dt^r}\right) = \frac{d^r w}{dt^r} \tag{12.7}$$

Suppose that we multiply (12.6) by 5, and also differentiate equation (12.6) once and multiply the result by 2. The two resulting equations are:

$$4 \times 5 \frac{d^2\eta}{dt^2} + 3 \times 5 \frac{d\eta}{dt} - 2 \times 5\eta = 5w \quad \text{(a)}$$

$$4 \times 2 \frac{d^2}{dt^2}\left(\frac{d\eta}{dt}\right) + 3 \times 2 \frac{d}{dt}\left(\frac{d\eta}{dt}\right) - 2 \times 2\left(\frac{d\eta}{dt}\right) = 2\frac{dw}{dt} \quad \text{(b)}$$

$$\tag{12.8}$$

Add these two equations to obtain

$$4\frac{d^2}{dt^2}\left(2\frac{d\eta}{dt} + 5\eta\right) + 3\frac{d}{dt}\left(2\frac{d\eta}{dt} + 5\eta\right) - 2\left(2\frac{d\eta}{dt} + 5\eta\right) = 2\frac{dw}{dt} + 5w \tag{12.9}$$

By comparing (12.9) with (12.5), we see that

$$y = 2\frac{d\eta}{dt} + 5\eta \tag{12.10}$$

We now write in (12.6) and (12.10)

$$\begin{aligned} x_1 &= \eta & \text{(a)} \\ x_2 &= \dot{x}_1 = \frac{d\eta}{dt} & \text{(b)} \\ \dot{x}_2 &= \frac{d^2\eta}{dt^2} = -\frac{3}{4}\frac{d\eta}{dt} + \frac{2}{4}\eta + \frac{w}{4} & \text{(c)} \end{aligned} \tag{12.11}$$

and these equations attain the form

$$\begin{aligned} \dot{x}_1 &= x_2 & \text{(a)} \\ \dot{x}_2 &= -\frac{3}{4}x_2 + \frac{2}{4}x_1 + \frac{1}{4}w & \text{(b)} \\ y &= 2x_2 + 5x_1 & \text{(c)} \end{aligned} \tag{12.12}$$

These equations are written in matrix form

$$\frac{d}{dt}\begin{bmatrix} x_1 \\ x_2 \end{bmatrix} = \begin{bmatrix} 0 & 1 \\ \frac{2}{4} & -\frac{3}{4} \end{bmatrix}\begin{bmatrix} x_1 \\ x_2 \end{bmatrix} + \begin{bmatrix} 0 \\ \frac{1}{4} \end{bmatrix} w \quad \text{(a)} \tag{12.13}$$

with the output vector

$$y = \begin{bmatrix} 5 & 2 \end{bmatrix}\begin{bmatrix} x_1 \\ x_2 \end{bmatrix} \quad \text{(b)}$$

∎

We follow the same procedure as in the foregoing example to reduce to normal form the general differential equation of the nth order; namely,

$$a_n \frac{d^n y}{dt^n} + a_{n-1} \frac{d^{n-1} y}{dt^{n-1}} + \cdots + a_0 y = b_m \frac{d^m w}{dt^m} + b_{m-1} \frac{d^{m-1} w}{dt^{m-1}} + \cdots + b_0 w \tag{12.14}$$

The result is the pair of equations

$$a_n \frac{d^n \eta}{dt^n} + a_{n-1} \frac{d^{n-1} \eta}{dt^{n-1}} + \cdots + a_0 \eta = w \quad \text{(a)}$$

$$y = b_m \frac{d^m \eta}{dt^m} + b_{m-1} \frac{d^{m-1} \eta}{dt^{m-1}} + \cdots + b_0 \eta \quad \text{(b)} \tag{12.15}$$

To express these results in normal form, we define the entities

$$x_1 = \eta$$

$$x_2 = \dot{x}_1 = \frac{d\eta}{dt}$$

$$x_3 = \dot{x}_2 = \frac{d^2 \eta}{dt^2} \tag{12.16}$$

$$x_n = \dot{x}_{n-1} = \frac{d^{n-1} \eta}{dt^{n-1}}$$

The final equation is obtained by combining these forms with (12.15a)

$$\dot{x}_n = -\frac{a_{n-1}}{a_n} x_{n-1} - \frac{a_{n-2}}{a_n} x_{n-2} - \cdots - \frac{a_0}{a_n} x_1 + \frac{w}{a_n} \tag{12.17}$$

Also, the expression for $y(t)$ is given by (12.15b) and is written

$$y = b_m x_{m+1} + b_{m-1} x_m + \cdots + b_0 x_1 \tag{12.18}$$

In matrix form, these expressions have the form

$$\frac{d\mathbf{x}}{dt} = \mathbf{A}\mathbf{x} + \mathbf{B}w \quad \text{(a)}$$

$$\mathbf{y} = \mathbf{C}\mathbf{x} \quad \text{(b)} \tag{12.19}$$

STATE VARIABLES AND STATE EQUATIONS

where

$$\mathbf{A} = \begin{bmatrix} 0 & 1 & 0 & \cdots & 0 \\ 0 & 0 & 1 & & 0 \\ \vdots & \vdots & \vdots & \cdots & \vdots \\ 0 & 0 & 0 & & 1 \\ -\dfrac{a_0}{a_n} & -\dfrac{a_1}{a_n} & -\dfrac{a_2}{a_n} & \cdots & -\dfrac{a_{n-1}}{a_n} \end{bmatrix} \quad \mathbf{B} = \begin{bmatrix} 0 \\ 0 \\ \vdots \\ 0 \\ \dfrac{1}{a_n} \end{bmatrix}$$

$$\mathbf{C} = \begin{bmatrix} b_0 & b_1 & b_2 & \cdots & b_m & 0 & \cdots & 0 \end{bmatrix}$$

This representation is also known as the **phase variable form.** A block diagram representation of the matrix state form is given in Figure 12.3a. Figure 12.3b is a signal flow graph portrayal of (12.15a) and (12.15b), where p is the derivative operator d/dt (often referred to as the Heaviside operator).

The notational simplicity of (12.19) is evident since a system of order n has the appearance of a first-order system. In addition to the compact notation, we will find that advantages exist when we undertake the solution of these matrix equations. If $m > n$, the system is said to be improper, and in this case derivatives of the input function will appear in the output state equations.

Figure 12.3
(a) Block diagram representation of the vector-matrix form of (12.19).
(b) Signal flow graph representation of (12.15).

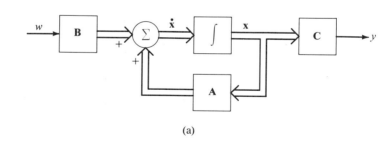

(a)

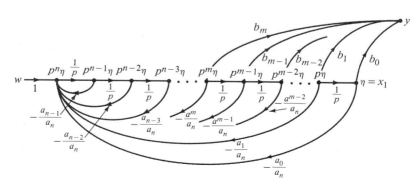

(b)

Another important form for the state equations somewhat like that given in (12.19) is the **first canonical form.** We first consider a third-order system and then extend the results to an nth-order system. We begin by considering the equation

$$a_3 \frac{d^3y}{dt^3} + a_2 \frac{d^2y}{dt^2} + a_1 \frac{dy}{dt} + a_0 y = b_3 \frac{d^3w}{dt^3} + b_2 \frac{d^2w}{dt^2} + b_1 \frac{dw}{dt} + b_0 w \tag{12.20}$$

We define the state variables as follows:

$$x_1 = a_3 y - b_3 w \tag{a}$$

$$x_2 = a_3 \frac{dy}{dt} - b_3 \frac{dw}{dt} + a_2 y - b_2 w \tag{b}$$

$$x_3 = a_3 \frac{d^2y}{dt^2} - b_3 \frac{d^2w}{dt^2} + a_2 \frac{dy}{dt} - b_2 \frac{dw}{dt} + a_1 y - b_1 w \tag{c}$$

(12.21)

This set can be written in the following form

$$x_1 = a_3 y - b_3 w \tag{a}$$

$$x_2 = \dot{x}_1 + a_2 y - b_2 w \tag{b}$$

$$x_3 = \dot{x}_2 + a_1 y - b_1 w \tag{c}$$

(12.22)

By differentiating (12.21c) and comparing it with (12.20), we obtain

$$\dot{x}_3 = -(a_0 y - b_0 w) \tag{12.23}$$

By rearrangement and combination of (12.22) and (12.23), we have

$$\dot{x}_1 = -a_2 y + x_2 + b_2 w = -a_2 \left(\frac{x_1}{a_3} + \frac{b_3}{a_3} w \right) + x_2 + b_2 w \tag{a}$$

$$\dot{x}_2 = -a_1 y + x_3 + b_1 w = -a_1 \left(\frac{x_1}{a_3} + \frac{b_3}{a_3} w \right) + x_3 + b_1 w \tag{b}$$

$$\dot{x}_3 = -\left[a_0 \left(\frac{x_1}{a_3} + \frac{b_3}{a_3} w \right) - b_0 w \right] \tag{c}$$

$$y = \frac{1}{a_3} (x_1 + b_3 w) \tag{d}$$

(12.24)

These equations can now be written in their standard first canonical form

$$\dot{x}_1 = \frac{1}{a_3} [-a_2 x_1 + a_3 x_2 + (a_3 b_2 - a_2 b_3) w] \tag{a}$$

$$\dot{x}_2 = \frac{1}{a_3} [-a_1 x_1 + a_3 x_3 + (a_3 b_1 - a_1 b_3) w] \tag{b}$$

(12.25)

STATE VARIABLES AND STATE EQUATIONS

$$\dot{x}_3 = \frac{1}{a_3}[-a_0 x_1 + (a_3 b_0 - a_0 b_3)w] \qquad \text{(c)}$$

$$y = \frac{1}{a_3}(x_1 + b_3 w) \qquad \text{(d)}$$

From these equations we can write the matrices

$$\mathbf{A} = \frac{1}{a_3}\begin{bmatrix} -a_2 & a_3 & 0 \\ -a_1 & 0 & a_3 \\ -a_0 & 0 & 0 \end{bmatrix} \qquad \mathbf{B} = \frac{1}{a_3}\begin{bmatrix} a_3 b_2 - a_2 b_3 \\ a_3 b_1 - a_1 b_3 \\ a_3 b_0 - a_0 b_3 \end{bmatrix} \qquad (12.26)$$

$$\mathbf{C} = \begin{bmatrix} \frac{1}{a_3} & 0 & 0 \end{bmatrix} \qquad \mathbf{D} = \begin{bmatrix} \frac{b_3}{a_3} \end{bmatrix}$$

We can readily extend the above equations into forms that apply for the nth-order differential equation

$$a_n \frac{d^n y}{dt^n} + a_{n-1}\frac{d^{n-1} y}{dt^{n-1}} + \cdots + a_0 y = b_n \frac{d^n w}{dt^n} + b_{n-1}\frac{d^{n-1}}{dt^{n-1}} + \cdots + b_0 w \qquad (12.27)$$

The nth-order first canonical form is written (paralleling 12.25)

$$\frac{d}{dt}\begin{bmatrix} x_1 \\ x_2 \\ \vdots \\ x_{n-1} \\ x_n \end{bmatrix} = \begin{bmatrix} -\frac{a_{n-1}}{a_n} & 1 & 0 & \cdots & 0 \\ -\frac{a_{n-2}}{a_n} & 0 & 1 & \cdots & 0 \\ \vdots & & & & \vdots \\ -\frac{a_1}{a_n} & 0 & 0 & \cdots & 1 \\ -\frac{a_0}{a_n} & 0 & 0 & \cdots & 0 \end{bmatrix}\begin{bmatrix} x_1 \\ x_2 \\ \vdots \\ x_{n-1} \\ x_n \end{bmatrix} + \begin{bmatrix} \frac{b_{n-1}}{a_n} - \frac{a_{n-1}b_n}{a_n^2} \\ \frac{b_{n-2}}{a_n} - \frac{a_{n-2}b_n}{a_n^2} \\ \vdots \\ \frac{b_1}{a_n} - \frac{a_1 b_b}{a_n^2} \\ \frac{b_0}{a_n} - \frac{a_0 b_n}{a_n^2} \end{bmatrix} a_n w$$

$$(12.28)$$

$$y = \begin{bmatrix} \frac{1}{a_n} & 0 & \cdots & 0 \end{bmatrix}\begin{bmatrix} x_1 \\ x_2 \\ \vdots \\ x_n \end{bmatrix} + \begin{bmatrix} \frac{b_n}{a_n} \end{bmatrix} w \qquad (12.29)$$

For the case of $b_n = b_{n-1} = \cdots = b_{m-1} = 0$, $m < n$, the block diagram representation of (12.28) is shown in Figure 12.4.

EXAMPLE 12.4
Write the differential equation of Example 12.3 in its first canonical form.

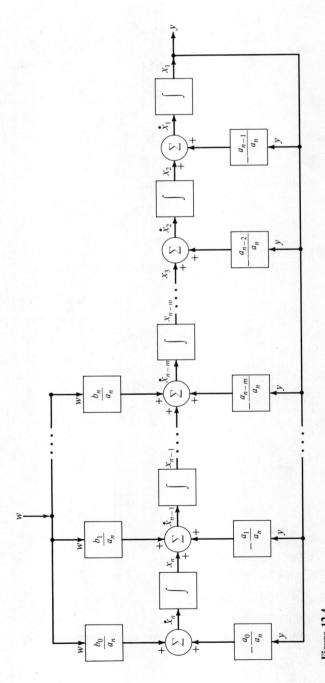

Figure 12.4
Block diagram representation of first canonical form.

Figure 12.5
Illustrating Example 12.4.

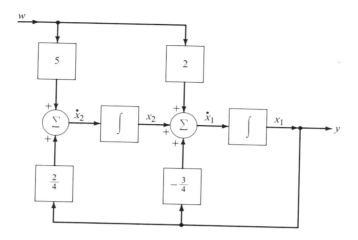

Solution: By comparing (12.5) with (12.28), we have the equivalent relationships:

$$a_n \triangleq a_2 = 4 \qquad a_{n-1} \triangleq a_1 = 3 \qquad a_{n-2} \triangleq a_0 = -2$$
$$b_n \triangleq b_2 = 0 \qquad b_{n-1} \triangleq b_1 = 2 \qquad b_{n-2} \triangleq b_0 = 5$$

By inspection, therefore, we write

$$\frac{d}{dt}\begin{bmatrix} x_1 \\ x_2 \end{bmatrix} = \begin{bmatrix} -\frac{3}{4} & 1 \\ -\frac{2}{4} & 0 \end{bmatrix} \begin{bmatrix} x_1 \\ x_2 \end{bmatrix} + \begin{bmatrix} \frac{2}{4} \\ \frac{5}{4} \end{bmatrix} 4w$$

$$y = \frac{1}{4}\begin{bmatrix} 1 & 0 \end{bmatrix} \begin{bmatrix} x_1 \\ x_2 \end{bmatrix}$$

The block diagram representation of this first canonical form expression is shown in Figure 12.5.

■ ■ ■

12-4 STATE VARIABLE TRANSFORMATIONS

We will now show that given one normal-form state equation representation of a system, we can find a new normal-form state equation representation in which the new variables are linear combinations of the original state variables. We begin with the state equations given by

$$\dot{\mathbf{x}} = \mathbf{A}\mathbf{x} + \mathbf{B}\mathbf{w} \quad \text{(a)}$$
$$\mathbf{y} = \mathbf{C}\mathbf{x} + \mathbf{D}\mathbf{w} \quad \text{(b)} \qquad\qquad (12.30)$$

Introduce a new set of state variables $\hat{x}_1, \hat{x}_2, \ldots, \hat{x}_n$, each of which is a linear combination of the original state variables x_j. That is,

$$\hat{x}_i = \sum_{j=1}^n m_{ij} x_j \qquad i = 1, 2, \ldots, n \quad \text{(a)} \tag{12.31}$$

which in vector form is

$$\hat{\mathbf{x}} = \mathbf{M}\mathbf{x} \qquad \text{(b)}$$

where $\mathbf{M}$ is a nonsingular matrix with matrix elements m_{ij}. Now assume that the elements of $\hat{\mathbf{x}}$ are linearly independent; therefore $\mathbf{M}$ is nonsingular and its inverse $\mathbf{M}^{-1}$ exists. Thus we can write

$$\mathbf{x} = \mathbf{M}^{-1}\hat{\mathbf{x}} \quad \text{(a)} \tag{12.32}$$
$$\dot{\mathbf{x}} = \mathbf{M}^{-1}\dot{\hat{\mathbf{x}}} \quad \text{(b)}$$

Combine these with (12.30)

$$\mathbf{M}^{-1}\dot{\hat{\mathbf{x}}} = \mathbf{A}\mathbf{M}^{-1}\hat{\mathbf{x}} + \mathbf{B}\mathbf{w} \tag{12.33}$$
$$\mathbf{y} = \mathbf{C}\mathbf{M}^{-1}\hat{\mathbf{x}} + \mathbf{D}\mathbf{w}$$

The first of these equations is premultiplied by $\mathbf{M}$, and this yields

$$\dot{\hat{\mathbf{x}}} = \mathbf{M}\mathbf{A}\mathbf{M}^{-1}\hat{\mathbf{x}} + \mathbf{M}\mathbf{B}\mathbf{w} \tag{12.34}$$

Thus the normal form in terms of the new state variables is

$$\dot{\hat{\mathbf{x}}} = \bar{\mathbf{A}}\hat{\mathbf{x}} + \bar{\mathbf{B}}\mathbf{w} \tag{12.35}$$
$$\mathbf{y} = \bar{\mathbf{C}}\hat{\mathbf{x}} + \bar{\mathbf{D}}\mathbf{w}$$

where

$$\bar{\mathbf{A}} = \mathbf{M}\mathbf{A}\mathbf{M}^{-1} \qquad \bar{\mathbf{B}} = \mathbf{M}\mathbf{B} \qquad \bar{\mathbf{C}} = \mathbf{C}\mathbf{M}^{-1} \qquad \bar{\mathbf{D}} = \mathbf{D}$$

■ ■ ■

12-5 SOLUTION OF THE CONTINUOUS-TIME STATE EQUATIONS: FORCE-FREE EQUATIONS

We will consider the solution of (12.1) in two steps:

$$\dot{\mathbf{x}}(t) = \mathbf{A}\mathbf{x}(t) + \mathbf{B}\mathbf{w}(t)$$
$$\mathbf{y}(t) = \mathbf{C}\mathbf{x}(t) + \mathbf{D}\mathbf{w}(t)$$

Initially we consider the force-free state equations (that is, the homogeneous equations)

$$\dot{\mathbf{x}} = \mathbf{A}\mathbf{x} \tag{12.36}$$

with known initial conditions $\mathbf{x}(0)$ at $t_0 = 0$. In this connection we examine the matrix function

STATE VARIABLES AND STATE EQUATIONS

$$\boxed{\Phi(t) = e^{\mathbf{A}t} = \mathbf{I} + \mathbf{A}t + \mathbf{A}^2 \frac{t^2}{2!} + \cdots = \sum_{k=0}^{\infty} \mathbf{A}^k \frac{t^k}{k!}} \qquad (12.37)$$

This function is called the **fundamental matrix,** and is often referred to as the **state transition matrix** of the system. It can be shown by a term-by-term differentiation of (12.37) that

$$\frac{d}{dt} e^{\mathbf{A}t} = \mathbf{A} e^{\mathbf{A}t} = e^{\mathbf{A}t} \mathbf{A} \qquad (12.38)$$

Then by direct substitution we find that the solution to (12.36) is

$$\mathbf{x}(t) = e^{\mathbf{A}t}\mathbf{x}(0) \qquad (12.39)$$

or expressed more generally

$$\boxed{\mathbf{x}(t) = \Phi(t)\mathbf{x}(0)} \qquad (12.40)$$

This result can be generalized to initial time $t = t_0$ instead of $t = 0$. The solution in this case is

$$\mathbf{x}(t) = e^{\mathbf{A}(t-t_0)}\mathbf{x}(t_0) \qquad (12.41)$$

or more generally

$$\boxed{\mathbf{x}(t) = \Phi(t - t_0)\mathbf{x}(t_0)} \qquad (12.42)$$

This can be easily proved by inserting $\mathbf{x}(t)$ of (12.41) into (12.36); the result is an identity. As defined, **the fundamental matrix relates the state at time t with that at time t_0.**

That a physical interpretation of (12.40) is possible provides some insight into the meaning of this expression. Specifically, consider a second-order system written explicitly as

$$\begin{bmatrix} x_1(t) \\ x_2(t) \end{bmatrix} = \begin{bmatrix} \varphi_{11}(t) & \varphi_{12}(t) \\ \varphi_{21}(t) & \varphi_{22}(t) \end{bmatrix} \begin{bmatrix} x_1(0) \\ x_2(0) \end{bmatrix} \qquad (12.43)$$

In order to find $\varphi_{11}(t)$, we must set $x_1(0) = 1$ and $x_2(0) = 0$; in this case $x_1(t) = \varphi_{11}(t)$. This shows that $\varphi_{11}(t)$ is the response at the output of integrator 1 to an impulse at the input of this integrator, when the initial condition to integrator 2 is zero. In general, $\varphi_{ij}(t)$ is the response at the output of the ith integrator to an impulse at the input of the jth integrator, with the initial conditions to all other integrators being zero.

EXAMPLE 12.5
A system is specified by the fundamental matrix $\exp[\mathbf{A}t]$ with

$$\mathbf{A} = \begin{bmatrix} 0 & 1 \\ -\frac{1}{2} & -1 \end{bmatrix}$$

Determine the response of the system to a specified initial condition vector $\mathbf{x}(0)$.

Solution: We solve this problem using the direct method of solution, which involves (12.37) and requires that we find $\mathbf{A}^2$, $\mathbf{A}^3$, and so on

$$\mathbf{A}^2 = \begin{bmatrix} -\frac{1}{2} & -1 \\ \frac{1}{2} & \frac{1}{2} \end{bmatrix} \qquad \mathbf{A}^3 = \begin{bmatrix} \frac{1}{2} & \frac{1}{2} \\ -\frac{1}{4} & 0 \end{bmatrix} \qquad \cdots$$

Hence we have

$$\mathbf{\Phi}(t) = e^{\mathbf{A}t} = \begin{bmatrix} 1 + 0t - \frac{\frac{1}{2}t^2}{2!} + \frac{\frac{1}{2}t^3}{3!} + \cdots & 0 + t - \frac{t^2}{2!} + \frac{\frac{1}{2}t^3}{3!} + \cdots \\ 0 - \frac{1}{2}t + \frac{\frac{1}{2}t^2}{2!} - \frac{\frac{1}{4}t^3}{3!} + \cdots & 1 - t + \frac{\frac{1}{2}t^2}{2!} + \frac{0t^3}{3!} + \cdots \end{bmatrix}$$

$$\triangleq \begin{bmatrix} \varphi_{11}(t) & \varphi_{12}(t) \\ \varphi_{21}(t) & \varphi_{22}(t) \end{bmatrix}$$

It is not possible to recognize a closed form expression corresponding to these series. Often it is possible to recognize a closed form expression corresponding to a series although solutions in series form are often useful. ■

EXAMPLE 12.6
Find the zero input response to the circuit shown in Figure 12.6a using the state space method.

Solution: By a straightforward application of the KVL we write, for the left-hand loop

$$v_1 - v_{R_1} - L\frac{di_L(t)}{dt} - R_2 i_L(t) = 0$$

But

$$v_{R_1} = v_1(t) - v_c(t) - v_2(t)$$

Combine these equations to write

$$\frac{di_L(t)}{dt} = -\frac{R_2}{L} i_L(t) + \frac{1}{L} v_c(t) + \frac{1}{L} v_2(t)$$

Figure 12.6
An electric network and its block diagram representation.

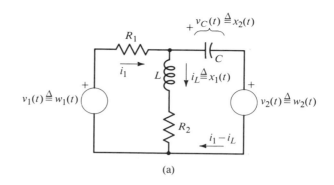

(a)

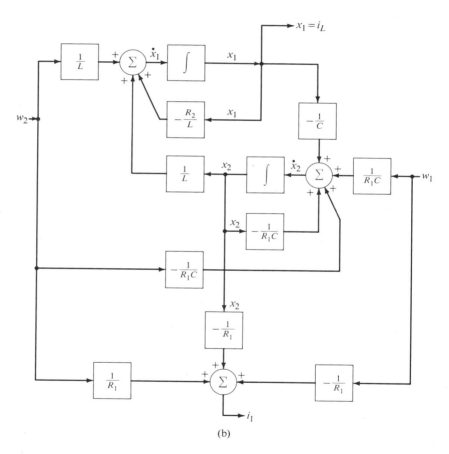

(b)

In state space form, this expression is written ($x_1 = i_L$ and $x_2 = v_c$)

$$\dot{x}_1(t) = -\frac{R_2}{L} x_1(t) + \frac{1}{L} x_2(t) + \frac{1}{L} w_2(t) \qquad (12.44)$$

For the outside loop, the voltage equation is

$$v_1(t) - i_1(t)R_1 - v_c(t) - v_2(t) = 0$$

However,

$$C\frac{dv_c(t)}{dt} = i_1(t) - i_L(t)$$

Combine these equations to get

$$v_1(t) - R_1\left[C\frac{dv_c(t)}{dt} + i_L(t)\right] - v_c(t) - v_2(t) = 0$$

or

$$\frac{dv_c(t)}{dt} = -\frac{1}{C}i_L(t) - \frac{1}{R_1 C}v_c(t) + \frac{1}{R_1 C}v_1(t) - \frac{1}{R_1 C}v_2(t)$$

In state variable form, this equation is

$$\dot{x}_2(t) = -\frac{1}{C}x_1(t) - \frac{1}{R_1 C}x_2(t) + \frac{1}{R_1 C}w_1(t) - \frac{1}{R_1 C}w_2(t) \qquad (12.45)$$

We combine (12.44) and (12.45) and write the result in matrix form. Thus

$$\begin{bmatrix}\dot{x}_1(t)\\\dot{x}_2(t)\end{bmatrix} = \begin{bmatrix}-\dfrac{R_2}{L} & \dfrac{1}{L}\\-\dfrac{1}{C} & -\dfrac{1}{R_1 C}\end{bmatrix}\begin{bmatrix}x_1(t)\\x_2(t)\end{bmatrix} + \begin{bmatrix}0 & \dfrac{1}{L}\\\dfrac{1}{R_1 C} & -\dfrac{1}{R_1 C}\end{bmatrix}\begin{bmatrix}w_1(t)\\w_2(t)\end{bmatrix} \qquad (12.46)$$

If we select the currents i_1 and i_L as the outputs, the output equations become

$$i_L(t) = x_1(t)$$

$$i_1(t) = -\frac{1}{R_1}x_2(t) - \frac{w_2(t)}{R_1} + \frac{w_1(t)}{R_1}$$

In matrix form these are

$$\begin{bmatrix}i_L(t)\\i_1(t)\end{bmatrix} = \begin{bmatrix}1 & 0\\0 & -\dfrac{1}{R_1}\end{bmatrix}\begin{bmatrix}x_1(t)\\x_2(t)\end{bmatrix} + \begin{bmatrix}0 & 0\\\dfrac{1}{R_1} & -\dfrac{1}{R_1}\end{bmatrix}\begin{bmatrix}w_1(t)\\w_2(t)\end{bmatrix} \qquad (12.47)$$

Equations (12.46) and (12.47) are included in the block diagram representation in Figure 12.6b.

Let us now choose the following numerical values for the circuit parameters:

$$R_2 = 0 \qquad L = 1\text{ H} \qquad C = 0.5\text{ F} \qquad \frac{1}{CR_1} = 3s^{-1}$$

We then proceed to find the quantities $\mathbf{A}, \mathbf{A}^2, \mathbf{A}^3, \mathbf{A}^4, \ldots$. These are

STATE VARIABLES AND STATE EQUATIONS

$$\mathbf{A} = \begin{bmatrix} 0 & 1 \\ -2 & -3 \end{bmatrix} \quad \mathbf{A}^2 = \begin{bmatrix} -2 & -3 \\ 6 & 7 \end{bmatrix}$$

$$\mathbf{A}^3 = \begin{bmatrix} 6 & 7 \\ -14 & -15 \end{bmatrix} \quad \mathbf{A}^4 = \begin{bmatrix} -14 & -15 \\ 30 & 31 \end{bmatrix} \cdots$$

The fundamental matrix has the form [from (12.37)]

$$\mathbf{\Phi}(t) = e^{\mathbf{A}t} = \begin{bmatrix} 1 - \dfrac{2t^2}{2!} + 6\dfrac{t^3}{3!} - 14\dfrac{t^4}{4!} \cdots & t - 3\dfrac{t^2}{2!} + 7\dfrac{t^3}{3!} - 15\dfrac{t^4}{4!} \cdots \\ -2t + 6\dfrac{t^2}{2!} - 14\dfrac{t^3}{3!} + 30\dfrac{t^4}{4!} \cdots & 1 - 3t + 7\dfrac{t^2}{2!} - 15\dfrac{t^3}{3!} + 31\dfrac{t^4}{4!} \cdots \end{bmatrix}$$

(12.48)

We cannot identify closed form expressions for the elements in this matrix. However, we can ascertain the closed form by proceeding in a classical way. That is, we can write from (12.46) for the force-free case and the specified R, L, and C elements values

$$\dot{x}_1(t) = x_2(t) \quad \text{(a)}$$
$$\dot{x}_2(t) = -2x_1(t) - 3x_2(t) \quad \text{(b)}$$

(12.49)

These expressions are combined to yield

$$\ddot{x}_1(t) + 3\dot{x}_1(t) + 2x_1(t) = 0 \tag{12.50}$$

The general solution to this differential equation is, employing elementary methods,

$$x_1(t) = C_1 e^{-t} + C_2 e^{-2t} \tag{12.51}$$

By introducing this expression into (12.49), we have

$$x_2(t) = -C_1 e^{-t} - 2C_2 e^{-2t} \tag{12.52}$$

These expressions are combined with (12.39) to specify the matrix $\mathbf{\Phi}$ that multiplies the initial state vector $\mathbf{x}(0)$ to give $\mathbf{x}(t)$. Thus we have

$$C_1 + C_2 = x_1(0)$$
$$-C_1 - 2C_2 = x_2(0)$$

From these we find

$$C_1 = 2x_1(0) + x_2(0)$$
$$C_2 = -x_1(0) - x_2(0)$$

Substitute these values in (12.51) and (12.52), with the result that

$$\begin{bmatrix} x_1(t) \\ x_2(t) \end{bmatrix} = \begin{bmatrix} 2e^{-t} - e^{-2t} & e^{-t} - e^{-2t} \\ -2e^{-t} + 2e^{-2t} & -e^{-t} + 2e^{-2t} \end{bmatrix} \cdot \begin{bmatrix} x_1(0) \\ x_2(0) \end{bmatrix} \tag{12.53}$$

■

EXAMPLE 12.7

Determine the solution for the system given in Example 12.5 using the eigenvalue method.

Solution: The solution follows from Appendix 2 at the end of the book. First we find the eigenvalues of **A** from the characteristic equation

$$\Delta(\lambda) = |\mathbf{A} - \lambda\mathbf{I}| = \begin{vmatrix} -\lambda & 1 \\ -\tfrac{1}{2} & -(1+\lambda) \end{vmatrix} = \lambda(\lambda+1) + \tfrac{1}{2} = \lambda^2 + \lambda + \tfrac{1}{2}$$

The eigenvalues are $\lambda_1 = -\tfrac{1}{2} + j\tfrac{1}{2}$, $\lambda_2 = -\tfrac{1}{2} - j\tfrac{1}{2}$. Also, by the Cayley-Hamilton theorem, the solution will be of the form

$$e^{\mathbf{A}t} = \gamma_0 \mathbf{I} + \gamma_1 \mathbf{A} = \begin{bmatrix} \gamma_0 & 0 \\ 0 & \gamma_0 \end{bmatrix} + \begin{bmatrix} \gamma_1 a_{11} & \gamma_1 a_{12} \\ \gamma_1 a_{12} & \gamma_1 a_{22} \end{bmatrix}$$

We can evaluate the constants γ_0 and γ_1 from the equations

$$e^{[-(1/2)+j(1/2)]t} = \gamma_0 + \gamma_1[-(1/2) + j(1/2)]$$
$$e^{[-(1/2)-j(1/2)]t} = \gamma_0 + \gamma_1[-(1/2) - j(1/2)]$$

By subtracting these equations, it is found after simple manipulations that

$$\gamma_1 = 2e^{-t/2} \sin\frac{t}{2}$$

Combine this result with either of the equations above. The result is found to be

$$\gamma_0 = e^{-t/2}\left(\cos\frac{t}{2} + \sin\frac{t}{2}\right)$$

Combine these values with the above expression for $\exp[\mathbf{A}t]$ to find

$$e^{\mathbf{A}t} = \begin{bmatrix} e^{-t/2}\left(\cos\dfrac{t}{2} + \sin\dfrac{t}{2}\right) & 2e^{-t/2}\sin\dfrac{t}{2} \\ -e^{-t/2}\sin\dfrac{t}{2} & e^{-t/2}\left(\cos\dfrac{t}{2} - \sin\dfrac{t}{2}\right) \end{bmatrix}$$

The final solution is then

$$\begin{bmatrix} x_1(t) \\ x_2(t) \end{bmatrix} = e^{-t/2} \begin{bmatrix} \left(\cos\dfrac{t}{2} + \sin\dfrac{t}{2}\right)x_1(0) + 2\sin\dfrac{t}{2}x_2(0) \\ -\sin\dfrac{t}{2}x_1(0) + \left(\cos\dfrac{t}{2} - \sin\dfrac{t}{2}\right)x_2(0) \end{bmatrix}$$

∎

12-6 PROPERTIES OF THE FUNDAMENTAL MATRIX

A number of properties of the fundamental matrix are important. These include the following:

STATE VARIABLES AND STATE EQUATIONS

1. The Derivative Property

$$\frac{\partial \mathbf{\Phi}(t - t_0)}{\partial t} = \mathbf{A}\mathbf{\Phi}(t - t_0) \tag{12.54}$$

Proof: Substitute $\mathbf{x}(t) = \mathbf{\Phi}(t - t_0)\mathbf{x}(t_0)$ into (12.36). This yields

$$\frac{\partial \mathbf{\Phi}(t - t_0)}{\partial t} \mathbf{x}(t_0) = \mathbf{A}\mathbf{\Phi}(t - t_0)\mathbf{x}(t_0)$$

2. Identity Property

$$\mathbf{\Phi}(t_0 - t_0) = \mathbf{\Phi}(0) = \mathbf{I} \tag{12.55}$$

Proof: Substitute $t = t_0$ into (12.42) to obtain

$$\mathbf{x}(t_0) = \mathbf{\Phi}(0)\mathbf{x}(t_0) \quad \text{or} \quad \mathbf{\Phi}(0) = \mathbf{I}$$

3. Initial Value Property

$$\left.\frac{\partial \mathbf{\Phi}(t - t_0)}{\partial t}\right|_{t=t_0} = \mathbf{A} \tag{12.56}$$

Proof: This result follows directly from (12.54) and (12.55).

4. Transition Property

$$\mathbf{\Phi}(t_2 - t_0) = \mathbf{\Phi}(t_2 - t_1)\mathbf{\Phi}(t_1 - t_0) \qquad t_0 \leq t_1 \leq t_2 \tag{12.57}$$

Proof: Equation 12.42 is valid for any t_0; from this we can write

$$\mathbf{x}(t) = \mathbf{\Phi}(t - t_0)\mathbf{x}(t_0) \quad \text{also} \quad \mathbf{x}(t_1) = \mathbf{\Phi}(t_1 - t_0)\mathbf{x}(t_0)$$

But we note that

$$\mathbf{x}(t_2) = \mathbf{\Phi}(t_2 - t_0)\mathbf{x}(t_0) = \mathbf{\Phi}(t_2 - t_1)\mathbf{x}(t_1) = \mathbf{\Phi}(t_2 - t_1)\mathbf{\Phi}(t_1 - t_0)\mathbf{x}(t_0)$$

which thus implies that

$$\mathbf{\Phi}(t_2 - t_0) = \mathbf{\Phi}(t_2 - t_1)\mathbf{\Phi}(t_1 - t_0)$$

5. Inverse Property

$$\mathbf{\Phi}(t_0 - t_1) = \mathbf{\Phi}^{-1}(t_1 - t_0) \tag{12.58}$$

Proof: Since $\mathbf{\Phi}^{-1}(t_1 - t_0)\mathbf{\Phi}(t_1 - t_0) = \mathbf{I}$ and also $\mathbf{\Phi}(t_0 - t_0) = \mathbf{I}$ by (12.57), we have

$$\mathbf{\Phi}^{-1}(t_1 - t_0)\mathbf{\Phi}(t_1 - t_0) = \mathbf{\Phi}(t_0 - t_0) = \mathbf{\Phi}(t_0 - t_1)\mathbf{\Phi}(t_1 - t_0)$$

This implies (12.58).

6. Separation Property

$$\boxed{\mathbf{\Phi}(t_1 - t_0) = \mathbf{\Phi}(t_1)\mathbf{\Phi}^{-1}(t_0)} \tag{12.59}$$

Proof: Based on the properties already considered, we can write

$$\mathbf{\Phi}(t_1 - t_0) = \mathbf{\Phi}(t_1 - 0)\mathbf{\Phi}(0 - t_0) = \mathbf{\Phi}(t_1)\mathbf{\Phi}^{-1}(t_0 - 0)$$
$$= \mathbf{\Phi}(t_1)\mathbf{\Phi}^{-1}(t_0)$$

EXAMPLE 12.8
Verify Properties 2, 3, and 4 for the fundamental matrix given in (12.53).

Solution: We perform the processes indicated. Thus we have for Property 2:

$$\mathbf{\Phi}(t = 0) = \begin{bmatrix} 2 - 1 & 1 - 1 \\ -2 + 2 & -1 + 2 \end{bmatrix} = \begin{bmatrix} 1 & 0 \\ 0 & 1 \end{bmatrix} = \mathbf{I}$$

For Property 3

$$\left.\frac{\partial \mathbf{\Phi}}{\partial t}\right|_{t=0} = \begin{bmatrix} -2e^{-t} + 2e^{-2t} & -e^{-t} + 2e^{-2t} \\ 2e^{-t} - 4e^{-2t} & e^{-t} - 4e^{-2t} \end{bmatrix} = \begin{bmatrix} 0 & 1 \\ -2 & -3 \end{bmatrix} = \mathbf{A}$$

For Property 4 when $t_0 = 0 \leq t_1 = 1 \leq t_2 = 2$, we obtain

$$\mathbf{\Phi}(2 - 0) = \mathbf{\Phi}(2 - 1)\mathbf{\Phi}(1 - 0)$$

$$= \begin{bmatrix} 2e^{-1} - e^{-2} & e^{-1} - e^{-2} \\ -2e^{-1} + 2e^{-2} & -e^{-1} + 2e^{-2} \end{bmatrix}$$

$$\times \begin{bmatrix} 2e^{-1} - e^{-2} & e^{-1} - e^{-2} \\ -2e^{-1} + 2e^{-2} & -e^{-1} + 2e^{-2} \end{bmatrix}$$

$$= \begin{bmatrix} 2e^{-2} - e^{-4} & e^{-2} - e^{-4} \\ -2e^{-2} + 2e^{-4} & -e^{-2} + 2e^{-4} \end{bmatrix}$$

■ ■ ■

12-7 COMPLETE SOLUTION OF THE CONTINUOUS-TIME STATE EQUATIONS

We now seek a solution to the general state equation given by (12.1)

$$\dot{\mathbf{x}}(t) = \mathbf{A}\mathbf{x}(t) + \mathbf{B}\mathbf{w}(t) \tag{12.60}$$

STATE VARIABLES AND STATE EQUATIONS

To accomplish this, multiply each term in this equation by the factor $\exp[-\mathbf{A}t]$. This gives

$$e^{-\mathbf{A}t}\dot{\mathbf{x}} = e^{-\mathbf{A}t}\mathbf{A}\mathbf{x} + e^{-\mathbf{A}t}\mathbf{B}\mathbf{w}$$

Now rearrange this equation to the form

$$e^{-\mathbf{A}t}\dot{\mathbf{x}} - e^{-\mathbf{A}t}\mathbf{A}\mathbf{x} = e^{-\mathbf{A}t}\mathbf{B}\mathbf{w}$$

which is

$$\frac{d}{dt}(e^{-\mathbf{A}t}\mathbf{x}) = e^{-\mathbf{A}t}\mathbf{B}\mathbf{w}$$

Multiply by dt and integrate over the interval t_0 to t. Also, to avoid confusion a change of variable is made. The result is

$$\int_{t_0}^{t} \frac{d}{d\tau}(e^{-\mathbf{A}\tau}\mathbf{x})\, d\tau = \int_{t_0}^{t} e^{-\mathbf{A}\tau}\mathbf{B}\mathbf{w}\, d\tau$$

which is

$$e^{-\mathbf{A}t}\mathbf{x}(t) - e^{-\mathbf{A}t_0}\mathbf{x}(t_0) = \int_{t_0}^{t} e^{-\mathbf{A}\tau}\mathbf{B}\mathbf{w}\, d\tau$$

Now premultiply all terms in this equation by $\exp[\mathbf{A}t]$, to get

$$\mathbf{x}(t) = e^{\mathbf{A}(t-t_0)}\mathbf{x}(t_0) + \int_{t_0}^{t} e^{\mathbf{A}(t-\tau)}\mathbf{B}\mathbf{w}(\tau)\, d\tau \tag{12.61}$$

When written in terms of the fundamental matrix, the expression is

$$\boxed{\mathbf{x}(t) = \mathbf{\Phi}(t-t_0)\mathbf{x}(t_0) + \int_{t_0}^{t} \mathbf{\Phi}(t-\tau)\mathbf{B}\mathbf{w}(\tau)\, d\tau} \tag{12.62}$$

The first term on the right represents the contribution to the total response arising from the free or natural response of the system due to initial conditions. The second term is the contribution due to the particular applied drivers or forcing functions, and appears in a matrix convolution integral. Be advised that obtaining a solution to this general matrix equation is usually a complicated task.

To find the corresponding output expression, we combine the general form

$$\mathbf{y}(t) = \mathbf{C}\mathbf{x}(t) + \mathbf{D}\mathbf{w}(t) \tag{12.63}$$

with the expression for $\mathbf{x}(t)$ given by (12.62). The result is

$$\mathbf{y}(t) = \mathbf{C}e^{\mathbf{A}(t-t_0)}\mathbf{x}(t_0) + \int_{t_0}^{t} \mathbf{C}e^{\mathbf{A}(t-\tau)}\mathbf{B}\mathbf{w}(\tau)\, d\tau + \mathbf{D}\mathbf{w}(t)$$

or

$$\boxed{\mathbf{y}(t) = \mathbf{C}e^{\mathbf{A}(t-t_0)}\mathbf{x}(t_0) + \int_{t_0}^{t} [\mathbf{C}e^{\mathbf{A}(t-\tau)}\mathbf{B} + \mathbf{D}\,\delta(t-\tau)]\mathbf{w}(\tau)\, d\tau} \tag{12.64}$$

Note that the entity $[Ce^{At}B + D\delta(t)]$ is the impulse response function $\mathbf{h}(t)$, with $\mathbf{D}$ denoting the paths straight through from input to output. To see this, we set the initial conditions equal to zero (that is, $\mathbf{x}(t_0) = \mathbf{0}$) with the result

$$\mathbf{h}(t) = \begin{cases} \mathbf{C}\mathbf{\Phi}(t)\mathbf{B} + \mathbf{D}\,\delta(t) & t \geq 0 \\ 0 & \text{otherwise} \end{cases} \tag{12.65}$$

If we define two functions as follows:

$$\begin{aligned}\mathbf{x}_1(t; t_0, \mathbf{0}, \mathbf{w}) &\triangleq \int_{t_0}^{t} \mathbf{\Phi}(t - \tau)\mathbf{B}\mathbf{w}(\tau)\,d\tau \quad \text{(a)} \\ \mathbf{x}_2(t; t_0, \mathbf{x}(t_0), \mathbf{0}) &\triangleq \mathbf{\Phi}(t - t_0)\mathbf{x}(t_0) \quad \text{(b)} \end{aligned} \tag{12.66}$$

then we observe that $\mathbf{x}_1(\cdot)$ is a solution to (12.1a) with zero initial state vector $\mathbf{x}(t_0) = \mathbf{0}$, and it is called the **zero-state response**. We also observe that $\mathbf{x}_2(\cdot)$ is the solution to the same equation with zero input, and it is referred to as the **zero-input response**. Therefore the output of any system is the sum of these two responses.

EXAMPLE 12.9
Find the state representation and the impulse response of the network shown in Figure 12.7.

Solution: The first step is to choose the state variables. Based on our prior discussion, we choose the voltages across the capacitors as the state variables. But since $q = Cv$, then for $C = 1$, $q = v$; we can then choose the charge q as the state variable, as shown in Figure 12.7. Moreover, since $C\,dv/dt = i_c$, we can use the mesh current equations to write a relation between the mesh currents and the state variables. Thus

$$\frac{dq_1}{dt} = i_1 - i_2$$

$$\frac{dq_2}{dt} = i_2$$

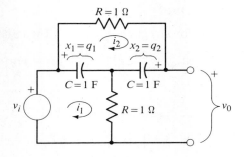

Figure 12.7 Illustrating Example 12.9.

STATE VARIABLES AND STATE EQUATIONS

From the mesh voltage equations for the two loops, we write

$$v_i - \frac{q_1}{1} - i_1 \cdot 1 = 0$$

$$\frac{-q_1}{1} + \frac{q_2}{1} + i_2 \cdot 1 = 0$$

Eliminate i_1 and i_2 from the foregoing equations and incorporate the state variable notation, with $v_i = w$, $y = v_o$; we find the following set

$$\frac{dx_1}{dt} = -2x_1 + x_2 + w$$

$$\frac{dx_2}{dt} = x_1 - x_2$$

or in matrix form

$$\frac{d}{dt}\begin{bmatrix} x_1 \\ x_2 \end{bmatrix} = \begin{bmatrix} -2 & 1 \\ 1 & -1 \end{bmatrix}\begin{bmatrix} x_1 \\ x_2 \end{bmatrix} + \begin{bmatrix} 1 \\ 0 \end{bmatrix} w$$

The output equation becomes $v_o = -q_1 + q_2 + v_i$, and its state representation is

$$y = -x_1 + x_2 + w = \begin{bmatrix} -1 & 1 \end{bmatrix}\begin{bmatrix} x_1 \\ x_2 \end{bmatrix} + w$$

The characteristic equation of $\mathbf{A}$ is

$$\begin{bmatrix} -2-\lambda & 1 \\ 1 & -1-\lambda \end{bmatrix} = \left(\lambda + \frac{3+\sqrt{5}}{2}\right)\left(\lambda + \frac{3-\sqrt{5}}{2}\right) = 0$$

with the eigenvalues

$$\lambda_1 = \frac{-3-\sqrt{5}}{2} = -2.618 \qquad \lambda_2 = \frac{-3+\sqrt{5}}{2} = -0.382$$

The Cayley-Hamilton theorem requires that we first find the γ's from the relations

$$e^{-2.618t} = \gamma_0 + (-2.618)\gamma_1$$

$$e^{-0.382t} = \gamma_0 + (-0.382)\gamma_1$$

Upon solving these equations, we obtain

$$\gamma_0 = 1.171e^{-0.382t} - 0.171e^{-2.618t}$$

$$\gamma_1 = 0.447(e^{-0.382t} - e^{-2.618t})$$

Therefore the expression for $\exp[\mathbf{A}t]$ is

$$e^{\mathbf{A}t} = \begin{bmatrix} 0.277e^{-0.382t} + 0.723e^{-2.618t} & 0.447(e^{-0.382t} - e^{-2.618t}) \\ 0.477(e^{-0.382t} - e^{-2.618t}) & 0.724e^{-0.382t} + 0.276e^{-2.618t} \end{bmatrix}$$

The impulse response, for $t > 0$, is

$$\mathbf{h}(t) = \mathbf{C}\boldsymbol{\Phi}(t)\mathbf{B} + \mathbf{D}\,\delta(t) = 0.2e^{-0.382t} - 1.2e^{-2.618t} \qquad \blacksquare$$

■ ■ ■

12-8 LAPLACE TRANSFORM SOLUTION TO THE STATE EQUATIONS

It is of analytic interest to examine the solution of the state equations for linear systems by Laplace transform techniques. The procedure is direct, but due account must be taken of the matrix nature of the equations.

We again begin with the general state equations

$$\dot{\mathbf{x}} = \mathbf{A}\mathbf{x} + \mathbf{B}\mathbf{w}$$
$$\mathbf{y} = \mathbf{C}\mathbf{x} + \mathbf{D}\mathbf{w} \tag{12.67}$$

We consider the first equation of this set and Laplace-transform it. There results [see also (6.19)]

$$s\mathbf{x}(s) - \mathbf{x}(0+) = \mathbf{A}\mathbf{x}(s) + \mathbf{B}\mathbf{w}(s) \tag{12.68}$$

where $\mathbf{x}(0+)$ is the initial state vector. This equation is rearranged to the form

$$[s\mathbf{I} - \mathbf{A}]\mathbf{x}(s) = \mathbf{x}(0+) + \mathbf{B}\mathbf{w}(s)$$

where $\mathbf{I}$ is the unit matrix. Premultiply this expression by $[s\mathbf{I} - \mathbf{A}]^{-1}$, with the result

$$\boxed{\mathbf{x}(s) = [s\mathbf{I} - \mathbf{A}]^{-1}\mathbf{x}(0+) + [s\mathbf{I} - A]^{-1}\mathbf{B}\mathbf{w}(s)} \tag{12.69}$$

This is conveniently written

$$\mathbf{x}(s) = \boldsymbol{\Phi}(s)\mathbf{x}(0+) + \boldsymbol{\Phi}(s)\mathbf{B}\mathbf{w}(s) \quad \text{(a)}$$

where $\tag{12.70}$

$$\boldsymbol{\Phi}(s) = [s\mathbf{I} - \mathbf{A}]^{-1} \quad \text{(b)}$$

$\boldsymbol{\Phi}(s)$ is called the **characteristic matrix** of the system. Figure 12.8 shows in block diagram representation a multiple-input, multiple-output system.

To find $\mathbf{x}(t)$ from this equation, the inverse Laplace transform is taken. Initially consider the case when the sources are zero. In this situation the function under review is

$$\mathbf{x}(s) = \boldsymbol{\Phi}(s)\mathbf{x}(0+) \tag{12.71}$$

The inverse transform is

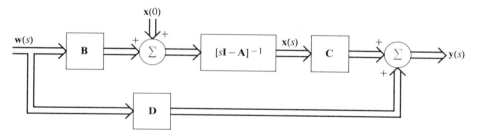

Figure 12.8
Block diagram representation of a multiple-input, multiple-output system.

$$\mathbf{x}(t) = \mathcal{L}^{-1}[\mathbf{\Phi}(s)]\mathbf{x}(0+) = \mathbf{\Phi}(t)\mathbf{x}(0+) \quad \text{(a)} \qquad (12.72)$$

where

$$\mathcal{L}^{-1}[\mathbf{\Phi}(s)] = \mathbf{\Phi}(t) = e^{\mathbf{A}t} \quad \text{(b)}$$

EXAMPLE 12.10
Repeat Example 12.5 using Laplace methods, where

$$\mathbf{A} = \begin{bmatrix} 0 & 1 \\ -\frac{1}{2} & -1 \end{bmatrix}$$

Solution: Evaluate the quantity

$$[s\mathbf{I} - \mathbf{A}]^{-1} = \begin{bmatrix} s & -1 \\ \frac{1}{2} & s+1 \end{bmatrix}^{-1}$$

But from the definition (see Appendix 2)

$$\mathbf{A}^{-1} = [a_{ij}]^{-1} = \frac{\text{transpose of cofactor } a_{ij}}{\det \mathbf{A}} = \frac{\text{cofactor or adjoint } a_{ji}}{\Delta(\mathbf{A})}$$

so that

$$[s\mathbf{I} - \mathbf{A}]^{-1} = \frac{\begin{bmatrix} s+1 & 1 \\ -\frac{1}{2} & s \end{bmatrix}}{s(s+1) + \frac{1}{2}} = \begin{bmatrix} \frac{s+1}{s^2+s+\frac{1}{2}} & \frac{1}{s^2+s+\frac{1}{2}} \\ \frac{-\frac{1}{2}}{s^2+s+\frac{1}{2}} & \frac{s}{s^2+s+\frac{1}{2}} \end{bmatrix}$$

The inverse transform of each term in this matrix is evaluated. These yield

$$\mathcal{L}^{-1}\left[\frac{s+1}{s^2+s+\frac{1}{2}}\right] = \mathcal{L}^{-1}\left[\frac{s+1}{(s+\frac{1}{2})^2+\frac{1}{2}^2}\right] = \mathcal{L}^{-1}\left[\frac{(s+\frac{1}{2})+\frac{1}{2}}{(s+\frac{1}{2})^2+\frac{1}{2}^2}\right]$$

$$= e^{-t/2}\cos\frac{t}{2} + e^{-t/2}\sin\frac{t}{2}$$

$$\mathscr{L}^{-1}\left[\frac{1}{s^2+s+\frac{1}{2}}\right] = 2e^{-t/2}\sin\frac{t}{2}$$

$$\mathscr{L}^{-1}\left[\frac{-\frac{1}{2}}{s^2+s+\frac{1}{2}}\right] = -e^{-t/2}\sin\frac{t}{2}$$

$$\mathscr{L}^{-1}\left[\frac{s}{s^2+s+\frac{1}{2}}\right] = \frac{(s+\frac{1}{2})-\frac{1}{2}}{s^2+s+\frac{1}{2}} = e^{-t/2}\cos\frac{t}{2} - e^{-t/2}\sin\frac{t}{2}$$

These elements of the matrix are the same as those obtained in Example 12.7. ∎

To find the complete solution, we write

$$\mathbf{x}(t) = \mathscr{L}^{-1}[\mathbf{\Phi}(s)\mathbf{x}(0+)] + \mathscr{L}^{-1}[\mathbf{\Phi}(s)\mathbf{B}\mathbf{w}(s)] \tag{12.73}$$

This is a generalization of the convolution integral in Laplace form, the result being

$$\mathbf{x}(t) = \mathbf{\Phi}(t)\mathbf{x}(0+) + \int_0^t \mathbf{\Phi}(t-\tau)\mathbf{B}\mathbf{w}(\tau)\,d\tau \quad \text{(a)} \tag{12.74}$$

since

$$\mathscr{L}^{-1}[\mathbf{\Phi}(s)] = \mathbf{\Phi}(t) \quad \text{(b)}$$

and

$$\mathscr{L}^{-1}[\mathbf{\Phi}(s)\mathbf{B}\mathbf{w}(s)] = \int_0^t \mathbf{\Phi}(t-\tau)\mathbf{B}\mathbf{w}(\tau)\,d\tau \quad \text{(c)}$$

Now consider the output vector, which is obtained by Laplace-transforming the second equation in (12.67). This gives

$$\mathbf{y}(s) = \mathbf{C}\mathbf{x}(s) + \mathbf{D}\mathbf{w}(s) \tag{12.75}$$

Combine this with (12.70) to write

$$\mathbf{y}(s) = \mathbf{C}\mathbf{\Phi}(s)\mathbf{x}(0+) + [\mathbf{C}\mathbf{\Phi}(s)\mathbf{B} + \mathbf{D}]\mathbf{w}(s) \tag{12.76}$$

This expression shows that the output is the sum of the responses due to the initial state vector and those due to the drivers. By taking the inverse transform of this expression, we have

$$\mathbf{y}(t) = \mathbf{C}\mathbf{\Phi}(t)\mathbf{x}(0+) + \int_0^t [\mathbf{C}\mathbf{\Phi}(t-\tau)\mathbf{B} + \mathbf{D}\,\delta(t-\tau)]\mathbf{w}(\tau)\,d\tau \tag{12.77}$$

in agreement with (12.64).

With zero initial conditions (12.76) becomes

$$\mathbf{y}(s) = [\mathbf{C}\mathbf{\Phi}(s)\mathbf{B} + \mathbf{D}]\mathbf{w}(s) = \mathbf{H}(s)\mathbf{w}(s) \tag{12.78}$$

where $\mathbf{H}(s)$ is the transfer matrix (function) [using (12.70b)]

$$\mathbf{H}(s) = \mathbf{C}\mathbf{\Phi}(s)\mathbf{B} + \mathbf{D} = \mathbf{C}[s\mathbf{I} - \mathbf{A}]^{-1}\mathbf{B} + \mathbf{D} \tag{12.79}$$

STATE VARIABLES AND STATE EQUATIONS

Note that since $y_i(s)$ from (12.78) is given by

$$y_i(s) = H_{i1}(s)w_1(s) + H_{i2}(s)w_2(s) + \cdots + H_{in}(s)w_n(s) \tag{12.80}$$

it is then evident that the element $H_{ij}(s)$ of the matrix $\mathbf{H}(s)$ is the transfer function between $w_j(t)$ and $y_i(t)$. Hence we may write

$$H_{ij}(s) = \mathscr{L}\{h_{ij}(t)\} \tag{12.81}$$

where $h_{ij}(t)$ is the response of the ith output to the jth input terminal.

EXAMPLE 12.11
Find the response i_1 of Example 12.6 for an input $v_2(t) = w_2(t) = \delta(t)$.

Solution: The characteristic matrix is

$$\mathbf{\Phi}(s) = [s\mathbf{I} - \mathbf{A}]^{-1} = \begin{bmatrix} s & -1 \\ 2 & s+3 \end{bmatrix}^{-1} = \frac{\begin{bmatrix} s+3 & 1 \\ -2 & s \end{bmatrix}}{(s+1)(s+2)}$$

$$= \begin{bmatrix} \dfrac{s+3}{(s+1)(s+2)} & \dfrac{1}{(s+1)(s+2)} \\ -\dfrac{2}{(s+1)(s+2)} & \dfrac{s}{(s+1)(s+2)} \end{bmatrix}$$

The inverse transform of this matrix becomes

$$\mathbf{\Phi}(t) = e^{\mathbf{A}t} = \begin{bmatrix} 2e^{-t} - e^{-2t} & e^{-t} - e^{-2t} \\ -2e^{-t} + 2e^{-2t} & -e^{-t} + 2e^{-2t} \end{bmatrix}$$

which is identical with (12.53).

The transfer matrix is given by (12.79). This is given by

$$\mathbf{H}(s) = \begin{bmatrix} 1 & 0 \\ 0 & -1.5 \end{bmatrix} \begin{bmatrix} \dfrac{s+3}{(s+1)(s+2)} & \dfrac{1}{(s+1)(s+2)} \\ \dfrac{-2}{(s+1)(s+2)} & \dfrac{s}{(s+1)(s+2)} \end{bmatrix} \times \begin{bmatrix} 0 & 1 \\ 3 & -3 \end{bmatrix}$$

$$+ \begin{bmatrix} 0 & 0 \\ -1.5 & 1.5 \end{bmatrix}$$

$$= \begin{bmatrix} \dfrac{3}{(s+1)(s+2)} & \dfrac{(s+3)-3}{(s+1)(s+2)} \\ -4.5\dfrac{s}{(s+1)(s+2)} - 1.5 & \dfrac{3+4.5s}{(s+1)(s+2)} + 1.5 \end{bmatrix}$$

Since $i_1 = y_2$ and $w_2 = \delta(t)$, we must find the inverse of the element $H_{22}(s)$, which is

$$h_{22}(t) = \mathcal{L}^{-1}\{H_{22}(s)\} = -1.5e^{-t} + 6e^{-2t} + 1.5\delta(t)$$

which specifies the current i_1 due to the impulse input $v_2(t)$. ∎

EXAMPLE 12.12
Find and plot the time response of the system given in Example 12.6 for $\mathbf{x}(0) = [-0.5 \quad 0]^T$ and $w_1 = u(t)$ and $w_2(t) = 0$.

Solution: Using the results of Example 12.11 for the characteristic matrix in (12.74), we write

$$\mathbf{x}(t) \triangleq \begin{bmatrix} x_1(t) \\ x_2(t) \end{bmatrix} = \begin{bmatrix} 2e^{-t} - e^{-2t} & e^{-t} - e^{-2t} \\ -2e^{-t} + 2e^{-2t} & -e^{-t} + 2e^{-2t} \end{bmatrix} \begin{bmatrix} -0.5 \\ 0 \end{bmatrix}$$

$$+ \int_0^t \begin{bmatrix} 2e^{-(t-\tau)} - 2e^{-2(t-\tau)} & e^{-(t-\tau)} - e^{-2(t-\tau)} \\ -2e^{-(t-\tau)} + 2e^{-2(t-\tau)} & -e^{-(t-\tau)} + 2e^{-2(t-\tau)} \end{bmatrix}$$

$$\times \begin{bmatrix} 0 & 1 \\ 1 & -1 \end{bmatrix} \begin{bmatrix} 1 \\ 0 \end{bmatrix} d\tau$$

$$= \begin{bmatrix} -e^{-t} + 0.5e^{-2t} \\ e^{-t} - e^{-2t} \end{bmatrix} + \begin{bmatrix} \int_0^t [e^{-(t-\tau)} - e^{-2(t-\tau)}] d\tau \\ \int_0^t [-e^{-(t-\tau)} + 2e^{-2(t-\tau)}] d\tau \end{bmatrix}$$

$$= \begin{bmatrix} \tfrac{1}{2} - 2e^{-t} + e^{-2t} \\ 2e^{-t} - 2e^{-2t} \end{bmatrix}$$

Figure 12.9a shows the time response of the state variable $x_1(t)$, which is the current through the inductor of Figure 12.6a. In the same figure is shown the time response of the state variable $x_2(t)$, which is the voltage across the capacitor of the circuit. We note that the **state space** is the coordinate system with axes being the state variables of the system under investigation. Figure 12.9b is the time response plotted as a trajectory in state space.

■ ■ ■

12-9 STATE RESPONSE TO PERIODIC INPUTS

We wish to examine the changes that occur in (12.62) or (12.74) when the input **w** is a periodic function; that is, when

$$\mathbf{w}(t) = \mathbf{w}(t + T)$$

Choose the interval $(0, T)$ and write (12.62) as

$$\mathbf{x}(t) = e^{\mathbf{A}t}\mathbf{x}(0) + \int_0^t e^{\mathbf{A}(t-\tau)}\mathbf{B}\mathbf{w}(\tau)\,d\tau \qquad (12.82)$$

Figure 12.9
(a) Time response of the state variables $x_1(t)$ and $x_2(t)$. (b) Time response as a trajectory in state space.

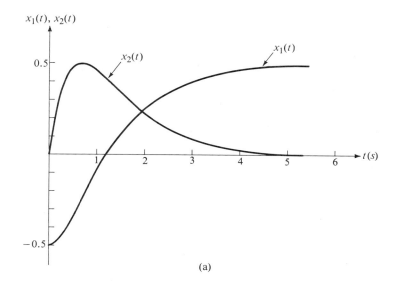

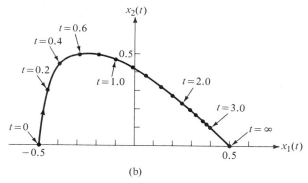

For the periodic case, we require that $x(0)$ at $t = 0$ is equal to the forced value of the function $\mathbf{x}(t)|_{t=T} = \mathbf{x}(T)$. This requires that

$$\mathbf{x}(T) = e^{\mathbf{A}T}\mathbf{x}(0) + \int_0^T e^{\mathbf{A}(T-\tau)}\mathbf{Bw}(\tau)\,d\tau = \mathbf{x}(0)$$

Solve this expression for $\mathbf{x}(0)$ to find

$$\mathbf{x}(0) = [\mathbf{I} - e^{\mathbf{A}T}]^{-1} \int_0^T e^{\mathbf{A}(T-\tau)}\mathbf{Bw}(\tau)\,d\tau \qquad (12.83)$$

Now combine this expression with (12.82) to find

$$\mathbf{x}(t) = e^{\mathbf{A}t}[\mathbf{I} - e^{\mathbf{A}T}]^{-1} \int_0^T e^{\mathbf{A}(T-\tau)}\mathbf{Bw}(\tau)\,d\tau + \int_0^t e^{\mathbf{A}(t-\tau)}\mathbf{Bw}(\tau)\,d\tau$$

or

$$\mathbf{x}(t) = e^{\mathbf{A}t}\left\{[\mathbf{I} - e^{\mathbf{A}T}]^{-1}e^{\mathbf{A}T}\int_0^T e^{-\mathbf{A}\tau}\mathbf{B}\mathbf{w}(\tau)\,d\tau + \int_0^t e^{-\mathbf{A}\tau}\mathbf{B}\mathbf{w}(\tau)\,d\tau\right\} \quad (12.84)$$

For the case when the input is sinusoidal of the form $w = \exp[j\omega t]$, the complete solution is given by (12.84) with $T = 2\pi$. This expression assumes the form

$$\mathbf{x}(t) = e^{\mathbf{A}t}\left\{[\mathbf{I} - e^{2\pi\mathbf{A}}]^{-1}e^{2\pi\mathbf{A}}\int_0^{2\pi} e^{-\mathbf{A}(-j\omega+1)\tau}\mathbf{B}\,d\tau + \int_0^t \mathbf{B}e^{-\mathbf{A}(-j\omega+1)\tau}\,d\tau\right\} \quad (12.85)$$

By carrying out the integrations and rearranging the results, we find that

$$\mathbf{x}(t) = -\frac{\mathbf{B}\mathbf{A}^{-1}e^{-\mathbf{A}t}}{(-j\omega+1)}\{e^{-\mathbf{A}(-j\omega+1)t}$$
$$+ [\mathbf{I} - e^{2\pi\mathbf{A}}]^{-1}e^{2\pi\mathbf{A}}(e^{-\mathbf{A}(-j\omega+1)2\pi} - 1) - 1\} \quad (12.86)$$

The interpretation of this complicated expression is not readily apparent. However, since the system is linear time-invariant, the output $\mathbf{y}(t)$ is of the form $\mathbf{y}(j\omega)\exp[j\omega t] = H(j\omega)\exp[j\omega t]$, and the steady-state solution for the state $\mathbf{x}(t)$ is $\mathbf{x}(j\omega)\exp[j\omega t]$. Understandable results are possible; that is, the state equation becomes

$$j\omega\mathbf{x}(j\omega)e^{j\omega t} = \mathbf{A}\mathbf{x}(j\omega)e^{j\omega t} + \mathbf{B}e^{j\omega t}$$

which leads to

$$[j\omega\mathbf{x}(j\omega) - \mathbf{A}\mathbf{x}(j\omega)]e^{j\omega t} = \mathbf{B}e^{j\omega t}$$

Cancel the common factor $\exp[j\omega t]$ that appears on both sides of the equation, and

$$[j\omega\mathbf{I} - \mathbf{A}]\mathbf{x}(j\omega) = \mathbf{B}$$

from which

$$\mathbf{x}(j\omega) = [j\omega\mathbf{I} - \mathbf{A}]^{-1}\mathbf{B} \quad (12.87)$$

This corresponds to (12.70) with zero initial state $\mathbf{x}(0+) = \mathbf{0}$. Now combine this with the second state equation, which becomes

$$\mathbf{y}(j\omega)e^{j\omega t} = \mathbf{C}\mathbf{x}(j\omega)e^{j\omega t} + \mathbf{D}e^{j\omega t}$$

When this is combined with (12.87), we obtain, upon cancelling the common factor $\exp[j\omega t]$

$$\mathbf{y}(j\omega) = \mathbf{C}[j\omega\mathbf{I} - \mathbf{A}]^{-1}\mathbf{B} + \mathbf{D} \quad (12.88)$$

This form corresponds to (12.76) with zero initial state $\mathbf{x}(0+) = \mathbf{0}$.

■ ■ ■

12-10 INITIAL STATE VECTORS AND INITIAL CONDITIONS

A knowledge of the initial state vector $\mathbf{x}(0+)$ is essential to the solution of the state equations. We will write $\mathbf{x}(0)$ for $\mathbf{x}(0+)$ for ease in writing. The initial state vector $\mathbf{x}(0)$ is found from the specified initial conditions that apply in any given case. The means for relating the physical initial conditions with the initial state vector is very important. We will next discuss this matter in terms of a particular example.

Consider a specific system described by the state equation

$$\begin{bmatrix} \dot{x}_1 \\ \dot{x}_2 \\ \dot{x}_3 \end{bmatrix} = \begin{bmatrix} -2 & +1 & 0 \\ -3 & 0 & +1 \\ 0 & 0 & -5 \end{bmatrix} \begin{bmatrix} x_1 \\ x_2 \\ x_3 \end{bmatrix} + \begin{bmatrix} 0 \\ 1 \\ 3 \end{bmatrix} w$$

$$[y] = \begin{bmatrix} 1 & 0 & 0 \end{bmatrix} \begin{bmatrix} x_1 \\ x_2 \\ x_3 \end{bmatrix}$$

(12.89)

subject to a set of specified initial conditions $y(0), \dot{y}(0), \ddot{y}(0)$. First expand the state equations to the set

$$\dot{x}_1 = -2x_1 + x_2 \quad \textbf{(a)}$$
$$\dot{x}_2 = -3x_1 + x_3 + w \quad \textbf{(b)}$$
$$\dot{x}_3 = -5x_3 + 3w \quad \textbf{(c)}$$
$$y = x_1 \quad \textbf{(d)}$$

(12.90)

The initial conditions translate by (12.90d) into the equivalent conditions

$$y(0) = x_1(0) \qquad \dot{y}(0) = \dot{x}_1(0) \qquad \ddot{y}(0) = \ddot{x}_1(0)$$ (12.91)

Combine these with (12.90a) to get

$$x_2(0) = \dot{x}_1(0) + 2x_1(0)$$
$$= \dot{y}(0) + 2y(0)$$

(12.92)

Further, by differentiating (12.90a) we then find that

$$\dot{x}_2(0) = \ddot{y}(0) + 2\dot{y}(0)$$ (12.93)

Combine this equation with (12.91) and apply this to (12.90b). The quantity w is omitted since it is not involved in the initial state. The result is

$$x_3(0) = \dot{x}_2(0) + 3x_1(0) = \ddot{y}(0) + 2\dot{y}(0) + 3y(0)$$ (12.94)

From these we write finally

$$x_1(0) = y(0)$$
$$x_2(0) = \dot{y}(0) + 2y(0)$$
$$x_3(0) = \ddot{y}(0) + 2\dot{y}(0) + 3y(0)$$

(12.95)

∎

B. DISCRETE TIME SYSTEMS

12-11 REPRESENTATION OF DISCRETE TIME SYSTEMS

A discrete time system can be represented by a set of equations that resemble (12.1) for linear continuous time systems. These equations are:

$$\begin{aligned} \mathbf{x}(k+1) &= \mathbf{A}\mathbf{x}(k) + \mathbf{B}\mathbf{w}(k) \quad &\text{(a)} \\ \mathbf{y}(k) &= \mathbf{C}\mathbf{x}(k) + \mathbf{D}\mathbf{w}(k) \quad &\text{(b)} \end{aligned} \quad (12.96)$$

where the matrices **A**, **B**, **C**, and **D** may depend on the discrete time variable k for time-varying systems. We will treat time-invariant systems in our study, and the elements of these matrices will be constant numbers.

EXAMPLE 12.13
Find the state-space representation of the discrete time system described by the difference equation

$$y(k) + \alpha_1 y(k-1) + \alpha_2 y(k-2) = \beta_0 w(k) \quad (12.97)$$

which is shown in Figure 12.10a.

Solution: To obtain a set of state variables from this difference equation, proceed in a manner that parallels our discussion in Section 12-3 and convert this difference equation into normal form. To do this we write

$$\begin{aligned} x_1(k) &= y(k-2) \\ x_2(k) &= x_1(k+1) = y(k-1) \\ x_2(k+1) &= y(k) = -\alpha_1 y(k-1) - \alpha_2 y(k-2) + \beta_0 w(k) \\ &= -\alpha_1 x_2(k) - \alpha_2 x_1(k) + \beta_0 w(k) \end{aligned} \quad (12.98)$$

We write these in matrix form

$$\begin{bmatrix} x_1(k+1) \\ x_2(k+1) \end{bmatrix} = \begin{bmatrix} 0 & 1 \\ -\alpha_2 & -\alpha_1 \end{bmatrix} \begin{bmatrix} x_1(k) \\ x_2(k) \end{bmatrix} + \begin{bmatrix} 0 \\ \beta_0 \end{bmatrix} w(k) \triangleq \mathbf{A}\mathbf{x}(k) + \mathbf{B}\mathbf{w}(k) \quad (12.99)$$

We write, from the third equation of (12.98), in matrix form

$$y(k) = \mathbf{C}\mathbf{x}(k) + \mathbf{D}w(k) \quad (12.100)$$

where

$$\mathbf{C} = [-\alpha_2 \; -\alpha_1] \quad \mathbf{D} = \beta_0$$

The vector matrix block diagram representation is shown in Figure 12.10b. ∎

STATE VARIABLES AND STATE EQUATIONS

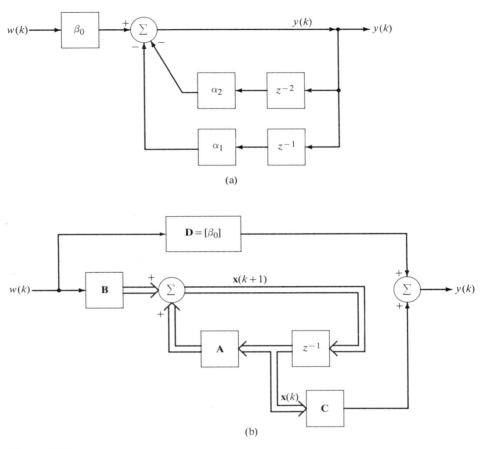

Figure 12.10
Illustrating Example 12.13.

In the general case, the difference equation description of a system will be of the form:

$$\alpha_n y(k+n) + \alpha_{n-1} y(k+n-1) + \cdots + \alpha_0 y(k)$$
$$= \beta_m w(k+m) + \beta_{m-1} w(k+m-1) + \cdots + \beta_0 w(k) \quad (12.101)$$

This equation is a recursive relation suitable for numerical calculation. It permits the calculation of $y(k+n)$ from a knowledge of all previous values of y. Thus one would begin with the initial state, say $y(0)$, and ascertain the successive values of y by taking into account the appropriate past values of y and w, as required by the equation. To discuss this process further, a change in notation is helpful. Suppose that we denote the latest value of y as $y(k)$; hence the previous values will be $y(k-1)$, $y(k-2)$, ... Correspondingly, the latest value of w will be $w(k)$ with previous values being $w(k-1)$, $w(k-2)$, ... Including this change of notation, (12.101) is written

$$\xi_0 y(k) + \xi_1 y(k-1) + \cdots + \xi_k y(k-n)$$
$$= \zeta_0 w(k) + \zeta_1 w(k-1) + \cdots + \zeta_m w(k-m) \tag{12.102}$$

where

$$\xi_i = \alpha_{n-i} \qquad \zeta_i = \beta_{m-i}$$

Equation (12.102) can be written in compact form

$$\sum_{j=0}^{n} \xi_j y(k-j) = \sum_{j=0}^{m} \zeta_j w(k-j) \tag{12.103}$$

The latest value $y(k)$ can be written explicitly as

$$y(k) = \sum_{j=0}^{m} \frac{\zeta_j}{\xi_0} w(k-j) - \sum_{j=1}^{n} \frac{\xi_j}{\xi_0} y(k-j)$$

which is written

$$\boxed{y(k) = \sum_{j=0}^{m} b_j w(k-j) - \sum_{j=1}^{n} a_j y(k-j)} \tag{12.104}$$

where

$$b_j = \frac{\zeta_j}{\xi_0} \qquad a_j = \frac{\xi_j}{\xi_0}$$

Clearly, if all of the a_j's are zero, the output $y(k)$ is a linear weighting of the previous $(k-1)$ samples of the input. This is a **nonrecursive** process.

To carry out the details of (12.104) for the present value of $y(k)$ requires that the prior m values of w and n values of y be available. Such a **recursive** process imposes certain memory demands, and if this calculation is being accomplished with a digital computer, these data would be stored in memory. It is possible to reduce the memory requirement by replacing this one recursion relation by two expressions, in a manner that parallels replacing the differential equation of (12.14) by the pair of equations of (12.15). We begin with the reduced equation

$$\eta(k) + a_1 \eta(k-1) + \cdots + a_n \eta(k-n) = w(k) \tag{12.105}$$

or equivalently

$$\eta(k) = w(k) - \sum_{j=1}^{n} a_j \eta(k-j) \tag{12.106}$$

Also, we can write a similar relation for $y(k)$ that combines the appropriate values of $w(k-j)$. This gives

$$y(k) = \sum_{j=0}^{m} b_j \eta(k-j) \tag{12.107}$$

STATE VARIABLES AND STATE EQUATIONS

Hence we see that (12.106) and (12.107) taken together replace the single equation (12.104). Observe that with these equations it is only necessary to store the m or n previous values of η, depending on which is greater.

To verify (12.106) and (12.107), let us investigate a second-order system of the form

$$y(k) + a_1 y(k-1) + a_2 y(k-2) = b_0 w(k) + b_1 w(k-1)$$

We begin by writing the equations

$$\eta(k) + a_1 \eta(k-1) + a_2 \eta(k-2) = w(k)$$

From this equation we write two related equations. In one we multiply by b_0, and in the other we multiply by b_1 and shift the order of the equation by one. The resulting two equations are

$$b_0 \eta(k) + b_0 a_1 \eta(k-1) + b_0 a_2 \eta(k-2) = b_0 w(k)$$
$$b_1 \eta(k-1) + b_1 a_1 \eta(k-2) + b_1 a_2 \eta(k-3) = b_1 w(k-1)$$

Add these two equations to obtain

$$b_0 \eta(k) + b_1 \eta(k-1) + a_1 [b_0 \eta(k-1) + b_1 \eta(k-2)]$$
$$+ a_2 [b_0 \eta(k-2) + b_1 \eta(k-3)] = b_0 w(k) + b_1 w(k-1)$$

Now observe that if we set

$$y(k) = b_0 \eta(k) + b_1 \eta(k-1)$$

in this equation, we obtain the original difference equation. Therefore, for the general case, (12.106) and (12.107) represent the original difference equation (12.104).

We can employ a procedure for writing the difference equation (12.101) in state form that parallels the procedure used in reducing (12.14) to state form. We begin with (12.101) and write the reduced equation

$$\alpha_n \eta(k+n) + \alpha_{n-1} \eta(k+n-1) + \cdots + \alpha_0 \eta(k) = w(k) \tag{12.108}$$

Now define the quantities

$$\begin{aligned}
x_1(k) &= \eta(k) \\
x_2(k) &= \eta(k+1) \quad = x_1(k+1) \\
x_3(k) &= \eta(k+2) \quad = x_2(k+1) \\
&\vdots \\
x_n(k) &= \eta(k+n-1) = x_{n-1}(k+1)
\end{aligned} \tag{12.109}$$

Combine these expressions with (12.108) to write

$$x_n(k+1) = \frac{1}{\alpha_n} [w(k) - \alpha_{n-1} x_n(k) - \alpha_{n-2} x_{n-1}(k) - \cdots - \alpha_0 x_1(k)]$$

$$\tag{12.110}$$

Consequently, (12.108) has now been replaced by the set of equations (12.109), which in turn is written in matrix form

$$\mathbf{x}(k+1) = \mathbf{A}\mathbf{x}(k) + \mathbf{B}w(k) \tag{12.111}$$

where

$$\mathbf{A} = \begin{bmatrix} 0 & 1 & 0 & \cdots & 0 & 0 \\ 0 & 0 & 1 & & 0 & 0 \\ \vdots & \vdots & \vdots & & \vdots & \vdots \\ 0 & 0 & 0 & & 0 & 1 \\ -\dfrac{\alpha_0}{\alpha_n} & -\dfrac{\alpha_1}{\alpha_n} & -\dfrac{\alpha_2}{\alpha_n} & & -\dfrac{\alpha_{n-2}}{\alpha_n} & -\dfrac{\alpha_{n-1}}{\alpha_n} \end{bmatrix} \quad \mathbf{B} = \begin{bmatrix} 0 \\ 0 \\ \vdots \\ 0 \\ \dfrac{1}{\alpha_n} \end{bmatrix}$$

$$\mathbf{x}(k) = \begin{bmatrix} x_1(k) \\ x_2(k) \\ \vdots \\ x_n(k) \end{bmatrix}$$

We now multiply (12.108) in succession by $\beta_0, \beta_1, \beta_2, \ldots, \beta_m$, respectively, and shift the successive expressions to the left by an amount equal to that indicated by the subscript of the β's. Add the resulting expressions and combine with (12.109) to yield

$$y(k) = \beta_m x_{m+1}(k) + \beta_{m-1} x_m(k) + \cdots + \beta_0 x_1(k) \tag{12.112}$$

which is

$$\mathbf{y}(k) = \mathbf{C}\mathbf{x}(k) \tag{12.113}$$

where $\mathbf{C}$ is the row matrix

$$\mathbf{C} = [\beta_0 \quad \beta_1 \quad \beta_2 \quad \beta_m \quad 0 \quad \cdots \quad 0]$$

This development shows that the phase variable state equation formulation for the nth-order difference equation is

$$\mathbf{x}(k+1) = \mathbf{A}\mathbf{x}(k) + \mathbf{B}w(k)$$
$$\mathbf{y}(k) = \mathbf{C}\mathbf{x}(k) \tag{12.114}$$

where $\mathbf{A}$, $\mathbf{B}$, and $\mathbf{C}$ are given above. The system realization is shown in Figure 12.11.

We again note that one can find many state variables appropriate to a given situation. In fact, the results of Section 12-4 appropriately modified will also be applicable to discrete state variables. We will show a number of examples for which different approaches to selecting state variables will be made.

STATE VARIABLES AND STATE EQUATIONS

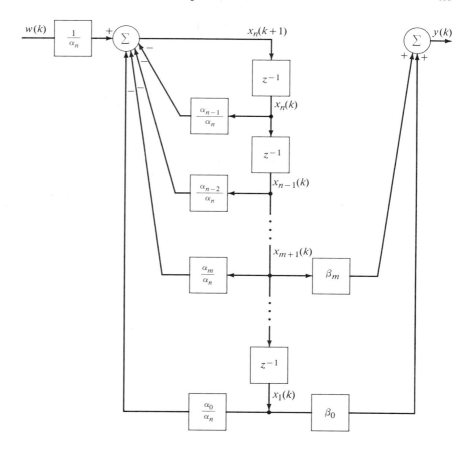

Figure 12.11
Block diagram representation of Equations (12.110) and (12.112).

EXAMPLE 12.14
Repeat Example 12.13 employing a different selection of states.

Solution: Begin with (12.97), which we write

$$y(k) = -\alpha_1 y(k-1) - \alpha_2 y(k-2) + \beta_0 w(k) \qquad (12.115)$$

We now define the state variable

$$x_2(k) = -\alpha_2 y(k-2) - \alpha_1 y(k-1) \qquad (12.116)$$

Combine this with (12.115) to get

$$y(k) = x_2(k) + \beta_0 w(k) \tag{12.117}$$

By increasing the index by 1, we can write (12.116) in the form

$$x_2(k + 1) = -\alpha_2 y(k - 1) - \alpha_1 y(k) \tag{12.118}$$

We now select as a second state variable

$$x_1(k) = -\alpha_2 y(k - 1) \tag{12.119}$$

By combining the foregoing expressions, we have

$$\begin{aligned} x_2(k + 1) &= x_1(k) - \alpha_1 y(k) = x_1(k) - \alpha_1 [x_2(k) + \beta_0 w(k)] \\ &= x_1(k) - \alpha_1 x_2(k) - \alpha_1 \beta_0 w(k) \end{aligned} \tag{12.120}$$

But from (12.119)

$$x_1(k + 1) = -\alpha_2 y(k) = -\alpha_2 x_2(k) - \alpha_2 \beta_0 w(k) \tag{12.121}$$

We combine (12.120) and (12.121) in matrix form, with the result

$$\begin{bmatrix} x_1(k+1) \\ x_2(k+1) \end{bmatrix} = \begin{bmatrix} 0 & -\alpha_2 \\ 1 & -\alpha_1 \end{bmatrix} \begin{bmatrix} x_1(k) \\ x_2(k) \end{bmatrix} + \begin{bmatrix} -\alpha_2 \beta_0 \\ -\alpha_1 \beta_0 \end{bmatrix} w(k) \tag{12.122}$$

which is, in concise form,

$$\mathbf{x}(k + 1) = \mathbf{A}\mathbf{x}(k) + \mathbf{B}w(k) \tag{12.123}$$

The output $y(k)$ specified by (12.117) is

$$y(k) = x_2(k) + \beta_0 w(k) = \mathbf{C}\mathbf{x}(k) + \mathbf{D}w(k) \quad \text{(a)} \tag{12.124}$$

where

$$\mathbf{C} = \begin{bmatrix} 0 & 1 \end{bmatrix} \quad \mathbf{D} = \begin{bmatrix} \beta_0 \end{bmatrix} \quad \text{(b)}$$

As an extension of this result, consider the general equation

$$\begin{aligned} y(k) &+ \alpha_1 y(k - 1) + \alpha_2 y(k - 2) + \cdots + \alpha_n y(k - n) \\ &= \beta_0 w(k) + \beta_1 w(k - 1) + \cdots + \beta_n w(k - n) \end{aligned} \tag{12.125}$$

By following the same procedure as above, we would obtain the first canonical form matrices:

$$\mathbf{A} = \begin{bmatrix} 0 & 0 & 0 & \cdots & 0 & 0 & -\alpha_n \\ 1 & 0 & 0 & & 0 & 0 & -\alpha_{n-1} \\ 0 & 1 & 0 & & 0 & 0 & -\alpha_{n-2} \\ \vdots & \vdots & \vdots & & \vdots & \vdots & \vdots \\ 0 & 0 & 0 & & 1 & 0 & -\alpha_2 \\ 0 & 0 & 0 & & 0 & 1 & -\alpha_1 \end{bmatrix} \quad \mathbf{B} = \begin{bmatrix} \beta_n - \alpha_n \beta_0 \\ \beta_{n-1} - \alpha_{n-1} \beta_0 \\ \beta_{n-2} - \alpha_{n-2} \beta_0 \\ \vdots \\ \beta_2 - \alpha_2 \beta_0 \\ \beta_1 - \alpha_1 \beta_0 \end{bmatrix} \tag{12.126}$$

$$\mathbf{C} = \begin{bmatrix} 0 & 0 & 0 & \cdots & 0 & 1 \end{bmatrix} \quad \mathbf{D} = \begin{bmatrix} \beta_0 \end{bmatrix}$$

Figure 12.12
Another realization form of the discrete systems described by the difference equation (12.125).

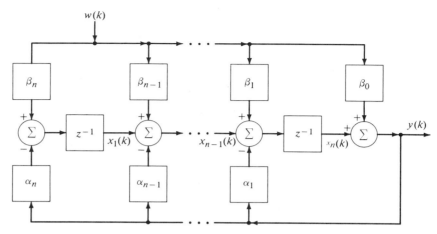

The realization of this system is shown in Figure 12.12. The reader can easily show that the realization represents (12.125) by means of a second-order example.

EXAMPLE 12.15
Obtain a set of state equations in phase variable form for the second order equation

$$y(k) + \alpha_1 y(k-1) + \alpha_2 y(k-2) = \beta_0 w(k) + \beta_1 w(k-1) + \beta_2 w(k-2) \tag{12.127}$$

Solution: Introduce a new variable $p(k)$, which is defined through the expression

$$w(k) = p(k) + \alpha_1 p(k-1) + \alpha_2 p(k-2) \tag{12.128}$$

Then it will be found that

$$y(k) = \beta_0 p(k) + \beta_1 p(k-1) + \beta_2 p(k-2) \tag{12.129}$$

The latter two equations are equivalent to (12.127), as can be verified by direct substitution, which leads to an identity.

Now we define the state variables

$$\begin{aligned} x_1(k) &= p(k-2) \quad \textbf{(a)} \\ x_2(k) &= p(k-1) \quad \textbf{(b)} \end{aligned} \tag{12.130}$$

From these we obtain the new set

$$\begin{aligned} x_1(k+1) &= p(k-1) = x_2(k) & \textbf{(a)} \\ x_2(k+1) &= p(k) = w(k) - \alpha_1 p(k-1) - \alpha_2 p(k-2) \\ &= w(k) - \alpha_1 x_2(k) - \alpha_2 x_1(k) & \textbf{(b)} \end{aligned} \tag{12.131}$$

CHAPTER 12

These states can be written in matrix form

$$\begin{bmatrix} x_1(k+1) \\ x_2(k+1) \end{bmatrix} = \begin{bmatrix} 0 & 1 \\ -\alpha_2 & -\alpha_1 \end{bmatrix} \begin{bmatrix} x_1(k) \\ x_2(k) \end{bmatrix} + \begin{bmatrix} 0 \\ 1 \end{bmatrix} w(k) \quad \text{(a)} \qquad (12.132)$$

so that

$$\mathbf{A} = \begin{bmatrix} 0 & 1 \\ -\alpha_2 & -\alpha_1 \end{bmatrix} \qquad \mathbf{B} = \begin{bmatrix} 0 \\ 1 \end{bmatrix} \quad \text{(b)}$$

The output equation is obtained using the difference equation specified by

$$y(k) - \beta_0 w(k) = (-\alpha_1 \beta_0 + \beta_1) p(k-1) + (-\alpha_2 \beta_0 + \beta_2) p(k-2)$$

from which

$$y(k) = (\beta_1 - \alpha_1 \beta_0) x_2(k) + (\beta_2 - \alpha_2 \beta_0) x_1(k) + \beta_0 w(k) \qquad (12.133)$$

This has the matrix representation

$$y(k) = \begin{bmatrix} \beta_1 - \alpha_1 \beta_0 & \beta_2 - \alpha_2 \beta_0 \end{bmatrix} \begin{bmatrix} x_1(k) \\ x_2(k) \end{bmatrix} + [\beta_0] w(k) \quad \text{(a)} \qquad (12.134)$$

from which, therefore

$$\mathbf{C} = \begin{bmatrix} \beta_1 - \alpha_1 \beta_0 & \beta_2 - \alpha_2 \beta_0 \end{bmatrix} \qquad \mathbf{D} = [\beta_0] \quad \text{(b)}$$

An extension of this procedure for the nth-order equation

$$\begin{aligned} y(k) + \alpha_1 y(k-1) + \cdots + \alpha_n y(k-n) \\ = \beta_0 w(k) + \beta_1 w(k-1) + \cdots + \beta_n w(k-n) \end{aligned} \qquad (12.135)$$

will lead to the standard first canonical form matrix equations with

$$\mathbf{A} = \begin{bmatrix} 0 & 1 & 0 & \cdots & 0 & 0 \\ 0 & 0 & 1 & \cdots & 0 & 0 \\ \vdots & \vdots & \vdots & & & \\ 0 & 0 & 0 & & 0 & 1 \\ -\alpha_n & -\alpha_{n-1} & & & -\alpha_2 & -\alpha_1 \end{bmatrix} \qquad \mathbf{B} = \begin{bmatrix} 0 \\ 0 \\ \vdots \\ 0 \\ 1 \end{bmatrix} \qquad (12.136)$$

$$\mathbf{C} = \begin{bmatrix} (\beta_n - \alpha_n \beta_0) & (\beta_{n-1} - \alpha_{n-1} \beta_0) & \cdots & (\beta_1 - \alpha_1 \beta_0) \end{bmatrix} \qquad \mathbf{D} = [\beta_0] \quad \blacksquare$$

EXAMPLE 12.16

Write the state-space representation of the finite impulse response (FIR) system (see Section 11-12)

$$y(k) = \beta_0 w(k) + \beta_1 w(k-1) + \beta_2 w(k-2) + \beta_3 w(k-3) \qquad (12.137)$$

Solution: A direct application of (12.136) yields

STATE VARIABLES AND STATE EQUATIONS

Figure 12.13
A second-order discrete system.

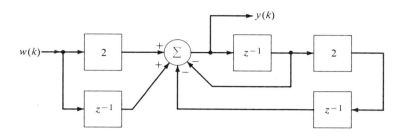

$$\begin{bmatrix} x_1(k+1) \\ x_2(k+1) \\ x_3(k+1) \end{bmatrix} = \begin{bmatrix} 0 & 1 & 0 \\ 0 & 0 & 1 \\ 0 & 0 & 0 \end{bmatrix} \begin{bmatrix} x_1(k) \\ x_2(k) \\ x_3(k) \end{bmatrix} + \begin{bmatrix} 0 \\ 0 \\ 1 \end{bmatrix} w(k)$$

$$\mathbf{y}(k) = \begin{bmatrix} \beta_3 & \beta_2 & \beta_1 \end{bmatrix} \begin{bmatrix} x_1(k) \\ x_2(k) \\ x_3(k) \end{bmatrix} + \beta_0 w(k)$$

EXAMPLE 12.17
Find the state-space description of the system shown in Figure 12.13.

Solution: The difference equation specified by Figure 12.13 is readily found to be

$$y(k) + y(k-1) + 2y(k-2) = 2w(k) + w(k-1)$$

If the states are chosen as described in (12.136), the results are

$$\mathbf{A} = \begin{bmatrix} 0 & 1 \\ -2 & -1 \end{bmatrix} \quad \mathbf{B} = \begin{bmatrix} 0 \\ 1 \end{bmatrix} \quad \mathbf{C} = \begin{bmatrix} -4 & -1 \end{bmatrix} \quad \mathbf{D} = \begin{bmatrix} 2 \end{bmatrix}$$

where $x_1(k) = p(k-2)$; $x_2(k) = p(k-1)$.

If the states are chosen as described in (12.126), the results are

$$\mathbf{A} = \begin{bmatrix} 0 & -2 \\ 1 & -1 \end{bmatrix} \quad \mathbf{B} = \begin{bmatrix} -4 \\ -1 \end{bmatrix} \quad \mathbf{C} = \begin{bmatrix} 0 & 1 \end{bmatrix} \quad \mathbf{D} = \begin{bmatrix} 2 \end{bmatrix}$$

where $x_1(k) = \eta(k-2)$; $x_2(k) = \eta(k-1)$.

EXAMPLE 12.18
Find the state-space description of the multiple-input, multiple-output system shown in Figure 12.14.

Solution: From the figure we obtain the equations

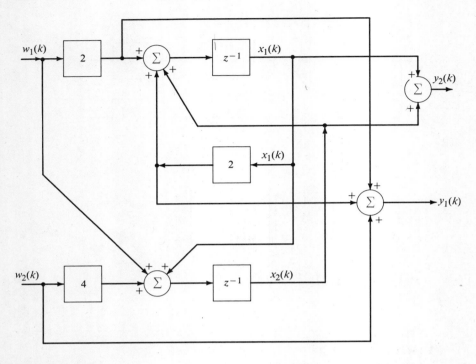

Figure 12.14
A multiple-input, multiple-output discrete system.

$$\begin{bmatrix} x_1(k+1) \\ x_2(k+1) \end{bmatrix} = \begin{bmatrix} 2 & 1 \\ 1 & 0 \end{bmatrix} \begin{bmatrix} x_1(k) \\ x_2(k) \end{bmatrix} + \begin{bmatrix} 2 & 0 \\ 1 & 4 \end{bmatrix} \begin{bmatrix} w_1(k) \\ w_2(k) \end{bmatrix}$$

$$\begin{bmatrix} y_1(k) \\ y_2(k) \end{bmatrix} = \begin{bmatrix} 2 & 0 \\ 1 & 1 \end{bmatrix} \begin{bmatrix} x_1(k) \\ x_2(k) \end{bmatrix} + \begin{bmatrix} 2 & 1 \\ 0 & 0 \end{bmatrix} \begin{bmatrix} w_1(k) \\ w_2(k) \end{bmatrix}$$

∎

12-12 SOLUTION TO DISCRETE TIME STATE EQUATIONS

We will initially address our attention to the solution of the homogeneous state equation

$$\mathbf{x}(k+1) = \mathbf{A}\mathbf{x}(k) \qquad (12.138)$$

For the case of an initial state $\mathbf{x}(k_0)$ at time k_0, the succeeding states of the system are obtained by a direct iteration process, as follows:

$$\mathbf{x}(k_0 + 1) = \mathbf{A}\mathbf{x}(k_0)$$

STATE VARIABLES AND STATE EQUATIONS

$$\mathbf{x}(k_0 + 2) = \mathbf{A}\mathbf{x}(k_0 + 1) = \mathbf{A}\mathbf{A}\mathbf{x}(k_0) = \mathbf{A}^2\mathbf{x}(k_0) \quad (12.139)$$
$$\vdots$$
$$\mathbf{x}(k_0 + n) = \mathbf{A}^n\mathbf{x}(k_0)$$

If we set $k_0 + n = k$ in this last iterative element, there results

$$\boxed{\mathbf{x}(k) = \mathbf{A}^{k-k_0}\mathbf{x}(k_0)} \quad (a) \quad (12.140)$$

and for $k_0 = 0$

$$\boxed{\mathbf{x}(k) = \mathbf{A}^k\mathbf{x}(0)} \quad (b)$$

The **state transition matrix** (also, the fundamental matrix), which is written $\mathbf{\Phi}(k, k_0)$, defines the $n \times n$ matrix such that $\mathbf{x}(k) = \mathbf{\Phi}(k, k_0)\mathbf{x}(k_0)$ is the solution of (12.138) with $\mathbf{x}(k_0)$ as the initial condition. Equations (12.140) are thus used to write

$$\boxed{\begin{aligned}\mathbf{\Phi}(k, k_0) &= \mathbf{\Phi}(k - k_0) = \mathbf{A}^{k-k_0} \quad (a) \\ \mathbf{\Phi}(k) &= \mathbf{A}^k \quad (b)\end{aligned}} \quad (12.141)$$

From these relations we can readily deduce the following properties of the state transition matrix:

1. $\mathbf{\Phi}(k + 1 - k_0) = \mathbf{A}\mathbf{\Phi}(k - k_0)$ (a)
2. $\mathbf{\Phi}(k_0 - k_0) = \mathbf{\Phi}(0) = \mathbf{I}$ (b)
3. $\mathbf{\Phi}(k_0 + 1 - k_0) = \mathbf{\Phi}(1) = \mathbf{A}$ (c)
4. $\mathbf{\Phi}(k_2 - k_0) = \mathbf{\Phi}(k_2 - k_1)\mathbf{\Phi}(k_1 - k_0)$ (d) (12.142)
5. $\mathbf{\Phi}(k_0 - k_1) = \mathbf{\Phi}^{-1}(k_1 - k_0)$ (e)
6. $\mathbf{\Phi}(k_1 - k_0) = \mathbf{\Phi}(k_1)\mathbf{\Phi}^{-1}(k_0)$ (f)

We now consider the solution of the general state equations given by (12.96) for the initial vector at $k = k_0$. We build up the solutions by noting that

$$\mathbf{x}(k_0 + 1) = \mathbf{A}\mathbf{x}(k_0) + \mathbf{B}\mathbf{w}(k_0)$$
$$\begin{aligned}\mathbf{x}(k_0 + 2) &= \mathbf{A}\mathbf{x}(k_0 + 1) + \mathbf{B}\mathbf{w}(k_0 + 1) \\ &= \mathbf{A}^2\mathbf{x}(k_0) + \mathbf{A}\mathbf{B}\mathbf{w}(k_0) + \mathbf{B}\mathbf{w}(k_0 + 1)\end{aligned}$$
$$\begin{aligned}\mathbf{x}(k_0 + 3) &= \mathbf{A}\mathbf{x}(k_0 + 2) + \mathbf{B}\mathbf{w}(k_0 + 2) \\ &= \mathbf{A}^3\mathbf{x}(k_0) + \mathbf{A}^2\mathbf{B}\mathbf{w}(k_0) + \mathbf{A}\mathbf{B}\mathbf{w}(k_0 + 1) + \mathbf{B}\mathbf{w}(k_0 + 2)\end{aligned}$$
$$\vdots$$

$$\mathbf{x}(k_0 + q) = \mathbf{A}^q \mathbf{x}(k_0) + \sum_{m=k_0}^{k_0+q-1} \mathbf{A}^{k_0+q-m-1} \mathbf{B} \mathbf{w}(m)$$

If we set $k_0 + q = k$, then the foregoing results take on the form

$$\mathbf{x}(k) = \mathbf{A}^{k-k_0} \mathbf{x}(k_0) + \sum_{m=k_0}^{k-1} \mathbf{A}^{k-m-1} \mathbf{B} \mathbf{w}(m) \qquad (12.143)$$

or equivalently

$$\mathbf{x}(k) = \mathbf{\Phi}(k - k_0) \mathbf{x}(k_0) + \sum_{m=k_0}^{k-1} \mathbf{\Phi}(k - m - 1) \mathbf{B} \mathbf{w}(m). \qquad (12.144)$$

This equation specifies the state vector at time $k \geq k_0$ and shows it to be the sum of two terms, one representing the contributions due to the initial state $\mathbf{x}(k_0)$ and the second being the contribution due to the input $\mathbf{w}(k)$ over the interval $(k_0, k - 1)$.

Introduce the value of $\mathbf{x}(k)$ of (12.144) into (12.96b) to obtain

$$\mathbf{y}(k) = \mathbf{C} \mathbf{\Phi}(k - k_0) \mathbf{x}(k_0) + \sum_{m=k_0}^{k-1} \mathbf{C} \mathbf{\Phi}(k - m - 1) \mathbf{B} \mathbf{w}(m) + \mathbf{D} \mathbf{w}(k)$$

$$(12.145)$$

When we set the initial conditions equal to zero $\mathbf{x}(0) = \mathbf{0}$, and we use an input $w(k) = \delta(k)$, the output is the impulse response function

$$\mathbf{h}(k) = \sum_{m=0}^{k-1} \mathbf{C} \mathbf{\Phi}(k - m - 1) \mathbf{B} \, \delta(m) + \mathbf{D} \, \delta(k) \qquad (12.146)$$

Since $\delta(k)$ is zero for $k \neq 0$, the second term is equal to $\mathbf{D}$ when $k = 0$. When $k \leq 0$, the sum has zero value. Hence the expression for $\mathbf{h}(k)$ is written

$$\mathbf{h}(k) = \begin{cases} \mathbf{D} & k = 0 \\ \sum_{m=0}^{k-1} \mathbf{C} \mathbf{\Phi}(k - m - 1) \mathbf{B} \, \delta(m) = \mathbf{C} \mathbf{\Phi}(k - 1) \mathbf{B} & k > 0 \\ \mathbf{O} & k < 0 \end{cases} \qquad (12.147)$$

where

$$\mathbf{\Phi}(k - 1) = \mathbf{A}^{k-1}$$

EXAMPLE 12.19

Deduce the solution to the system shown in Figure 12.15 for $k \geq 0$ and $w(k) = u(k)$.

Figure 12.15
A simple discrete system.

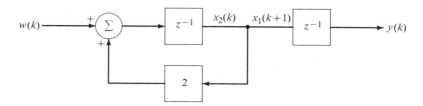

Solution: From the figure we obtain the following state equations

$$\begin{bmatrix} x_1(k+1) \\ x_2(k+1) \end{bmatrix} = \begin{bmatrix} 0 & 1 \\ 0 & 2 \end{bmatrix} \begin{bmatrix} x_1(k) \\ x_2(k) \end{bmatrix} + \begin{bmatrix} 0 \\ 1 \end{bmatrix} w(k)$$

$$y(k) = \begin{bmatrix} 1 & 0 \end{bmatrix} \begin{bmatrix} x_1(k) \\ x_2(k) \end{bmatrix}$$

The respective powers of the matrix **A** are found to be

$$\mathbf{A}^2 = \begin{bmatrix} 0 & 2 \\ 0 & 2^2 \end{bmatrix} \qquad \mathbf{A}^3 = \begin{bmatrix} 0 & 2^2 \\ 0 & 2^3 \end{bmatrix} \qquad \mathbf{A}^4 = \begin{bmatrix} 0 & 2^3 \\ 0 & 2^4 \end{bmatrix} \cdots$$

Combine these equations with (12.143) to write

$$\begin{bmatrix} x_1(k) \\ x_2(k) \end{bmatrix} = \begin{bmatrix} 0 & 2^{k-1} \\ 0 & 2^k \end{bmatrix} \begin{bmatrix} x_1(0) \\ x_2(0) \end{bmatrix} + \sum_{m=0}^{k-1} \begin{bmatrix} 0 & 2^{k-m-2} \\ 0 & 2^{k-m-1} \end{bmatrix} \begin{bmatrix} 0 \\ 1 \end{bmatrix} w(m)$$

$$= \begin{bmatrix} 0 & 2^{k-1} \\ 0 & 2^k \end{bmatrix} \begin{bmatrix} x_1(0) \\ x_2(0) \end{bmatrix} + \begin{bmatrix} 2^{k-2} + 2^{k-3} + \cdots + 2^{-1} \\ 2^{k-1} + 2^{k-2} + \cdots + 2^0 \end{bmatrix}$$

The output of the system is given by

$$y(k) = 2^{k-1} x_2(0) + \sum_{m=0}^{k-1} 2^{k-2} 2^{-m} = 2^{k-1} x_2(0) + 2^{k-2} \sum_{m=0}^{k-1} (\tfrac{1}{2})^m$$

$$= 2^{k-1} x_2(0) + 2^{k-2} \frac{1 - (\tfrac{1}{2})^k}{1 - \tfrac{1}{2}}$$

$$= 2^{k-1} x_2(0) + 2^{k-1} [1 - (\tfrac{1}{2})^k] \qquad k = 1, 2, 3, \ldots \qquad \blacksquare$$

Appendix 2 shows that if a matrix has distinct eigenvalues, we can always find a matrix **S** whose columns are the eigenvectors of **A** such that

$$\mathbf{S}^{-1} \mathbf{A} \mathbf{S} = \mathbf{\Lambda} \triangleq \begin{bmatrix} \lambda_1 & & & 0 \\ & \lambda_2 & & \\ & & \ddots & \\ 0 & & & \lambda_n \end{bmatrix} \qquad (12.148)$$

That is, such a similarity transformation diagonalizes the matrix **A**, the elements on the diagonal being the eigenvalues. From this equation we can also write

$$\mathbf{A} = \mathbf{S} \mathbf{\Lambda} \mathbf{S}^{-1}$$

This permits the ready determination of $\mathbf{A}^k$, since

$$\mathbf{A}^k = (\mathbf{S\Lambda S}^{-1})(\mathbf{S\Lambda S}^{-1})\cdots(\mathbf{S\Lambda S}^{-1}) = \mathbf{S\Lambda\Lambda}\cdots\mathbf{\Lambda S}^{-1} = \mathbf{S\Lambda}^k\mathbf{S}^{-1}$$

(12.149)

EXAMPLE 12.20
Find the solution to the system shown in Figure 12.16 for $k > 0$ and $w(k) = \delta(k)$.

Solution: The difference equation that describes this system is

$$y(k) + 4y(k-1) + 2y(k-2) = w(k)$$

Therefore from (12.126) with $\alpha_1 = 4$, $\alpha_2 = 2$, and $\beta_0 = 1$, we deduce the coefficient matrices

$$\mathbf{A} = \begin{bmatrix} 0 & -2 \\ 1 & -4 \end{bmatrix} \qquad \mathbf{B} = \begin{bmatrix} -2 \\ -4 \end{bmatrix} \qquad \mathbf{C} = \begin{bmatrix} 0 & 1 \end{bmatrix} \qquad \mathbf{D} = \begin{bmatrix} 1 \end{bmatrix}$$

The state equations are given by

$$\begin{bmatrix} x_1(k+1) \\ x_2(k+1) \end{bmatrix} = \begin{bmatrix} 0 & -2 \\ 1 & -4 \end{bmatrix}\begin{bmatrix} x_1(k) \\ x_2(k) \end{bmatrix} + \begin{bmatrix} -2 \\ -4 \end{bmatrix} w(k)$$

$$y(k) = \begin{bmatrix} 0 & 1 \end{bmatrix}\begin{bmatrix} x_1(k) \\ x_2(k) \end{bmatrix} \quad w(k)$$

The eigenvalues of $\mathbf{A}$ are found from the expression

$$|\mathbf{A} - \lambda\mathbf{I}| = \begin{vmatrix} 0 - \lambda & -2 \\ 1 & -4 - \lambda \end{vmatrix} = \lambda^2 + 4\lambda + 2 = 0$$

from which we find that

$$\lambda_1 = -2 - \sqrt{2} \qquad \lambda_2 = -2 + \sqrt{2}$$

Figure 12.16
Illustrating Example 12.20.

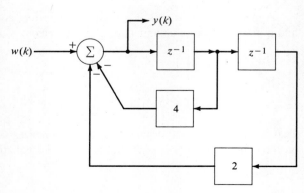

STATE VARIABLES AND STATE EQUATIONS

The eigenvector $\mathbf{e}_1$ corresponding to the eigenvalue $\lambda_1 = -2 - \sqrt{2}$ is determined from the equation

$$(\mathbf{A} - \lambda_1 \mathbf{I})\mathbf{e}_1 = \mathbf{0} \quad \text{or} \quad \begin{bmatrix} 2 + \sqrt{2} & -2 \\ 1 & -2 + \sqrt{2} \end{bmatrix} \begin{bmatrix} e_{11} \\ e_{12} \end{bmatrix} = \begin{bmatrix} 0 \\ 0 \end{bmatrix}$$

Thus the components of $\mathbf{e}_1$ must simultaneously satisfy the equations

$$(2 + \sqrt{2})e_{11} - 2e_{12} = 0$$
$$e_{11} + (-2 + \sqrt{2})e_{12} = 0$$

These lead to the values $e_{11} = 2/(2 + \sqrt{2})$ and $e_{12} = 1$; hence the vector $\mathbf{e}_1$ is

$$\mathbf{e}_1 = \begin{bmatrix} \dfrac{2}{2 + \sqrt{2}} \\ 1 \end{bmatrix}$$

In a similar way, we find the second eigenvector $\mathbf{e}_2$ to be

$$\mathbf{e}_2 = \begin{bmatrix} \dfrac{2}{2 - \sqrt{2}} \\ 1 \end{bmatrix}$$

Because the eigenvalues are distinct, we know from matrix theory that the eigenvectors are linearly independent. As a consequence, the transformation matrix $\mathbf{S}$ is nonsingular and is given by

$$\mathbf{S} = [\mathbf{e}_1 \quad \mathbf{e}_2] = \begin{bmatrix} \dfrac{2}{2 + \sqrt{2}} & \dfrac{2}{2 - \sqrt{2}} \\ 1 & 1 \end{bmatrix}$$

Since the input is a delta function, the output is the impulse response of the system, and this is given by (12.147). Further, the value of $\mathbf{\Phi}(k - 1)$ is given by [see (12.149)]

$$\mathbf{\Phi}(k - 1) = \mathbf{A}^{k-1} = \mathbf{A}^k \mathbf{A}^{-1} = \mathbf{S}\mathbf{\Lambda}^k \mathbf{S}^{-1} \mathbf{A}^{-1}$$

which is

$$\mathbf{\Phi}(k - 1) = \begin{bmatrix} \dfrac{2}{2 + \sqrt{2}} & \dfrac{2}{2 - \sqrt{2}} \\ 1 & 1 \end{bmatrix} \begin{bmatrix} (-2 - \sqrt{2})^k & 0 \\ 0 & (-2 + \sqrt{2})^k \end{bmatrix} \left(-\dfrac{1}{2\sqrt{2}}\right) \begin{bmatrix} 1 & \dfrac{2}{2 - \sqrt{2}} \\ -1 & \dfrac{2}{2 + \sqrt{2}} \end{bmatrix} \begin{bmatrix} 0 & -2 \\ 1 & -4 \end{bmatrix}$$

or

$$\mathbf{\Phi}(k - 1) = \dfrac{1}{\sqrt{2}} \begin{bmatrix} (-1)^k[(2 + \sqrt{2})^k - (2 - \sqrt{2})^k] & 2(-1)^k[(5 + 2\sqrt{2})(2 + \sqrt{2})^k + (3 + 2\sqrt{2})(2 - \sqrt{2})^{k-1}] \\ \tfrac{1}{2}(-1)^k[(2 + \sqrt{2})^{k+1} - (2 - \sqrt{2})^{k+1}] & (-1)^k[(5 + 2\sqrt{2})(2 + \sqrt{2})^k + (3 + 2\sqrt{2})(2 - \sqrt{2})^k] \end{bmatrix}$$

Therefore the solution is [see (12.147)]

$$h(k) = \begin{cases} 1 & k = 0 \\ -2\varphi_{21} - 4\varphi_{22} = -\dfrac{1}{\sqrt{2}}(-1)^k[(2+\sqrt{2})^{k+1}] - (2-\sqrt{2})^{k+1}] - \dfrac{4}{\sqrt{2}}(-1)^k \\ \qquad \times [(5+2\sqrt{2})(2+\sqrt{2})^k + (3+2\sqrt{2})(2-\sqrt{2})^k] & k > 1 \\ 0 & k < 0 \end{cases}$$

∎

Observe that in this section we used the direct multiplication method to find $\mathbf{A}^k$ and the diagonalization method when $\mathbf{A}$ had distinct eigenvalues. For the case when $\mathbf{A}$ has multiple eigenvalues, we use the basic matrix development (see Appendix 2), which states that if the degree of the minimal polynomial of a matrix $\mathbf{A}$ is m, any function $\mathbf{f(A)}$ can be expressed as a linear combination of the m linearly independent matrices $\mathbf{I}, \mathbf{A}, \mathbf{A}^2, \ldots, \mathbf{A}^{m-1}$; that is,

$$f(\lambda) \triangleq \mathbf{f(A)} = \gamma_{m-1}\mathbf{A}^{m-1} + \gamma_{m-2}\mathbf{A}^{m-2} + \cdots + \gamma_1\mathbf{A} + \gamma_0\mathbf{I} \triangleq \mathbf{g(A)} \quad (12.150)$$

If λ_i is a k-fold degenerate eigenvalue of $\mathbf{A}$, the algebraic functions $f(\lambda)$ and $g(\lambda)$ satisfy the k equations

$$\begin{aligned} f(\lambda)|_{\lambda=\lambda_i} &= g(\lambda)|_{\lambda=\lambda_i} \\ \dfrac{df(\lambda)}{d\lambda}\bigg|_{\lambda=\lambda_i} &= \dfrac{dg(\lambda)}{d\lambda}\bigg|_{\lambda=\lambda_i} \\ &\vdots \\ \dfrac{d^{k-1}f(\lambda)}{d\lambda^{k-1}}\bigg|_{\lambda=\lambda_i} &= \dfrac{d^{k-1}g(\lambda)}{d\lambda^{k-1}}\bigg|_{\lambda=\lambda_i} \end{aligned} \quad (12.151)$$

The following example illustrates the procedure.

EXAMPLE 12.21
Find the values of $\mathbf{A}^n$ and $e^{-\mathbf{A}}$, where

$$\mathbf{A} = \begin{bmatrix} 0 & -4 \\ 1 & 4 \end{bmatrix}$$

Solution: **a.** The matrix $\mathbf{A}$ has the double root $\lambda_1 = \lambda_2 = 2$, and is of the second order. Hence we write

$$\mathbf{g(A)} = \gamma_1 \mathbf{A} + \gamma_0 \mathbf{I}$$
$$g(\lambda) = \gamma_1 \lambda + \gamma_0$$

Therefore the coefficients γ_1 and γ_0 are found from the conditions

$$f(2) = g(2) \qquad \dfrac{df(\lambda)}{d\lambda}\bigg|_{\lambda=2} = \dfrac{dg(\lambda)}{d\lambda}\bigg|_{\lambda=2}$$

From these we have

STATE VARIABLES AND STATE EQUATIONS

$$\lambda^n|_{\lambda=2} = 2^n = 2\gamma_1 + \gamma_0$$

$$\frac{d\lambda^n}{d\lambda}\bigg|_{\lambda=2} = n \times 2^{n-1} = \gamma_1$$

from which $\gamma_1 = n \times 2^{n-1}$ and $\gamma_0 = 2^n(1-n)$. Therefore we find

$$\mathbf{A}^n = n \times 2^{n-1} \begin{bmatrix} 0 & -4 \\ 1 & 4 \end{bmatrix} + 2^n(1-n) \begin{bmatrix} 1 & 0 \\ 0 & 1 \end{bmatrix}$$

$$= \begin{bmatrix} (1-n)2^n & -2n2^n \\ n2^{n-1} & (n+1)2^n \end{bmatrix}$$

b. Similarly, from the expressions

$$e^{-\lambda}|_{\lambda=2} = 2\gamma_1 + \gamma_0$$

$$\frac{de^{-\lambda}}{d\lambda}\bigg|_{\lambda=2} = \gamma_1$$

we find that $\gamma_1 = -\exp[-2]$ and $\gamma_0 = 3\exp[-2]$. The function $\exp[-\mathbf{A}]$ is now written in the form

$$e^{-\mathbf{A}} = -e^{-2} \begin{bmatrix} 0 & -4 \\ 1 & 4 \end{bmatrix} + 3e^{-2} \begin{bmatrix} 1 & 0 \\ 0 & 1 \end{bmatrix} = \begin{bmatrix} 3e^{-2} & 4e^{-2} \\ -e^{-2} & -e^{-2} \end{bmatrix} \qquad \blacksquare$$

■ ■ ■

12-13 CONTINUOUS-TIME SYSTEMS WITH SAMPLED INPUTS

Suppose that the system inputs are approximated by step function approximations having transition times of fundamental period T. In essence, therefore, the inputs are replaced by stepwise approximations. At two successive time intervals, we obtain from (12.39), at $t = nT$,

$$\mathbf{x}(nT) = e^{\mathbf{A}nT}\mathbf{x}(0) \tag{12.152}$$

and for $t = (n+1)T$

$$\mathbf{x}[(n+1)T] = e^{\mathbf{A}(n+1)T}\mathbf{x}(0) \tag{12.153}$$

From these it follows immediately that

$$\mathbf{x}[(n+1)T] = e^{\mathbf{A}T}\mathbf{x}(nT) \tag{12.154}$$

For the complete solution given by (12.62) under these same sampled conditions, we readily find that, for $t = nT$,

$$\mathbf{x}(nT) = e^{\mathbf{A}nT}\mathbf{x}(0) + e^{\mathbf{A}nT}\int_0^{nT} e^{-\mathbf{A}\tau}\mathbf{B}\mathbf{w}(\tau)\,d\tau \tag{12.155}$$

and for $t = (n+1)T$

$$\mathbf{x}[(n+1)T] = e^{\mathbf{A}(n+1)T}\mathbf{x}(0) + e^{\mathbf{A}(n+1)T}\int_0^{(n+1)T} e^{-\mathbf{A}\tau}\mathbf{B}\mathbf{w}(\tau)\,d\tau \tag{12.156}$$

Now multiply (12.155) by $\exp[\mathbf{A}T]$ and subtract from (12.156). The result is

$$\mathbf{x}[(n+1)T] = e^{\mathbf{A}T}\mathbf{x}(nT) + e^{\mathbf{A}(n+1)T} \int_{nT}^{(n+1)T} e^{-\mathbf{A}\tau}\mathbf{B}\mathbf{w}(\tau)\,d\tau \qquad (12.157)$$

Suppose that $w(t)$ is stepwise constant during the interval $nT \leq t \leq (n+1)T$. The integrations can be carried out under these conditions. After some algebraic manipulation, the result is

$$\mathbf{x}[(n+1)T] = e^{\mathbf{A}T}\mathbf{x}(nT) + (e^{\mathbf{A}T} - 1)\mathbf{A}^{-1}\mathbf{B}\mathbf{w}(nT) \qquad (12.158)$$

Let us write this expression as

$$\mathbf{x}[(n+1)T] = \mathbf{F}\mathbf{x}(nT) + \mathbf{G}\mathbf{w}(nT) \quad (a) \qquad (12.159)$$

where

$$\mathbf{F} = e^{\mathbf{A}T} \qquad (b)$$

$$\mathbf{G} = (e^{\mathbf{A}T} - 1)\mathbf{A}^{-1}\mathbf{B} \qquad (c)$$

It is observed that both $\mathbf{F}$ and $\mathbf{G}$ are constants that can be evaluated at the start of a computation.

This development shows that a system described by (12.157) with piecewise constant inputs at regular sampling intervals is equivalent at the sampling instants to the system given by the difference equation (12.159a). This form is identical to (12.100a) with the solution given by (12.143).

■ ■ ■

12-14 Z-TRANSFORM SOLUTION TO DISCRETE TIME STATE EQUATIONS

Just as the Laplace transform provides a vehicle for solving the continuous-time (analog) state equations, we wish to show that the Z-transform provides a vehicle for solving discrete time (digital) state equations. We begin with the state equations

$$\mathbf{x}(k+1) = \mathbf{A}\mathbf{x}(k) + \mathbf{B}\mathbf{w}(k)$$
$$\mathbf{y}(k) = \mathbf{C}\mathbf{x}(k) + \mathbf{D}\mathbf{w}(k) \qquad (12.160)$$

where $\mathbf{A}$, $\mathbf{B}$, $\mathbf{C}$, and $\mathbf{D}$ are constant matrices. The Z-transform is taken of these equations, with the result

$$z\mathbf{x}(z) = \mathbf{A}\mathbf{x}(z) + \mathbf{B}\mathbf{w}(z) + z\mathbf{x}(0)$$
$$\mathbf{y}(z) = \mathbf{C}\mathbf{x}(z) + \mathbf{D}\mathbf{w}(z) \qquad (12.161)$$

From these we can write

$$\mathbf{x}(z) = [z\mathbf{I} - \mathbf{A}]^{-1}\mathbf{B}\mathbf{w}(z) + [z\mathbf{I} - \mathbf{A}]^{-1}z\mathbf{x}(0)$$
$$\mathbf{y}(z) = (\mathbf{C}[z\mathbf{I} - \mathbf{A}]^{-1}\mathbf{B} + \mathbf{D})\mathbf{w}(z) + \mathbf{C}[z\mathbf{I} - \mathbf{A}]^{-1}z\mathbf{x}(0) \qquad (12.162)$$

STATE VARIABLES AND STATE EQUATIONS

The fundamental matrix $\Phi(z)$ for the discrete system is defined by

$$\Phi(z) = \mathscr{Z}\{\Phi(k)\} = z[z\mathbf{I} - \mathbf{A}]^{-1} \quad (12.163)$$

Recall, however, that the fundamental matrix for the discrete time system was defined in (12.141) as

$$\Phi(k) = \mathbf{A}^k \quad (12.164)$$

It follows, therefore, that

$$\Phi(z) = \mathscr{Z}\{\Phi(k)\} = \mathscr{Z}\{\mathbf{A}^k\} = z[z\mathbf{I} - \mathbf{A}]^{-1} = [\mathbf{I} - z^{-1}\mathbf{A}]^{-1} \quad (12.165)$$

If we set $k_0 = 0$ in (12.145) and take its Z-transform, keeping in mind that the summation is a discrete convolution, we obtain the following equation

$$\mathbf{y}(z) = \mathbf{C}\Phi(z)\mathbf{x}(0) + [\mathbf{C}\Phi(z)z^{-1}\mathbf{B} + \mathbf{D}]\mathbf{w}(z)$$

Upon comparing this equation with (12.162), we observe that $\Phi(z)$ is defined by (12.163). A comparison of (12.165) with (12.70b) for the continuous system shows that the form of $\Phi(z)$ is similar to that for $\Phi(s)$, but it differs in form by the presence of the factor z in the expression for $\Phi(z)$.

The relation between the excitation and the response function, in the absence of initial conditions is, from (12.162)

$$\mathbf{y}(z) = (\mathbf{C}[z\mathbf{I} - \mathbf{A}]^{-1}\mathbf{B} + \mathbf{D})\mathbf{w}(z) = [\mathbf{C}\Phi(z)z^{-1}\mathbf{B} + \mathbf{D}]\mathbf{w}(z) \quad (12.166)$$

The Z-domain system transfer matrix is then given by

$$\mathbf{H}(z) = \mathbf{C}\Phi(z)z^{-1}\mathbf{B} + \mathbf{D} \quad (12.167)$$

An important feature of these equations is specifically noted: although there is not an exact analogy between the z-domain and s-domain forms for the corresponding fundamental matrices, the extra z in the definition for $\Phi(z)$ disappears in the expression for $\mathbf{H}(z)$.

The time sequence vector $\mathbf{y}(k)$ corresponding to $\mathbf{y}(z)$ in (12.166) will require inverting $\mathbf{y}(z)$. This involves the quantity $[z\mathbf{I} - \mathbf{A}]^{-1}$. Here, as noted in the discussion of the comparable form in Example 12.10,

$$[z\mathbf{I} - \mathbf{A}]^{-1} = \frac{\text{adjoint}[z\mathbf{I} - \mathbf{A}]}{\det(z\mathbf{I} - \mathbf{A})} \quad (12.168)$$

This is a matrix whose entries are rational functions of z, and it can be expanded into partial-fraction form. Thus, if $z = z_i$ is a zero of $\Delta(z) = \det(z\mathbf{I} - \mathbf{A})$ and $\mathbf{x}(0) = \mathbf{0}$, the expression for $y(z)$ contains terms of the form

$$\mathbf{y}_i(z) = \frac{\mathbf{w}_i}{z - z_i}\mathbf{w}(z) \quad (12.169)$$

where $\mathbf{w}_i$ is the residue matrix that corresponds to the pole specified by $z = z_i$. It can be verified that $\mathbf{y}_i(z)$ is the Z-transform of the sequence

$$\mathbf{y}(k) = \begin{cases} \mathbf{w}C_{n-1}^{k}\mathbf{A}^{k-n+1} & k \geq n-1 \\ 0 & k < n-1 \end{cases} \tag{12.170}$$

where the binomial constant C_{n-1}^{k} (often written $\binom{k}{n-1}$) is given by

$$C_{n-1}^{k} = \frac{k!}{(n-1)!(k-n+1)!} = \frac{k(k-1)\cdots(k-n+2)}{(n-1)!}$$

This specifies the solution to (12.166).

EXAMPLE 12.22
Suppose that the state-space equations for a given discrete time-invariant system have the form

$$\begin{bmatrix} x_1(k+1) \\ x_2(k+1) \end{bmatrix} = \begin{bmatrix} 0 & 1 \\ -2 & -3 \end{bmatrix} \begin{bmatrix} x_1(k) \\ x_2(k) \end{bmatrix} + \begin{bmatrix} 0 \\ 1 \end{bmatrix} w(k)$$

$$y(k) = \begin{bmatrix} 1 & 0 \end{bmatrix} \begin{bmatrix} x_1(k) \\ x_2(k) \end{bmatrix}$$

with $\mathbf{y}(0) = \mathbf{y}(1) = 1$. Find $y(k)$ for $w(k) = u(k)$.

Solution: We need the following quantities:

1. $[z\mathbf{I} - \mathbf{A}]^{-1} = \left(\begin{bmatrix} z & 0 \\ 0 & z \end{bmatrix} - \begin{bmatrix} 0 & 1 \\ -2 & -3 \end{bmatrix}\right)^{-1} = \dfrac{\begin{bmatrix} z+3 & 1 \\ -2 & z \end{bmatrix}}{z(z+3)+2}$

2. $w(z) = \mathscr{L}\{u(k)\} = \dfrac{z}{z-1}$

3. Initial conditions $x(0)$

From the state equations we have

$x_1(1) = x_2(0)$

$y(0) = x_1(0) = 1$

$y(1) = x_1(1) = 1$

so that $x(0) = \begin{bmatrix} 1 & 1 \end{bmatrix}^T$.

Introduce these quantities into (12.162), and we obtain

$$y(z) = \left(\begin{bmatrix} 1 & 0 \end{bmatrix} \frac{\begin{bmatrix} z+3 & 1 \\ -2 & z \end{bmatrix}}{z^2+3z+2} \begin{bmatrix} 0 \\ 1 \end{bmatrix}\right) \frac{z}{z-1} + \begin{bmatrix} 1 & 0 \end{bmatrix} \frac{\begin{bmatrix} z+3 & 1 \\ -2 & z \end{bmatrix}}{z^2+3z+2} \begin{bmatrix} z \\ z \end{bmatrix}$$

which can be rewritten

$$\frac{y(z)}{z} = \frac{1}{(z-1)(z+1)(z+2)} + \frac{z+4}{(z+1)(z+2)}$$

By employing known techniques for inverting this Z-transform expression, we obtain (see Table 9.3)

$$y(k) = \frac{5}{2}(-1)^k + \frac{1}{6}(1)^k - \frac{5}{3}(-2)^k$$

∎

■ ■ ■

†C. ADDITIONAL TOPICS

12.15 CONTROLLABILITY AND OBSERVABILITY OF LINEAR SYSTEMS

Implicit in our foregoing discussions has been the assumption that an input to a system will excite all of the natural modes of that system; that is, the response of the unforced system is assumed to be the superposition of the effects of all of the modes of the system. The concept of controllability of a given mode relates to whether or not that mode is coupled to the input variable. The question of observability, on the other hand, has to do with the existence of coupling of a given mode to the output variable. As a simple example of the ideas, refer to Figure 12.17. A simple calculation will show that the state variables v_c and i_L are dependent on the input excitation; hence the modes are controllable. If the current i is specified as the output; then (owing to the cancellation of the zero and pole of this network) the input admittance of the network $Y(s) = 1$. The modes are not coupled to the output, hence these modes are not observable.

The concepts of controllability and observability of linear systems play very important roles in modern control theory. Specifically, suppose that if unstable, unobservable modes exist in a feedback system, they will not be detected in the output. Also, if the system is not completely controllable, a control signal may be unable to affect unstable modes if they exist. Clearly, for a feedback system to stabilize a system, the design assumes both controllability and observability.

To discuss these concepts in some detail, we proceed in a formal way.

Definition 12.1 If all initial conditions $\mathbf{x}(t_0)$ [or $x(k_0)$ for discrete time systems] can be changed to $\mathbf{x}(t_1)$ [or $\mathbf{x}(k_1)$] at a finite time t_1 (or k_1) by an input

Figure 12.17
A network with unobservable modes.

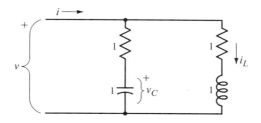

vector $\mathbf{w}$ defined in (t_0, t_1) [or in (k_0, k_1)], then the state $\mathbf{x}(t_1)$ of the system is said to be **controllable** at time t_0 (or k_0). The negative of this statement defines an **uncontrollable** system at t_0 (or k_0).

Definition 12.2 If the knowledge of the input $\mathbf{w}(t)$ [$\mathbf{w}(k)$ for discrete systems] the output $\mathbf{y}(t)$ [$\mathbf{y}(k)$] over an interval (t_0, t_1) [(k_0, k_1)] completely determines the state $\mathbf{x}(t_0)$, then the state $\mathbf{x}(t_1)$ [$\mathbf{x}(k_1)$] of the system is said to be **observable** at $t_0(k_0)$. The negative of this statement defines an **unobservable** system at $t_0(k_0)$.

EXAMPLE 12.23

Determine whether the system specified by the relationship

$$\dot{\mathbf{x}} = \begin{bmatrix} 0 & 1 \\ -2 & -3 \end{bmatrix} \mathbf{x} + \begin{bmatrix} a \\ b \end{bmatrix} w(t)$$

is controllable. Assume $\mathbf{x}(t_0) = 0$, $t_0 = 0$, $a = 1$, $b = -1$.

Solution: The solution of the state equation [see (12.61)] is

$$\mathbf{x}(t_1) = e^{\mathbf{A}t_1} \int_0^{t_1} e^{-\mathbf{A}\tau} \begin{bmatrix} +1 \\ -1 \end{bmatrix} w(\tau)\,d\tau$$

$$= e^{\mathbf{A}t_1} \int_0^{t_1} \begin{bmatrix} e^{-\mathbf{A}\tau} \\ -e^{-\mathbf{A}\tau} \end{bmatrix} w(\tau)\,d\tau = e^{\mathbf{A}t_1} \begin{bmatrix} f_1(t_1) \\ -f_1(t_1) \end{bmatrix} = \begin{bmatrix} f_1(t_1)e^{-t_1} \\ -f_1(t_1)e^{-t_1} \end{bmatrix}$$

This indicates that we cannot control $x_1(t_1)$ and $x_2(t_1)$ independently; hence the system is uncontrollable. This conclusion follows from the fact that when we set a value for $x_1(t_1)$, $x_2(t_1)$ is also determined. ■

EXAMPLE 12.24

A discrete system is described by the set of equations

$$\mathbf{x}(k+1) = \begin{bmatrix} 1 & 0 \\ -1 & -1 \end{bmatrix} \mathbf{x}(k) + \begin{bmatrix} 1 \\ 0 \end{bmatrix} \mathbf{w}(k)$$

Consider that the final state is $\mathbf{x}(k) = \mathbf{x}(2)$. Determine if the system is controllable.

Solution: For $k = 0$ we obtain

$$\mathbf{x}(1) = \begin{bmatrix} 1 & 0 \\ -1 & -1 \end{bmatrix} \mathbf{x}(0) + \begin{bmatrix} 1 \\ 0 \end{bmatrix} w(0) = \begin{bmatrix} x_1(0) + w(0) \\ -x_1(0) - x_2(0) \end{bmatrix}$$

and for $k = 1$

$$\mathbf{x}(2) = \begin{bmatrix} 1 & 0 \\ -1 & -1 \end{bmatrix} \mathbf{x}(1) + \begin{bmatrix} 1 \\ 0 \end{bmatrix} w(1) = \begin{bmatrix} x_1(0) + w(0) + w(1) \\ x_2(0) - w(0) \end{bmatrix}$$

Solve this latter expression for $w(0)$ and $w(1)$; we obtain the relation

$$\begin{bmatrix} x_1(2) - x_1(0) \\ x_2(2) - x_2(0) \end{bmatrix} = \begin{bmatrix} 1 & 1 \\ -1 & 0 \end{bmatrix} \begin{bmatrix} w(0) \\ w(1) \end{bmatrix}$$

We observe that the matrix multiplying w is nonsingular, which means that we can find unique values for $w(0)$ and $w(1)$. Thus the system is controllable. ∎

EXAMPLE 12.25

Consider the continuous time system described by the equations

$$\dot{\mathbf{x}} = \begin{bmatrix} 0 & 1 \\ 1 & 0 \end{bmatrix} \mathbf{x} \qquad y = \begin{bmatrix} a & b \end{bmatrix} \mathbf{x}$$

This system is shown in Figure 12.18. Under what values of a and b is the system observable?

Solution: We readily find that the response of this system is

$$y(t) = \begin{bmatrix} a & b \end{bmatrix} \begin{bmatrix} \cosh(t - t_0) & \sinh(t - t_0) \\ \sinh(t - t_0) & \cosh(t - t_0) \end{bmatrix} \begin{bmatrix} x_1(t_0) \\ x_2(t_0) \end{bmatrix}$$

$$= [a \cosh(t - t_0) + b \sinh(t - t_0)] x_1(t_0)$$
$$+ [a \sinh(t - t_0) + b \cosh(t - t_0)] x_2(t_0)$$

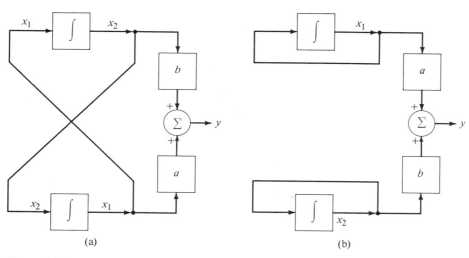

Figure 12.18
Illustrating Example 12.25.

We see from this equation that even if either a or b is zero, we have both $x_1(t_0)$ and $x_2(t_0)$ at the output. Therefore the system is observable.

We note that if the matrix $\mathbf{A}$ has the form

$$\mathbf{A} = \begin{bmatrix} 1 & 0 \\ 0 & 1 \end{bmatrix}$$

then if either a or b were zero, the system would be unobservable. Figure 12.18b shows this configuration. ∎

■ ■ ■

12-16 CONTROLLABILITY AND OBSERVABILITY USING A, B, AND C MATRICES

If we wish to ascertain controllability or observability of a system without specifically determining which of the modes are not controllable or observable, the following two theorems provide an answer in a relatively direct manner.

Theorem 12.1 Any time-invariant linear system described by Equation (12.1a) (analog) or (12.96a) (digital) is controllable if and only if the $n \times (np)$ **controllability matrix** $\mathbf{G}_c$ has rank n, where

$$\mathbf{G}_c = [\mathbf{B}, \mathbf{AB}, \mathbf{A}^2\mathbf{B}, \cdots \mathbf{A}^{n-1}\mathbf{B}] \tag{12.171}$$

$\mathbf{A}$ is $n \times n$, $\mathbf{B}$ is $n \times p$, $\mathbf{x}$ is an n-vector, and $\mathbf{w}$ is a p-vector.

EXAMPLE 12.26
Determine whether the systems of Examples 12.23 and 12.24 are controllable.

Solution: (a) Equation (12.171) deduced from the specifications in Example 12.23 yields the controllability matrix

$$\mathbf{G}_c = \begin{bmatrix} 1 & -1 \\ -1 & 1 \end{bmatrix}$$

Since the determinant of $\mathbf{G}_c$ is equal to zero, this indicates that the rank is less than 2. Hence the system is uncontrollable.

(b) In a similar way for Example 12.24 we obtain

$$\mathbf{G}_c = \begin{bmatrix} 1 & 1 \\ 0 & -1 \end{bmatrix}$$

The determinant in this case has the value -1. This indicates that the rank of $\mathbf{G}_c$ is 2; hence the system is controllable. ∎

Theorem 12.2 Any time-invariant linear system with p inputs and q outputs described by (12.1) or (12.96) is observable if and only if the $(nq \times n)$

observability matrix

$$\mathbf{G}_o = \begin{bmatrix} \mathbf{C} \\ \mathbf{CA} \\ \vdots \\ \mathbf{CA}^{n-1} \end{bmatrix} = [\mathbf{C}^T, \mathbf{A}^T\mathbf{C}^T, \mathbf{A}^{2T}\mathbf{C}^T, \ldots, \mathbf{A}^{n-1}\mathbf{C}^T] \qquad (12.172)$$

has rank n.

EXAMPLE 12.27

Ascertain if the system defined by

$$\dot{\mathbf{x}}(t) = \begin{bmatrix} -1 & -1 & 2 \\ 0 & 0 & 1 \\ -2 & -2 & -1 \end{bmatrix} \mathbf{x}(t) + \begin{bmatrix} 1 \\ 0 \\ 0 \end{bmatrix} w(t) \qquad y = \begin{bmatrix} 1 & 0 & 0 \end{bmatrix} \mathbf{x}(t)$$

is controllable and observable.

Solution: From (12.171) and (12.172), we find

$$\mathbf{G}_c = \begin{bmatrix} 1 & -1 & -3 \\ 0 & 0 & -2 \\ 0 & -2 & 4 \end{bmatrix} \qquad \mathbf{G}_o = \begin{bmatrix} 1 & 0 & 0 \\ -1 & -1 & 2 \\ -3 & -3 & -4 \end{bmatrix}$$

An evaluation will show that the determinants for $\mathbf{G}_c$ and $\mathbf{G}_o$ are different from zero. Therefore their ranks are 3, and the system is both controllable and observable.

■ ■ ■

12-17 STABILITY OF SYSTEMS

It is virtually impossible to ascertain the stability of a system under all possible input conditions by solving the differential equations describing the system. It is understandable that considerable attention has been given to developing general criteria to establish stability without resorting to the solution of the set of equations.

Because stability is not a unique property, a number of different definitions are available to appropriately define stability. When one proceeds from a state equation to the formulation of the system behavior, the physical idea of stability is closely related to a bounded system response to a sudden disturbance or input. Suppose that a system is disturbed and is displaced slightly from the equilibrium state. Several different behaviors are possible. If the system remains near the equilibrium state, the system is said to be stable. That is, if the system is relaxed and a bounded input is applied (with the output being bounded), the system is stable. This is referred to as **bounded-input, bounded-output** stability (bibo). If

Figure 12.19
Stability regions in state space.

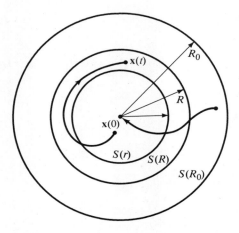

the system is displaced from the equilibrium state and tends to return to the equilibrium state, it is said to be **asymptotically stable**. An equilibrium state $\mathbf{x}_e$ is said to be asymptotically stable **in the large** if it is asymptotically stable for any initial state vector $\mathbf{x}(0)$ such that every motion converges to $\mathbf{x}_e$ as $t \to \infty$.

The foregoing discussion of stability can be given graphical display. Suppose that an equilibrium state exists at the origin, so that $\mathbf{x} = \mathbf{0}$. It may require a transformation of coordinates to refer the equations to the origin, but this can always be done. Now consider the norm from the origin, which is written

$$\|\mathbf{x}\| = (x_1^2 + x_2^2 + \cdots + x_n^2)^{1/2} \tag{12.173}$$

Let $S(R)$ be a spherical region of radius $R > 0$ about the origin, as illustrated in Figure 12.19. Clearly, $S(R)$ will contain those points $\mathbf{x}$ that satisfy the condition $\|\mathbf{x}\| < R$. Now consider a second region $S(r)$ within $S(R)$, as shown. The origin state is said to be stable if corresponding at each $S(R)$, there is an $S(r)$ such that a solution starting in $S(r)$ does not leave $S(R)$. Under these conditions with $\mathbf{x}(0)$ in $S(r)$, $\mathbf{x}(t)$ will remain within $S(R)$ for all $t > 0$. If there is a region $S(R_0)$ such that every solution starting in $S(R_0)$ approaches the origin as $t \to \infty$, the system is asymptotically stable. If R_0 is arbitrarily large, the solution is asymptotically stable in the large.

We can express the conditions for various stabilities in precise form.

12-17.a Zero Input Stability

Theorem 12.3 (a) The equilibrium state $\mathbf{x} = \mathbf{0}$ of $\dot{\mathbf{x}}(t) = \mathbf{A}\mathbf{x}(t)$ is stable if all eigenvalues λ_i of $\mathbf{A}$ have nonpositive real parts (that is, $\text{Re}\{\lambda_i\} \leq 0$) and all eigenvalues for which $\text{Re}\{\lambda_i\} = 0$ are simple zeros.

(b) The equilibrium state $\mathbf{x} = \mathbf{0}$ of $\dot{\mathbf{x}}(t) = \mathbf{A}\mathbf{x}(t)$ is asymptotically stable if all eigenvalues have negative real parts, $\text{Re}\{\lambda_i\} < 0$.

EXAMPLE 12.28
Determine whether the systems specified by the following **A** matrices are stable.

(a) $\mathbf{A} = \begin{bmatrix} 0 & 1 \\ 0 & 0 \end{bmatrix}$ (b) $\mathbf{A} = \begin{bmatrix} 0 & -2 \\ 0 & -1 \end{bmatrix}$ (c) $\mathbf{A} = \begin{bmatrix} 0 & 1 \\ -4 & -6 \end{bmatrix}$

Solution: (a) The eigenvalues are deduced from the relation $|\mathbf{A} - \mathbf{I}\lambda| = 0$. In this case we find that $\lambda^2 = 0$, which indicates that the system does not have simple zeros as its eigenvalues. Therefore, each system and its state $\mathbf{x} = \mathbf{0}$ is not stable.

(b) For the system defined by the given **A**, the eigenvalues follow from the equation $\lambda^2 + \lambda = 0$, which yields $\lambda_1 = 0$ and $\lambda_2 = -1$. Thus the system state $\mathbf{x} = \mathbf{0}$ is stable but is not asymptotically stable.

(c) In this case the eigenvalue polynomial is $\lambda^2 + 6\lambda + 4 = 0$, the resulting eigenvalues being $\lambda_1 = -3 - \sqrt{5} < 0$ and $\lambda_2 = -3 + \sqrt{5} < 0$. Therefore $\mathbf{x} = \mathbf{0}$ is asymptotically stable. ∎

Theorem 12.4 (a) The equilibrium state $\mathbf{x} = \mathbf{0}$ of $\mathbf{x}(k+1) = \mathbf{A}\mathbf{x}(k)$ is stable if all eigenvalues λ_i of **A** have magnitude less than or equal to one ($|\lambda_i| \leq 1$) and all eigenvalues for which $|\lambda_i| = 1$ are simple zeros of the minimal polynomial of **A**.

(b) The equilibrium state $\mathbf{x} = \mathbf{0}$ of $\mathbf{x}(k+1) = \mathbf{A}\mathbf{x}(k)$ is asymptotically stable if all eigenvalues λ_i of **A** have magnitude less than one ($|\lambda_i| < 1$).

EXAMPLE 12.29
Determine whether the systems specified by the following **A** matrices are stable

(a) $\mathbf{A} = \begin{bmatrix} 0 & 1 \\ 1 & 0 \end{bmatrix}$ (b) $\mathbf{A} = \begin{bmatrix} 0 & 2 \\ -1 & -1 \end{bmatrix}$

Solution: (a) The minimum polynomial is $\lambda^2 - 1 = 0$, from which we find $\lambda_1 = 1$ and $\lambda_2 = -1$. The system is stable.

(b) The minimum polynomial is $\lambda^2 + \lambda + 2 = 0$ from which $\lambda_{1,2} = -\frac{1}{2} \pm j\frac{1}{2}\sqrt{7}$. Here $|\lambda_1| = |\lambda_2| = 1.41 > 0$; the system is unstable. ∎

12-17.b Bounded-input, Bounded-output Stability In the foregoing we established stability criteria when the input vector $\mathbf{w}(t)$ was zero. We now consider stability of systems when an input vector is present and the system is in the relaxed state $\mathbf{x}(t_0) = \mathbf{0}$.

Theorem 12.5 A system specified by

$$\dot{\mathbf{x}}(t) = \mathbf{A}\mathbf{x}(t) + \mathbf{B}\mathbf{w}(t) \qquad \mathbf{y}(t) = \mathbf{C}\mathbf{x}(t) \qquad \mathbf{x}(t_0) = \mathbf{0}$$

is bibo stable if all poles of every element $H_{ij}(s)$ of the transfer matrix $\mathbf{H}(s) = \mathbf{C}\mathbf{\Phi}(s)\mathbf{B}$ have negative real parts.

The possibility exists for pole-zero cancellations in $H(s)$. In this case some components of $\mathbf{x}$ may be unbounded even though the output is bounded. Furthermore, it is quite possible that we may not control the unbounded components of $\mathbf{x}$ by using bounded inputs. As already noted such a system may not be completely controllable or observable.

We note that the system of Example 12.6 (further discussed in Example 12.11) is bibo stable since each element of its transfer matrix $\mathbf{H}(s)$ has negative real parts.

Theorem 12.6 The discrete system specified by

$$\mathbf{x}(k+1) = \mathbf{A}\mathbf{x}(k) + \mathbf{B}\mathbf{w}(k) \qquad \mathbf{y}(k) = \mathbf{C}\mathbf{x}(k) \qquad \mathbf{x}(k_0) = 0$$

is bibo stable if all poles of each element $H_{ij}(z)$ of the transfer matrix $\mathbf{H}(z) = \mathbf{C}\boldsymbol{\Phi}(z)z^{-1}\mathbf{B}$ have magnitudes less than unity.

Since pole-zero cancellations are possible for discrete systems, the comments above for continuous systems will be equally applicable for discrete systems.

■ ■ ■

12-18 STABILITY IN THE SENSE OF LIAPUNOV

The direct method of Liapunov, which is also known as the Liapunov second method, allows one to ascertain the stability of a system. A feature of this method, in common with the Routh-Hurwitz test, the Nyquist tests, and root-locus considerations, is that it permits a study of the stability of a dynamic system without requiring the solution of the vector equations of state.

The essence of the direct method of Liapunov is understood by considering the total energy of a system. It is a generalization of the idea that the total energy of a system is always decreasing in the neighborhood of an equilibrium state. To understand the ideas more clearly, consider the series RL circuit with constant excitation V when the solution for the current is given by the equation (see Chapter 2)

$$i = \frac{V}{R}(1 - e^{-Rt/L})$$

which shows that the system is stable. Let us consider the magnetic coenergy stored in the magnetic field of the inductor. This is

$$W_m = \frac{1}{2}Li^2 = \frac{L}{2}\left(\frac{V}{R}\right)^2 (1 - e^{-Rt/L})^2$$

which is a positive quantity. Now examine the time derivative of this function. It is

$$\frac{dW_m}{dt} = \frac{L}{2} 2\left[\frac{V}{R}\right]^2 (1 - e^{-Rt/L})\left(\frac{R}{L}e^{-Rt/L}\right) = \left[\frac{Re^{-Rt/L}}{1 - e^{-Rt/L}}\right]W_m$$

This shows that the time derivative of the energy is always decreasing.

STATE VARIABLES AND STATE EQUATIONS 677

The Liapunov generalization is essentially the following: Suppose that within some neighborhood $S(R)$ of the origin, a scalar function $V(\mathbf{x})$ of the state $\mathbf{x}(t)$ can be constructed such that $V(\mathbf{x})$ has continuous first partial derivatives and $V(\mathbf{x}) = 0$ when $\mathbf{x}$ equals the equilibrium state. Further, it is required that $V(\mathbf{x}) > 0$ for all $\mathbf{x}$'s other than the equilibrium state and that the time derivative of $V(\mathbf{x})$ is negative. Then the system represented by $V(\mathbf{x})$ is asymptotically stable. These results are specified by the Liapunov stability theorem.

Theorem 12.7 The given system is specified by $\dot{\mathbf{x}} = X(\mathbf{x})$, where $\dot{\mathbf{x}}_i$ is continuous in the state variable $\mathbf{x}_j$ for all $i, j = 1, 2, \ldots, n$. If there exists a $V(\mathbf{x})$ such that

    a. $V(\mathbf{x})$ has continuous first partial-derivatives with respect to $\mathbf{x}_i$.
    b. $V(\mathbf{x})$ is positive definite for all $\mathbf{x}_i > 0$ (see Appendix 2).
    c. $V(\mathbf{x}) \to \infty$ for $|\mathbf{x}| \to \infty$.

then

1. The system is stable with respect to $\mathbf{x}_i$ if there is a region $S(R)$ defined by $0 < \|\mathbf{x}_i\| < S(R)$, where $S(R)$ is a real positive constant such that in this region $-\dot{V}(\mathbf{x})$ is positive semidefinite; that is,

$$-\dot{V}(\mathbf{x}) = \sum_{i=1}^{n} \frac{\partial V}{\partial \mathbf{x}_i} \dot{\mathbf{x}}_i \geq 0 \qquad \text{for all } (\mathbf{x}_i, t > 0)$$

2. The system is asymptotically stable with respect to $\mathbf{x}_i$ if, in $S(R)$, $-\dot{V} > 0$, $\mathbf{x} \neq 0$; that is, $-\dot{V}$ is positive definite.

3. The system is asymptotically stable in the large if Condition 2 is satisfied and if $S(R)$ is the whole of state space.

This theorem gives a set of sufficiency conditions for stability; thus if we can find a positive definite function which is nonincreasing, the origin state is thereby proved to be stable. However, if some arbitrarily selected positive definite function $V(\mathbf{x})$ is increasing for some motion of the state vector, no conclusions can be drawn. This means that if a proposed Liapunov function fails to satisfy the requirements of the stability theorem, no valid conclusions are possible since another choice of Liapunov function may satisfy the stability requirements. Hence a difficulty in the use of the Liapunov stability theorem is in the construction of a suitable $V(\mathbf{x})$ for a given system. Formal methods exist for generating suitable Liapunov functions for linear systems, and certain classes of nonlinear problems lend themselves to formal procedures in finding suitable $V(\mathbf{x})$ functions. However, generating suitable $V(\mathbf{x})$ functions in the general nonlinear case is not yet possible.

EXAMPLE 12.30
A simple nonlinear system is described by the differential equation

$$\frac{d^2x}{dt^2} + a\frac{dx}{dt} + bx + cx^2 = 0 \qquad a > 0 \quad b > 0 \quad c > 0$$

Discuss the system stability in the neighborhood of $x = 0$ **(a)** from considerations of the differential equation and **(b)** by the Liapunov theorem.

Solution: (a) In the neighborhood of $x = 0$ the differential equation approximates to

$$\frac{d^2x}{dt^2} + a\frac{dx}{dt} + bx = 0$$

This is a simple second-order linear equation, which we know to be asymptotically stable.

(b) Transform the differential equation to normal form. This becomes

$$\dot{x} = y$$
$$\dot{y} = -ay - bx - cx^2$$

Define as a possible Liapunov function

$$V(x, y) = \tfrac{1}{2}(bx^2 + y^2 + 2\beta xy) \qquad \beta > 0$$

$V(x, y)$ is positive definite, as required. From this we write

$$\dot{V} = \frac{\partial V}{\partial x}\dot{x} + \frac{\partial V}{\partial y}\dot{y} = (bx + \beta y)\dot{x} + (y + \beta x)\dot{y}$$
$$= (bx + \beta y)y - (y + \beta x)(ay + bx + cx^2)$$
$$= -(a - \beta)y^2 - a\beta xy - cyx^2 - b\beta x^2 - c\beta x^3$$

Clearly, for $\beta < a$, $\dot{V}(x, y) < 0$. Thus, $V(x, y)$ is a Liapunov function, and by the Liapunov theorem, the state at the origin is asymptotically stable. ∎

■ ■ ■

12-19 GENERATING LIAPUNOV FUNCTIONS

It was noted above that an explicit method exists for generating a Liapunov function for a linear system. To see how this is accomplished, consider the system described by

$$\dot{\mathbf{x}} = \mathbf{A}\mathbf{x} \qquad (12.174)$$

Let us select as a Liapunov function

$$V(\mathbf{x}) = \mathbf{x}^T \mathbf{E} \mathbf{x} \qquad (12.175)$$

where $\mathbf{E}$ is a matrix to be determined. The time derivative of V is

$$\dot{V} = \mathbf{x}^T \mathbf{E} \dot{\mathbf{x}} + \dot{\mathbf{x}}^T \mathbf{E} \mathbf{x} \qquad (12.176)$$

STATE VARIABLES AND STATE EQUATIONS 679

Combine this equation with (12.174) to obtain

$$\dot{V} = \mathbf{x}^T \mathbf{E} \mathbf{A} \mathbf{x} + (\mathbf{A}\mathbf{x})^T \mathbf{E} \mathbf{x} \qquad (12.177)$$

But since $(\mathbf{A}\mathbf{x})^T = \mathbf{x}^T \mathbf{A}^T$, then

$$\dot{V} = \mathbf{x}^T (\mathbf{E}\mathbf{A} + \mathbf{A}^T \mathbf{E}) \mathbf{x} \qquad (12.178)$$

Since $V(\mathbf{x})$ is to be positive definite and for asymptotic stability $-\dot{V}$ must also be positive definite, we can define a vector $\mathbf{F}$ such that

$$\dot{V} = -\mathbf{x}^T \mathbf{F} \mathbf{x} \qquad (12.179)$$

Then

$$-\mathbf{F} = \mathbf{E}\mathbf{A} + \mathbf{A}^T \mathbf{E} \qquad (12.180)$$

Thus for asymptotic stability of a linear system, it is sufficient that $\mathbf{F}$ be positive definite. The necessity portion of the proof relies on several theorems that we have not discussed.

EXAMPLE 12.31

Consider a second-order system defined by

$$\frac{d}{dt}\begin{bmatrix} x_1 \\ x_2 \end{bmatrix} = \begin{bmatrix} a_{11} & a_{12} \\ a_{21} & a_{22} \end{bmatrix} \begin{bmatrix} x_1 \\ x_2 \end{bmatrix}$$

Specify the conditions on $\mathbf{A}$ and $\mathbf{E}$ to satisfy (12.180).

Solution: Since $\mathbf{F}$ is an arbitrary symmetric positive definite matrix, we choose it to be the unity matrix, $\mathbf{F} = \mathbf{I}$. Equation (12.180) becomes

$$\begin{bmatrix} e_{11} & e_{12} \\ e_{21} & e_{22} \end{bmatrix} \begin{bmatrix} a_{11} & a_{12} \\ a_{21} & a_{22} \end{bmatrix} + \begin{bmatrix} a_{11} & a_{21} \\ a_{12} & a_{22} \end{bmatrix} \begin{bmatrix} e_{11} & e_{12} \\ e_{21} & e_{22} \end{bmatrix} = \begin{bmatrix} -1 & 0 \\ 0 & -1 \end{bmatrix}$$

This matrix equation is expanded and rearranged to the following form

$$\begin{bmatrix} 2a_{11} & 2a_{21} & 0 \\ a_{12} & a_{11} + a_{22} & a_{21} \\ 0 & 2a_{22} & 2a_{22} \end{bmatrix} \begin{bmatrix} e_{11} \\ e_{12} \\ e_{22} \end{bmatrix} = \begin{bmatrix} -1 \\ 0 \\ -1 \end{bmatrix}$$

These three equations can be solved simultaneously to yield the elements of the matrix $\mathbf{E}$,

$$\begin{bmatrix} e_{11} & e_{12} \\ e_{21} & e_{22} \end{bmatrix} = \frac{-1}{2(\text{Tr } \mathbf{A}) \Delta(\mathbf{A})}$$

$$\times \begin{bmatrix} \Delta(\mathbf{A}) + a_{21}^2 + a_{22}^2 & -(a_{12}a_{22} + a_{21}a_{11}) \\ -(a_{12}a_{22} + a_{21}a_{11}) & \Delta(\mathbf{A}) + a_{11}^2 + a_{12}^2 \end{bmatrix}$$

where $\text{Tr } \mathbf{A} = a_{11} + a_{22}$ is the sum of the diagonal of matrix $\mathbf{A}$. Matrix $\mathbf{E}$ is positive definite if each element on the diagonal is positive. This requires that

$$e_{11} = -\frac{\Delta(\mathbf{A}) + a_{21}^2 + a_{22}^2}{2(\text{Tr }\mathbf{A})\,\Delta(\mathbf{A})} > 0 \qquad e_{22} = -\frac{\Delta(\mathbf{A}) + a_{11}^2 + a_{12}^2}{2(\text{Tr }\mathbf{A})\,\Delta(\mathbf{A})} > 0$$

$$e_{12} = \frac{a_{12}a_{22} + a_{21}a_{11}}{2(\text{Tr }\mathbf{A})\Delta(\mathbf{A})} > 0 \qquad \Delta(E) = \frac{(a_{11} + a_{22})^2 + (a_{12} - a_{21})^2}{2\text{Tr }\mathbf{A}\,(\Delta\mathbf{A})^2} > 0$$

Consider the expression for e_{11}. Since the numerator of the inequality is always positive, it requires that $\Delta(\mathbf{A})$ be positive; that is,

$$\Delta(\mathbf{A}) = a_{11}a_{22} - a_{12}a_{21} > 0$$

and consequently,

$$\text{Tr }\mathbf{A} = a_{11} + a_{22} < 0$$

These two expressions give the required conditions for the stability of the second-order system. ∎

The problem of establishing necessary and sufficient conditions for the stability of nonlinear systems is still only partially resolved. Lur'e and Letov have shown that for particular classes of nonlinear systems for which the system equations can be expressed in one of two canonical forms, a suitable form of a Liapunov function is known. This is an important step since many control system problems can be arranged in the prescribed canonical form. Other important methods exist for generating Liapunov functions and for studying the stability of nonlinear systems. Pursuing this matter is beyond the scope of this book.

REFERENCES

1. Gabel, R. A., and R. A. Roberts, *Signals and Linear Systems*. 2d ed. New York: Wiley, 1980.
2. Lewis, J. B. *Analysis of Linear Dynamic Systems*. Champaign, Ill.: Matrix, 1977.
3. Liapunov, A. "Problème general de la stabilité du movement." In *Annals of Mathematical Studies Number 17*, Princeton, N.J.: Princeton University Press, 1949.
4. Seely, S. *An Introduction to Engineering Systems*. New York: Pergamon Press, 1972.
5. Seely, S., and A. D. Poularikas. *Electrical Engineering—Introduction and Concepts*. Beaverton, Or.: Matrix, 1982.
6. Swisher, G. M. *Introduction to Linear Systems Analysis*. Champaign, Ill.: Matrix, 1976.

PROBLEMS

12-2.1 Deduce state models for the circuits shown in Figure P12-2.1.

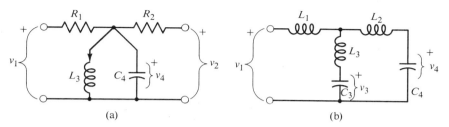

Figure P12-2.1

12-2.2 Deduce state models for the systems shown in Figure P12-2.2.

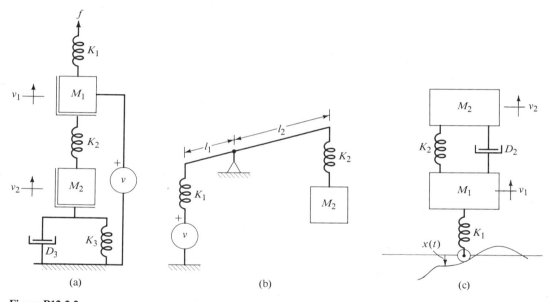

Figure P12-2.2

12-3.1 Write each of the following differential equations in normal form. Specify the **A** and **B** matrices for each. Show an appropriate signal flow graph for each ($p^n = d^n/dt^n$).
  a. $(p^2 + 5p + 2)y = w$
  b. $(3p^2 + 2p + 1)y = w + 2$
  c. $(p^3 + 2p^2 + 4p - 3)y = pu$

12-3.2 Deduce the **A, B, C, D** matrices from the normal form of the differential equation given. y denotes the output and $p^n = d^n/dt^n$

$(p^3 + 2p^2 + 3p + 1)y = (1 + p)x$

12-4.1 Refer to the circuit given in Figure P12-4.1.
  a. Write the differential equations for i_1 and i_2.
  b. Express these in normal form and deduce the appropriate **A** and **B** matrices.
  c. Deduce a state model for this network.

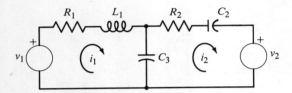

Figure P12-4.1

12-4.2 A system is defined by the state equations given in (12.30) where

$$\mathbf{A} = \begin{bmatrix} 0 & 1 \\ 2 & 3 \end{bmatrix} \quad \mathbf{B} = \begin{bmatrix} 0 \\ 1 \end{bmatrix} \quad \mathbf{C} = [2 \quad 1] \quad \mathbf{D} = [0]$$

  a. Sketch an SFG of the system.
  b. Suppose that the state vector is changed by the transformation matrix $\mathbf{M} = \begin{bmatrix} 1 & 0 \\ 0 & 2 \end{bmatrix}$. Sketch the SFG of the transformed system.
  c. Which system contains fewer integrators?

12-5.1 Determine expressions for $\mathbf{A}^k$ by induction by evaluating $\mathbf{A}^2$, $\mathbf{A}^3$ for the following:

  a. $\mathbf{A} = \begin{bmatrix} \alpha & 0 \\ 0 & \alpha \end{bmatrix}$ **b.** $\mathbf{A} = \begin{bmatrix} 0 & \beta \\ \beta & 0 \end{bmatrix}$ **c.** $\mathbf{A} = \begin{bmatrix} \alpha & 1 \\ 0 & \alpha \end{bmatrix}$

Verify each by means of Section A2-3 of Appendix 2.

12-5.2 Show that (12.41) follows from (12.39).

12-5.3 Find the closed form expression for $\exp[\mathbf{A}t]$ for each of the following:

  a. $\mathbf{A} = \begin{bmatrix} 2 & 1 \\ 4 & 2 \end{bmatrix}$ **b.** $\mathbf{A} = \begin{bmatrix} 3 & 0 \\ 2 & 2 \end{bmatrix}$ **c.** $\mathbf{A} = \begin{bmatrix} -1 & 1 \\ 0 & -1 \end{bmatrix}$ **d.** $\mathbf{A} = \begin{bmatrix} 0 & 1 \\ -2 & -3 \end{bmatrix}$

12-5.4 a. Determine the conditions on **A** and **B** (both are $n \times n$ matrices) for which $e^{\mathbf{A}+\mathbf{B}} = e^{\mathbf{A}} e^{\mathbf{B}}$. Hint: Compare $e^{\mathbf{A}+\mathbf{B}}$ and $e^{\mathbf{A}} e^{\mathbf{B}}$ on a term-by-term basis in the series expansion for the exponentials [see (12.37)].
  b. Use these results to show that $[e^{\mathbf{A}}]^{-1} = e^{-\mathbf{A}}$.

12-6.1 Verify the properties of the fundamental matrix

$$\Phi(t) = e^{-t/2} \begin{bmatrix} \cos\frac{t}{2} + \sin\frac{t}{2} & 2\sin\frac{t}{2} \\ -\sin\frac{t}{2} & \cos\frac{t}{2} - \sin\frac{t}{2} \end{bmatrix}$$

where

$$\mathbf{A} = \begin{bmatrix} 0 & -1 \\ -\frac{1}{2} & -1 \end{bmatrix}$$

12-7.1 The inputs to the system in Example 12.6 are $w_1(t) = u(t)$ and $w_2(t) = 0$. The initial conditions are specified by $\mathbf{x}(0) = [1 \quad 0]^T$. Find the outputs $\mathbf{y}(t) = [i_L(t) \quad i_1(t)]^T$, and the impulse response of the system.

12-8.1 Repeat Problem 12-5.3 using Laplace transform methods.

12-8.2 Determine the inverse of the matrix $[s\mathbf{I} - \mathbf{A}]$, where

$$\mathbf{A} = \begin{bmatrix} 3 & 2 & 7 \\ 1 & 5 & 6 \\ -4 & -1 & -3 \end{bmatrix}$$

  a. By direct expansion using Laplace methods.
  b. Using the algorithm given in Appendix 2.

12-8.3 Given a system matrix

$$\mathbf{A} = \begin{bmatrix} 0 & 1 \\ -2 & -3 \end{bmatrix}$$

  a. Find the eigenvalues of the system.
  b. Find a transformation matrix $\mathbf{M}$ such that $\mathbf{MAM}^{-1}$ is the diagonal matrix $\mathbf{\Lambda} = \begin{bmatrix} -1 & 0 \\ 0 & -2 \end{bmatrix}$

12-8.4 Refer to the circuit of Figure P12-8.4.
  a. Write the differential equation that relates i to v.
  b. Draw an SFG of this network.
  c. Deduce a state model from the differential equation.
  d. Deduce a state model from the circuit.
  e. Show that the eigenvalues obtained from (c) and those in (d) are the same.

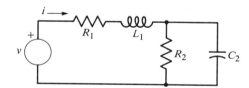

Figure P12-8.4

12-9.1 Determine the matrix $[j\omega\mathbf{I} - \mathbf{A}]^{-1}$ for each $\mathbf{A}$ in Problem 12-5.3.

12-11.1 Write the difference equation descriptions of the block diagram representations shown in Figure P12-11.1.

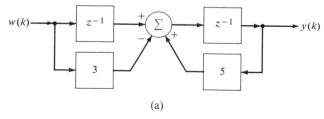

(a)

Figure P12-11.1a

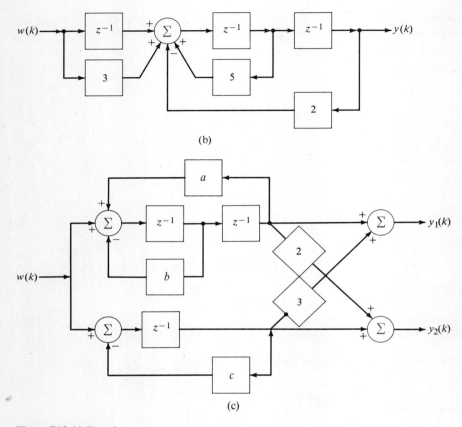

(b)

(c)

Figure P12-11.1b and c

12-11.2 A discrete time system is specified by the difference equation
$$2y(k+2) + 5y(k+1) + y(k) = w(k)$$
a. Express the equation in state form.
b. Draw a block diagram of the system.

12-11.3 A discrete time system is specified by the difference equation
$$3y(k+2) + 2y(k+1) + 5y(k) = 2w(k+1) + 3w(k)$$
a. Give an SFG representation of this equation.
b. Express the system description in state variable form.

12-11.4 A system is characterized by the simultaneous difference equations
$$2y_1(k+1) + 3y_1(k) - 5y_2(k) = w_1(k)$$
$$y_1(k) + 3y_2(k+2) + y_2(k) = w_2(k)$$
a. Express this system in state form.
b. Draw an SFG for the system.

12-11.5 Obtain a numerical solution of the state equations for k from 1 to 10. Sketch the results.
$$\begin{bmatrix} x_1(k+1) \\ x_2(k+1) \\ x_3(k+1) \end{bmatrix} = \begin{bmatrix} 2 & -1 & -1 \\ 1 & 0 & -1 \\ 0 & 0 & 1 \end{bmatrix} \begin{bmatrix} x_1(k) \\ x_2(k) \\ x_3(k) \end{bmatrix} \quad \text{for } \mathbf{x}(0) = \begin{bmatrix} 1 \\ 1 \\ 1 \end{bmatrix}$$

12-12.1 The system equations of a discrete system are

$$\begin{bmatrix} x_1(k+1) \\ x_2(k+1) \\ x_3(k+1) \end{bmatrix} = \begin{bmatrix} 0 & 0 & 4 \\ 1 & 0 & -8 \\ 0 & 1 & 5 \end{bmatrix} \begin{bmatrix} x_1(k) \\ x_2(k) \\ x_3(k) \end{bmatrix} + \begin{bmatrix} 4 \\ -8 \\ 5 \end{bmatrix} w(k)$$

$$y(k) = \begin{bmatrix} 0 & 0 & 1 \end{bmatrix} \begin{bmatrix} x_1(k) \\ x_2(k) \\ x_3(k) \end{bmatrix} + [1] w(k)$$

Determine the following expressions: (a) $f(A) = 2^A$; (b) $f(A) = A^k$ (c) $f(A) = e^{-A}$ (d) the impulse response of the system.

12-12.2 Find the expressions:
  a. $f(A) = A^k$
  b. $f(A) = e^{At}$
  c. $f(A) = e^{-2A}$ if

$$A = \begin{bmatrix} 0 & -2 \\ 1 & 3 \end{bmatrix}$$

12-14.1 The SFG of a specified discrete time system is given in Figure P12-14.1.
  a. Write the difference equation of the system.
  b. Deduce a state description.
  c. Determine the fundamental matrix $\Phi(z)$.

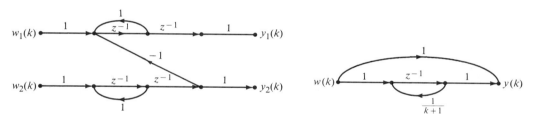

Figure P12-14.1

Figure P12-14.2

12-14.2 A system is described by the SFG shown in Figure P12-14.2.
  a. Determine the fundamental matrix of the system.
  b. Give the unit step response of the system.
  c. Specify the input-output transmittance $H(z)$.

12-16.1 Systems are characterized by the following matrix sets. Ascertain whether the systems are controllable and whether they are observable.

  a. $A = \begin{bmatrix} 1 & 1 \\ 0 & -1 \end{bmatrix}$ $B = \begin{bmatrix} 1 \\ -2 \end{bmatrix}$ $C = \begin{bmatrix} 1 & 0 \end{bmatrix}$

  b. $A = \begin{bmatrix} -2 & -2 & -1 \\ 0 & 0 & 1 \\ -1 & -1 & 2 \end{bmatrix}$ $B = \begin{bmatrix} 0 \\ 0 \\ 1 \end{bmatrix}$ $C = \begin{bmatrix} 1 & 0 & 0 \end{bmatrix}$

12-17.1 Consider the system given in the SFG of Figure P12-17.1.
  a. Deduce the state equations of the system.
  b. Determine the level of stability.

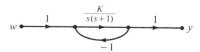

Figure P12-17.1

12-17.2 Ascertain the level of stability of systems described by

    a. $\dot{\mathbf{x}} = \begin{bmatrix} 1 & -1 \\ -1 & 1 \end{bmatrix} \mathbf{x} + \begin{bmatrix} 1 \\ 0 \end{bmatrix} \mathbf{w}$ **b.** $\mathbf{x}(k+1) = \begin{bmatrix} 1 & -\frac{1}{2} \\ -2 & 1 \end{bmatrix} \mathbf{x}(k) + \begin{bmatrix} 1 \\ 0 \end{bmatrix} \mathbf{w}(k)$

12-17.3 Systems are described by the following **A** matrices. Are they stable?

    a. $\mathbf{A} = \begin{bmatrix} 0 & 1 & 0 \\ 0 & 0 & 1 \\ 1.6 & -0.8 & 3 \end{bmatrix}$ **b.** $|s\mathbf{I} - \mathbf{A}| = 4s^3 + 21s^2 + 3s + 3 = 0$

12-18.1 A system is described by the state equations
$$\dot{x}_1 = x_2$$
$$\dot{x}_2 = -ax_2 - bx_1 - x_1^2 \qquad a > 0 \quad b > 0$$
Define as a possible Liapunov function
$$V(x_1, x_2) = \tfrac{1}{2}[bx_1^2 + 2\beta x_1 x_2 + x_2^2] \qquad \beta > 0$$
Is the origin stable? If so, establish the level of stability.

12-18.2 A system is described by the state equation
$$\frac{d}{dt}\begin{bmatrix} x_1 \\ x_2 \end{bmatrix} = \begin{bmatrix} 0 & 1 \\ -1 & 0 \end{bmatrix}\begin{bmatrix} x_1 \\ x_2 \end{bmatrix} - a\begin{bmatrix} x_1 & x_1 \\ x_2 & x_2 \end{bmatrix}\begin{bmatrix} x_1^2 \\ x_2^2 \end{bmatrix} \qquad a > 0$$
A possible Liapunov function is
$$V(x_1, x_2) = x_1^2 + x_2^2$$
Is the system stable? If so, establish the level of stability.

12-18.3 The SFG of a simple feedback system known to be asymptotically stable is given in Figure P12-17.1.
    a. Set up state equations for the system.
    b. Two possible Liapunov functions are suggested
$$V(x_1, x_2) = x_1^2 + x_2^2 \qquad V(x_1, x_2) = Kx_1^2 + (x_2 + ax_1)^2 \qquad a > 0$$
    Ascertain whether either is a suitable Liapunov function.

12-18.4 A system is described by the state equation
$$\frac{d}{dt}\begin{bmatrix} x_1 \\ x_2 \end{bmatrix} = \begin{bmatrix} 1 & -1 \\ -1 & 1 \end{bmatrix}\begin{bmatrix} x_1 \\ x_2 \end{bmatrix} + \begin{bmatrix} w \\ 0 \end{bmatrix}$$
Ascertain whether the system is stable.

12-18.5 Investigate the stability of the system described by the state equation
$$\frac{d}{dt}\begin{bmatrix} x_1 \\ x_2 \end{bmatrix} = \begin{bmatrix} 0 & 1 \\ -\alpha & -\beta \end{bmatrix}\begin{bmatrix} x_1 \\ x_2 \end{bmatrix} + \begin{bmatrix} w \\ 0 \end{bmatrix}$$
Select as a possible Liapunov function
$$V(x_1, x_2) = x_1^2 + x_2^2$$

12-18.6 An input-free system is characterized by the matrix
$$\mathbf{A} = \begin{bmatrix} 0 & 1 & 0 \\ 0 & 0 & 1 \\ 1.6 & -0.8 & 3 \end{bmatrix}$$
Apply the Liapunov test to determine whether or not this system is stable.

CHAPTER APPENDIXES

CHAPTER APPENDIX 1-1

THE DELTA FUNCTION AND ITS PROPERTIES

A1-1 DELTA FUNCTION AS A LIMIT OF SPECIAL SEQUENCES

One of the accepted definitions of the delta function is associated with sequences of functions whose limit functions behave like a delta function. Consider the function

$$f_\epsilon(t) = \frac{1}{\pi} \frac{\epsilon}{t^2 + \epsilon^2} \qquad \epsilon > 0 \tag{A1-1.1}$$

The form of this function is shown in Figure A1-1.1a for two different values of the parameter ϵ. It is apparent that as $\epsilon \to 0$, the height at $t = 0$ will become infinite. This is easily seen by applying l'Hospital's rule to the function. In addition

$$\lim_{\epsilon \to 0} \int_{-a}^{a} f_\epsilon(t)\,dt = \lim_{\epsilon \to 0} \frac{1}{\pi} \left[\tan^{-1} \frac{a}{\epsilon} - \tan^{-1} \frac{(-a)}{\epsilon} \right] = 1 \tag{A1-1.2}$$

Figure A1-1.1
(a) The function $f_\epsilon(t) = (\epsilon/\pi)/(t^2 + \epsilon^2)$ for two different values of ϵ.
(b) The derivative of $f_\epsilon(t)$ for two different values of ϵ.

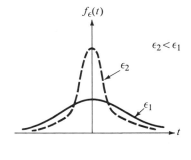

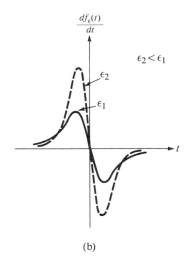

(a)

(b)

which shows that the limiting function satisfies the definition of the delta function, (1.15b). Another property follows from the fact that if in the limit $\epsilon \to 0$ $f_\epsilon(t) \to \delta(t)$, we have

$$\lim_{\epsilon \to 0} \frac{df_\epsilon(t)}{dt} = \lim_{\epsilon \to 0} \left(-\frac{2}{\pi} \frac{\epsilon t}{(t^2 + \epsilon^2)^2} \right) = \frac{d\delta(t)}{dt} \qquad \text{(A1-1.3)}$$

Figure A1-1.1b shows a graph of the derivative of $f_\epsilon(t)$ for two values of ϵ. From Figure A1-1.1a we also conclude that while the delta function is **even**, its derivative is **odd**.

Table A1-1.1 shows some additional sequences that represent delta functions.

Table A1-1.1 Sequences leading to delta functions

$\dfrac{1}{a} u(t) - \dfrac{1}{a} u(t - a)$	$a \to 0, \quad a \geq 0$
$ae^{-at} u(t)$	$a \to \infty, \quad a > 0$
$\dfrac{1}{2\sqrt{\pi a}} e^{-t^2/4a}$	$a \to 0, \quad a \geq 0$
$\dfrac{1}{\pi} \dfrac{\sin at}{t}$	$a \to \infty, \quad a > 0$

■ ■ ■

A1-2 PROPERTIES OF DELTA FUNCTIONS

Although the proofs of some of these properties are presented below, many of the special properties of the δ-function are presented without proof. Some important properties of the δ-function follow:

1. **Scaling Properties**

$$\delta\left(\frac{t - t_0}{a}\right) = |a| \delta(t - t_0) \qquad \text{(a)}$$

$$\delta(at - t_0) = \frac{1}{|a|} \delta\left(t - \frac{t_0}{a}\right) \qquad \text{(b)} \qquad \text{(A1-1.4)}$$

$$\delta(-t + t_0) = \delta(t - t_0) \qquad \text{(c)}$$

$$\delta(-t) = \delta(t) \qquad \text{(d)}$$

THE DELTA FUNCTION AND ITS PROPERTIES

EXAMPLE A1-1.1
Prove the equality specified in (A1-1.4a).

Solution: From the definition of the delta function given in (1.15), we can write

$$\int_{-\infty}^{\infty} f(t) \delta\left(\frac{t - t_0}{a}\right) dt = |a| \int_{-\infty}^{\infty} f(\tau a) \delta\left(\tau - \frac{t_0}{a}\right) d\tau$$

$$= |a| f(\tau a)|_{\tau = t_0/a} = |a| f(t_0)$$

where we have set $t = a\tau$. It follows therefore that

$$\int_{-\infty}^{\infty} f(t) |a| \delta(t - t_0) dt = |a| f(t_0)$$

The equality given in (A1-1.4a) follows by comparing the initial and final forms. ■

Observe that the scaling properties given by (A1-1.4c) and (A1-1.4d) show that the delta function is an even function. This property was also shown (see Figure A1-1) to be valid for the sequences of functions that in the limit behave exactly as the delta function.

2. **Integral Properties**

$$\int_{-\infty}^{\infty} A \delta(t - t_0) dt = A \quad \text{(a)}$$

$$\int_{-\infty}^{\infty} \delta(\tau - t_1) \delta(t - \tau - t_2) d\tau = \delta[t - (t_1 + t_2)]$$

$$= \delta(t - t_1) * \delta(t - t_2) \quad \text{(b)} \quad \text{(A1-1.5)}$$

$$\int_{-\infty}^{\infty} f(t - \tau) \delta(\tau) d\tau = f(t) * \delta(t) = f(t) \quad \text{(c)}$$

This latter expression has been written in convolution form, where the general form of the convolution of two functions is defined by

$$f_1(t) * f_2(t) \triangleq \int_{-\infty}^{\infty} f_1(\tau) f_2(t - \tau) d\tau = \int_{-\infty}^{\infty} f_1(t - \tau) f_2(\tau) d\tau$$

The importance of convolution is explored extensively in Chapter 2.

3. **Derivative Properties**

$$\frac{d^n \delta(t - t_0)}{dt^n} = 0 \quad t \neq t_0 \quad \text{(a)}$$

$$\int_{-\infty}^{\infty} f(t) \frac{d \delta(t - t_0)}{dt} dt = -\frac{df(t_0)}{dt} \quad \text{(b)}$$

$$\int_{-\infty}^{\infty} f(t) \frac{d^n \delta(t - t_0)}{dt^n} dt = (-1)^n \frac{d^n f(t_0)}{dt^n} \quad \text{(c)}$$

$$\int_{-\infty}^{\infty} \frac{d\,\delta(t)}{dt}\,dt = 0 \tag{d}$$

$$\frac{d\,\delta(t)}{dt} * f(t) = \int_{-\infty}^{\infty} f(t-\tau)\frac{d\,\delta(\tau)}{d\tau}\,d\tau = \frac{df(t)}{dt} \tag{e}$$

$$\frac{d^n\,\delta(-t)}{dt^n} = (-1)^n \frac{d^n\,\delta(t)}{dt^n} \tag{f}$$

$$t\,\frac{d\,\delta(t)}{dt} = -\delta(t) \tag{g}$$

$$t^2\,\frac{d\,\delta(t)}{dt} = 0 \tag{h}$$

$$f(t)\,\frac{d\,\delta(t)}{dt} = f(0)\,\frac{d\,\delta(t)}{dt} - \frac{df(0)}{dt}\,\delta(t) \tag{i}$$

(A1-1.6)

EXAMPLE A1-1.2
Prove (A1-1.6b).

Solution: Perform the following steps by a change of variable and integration by parts:

$$\int_{-\infty}^{\infty} f(t)\frac{d\,\delta(t-t_0)}{dt}\,dt = \int_{-\infty}^{\infty} f(\tau+t_0)\frac{d\,\delta(\tau)}{d\tau}\,d\tau$$

$$= f(\tau+t_0)\delta(\tau)\bigg|_{\tau=-\infty}^{\infty} - \int_{-\infty}^{\infty} \frac{df(\tau+t_0)}{d\tau}\delta(\tau)\,d\tau$$

$$= -\frac{df(\tau+t_0)}{d\tau}\bigg|_{\tau=0} \triangleq -\frac{df(t+t_0)}{dt}\bigg|_{t=0}$$

$$= \frac{-df(t_0)}{dt}$$

∎

EXAMPLE A1-1.3
Evaluate the function given to show that (A1-1.6i) is an identity.

$$\int_{-\infty}^{\infty} \varphi(t)f(t)\frac{d\,\delta(t)}{dt}\,dt$$

Solution: Proceed by integrating by parts. This leads to

$$\int_{-\infty}^{\infty} \varphi(t)f(t)\frac{d\,\delta(t)}{dt}\,dt = \varphi(t)f(t)\delta(t)\bigg|_{-\infty}^{\infty}$$

$$-\int_{-\infty}^{\infty} \delta(t)\left[\frac{d\varphi(t)}{dt}f(t) + \varphi(t)\frac{df(t)}{dt}\right]dt$$

THE DELTA FUNCTION AND ITS PROPERTIES

$$= -f(0)\frac{d\varphi(0)}{dt} - \frac{df(0)}{dt}\varphi(0)$$

Also, from (A1-1.6i)

$$\int_{-\infty}^{\infty} \varphi(t)f(0)\frac{d\,\delta(t)}{dt}dt - \int_{-\infty}^{\infty} \varphi(t)\frac{df(0)}{dt}\delta(t)\,dt = -f(0)\frac{d\varphi(0)}{dt} - \frac{df(0)}{dt}\varphi(0)$$

which shows that (A1-1.6i) is an identity. ∎

EXAMPLE A1-1.4
Prove (A1-1.6e).

Solution: Proceed in the manner shown:

$$\frac{d\,\delta(t)}{dt} * f(t) = \int_{-\infty}^{\infty} f(t-\tau)\frac{d\,\delta(\tau)}{d\tau}d\tau$$

$$= f(t-\tau)\delta(\tau)\Big|_{-\infty}^{\infty} - \int_{-\infty}^{\infty} \delta(\tau)\frac{df(t-\tau)}{d\tau}d\tau$$

$$= -\int_{\infty}^{-\infty} \delta(t-\lambda)\frac{df(\lambda)}{d\lambda}d\lambda = \int_{-\infty}^{\infty} \delta(t-\lambda)\frac{df(\lambda)}{d\lambda}d\lambda$$

$$= \int_{-\infty}^{\infty} \delta(\lambda-t)\frac{df(\lambda)}{d\lambda}d\lambda = \frac{df(t)}{dt}$$

where we have written $\tau = t - \lambda$. ∎

4. General Argument

$$\delta[\alpha(t)] = 0 \qquad \alpha(t) \neq 0 \qquad \text{(a)}$$

$$\int_{-\infty}^{\infty} f(t)\,\delta[\alpha(t)]\,dt = \sum_{n} \frac{f(t_n)}{\left|\dfrac{d\alpha(t_n)}{dt}\right|} \qquad \frac{d\alpha(t_n)}{dt} \neq 0 \quad \text{(b)} \qquad \text{(A1-1.7)}$$

where t_n is defined by $\alpha(t_n) = 0$, the points at which the function crosses the t-axis.

To see that (A1-1.7) is true, consider the particular function $\alpha(t) = -2t + 1$ as the independent variable of the delta function. This function and its inverse $t(x) = (1 - x)/2$ are plotted in Figures A1-1.2a and b within the region $[0, 1]$. It then follows that

$$\int_0^1 \delta(-2t+1)\,dt = \int_1^{-1} \delta(x)\frac{dx}{\left(\dfrac{dx}{dt}\right)} = -\int_{-1}^1 \delta(x)\frac{dx}{-2}$$

$$= \int_{-1}^1 \delta(x)\frac{dx}{|-2|} = \frac{1}{2} \triangleq \frac{1}{|\alpha^{(1)}(\frac{1}{2})|}$$

Figure A1-1.2
A delta function with a general argument.

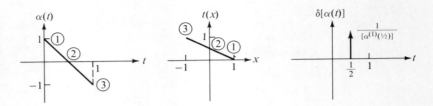

where $t = 1/2$ is the point at which $\alpha(t)$ crosses the t-axis and where $\alpha^{(1)}(t)$ indicates the first-order derivative of $\alpha(t)$. Observe from Figure A1-1.2b that the inverse function $t(x)$ has a negative slope also, but the limits reverse. Therefore we can reverse the order of integration and take the absolute value of the derivative. However, when the function is positive monotone, its inverse has the same slope as the function but the limits do not reverse. Hence (A1-1.7) is the correct expression for both cases.

It is further noted that since

$$\int_0^1 f(t) \tfrac{1}{2} \delta(t - \tfrac{1}{2})\, dt = \tfrac{1}{2} f(\tfrac{1}{2}) \triangleq \int_0^1 f(t)\, \delta(-2t + 1)\, dt$$

it means that $\delta[\alpha(t)]$ is equivalently

$$\delta[\alpha(t)] \triangleq \sum_n \frac{\delta(t - t_n)}{|\alpha^{(1)}(t_n)|} \qquad \text{(a)} \qquad \text{(A1-1.8)}$$

We also have

$$\frac{d\,\delta(t)}{dt} \triangleq \sum_n \frac{1}{\alpha^{(1)}(t)\,|\alpha^{(1)}(t)|} \frac{d\,\delta(t - t_n)}{dt} \qquad \text{(b)}$$

and

$$\delta(t^2 - a^2) = (2|a|)^{-1}[\delta(t - |a|) + \delta(t + |a|)] \qquad \text{(c)}$$

PROBLEMS

A1-2.1 Evaluate the following integrals

a. $\displaystyle\int_0^{10} (t^3 + 4t - 1)\,\delta(t - 1)\, dt$
b. $\displaystyle\int_0^{10} (t^3 + 4t - 1) \frac{d\,\delta(t - 2)}{dt}\, dt$
c. $\displaystyle\int_{-\infty}^{\infty} e^{-t} u(t) \frac{d\,\delta(t - 2)}{dt}\, dt$

d. $\displaystyle\int_{-3}^{3} e^{-t} \delta\!\left(\frac{t - 1}{2}\right) dt$
e. $\displaystyle\int_{-\infty}^{\infty} e^{-t^2}\,\delta(2t + 3)\, dt$

A1-2.2 Verify (A-1.6c)–(A1-1.6h).

A1-2.3 Evaluate the following integrals

a. $\displaystyle\int_{-\infty}^{\infty} p_4(t)\,\delta(\cos 3t)\, dt$
b. $\displaystyle\int_{-4}^{4} t\,\delta[\sin(t - 1)]\, dt$
c. $\displaystyle\int_{-\infty}^{\infty} p_{10}(t)\,\delta[\operatorname{sinc}_2(t)]\, dt$

CHAPTER APPENDIX 1-2

BINARY AND GRAY CODE RELATIONSHIPS

The Gray code is a unit distance code, which means that in going from the coding for any integer I to the coding for $I + 1$ only one bit changes regardless of the value of I. The four-bit Gray code and its conversion to binary are contained in Table A1-2.1.

Table A1-2.1 Gray Code and Conversion to Binary

I	Gray Code				Binary Code			
	g_8	g_4	g_2	g_1	b_8	b_4	b_2	b_1
0	0	0	0	0	0	0	0	0
1	0	0	0	1	0	0	0	1
2	0	0	1	1	0	0	1	0
3	0	0	1	0	0	0	1	1
4	0	1	1	0	0	1	0	0
5	0	1	1	1	0	1	0	1
6	0	1	0	1	0	1	1	0
7	0	1	0	0	0	1	1	1
8	1	1	0	0	1	0	0	0
9	1	1	0	1	1	0	0	1
10	1	1	1	1	1	0	1	0
11	1	1	1	0	1	0	1	1
12	1	0	1	0	1	1	0	0
13	1	0	1	1	1	1	0	1
14	1	0	0	1	1	1	1	0
15	1	0	0	0	1	1	1	1

Notice that in this table b_8 is the same as g_8; that is,

$$b_8 = g_8$$

Next, looking at b_4, it is seen that b_4 is the same as g_4 when $b_8 = 0$. However, if $b_8 = 1$ then b_4 is the same as the complement of g_4, designated by $\bar{g}_4$. That is

$$b_4 = \begin{cases} g_4 & \text{if } b_8 = 0 \\ \bar{g}_4 & \text{if } b_8 = 1 \end{cases}$$

This can be reexpressed using the Exclusive OR connective as

$$b_4 = g_4 \oplus b_8$$

Similarly, it is observed that b_2 is the same as g_2 if $b_4 = 0$, and is equal to g_2 if $b_4 = 1$. This leads to the Boolean equation

$$b_2 = g_2 \oplus b_4$$

A similar analysis leads to the final relation

$$b_1 = g_1 \oplus b_2$$

A gate network for accomplishing this conversion is shown in Figure A1-2.1. In this network it is implied that all inputs are available simultaneously and the outputs are generated simultaneously.

Another way of effecting the Gray-to-binary transformation is by starting from the most significant digit and then moving to the least significant digit. We set $b_i = g_i$ if the number of ones preceding g_i is even, and we set $b_i = \bar{g}_i$ (complement) if the number of ones preceding g_i is odd. The process is shown in Figure A1-2.2a. Figure A1-2.2b shows the converse or binary-to-Gray process.

The binary-to-Gray conversion is accomplished using the relation

$$g_i = b_i \oplus b_{i+1}$$

The most significant Gray digit is identical with the corresponding binary digit.

Figure A1-2.1 Gray-to-binary conversion network.

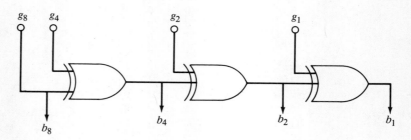

Figure A1-2.2 (a) Gray-to-binary conversion. (b) Binary-to-Gray conversion.

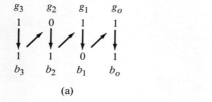
(a)

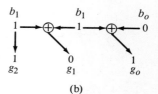
(b)

CHAPTER APPENDIX 4-1

BESSEL FUNCTIONS

A4-1 BESSEL FUNCTIONS

The Bessel function of the first kind of order n is given by the following expression, for positive integral n,

$$J_n(t) = \sum_{k=0}^{\infty} \frac{1}{k!} \frac{(-1)^k}{(k+n)!} \left(\frac{t}{2}\right)^{n+2k} \tag{A4-1.1}$$

with

$$J_n(0) = 0 \quad n \neq 0 \quad J_0(0) = 1$$

The following expressions relate different orders of Bessel functions

$$J_n(t) = (-1)^n J_{-n}(t) = J_n(-t) \quad \text{for } n = \text{integer} \quad \textbf{(a)}$$

$$\frac{2n}{t} J_n(t) = J_{n-1}(t) + J_{n+1}(t) \quad \textbf{(b)}$$

$$2 \frac{d}{dt} J_n(t) = J_{n-1}(t) - J_{n+1}(t) \quad \textbf{(c)} \tag{A4-1.2}$$

$$t \frac{d}{dt} J_n(t) = n J_n(t) - t J_{n+1}(t) \quad \textbf{(d)}$$

Important integral expressions involving Bessel functions include:

$$\int_0^t t J_n^2(\alpha t) \, dt = \frac{t^2}{2} \left[J_n^2(\alpha t) - J_{n+1}(\alpha t) J_{n-1}(\alpha t) \right]$$

$$\int_0^t t^n J_{n-1}(t) \, dt = t^n J_n(t)$$

Figure A4-1.1 shows the Bessel function of the two real variables n and t.

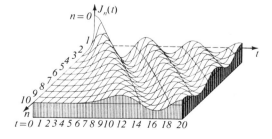

Figure A4-1.1
The Bessel functions $J_n(t)$ of the two variables t and n. (Reprinted, by permission, from Jahnke and Emde, *Tables of Functions*, 126.)

CHAPTER APPENDIX 7-1

THE ERROR FUNCTION

The **error function** is defined by the expressions

$$\text{erf}(t) = \frac{2}{\sqrt{\pi}} \int_0^t e^{-\tau^2} d\tau = \frac{2}{\sqrt{\pi}} \sum_{n=0}^{\infty} \frac{(-1)^n t^{2n+1}}{n!(2n+1)} \qquad \text{(A7-1.1)}$$

Further,

$$\int_0^{\infty} e^{-\tau^2} d\tau = \frac{\sqrt{\pi}}{2} \quad \text{(a)} \qquad \text{(A7-1.2)}$$

from which

$$\text{erf}(\infty) = 1 \quad \text{(b)}$$

Additionally, we easily find from (A7-1.1) that

$$\text{erf}(0) = 0 \quad \text{(a)}$$
$$\text{erf}(-t) = -\text{erf}(t) \quad \text{(b)} \qquad \text{(A7-1.3)}$$

A related function is the **complementary error function** which is defined by

$$\text{erfc}(t) = 1 - \text{erf}(t) = \frac{2}{\sqrt{\pi}} \int_t^{\infty} e^{-\tau^2} d\tau \qquad \text{(A7-1.4)}$$

Tables have been prepared which gives values of erf(t) for different values of t. Table A7-1.1 contains such values.

REFERENCES

1. Abramowitz, M., and I. A. Stegun. *Handbook of Mathematical Tables.* New York: Dover, 1965.
2. Korn, G. A., and T. M. Korn. *Mathematical Handbook for Scientists and Engineers.* New York: McGraw-Hill, 1961.
3. Gradshteyn, I. S., and I. M. Ryzhik. *Tables of Integrals, Series and Products.* New York: Academic Press, 1965.

THE ERROR FUNCTION

Table A7-1.1 $\text{erf}(t) = \dfrac{2}{\sqrt{\pi}} \int_0^t e^{-\tau^2} d\tau$

t	$\text{erf}(t)$	t	$\text{erf}(t)$
0.00	0.00000	0.65	0.64203
0.02	0.02256	0.70	0.67780
0.04	0.04511	0.75	0.71115
0.06	0.06762	0.80	0.74210
0.08	0.09008	0.85	0.77067
0.10	0.11246	0.90	0.79691
0.12	0.13476	0.95	0.82089
0.14	0.15695	1.00	0.84270
0.16	0.17901	1.05	0.86244
0.18	0.20093	1.10	0.88020
0.20	0.22270	1.15	0.89612
0.22	0.24429	1.20	0.91031
0.24	0.26570	1.25	0.92290
0.26	0.28690	1.30	0.93401
0.28	0.30788	1.35	0.94376
0.30	0.32863	1.40	0.95228
0.32	0.34913	1.45	0.95969
0.34	0.36936	1.50	0.96610
0.36	0.38933	1.55	0.97162
0.38	0.40901	1.60	0.97635
0.40	0.42839	1.65	0.98037
0.42	0.44747	1.70	0.98379
0.44	0.46622	1.75	0.98667
0.46	0.48465	1.80	0.98909
0.48	0.50275	1.85	0.99111
0.50	0.52050	1.90	0.99279
0.55	0.56332	1.95	0.99418
0.60	0.60386	2.00	0.99532

GENERAL APPENDIXES

GENERAL APPENDIX 1

FUNCTIONS OF A COMPLEX VARIABLE

The student who has studied Fourier transforms and Laplace transforms applied to network problems and, therefore, has become familiar with such term as system functions, proper fractions, poles, zeros, contour integrals, s-plane, and $H(s)$ plane has been introduced to many concepts from the theory of functions of a complex variable. The purpose of this Appendix is to place these concepts in the mathematical formalism of complex variable theory. The student should try to relate his previous studies to this more mathematical presentation.

■ ■ ■

A1-1 BASIC CONCEPTS

A complex variable z defined by

$$z = x + jy \tag{A1.1}$$

assumes certain values over a region R_z of the complex plane. If a complex quantity $W(z)$ is so connected with z that each z in R_z corresponds with one value of $W(z)$ in R_w, then we say that $W(z)$ is a single-valued function of z

$$W(z) = u(x, y) + jv(x, y) \tag{A1.2}$$

which has a **domain** R_z and a **range** R_w (see Figure A1.1). The function $W(z)$ can be **single-valued** or **multiple-valued**. Examples of single-valued functions include:

$$W = a_0 + a_1 z + a_2 z^2 + \cdots + a_n z^n \qquad n \text{ integral}$$

$$W = e^z$$

Examples of multi-valued functions are:

$$W = z^n \qquad n, \text{ not integral}$$

$$W = \log z$$

$$W = \sin^{-1} z$$

Definition A1.1 A function $W(z)$ is **continuous** at a point z of R_z if, for each number $\epsilon > 0$, there exists another number $\delta > 0$ such that whenever

Figure A1.1
Illustration of the range and domain of complex functions.

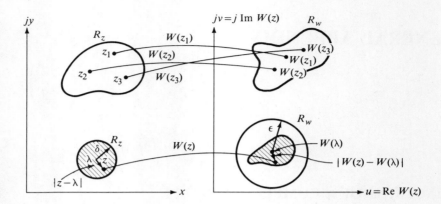

$$|z - \lambda| < \delta \quad \text{then} \quad |W(z) - W(\lambda)| < \epsilon \tag{A1.3}$$

The geometric representation of this equation is shown in Figure A1.1.

Definition A1.2 A function $W(z)$ is **analytic** at a point λ if, for each number $\epsilon > 0$ there exists another number $\delta > 0$ such that whenever

$$|z - \lambda| < \delta \quad \text{then} \quad \left| \frac{W(z) - W(\lambda)}{z - \lambda} - \frac{dW}{dz}\bigg|_{z=\lambda} \right| < \epsilon \tag{A1.4}$$

EXAMPLE A1.1
Show that the function $W(z) = e^z$ satisfies (A1.4)

Solution: From (A1.4) we obtain

$$\lim_{z \to \lambda} \frac{W(z) - W(\lambda)}{z - \lambda} = \lim_{z \to \lambda} \frac{e^z - e^\lambda}{z - \lambda} = \lim_{z \to \lambda} e^z \left[1 - \frac{(z - \lambda)}{2!} + \frac{(z - \lambda)^2}{3!} - \cdots \right]$$

$$= e^\lambda = \frac{de^z}{dz}\bigg|_{z=\lambda}$$

which proves the assertion. ∎

In this example we did not mention the direction from which z approaches λ. We might surmise from this that the derivative of our analytic function is independent of the path of z as it approaches the limiting point. However, this is not true in general. By setting $\lambda = z$ and $z = z + \Delta z$ in (A1.4), we obtain an alternative form of that equation; namely,

$$\frac{dW}{dz} = \lim_{\Delta z \to 0} \left\{ \frac{W(z + \Delta z) - W(z)}{\Delta z} \right\} \tag{A1.5}$$

For a function to possess a unique derivative, it is required that

FUNCTIONS OF A COMPLEX VARIABLE

$$\frac{dW}{dz} = \lim_{\Delta z \to 0} \frac{\Delta W}{\Delta z} = \lim_{\substack{\Delta x \to 0 \\ \Delta y \to 0}} \frac{\Delta u + j \Delta v}{\Delta x + j \Delta y}$$

But since

$$\Delta u = \frac{\partial u}{\partial x} \Delta x + \frac{\partial u}{\partial y} \Delta y$$

$$\Delta v = \frac{\partial v}{\partial x} \Delta x + \frac{\partial v}{\partial y} \Delta y$$

the unique derivative requirement becomes

$$\frac{dW}{dz} = \lim_{\substack{\Delta x \to 0 \\ \Delta y \to 0}} \frac{\left(\frac{\partial u}{\partial x} + j\frac{\partial v}{\partial x}\right)\Delta x + j\left(\frac{\partial v}{\partial y} - j\frac{\partial u}{\partial y}\right)\Delta y}{\Delta x + j \Delta y}$$

For this to be independent of how Δx and Δy approach zero (that is, for the derivative to be unique), it is necessary and sufficient that $\Delta x + j\Delta y$ cancel in the numerator and denominator. This requires that

$$\frac{dW}{dz} = \frac{\partial u}{\partial x} + j\frac{\partial v}{\partial y} = \frac{\partial v}{\partial x} - j\frac{\partial u}{\partial y}$$

This condition can be met if

$$\frac{\partial u}{\partial x} = \frac{\partial v}{\partial y} \qquad \frac{\partial v}{\partial x} = -\frac{\partial u}{\partial y} \qquad \text{(A1.6)}$$

These are the **Cauchy-Riemann** differential equations. If the function satisfies these equations, it possesses a unique derivative.

Integration of a complex function is defined in a manner like that for a real function, except for the important difference that the path of integration as well as the end points must be specified. A number of important theorems relate to integration, as we will discuss later.

Recall that the real integral $\int_A^B f(x)dx$ means that the X-axis is broken into tiny elements Δx from A to B, each element is multiplied by the mean value of $f(x)$ in that element, and then the sum of all such products from A to B is taken as $\Delta x \to 0$. The same general procedure is used to define the integral in the complex plane. Instead of being restricted to the X-axis, the path of integration can be anywhere in the z-plane—for example, the arc ACB in Figure A1.2. This arc is broken into n elements Δz_s and the corresponding mean value of $W(z)$ over each element is written W_s. Now form the sum $\sum_{s=1}^{n} W_s \Delta z_s$ over all values of s from A to B, and take the limit $\Delta z_s \to 0$, $n \to \infty$. This limit, if it exists, is the integral

$$I = \int_A^B W(z)\,dz \qquad \text{(A1.7)}$$

Figure A1.2
The path of integration in the complex plane.

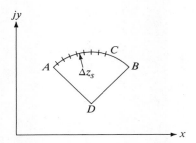

The only innovation introduced here is that the path over which the integral is to be taken must be specified.

EXAMPLE A1.2
Evaluate the integral in (A1.7) for the function $W = 1/z$ over the semicircles shown in Figure A1.3.

Solution: Refer first to Figure A1.3a, and introduce the polar coordinates

$$z = re^{j\theta} \qquad dz = jre^{j\theta}\, d\theta$$

Then

$$\int W\, dz = \int \frac{dz}{z} = \int j\, d\theta$$

Over the path ACB, θ varies between 0 and π, and the integral equals $j\pi$. Over the path ADB, θ varies between 0 and $-\pi$, and the integral equals $-j\pi$. Thus, although the end points are the same, the integrals over the two paths are different. (The fact that one integral is numerically the negative of the other has no general significance.)

In evaluating the real integral by starting at A and integrating to B and then back to A, the result will be zero since the integral from A to B is the negative of the integral from B to A. The same result is not necessarily true for complex

Figure A1.3
Integral of the function $W = 1/z$ over two paths.

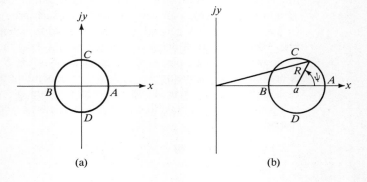

(a)   (b)

variables, unless the path from A to B coincides with the path from B to A. In the present complex integral, the integration from A to B via C and then back to A via D yields $j\pi - (-j\pi) = 2j\pi$ and not zero.

Now consider the integration over the semicircle displaced from the origin, as shown in Figure A1.3b. Introduce the coordinates

$$z = a + Re^{j\psi} \qquad dz = jRe^{j\psi} d\psi$$

Then

$$\int_{ACBDA} \frac{dz}{z} = \int_0^{2\pi} \frac{jRe^{j\psi}}{a + Re^{j\psi}} d\psi = \ln(a + Re^{j\psi})\Big|_0^{2\pi} = \ln z \Big|_A^A = 0$$

The results of these calculations point up the fact that the two paths possess different features. The difference is that in Figure A1.3a the path encloses a singularity (the function becomes infinite) at the origin, whereas the path in Figure A1.3b does not enclose the singularity and $W = 1/z$ is everywhere analytic in the region of integration and on the boundary.

It is easily shown that the integrals of the functions $W(z) = 1/z^2$, $W(z) = 1/z^3, \ldots, W(z) = 1/z^n$ around a contour encircling the origin of the coordinate axis are each equal to zero; that is,

$$\oint \frac{1}{z^2} dz = \oint \frac{1}{z^3} dz = \cdots = \oint \frac{1}{z^n} dz = 0 \qquad \text{(A1.8)}$$

∎

EXAMPLE A1.3

Find the value of the integral $\int_0^{z_0} z \, dz$ from the point $(0, 0)$ to $(2, j4)$.

Solution: Since z is an analytic function along any path, then

$$\int_0^{z_0} z \, dz = \frac{z^2}{2}\Big|_0^{2+j4} = -6 + j8$$

Equivalently, we could write

$$\int_0^{z_0} z \, dz = \int_0^2 x \, dx - \int_0^4 y \, dy + j\int_0^4 dy = \frac{x^2}{2}\Big|_0^2 - \frac{y^2}{2}\Big|_0^4 + jxy\Big|_0^4$$

$$= 2 - \frac{16}{2} + j2 \times 4 = -6 + j8$$

∎

We now consider a very important theorem, and this is often referred to as the **Principal Theorem of Complex Function Theory**. This is the **Cauchy First Integral Theorem**.

Theorem A1.1 Given a region of the complex plane within which $W(z)$ is analytic and **any** closed curve C that lies entirely within this region, then

$$\oint_C W(z) \, dz = 0 \qquad \text{(A1.9)}$$

Figure A1.4
A general configuration in the proof of the Cauchy First Integral Theorem.

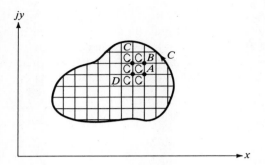

The integration over a closed path is called a contour integral. Also, by convention the positive direction of integration is taken so that when traversing the contour, the enclosed region is always to the left.

Proof: Refer to Figure A1.4, which shows a closed contour in the z-plane. The area enclosed by the path of integration C is broken up into tiny sections by drawing lines parallel to the coordinate axes. If the integration is carried out over each section in the counterclockwise direction shown by the arrows, and all of these integrals are added, the result will be the integral over C. The reason for this is that the integrals over the sides common to two adjacent sections cancel since each side is traversed twice and in opposite directions. All that remains is the integral over the periphery C. Since

$$z = x + jy \quad \text{and} \quad W(z) = u(x, y) + jv(x, y)$$

then

$$\int W \, dz = \int (u + jv)(dx + j\,dy) = \int (u\,dx - v\,dy) + j(v\,dx + u\,dy) \quad \text{(A1.10)}$$

The value of this integral over an elementary section such as $ABCD$ is first considered. The first term in this integral over the segments is

path AB	$u(x, y)\,\Delta x$	since $\Delta y = 0$
BC	$-v(x + \Delta x, y)\,\Delta y$	$\Delta x = 0$
CD	$-u(x, y + \Delta y)\,\Delta x$	$\Delta y = 0$
DA	$v(x, y)\,\Delta y$	$\Delta x = 0$

The sum is

$$-[u(x, y + \Delta y) - u(x, y)]\Delta x - [v(x + \Delta x, y) - v(x, y)]\Delta y$$

But by the definition of a partial derivative

$$\frac{\partial u}{\partial y} = \lim_{\Delta y \to 0} \left[\frac{u(x, y + \Delta y) - u(x, y)}{\Delta y} \right]$$

Also

FUNCTIONS OF A COMPLEX VARIABLE

$$\frac{\partial v}{\partial x} = \lim_{\Delta x \to 0} \left[\frac{v(x + \Delta x, y) - v(x, y)}{\Delta x} \right]$$

Hence, in the limit of infinitely small sections, the integral around $ABCD$ becomes $[-(\partial u/\partial y) - (\partial v/\partial x)]\, dx\, dy$, which vanishes in accordance with the Cauchy-Riemann equations. In a similar manner, the second integral term in (A1.10) can also be shown to be zero. Thus over the entire contour the integral vanishes, as specified by (A1.9).

Note specifically that the proof depends on the fact that everywhere within C the Cauchy-Riemann equations are satisfied; that is, $W(z)$ possesses a unique derivative at all points of the path.

a. First Corollary. If contour C_2 completely encloses C_1, and if $W(z)$ is analytic in the region between C_1 and C_2 and also on C_1 and C_2, then

$$\oint_{C_1} W\, dz = \oint_{C_2} W\, dz \tag{A1.11}$$

Proof: Refer to Figure A1.5, which shows the two contours C_1 and C_2 and two connecting lines DE and GA. In the region enclosed by the contour $ABDEFGA$, the function $W(z)$ is everywhere analytic, and $\oint W\, dz = 0$ over the path. This means that

$$\int_{ABD} + \int_{DE} + \int_{EFG} + \int_{GA} = 0 \tag{A1.12}$$

where $W(z)\, dz$ is to be understood after each integral sign. Now allow A to approach D, and G to approach E, so that DE coincides with AG. Then

$$\int_{DE} = \int_{AG} = -\int_{GA}$$

Also

$$\int_{ABD} = -\int_{C_1} \tag{A1.13}$$

and

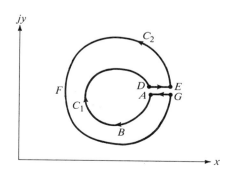

Figure A1.5 To prove the First Corollary.

$$\int_{EFG} = \int_{C_2}$$

where strict attention has been paid to the convention given in the determination of the positive direction of integration around a contour. Combine (A1.12) and (A1.13) so that

$$-\int_{C_1} + \int_{C_2} = 0$$

or

$$\int_{C_1} W\,dz = \int_{C_2} W\,dz$$

which was to be proved.

This is an important theorem since it allows the evaluation around one contour by replacing that contour with a simpler one, the only restriction being that in the region between the two contours the integrand must be regular. It does not require that the function $W(z)$ be analytic within C_1.

b. Second Corollary. If $W(z)$ has a finite number n of isolated singularities within a region G bounded by curve C, then

$$\oint_C W\,dz = \sum_{n=1}^{s} \oint_{C_s} W\,dz \qquad (A1.14)$$

where C_s is any contour surrounding the sth singularity.

Proof: Refer to Figure A1.6. The proof for this case is evident from the manner in which the First Corollary was proved.

c. Third Corollary. The integral $\int_A^B W\,dz$ depends only upon the end points A and B (refer to Figure A1.2) and does not depend upon the path of integration, provided that this path lies entirely within the region in which $W(z)$ is analytic.

Proof: Consider $ACBDA$ of Figure A1.2 as a contour that encloses no singularity of W. Then

$$\oint_C = 0 = \int_{ADB} + \int_{BCA}$$

Figure A1.6
A contour enclosing n isolated singularities.

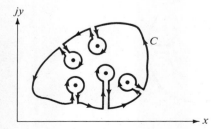

FUNCTIONS OF A COMPLEX VARIABLE

Figure A1.7
To prove the Cauchy Second Integral Theorem.

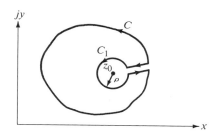

or

$$\int_{ADB} = \int_{ACB} \tag{A1.15}$$

Hence the integral is the same whether taken over path D or C, and thus is independent of the path and depends only upon the end points A and B.

Theorem A1.2 The Cauchy Second Integral Theorem If $W(z)$ is the function $W(z) = f(z)/(z - z_0)$ and the contour encloses the singularity at z_0, then

$$\oint_C \frac{f(z)}{z - z_0} dz = j2\pi f(z_0) \tag{A1.16}$$

Proof: Refer to Figure A1.7. Begin with Corollary 2 and draw a circle C_1 about the point z_0. Then

$$\int_C \frac{f(z)}{z - z_0} dz = \int_{C_1} \frac{f(z)}{z - z_0} dz \tag{A1.17}$$

Let $z' = z - z_0 = \rho e^{j\theta}$, which permits writing

$$\int_{C_1} \frac{f(z)}{z - z_0} dz = \int_0^{2\pi} \frac{f(z' + z_0)}{\rho e^{j\theta}} j\rho e^{j\theta} d\theta = j \int_0^{2\pi} f(z' + z_0) d\theta$$

In the limit as $\rho \to 0$, $z' \to 0$, and

$$j \int_0^{2\pi} f(z' + z_0) d\theta \big|_{\lim_{\rho \to 0}} = 2\pi j f(z_0)$$

Combine this with (A1.14) to find

$$\int_C \frac{f(z)}{z - z_0} dz = 2\pi j f(z_0)$$

which proves the theorem.

■ ■ ■

A1-2 SEQUENCES AND SERIES

Consider a sequence of numbers, such as those that might arise in connection with the Z-transform. Suppose that the sequence of complex numbers is given as $z_0, z_1, z_2, \ldots$.

The sequence of complex numbers is said to **converge** to the limit L; that is,

$$\lim_{n \to \infty} z_n = L$$

if, for every positive δ there exists an integer N such that

$$|z_n - N| < \delta \qquad \text{for all } n > N$$

That is, a convergent sequence is one whose terms approach arbitrarily close to the limit N as n increases. If the series does not converge, it is said to **diverge**.

The sum of an infinite sequence of complex numbers $z_0, z_1, z_2, \ldots$ is given by

$$S = z_0 + z_1 + z_2 + \cdots = \sum_{n=0}^{\infty} z_n \tag{A1.18}$$

Consider the partial sum sequence of n terms, which is designated S_n. The infinite series converges to the sum S if the partial sum sequence S_n converges to S. That is, the series converges if for

$$S_n = \sum_{n=0}^{n} z_n$$

$$\lim_{n \to \infty} S_n = S \tag{A1.19}$$

When the partial sum S_n diverges, the series is said to diverge.

The series (A1.19) is **absolutely convergent** if the series $\sum_{n=0}^{\infty} |z_n|$ converges. Clearly, any series that is absolutely convergent also converges in the sense of (A1.19).

Two important tests for series convergence will be given without proof.

Ratio Test: Consider the ratio of successive elements of the series

$$\lim_{n \to \infty} \left| \frac{z_{n+1}}{z_n} \right|$$

If this ratio converges to L, then the series (A1.18) is absolutely convergent if $L < 1$ and is divergent if $L > 1$. More generally, if this ratio never exceeds a number m for n sufficiently large, then the series is absolutely convergent for $m < 1$ and is divergent if $m > 1$.

Root Test: Consider the sequence

$$r_n = \sqrt[n]{|z_n|}$$

If this sequence converges to L as n approaches infinity, then the series (A1.18) converges absolutely if $L < 1$ and diverges if $L > 1$. More generally, if for sufficiently large n, r_n never exceeds a number N, then the series converges absolutely if $N < 1$ and diverges if $N > 1$.

EXAMPLE A1.4

Consider the series $\sum_{n=0}^{\infty} a^n$ where a is a complex number. Determine whether this series converges or diverges.

Solution: By the ratio test, with $z_n = a^n$

$$\left|\frac{z_{n+1}}{z_n}\right| = \left|\frac{a^{n+1}}{a^n}\right| = |a|$$

Thus $L = |a|$ and the given series is absolutely convergent if $|a| < 1$, and it diverges if $|a| > 1$.

By the root test, we have

$$\sqrt[n]{|z_n|} = \sqrt[n]{|a^n|} = |a|$$

Hence the same conclusion is reached using either test. ∎

■ ■ ■

A1-3 POWER SERIES

A series of the form

$$W = a_0 + a_1(z - z_0) + a_2(z - z_0)^2 + \cdots = \sum_{n=0}^{\infty} a_n(z - z_0)^n \quad \text{(A1.20)}$$

where the coefficients a_n are given by

$$a_n = \frac{1}{n!} \left.\frac{d^n w}{dz^n}\right|_{z=z_0} \quad \text{(A1.21)}$$

is called a **Taylor** series that is expanded about the point $z = z_0$, where z_0 is a complex constant. That is, the Taylor series expands an analytic function as an infinite sum of component functions. More precisely, the Taylor series expands a function $W(z)$, which is analytic in the neighborhood of the point $z = z_0$, into an infinite series whose coefficients are the successive derivatives of the function at the given point. However, we know that the definition of a derivative of any order does not require more than the knowledge of the function in an arbitrarily small neighborhood of the point $z = z_0$. This means, therefore, that the Taylor series indicates that the shape of the function at a finite distance z_0 from the point z is determined by the behavior of the function in the infinitesimal vicinity of $z = z_0$. Thus the Taylor series implies that an analytic function has a very strong interconnected structure, and that by studying the function in a small vicinity of the point $z = z_0$, we can precisely predict what happens at the point $z = z_0 + \Delta z_0$, which is a finite distance from the point of study.

If $z_0 = 0$, the expansion is said to be about the origin and is called a **Maclaurin** series.

A power series of negative powers of $(z - z_0)$,

$$W = a_0 + a_1(z - z_0)^{-1} + a_2(z - z_0)^{-2} + \cdots \tag{A1.22}$$

is called a negative power series.

We first focus our attention on the positive power series (A1.20). Clearly, this series converges to a_0 when $z = z_0$. To ascertain whether it converges for other values of z, we write:

Theorem A1.3 A positive power series converges absolutely in a circle of radius R^+ centered at z_0 where $|z - z_0| < R^+$; it diverges outside of this circle where $|z - z_0| > R^+$.

The value of R^+ may be zero, a positive number, or ∞. If $R^+ = 0$, the series converges only at $z = z_0$. If $R^+ = \infty$, the series converges everywhere. The radius R^+ is found from the relation

$$R^+ = \lim_{n \to \infty} \left| \frac{a_n}{a_{n+1}} \right| \quad \text{if the limit exists} \tag{A1.23}$$

or by

$$R^+ = \lim_{n \to \infty} \frac{1}{\sqrt[n]{|a_n|}} \quad \text{if the limit exists} \tag{A1.24}$$

Proof: For a fixed value z, apply the ratio test, where

$$z_n = a_n(z - z_0)^n$$

That is,

$$\left| \frac{z_{n+1}}{z_n} \right| = \left| \frac{a_{n+1}(z - z_0)^{n+1}}{a_n(z - z_0)^n} \right| = \left| \frac{a_{n+1}}{a_n} \right| |z - z_0|$$

For the power series to converge, the ratio test requires that

$$\lim_{n \to \infty} \left| \frac{a_{n+1}}{a_n} \right| |z - z_0| < 1$$

or

$$|z - z_0| < \lim_{n \to \infty} \left| \frac{a_n}{a_{n+1}} \right| = R^+$$

That is, the power series converges absolutely for all z that satisfy this inequality. It diverges for all z for which

$$|z - z_0| > R^+$$

The value of R^+ specified by (A1.24) is deduced by applying the root test. Figure A1.8 shows the region of convergence for a positive power series.

Figure A1.8
(a) Convergence region of a positive power series.
(b) Convergence region of a negative power series.

EXAMPLE A1.5
Determine the region of convergence for the power series

$$W = \frac{1}{1+z} = 1 - z + z^2 - z^3 + \cdots$$

Solution: We have $a_n = (-1)^n$, from which

$$R^+ = \lim_{n \to \infty} \left| \frac{(-1)^n}{(-1)^{n+1}} \right| = |1|$$

This series converges for all z for which

$$|z| < 1$$

Hence this expansion converges for any value of z within a circle of unit radius about the origin. Note that there will always be at least one singular point of W on the circle of convergence. In the present case, the point $z = -1$ is a singular point. ∎

EXAMPLE A1.6
Determine the region of convergence for the power series

$$W = e^z = 1 + z + \frac{z^2}{2!} + \frac{z^3}{3!} + \cdots = \sum_{n=1}^{\infty} \frac{1}{n!} z^n$$

Solution: We have $a_n = \frac{1}{n!}$ from which

$$R^+ = \lim_{n \to \infty} \left| \frac{(n+1)!}{n!} \right| = \lim_{n \to \infty} (n+1) = \infty$$

The circle of convergence is specified by $R^+ = \infty$; hence e^z converges for all finite values of z. ∎

Theorem A1.4 A negative power series (A1.22) converges absolutely outside a circle of radius R^- centered at z_0, where $|z - z_0| > R^-$; it diverges inside of this circle, where $|z - z_0| < R^-$.

The radius R^- is determined from

$$R^- = \lim_{n \to \infty} \left| \frac{a_{n+1}}{a_n} \right| \quad \text{if this limit exists} \tag{A1.25}$$

or by

$$R^- = \lim_{n \to \infty} \sqrt[n]{|a_n|} \quad \text{if this limit exists} \tag{A1.26}$$

Proof: The proof of this theorem parallels that for Theorem A1.3. The region of convergence for a negative power series is shown in Figure A1.8b.

If a function has a singularity at $z = z_0$, it cannot be expanded in a Taylor series about this point. However, if one deletes the neighborhood of z_0 so that W is analytic in the deleted neighborhood of z_0, it can be expressed in the form of a **Laurent** series. The Laurent series is written

$$W = \cdots + \frac{a_{-2}}{(z - z_0)^2} + \frac{a_{-1}}{(z - z_0)} + a_0 + a_1(z - z_0) + a_2(z - z_0)^2 + \cdots$$

$$= \sum_{n=-\infty}^{\infty} a_n(z - z_0)^n \tag{A1.27}$$

If a circle is drawn about the point z_0 such that the nearest singularity of W (outside of z_0 itself) lies on this circle, then (A1.27) defines an analytic function everywhere within this circle except at its center. The portion $\sum_{n=0}^{\infty} a_n(z-z_0)^n$ is regular at $z = z_0$. The portion $\sum_{n=-1}^{-\infty} a_n(z - z_0)^n$ is not regular and is called the principal part of $W(z)$ at $z = z_0$.

The region of convergence for the positive series part of the Laurent series is of the form

$$|z - z_0| < R^+ \tag{A1.28}$$

while that for the principal part is given by

$$|z - z_0| > R^- \tag{A1.29}$$

The evaluation of R^+ and R^- proceed according to the methods already discussed. Hence the region of convergence of the Laurent series is given by those points common to (A1.28) and (A1.29) or for

$$R^- < |z - z_0| < R^+ \tag{A1.30}$$

If $R^- > R^+$, the series converges nowhere. The annular region of convergence for a typical Laurent series is shown in Figure A1.9.

EXAMPLE A1.7
Consider the Laurent series

Figure A1.9
The annular region of convergence for the Laurent series.

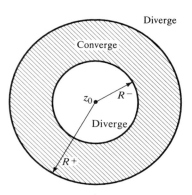

$$W(z) = \sum_n a_n z^n$$

where

$$a_n = \begin{cases} (\tfrac{1}{3})^n & \text{for } n = 0, 1, 2, \ldots \\ 2^n & \text{for } n = -1, -2, -3, \ldots \end{cases}$$

Determine the region of convergence.

Solution: By (A1.28) and (A1.23), we have

$$R^+ = 3$$

By (A1.29) and (A1.25), we have

$$R^- = 2$$

Hence this series converges for all z for which

$$2 < |z| < 3 \qquad\blacksquare$$

No convenient expression exists for obtaining the coefficients of the Laurent series. However, since there is only one Laurent expansion for a given function, the resulting series, however derived, will be the appropriate one. For example,

$$e^{1/z} = 1 + \frac{1}{z} + \frac{1}{2!z^2} + \frac{1}{3!z^3} + \cdots \qquad \text{(A1.31)}$$

is obtained by replacing z by $1/z$ in the Maclaurin expansion of e^z. Note that in this case the coefficients of all positive powers of z in the Laurent expansion are zero. As a second illustration, consider the function $W = (\cos z)/z$. This is found by dividing the Maclaurin series for $\cos z$ by z, with the result

$$\frac{\cos z}{z} = \frac{1}{z}\left(1 - \frac{z^2}{2!} + \frac{z^4}{4!} - \cdots\right) = \frac{1}{z} - \frac{z}{2!} + \frac{z^3}{4!} - \cdots \qquad \text{(A1.32)}$$

In this case the Laurent expansion includes only one term $1/z$ in descending powers of z, but an infinite number of terms in ascending powers of z. That is, $a_{-1} = 1$ and $a_{-n} = 0$ if $n \neq 1$.

■ ■ ■

A1-4 ANALYTIC CONTINUATION

The Taylor theorem shows that if a function $f(z)$ is given by a power series in z, it can also be represented as a power series in $z - z_0 = f[(z - z_0) + z_0]$ where z_0 is any point within the original circle of convergence, and this series will converge within any circle about z_0 that does not pass beyond the original circle of convergence. Actually, it may converge within a circle that does pass beyond the original circle of convergence. Consider, for example, the function

$$f(z) = 1 + z + z^2 + \cdots = \frac{1}{1-z} \quad \text{for } |z| < 1$$

Choose $z_0 = \frac{1}{2}j$, and the Taylor expansion of

$$f(z) = \frac{1}{1 - [(z - \frac{1}{2}j) + \frac{1}{2}j]} = \frac{1}{(1 - \frac{1}{2}j) - (z - \frac{1}{2}j)} = \frac{1}{(1 - \frac{1}{2}j) - z'}$$

in powers of $z' = z - \frac{1}{2}j$ is

$$f(z) = \frac{1}{1 - \frac{1}{2}j} + \frac{z'}{(1 - \frac{1}{2}j)^2} + \frac{z'^2}{(1 - \frac{1}{2}j)^3} + \cdots$$

This series must converge and be equal to the original function if $|z'| < \frac{1}{2}$, since j is the point of the circle $|z| = 1$ nearest to $\frac{1}{2}j$, a requirement of Taylor's theorem. Actually this series converges if $|z'| < |1 - \frac{1}{2}j| = \frac{1}{2}\sqrt{5}$.

Suppose that the considered series represented no previously known function. In this case the new Taylor series would define values of an analytic function over a range of z where no function is defined by the original series. Then we can extend the range of definition by taking a new Taylor series about a point in the new region. This process is called **analytic continuation.** In practice when continuation is required, the direct use of the Taylor series is laborious and is seldom used. Of more convenience is the following theorem.

Theorem A1.5 If two functions $f_1(z)$ and $f_2(z)$ are analytic in a region D and equal in a region D' within D, they are equal everywhere in D.

Proof: Let z_0 be any ordinary point in D' and Z any other point in D. Then $f_1(z) = f_2(z)$ within any circle about z_0 that does not reach a singularity or the boundary. Now suppose that z_0 and Z are connected by a curve of finite length in D not reaching the boundary or passing through a singularity. Then the distances between points on the curve and the singularities have a positive lower bound δ, so also do the distances between points on the curve

FUNCTIONS OF A COMPLEX VARIABLE

and the boundary. Thus we can choose points $z_1, z_2, \ldots, z_n = Z$ such that $|z_r - z_{r-1}| < \delta$, with n finite. But z_1 lies within the circle of convergence of the series representing $f_1(z)$ and $f_2(z)$ in powers of $z - z_0$, and both functions have the same Taylor series in $z - z_1$ and z_2 is within its circle of convergence. By proceeding in this way, we can show that both functions have the same Taylor series in $(z - z_{n-1})$, the circle of convergence of which includes Z. Hence $f_1(Z) = f_2(Z)$.

■ ■ ■

A1-5 SINGULARITIES OF A COMPLEX FUNCTION

A singularity has already been defined as a point at which a function ceases to be analytic. Thus a discontinuous function has a singularity at the point of discontinuity, and a multivalued function has a singularity at a branch point. There are two important classes of singularities that a continuous, single-valued function may possess.

Definition A1.3 A function has an **essential singularity** at $z = z_0$ if its Laurent expansion about the point z_0 contains an infinite number of terms in inverse powers of $(z - z_0)$.

Definition A1.4 A function has a **nonessential singularity** or **pole of order** m if its Laurent expansion can be expressed in the form

$$W(z) = \sum_{n=-m}^{\infty} a_n(z - z_0)^n \tag{A1.33}$$

Note that the summation extends from $-m$ to ∞ and not from $-\infty$ to ∞; that is, the highest inverse power of $(z - z_0)$ is m.

An alternative definition that is equivalent to this but somewhat simpler to apply is the following: If $\lim_{z \to z_0}[(z - z_0)^m W(z)] = c$, a nonzero constant (here m is a positive number), then $W(z)$ is said to possess a pole of order m at z_0.

We illustrate these definitions with a number of examples.

1. $e^{1/z}$ [see (A1.31)] has an essential singularity at the origin.
2. $\cos z/z$ [see (A1.32)] has a pole of order 1 at the origin.
3. Consider the function

$$W = \frac{e^z}{(z-4)^2(z^2+1)}$$

Note that functions of this general type exist frequently in the Laplace inversion integral. Since e^z is regular at all finite points of the z-plane, the singularities of W must occur at the points for which the denominator vanishes; that is, for

$$(z-4)^2(z^2+1) = 0 \quad \text{or} \quad z = 4, +j, -j$$

By the alternate definition of A1.4, it is easily shown that W has a second-order pole at $z = 4$, and first-order poles at the two points $\pm j$. That is,

$$\lim_{z \to 4} (z - 4)^2 \left[\frac{e^z}{(z - 4)^2(z^2 + 1)} \right] = \frac{e^4}{17} \neq 0$$

$$\lim_{z \to j} (z - j) \left[\frac{e^z}{(z - 4)^2(z^2 + 1)} \right] = \frac{e^j}{(j - 4)^2 2j} \neq 0$$

4. An example of a function with an infinite number of singularities occurs in heat flow, wave motion, and similar problems. The function involved is

$$W = \frac{1}{\sinh az}$$

The singularities in this function occur when $\sinh az = 0$ or $az = js\pi$, where $s = 0, \pm 1, \pm 2, \ldots$. That each of these is a first-order pole follows from

$$\lim_{z \to j(s\pi/a)} \left(z - j\frac{s\pi}{a} \right) \frac{1}{\sinh az} = \frac{0}{0}$$

This can be evaluated in the usual manner by differentiating numerator and denominator (l'Hospital's rule) to find

$$\lim_{z \to j(s\pi/a)} \frac{1}{a \cosh az} = \frac{1}{a \cosh js\pi} = \frac{1}{a \cos s\pi} \neq 0$$

■ ■ ■

A1-6 THEORY OF RESIDUES

It has already been shown that the contour integral of any function that encloses no singularities of the integrand will vanish. Now our purpose is to examine the integral, the path of which encloses one singularity, say at $z = z_0$. The Laurent expansion of such a function is

$$W = \sum_{n=-\infty}^{\infty} a_n(z - z_0)^n$$

and so

$$\oint_C W\, dz = \sum_{n=-\infty}^{\infty} a_n \oint_C (z - z_0)^n\, dz$$

But by (A1.12) each term in the sum vanishes except that for $n = -1$, with

$$\oint_C (z - z_0)^{-1}\, dz = 2\pi j$$

It then follows that

$$\oint_C W\, dz = \sum_{n=-\infty}^{\infty} \oint (z - z_0)^n\, dz = 2\pi j a_{-1} \tag{A1.34}$$

FUNCTIONS OF A COMPLEX VARIABLE

Because the integral $(1/2\pi j) \oint_C W\, dz$ will appear frequently in subsequent applications, it is given a name; it is called the **residue of W** at z_0 and is abbreviated $\text{Res}(W)_{z_0}$.

From the second corollary (A1.13), it follows that if W has n isolated singularities within C, then

$$\frac{1}{2\pi j} \oint_C W\, dz = \sum_{s=1}^{n} \frac{1}{2\pi j} \oint_{C_s} W\, dz = \sum_{s=1}^{n} \text{Res}_s(W) \tag{A1.35}$$

or, in words, $(1/2\pi j) \oint_C W\, dz$ equals the sum of the residues within C. Observe that to evaluate integrals in the complex plane, it is only necessary to find the residues at the singularities of the integrand within the contour. One obvious way of doing this is [see (A1.34)] to find the coefficient a_{-1} in the Laurent expansion about each singularity. However, this is not always an easy task; it is accomplished by expanding the integrand in a partial-fraction expansion, as discussed in Section 6-7.

Several theorems exist that make evaluating residues relatively easy. We introduce these.

Theorem A1.6 If the $\lim_{z \to z_0} [(z - z_0)W]$ is finite, this limit is the residue of W at $z = z_0$. If the limit is not finite, then W has a pole of at least second order at $z = z_0$ (it may possess an essential singularity here). If the limit is zero, then W is regular at $z = z_0$.

Proof: Suppose that the function is expanded into the Laurent series

$$W = \frac{a_{-1}}{z - z_0} + a_0 + a_1(z - z_0) + a_2(z - z_0)^2 + \cdots$$

Then the expression

$$\lim_{z \to z_0} [(z - z_0)W] = \lim_{z \to z_0} [a_{-1} + a_0(z - z_0) + a_1(z - z_0)^2 + \cdots]$$
$$= a_{-1}$$

This proves the theorem.

This process was previously used to ascertain whether or not a function had a first-order pole at $z = z_0$. Thus referring back to the examples in Section A1-5 we have

$$\text{Res}\left(\frac{\cos z}{z}\right)_0 = 1$$

$$\text{Res}\left[\frac{e^z}{(z-4)^2(z^2+1)}\right]_j = \frac{e^j}{(j-4)^2 \cdot 2j}$$

$$\text{Res}\left[\frac{1}{\sinh az}\right]_{j(s\pi/a)} = \frac{1}{a \cos s\pi}$$

Many of the singularities that arise in system function studies are first-order poles. The evaluation of the integral is relatively direct.

EXAMPLE A1.8
Evaluate the following integral

$$\frac{1}{2\pi j}\oint_C \frac{e^{zt}}{(z^2 + \omega^2)}\,dz$$

when the contour C encloses both first-order poles at $z = \pm j\omega$. Note that this is precisely the Laplace inversion integral of the function $1/(z^2 + \omega^2)$.

Solution: This involves finding the following residues

$$\left[\frac{e^{zt}}{z^2 + \omega^2}\right]_{j\omega} = \frac{e^{j\omega t}}{2j\omega}$$

$$\left[\frac{e^{zt}}{z^2 + \omega^2}\right]_{-j\omega} = -\frac{e^{-j\omega t}}{2j\omega}$$

Hence

$$\frac{1}{2\pi j}\oint_C \frac{e^{zt}}{z^2 + \omega^2}\,dz = \sum \text{Res} = \left(\frac{e^{j\omega t} - e^{-j\omega t}}{2j\omega}\right) = \frac{\sin \omega t}{\omega} \qquad\blacksquare$$

A slight modification of the method for finding residues of simple poles

$$\text{Res } W(z_0) = \lim_{z \to z_0} \left[(z - z_0)W\right] \qquad (A1.36)$$

makes the process even simpler. This is specified by the theorem:

Theorem A1.7 Suppose that $f(z)$ is analytic at $z = z_0$ and suppose that $g(z)$ is divisible by $z - z_0$ but not by $(z - z_0)^2$. Then

$$\text{Res}\left[\frac{f(z)}{g(z)}\right]_{z_0} = \frac{f(z_0)}{g'(z_0)} \qquad \text{where } g'(z) = \frac{dg}{dz} \qquad (A1.37)$$

Proof: Write the relation

$$(z - z_0)h(z) = g(z)$$

then

$$g'(z) = (z - z_0)h'(z) + h(z)$$

so that for $z = z_0$, $g'(z_0) = h(z_0)$. Then we have

$$\text{Res}\left[\frac{f(z)}{g(z)}\right]_{z_0} = \lim_{z \to z_0}\left[(z - z_0)\frac{f(z)}{g(z)}\right] = \lim_{z \to z_0}\left[\frac{f(z)}{h(z)}\right] = \frac{f(z_0)}{h(z_0)} = \frac{f(z_0)}{g'(z_0)}$$

which is the given result.

FUNCTIONS OF A COMPLEX VARIABLE

In reality, this theorem has already been used in the evaluation of $\text{Res}(1/\sinh az)_{j(s\pi/a)}$. Here $f(z) = 1$, $g(z) = \sinh az$, and $g'(z) = a \cosh az$.

As a second illustration, consider the previously used function

$$W = \frac{e^z}{(z-4)^2(z^2+1)}$$

here we take

$$f(z) = \frac{e^z}{(z-4)^2} \qquad g(z) = z^2 + 1$$

thus $g'(z) = 2z$ and the previous result follows immediately with

$$\text{Res}\left[\frac{e^z}{(z-4)^2(z^2+1)}\right] = \frac{e^j}{(j-4)^2 \cdot 2j}$$

Equation (A1.36) permits a simple proof of the Cauchy Second Integral Theorem, (A1.16). This involves choosing $g(z) = (z - z_0)$ in the integral

$$\frac{1}{2\pi j}\oint_C \frac{f(z)}{z-z_0}\,dz = \frac{f(z_0)}{1} = f(z_0) \tag{A1.38}$$

Suppose that (A1.38) is differentiated $n - 1$ times with respect to z_0. Then we write

$$\frac{d^{n-1}f(z_0)}{dz_0^{n-1}} = f^{(n-1)}(z_0) = \frac{(n-1)!}{2\pi j}\oint_C \frac{f(z)}{(z-z_0)^n}\,dz \tag{A1.39}$$

This specifies any-order derivative of a complex function expressed as a contour integral.

Our discussion thus far has concentrated on finding the residue of a first-order pole. However, (A1.39) permits finding the residue of a pole of any order. If, for example, $W = [f(z)/(z-z_0)^n]$, then evidently W has a pole of order n at $z = z_0$ since $f(z)$ is analytic at $z = z_0$. Then $f(z) = (z-z_0)^n W$, and (A1.39) becomes

$$\text{Res}(W)|_{z_0} = \frac{1}{2\pi j}\oint_C W\,dz = \frac{1}{(n-1)!}\frac{d^{n-1}}{dz^{n-1}}[(z-z_0)^n W]_{z_0} \tag{A1.40}$$

This result was developed differently as (6.82) from considerations of the partial-fraction expansion.

EXAMPLE A1.9
Evaluate the residue at the second-order pole at $z = 4$ of the previously considered function

$$W = \frac{e^z}{(z-4)^2(z^2+1)}$$

Solution: It follows from (A1.40) that

$$\operatorname{Res} W\Big|_{z=4} = \frac{1}{1!}\frac{d}{dz}\left[\frac{e^z}{z^2+1}\right]_{z=4} = \left[\frac{(z^2+1)e^z - 2ze^z}{(z^2+1)^2}\right]_4 = \frac{9e^4}{289}$$

■

EXAMPLE A1.10
Evaluate the residue at the third-order pole of the function

$$W = \frac{e^{zt}}{(z+1)^3}$$

Solution: A direct application of (A1.40) yields

$$\operatorname{Res} W\Big|_{-1} = \frac{1}{2!}\frac{d^2}{dz^2}(e^{zt})\Big|_{-1} = \frac{1}{2}\frac{d}{dz}(te^{zt})\Big|_{-1} = \frac{1}{2}t^2 e^{zt}\Big|_{-1} = \frac{1}{2}t^2 e^{-t}$$

■

Note that the process of finding the residue of a function at a higher-order pole can become very involved and cumbersome, although the process is straightforward in principle.

There is no simple way of finding the residue at an essential singularity. The Laurent expansion must be found and the coefficient a_{-1} is thereby obtained. For example, from (A1.31) it is seen that the residue of $\exp[1/z]$ at the origin is unity. Fortunately, an essential singularity seldom arises in practical applications.

■ ■ ■

A1-7 AIDS TO INTEGRATION

The following three theorems will substantially simplify the evaluation of certain integrals in the complex plane. Examples will be found in later applications.

Theorem A1.8 If AB is the arc of a circle of radius $|z| = R$ for which $\theta_1 \leq \theta \leq \theta_2$ and if $\lim_{R\to\infty}(zW) = k$, a constant which may be zero, then

$$\lim_{R\to\infty}\int_{AB} W\,dz = jk(\theta_2 - \theta_1) \tag{A1.41}$$

Proof: Let $zW = k + \epsilon$, where $\epsilon \to 0$ as $R \to \infty$. Then

$$\int_{AB} W\,dz = \int_{AB}\frac{k+\epsilon}{z}dz = (k+\epsilon)\int_{\theta_1}^{\theta_2} j\,d\theta = (k+\epsilon)j(\theta_2 - \theta_1)$$

In carrying out this integration, the procedure employed in Example A1.2 is used. In the limit as $R \to \infty$, (A1.41) follows.

This theorem can be shown to be valid even if there are a finite number of points on the arc AB for which $\lim_{R\to\infty}(zW) \neq k$, provided only that the limit remains finite for finite R at these points. This theorem can also be proved

Figure A1.10 For Theorem A1.9 proof.

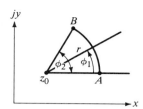

true if we choose $\lim_{R \to \infty} (z - a)W(z) = k$ when the integral is taken around the arc $\theta_1 \leq \arg(z - a) \leq \theta_2$ of the circle $|z - a| = r$.

Theorem A1.9 If AB is the arc of a circle of radius $|z - z_0| = r$ for which $\varphi_1 \leq \varphi \leq \varphi_2$ (as shown in Figure A1.10) and if $\lim_{z \to z_0} [(z - z_0)W] = k$, a constant which may be zero, then

$$\lim_{r \to 0} \int_{AB} W \, dz = jk(\varphi_2 - \varphi_1) \tag{A1.42}$$

where r and φ are introduced polar coordinates, with the point $z = z_0$ as origin.

Proof: The proof of this theorem follows along similar lines to that of Theorem A1.8.

Note specifically that Theorem A1.8 will allow the evaluation of integrals over infinitely large arcs, whereas Theorem A1.9 will allow the evaluation over infinitely small arcs.

Theorem A1.10 If the maximum value of W along a path C (not necessarily closed) is M, the maximum value of the integral of W along C is Ml, where l is the length of C. When expressed analytically, this specifies that

$$\left| \int_C W \, dz \right| \leq Ml \tag{A1.43}$$

Proof: The proof of this theorem is very simple if recourse is made to the definition of an integral. Thus from Figure A1.2

$$\int_C W \, dz = \lim_{\substack{\Delta z_s \to 0 \\ n \to \infty}} \sum_{s=1}^{n} W_s \Delta z_s \leq M \lim_{\substack{\Delta z_s \to 0 \\ n \to \infty}} \sum_{s=1}^{n} \Delta z_s = Ml$$

■ ■ ■

A1-8 EVALUATION OF DEFINITE INTEGRALS

The principles discussed above find considerable applicability in the evaluation of certain definite real integrals. This is a common application of the developed theory, as it is often extremely difficult to evaluate some of these real integrals by other methods. We employed such methods in the evaluation of Fourier integrals. In practice, the given integrand is replaced by a complex function that yields the specified integrand in its appropriate limit. The integration is then carried out in the complex plane, with the real integral being extracted for the total result. Several examples show this procedure.

EXAMPLE A1.11
Consider the evaluation of the real integral

$$\int_{-\infty}^{\infty} \frac{\cos x}{x^2 + a^2} \, dx$$

where a is a real positive constant.

Solution: In place of the given integral, the following integral will be evaluated along the path indicated in Figure A1.11.

$$I = \oint_C \frac{e^{jz}}{z^2 + a^2} \, dz$$

We carry out the integration along the indicated path; we must examine

$$I = \lim_{R \to \infty} \left[\int_{\text{semicircle}} + \int_{-R \atop \text{along real axis}}^{R} \right] = 2\pi j \, \text{Res} \, W \big|_{ja}$$

The only singularity enclosed is that at $z = ja$; the residue there is, by (A1.36), $e^{-a}/2ja$. Now along the semicircle

$$\lim_{R \to \infty} \left[\frac{ze^{jz}}{z^2 + a^2} \right] = \lim_{R \to \infty} \frac{e^{jR\cos\theta} e^{-R\sin\theta}}{Re^{j\theta}} = 0$$

Hence the integral along the semicircle vanishes as R approaches infinity. There remains, therefore, the integral along the real axis, so that

$$\frac{2\pi j e^{-a}}{2ja} = \int_{-\infty}^{\infty} \frac{e^{jx}}{x^2 + a^2} \, dx = \int_{-\infty}^{\infty} \left[\frac{\cos x}{x^2 + a^2} + j \frac{\sin x}{x^2 + a^2} \right] dx$$

Figure A1.11
The path for the contour integral in Example A1.11.

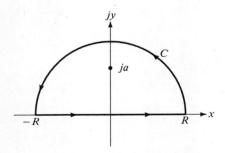

FUNCTIONS OF A COMPLEX VARIABLE

Equating real and imaginary parts of each side gives

$$\int_{-\infty}^{\infty} \frac{\cos x}{x^2 + a^2} dx = \frac{\pi}{a} e^{-a}$$

and

$$\int_{-\infty}^{\infty} \frac{\sin x}{x^2 + a^2} dx = 0$$

The last equation is obviously true since the integrand is an odd function of x; hence the sum is zero. ■

EXAMPLE A1.12
Evaluate the integral

$$\int_0^{\infty} \frac{\sin x}{x} dx = \frac{\pi}{2}$$

Solution: Proceed by considering the contour integral

$$I = \int_C \frac{e^{jz}}{z} dz$$

along the path illustrated in Figure A1.12. Since there are no singularities of the integrand enclosed by this contour, the value of the total contour integral is zero. Therefore with obvious notation

$$I = \oint_C = \int_{\text{large semicircle}} + \int_{\text{small semicircle}} + \int_{-R}^{-r} + \int_r^R = I_1 + I_2 + I_3 + I_4 = 0$$

Each integral is considered in turn. By (A1.41)

$$\lim_{R \to \infty} e^{jR \cos \theta} e^{-R \sin \theta} = 0$$

for all values of θ in the first and second quadrants, exclusive of the end points $\theta = 0$ or π. However, at these points this limit is finite—in fact, its maximum value is unity. Hence by (A1.41) $I_1 = 0$.

Figure A1.12
The contour for Example A1.12.

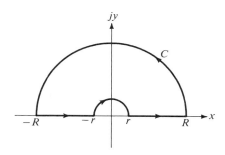

In estimating the maximum value of a function, remember that the absolute value of $\exp[j\psi]$, where ψ is real, is always unity. The integral I_2 can be evaluated using (A1.42). Thus

$$\lim_{r \to 0} e^{jr\cos\varphi} e^{-r\sin\varphi} = 1$$

Hence $I_2 = -j\pi$, since $\varphi_1 = \pi$ and $\varphi_2 = 0$ in (A1.42).

As $R \to \infty$ and $r \to 0$, $I_3 + I_4$ become $\int_{-\infty}^{\infty}$, and integral I becomes

real axis

$$\int_{-\infty}^{\infty} \frac{e^{jx}}{x}\, dx = j\pi = \int_{-\infty}^{\infty}\left(\frac{\cos x}{x} + j\frac{\sin x}{x}\right)dx$$

or

$$\int_{-\infty}^{\infty} \frac{\cos x}{x}\, dx = 0$$

This is an expected result since the integrand is an odd function of x, and

$$\int_{-\infty}^{\infty} \frac{\sin x}{x}\, dx = \pi$$

This integrand is an even function of x and the integral from $-\infty$ to 0 is the same as that from 0 to ∞, so $\int_0^{\infty} = \frac{1}{2}\int_{-\infty}^{\infty}$. Therefore it follows that

$$\int_0^{\infty} \frac{\sin x}{x}\, dx = \frac{\pi}{2}$$

Two points must be noted specifically. First, observe the way that the tiny hook around the singularity at the origin excluded the singularity. In this case the residue was zero. If the hook had been taken to include the singularity, the residue would have been $2j\pi$, but the integral around the hook would have been $+j\pi$. The end result would have been the same. ∎

■ ■ ■

A1-9 PRINCIPAL VALUE OF AN INTEGRAL

Refer to the limiting process employed in Example A1.12, which was written

$$\lim_{R \to \infty} \int_{-R}^{R} \frac{e^{jx}}{x}\, dx = j\pi$$

The limit is called the **Cauchy principal value** of the integral in the equation

$$\int_{-\infty}^{\infty} \frac{e^{jx}}{x}\, dx = j\pi$$

In general, if $f(x)$ becomes infinite at a point $x = c$ inside the range of integration, and if

$$\lim_{\epsilon \to 0} \int_{-R}^{R} f(x)\, dx = \lim_{\epsilon \to 0}\left[\int_{-R}^{c-\epsilon} f(x)\, dx + \int_{c+\epsilon}^{R} f(x)\, dx\right]$$

FUNCTIONS OF A COMPLEX VARIABLE

and if the separate limits on the right also exist, then the integral is convergent and the integral is written $\dashint$ where the - drawn through the integral sign denotes the principal value. Whenever each of the integrals

$$\int_{-\infty}^{0} f(x)\,dx \qquad \int_{0}^{\infty} f(x)\,dx$$

has a value, here for $R \to \infty$, the principal value is the same as the integral. For example, if $f(x) = x$, the principal value of the integral is zero, although the value of the integral itself does not exist.

As another example, consider the integral

$$\int_{a}^{b} \frac{dx}{x} = \log\frac{b}{a}$$

If a is negative and b is positive, the integral diverges at $x = 0$. However, we can still define

$$\dashint_{a}^{b} \frac{dx}{x} = \lim_{\epsilon \to 0}\left[\int_{a}^{-\epsilon}\frac{dx}{x} + \int_{\epsilon}^{b}\frac{dx}{x}\right] = \lim_{\epsilon \to 0}\left(\log\frac{\epsilon}{-a} + \log\frac{b}{\epsilon}\right)$$
$$= \log\frac{b}{-a} = \log\frac{b}{|a|}$$

This principal value integral is unambiguous. The condition that the same value of ϵ must be used on both sides is essential; otherwise the limit could be almost anything by taking the first integral from a to $-\epsilon$ and the second from κ to b and making ϵ and κ tend to zero in a suitable ratio.

If the complex variable were used, we could complete the path by a semicircle (hook) from $-\epsilon$ to $+\epsilon$ about the origin, either above or below the real axis. If the upper semicircle were chosen, there would be a contribution $-j\pi$, whereas if the lower semicircle were chosen, the contribution to the integral would be $+j\pi$. Thus according to the path permitted in the complex plane (branch cuts, discussed in the next section, could impose restrictions on the path), we should have

$$\int_{a}^{b} \frac{dz}{z} = \log\frac{b}{|a|} \pm j\pi$$

The principal value is the mean of these alternatives.

If a path in the complex plane passes through a simple pole a, we can define a principal value of the integral along the path by using a hook of small radius ϵ about a and then making ϵ tend to zero, as already discussed. If we change the variable z to ζ and $dz/d\zeta$ is finite and not equal to zero at the pole, this procedure will define an integral in the ζ-plane, but the values of the integrals will be the same. Suppose that the hook in the z-plane cuts the path at $a - \epsilon$ and $a + \epsilon'$, where $|\epsilon| = |\epsilon'|$, and in the ζ-plane the hook cuts the path at $\alpha - \kappa$ and $\alpha + \kappa'$. Then if κ and κ' tend to zero so that $\epsilon/\epsilon' \to 1$, κ and κ' will tend to 0 so that $\kappa/\kappa' \to 1$.

Figure A1.13
The contour for integral I.

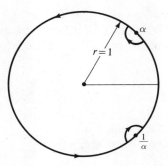

To illustrate this discussion, suppose we want to evaluate the integral

$$I = \int_0^\pi \frac{d\theta}{a - b\cos\theta}$$

where a and b are real and $a > b > 0$. A change of variable by writing $z = \exp[j\theta]$ transforms this integral to (where a new constant, α, is introduced)

$$I = \int_0^\pi \frac{2e^{j\theta}\,d\theta}{2ae^{j\theta} - b(e^{2j\theta} + 1)} = -\frac{1}{j}\int_C \frac{2\,dz}{bz^2 - 2az + b}$$

$$= -\frac{1}{j}\int_C \frac{2\,dz}{b(z-\alpha)(z-1/\alpha)}$$

where the path of integration is around the unit circle. Because the contour would pass through the poles, hooks are used to isolate the poles, as shown in Figure A1.13. Since no singularities are enclosed by the path, the integral is zero. The contributions of the hooks are $-j\pi$ times the residue, where the residues are

$$-\frac{1}{j}\frac{\frac{2}{b}}{\alpha - \frac{1}{\alpha}} \qquad -\frac{1}{j}\frac{\frac{2}{b}}{\frac{1}{\alpha} - \alpha}$$

These are equal and opposite and cancel each other. Therefore the principal value of the integral around the unit circle is zero.

This approach for defining a principal value succeeds only at simple poles.

■ ■ ■

A1-10 BRANCH POINTS AND BRANCH CUTS

The singularities that have been considered are those points at which $|W(z)|$ ceases to be finite. At a branch point, $|W(z)|$ may be finite, but $W(z)$ is not single-valued, and hence is not regular. One of the simplest functions with these properties is

$$W_1(z) = z^{1/2} = \sqrt{r}\, e^{j\theta/2} \tag{A1.44}$$

which takes on two values for each value of z, one the negative of the other depending on the choice of θ. This follows since we can write an equally valid form for $z^{1/2}$ as

$$W_2(z) = \sqrt{r}\, e^{j(\theta + 2\pi)/2} = -\sqrt{r}\, e^{j\theta/2} = -W_1(z) \tag{A1.45}$$

Clearly, $W_1(z)$ is not continuous at points on the positive real axis since

$$\lim_{\theta \to 2\pi} (\sqrt{r}\, e^{j\theta/2}) = -\sqrt{r}$$

while

$$\lim_{\theta \to 0} (\sqrt{r}\, e^{j\theta/2}) = \sqrt{r}$$

Hence $W'(z)$ does not exist when z is real and positive. However, the branch $W_1(z)$ is analytic in the region $0 \leq \theta \leq 2\pi$, $r \to 0$. That part of the real axis where $x \geq 0$ is called a **branch cut** for the branch $W_1(z)$, and the branch is analytic except at points on this cut. Hence the cut is a boundary introduced so that the corresponding branch is single-valued and analytic throughout the open region bounded by the cut.

Suppose that we consider the function $W(z) = z^{1/2}$ and contour C, as shown in Figure A1.14a, which encloses the origin. Clearly, after one complete circuit in the positive direction enclosing the origin, θ is increased by 2π, giving a value of $W(z)$ that changes from $W_1(z)$ to $W_2(z)$; that is, the function has changed from one branch to a second branch. To avoid this and to make the function analytic, the contour C is replaced by a contour Γ, which consists of a small circle γ surrounding the branch point, a semi-infinite cut connecting γ and C, and C itself (as shown in Figure A1.14b). Such a contour, which avoids crossing the branch cut, ensures that $W(z)$ is single-valued. Since $W(z)$ is single valued and excludes the origin, we would write for this composite contour C

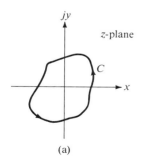
(a)

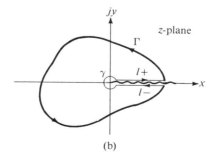
(b)

Figure A1.14
The modifications in a contour to include a branch cut.

$$\oint_C W\,dz = \int_\Gamma + \int_{l-} + \int_\gamma + \int_{l+} = 2\pi j \sum \text{Res} \tag{A1.46}$$

Evaluating the function along the various segments of C then proceeds as before.

EXAMPLE A1.13

If $0 < a < 1$, show that

$$\int_0^\infty \frac{x^{a-1}}{1+x}\,dx = \frac{\pi}{\sin a\pi}$$

Solution: Consider the integral

$$\oint_C \frac{z^{a-1}}{1+z}\,dz = \int_\Gamma + \int_{l-} + \int_\gamma + \int_{l+} = I_1 + I_2 + I_3 + I_4 = \sum \text{Res}$$

which we will evaluate using the contour shown in Figure A1.15. Under the conditions

$$\left|\frac{z^a}{1+z}\right| \to 0 \quad \text{as} \quad |z| \to 0 \qquad \text{if } a > 0$$

$$\left|\frac{z^a}{1+z}\right| \to 0 \quad \text{as} \quad |z| \to \infty \qquad \text{if } a < 1$$

the integrals become

By (A1.41) $\int_\Gamma \to 0 \qquad \int_{l-} = -e^{2\pi ja} \int_0^\infty$

By (A1.42) $\int_\gamma \to 0 \qquad \int_{l+} = 1 \int_0^\infty$

Thus

$$(1 - e^{2\pi ja}) \int_0^\infty \frac{x^{a-1}}{1+x}\,dx = 2\pi j \sum \text{Res}$$

Figure A1.15
The contour for Example A1.13.

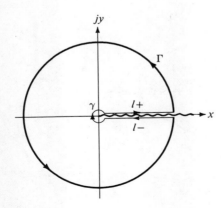

Further, the residue at the simple pole $z = -1$, which is enclosed, is

$$\lim_{z=e^{j\pi}} (1+z) \frac{z^{a-1}}{1+z} = e^{j\pi(a-1)} = -e^{j\pi a}$$

Therefore

$$\int_0^\infty \frac{x^{a-1}}{1+x} dx = 2\pi j \frac{e^{j\pi a}}{e^{j2\pi a} - 1} = 2\pi j \frac{1}{e^{j\pi a} - e^{-j\pi a}} = \frac{\pi}{\sin \pi a}$$

■ ■ ■

A1-11 INTEGRAL OF LOGARITHMIC DERIVATIVE

Of importance in the study of mapping from the z-plane to the $W(z)$-plane is the integral of the logarithmic derivative. Consider therefore the function

$$F(z) = \log W(z) \tag{A1.47}$$

Then

$$\frac{dF(z)}{dz} = \frac{1}{W(z)} \frac{dW(z)}{dz} = \frac{W'(z)}{W(z)}$$

The function to be examined is the following:

$$\int_C \frac{dF(z)}{dz} dz = \int_C \frac{W'(z)}{W(z)} dz \tag{A1.48}$$

The integrand of this expression will be analytic within the contour C except for points at which $W(z)$ is either zero or infinite.

Suppose that $W(z)$ has a pole of order n at z_0. This means that $W(z)$ can be written

$$W(z) = (z - z_0)^n g(z) \tag{A1.49}$$

with n positive for a zero and n negative for a pole. We differentiate this expression to get

$$W'(z) = n(z - z_0)^{n-1} g(z) + (z - z_0)^n g'(z)$$

and so

$$\frac{W'(z)}{W(z)} = \frac{n}{z - z_0} + \frac{g'(z)}{g(z)} \tag{A1.50}$$

For n positive, $W'(z)/W(z)$ will possess a pole of order one. Similarly, for n negative $W'(z)/W(z)$ will possess a pole of order one, but with a negative sign. Thus for the case of n positive or negative, the contour integral in the positive sense yields

$$\int_C \frac{W'(z)}{W(z)} dz = \pm \int_C \frac{n}{z - z_0} dz + \int_C \frac{g'(z)}{g(z)} dz \tag{A1.51}$$

But since $g(z)$ is analytic at the point z_0, then $\int_C [g'(z)/g(z)]\, dz = 0$, and by (A1.34)

$$\int_C \frac{W'(z)}{W(z)}\, dz = \pm 2\pi j n \tag{A1.52}$$

Thus the existence of a zero of $W(z)$ introduces a contribution $2\pi j n_z$ to the contour integral, where n_z is the multiplicity of the zero of $W(z)$ at z_0. Clearly, if a number of zeros of $W(z)$ exist, the total contribution to the contour integral is $2\pi j N$, where N is the weighted value of the zeros of $W(z)$ (weight 1 to a first-order zero, weight 2 to a second-order zero, and so on).

For the case where n is negative, which specifies that $W(z)$ possesses a pole of order n at z_0, then in (A1.52) n is negative and the contribution to the contour integral is now $-2\pi j n_p$ for each pole of $W(z)$; the total contribution is $-2\pi j P$, where P is the weighted number of poles. Clearly, since both zeros and poles of $F(z)$ cause poles of $W'(z)/W(z)$ with opposite signs, then the total value of the integral is

$$\int_C \frac{W'(z)}{W(z)}\, dz = \pm 2\pi j (N - P) \tag{A1.53}$$

Note further that

$$\int_C W'\, dz = \int_C \frac{dW(z)}{dz}\, dz = \int d[\log W(z)]$$

$$= \int d[\log |W(z)| + j \arg W(z)]$$

$$= \log W(z)\Big|_0^\pi + j[\arg W(2\pi) - \arg W(0)]$$

$$= 0 + j[\arg W(2\pi) - \arg W(0)]$$

so that

$$[\arg W(0) - \arg W(2\pi)] = 2\pi (N - P) \tag{A1.54}$$

This relation can be given simple graphical interpretation. Suppose that the function $W(z)$ is represented by its pole and zero constellation on the z-plane. As z traverses the prescribed contour on the z-plane, $W(z)$ will move on the $W(z)$-plane according to its functional dependence on z. But the left-hand side of this equation denotes the total change in the phase angle of $W(z)$ as z travels around the complete contour. Therefore the number of times that the moving point representing $W(z)$ revolves around the origin in the $W(z)$-plane as z moves once around the specified contour is given by $N - P$.

The foregoing is conveniently illustrated graphically. Figure A1.16a shows the prescribed contour in the z-plane, and Figure A1.16b shows a possible form for the variation of $W(z)$. For this particular case, the contour in the z-plane encloses one zero and no poles; hence $W(z)$ encircles the origin once in the clockwise direction in the $W(z)$-plane.

Figure A1.16
Mapping a contour in the z-plane into the contour in the $f(z)$-plane for the specified mapping function $f(z)$.

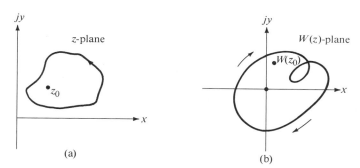

Note that corresponding to a point z_0 within the contour in the z-plane, the point $W(z_0)$ is mapped inside the $W(z)$ contour in the $W(z)$-plane. In fact, every point on the inside of the contour in the z-plane maps onto the inside of the $W(z)$ contour in the $W(z)$-plane (for single-valued functions). Clearly, there is one point in the z-plane that maps into $W(z) = 0$, the origin.

On the other hand, if the contour includes a pole but no zeros, it can be shown by a similar argument that any point in the interior of the z-contour must correspond to a corresponding point outside of the $W(z)$ contour in the $W(z)$-plane. This is manifested by the fact that the $W(z)$ contour is traversed in a counterclockwise direction. With both zeros and poles present, the situation depends on the values of N and P.

Of special interest is the locus of a network function that contains no poles in the right-half plane or on the $j\omega$-axis. In this case the frequency locus is completely traced as z varies along the ω-axis from $-j\infty$ to $+j\infty$. To show this, since $W(z)$ is analytic along this path, $W(z)$ can be written for the neighborhood of a point z_0 in a Taylor series

$$W(z) = \alpha_0 + \alpha_1(z - z_0) + \alpha_2(z - z_0)^2 + \cdots$$

For the neighborhood $z \to \infty$, we examine $W(z')$, where $z' = 1/z$. Since $W(z)$ does not have a pole at $z \to \infty$, then $W(z')$ does not have a pole at $z' = 0$. Therefore we can expand $W(z')$ in a Maclaurin series

$$W(z') = \alpha_0 + \alpha_1 z' + \alpha_2 (z')^2 + \cdots$$

which means that

$$W(z) = \alpha_0 + \frac{\alpha_1}{z} + \frac{\alpha_2}{z^2} + \cdots$$

But as $z \to \infty$, $W(\infty) \to \infty$. In a real network function when z^* is written for z, then $W(z^*) = W^*(z)$. This condition requires that $\alpha_0 = a_0 + j \cdot 0$ be a real number irrespective of how $z \to \infty$; that is, as $z \to \infty$, $W(z) \to$ a fixed point in the $W(z)$-plane.

Figure A1.17 The path for a stable network function.

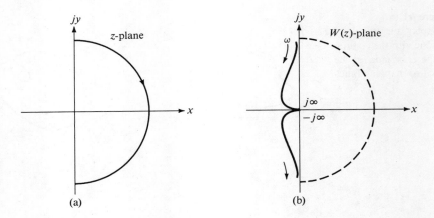

This shows that as z varies around the specified contour in the z-plane, $W(z)$ varies from $W(-j\infty)$ to $W(j\infty)$ as z varies from $-j\infty$ to $+j\infty$. However, $W(-j\infty) = W(j\infty)$, from the above, which thereby shows that the locus is completely determined. This is illustrated in Figure A1.17.

REFERENCES

1. Churchill, R. V. *Introduction to Complex Variables and Applications.* New York: McGraw-Hill, 1948.
2. Irving, J., and N. Mullineux. *Mathematics in Physics and Engineering.* New York: Academic Press, 1959.
3. Whittaker, E. T., and G. N. Watson. *A Course in Modern Analysis.* Cambridge University Press, 4th ed., 1927.

PROBLEMS

A1-1.1 Prove that $\int_C z^n \, dz = 0$ unless $n = -1$.

A1-1.2 Prove that $\int_C dz/(z-a)^n = 0$ unless $n = +1$.

A1-1.3 Find the value of $(1/2\pi j) \int_C [W(z)/(z-z_0)] \, dz$ in the following cases:
  a. $W(z)$ is regular in the finite part of the z-plane and $z = z_0$ lies inside C.
  b. $W(z)$ is regular at all points of the circle γ; $|z - z_0| = \rho$, and all points of C are internal points of γ.
  c. $W(z)$ has a simple pole at $z = \beta$, which lies inside of C.
  d. $W(z)$ has a simple pole at $z = \beta$, which lies outside of C.
  e. By taking C to be the unit circle with center at the origin, $W(z) = \cos z$, and choosing a suitable value of z_0, show that
  $$\int_0^\pi \cos(\cos \theta) \cosh(\sin \theta) \, d\theta = \pi$$

FUNCTIONS OF A COMPLEX VARIABLE

A1-3.1 Obtain the Taylor series expansion

$$\frac{z-1}{z^2} = \sum_{n=0}^{\infty} (-1)^n (n+1)(z-1)^{n+1} \qquad |z-1| < 1$$

A1-3.2 Obtain the Maclaurin series expansion

$$\frac{z+1}{z-1} = -1 - 2\sum_{n=1}^{\infty} z^n \qquad |z| < 1$$

A1-3.3 Expand $1/(2z-1)(z-2)^2$ as a Laurent series about the points $z = 1/2$ and $z = 2$. Specify the coefficients a_{-1} in each case.

A1-3.4 Show that the first four terms of the Laurent expansion of the following function are,

$$\frac{e^z}{z(z^2+1)} = \frac{1}{z} + 1 - \frac{z}{2} - \frac{5z^2}{6} + \cdots$$

A1-3.5 Determine the convergence or divergence properties of the following series:

a. $\displaystyle\sum_{k=1}^{\infty} \frac{1}{k^2+1}$ b. $\displaystyle\sum_{k=1}^{\infty} \left(\frac{2}{z}\right)^k$ c. $\displaystyle\sum_{k=1}^{\infty} \frac{1}{2,4,6,\ldots,(2k+2)}$

A1-6.1 Find the residues of the following functions

a. $\dfrac{z+1}{z(z-2)}$ b. $\dfrac{z^2+1}{z(z^2+3z+2)}$ c. $\dfrac{e^{2z}}{(z-1)^2}$ d. $\dfrac{1+e^z}{\sin z + z \cos z}$ at $z = 0$

A1-6.2 Find the value of the integral $\int_C dz/z^3(z+4)$ when
  a. C is a contour that includes the entire z-plane.
  b. C is a circle with $|z| = 2$.   Ans. $j\pi/32$
  c. C is a circle with $|z+2| = 3$ Ans. 0

A1-6.3 If C is a unit circle about the origin, evaluate the integrals

a. $\displaystyle\int_C \frac{e^{-z}}{z^2} dz$ b. $\displaystyle\int_C \frac{dz}{z \sin z}$ c. $\displaystyle\int_C z e^{1/z} dz$ Ans. (a.) $-2\pi j$ (b.) 0 (c.) $j\pi$

A1-8.1 Use the residue theory to evaluate the following real integrals. Choose suitable contours

a. $\displaystyle\int_0^\infty \frac{dx}{x^4+1}$ b. $\displaystyle\int_0^\infty \frac{\cos x}{(x^2+1)^2} dx$ c. $\displaystyle\int_{-\infty}^\infty \frac{\sin x}{x^2+4x+5} dx$ Ans. (a.) $\dfrac{\pi\sqrt{2}}{4}$ (b.) $\dfrac{\pi}{2e}$ (c.) $\dfrac{-\pi \sin 2}{e}$

A1-8.2 Evaluate

a. $\displaystyle\int_0^{2\pi} \frac{d\theta}{5+3\cos\theta}$ b. $\displaystyle\int_0^{2\pi} \frac{2\cos^2 3\theta}{5-4\cos 2\theta} d\theta$ Ans. (a.) $\dfrac{\pi}{2}$ (b.) $\dfrac{3\pi}{4}$

A1-8.3 Show that

$$\int_0^\infty \frac{\sin ax}{e^{2\pi x}+1} dx = \frac{1}{2a} - \frac{1}{4 \sinh \dfrac{a}{2}}$$

GENERAL APPENDIX 2

MATRICES

A2-1 INTRODUCTION

We find extensive use of matrices in representing systems of equations with both continuous and discrete variables. They are particularly important in studies involving state variables.

To introduce the concept of a matrix, consider the following set of linear algebraic equations:

$$a_{11}x_1 + a_{12}x_2 + \cdots + a_{1n}x_n = w_1$$
$$a_{21}x_1 + a_{22}x_2 + \cdots + a_{2n}x_n = w_2$$
$$\vdots \qquad \vdots \qquad \cdots \qquad \vdots \qquad \vdots$$
$$a_{n1}x_1 + a_{n2}x_2 + \cdots + a_{nn}x_n = w_n$$

Here the x's may denote the state variables of a system and the w's may be the excitation (input) functions. The a's are linear coefficients defined by the system. In matrix form this set of equations is written

$$\begin{bmatrix} a_{11} & a_{12} & \cdots & a_{1n} \\ a_{21} & a_{22} & & a_{2n} \\ \vdots & \vdots & & \vdots \\ a_{n1} & a_{n2} & & a_{nn} \end{bmatrix} \cdot \begin{bmatrix} x_1 \\ x_2 \\ \vdots \\ x_n \end{bmatrix} = \begin{bmatrix} w_1 \\ w_2 \\ \vdots \\ w_n \end{bmatrix}$$

This is usually written more simply as

$$\mathbf{Ax} = \mathbf{w} \quad \text{or} \quad \sum_{j=1}^{n} a_{ij} x_j = w_i \qquad i = 1, 2, \ldots, n$$

where

$$\mathbf{A} = \begin{bmatrix} a_{11} & a_{12} & \cdots & a_{1n} \\ a_{21} & a_{22} & & a_{2n} \\ \vdots & \vdots & & \vdots \\ a_{n1} & a_{n2} & & a_{nn} \end{bmatrix} \qquad \mathbf{x} = \begin{bmatrix} x_1 \\ x_2 \\ \vdots \\ x_n \end{bmatrix} \qquad \mathbf{w} = \begin{bmatrix} w_1 \\ w_2 \\ \vdots \\ w_n \end{bmatrix}$$

MATRICES

Observe that the matrices as defined here are arrays of coefficients; there is no numerical value associated with any of these arrays.

The matrix **A** comprises the coefficients a_{ij}, where i refers to the row and j refers to the column; hence a_{ij} is the coefficient belonging to the ith row and jth column. Clearly, matrices may be square, they may be rectangular, they may be columnar, or they may be row. The relationship among the number of rows and columns in matrices is important in certain operations. We wish to examine some of these relations.

■ ■ ■

A2-2 DEFINITIONS

A matrix will ordinarily be defined by the character of the array, the elements of the array, or some special property of the array. The important definitions follow:

  a. Elements of a matrix **A**: the coefficients a_{ij}.
  b. Order of a matrix: a matrix of m rows and n columns is called an $m \times n$ matrix.
  c. Square matrix: a matrix of order $m \times m$.
  d. Column matrix: a matrix containing only one column; hence of order $m \times 1$.
  e. Row matrix: a matrix containing only one row; it is of order $1 \times m$.
  f. Real matrix: all elements of the matrix are real.
  g. Complex matrix: some or all elements of the matrix are algebraically complex.
  h. Zero or null matrix: all elements of the matrix are zero.
  i. Sparse matrix: many of the elements of the matrix are zero (no quantitative measure is given).
  j. Diagonal matrix: a square matrix with nonzero elements only along the diagonal.
  k. Unit matrix: a diagonal matrix whose elements are unity.
  l. Transposed matrix: A transposed matrix $\mathbf{A}^T$ has rows and columns that are interchanged with the rows and columns of the original matrix **A**; that is, if $\mathbf{A} = [a_{ij}]$, then $\mathbf{A}^T = [a_{ji}]$.
  m. Complex conjugate of a matrix: $\mathbf{A}^*$ is the complex conjugate of matrix **A** if each element of $\mathbf{A}^*$ is the complex conjugate of the corresponding element of **A**. Clearly, for real matrices $\mathbf{A} = \mathbf{A}^*$.
  n. Determinant of a matrix: the determinant that has the same elements as the matrix.
  o. Trace of a matrix: the sum of the diagonal elements of the matrix.
  p. Rank of a matrix: the number of dimensions spanned by the rows (or columns) considered as n (or m) vectors. The rank of an $n \times n$ nonsingular matrix is n. The rank of an $n \times n$ singular matrix is less than n. Equivalently, we say that an $n \times n$ matrix **A** is of rank $r < n$ if there exists a

submatrix $r \times r$ of **A** that is nonsingular (has a nonzero determinant) and any other submatrix of order higher than r is singular (has a zero determinant).

q. Cofactor of the element a_{ij}: the signed minor of a_{ij}; that is, cofactor $a_{ij} = (-1)^{i+j}$ minor a_{ij}, where minor a_{ij} is the determinant of the matrix formed by omitting the ith row and jth column of the original matrix **A**.

r. Adjoint of matrix **A**: the transposed cofactor of **A**.

s. Nonsingular matrix: A square matrix whose rows or columns are linearly independent. In this case $\det|\mathbf{A}| \neq 0$.

Additional definitions will be introduced as details of matrix operations are discussed.

■ ■ ■

A2-3 MATRIX ALGEBRA

We will now discuss some of the more important algebraic operations involving matrices.

a. Equality of matrices: Two matrices are equal if and only if each of their corresponding elements are equal. This prescribes the requirement that for two matrices to be equal, they must contain the same number of rows and columns.

b. Addition of matrices: The sum of two matrices **A** and **B**, both of the same order, is

$$\mathbf{A} + \mathbf{B} = [a_{ij} + b_{ij}] \quad \begin{matrix} i = 1, 2, \ldots, m \\ j = 1, 2, \ldots, n \end{matrix} \tag{A2.1}$$

For example, we have

$$\begin{bmatrix} 2 & 3 \\ 4 & -3 \end{bmatrix} + \begin{bmatrix} 1 & -2 \\ 2 & -4 \end{bmatrix} = \begin{bmatrix} 3 & 1 \\ 6 & -7 \end{bmatrix}$$

c. Subtraction of matrices (a corollary of b): The difference of two $m \times n$ matrices is another $m \times n$ matrix, the elements of which are the difference of the corresponding elements of the two matrices.

For example

$$\begin{bmatrix} 2 & 3 \\ 4 & -3 \end{bmatrix} - \begin{bmatrix} 1 & -2 \\ 2 & -4 \end{bmatrix} = \begin{bmatrix} 1 & 5 \\ 2 & 1 \end{bmatrix}$$

d. Matrix multiplication: The product of matrix **A** of m rows and n columns and matrix **B** of n rows and p columns is matrix **C** of m rows and p columns, with the elements as follows:

$$\mathbf{C} = \mathbf{AB} \tag{A2.2}$$

where

MATRICES

$$c_{ij} = \sum_{k=1}^{n} a_{ik} b_{kj} \quad \begin{array}{l} i = 1, 2, \ldots, m \\ j = 1, 2, \ldots, p \end{array}$$

For example

$$\begin{bmatrix} a_1 & a_2 \\ b_1 & b_2 \end{bmatrix} \begin{bmatrix} c_1 & c_2 \\ d_1 & d_2 \end{bmatrix} = \begin{bmatrix} a_1 c_1 + a_2 d_1 & a_1 c_2 + a_2 d_2 \\ b_1 c_1 + b_2 d_1 & b_1 c_2 + b_2 d_2 \end{bmatrix}$$

$[2 \times 2] \quad [2 \times 2] \qquad\qquad [2 \times 2]$

$$\begin{bmatrix} a_1 & a_2 \\ b_1 & b_2 \end{bmatrix} \begin{bmatrix} c_1 \\ d_1 \end{bmatrix} = \begin{bmatrix} a_1 c_1 + a_2 d_1 \\ b_1 c_2 + b_2 d_1 \end{bmatrix}$$

$[2 \times 2] \quad [2 \times 1] \qquad [2 \times 1]$

e. Conformable matrices: Two matrices that can be multiplied together are said to be conformable. This means that matrix **AB** might exist, whereas matrix **BA** might not exist; that is, **AB** might satisfy condition d, whereas **BA** might not do so. In general $\mathbf{AB} \neq \mathbf{BA}$.

f. Commutable matrices: Two matrices **A** and **B** are commutative if $\mathbf{AB} = \mathbf{BA}$.

g. Matrix division: Matrix division is not a defined operation. The equivalent operation involving the inverse of a matrix is defined.

h. Inverse of a matrix: The inverse $\mathbf{A}^{-1}$ of a square matrix **A** is defined by the relation

$$\mathbf{A}^{-1}\mathbf{A} = \mathbf{A}\mathbf{A}^{-1} = \mathbf{I} \quad \text{(the unit matrix)} \tag{A2.3}$$

This relation assumes that corresponding to a square matrix **A**, $\mathbf{A}^{-1}$ exists. When the inverse does not exist, **A** is said to be a singular matrix. In detail, to find the inverse of a square matrix

$$\mathbf{A} = [a_{ij}]$$

then

$$\mathbf{A}^{-1} = [a_{ij}]^{-1} = \frac{\text{transposed cofactor}[a_{ij}]}{\Delta(\mathbf{A})} = \frac{\text{cofactor}[a_{ji}]}{\Delta(\mathbf{A})} \tag{A2.4}$$

where $\Delta(\mathbf{A})$ is the determinant of the matrix, and the transposed cofactor is the adjoint of matrix **A**. If $\Delta(\mathbf{A}) = 0$, the matrix is singular and the inverse does not exist.

For example, to find the inverse of the matrix **A** given by

$$\mathbf{A} = \begin{bmatrix} 2 & 3 & 1 \\ 1 & 0 & 0 \\ 2 & 1 & 1 \end{bmatrix}$$

we must find the following

$$\underbrace{\begin{bmatrix} 0 & -1 & 1 \\ -2 & 0 & 4 \\ 0 & +1 & -3 \end{bmatrix}}_{\text{cofactor matrix}}; \underbrace{\begin{bmatrix} 0 & -2 & 0 \\ -1 & 0 & +1 \\ 1 & 4 & -3 \end{bmatrix}}_{\text{transpose cofactor matrix}}; \underbrace{\begin{bmatrix} 0 & 1 & 0 \\ \frac{1}{2} & 0 & -\frac{1}{2} \\ -\frac{1}{2} & -2 & \frac{3}{2} \end{bmatrix}}_{\substack{\text{transpose cofactor} \\ \text{divided by} \\ \Delta(\mathbf{A}) = -2}} = \mathbf{A}^{-1}$$

$$\underbrace{\begin{vmatrix} 2 & 3 & 1 \\ 1 & 0 & 0 \\ 2 & 1 & 1 \end{vmatrix} = -2}_{\text{determinant } \Delta(\mathbf{A})}$$

i. *Partitioning of matrices:* When matrices have a large number of rows and columns, it is often desirable to partition them into smaller sections or submatrices. Partitioning is best illustrated by an example. Suppose that a matrix **A** is partitioned into two parts as shown by the broken line, and **B** is partitioned into three parts, also as shown. The resulting matrix is

$$\left[\frac{\mathbf{A}_1}{\mathbf{A}_2}\right][\mathbf{B}_1 \mid \mathbf{B}_2 \mid \mathbf{B}_3] = \begin{bmatrix} \mathbf{A}_1\mathbf{B}_1 & \mathbf{A}_1\mathbf{B}_2 & \mathbf{A}_1\mathbf{B}_3 \\ \mathbf{A}_2\mathbf{B}_1 & \mathbf{A}_2\mathbf{B}_2 & \mathbf{A}_2\mathbf{B}_3 \end{bmatrix} \quad (A2.5)$$

The matrix product **AB** is thus evaluated as though the submatrices are ordinary matrix elements. For partitioning to be possible, the submatrices must be conformable.

j. *Inverse of product of matrices:* Consider the product of two conformable nonsingular matrices. From the fact that

$$(\mathbf{AB})(\mathbf{AB})^{-1} = \mathbf{I}$$

and also the fact that

$$\mathbf{ABB}^{-1}\mathbf{A}^{-1} = \mathbf{AA}^{-1} = \mathbf{I}$$

it then follows that

$$(\mathbf{AB})^{-1} = \mathbf{B}^{-1}\mathbf{A}^{-1} \quad (A2.6)$$

That is, the inverse of the product of two nonsingular matrices of the same order is the product of the inverse matrices in reverse order.

k. *Transpose of a product:* Consider two conformable matrices **A** and **B**. A typical term of their product is

$$[\mathbf{AB}]^T = [ab]_{ij}^T = [ab]_{ji}$$

$$= \sum_k a_{jk} b_{ki} = \sum_k b_{ik}^T a_{kj}^T = [b^T a^T]_{ij}$$

MATRICES

which shows that

$$[\mathbf{AB}]^T = \mathbf{B}^T\mathbf{A}^T \tag{A2.7}$$

That is, the transpose of a matrix product is the product of their transposes in reverse order.

l. Quadratic forms: Consider the function

$$Q = \mathbf{A}^T\mathbf{P}\mathbf{A} \tag{A2.8}$$

The entity on the right is known as a congruent transformation of $\mathbf{P}$. Specifically, suppose that the matrices are (where $p_{12} = p_{21}$, $p_{13} = p_{31}$, and $p_{23} = p_{32}$)

$$\mathbf{A} = \begin{bmatrix} a_1 \\ a_2 \\ a_3 \end{bmatrix} \qquad \mathbf{P} = \begin{bmatrix} p_{11} & p_{12} & p_{13} \\ p_{21} & p_{22} & p_{23} \\ p_{31} & p_{32} & p_{33} \end{bmatrix} \qquad \mathbf{A}^T = \begin{bmatrix} a_1 & a_2 & a_3 \end{bmatrix}$$

Upon expansion of (A2.8), there results

$$Q = p_{11}a_1^2 + p_{22}a_2^2 + p_{33}a_3^2 + 2p_{12}a_1a_2 + 2p_{32}a_2a_3 + 2p_{31}a_3a_1$$

This expression is homogeneous of the second degree in the variables a_1, a_2, a_3.

When $Q = \mathbf{A}^T\mathbf{P}\mathbf{A}$ is greater than zero for $\mathbf{A} \neq 0$, the quadratic form is called positive definite (it is customary to say that $\mathbf{P}$ is positive definite in this case). If Q is less than zero for $\mathbf{A} \neq 0$, the quadratic is negative definite. If Q is essentially positive but may become negative, it is called positive semidefinite. For example

$$x_1^2 + 2x_1x_2 + 2x_2^2 \qquad \text{positive definite}$$

$$-(x_1^2 + 2x_1x_2 + 2x_2^2) \qquad \text{negative definite}$$

$$x_1^2 + 2x_1x_2 + x_2^2 \qquad \text{positive semidefinite}$$

$$-(x_1^2 + 2x_1x_2 + x_2^2) \qquad \text{negative semidefinite}$$

■ ■ ■

A2-4 FUNCTIONS OF A MATRIX

We consider the following set of algebraic equations

$$\mathbf{A}\mathbf{x} = \lambda\mathbf{x} \tag{A2.9}$$

where $\mathbf{A}$ = a known real square matrix of order $n \times n$

$\mathbf{x}$ = an unknown column vector

λ = a scalar parameter (it is often the complex number s)

Here, of course, the vectors $\mathbf{x}$ and $\lambda\mathbf{x}$ have the same direction. An important question is—are there any directions that are left invariant by the transformation defined by $\mathbf{A}$? A nontrivial solution of (A2.9) exists only for certain specific values of the scalar parameter λ. To find these, write (A2.9) in the form

$$(\mathbf{A} - \lambda\mathbf{I})\mathbf{x} = \mathbf{0} \tag{A2.10}$$

This expression has a nontrivial solution **x** if and only if

$$\det|\mathbf{A} - \lambda\mathbf{I}| = 0 \tag{A2.11}$$

Now expand the determinant, which is written explicitly,

$$\det|\mathbf{A} - \lambda\mathbf{I}| = \begin{vmatrix} a_{11} - \lambda & a_{12} & \cdots & a_{1n} \\ a_{21} & a_{22} - \lambda & & a_{2n} \\ \vdots & \vdots & & \vdots \\ a_{n1} & a_{n2} & & a_{nn} - \lambda \end{vmatrix} = 0$$

Inspection of this determinant shows it to be of degree n in λ. The expansion of this determinant has the form

$$\lambda^n - (a_{11} + a_{22} + \cdots + a_{nn})\lambda^{n-1} + \cdots + (-1)^n|\mathbf{A}| = 0 \tag{A2.12}$$

where $|\mathbf{A}| = \Delta(\mathbf{A})$ is the determinant of **A**. This equation is called the **characteristic equation** of the matrix **A**. The roots of this equation are called the **eigenvalues** of matrix **A**.

An equation of degree n has n roots, some of which may be repeated and some or all of which may be real or complex. If the coefficients a are all real, the complex roots will occur in conjugate pairs. If we write the eigenvalues of **A** as $\lambda_1, \lambda_2, \ldots, \lambda_k$, the determinant can be written

$$g(\lambda) = |\mathbf{A} - \lambda\mathbf{I}| = (\lambda - \lambda_1)^{r_1}(\lambda - \lambda_2)^{r_2} \cdots (\lambda - \lambda_k)^{r_k} \tag{A2.13}$$

where the r_i specify the order of root λ_i. Note specifically that when **A** is symmetric, the eigenvalues are real.

Corresponding to each eigenvalue there exists a nonzero column vector **x** that satisfies (A2.10). $\mathbf{x}_i$ is the **eigenvector** (column) corresponding to the eigenvalue λ_i.

EXAMPLE A2.1
Find the eigenvalues and eigenvectors of the matrix

$$\mathbf{A} = \begin{bmatrix} 1 & 2 \\ 4 & 3 \end{bmatrix}$$

Solution: By (A2.11) we obtain

$$\begin{vmatrix} 1 - \lambda & 2 \\ 4 & 3 - \lambda \end{vmatrix} = \lambda^2 - 4\lambda - 5 = 0$$

From this polynomial we find the two roots to be: $\lambda_1 = -1$, $\lambda_2 = 5$. Further, since (A2.10) must be satisfied by each eigenvalue λ_i, we obtain

$$\begin{bmatrix} 1 + 1 & 2 \\ 4 & 3 + 1 \end{bmatrix} \begin{bmatrix} x_{11} \\ x_{12} \end{bmatrix} = \begin{bmatrix} 0 \\ 0 \end{bmatrix} \qquad \begin{bmatrix} 1 - 5 & 2 \\ 4 & 3 - 5 \end{bmatrix} \begin{bmatrix} x_{21} \\ x_{22} \end{bmatrix} = \begin{bmatrix} 0 \\ 0 \end{bmatrix}$$

The first of these yields the relations

MATRICES

$$2x_{11} + 2x_{12} = 0 \quad \text{and} \quad 4x_{11} + 4x_{12} = 0$$

Thus, corresponding to the eigenvalue $\lambda_1 = -1$, the eigenvector is

$$\mathbf{x}_1 = [x_{11} \quad x_{12}]^T \qquad \mathbf{x}_1 = \begin{bmatrix} x_{11} \\ x_{12} \end{bmatrix} = \begin{bmatrix} 1k \\ -1k \end{bmatrix}$$

where k is any positive constant; for example, $k = 1$.

From the second relation, we obtain

$$-4x_{21} + 2x_{22} = 0 \quad \text{and} \quad 4x_{21} - 2x_{22} = 0$$

Thus, corresponding to the eigenvalue $\lambda_2 = 5$, the eigenvector is

$$\mathbf{x}_2 = [x_{21} \quad x_{22}]^T \qquad \mathbf{x}_2 = \begin{bmatrix} x_{21} \\ x_{22} \end{bmatrix} = \begin{bmatrix} 1k \\ 2k \end{bmatrix}$$

where again k is any positive constant. ∎

If we create a matrix $\mathbf{E}$ whose columns are the eigenvectors of matrix $\mathbf{A}$, then the following condition holds

$$\mathbf{D} = \mathbf{E}^{-1}\mathbf{A}\mathbf{E} \tag{A2.14}$$

where $\mathbf{D}$ is a diagonal matrix. For example, for the matrix $\mathbf{A}$ in Example A2.1

$$\mathbf{D} = \frac{1}{3}\begin{bmatrix} 2 & -1 \\ 1 & 1 \end{bmatrix}\begin{bmatrix} 1 & 2 \\ 4 & 3 \end{bmatrix}\begin{bmatrix} 1 & 1 \\ -1 & 2 \end{bmatrix} = \frac{1}{3}\begin{bmatrix} -3 & 0 \\ 0 & 15 \end{bmatrix} = \begin{bmatrix} -1 & 0 \\ 0 & 5 \end{bmatrix} \tag{A2.15}$$

Observe that the diagonal elements of $\mathbf{D}$ are the eigenvalues of $\mathbf{A}$. If we premultiply (A2.14) by $\mathbf{E}$ and postmultiply by $\mathbf{E}^{-1}$, we obtain an equivalent relation

$$\mathbf{A} = \mathbf{E}\mathbf{D}\mathbf{E}^{-1} \tag{A2.16}$$

We can use this relation to show that the nth power of $\mathbf{A}$ is given by

$$\mathbf{A}^n = \mathbf{E}\mathbf{D}^n\mathbf{E}^{-1} \tag{A2.17}$$

To prove this assertion, consider $\mathbf{A}^3$, which we write

$$\mathbf{A}^3 = \mathbf{E}\mathbf{D}\underbrace{\mathbf{E}^{-1}\mathbf{E}}_{\mathbf{I}}\mathbf{D}\underbrace{\mathbf{E}^{-1}\mathbf{E}}_{\mathbf{I}}\mathbf{D}\mathbf{E}^{-1} = \mathbf{E}\mathbf{D}^3\mathbf{E}^{-1}$$

Similarly, we have

$$\mathbf{A}^{1/2} = \mathbf{E}\mathbf{D}^{1/2}\mathbf{E}^{-1} \tag{A2.18}$$

This equation is valid since we have

$$\mathbf{A}^{1/2}\mathbf{A}^{1/2} = \mathbf{E}\mathbf{D}^{1/2}\mathbf{E}^{-1}\mathbf{E}\mathbf{D}^{1/2}\mathbf{E}^{-1} = \mathbf{E}\mathbf{D}\mathbf{E}^{-1} = \mathbf{A}$$

It is not always possible to diagonalize a matrix. However, if we can find n linearly independent eigenvectors corresponding to an $n \times n$ matrix, the diagonalization is possible, with the diagonal elements equal to the eigenvalues of the matrix. The transformation vector $\mathbf{E}$ is known as a **similarity** transformation.

Often it is necessary to find the inverse of the matrix function

$$\Phi(s) = [sI - A]^{-1} = \frac{\text{adjoint } [sI - A]}{\Delta(s)} = \frac{\Delta_a(s)}{\Delta(s)} \tag{A2.19}$$

This matrix function arises in connection with the solution of the state equations of a linear system (see Chapter 12). Here, as specified by (A2.4), we have $\Delta_a(s) = [sI - A]_{ji} = [sI - A]_{ij}^T$, which is the transpose of the cofactor of the ijth element of $[sI - A]$. Also, $\Delta(s) = |sI - A|$ is the determinant of $[sI - A]$ and is precisely of the form given in (A2.12). Of course, evaluating the roots of this characteristic equation to find the eigenvalues is quite laborious for $n > 3$. Also, expanding $\Delta_a(s)$ is a complicated endeavor. The situation is eased somewhat if we express $\Phi(s)$ explicitly as

$$\Phi(s) = \frac{\Delta_a(s)}{\Delta(s)} = \frac{F_0 s^{n-1} + F_1 s^{n-2} + \cdots + F_{n-1}}{s^n + a_1 s^{n-1} + \cdots + a_n} \tag{A2.20}$$

where $F_0, F_1, F_2, \ldots$ are $n \times n$ matrices. This expression arises from the fact that each element of the adjoint matrix in the numerator of (A2.19) is a polynomial of degree $n - 1$. The matrices F_k and the coefficients a_k can be found by means of the algorithm

$$a_k = -\frac{1}{k} \text{Tr}(AF_{k-1}) \qquad k = 1, 2, \ldots, n$$

$$F_k = AF_{k-1} + a_k I \qquad k = 1, 2, \ldots, (n-1)$$

where Tr denotes the trace of the matrix. Note that the last relation is F_{n-1} and for $k = n$, $F_n = 0$; therefore $F_n = AF_{n-1} + a_n I = 0$.

■ ■ ■

A2-5 CAYLEY-HAMILTON THEOREM

The Cayley-Hamilton theorem is an important result in matrix theory. The theorem follows:

Theorem A2.1 Every $n \times n$ matrix satisfies its own characteristic equation. Specifically, if we write (A2.12) in the form

$$g(\lambda) = |A - \lambda I| = \lambda^n + b_{n-1} \lambda^{n-1} + \cdots + b_0 = 0 \tag{A2.21}$$

then the Cayley-Hamilton theorem specifies that

$$g(A) = A^n + b_{n-1} A^{n-1} + \cdots + b_1 A + b_0 I = 0 \tag{A2.22}$$

This theorem is of considerable value when calculating various functions of matrix A.

EXAMPLE A2.2
Show that the matrix A of Example 2.1 satisfies the Cayley-Hamilton theorem.

Solution: Apply (A2.22) to the results of Example 2.1 to write

$$\begin{bmatrix} 1 & 2 \\ 4 & 3 \end{bmatrix}^2 - 4 \begin{bmatrix} 1 & 2 \\ 4 & 3 \end{bmatrix} - 5 \begin{bmatrix} 1 & 0 \\ 0 & 1 \end{bmatrix} = \begin{bmatrix} 9-4-5 & 8-8 \\ 16-16 & 17-12-5 \end{bmatrix}$$
$$= \begin{bmatrix} 0 & 0 \\ 0 & 0 \end{bmatrix}$$

We note that (A2.22) can be written

$$\mathbf{g(A)} = (\mathbf{A} - \lambda_1 \mathbf{I})(\mathbf{A} - \lambda_2 \mathbf{I}) \cdots (\mathbf{A} - \lambda_n \mathbf{I}) = 0 \tag{A2.23}$$

and for the present example

$$\begin{bmatrix} 1+1 & 2 \\ 4 & 3+1 \end{bmatrix} \begin{bmatrix} 1-5 & 2 \\ 4 & 3-5 \end{bmatrix} = \begin{bmatrix} -8+8 & 4-4 \\ -16+16 & 8-8 \end{bmatrix} = \begin{bmatrix} 0 & 0 \\ 0 & 0 \end{bmatrix}$$

This form shows the validity of (A2.23). ∎

By means of the Cayley-Hamilton theorem, it is possible to reduce a polynomial of the nth-order matrix $\mathbf{A}$ to a polynomial whose highest degree in $\mathbf{A}$ is $n - 1$. This follows from (A2.22), which is rewritten

$$\mathbf{A}^n = -b_{n-1}\mathbf{A}^{n-1} - \cdots - b_1 \mathbf{A} - b_0 \mathbf{I} \tag{A2.24}$$

To extend this result, multiply this expression by $\mathbf{A}$. This yields

$$\mathbf{A}^{n+1} = -b_{n-1}\mathbf{A}^n - \cdots - b_1 \mathbf{A}^2 - b_0 \mathbf{A} \tag{A2.25}$$

Now substitute (A2.24) in this expression, with the result

$$\mathbf{A}^{n+1} = -b_{n-1}(-b_{n-1}\mathbf{A}^{n-1} - \cdots - b_1 \mathbf{A} - b_0 \mathbf{I}) - \cdots - b_1 \mathbf{A}^2 - b_0 \mathbf{A}$$
$$= (b_{n-1}^2 - b_{n-2})\mathbf{A}^{n-1} + (b_{n-1}b_{n-2} - b_{n-3})\mathbf{A}^{n-2}$$
$$+ \cdots + b_{n-1}b_0 \mathbf{I} \tag{A2.26}$$

Observe that this expresses $\mathbf{A}^{n+1}$ in terms of $\mathbf{A}^{n-1}, \mathbf{A}^{n-2}, \ldots, \mathbf{A}$ and $\mathbf{I}$. This process can be continued to prove that $\mathbf{A}$ to any power can be represented as the weighted sum of matrices involving $\mathbf{A}$ to powers not exceeding $n - 1$. As a result, functions of matrices that can be expressed in power series form, say

$$\mathbf{f(A)} = \zeta_0 \mathbf{I} + \zeta_1 \mathbf{A} + \cdots + \zeta_n \mathbf{A}^n + \cdots = \sum_{k=0}^{\infty} \zeta_k \mathbf{A}^k \tag{A2.27}$$

can be represented as

$$\mathbf{f(A)} = \gamma_0 \mathbf{I} + \gamma_1 \mathbf{A} + \cdots + \gamma_{n-1} \mathbf{A}^{n-1} = \sum_{k=0}^{n-1} \gamma_k \mathbf{A}^k \tag{A2.28}$$

The ζ-factors are functions of the γ-factors and n. The evaluation of the γ-factors can be accomplished as in (A2.26), although this is not a very convenient procedure.

For a more convenient method for finding the γ-factors, we retrace the above steps beginning with (A2.21), rather than with (A2.22). This essentially

APPENDIX 2

substitutes λ for $\mathbf{A}$ in (A2.25) through (A2.28). That is, we write polynomials of λ in terms of $\lambda, \lambda^2, \ldots, \lambda^{n-1}$, with the explicit form

$$\mathbf{f(A)} \triangleq f(\lambda) = \gamma_0 + \gamma_1 \lambda + \cdots + \gamma_{n-1} \lambda^{n-1} = \sum_{k=0}^{n-1} \gamma_k \lambda^k \triangleq \mathbf{g(A)} \qquad \text{(A2.29)}$$

Now we make use of the fact that this expression is valid for any λ that is a solution of the characteristic equation—that is, for any eigenvalue of the matrix $\mathbf{A}$. When the eigenvalues are distinct, (A2.29) yields n equations in n unknowns

$$\begin{aligned} f(\lambda_1) &= \gamma_0 + \gamma_1 \lambda_1 + \cdots + \gamma_{n-1} \lambda_1^{n-1} \\ f(\lambda_2) &= \gamma_0 + \gamma_1 \lambda_2 + \cdots + \gamma_{n-1} \lambda_2^{n-1} \\ f(\lambda_n) &= \gamma_0 + \gamma_1 \lambda_n + \cdots + \gamma_{n-1} \lambda_n^{n-1} \end{aligned} \qquad \text{(A2.30)}$$

This set of equations can be solved for the coefficients $\gamma_0, \gamma_1, \ldots, \gamma_{n-1}$. Since these values of γ are precisely those in (A2.28), the problem is solved.

For the case when λ_i denotes k-fold degenerate eigenvalues of $\mathbf{A}$, the functions $f(\lambda)$ and $g(\lambda)$ satisfy the following equations

$$\begin{aligned} f(\lambda)\big|_{\lambda=\lambda_i} &= g(\lambda)\big|_{\lambda=\lambda_i} \\ \frac{df(\lambda)}{d\lambda}\bigg|_{\lambda=\lambda_i} &= \frac{dg(\lambda)}{d\lambda}\bigg|_{\lambda=\lambda_i} \\ \frac{d^2 f(\lambda)}{d\lambda^2}\bigg|_{\lambda=\lambda_i} &= \frac{d^2 g(\lambda)}{d\lambda^2}\bigg|_{\lambda=\lambda_i} \\ &\vdots \\ \frac{d^{k-1} f(\lambda)}{d\lambda^{k-1}}\bigg|_{\lambda=\lambda_i} &= \frac{d^{k-1} g(\lambda)}{d\lambda^{k-1}}\bigg|_{\lambda=\lambda_i} \end{aligned} \qquad \text{(A2.31)}$$

EXAMPLE A2.3

Evaluate $\mathbf{f(A)} = \mathbf{A}^k$, for $\mathbf{A} = \begin{bmatrix} 2 & 0 \\ 1 & 1 \end{bmatrix}$

Solution: The characteristic equation is

$$g(\lambda) = |\mathbf{A} - \lambda \mathbf{I}| = \begin{vmatrix} 2-\lambda & 0 \\ 1 & 1-\lambda \end{vmatrix} = 0$$

From this

$$g(\lambda) = (2-\lambda)(1-\lambda) = 0 \qquad \text{with eigenvalues } \lambda_1 = 2 \quad \lambda_2 = 1$$

Now use (A2.30) to write

$$2^k = \gamma_0 + \gamma_1 \cdot 2$$
$$1^k = \gamma_0 + \gamma_1 \cdot 1$$

MATRICES

Subtract and solve for γ_1. The result is

$$\gamma_1 = 2^k - 1^k$$
$$\gamma_0 = 2 \times 1^k - 2^k$$

Using these results, we have (see A2.28)

$$\mathbf{f(A)} = \mathbf{A}^k = \gamma_0 \mathbf{I} + \gamma_1 \mathbf{A}$$

$$= \begin{bmatrix} 2 - 2^k & 0 \\ 0 & 2 - 2^k \end{bmatrix} + \begin{bmatrix} (2^k - 1) \times 2 & 0 \\ 2^k - 1 & 2^k - 1 \end{bmatrix} = \begin{bmatrix} 2^k & 0 \\ 2^k - 1 & 1 \end{bmatrix} \quad \blacksquare$$

EXAMPLE A2.4

Find the function $\mathbf{f(A)} = \exp[-\mathbf{A}t]$, given

$$\mathbf{A} = \begin{bmatrix} 1 & 0 & 4 \\ 0 & 2 & 0 \\ 0 & 1 & 2 \end{bmatrix}$$

Solution: The eigenvalues of $\mathbf{A}$ are found to be $\lambda_1 = 1$ with multiplicity 1, and $\lambda_2 = 2$ with multiplicity 2. From (A2.23) we write

$$\mathbf{f(A)} = e^{-\mathbf{A}t} \quad \text{or} \quad f(\lambda) = e^{-\lambda t}$$
$$\mathbf{g(A)} = \gamma_2 \mathbf{A}^2 + \gamma_1 \mathbf{A} + \gamma_0 \mathbf{I} \quad \text{or} \quad g(\lambda) = \gamma_2 \lambda^2 + \gamma_1 \lambda + \gamma_0$$

The coefficients γ_2, γ_1, and γ_0 are found from the conditions

$$f(\lambda)|_{\lambda=1} = g(\lambda)|_{\lambda=1}$$
$$f(\lambda)|_{\lambda=2} = g(\lambda)|_{\lambda=2}$$
$$\left.\frac{df(\lambda)}{d\lambda}\right|_{\lambda=2} = \left.\frac{dg(\lambda)}{d\lambda}\right|_{\lambda=2}$$

These yield, respectively, the relations

$$e^{-t} = \gamma_2 + \gamma_1 + \gamma_0$$
$$e^{-2t} = 4\gamma_2 + 2\gamma_1 + \gamma_0$$
$$-te^{-2t} = 4\gamma_2 + \gamma_1 + 0\gamma_0$$

The solution of this system of equations yields

$$\gamma_2 = e^{-t} - (te^{-2t} + e^{-2t})$$
$$\gamma_1 = (4 + 3t)e^{-2t} - 4e^{-t}$$
$$\gamma_0 = 4e^{-t} - (2t + 3)e^{-2t}$$

The matrix $\exp[-\mathbf{A}t]$ is given by

$$e^{-\mathbf{A}t} = \gamma_2 \mathbf{A}^2 + \gamma_1 \mathbf{A} + \gamma_0 \mathbf{I}$$

$$= \gamma_2 \begin{bmatrix} 1 & 4 & 12 \\ 0 & 4 & 0 \\ 0 & 4 & 4 \end{bmatrix} + \gamma_1 \begin{bmatrix} 1 & 0 & 4 \\ 0 & 2 & 0 \\ 0 & 1 & 2 \end{bmatrix} + \gamma_0 \begin{bmatrix} 1 & 0 & 0 \\ 0 & 1 & 0 \\ 0 & 0 & 1 \end{bmatrix}$$

$$= \begin{bmatrix} (\gamma_2 + \gamma_1 + \gamma_0) & 4\gamma_2 & 12\gamma_2 + 4\gamma_1 \\ 0 & 4\gamma_2 + 2\gamma_1 + \gamma_0 & 0 \\ 0 & 4\gamma_2 + \gamma_1 & 4\gamma_2 + 2\gamma_1 + \gamma_0 \end{bmatrix}$$ ∎

EXAMPLE A2.5

Find $\mathbf{A}^{-1}$ for the matrix of Example A2.3 using the Cayley-Hamilton theorem.

Solution: Since $\mathbf{A}$ satisfies its own characteristic equation, then

$$\mathbf{A}^2 - 3\mathbf{A} + 2\mathbf{I} = 0$$

from which, by premultiplying each term by $\mathbf{A}^{-1}$,

$$\mathbf{A} - 3\mathbf{I} + 2\mathbf{A}^{-1} = 0.$$

Therefore

$$\mathbf{A}^{-1} = -\tfrac{1}{2}\mathbf{A} + \tfrac{3}{2}\mathbf{I} = \tfrac{1}{2}\begin{bmatrix} 2 & 0 \\ 1 & 1 \end{bmatrix} + \tfrac{3}{2}\begin{bmatrix} 1 & 0 \\ 0 & 1 \end{bmatrix} = \begin{bmatrix} \tfrac{1}{2} & 0 \\ -\tfrac{1}{2} & 1 \end{bmatrix}$$ ∎

A second method for finding functions of a matrix, which often possesses some advantages over the method discussed above, is based on the **spectral decomposition** of a matrix. In this method an $n \times n$ matrix is represented by n **constituent** matrices $\mathbf{M}_1, \mathbf{M}_2, \ldots, \mathbf{M}_n$ by the expression

$$\mathbf{A} = \lambda_1 \mathbf{M}_1 + \lambda_2 \mathbf{M}_2 + \cdots + \lambda_n \mathbf{M}_n = \sum_{k=1}^{n} \lambda_k \mathbf{M}_k \qquad (A2.32)$$

where λ_k, $k = 1, 2, \ldots, n$, are the distinct eigenvalues of $\mathbf{A}$. The case for repeated roots will be considered below. The constituent matrices $\mathbf{M}_k$ have the following properties:

$$\mathbf{M}_i \mathbf{M}_j = \begin{cases} 0 & i \neq j \\ \mathbf{M}_i & i = j \end{cases} \quad (a)$$

$$\sum_{k=i}^{n} \mathbf{M}_k = \mathbf{I} \quad (b) \qquad (A2.33)$$

$$\mathbf{A}\mathbf{M}_k = \mathbf{M}_k \mathbf{A} = \lambda_k \mathbf{M}_k \quad (c)$$

$$\mathbf{M}_k \text{ have the rank 1} \quad (d)$$

On the basis of (A2.32) and our previous discussion, it follows that a function of a matrix $\mathbf{f}(\mathbf{A})$ can be written

$$\mathbf{f}(\mathbf{A}) = \sum_{k=1}^{n} f(\lambda_k)\mathbf{M}_k \qquad (A2.34)$$

MATRICES

To examine the application of this form, suppose that $\mathbf{A}$ is a 2×2 matrix with eigenvalues λ_1 and λ_2. We write

$$\mathbf{A}^k = \lambda_1^k \mathbf{M}_1 + \lambda_2^k \mathbf{M}_2 \tag{A2.35}$$

Specifically for $k = 0$

$$\mathbf{I} = \mathbf{M}_1 + \mathbf{M}_2 \quad \text{(a)} \tag{A2.36}$$

For $k = 1$

$$\mathbf{A} = \lambda_1 \mathbf{M}_1 + \lambda_2 \mathbf{M}_2 \quad \text{(b)}$$

From these, by multiplying the first by λ_1 and then subtracting from the second,

$$\mathbf{A} - \lambda_1 \mathbf{I} = (\lambda_2 - \lambda_1)\mathbf{M}_2$$

from which

$$\mathbf{M}_2 = \frac{\mathbf{A} - \lambda_1 \mathbf{I}}{\lambda_2 - \lambda_1} \quad \text{(a)} \tag{A2.37}$$

and similarly

$$\mathbf{M}_1 = \frac{\mathbf{A} - \lambda_2 \mathbf{I}}{\lambda_1 - \lambda_2} \quad \text{(b)}$$

The case for repeated eigenvalues proceeds in a slightly different manner. The representation for an $n \times n$ matrix now assumes the form

$$\mathbf{A} = \sum_{k=1}^{p} (\lambda_k \mathbf{M}_k + \mathbf{N}_k) \qquad p \le n \tag{A2.38}$$

instead of that given by (2.32). The matrix $\mathbf{N}_k$ is such that if the multiplicity of λ_k is r, then

$$\mathbf{N}_k^r = 0 \tag{A2.39}$$

The matrices $\mathbf{M}_k$ and $\mathbf{N}_k$ satisfy the following properties

$$\mathbf{M}_i \mathbf{M}_j = \begin{cases} 0 & i \ne j \\ \mathbf{M}_i & i = j \end{cases} \quad \text{(a)}$$

$$\mathbf{M}_i \mathbf{N}_j = \mathbf{N}_j \mathbf{M}_i = \begin{cases} 0 & i \ne j \\ \mathbf{N}_j & i = j \end{cases} \quad \text{(b)} \tag{A2.40}$$

$$\sum_{k=1}^{p} \mathbf{M}_k = \mathbf{I} \quad \text{(c)}$$

$$\mathbf{N}_k^r = 0 \qquad \text{where } r \text{ is the multiplicity of } \lambda_k \quad \text{(d)}$$

Functions of a matrix are calculated using the expression

$$\mathbf{f}(\mathbf{A}) = \sum_{k=1}^{p} \left[f(\lambda_k) \mathbf{M}_k + \sum_{m=1}^{r_k} \left[\frac{f^{m-1}(\lambda_k) \mathbf{N}_k^{m-1}}{(m-1)!} \right] \right] \tag{A2.41}$$

The second summation provides for the multiple eigenvalues. ∎

EXAMPLE A2.6

Determine $\mathbf{A}^k$ for the following 2×2 matrix

$$\mathbf{A} = \begin{bmatrix} 2 & 0 \\ 1 & 2 \end{bmatrix}$$

Solution: The characteristic equation is $g(\lambda) = (2 - \lambda)^2 = 0$, which has two repeated roots $\lambda_1 = \lambda_2 = 2$. In this case (A2.38) becomes

$$\mathbf{A} = 2\mathbf{M}_1 + \mathbf{N}_1$$

The matrix function $\mathbf{A}^k$ is then

$$\mathbf{f}(\mathbf{A}) = \mathbf{A}^k = 2^k \mathbf{M}_1 + k 2^{k-1} \mathbf{N}_1$$

For the specific values of k in this expression

$$k = 0 \quad \mathbf{I} = \mathbf{M}_1$$
$$k = 1 \quad \mathbf{A} = 2\mathbf{M}_1 + \mathbf{N}_1$$

From these we have that

$$\mathbf{M}_1 = \mathbf{I}$$
$$\mathbf{N}_1 = \mathbf{A} - 2\mathbf{I}$$

The result is

$$\mathbf{A}^k = 2^k \mathbf{I} + k 2^{k-1}(\mathbf{A} - 2\mathbf{I})$$
$$= \begin{bmatrix} 2^k & 0 \\ 0 & 2^k \end{bmatrix} + k \cdot 2^{k-1} \begin{bmatrix} 0 & 0 \\ 1 & 0 \end{bmatrix}$$
$$= \begin{bmatrix} 2^k & 0 \\ k \cdot 2^{k-1} & 2^k \end{bmatrix}$$

The reader should verify this result by following the procedure in Example A2.4. ■

REFERENCES

1. Franklin, J. N. *Matrix Theory*. Englewood Cliffs, N.J.: Prentice-Hall, 1968.
2. Perlis, S. *Theory of Matrices*. Reading, Mass.: Addison-Wesley, 1952.

ANSWERS

CHAPTER 1

1-2.3 b. nonperiodic   **c.** periodic; $\omega_0 = 4\pi \times 10^{-3}$ rad/s   **f.** periodic; $\omega_0 = \frac{1}{6}$ rad/s

1-3.1 c.   **1-3.2 c.**

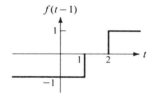
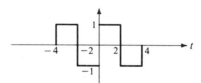

1-3.4 d.   **1-3.5 b.** $f(t) = 2\Lambda(t)(-\operatorname{sgn} t)$

1-3.7 a.   **1-3.8 d.**

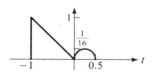
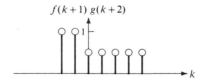

1-3.10 d.   **i.**

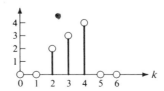

1-5.3 $\varphi_n(t) = \dfrac{1}{\sqrt{T}} e^{jn\omega_0 t}$ **1-5.4 b.** $c_0 = 0, c_1 = \sqrt{\dfrac{6}{25}}$

1-6.1 b. $\text{wal}_p(6, t) = \text{wal}_w(4, t)$ **1-6.2** $c_0 = 0.625, c_1 = -0.375, c_2 = -0.0625$

CHAPTER 2

2-2.2 $\frac{3}{2}(e^2 - 1)e^{-t} - 3$ **2-4.1 a.**

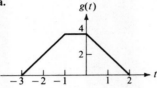

2-4.2 b.   d.   h.

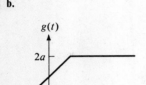 

j. $\delta(t)$

2-4.5 b.   **2-4.6 b.**

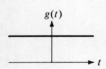 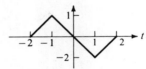

2-6.2 a. $g(t) = f(t) \star h(t) = \begin{cases} 0 & t < -1 \\ 1 - e^{-1}e^{-t} & -1 \leq t \leq 1 \\ e^{-t}(e - e^{-1}) & t \geq 1 \end{cases}$

c. $g(t) = f(t) \star h(t) = \begin{cases} \frac{1}{2}e^{-t} & t \geq 0 \\ \frac{1}{2}e^{t} & t \leq 0 \end{cases}$

2-7.1 a.

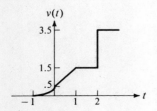

ANSWERS

2-8.1 $h(t) = \begin{cases} \dfrac{1}{RC} e^{-(t/RC)} & t \geq 0 \\ 0 & t < 0 \end{cases}$

2-8.2 $v_0(t) = \begin{cases} 0 & t \leq 0 \\ \dfrac{1}{R}(1 - e^{-(R/L)t}) & 0 \leq t \leq 2a \\ \dfrac{1}{R} e^{-(R/L)t}(e^{(R/L)2a} - 1) & 2a \leq t < \infty \end{cases}$

2-8.3 $h(t) = e^{-at}(k_1 \cos \sqrt{b^2 - a^2}\, t + k_2 \sin \sqrt{b^2 - a^2}\, t) \qquad t \geq 0 \quad k_1, k_2$ are constants

$a = \dfrac{R}{2L} \qquad b = \dfrac{1}{\sqrt{LC}}, \qquad \text{assume} \quad b > a$

Applying initial conditions: $h(t) = \dfrac{1}{L\sqrt{b^2 - a^2}} e^{-at} \sin \sqrt{b^2 - a^2}\, t$

2-8.4 b.

$M_1 \dfrac{dv_1}{dt} + D(v_1 - v_2) = f(t)$

$M_2 \dfrac{dv_2}{dt} + D(v_2 - v_1) + K \displaystyle\int v_2\, dt = 0$

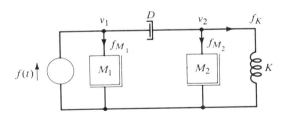

2-8.5 $v(t) = \begin{cases} 0 & t \leq 0 \\ \dfrac{t}{2} + \dfrac{1}{4} e^{-2t} - \dfrac{1}{4} & 0 \leq t \leq 2 \\ \left(\dfrac{3}{4} e^4 + \dfrac{1}{4}\right) e^{-2t} & 2 \leq t < \infty \end{cases}$

2-9.1 a. $\mathscr{C} = \dfrac{1}{L} \displaystyle\int h\, dt$

2-10.2 b. Linear, time-invariant

2-10.4 a. $H(\omega) = \dfrac{1}{R + \dfrac{1}{j\omega C}}$

d. $H(\omega) = \dfrac{1}{j\omega J + D + \dfrac{K}{j\omega}}$

2-10.5 b. $y(t) = \underbrace{e^{-t}}_{z_i} + \underbrace{t - 1 + e^{-t}}_{z_s} = \underbrace{2e^{-t}}_{\text{transient}} + \underbrace{(t - 1)}_{\text{steady state}}$

2-10.6 $y(t) = y_{z_i} + y_{z_s} = (8e^{-2t} - 6e^{-t}) + (7e^{-2t} - 5e^{-t}) = 15e^{-2t} - 11e^{-t}$

2-11.2 a. $f(t) \rightarrow \boxed{[1 + \mathcal{O}]^{-1}\mathcal{O}} \rightarrow g(t)$

b. $f(t) \rightarrow \boxed{[I - [I - \mathcal{O}_1]^{-1}\mathcal{O}_2\mathcal{O}_3]^{-1}[I - \mathcal{O}_1]^{-1}\mathcal{O}_2} \rightarrow g(t)$

2-11.3 c. $f(t) \rightarrow \boxed{a} \rightarrow g(t)$, feedback $\boxed{\dfrac{1}{a}} \rightarrow f(t)$

2-11.4 $f(t) \rightarrow \boxed{\dfrac{\mathcal{O}_1\mathcal{O}_2\mathcal{O}_3\mathcal{O}_4}{1 - \mathcal{O}_3\mathcal{O}_4\mathcal{H}_1 - \mathcal{O}_2\mathcal{O}_3\mathcal{H}_2 + \mathcal{O}_1\mathcal{O}_2\mathcal{O}_3\mathcal{O}_4\mathcal{H}_3}} \rightarrow y(t)$

2-12.1 a. $\dfrac{dx}{dt} = ax + bf$

$y = x$

2-12.2 $[\dot{\mathbf{x}}] = \begin{bmatrix} 0 & 1 \\ -\dfrac{k}{J} & 0 \end{bmatrix} [\mathbf{x}] + \begin{bmatrix} 0 \\ \dfrac{1}{J} \end{bmatrix} \mathcal{T} \qquad x_1 = \theta \quad x_2 = \omega$

2-12.5 $[\dot{\mathbf{x}}] = \begin{bmatrix} -\dfrac{R}{L} & -\dfrac{Bl}{L} \\ \dfrac{Bl}{M} & -\dfrac{D}{M} \end{bmatrix} [\mathbf{x}] + \begin{bmatrix} 0 \\ 1 \end{bmatrix} f_a \qquad x_1 = i \quad x_2 = v$

2-12.6 c. $[\dot{\mathbf{x}}] = \begin{bmatrix} 0 & -1 \\ 1 & -2 \end{bmatrix}[\mathbf{x}] + \begin{bmatrix} 5 \\ 2 \end{bmatrix} w$

$y = [0 \ 1][\mathbf{x}]$

CHAPTER 3

3-1.1 $\dfrac{T_1}{T_2} = \dfrac{n_2}{n_1}$ = rational number **3-1.2 a.** $\dfrac{n2\pi}{3}$ **b.** π

3-2.1 b.

	1st harmonic	3rd harmonic	5th harmonic ...
Amplitude Spectrum	$\dfrac{8}{\pi}$	$\dfrac{8}{3\pi}$	$\dfrac{8}{5\pi}$
Phase Spectrum	$\dfrac{\pi}{4}$	$\dfrac{3\pi}{4}$	$\dfrac{5\pi}{4}$

3-2.2 Delta functions at $t = n$ $n = 0, \pm 1, \pm 2, \ldots$

3-2.4 b. $f(t) = \sin \dfrac{2\pi}{t}$ **3-2.5 a.** Problem 3-2.2   **3-3.1** 2.72 Joules

3-3.3 a. B_n's   **b.** A_n's   **d.** B_n's

3-3.4 b.

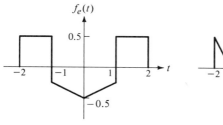 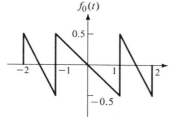

3-3.9 a. Integrate Equation (3.25) from $-T/2$ to $T/2$
  b. Denote the sum by S and multiply both sides by $2\sin(t/2)$. Combine trigonometric terms.

3-3.10 a.

	Zero Axis ($B_n = 0$)	Zero Point ($A_n = 0$)	Half Wave (no even)	$A_0 = 0$
a)	✓			
b)	✓		✓	✓
c)				
d)				✓

3-4.1 $f(x) = 1 + \dfrac{8}{\pi^2}\left[-\cos x + \dfrac{1}{3^2}\cos 3x - \dfrac{1}{5^2}\cos 5x + \dfrac{1}{7^2}\sin 7x \cdots\right]$

3-4.2 Only the phase spectrum is affected.

3-5.1 $v_0(t) = 1 + 2\left[\dfrac{1}{\sqrt{(RC)^2+1}}\cos(t - \tan^{-1} RC) + \dfrac{1}{\sqrt{(2RC)^2+1}}\cos(2t - \tan^{-1} 2RC)\right.$
$\left. + \dfrac{1}{\sqrt{(3RC)^2+1}}\cos(3t - \tan^{-1} 3RC)\right]$

3-5.4 The expression (a)

CHAPTER 4

4-1.2 b. $F(\omega) = \dfrac{E}{\omega}\left[\dfrac{1-\cos \omega T}{j} + \sin \omega T\right]$

4-1.3 b. $R(\omega) = \dfrac{4 \sin 4\omega}{\omega}$ $X(\omega) = \dfrac{1}{\omega}\left[4\cos 4\omega - \dfrac{1}{\omega}\sin 4\omega\right]$

4-2.3 a. $F(\omega) = e^{-j\omega}\dfrac{2\sin 2\omega}{\omega}$

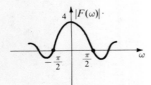
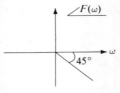

c. $F(\omega) = 0.5\mathscr{F}\{p_{0.5}(t)\} + 2\mathscr{F}\{p_{0.5}(t-2)\}\mathscr{F}\{p_{0.5}(t-2)\}$

g. $F(\omega) = \pi\displaystyle\int_{-\infty}^{\infty} p_2(\tau)p_1(\omega-\tau)\,d\tau$

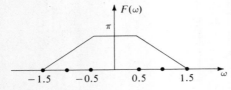

h. $F(\omega) = \dfrac{2\sin(\omega-2)}{\omega-2}$

j. $F(\omega) = \dfrac{2}{3}\dfrac{\sin 2\left(\dfrac{\omega}{3}-1\right)}{\left(\dfrac{\omega}{3}-1\right)}$

4-2.5 a. $F(\omega) = \pi e^{-|\omega|}$ **b.** $F(\omega) = \pi p(\omega)$ $-1 < \omega < 1$ **c.** $|a|e^{-jt_0\omega}$ **d.** $\displaystyle\lim_{n\to\infty} F(\omega) = 1$

4-2.7 a. $f(-t)$ **b.** shifted $f(t)$ by a.

4-2.8 $V(\omega) = V_c\pi\,\delta(\omega - \omega_c) + V_c\pi\,\delta(\omega + \omega_c) + mV_c\left[\dfrac{\sin(\omega-\omega_c)}{\omega-\omega_c} + \dfrac{\sin(\omega+\omega_c)}{\omega+\omega_c}\right]$

4-2.11 a. $G(\omega) = H(\omega)F(\omega) = \dfrac{1}{4+j\omega}\dfrac{1}{2+j\omega}$ **b.** $G(\omega) = H(\omega)F(\omega) = \dfrac{2}{(4-\omega^2+j5\omega)}\dfrac{1}{2+j\omega}$

4-2.12 $R_{g,f}(\omega) = G(\omega)F^*(\omega)$ $R_{f,f}(\omega) = |F(\omega)|^2$ $R_{g,g}(\omega) = |H(\omega)|^2 R_{f,f}(\omega)$

4-2.13 a. $H(\omega) = \dfrac{j\omega C}{1 - LC\omega^2}$ **c.** $H(\omega) = \dfrac{1}{K + j\omega D}$

4-2.14 b. $F(\omega) = \pi e^{-\omega^2/2}$ **4-2.18** Attenuates high and low frequencies

4-3.1 a. $F(\omega) = 2\pi\,\delta(\omega) + \dfrac{1}{1+(\omega-\omega_0)^2} + \dfrac{1}{1+(\omega+\omega_0)^2}$

ANSWERS

c. $F(\omega) = \dfrac{\pi}{2}\delta(\omega - \omega_0) + \dfrac{1}{2j}\dfrac{1}{(\omega - \omega_0)} + \dfrac{\pi}{2}\delta(\omega + \omega_0) + \dfrac{1}{2j}\dfrac{1}{(\omega + \omega_0)}$

d. $F(\omega) = \dfrac{1}{2[1 + j(\omega - \omega_0)]} + \dfrac{1}{2[1 + j(\omega + \omega_0)]}$

e. $F(\omega) = \pi[\delta(\omega - \omega_0) + \delta(\omega + \omega_0)] + \pi[\delta(\omega - \omega_0 - \omega_m) + \delta(\omega - \omega_0 + \omega_m)$
$\quad + \delta(\omega + \omega_0 - \omega_m) + \delta(\omega + \omega_0 + \omega_m)]$

4-3.2 a. $F(\omega) = 4\pi e^{-j\pi/2} \displaystyle\sum_{n=-\infty}^{\infty} \dfrac{[1 - \cos n\pi]}{n\pi}\delta(\omega - n\pi)$ **4-4.2** $F_t(\omega) = 2[Si(\omega + 1) - Si(\omega - 1)]$

4-7.1 complex envelope = $p_T\left(t - \dfrac{T}{2}\right)$ **4-7.2** $\hat{f}(t) = \dfrac{1}{\pi}\ln\left(\dfrac{T - a - t}{T + a - t}\right)$ **4-8.2** $T = \dfrac{2\pi}{10^7}$

4-8.3 For $a_1 = V_c$ and $2a_2/a_1 = m$ the form of the AM signal
$$v_0(t) = V_c v_m(t) + a_2 v_m^2(t) + a_2 \cos^2 \omega_c t + \underbrace{V_c[1 + mv_m(t)]\cos \omega_c t}_{\text{AM signal}}$$

4-8.5 $T \leq \dfrac{1}{2f_b}$ **4-9.2** FM and AM have same sideband components but they differ in phase.

4-9.4 ⑤ $v_0(t) = Akv_m(t)$

CHAPTER 5

5-1.1 a. distortionless but not realizable   **c.** nonlinear   **5-1.2 b.** $-3u(t - 2) - 3\delta(t - 5)$

5-1.3 $C = 0.0968\ \mu F$ **5-1.4 a.** $H_a(\omega) = 1 - H(\omega)$ **b.** $I_a|\omega| = F(\omega) - G(\omega)$

5-2.2 b. $\pi + \dfrac{2}{\pi}\cos \pi t$ **c.** π

5-2.5 c. $\dfrac{H_0}{\pi}\dfrac{\sin(t + t_0)}{t + t_0}$ **d.** $-\dfrac{H_0}{\pi}\dfrac{\sin \omega_0 t}{t}$

e. $G(\omega) = H_0 e^{j0.1 \sin \omega} \doteq H_0(1 + j0.1 \sin \omega);\ g(t) = \dfrac{H_0}{\pi}\dfrac{\sin \omega_0 t}{t} + \dfrac{0.1 H_0}{2\pi}\dfrac{\sin(t + 1)}{t + 1} - \dfrac{0.1 H_0}{2\pi}\dfrac{\sin(t - 1)}{t - 1}$

5-2.6 Similar to Example 5.2

5-2.7 a.   **b.** Same as (a)   **(c)** same as (a) but reflected down with respect to 0 dB line

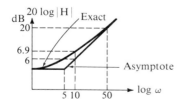

5-2.8 a. low-pass filter   **b.** high-pass filter

ANSWERS

5-3.1 a. 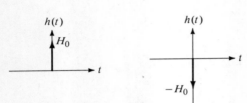 b. c. 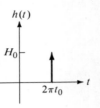 d.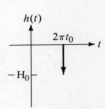

5-4.1 $h(t) = \dfrac{2}{\pi} \dfrac{1}{1 + (t - t_0)^2} \cos \omega_c t$

5-4.3 $h_1(t) = \mathscr{F}^{-1}\{H_0(\omega)\}$ = amplitude filter
$h_2(t) = \mathscr{F}^{-1}\{e^{-j\theta_h(\omega)}\}$ = phase filter $\quad h_3(t) = \mathscr{F}^{-1}\{e^{-j\omega t_0}\} = \delta(t - t_0)$ = phase filter

5-4.4 $H(\omega) = \dfrac{1}{1 + \dfrac{N(\omega)}{F(\omega)}}$ **5-4.5** The output is in the form of its spectrum function, $F(t)$.

5-4.6 $g(t) = \dfrac{A \sin 2t\tau}{t}$ implies pulse compression **5-4.7** $g(t) = H_0 R_{ff}(t - t_0)$ **5-6.2** $\varphi = \pm \dfrac{\pi}{2} \Rightarrow$ addition

5-6.4 $\tau(\omega_{y_1}) = [1 - p_\epsilon(\omega_{y_1} - \omega_{y_0}) - p_\epsilon(\omega_{y_1} + \omega_{y_0})]$

CHAPTER 6

6-3.1 b. $\dfrac{s^2 + 2s + 2}{s^3} \qquad \dfrac{s^2 + 8}{s(s^2 + 4^2)} \qquad \dfrac{2\omega s}{(s^2 + \omega^2)^2} \qquad \dfrac{1}{\omega} \tan^{-1} \dfrac{\omega}{s}$

6-3.2 a. $\dfrac{2}{s} + \dfrac{3}{s^2} \quad \sigma > 0$ b. $\dfrac{4}{0.1 + s} \quad \sigma > -0.1$ c. $\dfrac{2}{s^2 - 4} \quad \sigma > 2$ d. $\dfrac{1}{s} + \dfrac{1}{s^2 + 1} \quad \sigma > 0$

6-3.3 a. $\dfrac{2}{s} - \dfrac{48}{s^4}$ b. $\dfrac{s^2 - 4}{(s^2 + 4)^2}$ c. $\dfrac{s + 1}{s(s + 2)}$ d. $\dfrac{s - 1}{(s - 1)^2 + 1}$ e. $\dfrac{s - 1}{s^2 + 1}$

6-4.1 f. $\dfrac{1}{s(s^2 + 1)}$ **6-4.2** a. $\dfrac{A}{s} \tanh \dfrac{Ts}{4}$ b. $\dfrac{2A}{T} \dfrac{1}{s^2} \tanh \dfrac{Ts}{2}$ **6-4.3**

	Initial	Final
a.	1	0
b.	∞	0
c.	1	0
d.	1	0

6-4.4 a. $\dfrac{e^{-2(s+1)}}{s + 1} + sG(s) - g(0+)$ b. $\dfrac{1}{2\pi j} \displaystyle\int_{2-j\infty}^{2+j\infty} \dfrac{1}{s + 2} G(s - z)\,dz + \dfrac{1}{s(s + 1)} + \dfrac{1}{s}$

6-4.6 a. $F(s) = F(-s)$ b. $F(s) = -F(-s)$

6-5.1 a. $H(s) = \dfrac{1}{1 + RCs}$ b. $H(s) = \dfrac{1}{Ms + D_1 + D_2}$ **6-5.2** $H(s) = \dfrac{(2\pi rNB)R}{(2\pi rNB)^2 + (Ms + D)(Ls + R)}$

6-5.3 a. $H(s) = \dfrac{R_2(R_1 Cs + 1)}{R_1 R_2 Cs + R_1 + R_2}$

ANSWERS

6-5.4 a. $H(s) = \dfrac{H_1}{1 - H_2 + H_1 H_3}$ **b.** $H(s) = \dfrac{H_1 H_2 (1 - H_3)}{(1 - H_2) + H_1 H_2 (H_4 + H_5)(1 - H_3)}$

6-6.1 a. **b.** **c.**

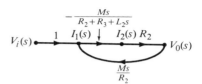

6-6.2 a. $\dfrac{x_5}{x_1} = \dfrac{ac}{1 - ab - cd + abcd}$ **b.** $\dfrac{x_6}{x_1} = \dfrac{acd}{1 - ab - de + abde}$ **c.** $\dfrac{x_5}{x_1} = \dfrac{abdf}{1 - bc - de}$

d. $\dfrac{x_6}{x_1} = \dfrac{abdfh}{1 - bc - de - fg + bcfg}$ **e.** $\dfrac{x_6}{x_1} = \dfrac{acdeg}{1 - cdb - fde}$

f. $\dfrac{x_8}{x_1} = \dfrac{ailm((1 - cd - ef) + bceg(1 - hi - kl) + ajeg(1 - kl)}{1 - (hi + lk + cd + ef) + (hicd + hief + lkcd + lkef)}$

6-7.1 a. $f_1(t) = u(t) - u(t - 2)$ **c.** $f_3(t) = \dfrac{t^3}{3!} e^{-2t}$ **e.** $f_5(t) = u(t) + r(t)$

f. $f_6(t) = u(t) - e^{-1.5t}[\cos 1.66t - 1.51 \sin 1.66t]$

6-7.2 a. $\dfrac{a}{b} \sin bt$ **c.** $\sin t - t \cos t$ **e.** $2(e^t - \cos t - \sin t)$

6-8.2

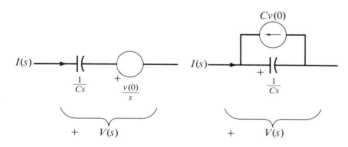

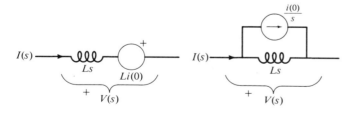

6-9.1 a. $y(t) = 10e^{-t} - 5e^{-2t}$ **b.** $y(t) = e^{-t} - e^{-2t}$ **6-9.2 a.** $i(t) = -\dfrac{10}{\sqrt{15}} e^{-t/4} \sin \dfrac{\sqrt{15}t}{4} + 5 \sin t$

6-9.5 $v_0(t) = t^2 e^{-2t}$
6-9.7 a. $h(t) = 1.077e^{-3.732t} - 0.077e^{-0.268t}$ **c.** $h(t) = 2e^{-t}(\tfrac{1}{2} \cos 3t - \tfrac{1}{6} \sin 3t)$

6-9.8 a. $v_0(t) = \dfrac{4e^{-t/2}}{\sqrt{7}} \sin\left(\dfrac{\sqrt{7}}{2} t\right)$ **6-9.9** $v_2(t) = \dfrac{2}{\sqrt{3}} e^{-t/2} \sin \dfrac{\sqrt{3}}{2} t$ **6-9.10** $v_0(t) = -4e^{-t} + \dfrac{14}{3} e^{-7t/6}$

6-10.1 a. $H(s) = \dfrac{2s + 1}{s^2 + 3s + 7}$ $h(t) = 2e^{-1.5t}[\cos 2.18t - 0.46 \sin 2.18t]$

CHAPTER 7

7-1.3 a.

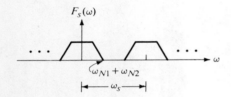

b.

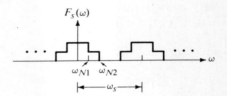

7-2.5 $T_s = \dfrac{\pi}{1.8}$ **7-2.9** $\omega_N = 1.496$

7-2.10

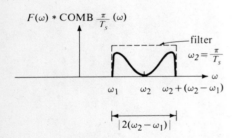

7-2.11 Filter bandwidth $2\omega_1$ from ω_0 to $\omega_0 + 2\omega_1$ **7-3.2** 18 kHz

CHAPTER 8

8-1.1 a. L, S   **b.** T.I., L, S, C   **c.** T.I., L, S   **d.** M, L, C   **8-2.1** 407.6

8-3.2 a. $y(k) + 2y(k-1) = 6x(k-1)$ **b.** $y(k) + \dfrac{5}{2} y(k-1) = 3x(k-1) + 5x(k-1)$

   c. $y(k) + k_1 y(k-1) = k_3 x(k) + (k_1 k_2 + k_1 k_3) x(k-1)$
   d. $y(k) + 15y(k-3) = 5x(k-1)$ **e.** $y(k) + 5y(k-4) = x(k) + 5x(k-2)$

8-3.3 a. $v(k) = \dfrac{a}{D}\left(1 - \left(\dfrac{M}{M+D}\right)^{k+1}\right)$ **8-3.4** For $T = 0.2$ $y(1.0) = 0.96 \doteq 1$ exact

8-3.6 a. $y(k) = y(-1) + (1+k)^2$ **b.** $y(k) = y(-1) + \dfrac{1+k}{2}k$

c. $y(k) = 3^k(y(0)+1) - 1$ **d.** $y(k) = (k+1)!y(-1)$

8-3.7

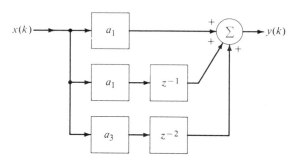

8-3.8 a. $v_0(0) = \dfrac{1}{3}$ $v_0(1) = \dfrac{1}{9}$ $v_0(2) = \dfrac{1}{27}$ $v_0(3) = \dfrac{1}{81} \cdots \to 0$

b. $\omega(0) = \dfrac{1}{6}$ $\omega(1) = \dfrac{2}{9}$ $\omega(2) = \dfrac{13}{54}$ $\omega(3) = \dfrac{20}{81} \cdots \to \dfrac{1}{4}$

8-5.1 a. $y(k) = C_1 2^k + C_2(-2)^k + \dfrac{2}{5}\cos\dfrac{k\pi}{2}$ **b.** $y(k) = C_1 5^k + C_2 2^k - 9 \times 3^k$

c. $y(k) = C_1 5^k + C_2 2^k - 2k 2^k$ **d.** $y(k) = C_1(-3)^k + C_2(-2)^k + \dfrac{7}{74}\cos\dfrac{k\pi}{2} - \dfrac{5}{74}\sin\dfrac{k\pi}{2}$

8-5.3 $v_c(k) = 1.08 \times 0.632^k \cos(37.74k - 22.8) + 1.045 \times 0.9^k$

8-5.5 a. $y_p(k) = -3 \times 2^k$ **b.** $y_p(k) = Ak^4 + Bk^3 + Ck^2$ **c.** $y_p(k) = Ak^3 2^k + Bk^2 2^k + C(-2)^k$

8-5.6 $y_p(k) = \dfrac{1}{7}\cos\dfrac{k\pi}{2}$

8-5.7 c.

8-5.8 a. Differentiation **b.** Double differentiation

8-6.1 a. $H(e^{j\omega}) = \dfrac{1}{5 - e^{-j\omega}} = \dfrac{1}{\sqrt{(5-\cos\omega)^2 + \sin^2\omega}} e^{-j\tan^{-1}(\sin\omega/(5-\cos\omega))}$

8-6.2 a. Low-pass **b.** High-pass **8-6.3** Same response as Example 8.15 **8-6.4** $h(k) = \dfrac{1}{\pi^2 k^2}(\cos\pi k - 1)$

8-6.5 $h(k) = 2\delta(k) + 4\delta(k-1) - 3\delta(k-2)$ **8-6.6** $H(e^{j0.5\omega}) = \dfrac{1}{1 - 2e^{j0.5\omega}}$

8-6.7 Nyquist frequency = 50 Hz

CHAPTER 9

9-1.1 a. $X(z) = \dfrac{z^3 + z^2 + z + 1}{z^3}$ **b.** $X(z) = \dfrac{z^3 + z^2 - z - 1}{z^3}$ **d.** $X(z) = \dfrac{z^4 + 2z^3 + 3z^2 + 2z^1 + 1}{z^5}$

9-1.2 a. $F(z) = \dfrac{2z}{2z - 1}$ **b.** $F(z) = \dfrac{a(1 + a) - z}{z(z - a)}$ **c.** $F(z) = T^2 \dfrac{z(z + 1)}{(z - 1)^3}$

9-1.3 a. $\{1, 4, 13, 40 \ldots\}$ **b.** $\{0, 1, 8, 28 \ldots\}$

9-2.1 a. $Y(z) = \dfrac{z^2}{(z - 1)^2}$ $|z| > 1$ **b.** $Y(z) = \dfrac{z(z + 1)}{(z - 1)^3}$ $|z| > 1$ **c.** $Y(z) = \dfrac{2z^2}{z^2 - a^2}$ $|z| > a$

d. $Y(z) = \dfrac{z}{z - e^{j\theta}}$ $|z| > 1$

9-2.2 $Y(z) = \dfrac{z(z - a\cos b)}{(z - ae^{jb})(z - ae^{-jb})}$ $|z| > a$

9-2.3 a. $Y(z) = 2\dfrac{z^5 - 1}{z^4(z - 1)} - \dfrac{z^5 - 3^5}{z^4(z - 3)} + \dfrac{z}{(z - 3)}$ $|z| > 3$ **b.** $Y(z) = \dfrac{2z\left(z - \dfrac{7}{24}\right)}{\left(z - \dfrac{1}{3}\right)\left(z - \dfrac{1}{4}\right)}$ $|z| > \dfrac{1}{3}$

9-2.4 $Y(z) = \dfrac{z^6 - 2^6}{z^5(z - 2)}$ $|z| > 0$

9-2.7 a. $F(z) = \dfrac{z^2 - z\cos\omega T}{z^2 - 2z\cos\omega T + 1}$ **b.** $F(z) = \dfrac{a^T z \sin\omega T}{z^2 - 2za^T \cos\omega T + a^{2T}}$

e. $F(z) = \dfrac{z(z - e^{-\alpha T}\cos\omega T)}{z^2 - 2z e^{-\alpha T}\cos\omega T + e^{-2\alpha T}}$ **f.** $F(z) = \dfrac{z}{z - a}$

9-3.2 Due to shifting, a pole appears at the origin

9-3.4 $Y_{(1)}(z) = \dfrac{z + 2}{z^2}$ $Y(z) = \dfrac{z^2(z + 2)}{z^4 - 1}$

9-3.7 $Y(z) = \dfrac{1}{(1 - z^{-1})(1 - 0.5z^{-1})}\left[\dfrac{1}{1 - 0.8z^{-1}} + \dfrac{1}{1 + 0.8z^{-1}}\right]$

9-3.8 $v(k) = -1.57(0.5)^{k+1} + 0.78e^{-0.2k}$ **9-3.12** $Y(z) = \dfrac{z}{z + 2}$

9-3.13 a. $\mathscr{L}\{e^{-\alpha k}u(k)\} = \dfrac{z}{z - e^{-\alpha}} \Rightarrow \mathscr{L}\{a^k e^{-\alpha k}u(k)\} = \dfrac{a^{-1}z}{a^{-1}z - e^{-\alpha}}$

d. $\mathscr{L}\{e^{-\alpha k}\sin k\omega\, u(k)\} = \dfrac{ze^{-\alpha}\sin\omega}{z^2 - 2e^{-\alpha}z\cos\omega + e^{-2\alpha}} \Rightarrow$

$\mathscr{L}\{a^k e^{-\alpha k}\sin k\omega\, u(k)\} = \dfrac{a^{-1}ze^{-\alpha}\sin\omega}{a^{-2}z^2 - 2e^{-\alpha}a^{-1}z\cos\omega + e^{-2\alpha}}$

ANSWERS

9-3.17 **a.** $Y(z) = \dfrac{Tz}{(z-1)^2}$ **b.** $Y(z) = \dfrac{T^2 z(z+1)}{(z-1)^3}$

c. $Y(z) = \dfrac{z^2(T^2 + T) + z(T^2 - T)}{(z-1)^3}$ **d.** $Y(z) = \dfrac{(T^2 - T)z^2 + z(T^2 + T)}{(z-1)^3}$

9-3.20 $h(k) = \dfrac{1}{3}\left(\dfrac{1}{1.5}\right)^k$ $k \geq 0$ $v_0(k0.5) = \sum_{n=0}^{k} h((k-n)0.5)[u(k0.5) - u((k-5)0.5)]$

9-4.1 **a.** $f(k) = -u(k) + 2^{k+1}$ **b.** $f(k) = k\left(\dfrac{1}{2}\right)^{k+1}$ **c.** $f(k) = \delta(k+1) - \dfrac{1}{2}(-1)^k - \dfrac{3}{2}(-3)^k$

9-5.1 **a.** $y(k) = -1 + (2)^k$ **b.** $y(k) = -2 - k + 2^{k+1}$ **c.** $y(k) = 1 + k 2^{k+1}$

9-5.2 $f(k-1) = \dfrac{3}{10} 2^{k-1} - \dfrac{9}{2}(-2)^{k-1} + \dfrac{26}{5}(-3)^{k-1}$ $k \geq 1$ **9-5.3** $f(k) = \sin \alpha k$ $k = 0, 1, 2, \ldots$

9-5.4 **a.** $f(k) = -3\delta(k) - 3\delta(k-1) - 2\delta(k-2)$ **9-6.2** $y(k) = 0.6u(k) + 0.2(-2.617)^k + 0.2(-0.383)^k$

9-6.3 **a.** $H(z) = \dfrac{1}{z^2 + a}$ **b.** $H(z) = \dfrac{C}{z^2 - az - b}$ **c.** $H(z) = \dfrac{z}{z^4 - az^2 - b}$

9-6.5 **b.**

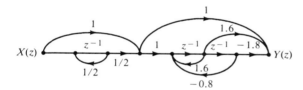

9-6.6 **a.** $x(k) \to \cdots \to y(k)$ **c.** $H(z) = \dfrac{1}{z^2 + 3z + 5}$

9-7.1 **a.** No poles   **b.** Pole at $z = 0$ **c.** Poles at a and b **d.** No poles

9-7.2 **a.** Dominant part $0.286 \dfrac{z}{z - 0.8}$; nondominant part $0.714 \dfrac{z}{z - 0.1}$

9-8.1 $|H(e^{j\omega})| = 1.12 \tan 2\omega$

9-9.2 $v(k) = \dfrac{v}{\left(\dfrac{3+\sqrt{5}}{2}\right)^n - \left(\dfrac{3-\sqrt{5}}{2}\right)^n} \left[\left(\dfrac{3+\sqrt{5}}{2}\right)^n \left(\dfrac{3-\sqrt{5}}{2}\right)^k - \left(\dfrac{3-\sqrt{5}}{2}\right)^n \left(\dfrac{3+\sqrt{5}}{2}\right)^k\right]$

9-9.4 **a.** $y(k) = y(0)(-1)^k - (1)^k$ $k = 0, 1, 2, \ldots$

b. $y(k) = y(0)(-1)^k + \dfrac{2}{\sqrt{2+\sqrt{2}}}\left(\dfrac{\sqrt{2}}{2}\right)^k \sin(\theta k + \theta_1)$ $\theta = \dfrac{\pi}{4}$ $\theta_1 = \tan^{-1} \dfrac{\sqrt{2}}{2+\sqrt{2}}$

CHAPTER 10

10-1.6 $F(n\Omega) = e^{j(\omega_0 - \Omega n)TN} \sin\left[\dfrac{(\omega_0 - \Omega n)T(N-1)}{2}\right] / \sin\left[\dfrac{(\omega_0 - \Omega n)T}{2}\right]$

10-1.7 b. $F_2\left(0\dfrac{\pi}{2}\right) = 20 \quad F_2\left(1\dfrac{\pi}{2}\right) = 4 - j4 \quad F_2\left(2\dfrac{\pi}{2}\right) = 4 \quad F_2\left(3\dfrac{\pi}{2}\right) = 4 - 4j \cdots$

10-2.10 $F(n\Omega) = e^{-j(8\pi/10)n} \dfrac{\sin\left(\dfrac{6\pi n}{10}\right)}{\sin\left(\dfrac{2\pi n}{10}\right)}$ **10-2.12 b.** $F(n) = \sum_{k=0}^{N-1} f_i(k) \sin nk + j \sum_{k=0}^{N-1} f_i(k) \cos nk$

10-2.13 $\{0, 1, 2, 3, 4, 4, 4, 3, 2, 1\}$ **10-2.14** $\begin{bmatrix} y(0) \\ y(1) \\ y(2) \end{bmatrix} = \begin{bmatrix} f(0) & f(1) & f(2) \\ f(1) & f(2) & f(0) \\ f(2) & f(0) & f(1) \end{bmatrix} \begin{bmatrix} y(0) \\ y(2) \\ y(1) \end{bmatrix}$

10-2.17 $\sum_0^3 f^2(kT) = 1.35; \quad \dfrac{1}{4}[(1.911)^2 + (0.735^2 + 0.378^2) + 0.615^2 + (0.737^2 + 0.378^2)] = 1.3499 = \dfrac{1}{N}\sum_0^3 |F(n)|^2$

10-2.18

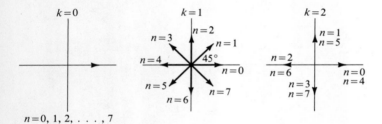

$n = 0, 1, 2, \ldots, 7$

10-4.1 a. Eight multiplication and addition   **b.** Twelve multiplication and addition

10-5.1 $F(n_3, n_2, n_1, n_0) = \sum_{k_0=0}^{1} \sum_{k_1=0}^{1} \sum_{k_2=0}^{1} \sum_{k_3=0}^{1} f_0(k_3, k_2, k_1, k_0) W^{8n_0 k_3} W^{(2n_1 + n_2)4k_2} W^{(4n_2 + 2n_1 + n_0)2k_1} W^{(8n_3 + 4n_2 + n_0)k_0}$

10-8.1 $a_{w71} = (-1)^{1.0 + 1[0+0] + 1[0+1]} = (-1)^1 = -1$

10-8.2 a. $\{F_1(n)\} = \{6, -4, 0, -2\}$ **b.** $\{F_2(n)\} = \{14, -8, 0, -4, 0, 7, 0, -2\}$

10-9.1 $F(r = 2) = [6, -2, -4, 0]^T$ **10-9.2** $a_{h31} = (-1)^{1\cdot1 + 1\cdot0 + 0\cdot0} = -1$

CHAPTER 11

11-2.1 a. $s = -0.7987 \quad s = \dfrac{-0.7987(1 \pm j\sqrt{3})}{2}$ **b.** 60 dB/decade

11-2.2 $H(s) = \dfrac{1}{(s+1)(s^2 + s + 1)} \quad s_1 = e^{j4\pi/6} \quad s_2 = e^{j\pi} \quad s_3 = e^{j8\pi/6} \quad s_4 = e^{j10\pi/6} \quad s_5 = e^{j2\pi} \quad s_6 = e^{j14\pi/6}$

ANSWERS

11-2.3 $H(s) = \dfrac{1}{s^3 + 2s^2 + 2s + 1}$ **11-3.2** $n = 3$

11-3.3 $\epsilon = 0.765$ $n = 3$ poles: $-0.369, -0.184 \pm j0.923$

11-3.5 a. $n = 17$ **b.** $n = 9$ **c.** $n = 8$ **11-9.1** $H(z) = \dfrac{0.449z^{-1}}{(1 - 0.28z^{-1})^2}$

11-9.4 $H(z) = \dfrac{1.358z}{z^2 - 0.1329z + 0.4913}$

11-10.2 $z|_{s=1} = 1.105$ $z|_{s=1+j(\pi/0.1)} = -0.436 + j0.93$ $z|_{s=1-j(\pi/0.1)} = -0.436 - j0.93$

$z|_{s=10} = 3$ $z|_{s=+\infty} = -1$

11-11.1 $|H(e^{j\omega t})| = \dfrac{\sin\dfrac{21}{2}\omega t}{\sin\dfrac{\omega t}{2}}$

11-14.2

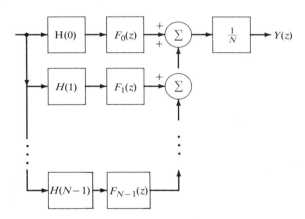

CHAPTER 12

12-2.1 a. $\dfrac{d}{dt}\begin{bmatrix} i_3 \\ v_4 \end{bmatrix} = \begin{bmatrix} 0 & \dfrac{1}{L} \\ -\dfrac{1}{C_4} & -\dfrac{1}{C_4}\left(\dfrac{1}{R_1} + \dfrac{1}{R_2}\right) \end{bmatrix} \begin{bmatrix} i_3 \\ v_4 \end{bmatrix} + \begin{bmatrix} 0 \\ \dfrac{1}{C_4 R_1} v + \dfrac{1}{C_4 R_2} v_2 \end{bmatrix}$

12-2.2 a. $\dfrac{d}{dt}\begin{bmatrix} f_2 \\ f_3 \\ v_2 \end{bmatrix} = \begin{bmatrix} 0 & 0 & -K_2 \\ 0 & 0 & K_3 \\ \dfrac{1}{M_2} & -\dfrac{1}{M_2} & -D_3 \end{bmatrix} \begin{bmatrix} f_2 \\ f_3 \\ v_2 \end{bmatrix} + \begin{bmatrix} K_2 \\ 0 \\ 0 \end{bmatrix} v$

12-3.1 a.

$$\begin{bmatrix} \dot{x}_1 \\ \dot{x}_2 \end{bmatrix} = \begin{bmatrix} 0 & 1 \\ -2 & -5 \end{bmatrix} \begin{bmatrix} x_1 \\ x_2 \end{bmatrix} + \begin{bmatrix} 0 \\ 1 \end{bmatrix} w \qquad \begin{array}{l} x_1 = y \\ \dot{x}_1 = x_2 = py \end{array}$$

b.

$$\begin{bmatrix} \dot{x}_1 \\ \dot{x}_2 \end{bmatrix} = \begin{bmatrix} 0 & 1 \\ -\frac{1}{3} & -\frac{2}{3} \end{bmatrix} \begin{bmatrix} x_1 \\ x_2 \end{bmatrix} + \begin{bmatrix} 0 \\ \frac{1}{3}(w+2) \end{bmatrix} \qquad \begin{array}{l} x_1 = y \\ \dot{x}_1 = x_2 = py \end{array}$$

12-3.2 $\mathbf{A} = \begin{bmatrix} 0 & 1 & 0 \\ 0 & 0 & 1 \\ -1 & -3 & -2 \end{bmatrix} \qquad \mathbf{B} = \begin{bmatrix} 0 \\ 0 \\ 1 \end{bmatrix} \qquad \mathbf{C} = \begin{bmatrix} 1 & 1 & 0 \end{bmatrix}$

12-4.1 a. $(a_1 p^3 + a_2 p^2 + a_3 p + a_4) i_1 = (b_0 p^2 + b_1 p) v_1 - b_2 p v_2 \qquad a_1 = R_2 L_1 \qquad a_2 = R_1 R_2 + \dfrac{L_1}{C}$

$a_3 = \dfrac{R_2}{C_3} + \dfrac{R_1}{C_2} \qquad a_4 = \dfrac{1}{C_3}\left(\dfrac{1}{C} - \dfrac{1}{C_3}\right) \qquad b_0 = R_2, \quad b_1 = \dfrac{1}{C} \qquad b_2 = \dfrac{1}{C_3}$

$(a_1 p^3 + a_2 p^2 + a_3 p + a_4) i_2 = b_2 p v_1 - (L_1 p^3 + R_1 p^2 + b_2 p) v_2$

c. $\dfrac{d}{dt}\begin{bmatrix} i_1 \\ v_{C_2} \\ v_{C_3} \end{bmatrix} = \begin{bmatrix} -\dfrac{R_1}{L_1} & 0 & -\dfrac{1}{L_1} \\ 0 & -\dfrac{1}{R_2 C_2} & \dfrac{1}{R_2 C_2} \\ \dfrac{1}{C_3} & \dfrac{1}{R_2 C_3} & -\dfrac{1}{R_2 C_3} \end{bmatrix} \begin{bmatrix} i_1 \\ v_{C_2} \\ v_{C_3} \end{bmatrix} + \begin{bmatrix} \dfrac{v_1}{L_1} \\ -\dfrac{v_2}{R_2 C_2} \\ \dfrac{v_2}{R_2 C_3} \end{bmatrix}$

12-4.2 a. $\begin{bmatrix} \dot{x}_1 \\ \dot{x}_2 \end{bmatrix} = \begin{bmatrix} 0 & 1 \\ 2 & 3 \end{bmatrix} \begin{bmatrix} x_1 \\ x_2 \end{bmatrix} + \begin{bmatrix} 0 \\ 1 \end{bmatrix} w \qquad y = \begin{bmatrix} 2, & 1 \end{bmatrix} \begin{bmatrix} x_1 \\ x_2 \end{bmatrix}$

c. Same number of integrators

12-5.1 a. $\mathbf{A}^k = \begin{bmatrix} \alpha^k & 0 \\ 0 & \alpha^k \end{bmatrix}$ **b.** $\mathbf{A}^k = \begin{bmatrix} 0 & \beta^k \\ \beta^k & 0 \end{bmatrix}$ **c.** $\mathbf{A}^k = \begin{bmatrix} \alpha^k & k\alpha^{k-1} \\ 0 & \alpha^k \end{bmatrix}$

ANSWERS

12-5.3 a. $e^{\mathbf{A}t} = \begin{bmatrix} e^{2t}\cosh 2t & \dfrac{e^{2t}}{2}\sinh 2t \\ 2e^{2t}\sinh 2t & e^{2t}\cosh 2t \end{bmatrix}$ **b.** $e^{\mathbf{A}t} = \begin{bmatrix} e^{3t} & 0 \\ 2(e^{3t}-e^{2t}) & e^{2t} \end{bmatrix}$

d. $e^{\mathbf{A}t} = \begin{bmatrix} (2e^{-t}-e^{-2t}) & (e^{-t}-e^{-2t}) \\ -2(e^{-t}-e^{-2t}) & (-e^{-t}+2e^{-2t}) \end{bmatrix}$

12-7.1 $i_L = -e^{-t} + \tfrac{1}{2}e^{-2t} + \tfrac{3}{2}$ $\quad t > 0$ $\qquad i_1 = -1.5\,e^{-t}$ $\quad t > 0$

12-8.2 $[s\mathbf{I}-\mathbf{A}]^{-1} = \dfrac{1}{s^3 - 5s^2 + 23s - 64}\begin{bmatrix} (s^2-2s-9) & (2s-1) & (7s-23) \\ s-21 & s^2+19 & 6s-11 \\ -4s+19 & -s-5 & s^2-8s+13 \end{bmatrix}$ **12-8.3** $m = \begin{bmatrix} 1 & -\tfrac{1}{2} \\ -1 & 1 \end{bmatrix}$

12-8.4 c. $p\begin{bmatrix} x_1 \\ x_2 \end{bmatrix} = \begin{bmatrix} 0 & 1 \\ -\dfrac{a_2}{a_0} & -\dfrac{a_1}{a_0} \end{bmatrix}\begin{bmatrix} x_1 \\ x_2 \end{bmatrix} + \begin{bmatrix} 0 \\ \dfrac{v}{a_0} \end{bmatrix}$ $\quad i = \begin{bmatrix} 1 & b_0 \end{bmatrix}\begin{bmatrix} x_1 \\ x_2 \end{bmatrix}$ $\quad a_0 = R_2 C_2 L_1 \quad a_1 = (L_1 + R_1 R_2 C_2)$

$a_2 = (R_1 + R_2) \quad b_0 = (R_2 C_2 p + 1) \quad p = \dfrac{d}{dt}$

d. $p\begin{bmatrix} i \\ v_c \end{bmatrix} = \begin{bmatrix} -\dfrac{R_1}{L_1} & -\dfrac{1}{L_1} \\ \dfrac{1}{C_2} & -\dfrac{1}{C_2 R_2} \end{bmatrix}\begin{bmatrix} i \\ v_c \end{bmatrix} + \begin{bmatrix} \dfrac{v}{L_1} \\ 0 \end{bmatrix}$

12-9.1 b. $\begin{bmatrix} \dfrac{1}{j\omega - 3} & 0 \\ \dfrac{2}{(j\omega-2)(j\omega-3)} & \dfrac{1}{j\omega-2} \end{bmatrix}$ **d.** $\dfrac{\begin{bmatrix} j\omega+3 & 1 \\ 2 & j\omega \end{bmatrix}}{(j\omega+2)(j\omega+1)}$

12-11.1 a. $y(k+1) - 5y(k) = -3w(k) + w(k-1)$ **b.** $y(k+2) - 5y(k+1) + 2y(k) = 3w(k) + w(k-1)$

12-11.2 a. $\begin{bmatrix} x_1(k+1) \\ x_2(k+1) \end{bmatrix} = \begin{bmatrix} 0 & 1 \\ -\tfrac{1}{2} & -\tfrac{5}{2} \end{bmatrix}\begin{bmatrix} x_1(k) \\ x_2(k) \end{bmatrix} + \begin{bmatrix} 0 \\ \tfrac{1}{2}w(k) \end{bmatrix}$ $\quad y(k) = \begin{bmatrix} 1 & 0 \end{bmatrix}\begin{bmatrix} x_1(k) \\ x_2(k) \end{bmatrix}$

12-12.1 a. $\gamma_2 + \gamma_1 + \gamma_0 = 2^1 \qquad 2^{\mathbf{A}} = \gamma_2 \mathbf{A}^2 + \gamma_1 \mathbf{A} + \gamma_0 \mathbf{I}$
$4\gamma_2 + 2\gamma_1 + \gamma_0 = 2^2$
$4\gamma_2 + \gamma_1 + 0\gamma_0 = 2.77$

b. $\gamma_2 + \gamma_1 + \gamma_0 = 1 \qquad \mathbf{A}^k = \gamma_2 \mathbf{A}^2 + \gamma_1 \mathbf{A} + \gamma_0 \mathbf{I}$
$4\gamma_2 + 2\gamma_1 + \gamma_0 = 2^k$
$4\gamma_2 + \gamma_1 + 0\gamma_0 = k2^{k-1}$

d. $h(k) = 4[1 - 2^{k-1} + (k-1)2^{k-2}] - 8[5(1 - 2^{k-1} + (k-1)2^{k-2}) - 4 + 2^{k+1} - 3(k-1)2^{k-2}]$
$\qquad + 5[17(1 - 2^{k-1} + (k-1)2^{k-2}) + 5(-4 + 4\cdot 2^{k-1} - 3(k-1)2^{k-2})$
$\qquad + 4 - 2^{k-1}\cdot 3 + 2(k-1)2^{k-2}] \quad k > 0 \qquad h(k) = 0 \quad \text{for } k < 0$

12-12.2 a. $\mathbf{A}^k = \begin{bmatrix} 2 - 2^k & -2^{k+1} + 2 \\ 2^k - 1 & 2^{k+1} - 1 \end{bmatrix}$ **b.** $e^{\mathbf{A}t} = k > 0 \begin{bmatrix} 2e^t - e^{2t} & 2(e^t - e^{2t}) \\ e^{2t} - e^t & 2e^{2t} - e^t \end{bmatrix}$

12-14.1 a. $\begin{bmatrix} Y_1(z) \\ Y_2(z) \end{bmatrix} = \begin{bmatrix} \dfrac{1}{z(z-1)} & \dfrac{1}{z^2(z-1)^2} \\ 0 & \dfrac{1}{z(z-1)} \end{bmatrix}\begin{bmatrix} W_1(z) \\ W_2(z) \end{bmatrix}$

c. $\Phi(z) = [\mathbf{I} - z^{-1}\mathbf{A}]^{-1} = \begin{bmatrix} 1 & -z^{-1} & 0 & 0 \\ 0 & 1-z^{-1} & z^{-1} & 0 \\ 0 & 0 & 1 & -z^{-1} \\ 0 & 0 & 0 & 1-z^{-1} \end{bmatrix}^{-1}$

12-14.2 a. $\Phi(k) = \mathbf{A}^k = \left(\dfrac{1}{k+1}\right)^k$

12-16.1 a. Uncontrollable, observable   **b.** Uncontrollable, observable

12-17.1 Bounded-input, bounded-output

12-17.2 Both unstable

12-17.3 a. Stable   **b.** Stable

12-18.1 For $\alpha > \beta$, $\dot{V}$ is negative and the origin is asymptotically stable

12-18.2 $\dot{V}$ is negative and the origin is asymptotically stable

12-18.4 System is unstable   **12-18.6** System is unstable

Appendix A1

A1-2.1 a. 4   **b.** 16   **c.** e^{-2} **d.** $2e^{-1}$ **e.** $\tfrac{1}{2}e^{-9/4}$

A1-2.3 a. 3   **b.** -1.712 **c.** 10.5π

INDEX

Absolute convergence, 713
  of Z-transform, 441
Admittance, indicial, 50
Aliasing, in sampling, 373
  elimination using bilinear transformation, 588
Amplitude modulation (*see* Modulation, amplitude)
Amplitude spectrum, continuous, 5
  discrete, 512
  Fourier series, 133
Analog to digital converter, 17
Analytic continuation, 718
Analytic function, 704
Aperture function, 85
Approximation:
  differential equation by difference equation, 410
  least square of Fourier series, 142
  in the mean, 23
  using interpolation formulas, 37
  using orthogonal functions, 19
Asymptotic stability, 347, 674
  existence of Liapunov function for, 677
Autocorrelation, 64
  Fourier transform of, 182
Auxiliary equation, of difference equation, 413

Bandlimited signals, 151
  sampling theorem for, 151, 370
Basis functions, 19
  complete, 20
  incomplete, 20
  orthonormal, 21
    Bessel functions, 227, 645, 697
    Chebyshev polynomials, 34, 569
    Hermite functions, 34
    Laguerre functions, 34
    Legendre polynomials, 32
    Walsh functions, 27, 547

Bessel function, 21, 697
  inequality, 24
  integral form, 227
  series expansion, 697
Bilinear transformation, 499, 588
Binary code, 695
Block diagram, 306
  reduction of, 306
  in system description, 308
Bode plot, 251
Branch:
  cuts, 352, 730
  points, 352, 730
Butterworth filter:
  analog, 562
  digital, 582, 592

Canonic form:
  state equations, 114, 623, 624
Casoration, 412
Cauchy:
  first integral theorem, 207, 707
  principle value, 728
  Riemann differential equations, 705
  second integral theorem, 711
Causal system, 93
Cayley-Hamilton theorem, 746
  matrix functions using, 639
Characteristic matrix, 641
Characteristic roots (*see* also Eigenvalues):
  determining stability, 347, 480
  of a matrix, 641
Chebyshev filter:
  analog, 568
  digital, 586
Comb function, 14
  Fourier transform of, 192
Complete solution:
  difference equation (*see* Z-transform)

Complete solution (*continued*)
  differential equation (*see* Laplace transform)
  state equations, 636
Complex Fourier series, 130
Complex frequency, 290
Complex variables, functions of, 703
  analytic continuation, 718
  basic concepts, 703
  branch points and branch cuts, 730
  Contour integration, 207
    evaluation of definite integrals, 725
  logarithmic derivative, integration of, 733
  principal value of an integral, 213, 728
  power series, 711
    convergence, regions of, 711
    convergence, test for, 711
    Laurent, 716
    Maclaurin, 713
    Taylor, 713
  residues, theory of, 207,
  singularities, 719
Conversation:
  charge, 335
  flux linkages, 335
Continuous time systems:
  impulse response of, from differential equations, 76
  from state variable description, 638
  stability of, 93, 344
  state variable description, 116, 615
  transfer function of (*see* System function)
  transform analysis of (*see* Fourier integral transform; Laplace transform)
  Superposition integral, 48
Contour integration, 207, 304, 350
Controllability:
  definition, 669
  using state matrix, 672

INDEX

Convergence:
  of Fourier series, 131
  of Laplace integral, 292, 357
  Z-transform, 440, 489
Convergence region (*see* Region of convergence)
Convolution:
  circular or periodic, 63, 144, 518
  in continuous time systems, 46, 178
  frequency, 521
  integral, 13, 52
  matrix:
    state, 637
    integral, 637
Convolution integral, 13, 53, 56, 87, 427
  in discrete Fourier transform, 518
  in Fourier series, 144
  in Fourier transforms, 302
  in Laplace transforms, 302
  in Z-transform, 456
Correlation:
  autocorrelation, 64
  cross-correlation, 66
Correlator, optical, 274
Criterions for stability (*see* Stability)
Critical frequencies:
  poles and zeros, 304
Cross-correlation, 64
  Fourier transform of, 182

D'Alembert's principle, 77
Decomposition of signals (*see* Fourier series; Singularity function)
Delay operations, 406
Delta function:
  definition, 10, 689
  properties of, 690
  transform of:
    discrete Fourier transform, 524
    Fourier transform of, 190
    Z-transform, 443
Delta sampling, 370
Demodulation, 221
Difference equations, 392
  approximation of differential equation, 410
  block diagram, 306
  characteristic equation of, 413
  characteristic of systems by, 394
  first order, 395
  higher order, 410
  instability, 400
  model for discrete time systems, 397
  normal form, 651
  solutions of:
    analytic, 401
    direct, 397
    undetermined coefficients, 418
    using state variables, 658
    using Z-transforms, 474
  state variable representation, 648

Differential equations:
  normal form of, 620
  signal flow graph representation, 321
  solution of:
    using convolution, 56
    using Fourier integral, 178
    using Laplace transform, 306, 336
Digital filters, 579
  Butterworth design, 592
  discrete Fourier transform as, 606
  FIR:
    design by discrete Fourier transform, 600
    design by Fourier series method, 394, 599
    use of window functions, 602
  frequency response of, 561
  IIR, frequency transformation of, 394, 609
  invariant impulse response method, 579
  folding and bilinear transformation, 588
  warping, 590
  maximally flat, 563
  pass band, 427
Digital conditions:
  in Fourier series, 130
  in Fourier transform, 164
  in Laplace transform, 293
Discrete Fourier transform, 510
  convolution, periodic, 518
  definition of, 511
  as a digital filter, 606
  evaluation of, 514
  fast Fourier transform (*see* Fast Fourier transform)
  inverse of, 511
  leakage, 543
  properties of, 512
  table of, 531
  relation to:
    Fourier integral, 526
    Fourier series, 498
    Walsh function, 546
    Z-transform, 511, 528
  windows in, 543
Discrete time systems, 392
  delay operator, 406
  difference equation of, 393
  equivalent for continuous time system, 397
  frequency response, 423
  fold-over frequency, 426
  fundamental matrix, 659
    properties of, 659
  impulse response of:
    from state variable representation, 660
    properties of, 392
  solution, 397, 401, 485
  to homogeneous difference equation, 402

Discrete time system (*continued*)
  stability of, 400, 499
  state variable representation, 648
  solutions of, 658
  system function, 470
  Z-transform analysis of, 452
Discrete transforms, 510
  discrete Fourier (*see* Discrete Fourier transform)
  fast Fourier (*see* Fast Fourier transform)
    Cooley-Turkey, 532
    Sande, 540
    Walsh-Hadamard, 552
    data flow graph, 553

Eigenfunction, 22, 96
Eigenvalue:
  definition of, 96
  Hermitian, 97
Elements, system:
  electrical, 67
  mechanical, rotational, 71
  mechanical, translation, 69
  optical, 74
Energy signal, 183
Equilibrium state, 673
  stability of (*see* Stability)
Equivalence:
  zero input, 98
  zero state, 99
Error function, 382, 698
Even function, 167, 517
Exp (At) (*see also* State, fundamental matrix), 629
  properties of, 634

Fast Fourier transform, 531
  base-2 algorithm, 537
  and convolution, 542
  data flow graph for, 535
  decimation in frequency: Sande, 540
  decimation in time:
    Cooley-Turkey, 532
    matrix approach, 534
  leakage in, 543
Filter:
  approximation to prescribed function, Butterworth, 562, 582
  approximation to prescribed function, Chebyshev, 568, 586
  elliptic, 576
  variations in pass band, 569
  classical types, 560
  digital (*see* Digital filters)
  distortionless, 243
  finite impulse response (FIR), 477, 599, 656
  ideal, 561
    high pass, 254
    low pass, 244, 562

INDEX

Filter (*continued*)
  infinite impulse response (IIR), 477
  frequency transformations, 609
  narrow-band, 219
  phase characteristic, 577
  quadrature, 213
Filter transformations, 577, 578, 609
Final value theorem:
  for Laplace transform, 298
  for Z-transform, 456
FIR filter (*see* Filter)
Flow graph, signal (*see* Signal flow graph)
Flow graph, data (*see* Fast Fourier transform)
Flux linkage, conservation of, 335
Focal plane, 203
Folding (*see* Convolution)
Fourier integral transform, 163
  and complex function theory, 211
  convolution:
    frequency, 179
    time, 178
  definition of, 163
  delta function, 190
  Gibbs phenomenon, 195
  interpretation of, 169
  inversion of, 163
  modulation property, 175
  of optical systems, 203
  Parseval's theorem, 182
  of periodic functions, 194
  power density, 182
  properties of, 166, 170
  relation to:
    Discrete Fourier transform, 526
    Hilbert transform (*see* Hilbert transform)
    Laplace transform, 291
    Z-transform, 497
  sampling theorem, 370
  system function, 177
  table of, 196
  window functions, 201
Fourier kernel, 152
Fourier series, 128
  choice of origin, 145
  convergence of, 131
  convolution, 144
  Dirichlet conditions, 130
  discrete, 528
  energy relation, 135
  exponential, 130
  Fejer factor, 155
  finite signals, 139
  Gibbs phenomenon, 152
  harmonics in, 139
  impulse train, 133
  Lanczos factor, 154
  least squares approximation, 142
  line spectrum, 133
  partial sums, 140
  phase, 133

Fourier series (*continued*)
  product of functions, 143
  properties of, 134
    Parseval's theorem, 134
  spectrum, amplitude, 133
  symmetry, effects of, 135
  table of, 000
  transform pairs, 131
  trigonometric, 4, 131
Frequency, complex, 292
Frequency convolution, 521
Frequency modulation (*see* Modulation, frequency)
Frequency response, 480
Function, analytic, 704
  Bassel, 227
  Butterworth, 562
  Chebyshev, 34, 569
  Comb, 14
  complex variable (*see* Complex variables, functions of)
  delta, 10, 687
  error, 382
  exponential order, 293
  Green's, 89
  Hermetian, 97
  Hermite, 34
  impulse (*see also*, Delta function), 689
  interpolation in sampling, 372
  Laguerre, 34
  Legendre, 32
  maximally flat, 563
  multiple-valued, 703
  ramp, 7
  sampled, 14
  signum, 9
  single-valued, 703
  singularity, 7
  transfer (*see* System function)
  unit step, 7
  Walsh, 27, 546
  weighting, 602
  window (*see* Window function)
Fundamental matrix, 629
  calculation of,
    discrete time system, 659
    Laplace transform of, 640
  properties of, 634
  Z-transform of, 464

Gain, signal flow graph, 323
Generalized function, 11, 164
Generating function for Z-transform, 438
Geometric series:
  use in Z-transform, 466
Gibbs phenomenon, 152, 195
Global stability (*see also* Stability), 674
Gram-Schmidt orthogonalization, 45
Graph, signal flow (*see* Signal flow graph)

Gray code, 695
Green's function, 89

Hadamard transform, 550
  data graph for, 553
  fast, 552
Walsh function, 550
Hilbert transform, 169, 212
  analytic signal, 216
  modulation, 215
  properties of, 214
  quadrature filter, 213
Homogeneity property, 94
Homogeneous solution:
  discrete time systems, 402
Hurwitz, polynomial, 565
  stability test (*see* Stability, Routh-Hurwitz test)

Improper fraction, 331
Impulse function (*see also* Delta function):
  Fourier transform of, 190
  Laplace transform of, 345
  replacement of initial conditions by, 336
  resolution of signal into, 52
  strength of, 11
  Z-transform of, 471
Impulse invariant response method, 579
Impulse response, 76
  calculation of:
    from difference equation, 471
    from differential equation, 76
  relation to step response, 52
  by state variable method:
    continuous time, 638
    discrete time, 660
  by transform methods:
    Laplace. 345
Impulse train:
  Fourier transform of, 192
  use in sampling theory, 370
Indicial admittance, 50
  response, 50
Inequality:
  Bessel, 24
  Schwartz, 64
Infinite impulse response (IIR) filter (*see* Filter)
Initial conditions, Laplace transform of, 334
  replacement of:
    in charged capacitors, 336
    in fluxed inductors, 336
  use in Z-transforms, 457
Initial state, 647
  relaxed, 334
Initial value theorem:
  in Laplace transforms, 305
  in Z-transforms, 455

INDEX

Input-output relation (*see* System function)
Integral, convolution, 13, 52
  principal value of, 212, 723
  superposition, 48
Interpolating function in sampling theory, 372
Inverse matrix (*see* Matrix, properties of)
Inverse transforms:
  discrete Fourier, 511
  Fourier integral, 163
  Hilbert, 169
  Laplace, 328, 350
  Z-transform, 463

Jordan, lemma, 209
  matrix, 000

Lagrange interpolation formula, 39
Laplace transform, 290
  bilateral, 292, 356
  convergence:
    abscissa of, 291
    region of, 292, 357
  convolution theorem, 302
  defining integral, 291
  expansion theorem, 331
  of fundamental matrix, 629
  initial conditions, 299, 301, 334
  inversion integral, 291, 328, 350
    contour integration techniques, 304, 356
    partial fraction expansion, 330
  linear time invariant systems, 306
  one-sided, 293
  problem solving using, 306, 336
  properties of, 296
  s-plane, 293
  stability, 344
  table of, 295
  two-sided, 293
  state equations, 640
Laurent series, 716
Least squares approximation of Fourier series, 142
Liapunov function, 677
  generating, 678
  stability theorem, 677
Linear difference equations (*see* Difference equations)
Linear differential equations (*see* Differential equations)
Linear filters, response of (*see* Filters)
Linear system, definition of, 242
Linear time invariant system, 306
Long division method of inverting Z-transforms, 465
Low-pass filter, continuous time, 562
  digital, 579
  effect on periodic input function, 148

Magnification, 87
Magnitude and phase plots, 140
Matrix, 738
  algebra of, 740
  Cayley-Hamiltion theorem, 746
  characteristic equation of, 744
  definitions, 739
  discrete time transfer, 470
  eigenvalues of, 744
  functions of, 743
  fundamental (*see* Fundamental matrix)
  non-singular, 740
  positive definite, 738
  properties of, 739
  quadratic form,
  state transition, 629, 659
  transformation similarity, 661
  transpose function, 38, 641
Maximally flat function, 563
Mean square error, 23
  in Fourier series approximation, 142
  minimization, 23
Memoryless system, 94, 616
Models and modeling, 66
  electrical elements, 67
  mechanical elements:
    rotational, 71
    translational, 69
  optical systems, 74
Modulation:
  amplitude, 16, 215, 219, 383
  frequency, 223, 225
    deviation ratio, 226
    instantaneous frequency, 224
    narrow-band, 226
    spectral distribution, 229
  light, 231
  modulation index, 220
    double-sideband suppressed carrier, 220
    single sideband, 216, 220
  phase, 230
  pulse,
    amplitude, 383
    code, 383
  theorem for Fourier transforms, 175
  time division, 383
Moment theorem,
Multiplexing:
  frequency division, 223
  time division, 383

Negative time functions:
  in Laplace transforms, 356
Network, graph, 316
  stability (*see* Stability)
Network behavior:
  calculation of fundamental matrix, 629
  convolution integral, 44, 178
  principle of superposition, 48

Network behavior (*continued*)
  step and impulse response, 53
  zero input, zero state, 98
Network function (*see* System function)
Newton law, 69
Non-causal functions, 93
Nonlinear system, definition of, 48
Nonperiodic signals, 7
Nonrecursive process, 477
  matrix, 674
Normal form of differential equations, 620
  procedure for determining, 621
Normalization, 729
  interval, 372

Observability, definition, 670
  using matrix, 672
Odd function, 167, 517
One sided Laplace transform, 293
One sided sequence, 438
  Z-transform of, 439
Operator, representation of systems, 47
  system, 87, 101
Optical filters:
  hologram, 279, 281
  operations with, 261
  matched, 282
  phase, 279
  processing of signals, 273
Optical systems, 74, 261
  convolution, 274
  correlation, 274
  diffraction-limited, 268
  filters, 277, 279
  Fourier transform plane, 275
  Fourier transforms in, 203, 274
  frequency response, 267, 268
  hologram, 281
  impulse response, 85
    coherent source, 261
    incoherent source, 261
  signal processing, optical filters, 261
  spread function, 262
  system function, 261
    coherent, 265
    cohenrent transfer function (CTF), 266
    incoherent, 263
    modulation transfer function (MTF), 264
    optical transfer function (OTF), 264
    space-invariant, 267
  vander Lugt filter, 279
Origin, stability of, 674
Orthogonal functions (*see also* Basis functions), 20, 32
Orthogonality:
  of basis functions, 21
  of exponentials, 131
  of sinusoids, 131

INDEX

Orthogonal vectors, 20
Orthonormal functions, 21
Output controllability, 672

Paley ordered Walsh functions, 30, 559
Parseval's theorem, 24, 182
   discrete Fourier transform, 522
   Fourier integral, 182
   Fourier series, 134
Partial fraction expansion:
   in contour integrals, 725
   in Laplace transform, 330
   relation to residue theory, 720
   in Z-transforms, 487
Particular response, state variable formulation, 628
Partitioning of matrix, 000
Periodic function, Fourier series, 130
   Fourier transform of, 194
   Laplace transform of,
Phase, of network function, 243
Phase spectrum, 5, 133
   Butterworth filter, 566
Physical systems, elements of (see Elements, system):
Piecewise constant signal, and superposition integral, 50
Poisson sum formula, 193
Pole, dominant, 478
   location and stability, 347, 480
   multiple order, 332, 494
   of network functions, 304, 332, 477
Polynomials, Butterworth, 565
Positive definite function, 677
Positive semi-definite matrix, 677
Positive time function, 93
Power density, 182
Power series (see also Series), 708
   convergence, 709
Power spectrum, 186
Principal value of integral, 213
Prewarping with bilinear transformation, 595
Projection theorem, 26
Proper fraction, 331
Pulse modulation (see Modulation, pulse)
Pulse train, Fourier series of, 133
   least squares approximation, 142
   spectrum of, 133

Quadrature filter, 213

Radius of convergence, 713
   gyration, 72
   of Z-transform, 440, 489
Ramp function, 7
Rational function, 331
   inverse Laplace transform of, 330
   inverse Z-transform of, 467, 490
Rectangular aperature, 206
Rectangular pulse, spectrum of, 172

Recursive filter, design by bilinear method, 588
   design by impulse invariant method, 579
Region of convergence, Laplace transform, 292
   Z-transform:
      one-sided, 440
      two-sided, 445
Residue, 207, 332, 493, 720
Resistor ladder network, 488
Resolution of continuous time functions:
   discrete time sequences, Z-transform, 438
   by impulses, 52
   by step function, 52
Response, forced:
   impulse (see Impulse response)
   indicial, 50
   state, discrete time, 648
      forced, 660
      force free, 659
      initial state, 647
   state, continuous time, 636
      forced, 636
      force free, 628
      initial state, 647
   step, 7
      impulse, 52
      step, 50
   zero input, 55, 98, 638
   zero state, 55, 99, 638
Response function (see system function)
Riemann (see Cauchy: Riemann differential equations)
Rise time, 246
Roots of characteristic equation, 641

s-plane, 293
Sample and hold, 385
Sampled signals, 366
   Fourier transform of, 367
   reconstruction from, 380
Sampling, aliasing, 373, 379
   by delta function, 370
   frequency (rate), 377
   interval, 366, 372
   Nyquist, 372
   by rectangular pulse train, 379
      flat top, 380
   time, 370
Sampling theorem, 151, 370
Schwartz inequality, 64
s-domain and stability, 347
Second method of Liapunov, 677
Second order difference equation (see Difference equations)
Sequence, weighting, 602
   Z-transform of, 438
Sequency, 27, 547

Series, 706, 711
   convergence, 707, 712
   Fourier (see Fourier series)
   Laurent, 716
   Maclaurin, 708, 713
   Taylor, 498, 713
Shift operator, 87, 101
Signum function, 8
   Fourier transform of, 190
Signal, complex amplitude of, 218
   analytic (preenvelope), 216
   discrete time (see also Sampled signals) 5
   energy, 25
   Fourier integral representation, 163
   Fourier series representation, 128
   nonperiodic, 7
      delta function (Dirac delta), 10
      Gaussian, 10
      ramp, 7
      rectangular pulse, 8
      signum, 9
      sinc, 9
      step, 7
      triangular, 9, 37
   periodic, 3
      sinusoid, 4
   polynomial interpolation, 37
   power, 25
   representation by orthogonal functions, 19
   sampled, 366
   stepwise approximation, 50
Signal flow graph, 316
   algebra of, 318
   interaction, 323
   Mason rule, 324
   nodes, 316
   properties of, 321
   rules for drawing, 318
   transmittance, 316, 321
Similarity transformation, 661
Sinc(x), definition of, 8
Sine integral, 200
Single side-band modulation, 215
Singularity function, 7, 719
Solution:
   difference equation (see Difference equations)
   difference equations (see Differential equations)
   state equations (see State, solutions of)
Spectrum, amplitude, 5
   amplitude-modulated signal, 216
   frequency modulated signal, 229
   phase, 5
   of spatial signals, 203
Spread function, 86, 89
Square wave, Fourier series representation, 133
   Laplace transform, 359

INDEX

Stability, 673
  asymptotic, 347, 674
  bounded-input bounded-output, 93, 673, 675
  criterions for, 348, 350
  equilibrium state, 674
  in the large, 674
  in Liapunov sense, 676
  of origin, 674
  relation to pole location, 347, 480
  zero input, 674
State:
  characteristic matrix, 640
  controllability, 672
  definition, 105
  equations, 105, 616
    block diagram, 623
    differential equation, 113, 615
    discrete time, 648
    signal flow graph, 623
  fundamental matrix (see also Fundamental matrix) 629
    properties of, 634
  impulse response, 345, 660
  matrix description, 106, 617
  observability (see Observability)
  solutions:
    continuous time, force free, 628
    continuous time, complete, 636
    discrete time, 658
    discrete time, impulse response, 660
    eigenvalue method, 634
    impulse, 638
    Laplace methods, 640
    periodic inputs, 644
    sampled inputs, 665
  space, 106
  stability, 344, 499, 673
    Liapunov function, 677
  transition matrix, 629, 659
  variables, selection of, 105, 106, 616
    transformation of, 627
  vector, 616
    initial state, 647
State equations, 106
  first canonical form, 114, 624
  phase variable form, 623
  solutions of (see State, solutions)
  transformation, 627
Step function, 7
Step response, related to impulse response, 52
Superposition integral, 48
  in terms of impulse response, 52
  in terms of step response, 50
Superposition, and convolution, 53
Symmetry property, Fourier transform, 170
System equations, normal form of, 620
System function, 57
  block diagram, 101

System function (continued)
  continuous time system, 48
  discrete system, 475
  frequency response, 177
  fundamental matrix, 629
  matrix, 642
  relation to impulse response, 345
  roots of, 347
  signal flow graph, 316
  state models, continuous time, 642
    discrete time, 667
Systems, block diagram for, 306
  cascade, 101
  causal, 93
  classification, 90
  continuous time (see Continuous time systems)
  controllable (see Controllability)
  discrete time (see Discrete time systems)
  distributed, 48
  linear time invariant, 48, 98
  linear time variant, 48
  lumped, 46
  memory of, 93
  nonlinear, 48, 91
  observable (see Observability)
  parallel, 102
  periodic inputs, 148
  signal flow graph for, 318
  stability of (see Stability)
  state representation (see State)

Taylor series, 498
Time convolution, Fourier transform, 178
Time division multiplexing, 383
Time invariance, 48
Time signals, sampling of, 370
Transfer function (see System function):
  coherent, 265
  optical, 264
Transfer matrix (see Fundamental matrix)
Transform:
  discrete Fourier (see Discrete Fourier transform)
  fast Fourier (see Fast Fourier transform)
  Fourier (see Fourier integral transform)
  Hadamard (see Hadamard transform)
  Hilbert (see Hilbert transform)
  Laplace (see Laplace transform)
  Z (see Z-transform)
Transform pairs, tables of:
  Fourier, 196
  Laplace, 295
  Z-, 464
  state space, 660
  transform methods, expansion theorem (see Laplace transform; Z-transform)

Transition matrix (see Fundamental matrix)
Transmission function (see system function):
Transmission without distortion, 149
Trigonometric form of Fourier series, 4, 131
  conversion from exponential, 131
Truncated series:
  Fourier, 139
  minimum mean square error, 142
Two-sided Laplace transform, 292
  convergence of, 292
  inverse, 356
Two-sided sequence, Z-transform, 461, 489

Underdamped response, 83
Undetermined coefficients, method of, 418
Unit delay operator, 406
Unit impulse (see Delta function):
Unit step function, 7
  Laplace transform of, 294
Unit step response, indicial admittance, 48
Unit step sequence, 439
  Z-transform of, 439
Unstable system (see Stability)

Variables, across, 68
  through, 68
Vectors, basis set, 19
Vandermonde matrix, 38

Waveform (see Signal)
Walsh functions, 27, 547
  Hadamard, 31, 550
  Paley, 30
Walsh transform, 546
  evaluation of, 549
  transformation kernels, 546
Warping, with bilinear transformation, prewarping, 595
Weighting sequence, 602
Window function, 152, 201, 544, 603
Wronskian (Casoratian), 412

Zero input responses, 98, 638
Zero state response, 99, 638
Z-transform, 438
  convergence:
    of one-sided, 440
    of two-sided, 489
  convolution, 456
    bilateral, 461
  definition:
    one-sided, 439
    two-sided, 445
  discrete Fourier transform, relation to, 511

Z-transform (*continued*)
   Fourier and Laplace transform, relation to, 497
   inversion, direct division, 465
      inversion integral, 463, 493
         partial fraction expansion, 467, 490
   matrix inversion, 667

Z-transform (*continued*)
   problem solving using, 485
   properties of, 449, 462
   solution of difference equations, 485, 666
   state variable application, 666

Z-transform (*continued*)
   system function, 470, 667
   table of, 464
   Taylor and Fourier series, 498
   transfer function, 470
   two-sided, 461, 489

Table 4.2 (*continued*)

$$f(t) = \frac{1}{2\pi} \int_{-\infty}^{\infty} F(\omega)e^{j\omega t}\, d\omega \qquad\qquad F(\omega) = \int_{-\infty}^{\infty} f(t)e^{-j\omega t}\, dt$$

$f(t)$	$F(\omega)$
$f(t) = \begin{cases} A & (T-a) < |t| < (T+a) \\ 0 & \text{otherwise} \end{cases}$	$F(\omega) = 4A\,\dfrac{\cos T\omega \sin a\omega}{\omega}$
$f(t) = \begin{cases} A\cos\omega_0 t & |t| \le a \\ 0 & \text{otherwise} \end{cases}$	$F(\omega) = A\left[\dfrac{\sin a(\omega-\omega_0)}{\omega-\omega_0} + \dfrac{\sin a(\omega+\omega_0)}{\omega+\omega_0}\right]$
$f(t) = A\,\delta(t)$	$F(\omega) = A$
$f(t) = \begin{cases} A & t > 0 \\ 0 & \text{otherwise} \end{cases}$	$F(\omega) = A\left[\pi\delta(\omega) - j\dfrac{1}{\omega}\right]$
$f(t) = \begin{cases} A & t > 0 \\ 0 & t = 0 \\ -A & t < 0 \end{cases}$	$F(\omega) = -j2A\,\dfrac{1}{\omega}$

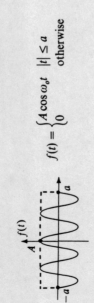

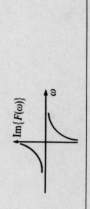

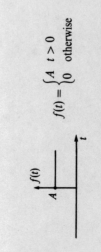

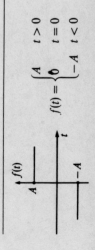